DR. AI   D1651004

Annual Review of Neuroscience

EDITORIAL COMMITTEE (2002)

DAVID J. ANDERSON
W. MAXWELL COWAN
STEVEN E. HYMAN
THOMAS M. JESSELL
WILLIAM T. NEWSOME
CARLA J. SHATZ
CHARLES F. STEVENS
MARC TESSIER-LAVIGNE
HUDA Y. ZOGHBI

RESPONSIBLE FOR THE ORGANIZATION OF VOLUME 25 (EDITORIAL COMMITTEE, 1999)

DAVID J. ANDERSON
W. MAXWELL COWAN
MARY E. HATTEN
STEVEN E. HYMAN
WILLIAM T. NEWSOME
CLIFFORD B. SAPER
DENNIS J. SELKOE
CARLA J. SHATZ
ERIC M. SHOOTER
CHARLES F. STEVENS
RICHARD F. THOMPSON

Production Editors: CLAIRE INSEL, JENNIFER E. MANN
Bibliographic Quality Control: MARY A. GLASS
Color Graphics Coordinator: EMÉ O. AKPABIO
Electronic Content Coordinator: SUZANNE K. MOSES
Subject Indexer: SUZANNE COPENHAGEN

Annual Review of Neuroscience

VOLUME 25, 2002

W. MAXWELL COWAN, *Editor*
Bethesda, Maryland

STEVEN E. HYMAN, *Associate Editor*
Harvard University

THOMAS M. JESSELL, *Associate Editor*
Columbia University

CHARLES F. STEVENS, *Associate Editor*
Salk Institute for Biological Studies

www.annualreviews.org   science@annualreviews.org   650-493-4400

ANNUAL REVIEWS
4139 El Camino Way • P.O. BOX 10139 • Palo Alto, California 94303-0139

ANNUAL REVIEWS
Palo Alto, California, USA

COPYRIGHT © 2002 BY ANNUAL REVIEWS, PALO ALTO, CALIFORNIA, USA. ALL RIGHTS RESERVED. The appearance of the code at the bottom of the first page of an article in this serial indicates the copyright owner's consent that copies of the article may be made for personal or internal use, or for the personal or internal use of specific clients. This consent is given on the condition that the copier pay the stated per-copy fee of $14.00 per article through the Copyright Clearance Center, Inc. (222 Rosewood Drive, Danvers, MA 01923) for copying beyond that permitted by Section 107 or 108 of the US Copyright Law. The per-copy fee of $14.00 per article also applies to the copying, under the stated conditions, of articles published in any *Annual Review* serial before January 1, 1978. Individual readers, and nonprofit libraries acting for them, are permitted to make a single copy of an article without charge for use in research or teaching. This consent does not extend to other kinds of copying, such as copying for general distribution, for advertising or promotional purposes, for creating new collective works, or for resale. For such uses, written permission is required. Write to Permissions Dept., Annual Reviews, 4139 El Camino Way, P.O. Box 10139, Palo Alto, CA 94303-0139 USA.

International Standard Serial Number: 0147-006X
International Standard Book Number: 0-8243-2425-0

All Annual Reviews and publication titles are registered trademarks of Annual Reviews.

∞ The paper used in this publication meets the minimum requirements of American National Standards for Information Sciences—Permanence of Paper for Printed Library Materials, ANSI Z39.48-1992.

Annual Reviews and the Editors of its publications assume no responsibility for the statements expressed by the contributors to this *Annual Review*.

Typeset by TechBooks, Fairfax, VA
Printed and Bound in the United States of America

Annual Review of Neuroscience
Volume 25, 2002

CONTENTS

THE HUMAN GENOME PROJECT AND ITS IMPACT ON PSYCHIATRY, *W. Maxwell Cowan, Kathy L. Kopnisky, and Steven E. Hyman*	1
AUDITORY SYSTEM DEVELOPMENT: PRIMARY AUDITORY NEURONS AND THEIR TARGETS, *Edwin W. Rubel and Bernd Fritzsch*	51
AMPA RECEPTOR TRAFFICKING AND SYNAPTIC PLASTICITY, *Roberto Malinow and Robert C. Malenka*	103
MOLECULAR CONTROL OF CORTICAL DENDRITE DEVELOPMENT, *Kristin L. Whitford, Paul Dijkhuizen, Franck Polleux, and Anirvan Ghosh*	127
FUNCTIONAL MRI OF LANGUAGE: NEW APPROACHES TO UNDERSTANDING THE CORTICAL ORGANIZATION OF SEMANTIC PROCESSING, *Susan Bookheimer*	151
INTENTIONAL MAPS IN POSTERIOR PARIETAL CORTEX, *Richard A. Andersen and Christopher A. Buneo*	189
BEYOND PHRENOLOGY: WHAT CAN NEUROIMAGING TELL US ABOUT DISTRIBUTED CIRCUITRY? *Karl Friston*	221
TRANSCRIPTIONAL CODES AND THE CONTROL OF NEURONAL IDENTITY, *Ryuichi Shirasaki and Samuel L. Pfaff*	251
THE ROLE OF HYPOCRETINS (OREXINS) IN SLEEP REGULATION AND NARCOLEPSY, *Shahrad Taheri, Jamie M. Zeitzer, and Emmanuel Mignot*	283
A DECADE OF MOLECULAR STUDIES OF FRAGILE X SYNDROME, *William T. O'Donnell and Stephen T. Warren*	315
CONTEXTUAL INFLUENCES ON VISUAL PROCESSING, *Thomas D. Albright and Gene R. Stoner*	339
LARGE-SCALE SOURCES OF NEURAL STEM CELLS, *David I. Gottlieb*	381
SCHIZOPHRENIA AS A DISORDER OF NEURODEVELOPMENT, *David A. Lewis and Pat Levitt*	409
THE CENTRAL AUTONOMIC NERVOUS SYSTEM: CONSCIOUS VISCERAL PERCEPTION AND AUTONOMIC PATTERN GENERATION, *Clifford B. Saper*	433
THE ROLE OF NOTCH IN PROMOTING GLIAL AND NEURAL STEM CELL FATES, *Nicholas Gaiano and Gord Fishell*	471

MULTIPLE SCLEROSIS: DEEPER UNDERSTANDING OF ITS
PATHOGENESIS REVEALS NEW TARGETS FOR THERAPY,
*Lawrence Steinman, Roland Martin, Claude Bernard, Paul Conlon,
and Jorge R. Oksenberg* 491

WIRED FOR REPRODUCTION: ORGANIZATION AND DEVELOPMENT OF
SEXUALLY DIMORPHIC CIRCUITS IN THE MAMMALIAN FOREBRAIN,
Richard B. Simerly 507

CENTRAL NERVOUS SYSTEM DAMAGE, MONOCYTES AND
MACROPHAGES, AND NEUROLOGICAL DISORDERS IN AIDS,
Kenneth C. Williams and William F. Hickey 537

LEARNING AND MEMORY FUNCTIONS OF THE BASAL GANGLIA,
Mark G. Packard and Barbara J. Knowlton 563

INDEXES
　Subject Index 595
　Cumulative Index of Contributing Authors, Volumes 16–25 603
　Cumulative Index of Chapter Titles, Volumes 16–25 607

ERRATA
　An online log of corrections to *Annual Review of Neuroscience* chapters
　(if any, 1997 to the present) may be found at http://neuro.annualreviews.org/

Related Articles

From the ***Annual Review of Biochemistry***, Volume 71 (2002)

Metabolism and the Control of Circadian Rhythms, Jared Rutter, Martin Reick, and Steven L. McKnight

Lipoprotein Receptors in the Nervous System, Joachim Herz and Hans H. Bock

From the ***Annual Review of Biomedical Engineering***, Volume 3 (2001)

Visual Prostheses, Edwin M. Maynard

Micro- and Nanomechanics of the Cochlear Outer Hair Cell, W. E. Brownell, A. A. Spector, R. M. Raphael, and A. S. Popel

From the ***Annual Review of Biomedical Engineering***, Volume 4 (2002)

Peptide Aggregation in Neurodegenerative Disease, Regina Murphy

From the ***Annual Review of Cell and Developmental Biology***, Volume 17 (2001)

Molecular Bases of Circadian Rhythms, Stacey L. Harmer, Satchidananda Panda, and Steve A. Kay

Early Eye Development in Vertebrates, Robert L. Chow and Richard A. Lang

Stem and Progenitor Cells: Origins, Phenotypes, Lineage Commitments, and Transdifferentiations, Irving L. Weissman, David J. Anderson, and Fred Gage

From the ***Annual Review of Genetics***, Volume 35 (2001)

Identification of Epilepsy Genes in Human and Mouse, Miriam H. Meisler, Jennifer Kearney, Ruth Ottman, and Andrew Escayg

Molecular Genetics of Hearing Loss, Christine Petit, Jacqueline Levilliers, and Jean-Pierre Hardelin

From the ***Annual Review of Medicine***, Volume 53 (2002)

Positron Emission Tomography Scanning: Current and Future Applications, Johannes Czernin and Michael E. Phelps

Attention Deficit/Hyperactivity Disorder Across the Lifespan, Timothy E. Wilens, Joseph Biederman, and Thomas J. Spencer

The Expanding Pharmacopoeia for Bipolar Disorder, Philip B. Mitchell and Gin S. Malhi

Multiple Sclerosis, B. Mark Keegan and John H. Noseworthy

From the ***Annual Review of Physiology***, Volume 64 (2002)

A Hundred Years of Sodium Pumping, Ian M. Glynn

Potassium Channel Ontogeny, Carol Deutsch

G Proteins and Pheromone Signaling, Henrik G. Dohlman

G Proteins and Phototransduction, Vadim Y. Arshavsky, Trevor D. Lamb, and Edward N. Pugh, Jr.

G Proteins and Olfactory Signal Transduction, Gabriele V. Ronnett and Cheil Moon

Calmodulin as an Ion Channel Subunit, Yoshiro Saimi and Ching Kung

Structure and Function of Dendritic Spines, Esther A. Nimchinsky, Bernardo L. Sabatini, and Karel Svoboda

Short-Term Synaptic Plasticity, Robert S. Zucker and Wade G. Regehr

From the ***Annual Review of Psychology***, Volume 52 (2001)

Episodic Memory: From Mind to Brain, Endel Tulving

Genetic Contributions to Addiction, John C. Crabbe

Depression: Perspectives from Affective Neuroscience, Richard J. Davidson, Diego Pizzagalli, Jack B. Nitschke, and Katherine Putnam

ANNUAL REVIEWS is a nonprofit scientific publisher established to promote the advancement of the sciences. Beginning in 1932 with the *Annual Review of Biochemistry*, the Company has pursued as its principal function the publication of high-quality, reasonably priced *Annual Review* volumes. The volumes are organized by Editors and Editorial Committees who invite qualified authors to contribute critical articles reviewing significant developments within each major discipline. The Editor-in-Chief invites those interested in serving as future Editorial Committee members to communicate directly with him. Annual Reviews is administered by a Board of Directors, whose members serve without compensation.

2002 Board of Directors, Annual Reviews

Richard N. Zare, *Chairman of Annual Reviews*
    Marguerite Blake Wilbur, Professor of Chemistry, Stanford University
John I. Brauman, *J. G. Jackson–C. J. Wood Professor of Chemistry, Stanford University*
Peter F. Carpenter, *Founder, Mission and Values Institute*
W. Maxwell Cowan, *Bethesda, Maryland*
Sandra M. Faber, *Professor of Astronomy and Astronomer at Lick Observatory,*
    University of California at Santa Cruz
Susan T. Fiske, *Professor of Psychology, Princeton University*
Eugene Garfield, *Publisher,* The Scientist
Samuel Gubins, *President and Editor-in-Chief, Annual Reviews*
Daniel E. Koshland, Jr., *Professor of Biochemistry, University of California at Berkeley*
Joshua Lederberg, *University Professor, The Rockefeller University*
Sharon R. Long, *Professor of Biological Sciences, Stanford University*
J. Boyce Nute, *Palo Alto, California*
Michael E. Peskin, *Professor of Theoretical Physics, Stanford Linear Accelerator Ctr.*
Harriet A. Zuckerman, *Vice President, The Andrew W. Mellon Foundation*

Management of Annual Reviews

Samuel Gubins, President and Editor-in-Chief
Richard L. Burke, Director for Production
Paul J. Calvi, Jr., Director of Information Technology
Steven J. Castro, Chief Financial Officer
John W. Harpster, Director of Sales and Marketing

Annual Reviews of

Anthropology
Astronomy and Astrophysics
Biochemistry
Biomedical Engineering
Biophysics and Biomolecular
    Structure
Cell and Developmental
    Biology
Earth and Planetary Sciences
Ecology and Systematics
Energy and the Environment
Entomology

Fluid Mechanics
Genetics
Genomics and Human Genetics
Immunology
Materials Research
Medicine
Microbiology
Neuroscience
Nuclear and Particle Science
Nutrition
Pharmacology and Toxicology
Physical Chemistry

Physiology
Phytopathology
Plant Biology
Political Science
Psychology
Public Health
Sociology

SPECIAL PUBLICATIONS
Excitement and Fascination of
    Science, Vols. 1, 2, 3, and 4

THE HUMAN GENOME PROJECT AND ITS IMPACT ON PSYCHIATRY*

W. Maxwell Cowan, Kathy L. Kopnisky, and Steven E. Hyman

National Institute of Mental Health, Bethesda, Maryland 20892;
email: mcowan1@mail.nih.gov; kkopnisk@mail.nih.gov; shyman@mail.nih.gov

Key Words schizophrenia, bipolar disorder, autism, linkage analysis, SNPs, human genome

■ **Abstract** There has been substantial evidence for more than three decades that the major psychiatric illnesses such as schizophrenia, bipolar disorder, autism, and alcoholism have a strong genetic basis. During the past 15 years considerable effort has been expended in trying to establish the genetic loci associated with susceptibility to these and other mental disorders using principally linkage analysis. Despite this, only a handful of specific genes have been identified, and it is now generally recognized that further advances along these lines will require the analysis of literally hundreds of affected individuals and their families. Fortunately, the emergence in the past three years of a number of new approaches and more effective tools has given new hope to those engaged in the search for the underlying genetic and environmental factors involved in causing these illnesses, which collectively are among the most serious in all societies. Chief among these new tools is the availability of the entire human genome sequence and the prospect that within the next several years the entire complement of human genes will be known and the functions of most of their protein products elucidated. In the meantime the search for susceptibility loci is being facilitated by the availability of single nucleotide polymorphisms (SNPs) and by the beginning of haplotype mapping, which tracks the distribution of clusters of SNPs that segregate as a group. Together with high throughput DNA sequencing, microarrays for whole genome scanning, advances in proteomics, and the development of more sophisticated computer programs for analyzing sequence and association data, these advances hold promise of greatly accelerating the search for the genetic basis of most mental illnesses while, at the same time, providing molecular targets for the development of new and more effective therapies.

---

*The U.S. Government has the right to retain a nonexclusive, royalty-free license in and to any copyright covering this paper.

INTRODUCTION

It is not generally appreciated, even among neuroscientists, that psychiatric disorders are among the most widespread and, from the point of view of health care, among the costliest of all illnesses in developed societies. Since they are not usually listed among the major causes of death (although suicide is an important factor in some mental illnesses), mental illnesses rarely receive the attention given to, say, heart disease, cancer, stroke, or AIDS. However, in terms of their overall prevalence, the long-sustained suffering they cause, and the economic burden they represent, they clearly exceed most other forms of ill health. Certainly no other disorder has a lifetime prevalence of over 30%, as is estimated to be the case for mental illnesses in the United States. Current estimates suggest that the collective cost of these disorders is on the order of $400 billion per year or roughly one third of the total health care budget. This is to say nothing of the suffering of those who are afflicted and the burden their illnesses impose on their families.

The following figures provide some indication of the magnitude of the problem in the United States. On the basis of a large epidemiological study involving some 20,000 individuals in five representative cities in 1991, it was estimated that the lifetime prevalence of mental illness was 32% and that in the year preceding the study, as much as 20% of the population was affected (Robins & Regier 1991). More recent estimates of the numbers in the United States affected by major psychiatric disorders indicate that in a given year as many as 44.3 million adults (18 and over) suffer from a diagnosable illness. Of these, about 19 million have some form of depression, with roughly 10 million having major depressive disorder (MDD) and a further 2.3 million with bipolar disorder [BPD (manic depression)]. Over 2.4 million have schizophrenia and about 19 million suffer from an anxiety disorder. More than 4 million have Alzheimer's disease (AD) and 1 in a 1000 children are thought to suffer from autism or a related developmental disorder (see Choi 2000, U.S. Department of Health and Human Services 1999, and www.nimh.nih.gov/publicat/numbers.cfm for details).

The World Health Organization, in conjunction with the World Bank and scientists at the Harvard School of Public Health, have introduced a new and more useful measure of the impact of various mental illnesses and other disorders called the *Disability Adjusted Life Year* (DALY). This measures the years of healthy life lost to mortality or disability (Murray & Lopez 1996, Hyman 2000, Michaud et al. 2001). By this measure, psychiatric disorders are responsible in the United States for a significant proportion of the years lost to disability and, when mortality is excluded, they account for seven of the 10 principal causes of disability (see Table 1). If mortality figures are included, mental illnesses rank second only to cardiovascular disorders. Several reasons are cited for the high disability figures for mental illness. Chief among these are: (*a*) they frequently begin early in life; (*b*) they are usually either chronically persistent or recur at intervals throughout the patient's life; (*c*) because they are prolonged, they nearly all result in frequent absences from the work force with consequent loss of earnings, and (*d*) they are the major cause of suicide; currently in the United States, suicide is the third leading cause

TABLE 1 Rank order of causes of disability in the U.S. and other developed nations[a]

		Percent of total YLDs
1	Unipolar major depression	14.3
2	Alcoholism	9.6
3	Osteoarthritis	5.8
4	Dementia & degen. CNS	5.1
5	Schizophrenia	4.7
6	Bipolar disorder	3.6
7	Cerebrovascular	3.3
8	Diabetes	3.2
9	Obsessive compulsive disorder	3.1
10	Drug use	3.0

[a]The data shown are for individuals over the age of five, expressed as a percent of the total years lost to disability (YLD) (with permission from Global Burden of Disease, Murray & Lopez 1996).

of death among teenagers and young adults and overall ranks eighth among the principal causes of mortality.

Considerable progress has been made in the past 20 years in our ability to diagnose more accurately and to treat more effectively many of these disorders, but, with few exceptions, we are still woefully ignorant of their underlying causes. Fortunately, advances in human genetics have opened up promising new lines for exploring the biological basis of the major cognitive and affective disorders and certain psychoses. The fact that the public at large is coming to accept that these various conditions are brain disorders is one of the most hopeful advances so far. By removing much of the mystery—and to some degree the stigma—associated with mental illness, we are paving the way for real understanding of the genetic, developmental, and environmental factors involved in the most serious psychiatric disorders.

Most serious mental illnesses run in families. Given the extraordinary difficulty of understanding the abnormalities of brain function that underlie mental illness, using genetics as a tool to approach their pathophysiology has seemed extremely attractive. Data collected over the past 25 years from studies of families with one or more affected members, and from the analysis of twins and adoptees, have established beyond doubt that schizophrenia, manic-depressive disorder, autism, and several other illnesses have a large genetic component. During the last two decades large numbers of genetic loci believed to be linked to susceptibility to these disorders have been identified and many possible candidate genes have been suggested. However, with the notable exception of AD—which is often thought of as a neurological disorder, although the cognitive impairment and dementia that characterize AD clearly fall within the bounds of psychiatry—no specific gene defects have yet been identified for any of the major psychiatric disorders. In large part this is attributable to the complexity of the disorders themselves and to the heterogeneity of the populations that have been available for study; but the fact that none of the disorders follows a simple mendelian pattern of inheritance and instead involves multiple interacting genes has made genetic analysis extremely difficult. We consider some of these difficulties later, but here we should emphasize that without the extensive DNA sequence information that has recently become available, it would be virtually impossible to progress from a locus identified by traditional linkage methods to the discovery of the relevant gene among the scores that are likely to be encompassed within the locus. In a few instances in which the locus included a gene that appeared from other evidence to be a likely candidate (such as the gene for one or another neurotransmitter receptor), close inspection has generally failed to yield a positive identification. The completion of the "first draft" of the human genome sequence and the availability of high throughput DNA sequencing should improve this situation dramatically, and we are confident that the search for the genes associated with most mental illnesses will now move forward at a greatly accelerated pace.

The announcement that the human genome sequence would shortly be completed was widely publicized at the time. By contrast the events that led up to this important development in biomedical research are not well known. Since we believe it is likely to have a major impact on psychiatry—initially by facilitating the diagnosis of mental illnesses and later by providing new molecular targets for therapy—we have provided a brief account of the history of the human genome project (HGP) in Appendix I.

THE GENETICS OF MENTAL ILLNESS PRIOR TO THE COMPLETION OF THE HUMAN GENOME PROJECT

The introduction to the fourth edition of the *Diagnostic and Statistical Manual of Mental Disorders* (DSM-IV), published by the American Psychiatric Association, points out the problematic nature of the term mental disorders. As the DSM-IV notes, this term implies that there is a distinction between mental and physical disorders. This archaic notion is wrong: Mental illnesses are as much disorders of the brain as those that are commonly referred to as neurological. As such, our ultimate understanding of them is to be sought in the rigorous analysis of the development and complex functioning of those systems of the brain that underlie cognition, emotion, and behavioral control. In this regard, the identification of genes involved in the development, maintenance, and plasticity of the brain will provide essential tools for the elucidation of both normal function and for what goes awry in mental disorders.

The DSM-IV lists operationalized diagnostic criteria for a very large number of mental disorders. These sets of criteria, based on symptoms and signs, have successfully elevated psychiatric diagnosis above its status of only three decades ago, which was roughly equivalent to the biblical story of human language after the fall of the tower of Babel. Despite the DSM's great success in improving the reliability of diagnosis (i.e., the ability of two clinical observers to agree), the validity of the criteria remains in question.

One of the seminal studies in the development of current psychiatric diagnoses was that of Robins & Guze (1970), who focused mainly on schizophrenia. They argued that diagnosis should be based on clinical description (which would identify correlated symptom clusters), laboratory studies, delineation of one disorder from another, family studies, and long-term follow-up studies to demonstrate the stability of diagnosis. From the perspective of three decades later, it is clear that this very reasonable model does not actually converge on clear diagnostic entities. For example, in the case of schizophrenia, if one focuses on diagnostic stability over the life of the patient, one would include (as the DSM-IV does) a requirement for chronic symptoms prior to making a definitive diagnosis. Unfortunately, when one observes the segregation of schizophrenia-like symptoms in families, this feature does not always breed true. Similarly the delineation of diagnoses within the DSM-IV has led to a situation in which many patients qualify for multiple diagnoses. Many similar observations confirm that the Robins & Guze model has important shortcomings. With the recognition, however, that mental disorders arise from complex gene-gene, gene-environment interactions, and with the hypothesis that no single gene may be both necessary and sufficient for a mental disorder phenotype, it is not surprising that like most genetically complex medical disorders, mental illnesses exhibit a great deal of clinically significant heterogeneity even within families.

Ultimately, the correlation of genotypes and phenotypes in psychiatry will be an iterative process. For now, what is critical is to avoid reification of DSM-IV diagnoses that almost certainly do not map onto actual disease entities. Having said that, it should also be acknowledged that current diagnostic schemes are not hopelessly wrong, as they yield high recurrence risk ratios in family studies of many mental disorders. The upshot is that at present, investigators have as much or greater difficulty in selecting phenotypes for genetic analysis as they have in making sense of the genotypes.

An additional difficulty—which is unique to the study of mental illness—is our limited understanding of the neural systems that determine what are loosely referred to as higher brain functions. Since it is the operations of these systems that are specifically disturbed in mental illnesses, we cannot look for real progress until we know a good deal more about the way these neural systems function and how disruption of their component parts can lead to the breakdown of the systems as a whole. Fortunately, considerable progress is now being made in this direction by the application of the new methods and tools of molecular, cognitive, and computational neuroscience. While the emergence of neuroscience as an intellectual discipline is widely acknowledged as one of the major scientific developments in the past half-century, progress in understanding such complex neural phenomena as language, perception, emotion, and executive function has been steady but relatively slow, and the even higher-level phenomena of consciousness and thought have so far eluded our grasp. It is thus understandable that we continue to have such difficulty in coming to grips with the disturbances in these activities that lie at the core of mental illness.

In recent years the most significant advances in neuroscience have come from the application of molecular genetic methods to the nervous system. Already this has led to the elucidation of the fundamental mechanisms involved in the propagation of nerve impulses, in synaptic transmission, and in early neural development. At a different level, the application identified the underlying genetic mutations responsible for a number of neurological disorders like Huntington's disease, the familial forms of Parkinson's disease and amyotrophic lateral sclerosis, fragile X syndrome, the spinocerebellar ataxias, and a new class of disorders, the channelopathies. The discovery of the genes associated with these disorders occurred fairly quickly (although it took almost 10 years from the identification of the locus for Huntington's disease to the discovery of the gene itself). Since no similar progress has been made in the elucidation of the genes involved in mental illness, it is important to point out that there are several reasons why progress in understanding the neurological disorders occurred so rapidly. First, the pattern of segregation in these neurological disorders followed classical mendelian patterns of inheritance and were either autosomal dominant, autosomal recessive, or X linked. In each pedigree studied, the disorder involved only a single gene. Second, a number of the disorders shared a common feature, namely the presence of an expanded trinucleotide sequence within or adjacent to the coding region; after the initial identification of such repeats in fragile X, a genome-wide search for others quickly led to the discovery of similar repeats in several other disorders including Huntington's disease. It also led to the elucidation of the clinically recognized phenomenon of anticipation, which is the tendency for the disorder to occur earlier and to be more serious in succeeding generations. Third, in the case of channelopathies, the genomic and expressed (cDNA) sequences of many of the ion channels involved in nerve signaling and in muscle contraction were already known. With the normal sequences in hand, and with a great deal of information available about the physiological functions of the ion channels, it was a relatively small step to the identification of the mutations responsible for the various disorders.

The situation is considerably more difficult in the case of the major psychiatric disorders. None of them follows a simple mendelian pattern of inheritance. Instead, the available evidence suggests that they are likely to be polygenic and to involve mutations or polymorphisms in several genes. In addition, they are likely to be multifactorial, involving both environmental and genetic factors. This by itself substantially increases the difficulty of elucidating the etiology of these illnesses (as it does for other multigenic disorders such as hypertension and diabetes). As Hyman wrote a couple of years ago: No genetically complex human disorder has yet been satisfactorily solved (1999). What was true in 1999 is, regrettably, still true in 2001. However, there have been a number of significant developments in genetics during the past few years that hold great promise for the identification of the genes associated with mental illness, and beyond this, for the development of more effective treatments. Among these developments, we believe the sequencing of the human genome and the availability of techniques for whole genome association studies are likely to be the most important.

In the past six or seven years the number of reports of the association of various mental illnesses with certain candidate genes or specific chromosomal loci has changed from a slow trickle to almost a flood. Hardly a month goes by without the appearance of one or more papers claiming to have identified a new locus for a specific psychiatric disorder or reporting a failure to corroborate the findings of an earlier study. It is not our purpose to summarize this extensive literature; rather, we focus attention on the three disorders that have been most intensively studied: schizophrenia, BPD, and autism. These examples serve to illustrate just how difficult it is to identify the genetic basis of a disorder when the illness itself is poorly defined and possibly heterogeneous in origin and when multiple genes are involved. While focusing on the possible genetic factors involved in these three illnesses, we are not unaware that stochastic developmental and environmental influences are undoubtedly also involved in their pathogenesis. We accept as a truism that all disorders are the product of the interaction of the patient's genetic background and the multitude of environmental experiences s/he has had from the moment of conception and throughout life. Since our focus is almost exclusively genetic, we are not considering other approaches to the study of these disorders while recognizing that they have contributed significantly to our understanding of the disorders we discuss.

SCHIZOPHRENIA

Schizophrenia is one of the most devastating brain disorders, affecting about 1% of the world's population and seriously disabling the majority of its victims. It is characterized by a number of positive symptoms and signs such as delusions, hallucinations, disorganized speech, and aberrant behavior, together with a variety of negative features including loss of affect, diminished motivation, and cognitive deficits. A number of subtypes of the disorder are recognized in the DSM-IV classification (paranoid, catatonic, and disorganized types) but these do not breed true and are of questionable status. There are also a number of associated illnesses such as schizoaffective, schizophreniform, and delusional disorders. In selecting patients for genetic analysis, these have not always been adequately distinguished. Relatives of individuals with schizophrenia have an increased risk of a broader category of illness currently classified as schizotypal personality disorder, in which florid positive symptoms are lacking. Schizophrenia commonly manifests itself during adolescence or in early adult life and occurs at about the same frequency in women and men. Until the introduction of chlorpromazine in the 1950s and later a wide range of other antipsychotic drugs, schizophrenia, and associated disorders had generally proved refractory to treatment. Indeed, until the late 1970s as many as one third of all mental hospital beds were occupied by patients suffering from schizophrenia or a related disorder.

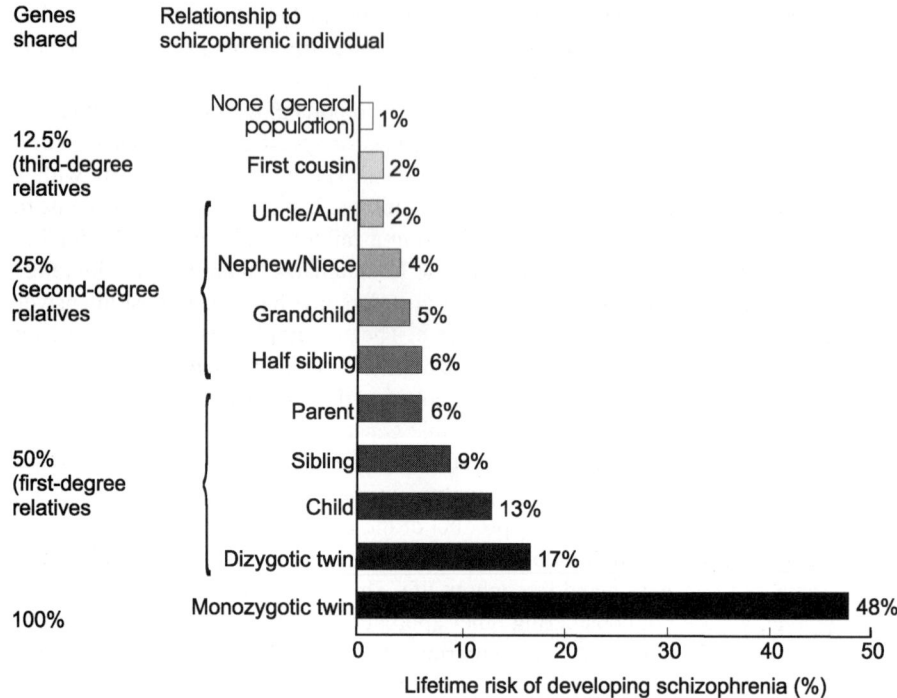

Figure 1 The lifetime risk of developing schizophrenia is correlated with genetic relatedness to an individual with schizophrenia. (From Gottesman 1991.)

Several different lines of evidence (in addition to its worldwide frequency) have indicated that schizophrenia is a genetic disorder. Its worldwide frequency and the striking similarity of symptoms and course of the disorder across cultures had long ago established that schizophrenia is not culture bound. Kallman's twin studies were the first to establish that the concordance rate for schizophrenia among monozygotic (MZ) twins was on the order of 50% but only 5%–15% among dizygotic twins and other siblings (Kallman 1938); for a review of more recent twin studies, see Cardno & Gottesman (2000). Gottesman (1991) later showed that the risk of family members developing the disorder increases significantly with the degree of their relationship to an affected individual (see Figure 1). Adoption studies (in which children, one of whose biological parents suffered from schizophrenia or a related disorder, were adopted shortly after birth and brought up by different adoptive parents) strongly suggest that these risks are due to genetic rather than environmental factors (Kety et al. 1976, Kendler & Gruenberg 1984). More recent evidence based on analyses of chromosomal aberrations, candidate gene approaches, and linkage studies have been thoughtfully summarized by Baron (2001). In what follows we have drawn heavily on Baron's review.

Chromosomal Aberrations

There is a fairly extensive body of work on the relation of chromosomal aberrations to schizophrenia, ranging from reports of partial trisomies to findings of various chromosomal translocations, inversions, deletions, certain fragile sites, and sex aneuploidies. (See DeLisi et al. 1994b for a review.) In retrospect, it seems reasonable to conclude with Baron (2001) that at present the evidence for the direct implication of most of these conditions in schizophrenia is equivocal, given the high incidence of the disorder in all populations studied and the small numbers of subjects with chromosomal changes that have been studied.

One notable exception to this general conclusion is the recent report from Porteous's group in Edinburgh of a balanced (1; 11) (q42.1; q14.3) translocation that has been found to segregate with schizophrenia and related illnesses in a large Scottish family with a maximum lod score of 6.0 (Millar et al. 2000). While there is a dearth of genes on chromosome 11 at the site of the translocation, there are several genes in this region of chromosome 1. Two, in particular, are of interest: they have been provisionally named DISC 1 and 2 (DISC for "disrupted in schizophrenia"). DISC 1 encodes a large novel protein with no apparent homology in the available databases; however, it is predicted to have a helical C-terminal domain that may form coiled-coil interactions. The closest similarities to other proteins include certain proteins involved in axonal transport and guidance, which may indicate that it has a role in neural development. DISC 2 specifies a noncoding RNA molecule that is antisense to DISC 1 and (by analogy with other such RNAs) may be critically involved in the regulation of DISC 1. The disruption of one or both of these two genes in this large family with many affected members strongly suggests that they are implicated (together no doubt with other genes) in the pathogenesis of schizophrenia and possibly other psychiatric disorders. It is interesting to note also that although linkages to other regions of chromosome 1 have been found in patients suffering from schizophrenia (at 1q32-41) and BPD (at 1q32—see below), to date there have been no linkages reported for the region of the translocation (1q14.3).

Candidate Genes

Attempts to short-circuit the effort required for linkage studies have generally involved looking for mutations or polymorphisms in a number of putative candidate genes. Most of these efforts have been directed toward the genes responsible for the biosynthesis, release, and re-uptake of specific neurotransmitters or their cognate receptors. Particular attention has been focused on dopaminergic and serotonergic transmission, following the discovery that many of the most effective psychotropic drugs act upon one or another aspect of transmission in neural systems that utilize these transmitters. While on the surface this approach seems rather straightforward, it may be fatally flawed by the lack of evidence tying these neurotransmitter systems to pathophysiology. It is further complicated by the existence of several pharmacologically distinct receptors for both of these transmitters (currently, 5 are known for dopamine and no fewer than 13 for serotonin) and by the fact that some antipsychotic drugs appear to act on more than one type of receptor. Moreover, there is no reason why a putative signaling abnormality in these pathways would necessarily involve the receptors or transporters just because these molecules are drug targets. An added complication is that the transporters involved in the reuptake of these transmitters have only recently been identified and there is still some uncertainty about their selectivity. Leaving these difficulties aside, there are several reports (some based on quite large numbers of subjects) of attempts to relate allelic variations in the dopamine receptors DRD2 and DRD3 (Spurlock et al. 1998) or the serotonin receptor 5H2-T2a (Williams et al. 1996), and for association with trinucleotide repeat size as a cause of the anticipation seen in some families with schizophrenia (Sasaki et al. 1996). Based on subject selection and power, however, none of these studies has yielded really convincing evidence for such involvement.

Linkage Studies

Because of the limitations of these other approaches and pending the development of tools for whole genome association studies, linkage studies have been the method of choice for analyzing the genetics of schizophrenia. Stated simply, linkage analysis examines the tendency of genes at specific loci to be inherited together with known markers because of their proximity on a chromosome. They are therefore potentially powerful in detecting the cosegregation of a known genetic marker and a disease phenotype (see Appendix II). To date, such analyses have identified more than a dozen chromosomal regions potentially linked with schizophrenia. Unfortunately, most of the reported linkages have failed to be corroborated in other family pedigrees, so at present it is difficult to know whether they are real or instead due to small gene effects that are more relevant in certain pedigrees (in particular those in which they were first identified). In light of the current view that schizophrenia is multifactorial in nature, it is conceivable that many of the familial linkages reported have correctly identified a different susceptibility locus and that collectively certain genes at the various loci contribute in some way to the schizophrenic phenotype.

Table 2 summarizes the results of a number of the linkage studies that have been carried out over the past seven years, together with the published lod scores. For the sake of brevity we discuss only three of these studies; these serve to illustrate this approach and the types of results it can yield. For a complete account of the linkage studies bearing on schizophrenia carried out prior to 2000, reference should be made to Riley & McGuffin's (2000) excellent review.

The first study we consider is by Brzustowicz et al. (1999). Their analysis was based on a genome-wide scan in 21 Canadian families that included 285 individuals. The families were of either Celtic or German descent and the members with schizophrenic illness appeared to be segregating in a unilineal, autosomal

TABLE 2 Suggestive genetic linkages for schizophrenia[a]

Chromosome	Marker	Pedigree	Model/disease definition/lod[b]	First author/year
1q21-q22	D1S1653-D1S1679	Canadian	Recessive/narrow/lod 6.5	Brzustowicz 2000
1q32-q41	D1S2891	Finnish	Dominant/narrow/lod 3.82	Hovatta 1999
4q31	D4S1586	Finnish	Dominant/narrow/lod 2.74	Hovatta 1999
5p14.1-13.1	D5S111	P. Rican	Dominant/broad/lod 4.37	Silverman 1996
5q21-q31	D5S804	Irish	Recessive/narrow/lod 3.35	Straub 1997
5q31	D5S399	Germ/Israeli	—/ASP/lod 1.8	Schwab 1997
6p24-p22	D6S296	Irish	Additive/intermediate/lod 3.51	Straub 1995
	D6S274	German &	—/ASP/lod 2.2	Schwab 1995
	D6S274	Israeli	—/ASP/lod 2.2	Schwab 2000
	D6S274	Euro/Canada/Taiwanese & U.S.	Other marginal linkages	Moises 1995
				Antonarakis 1995
6q21-22.3	D6S416	U.S.	—/ASP/MLS 3.06	Cao 1997
	D6S416	U.S.	—/ASP/MLS 3.11	Cao 1997
	D6S424–	U.S./Australia	—/ASP/lod 3.82	Martinez 1999
8p22-p21	D8S1771	U.S.	—/narrow/lod 3.64	Blouin 1998
	D8S1771		Dominant/broad/lod 6.17	Pulver 2000
	D8S1715-	Irish	Dominant/broad/lod 2.34	Kendler 1996
	D8S136	Canadian	Dominant/narrow/lod 3.49	Brzustowicz 1999
9q21-22	D9S922	Finnish	Dominant/narrow/lod 1.95	Hovatta 1999
	D9S257	U.S./Australia	Other marginal linkages	Moises 1995, Levinson 1998
10p11-15	D10S1423	U.S.	—/ASP/lod 3.4	Faraone 1998
	D10S582	U.S.	—/ASP/lod 3.2	Faraone 1998
	D10S1714	Germ/Israeli	—/ASP/lod 3.2	Schwab 1998b
	D10S2443	Irish	Recessive/intermediate/Hlod 1.95	Straub 1998
	D10S674	Irish	Recessive/intermediate/Hlod 3.2	Straub 1998
13q14-q32	D13S174	U.S.	—/narrow/lod 4.18	Blouin 1998
	D13S793	Canadian	Recessive/broad/lod 3.92	Brzustowicz 1999
		UK, Japan, Taiwanese	Other marginal linkages	Lin 1997, Shaw 1998, Antonarakis 1996, Kalsi 1996
15q13-q15	D15S1012	German	Dominant/catatonia/lod 3.57	Stober 2000
		U.S.	Other marginal linkages	Coon 1994b, Leonard 1998

(Continued)

TABLE 2 (*Continued*)

Chromosome	Marker	Pedigree	Model/disease definition/lod[b]	First author/year
17q21-22	D17S934	U.S.	Dominant/ schizophrenia-like psychosis/MLS lod 5.0	Bird 1997
22q12-q13.1	IL2RB	U.S.	Dominant/intermediate/ lod 2.82	Pulver 1994
22q11-13		U.S.	—/ASP/P = 0.009 Other marginal linkages	Lasseter 1995 Polymeropoulos 1994, Coon 1994a,b, Blouin 1998, Stober 2000
Xp11.4-p11.3	MAOB DXS7	Finnish	Recessive/narrow/lod 2.01 Other marginal linkages	Hovatta 1999 DeLisi 1994, Dann 1997

[a]Adapted, in part, from Baron (2001), Riley & McGuffin (2000), and Thaker & Carpenter (2001).
[b]lod = logarithm of the odds; ASP = affected sib pair; MLS = maximum lod score.

dominant manner. Using both recessive and dominant models of inheritance with broad and narrow definitions of schizophrenia, the researchers observed the most significant chromosomal linkage to markers on chromosome 13 at 13q32. The highest lod score at this locus, obtained using a recessive/broad definition model, was 3.92. Positive lod scores were also obtained, using all models, for a locus on chromosome 8p at 8p21. At the time these were among the strongest linkages identified on these two chromosomes and were within less than 5 cM (or roughly 5 Mbp) from the best previous linkages, reported by Blouin et al. in 1998, using a narrow definition of the disorder.

The following year, Brzustowicz and her colleagues identified another major susceptibility locus for familial schizophrenia, this time on chromosome 1 at 1q21-22 in the same group of Canadian families (but now expanded to 22 included families) (Brzustowicz et al. 2000). On average, 3.6 individuals with schizophrenia or schizoaffective disorder were studied in each family; the affected individuals spanned three generations in more than a quarter of the families, and family histories suggestive of schizophrenia spanning three to four generations were recorded in almost half the families.

Parametric linkage analyses, using four models (dominant or recessive, broad or narrow definition) were conducted and yielded a lod score of 5.79 with a marker that mapped to chromosome 1q22 under the narrow definition and a recessive mode of inheritance. Multipoint analysis with other markers in the region gave a maximum lod score of 6.5, with an estimated 75% of families linked to this locus. This is the highest recorded lod score among the studies we have reviewed, and, as the authors conclude, "The magnitude of [this] linkage, coupled with [its] clear localization .... should facilitate efforts to positional clone this susceptibility gene" (Evans et al. 2001). This effort is currently under way.

Several other groups of investigators have found linkages for schizophrenia on chromosome 13, specifically at 13q14.1-32. The abundance of these linkages is highly suggestive, especially since so many different pedigrees were studied, including families from the United States, Japan, Canada, Taiwan, Iceland, England, and Ireland. The reported lod scores for these linkages range from 1.4 to 4.5, depending upon the model used and the narrowness of the disease definition (see Riley & McGuffin 2000 and Figure 2). This linkage is of particular interest because the serotonin 5HT2A receptor gene is located in this chromosomal region (Williams et al. 1997). However, at least two other groups have failed to find any linkage associated with chromosome 13 (Jensen et al. 1998) or with the region 13q14.1-32 (Barden & Morissette 1999). Moreover, when the data from all the studies that have shown linkages to this region are pooled, the lod scores do not

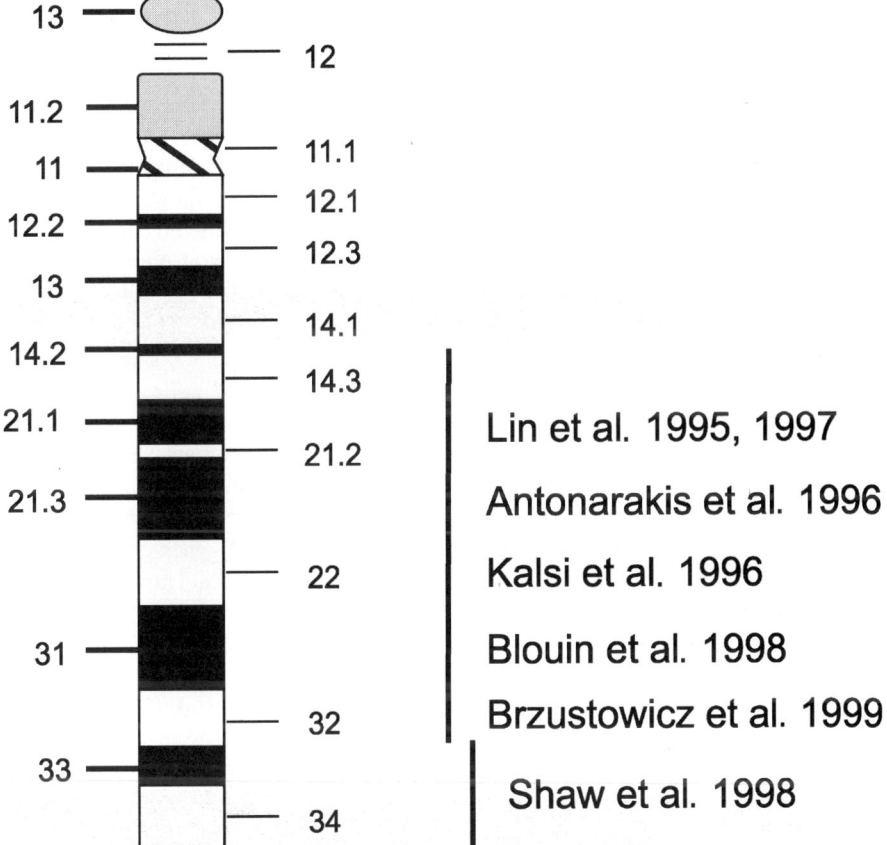

Figure 2   Location of positive linkages to schizophrenia on chromosome 13. (From Riley & McGuffin 2000.)

improve. This may be contrasted with the findings on the relationship of APOE4 to late onset AD, where a similar meta-analysis added further support to the view that this allele is an important risk factor for the disease.

A quite different set of linkages was reported by Hovatta et al. (1999), based on an extensive three-stage screen of schizophrenia in an isolated Finnish population of 18,000, whose roots extend back to the end of the seventeenth century. The lifetime risk of schizophrenia among this population is estimated to be 3.2%. No fewer than 365 families with at least one affected member and 69 families with at least two affected siblings were examined. Using both dominant and recessive models, the highest pairwise lod score observed in the scan was 3.82 at 1q32.2-q41. In the same region they observed a 6.6 cM haplotype segregating in four core families; since this was also present in healthy individuals, it appears that if there is a schizophrenia-associated gene at this locus, it must be nonpenetrant in these individuals. A second promising linkage was observed at 4q31 with a lod score of 2.74, using a narrow disease definition and a dominant model; sib pair analysis of this region also yielded a significant lod score of 2.09. Two other loci identified in this screen were at Xp11.4-p11.3 and 9q21, where the lod scores were 2.01 and 1.95 respectively. The locus on the X chromosome is particularly interesting because it encompasses the location of the gene for monoamine oxidase B that had earlier been found by DeLisi and colleagues in (1994a). However, since dideoxy fingerprinting of the *MAO B* gene in 100 male patients with schizophrenia showed no sequence variation (Sobell et al. 1997), this association must be considered equivocal.

As we indicated earlier, the availability of large numbers of human expressed sequence tags (ESTs) and of many complete genes has made it possible to examine gene expression on an unprecedented scale using microarrays. This approach has recently been applied to postmortem tissue from the prefrontal cortex of matched pairs of schizophrenic and control subjects by Levitt and colleagues (Mirnics et al. 2000). In an array of over 3700 ESTs, 4.8% were found to be differentially expressed in the prefrontal cortex from the schizophrenic subjects; 2.6% were expressed at higher than control levels and 2.2% at lower levels. The most striking changes were detected in a group of genes involved in presynaptic secretory machinery like NSF and SYN2. Within this group, 41 of 62 genes examined were altered in prefrontal tissue from the schizophrenic subjects. These findings are consistent with some of the linkage studies alluded to above (e.g., loci 22q11-13 and 17q21). It is also noteworthy that deletions in the 22q11-12 region (which includes several synapse-related genes) have been found to be associated with a 25-fold increase in the prevalence of schizophrenia (Murphy et al. 1999). Similarly, the locus at 17q21, which encompasses several presynaptic secretory-associated genes including those for NSF and ATPase, has been reported in one study to be linked to a non-Alzheimer's form of late onset schizophrenia-like psychosis followed by dementia, with a lod score of 5.0 (Bird et al. 1997).

Studies of MZ twins have shown a striking concordance (0.7%) for the age of onset of schizophrenia (Kendler et al. 1987, Cannon et al. 1998), which suggests that this phenotypic variable is also under genetic control. The search for possible linkages associated with the age of onset (or, rather, the age at the time of the first psychiatric contact) has so far been equivocal. Cardno et al. (2001) found six maximum likelihood lod scores of 1.5 or greater, but only one—at chromosome 17q—met the criteria for suggestive linkage in that it would be expected to occur only once or less per genome scan.

The disturbances in cognitive function, which are such a striking—and often dominant—feature of schizophrenia, have in the past eight years focused attention on the possibility that the disorder may be due to dysfunction of the frontal lobe and more specifically the dorsolateral prefrontal cortex (DLPC). Several experimental studies in monkeys have documented the role of this area in "working memory" (i.e., the ability to hold information transiently in consciousness so as to direct an appropriate response). At the same time, clinical studies have demonstrated a reduction in regional blood flow in the DLPC in schizophrenic subjects, and often an overall shrinkage of the frontal lobe associated with increased neuronal packing density in the DLPC (Goldman-Rakic 1994, Busatto et al. 1994, Harrison 1999, Callicott et al. 1999, Fuster 2000; and see Weinberger et al. 2001 for review).

In light of the large body of evidence implicating changes in dopaminergic activity in schizophrenia, Weinberger and colleagues have drawn attention to the possible involvement of a distinctively human single nucleotide polymorphism (SNP) in the gene that encodes the enzyme catechol-o-methyl transferase (COMT), which catalyzes the methylation of dopamine and other catecholamines. The polymorphism in question results in a valine to methionine change in the enzyme, with a consequent reduction of about 75% in its activity. Weinberger's studies, which were predicated on the notion that the *val* allele is likely to result in increased dopamine inactivation and hence compromised DLPC function, certainly suggest that COMT is involved in schizophrenia. However, as the authors point out: "By themselves [their findings] provide weak statistical evidence for the COMT *val* allele increasing the risk for schizophrenia" (Egan et al. 2001). Moreover, in an earlier study of the possible role of COMT, Karayiorgou and her colleagues (1998) had found no mutations in the COMT gene, and on this basis had ruled out a major role for the enzyme in the pathogenesis of the disorder but could not exclude the possibility that the val-met variant plays a minor role. It is noteworthy, however, that COMT maps to the region of chromosome 22q11, which is deleted in patients with velo-cardio-facial syndrome, many of whom have a history of psychosis, including especially schizophrenia (Murphy et al. 1999).

BIPOLAR DISORDER

Bipolar disorder (BPD) (or manic depressive illness as it used to be called) is one of the two most serious of the mood disorders. It is marked by recurrent episodes of depression and mania that tend to increase with age, and in perhaps 20% of cases, symptoms become chronic. Of the two types of BPD identified in the DSM-IV, bipolar I disorder is the more commonly recognized form of the illness, characterized by periods of depression of varying severity and one or more manic episodes. In bipolar II disorder on the other hand, periods of depression are associated with one or more bouts of hypomania that may or may not cause impairment. For the present, we consider the two types together since in most genetic studies of BPD, they have not been distinguished.

That BPD runs in families has long been acknowledged. Twin and adoption studies have taken this a good deal further by indicating the high degree of heritability (on the order of 0.8) of the disorder (Tsuang & Faraone 1990). This may be compared with the observed heritability of the other serious mood disorder, major depression, which is on the order of 0.4 (Sullivan et al. 2000). However, major depression is far more widespread, and current projections suggest that 20 years from now it will be second only to ischemic heart disease among the leading causes of disease burden worldwide (Murray & Lopez 1996).

Beginning with a highly publicized study of BPD among the Old Order Amish population, which reported linkage to a locus on chromosome 11 (at 11p15; Egeland et al. 1987), there have been several attempts to identify specific loci associated with BPD. Among these we consider only a few that have attracted the most attention. The first of these was the study by Kelsoe et al. (1989), which re-examined and extended the findings on the Old Order Amish population. In the original study a linkage to two markers (HRAS1 and INS) seemed highly suggestive with a lod score of 4.1 (for HRAS1) and 2.6 (for INS). But coincident with the report by Egeland and coworkers, two other studies, published in the same issue of *Nature*, found no indication of linkage to chromosome 11. The finding that two previously asymptomatic members of the original pedigree had developed serious mood disorders led to a re-examination of the data. This resulted in a marked drop in the lod score (to below 2 in the case of the linkage to HRAS1), and when other branches of the original pedigree were examined, the findings were strongly negative. To the consternation of many in the field, the original linkage had to be abandoned (Table 3).

A similar sense of disappointment followed the finding that a previously reported linkage of BPD to two markers on the X-chromosome, which had been based on the study of several large non-Ashkenazic families in Israel (Baron et al. 1987), could not be substantiated when other similar pedigrees were examined (Baron et al. 1993).

In 1997 the NIMH Genetics Initiative Bipolar Group reported the findings of a large-scale study of 97 families that included data from 232 individuals with BPD, 72 with BPD II, 88 with unipolar recurrent depression, and 32 with schizoaffective/bipolar illness. This study was unable to confirm any of the previously reported linkages and, at the same time, had to conclude that none of the linkages that appeared to be suggestive in their own analysis met the strict criteria that Lander & Kruglyak (1995) had advanced as a threshold for significance. Nor could they effectively exclude other areas, given the most plausible assumption regarding complex inheritance (Nurnberger, Jr. et al. 1997). In particular, they found only

TABLE 3 Suggestive genetic linkages for bipolar disorder[a]

Chromosome	Marker	Pedigree	Model/disease definition/lod	First author/year
1q25-32	GATA124	U.S.	—/BP & SA/lod 2.67	Detera-Wadleigh 1999
	F08		Dominant/BP & SA/lod 2.37	Detera-Wadleigh 1999
3q	D3S2403 &	U.S.	—/BP & SA & RUP/P = 0.029	Edenberg 1997
	D3S3038		—/BP & SA & RUP/P = 0.020	Edenberg 1997
4p16	D4S394	Scottish	Dominant/BP & RUP/lod 4.1	Blackwood 1996
	D4S394	Danish	Recessive/—/lod 2.0	Ewald 1998b
11p15	HRAS &	Amish	Dominant/BP & SA/lod 4.08	*Egeland 1987
	Insulin		Dominant/BP & SA/lod 2.36	**Kelsoe 1989
12q23-24	D12S1639	Danish	Dominant/—/lod 3.37	Ewald 1998a
	D12S78	Quebec	Recessive/BP & SA & RUP/lod 2.87	Morissette 1999
13q32	D13S1271-	U.S.	—/BP & SA/lod 3.4	Detera-Wadleigh 1999
	S779		Recessive/BP & SA/lod 2.06	
18p11.2	D18S21	Caucasian & European	—/BP & SA & RUP/P = 0.0004	Berrettini 1994
	D18S37	U.S.	—/BP & RUP/P = 0.0006	Stine 1995
	D18S37		—/broad/58% allele sharing	Lin & Bale 1997
	D18S53	Germ/Israeli	—/BP & Schz & SA & RUP/lod 3.1	Schwab 1998b
	D18S453	German	Recessive/pat. BPI/lod 1.91	Nothen 1999
	S1150-S71	U.S.	—/BP & SA/lod 2.32	Detera-Wadleigh 1999
18q21-22	D18S41	U.S.	—/pat BP & RUP/lod 3.51	Stine 1995
18q22-23	D18S70	Costa Rican	Dominant/BPI/lod 4.06	Freimer 1996
	D18S55...	Costa Rican	Dominant/BPI & SA/lod 1.3-2.59	McInnis 1996
	D18S55	U.S.	—/BP & SA/lod 2.2	McMahon 1997
18q22-23	D18S554	German	Recessive/BPI/lod 1.26	Nothen 1999
21q22.3	PFKL	U.S. & Israeli	Dominant/BP & RUP/lod 3.41	Straub 1994
	S267-S212	U.S.	Recessive/BP & MD/lod 1.79	Detera-Wadleigh 1996
	S1254-55	U.S.	—/BP & SA & RUP/P ≤ 0.05	Detera-Wadleigh 1997
	D21S1260	U.S. & Israeli	Dominant/BP & SA & RUP/lod 3.35	Aita 1999
22q11-13	D22S303	N. American	Dominant/BP & SA & RUP/lod 1.68	Lachman 1997
	D22S533	U.S.	—/BP & SA & RUP/lod 2.46	Edenberg 1997
22q12	D22S278	N.American	Dominant/BP/lod 3.84	Kelsoe 2001
Xq28	CB		—/MDI psychosis/lod 3.3	Winokur 1969
	G6PD & CB	Brussels	—/MDI psychosis/lod 3.3	Mendlewicz 1979, 1980, 1987
	G6PD/F9	Persian	—/MDI psychosis/lod 3.3	Baron 1977

(Continued)

TABLE 3 (*Continued*)

Chromosome	Marker	Pedigree	Model/disease definition/lod	First author/Year
	G6PD & CB	Sardenian	Dominant/BP & SA & RUP/ lod 1.47	Del Zompo 1984
	G6PD & CB	Israeli	Dominant/BP & SA & MD/ lod 7.0-9.17	*Baron 1987
	G6PD & CB	Israeli	Dominant/BP & SA & MD/ lod < 0.0	**Baron 1993
	G6PD & CB	Eur & Jewish	Dominant/BP & SA & MD/ lod < 0.0	#Berrettini 1990
Xq27-28	F9-CB	U.S.	Dominant/BP & SA & MD/ lod < 0.0	#Gejman 1990
Xq26-28	DXS1047	U.S.	—/BP & SA/lod 1.34	Stine 1997
Xq24-27.1	DXS994	Finland	SML/BP & SA & NOS/ lod 3.54	Pekkarinen 1995
Xp22	DXS989	U.S.	—/BP & SA/lod 0.94	Stine 1997

[a]Symbols and abbreviations: ** = not replicated in the same/original * pedigree; # = conflicting results compared with other studies; BPI = bipolar disorder I; BPII = bipolar disorder II; BP = BPI + BPII; RUP = recurrent unipolar depression; SA = schizoaffective disorder; Schz = schizophrenia; pat = paternal; MD = major depression; CB = color blindness; NOS = not otherwise specified; SML = single major locus; lod = logarithm of the odds.

weak evidence for linkage to chromosome 18p (for which others had put forth fairly strong evidence). With these caveats, the NIMH group concluded that their study provided evidence, not strong but suggestive, for linkages to several regions on a number of chromosomes including 21q21, Xq26, 16p, 7q, 6, 10, and 22q, some of which had been reported earlier by others (Straub et al. 1993, Pekkarinen et al. 1995, Ewald et al. 1995, Detera-Wadleigh et al. 1994).

In 1994 Berretini and colleagues, using affected sibling pair and affected pedigree member methods, had provided the first evidence for a locus on chromosome 18 that showed significant linkage to BPD in 22 Caucasian kindreds of European ancestry (Berrettini et al. 1994). This linkage was confirmed the following year by Stine et al. (1995), who also found evidence that the linkage was strongest in families with paternally transmitted illness. The combined lod scores for all the families involved were not significant, but in some kindreds, they reached as high as 2.38. The pericentromeric region of the chromosome implicated spans about 50 cM and probably contains as many as 50 genes. Berrettini and colleagues paid particular attention to two genes they thought might be involved in susceptibility to BPD: a corticotropin receptor gene and the gene for the α subunit of a GTP-binding protein involved in transducing neurotransmitter receptor action (which was thought also to be inhibited by lithium). Unfortunately, as we have seen, this particular linkage had not been corroborated in the NIMH Collaborative Study. However, some other studies that have examined all the available data from a total of 382 affected siblings, using two nonparametric approaches, have been more positive (see Lin & Bale 1997).

In 1986 a large group of collaborators from the University of California, San Francisco, three universities in the Netherlands, and the University of Costa Rica reported the results of a complete genome scan for genes that predispose to severe BPD in two families from the Central Valley of Costa Rica, where much of the population derives from a few founders in the sixteenth to eighteenth centuries (McInnes et al. 1996). Looking for linkage to some 473 microsatellite markers, they covered about 94% of the genome. Linkages that had significant lod scores were found at chromosomes 18q, 18p, and 11p, and suggestive linkages were observed in one or both families on chromosomes 1, 2, 3, 4, 5, 7, 13, 15, 16, and 17. The extension of the linkage on chromosome 18 found in this study is surprising when compared to earlier analyses and clearly requires confirmation. As yet, no such confirmation has been forthcoming. Indeed, when this (and other issues) was re-examined in 1998 by Baron and colleagues in a study that involved more than 1000 genotyped individuals in 53 unilineal multiplex pedigrees using 10 highly polymorphic markers, neither parametric nor nonparametric analysis yielded evidence for significant linkage between this region of chromosome 18 and BPD (Knowles et al. 1998).

This negative conclusion was confirmed two years later when Botstein and a large group of collaborators from Stanford, MIT, Johns Hopkins, and Cold Spring Harbor carried out a full-genome scan for linkage in 50 families segregating the BPD phenotype (Friddle et al. 2000). In addition to conventional linkage analysis, this group applied a "simultaneous-search algorithm" to the data in an attempt to circumvent the problem of genetic heterogeneity within the families studied. The results of the study were disappointing but clear-cut: No single locus or pair of loci was found that could account for the disorder in a substantial fraction of the families. As the authors discuss, their failure to identify significant loci could be due to the extreme heterogeneity in BPD. If a single dominant gene causes the disease (as some studies suggest, e.g., Blackwood et al. 1996), there must be many other such genes segregating in the population. Alternatively, there may be a number of different, relatively common disease alleles that, when combined in a single individual, predispose to BPD. Their final conclusion probably speaks for all such linkage studies for BPD: "We conclude that any underlying genetic etiology of BP disorders is too complex to be resolved in genome scans with single-major-locus-assumptions and in linkage analyses involving fewer than many hundreds of multiplex families."

Considering the long litany of uncorroborated studies of linkage as applied to BPD, we feel much as Risch & Botstein felt in 1996 when they wrote:

"In no field has the difficulty [of mapping inherited disease genes on to the human genome by linkage analysis] been more frustrating than in psychiatric genetics. Manic depression (bipolar illness) provides a typical case in point. Indeed, one might argue that the recent history of genetic linkage studies for this disease is rivaled only by the course of the illness itself. The euphoria of linkage findings being replaced by the dysphoria of non-replication has become a regular pattern, creating a roller coaster-type existence for many psychiatric genetics practitioners as well as their interested observers."

Not surprisingly numerous attempts have been made to identify candidate genes associated with BPD. Among the genes that have been examined are those for tyrosine hydroxylase, the dopamine receptors (DRD1, DRD2, DRD3, and DRD4),

FMR1, monoamine oxidase A&B, the ciliary neurotrophic factor (CNTF), the 5HT1A receptor, and the 5HT transporter (see Sanders et al. 1999). Since nearly all the studies that have suggested such causal associations have failed to be corroborated by later studies, at present we should probably suspend judgment on their validity.

Before leaving this review of BPD we should comment briefly on the genetic evidence for unipolar depression, or major depressive disorder (MDD) as it is now known. MDD and BPD are clearly related in some way; just as clearly, however, they are not the same condition. There is considerable evidence that the rates of MDD are significantly elevated among the relatives of individuals suffering from BPD. On the other hand, MZ twins show considerably greater concordance for one or the other disorder. This has led most investigators to conclude that the two disorders share some underlying genetic factor(s) but differ in other equally important (or even more important) factors (see Tsuang & Faraone 1990, Fava & Kendler 2000).

After reviewing most of the published data bearing on the genetics of MDD, Sullivan and colleagues identified five family studies that met their stringent inclusion criteria. The odds ratio of 2.84 for subjects with major depression versus first-degree relatives in these studies was found to be homogeneous across all five families. No adoption study met their fairly stringent criteria, which suggests that familial aggregation is probably due to additive genetic effects. Surprisingly, this metaanalysis revealed no substantial difference in heritability between males and females, although it is well-known from clinical studies that females are roughly twice as likely as males to suffer from MDD (Blazer et al. 1994, Fava & Kendler 2000).

AUTISM AND RETT SYNDROME

Autism is the best known of the broad class of pervasive developmental disorders recognized in DSM-IV that includes Rett syndrome, childhood disintegrative disorders, Asperger's syndrome, and a miscellaneous group of "developmental disorders not otherwise specified."

The most striking clinical features of children affected with autism are their restlessness and distraction, their lack of social interaction, their difficulty with language, and their repetitive and stereotyped motor behaviors. The disorder usually manifests itself before the age of three and commonly persists into adult life. Its prevalence in developed countries is on the order of 1 in 2500, with boys generally outnumbering girls by about 3:1. In about 25% of children with autism there is an associated neurological disorder like tuberous sclerosis or fragile X mental retardation. Most of the clinical signs of autism are recognizable early in life, but there are reports of children who appear to have developed fairly normally for up to five or six years of age before developing the characteristic movement disorder and regressing socially and linguistically. It is noteworthy that three quarters of the affected children suffer some degree of mental retardation (with IQs in the range of 35–50) and almost a third have epilepsy.

That there is a strong genetic component to the disorder was clearly borne out by twin and family studies. Monozygotic twins have a concordance rate of greater than 60%; this is in striking contrast to the low frequency of the disorder among same sex dizygotic twins (Folstein & Rutter 1977, Bailey et al. 1995). Family studies have shown also that there is a fairly high incidence of social and communicative impairment among the relatives of children with autism compared to those of other (control) children. Indeed, according to Bolton et al. (1994), the risk to siblings of individuals with idiopathic autism is said to be 75–150 times as great as in the population as a whole. As Lamb et al. (2000) point out in their excellent review of genetic studies of autism, the available evidence suggests that the disorder probably involves changes at several genetic loci that act epistatically. Estimates of the numbers of genes involved range from 3 (Pickles et al. 1995) to more than 15 (Risch et al. 1999). In addition, the association with fragile X and tuberous sclerosis points to genetic heterogeneity within the disorder.

As in the case of schizophrenia, autism has been associated with a variety of chromosomal abnormalities. According to Lamb et al. (2000), abnormalities in the form of interstitial or terminal deletions, balanced and unbalanced translocations, and inversions have been reported for every chromosome with the exception of chromosomes 14 and 20. Some of these chromosomal abnormalities involve well-known fragile sites, and nearly all arise de novo in the affected children. Changes in chromosome 15 have received particular attention; they often take the form of duplications, especially a maternally derived interstitial duplication of the region 15q11-q13 (Repetto et al. 1998). This is the same region in which deletions have been found in Prader-Willi and Angelman syndromes, and it is probably not without significance that patients with Angelman syndrome display many of the same behavioral disorders seen in autism. As recently reported, Angelman syndrome most frequently results from maternal microdeletions of the 15q11-q13 region, which contains the UBE 3A gene that codes for the E6-AP ubiquitin-protein ligase. However, no evidence has been found for a functional mutation in UBE3A in autism (Veenstra-VanderWeele et al. 1999), but other steps in the protein degradation pathway may be implicated. Among other genes located in this region that have also been thought to be involved in autism are those for the $\alpha 5$, $\beta 3$, and $\gamma 3$ subunits of the GABA receptor. There is one report suggesting a possible association of the $\beta 3$ subunit with autism (Cook et al. 1998) but this has not yet been confirmed. There has similarly been no confirmation of the suggested involvement of the serotonin transporter gene or for two serotonin receptor types (5-HT1 and H-HT2A).

Table 4, which is adapted from the review by Lamb et al. (2000), summarizes the results of four genome-wide linkage studies aimed at identifying loci associated with susceptibility to autism. The strongest evidence for such linkages are those found in the Collaborative Linkage Study of Autism (1999), which focused on chromosome 13, and those reported by the International Molecular Genetic Study of Autism Consortium (1998) and the Collaborative Linkage Study of Autism, which reported a linkage to chromosome 7. By contrast, the lack of significant linkage to chromosome 15 is surprising in light of the evidence from the chromosomal

TABLE 4 Suggestive genetic linkages for autism[a]

Chromosome	Marker	# and pedigree origin	Model & lod/p	First author/year
1p	~D1S1675	90 multiplex U.S.	MMLS 2.15	Risch 1999
2q	D2S2188	83 multiplex	MMLS 3.74	IMGSAC 2001
6q	D6S283-D6S261	51 multiplex	MMLS 2.23	Philippe 1999
7q	D7S530-	99 multiplex	MMLS 2.53	IMGSAC 1998
	D7S684	56 families U.K.	MMLS 3.55	IMGSAC 1998
	D7S684	51 multiplex	MMLS 0.83	Philippe 1999
	D7S477	83 multiplex	MMLS 3.2	IMGSAC 2001
	D7S495	76 multiplex	MLS/het 1.47	Ashley-Koch 1999
	D7S2527	76 multiplex	MMLS 1.77	Ashley-Koch 1999
	D7S640	76 multiplex	NPL score 2.01	Ashley-Koch 1999
	D7S1804	90 multiplex U.S.	MMLS 0.93	Risch 1999
	D7S1813	75 families U.S.	MMLS 2.2	CLSA 1999
13q	D13S800	75 families U.S.	MMLS/het 3.0	CLSA 1999
	D13S217-D13S1229	75 families U.S.	MMLS/het 2.3	CLSA 1999
	D13S800	90 multiplex U.S.	MMLS 0.68	Risch 1999
15q11-13	GABRB3	138 families	P = 0.0014	Cook 1998
	D15S118	51 multiplex	MMLS 1.10	Philippe 1999
	D15S217	63 multiplex	NPL Z 1.37	Bass 1999
	D15S975	75 families U.S.	MMLS/het 0.51	CLSA 1999
16p	D16S407-	99 families	MMLS 1.51	IMGSAC 1998
	D16S3114	56 families U.K.	MMLS 1.97	IMGSAC 1998
17p11.2	D17S1876	90 multiplex U.S.	MMLS 1.21	Risch 1999
18q	D18S68	51 multiplex	MMLS 0.62	Philippe 1999
	D18S878	90 multiplex U.S.	MMLS 1.0	Risch 1999
19p	D19S226	51 multiplex	MMLS 1.37	Philippe 1999
	D19S221-D19S49	99 families	MMLS 0.99	IMGSAC 1998
Xq	DXS424	38 multiplex	Max lod 1.24	Hallmayer 1996
	DXS1047	110 multiplex	MMLS 2.67	Liu 2001

[a] MMLS = maximum multipoint lod score; lod = logarithm of the odds; het = heterogeneity; NPL = nonparametric linkage.

abnormalities considered above. Among the genome scanning studies we have reviewed, only that by Philippe et al. (1999) identified a potential susceptibility locus on chromosome 15 (near q11-q15, overlapping the region involved in the Prader-Willi and Angelman syndromes). The significance of this finding is unclear, however, given that the maximum lod score reported was only 1.10. Perhaps equally surprising, considering the higher incidence of autism in boys and the frequent association of the disorder with the fragile X syndrome, is that none of the studies has reported linkage to the X chromosome.

The linkage to chromosome 13 at marker D13S800 was the strongest observed in the 1999 collaborative study, with a maximum lod score of 3.0. The International Consortium had earlier reported a weak linkage to the same chromosome, but at a distance of some 20 cM from marker D13S800. Since peaks around a true locus can vary by as much as 15 cM in either direction (Hauser et al. 1996), it is possible that those two sets of observations represent the same underlying locus as noted by the authors of the Collaborative Linkage Study (1999). The rather wide scatter of the observed linkages to chromosome 7, illustrated in Figure 3, is also noteworthy. Together the linkages cover the region from about 7q20 to 31.3. Whether or not this is because several different genes on chromosome 7 are involved in autism is unclear at this time, but it does suggest that sequencing the entire

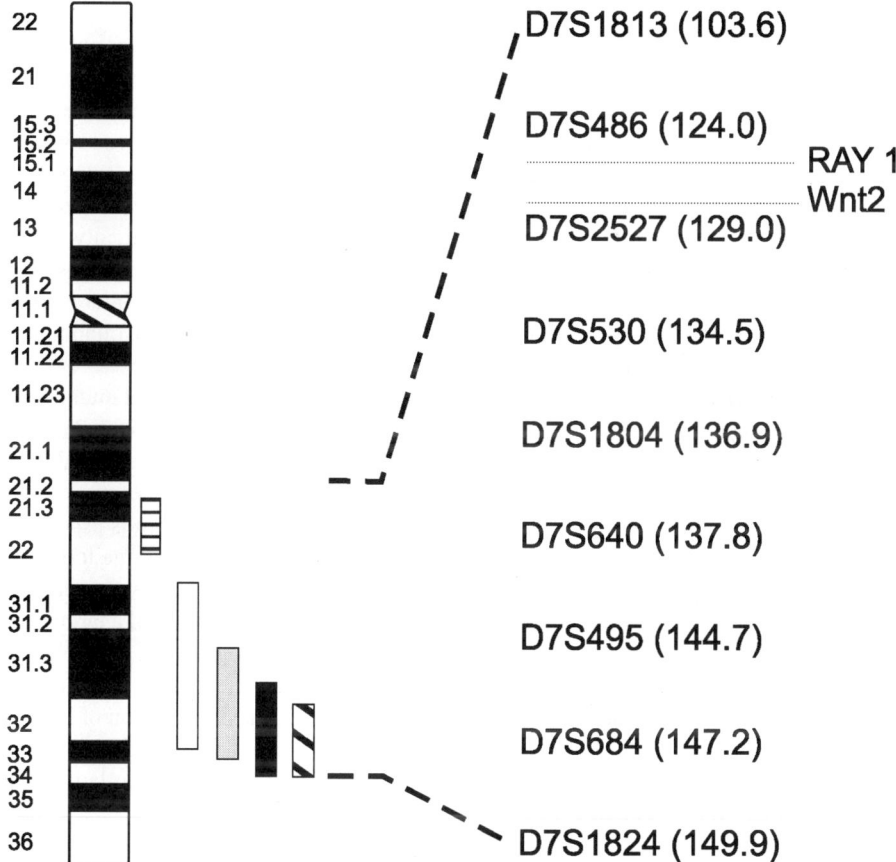

Figure 3 Chromosome 7 ideogram showing locations of markers positively linked to autism. The horizontally striped rectangle represents the CLSA study (1999); open rectangle, Philippe et al. (1999); dotted rectangle, Ashley-Koch et al. (1999); black rectangle, IMGSAC (1998); and the diagonally striped rectangle, Risch et al. (1999). (From Lamb et al. 2000.)

region for significant mutations or polymorphisms will be required, and this is not a trivial task. This is borne out by the study of Vincent et al. (2000), who, having discovered a novel gene at 7q31 (*RAY 1*) that was interrupted by a translocation breakpoint in an autistic patient, sequenced the entire coding region covering 16 exons. Mutation screening of this region in 27 unrelated autistic individuals failed to identify any phenotype-specific variations. However, as was pointed out in the International Consortium's study (1998), there are several other genes in this region that might be involved, including a G-protein coupled peptide receptor (*GPR 37*), a protein tyrosine phosphatase receptor (*PTPR11*), an ephrin tyrosine kinase receptor (*EPHB6*), and the muscarinic acetylcholine receptor (*GRM8*). Moreover, there is evidence that the gene responsible for a specific speech and language disorder (*SPCH1*) is located in the same general area (Fisher et al. 1998), which may indicate that autism and this speech disorder have an overlapping genetic etiology.

The most extensive genome screen for autism susceptibility loci yet published is that of Liu et al. (2001). This study was based on an analysis of some 335 microsatellite markers genotyped in 110 multiplex families; each included two affected siblings of whom at least one was autistic and the other diagnosed as having either pervasive developmental disorders or Asperger's syndrome. Suggestive linkages with lod scores that reached Lander & Kruglyak's (1995) threshold for schizophrenia were observed at chromosomes 5q and 19p, and statistically a third linkage at Xq was also considered significant. Other more equivocal linkages were observed on chromosomes 2, 3, 4, 8, 10, 11, 12, 15, 16, 18, and 20. Interestingly no linkage was reported for chromosomes 7 and 13 as found in the Collaborative Linkage Study (1999).

In a recent study, Wassink et al. (2001) report the finding of a point mutation in the coding regions of a *WNT2* gene (that is also located in this region—7q31-33— adjacent to *RAY1*) in two families in which one parent and a number of siblings were autistic. In a much larger group of autistic children SNP was found in the noncoding region of the gene. The frequency of this polymorphism is especially high in the individuals with the most serious linguistic impairment; the lod score in these cases was 3.7. This study is especially interesting since it was prompted by more than just the recognition that the *WNT-2* gene—a member of a large gene family involved in early development, including the development of the CNS—is located in the 7q31-33 region. The study also came about because of an earlier report that mice whose disheveled gene (*Dvl 1*, which acts downstream of *WNT2*) had been knocked out show a curious lack of social interaction: They do not groom each other, they do not huddle together when sleeping, and they care little for their offspring (Lijam 1997).

Until this latest study, progress in elucidating the genetic basis of autism—as in other multigenic disorders—had been relatively slow. This has not been the case in Rett syndrome, a single gene disorder that appears behaviorally to be quite similar to autism (Amir & Zoghbi 2000, Shahbazian & Zoghbi 2001). Rett syndrome is the leading cause of inherited mental retardation in females. It is characterized by a wide range of neurological and behavioral disorders: from mild learning disability to severe mental retardation, and from incessant hand wringing and clapping to marked ataxia, seizures, and neonatal encephalopathy. Zoghbi and her colleagues have recently identified the principal cause of the syndrome, namely mutations in the X-linked gene that encodes the methyl-CpG-binding protein 2 (MeCP2). MeCP2 is a transcriptional repressor that binds to methylated CpG dinucleotides throughout the genome, and the observed mutations in the gene result in a loss of function. Favorable X inactivation, on the other hand, may result in essentially no behavioral phenotype, but the few males born into Rett's kindreds have a severe neonatal encephalopathy and they rarely survive beyond the second year. While the identification of the gene involved in Rett syndrome opens up the possibility of identifying the genes whose expression is misregulated in the absence of a functional MeCP2, this discovery is rightly considered one of the most striking recent advances in the genetics of mental illness.

DISCUSSION

Despite all the effort that has gone into the search for candidate genes and genetic loci for schizophrenia and BPD, the available literature reveals little or no consensus and on reviewing it, one is left with an abiding sense of disappointment and frustration. It is not that there are too few loci or suggested candidate genes. If anything, there are too many. As Ostrander & Stanford (2000) have stated in a quite different context: "Too many loci, too few genes." Moreover, as we have pointed out above, relatively few of the suggested loci have been corroborated, and only a handful have achieved an acceptable degree of certainty (Lander & Kruglyak 1995).

Does this mean that little or no progress has been made? Or are we to conclude that the sheer complexity of such multigenic disorders is beyond our ability at this time to identify the likely locations of the many genes involved? Or are the populations of subjects available for study so heterogeneous that it will be impossible in the near future to recognize the significant loci by linkage analysis, as Botstein and colleagues (Friddle et al. 2000) have concluded? Or should we proceed on the assumption that many of the relevant loci have in fact been identified, and using the new tools that are now available (and those that will shortly become available) try to identify mutations or polymorphisms in some of the genes that lie within the identified loci?

This last possibility might have been considered utopian only two or three years ago, but several recent developments give us reason to be hopeful. The first development is, of course, the near completion of the human genome sequence. As the sequence is annotated and all the genes are identified (which may take another two or three years), the possibility of identifying the genes most likely to be implicated within any given locus should be within our reach. The second significant development is the generation of large numbers (now numbering in the millions) of SNPs. These closely spaced markers should greatly facilitate the future identification of disease-associated loci and, in the present context, narrow down relevant regions within the loci already identified. A particular SNP may itself be the genomic variation that produces disease risk within a candidate gene. However, such a direct association scenario is not unlikely to be common because it would require prior pathophysiologic knowledge implicating genes as candidates. At this time, the greater utility of SNPs will be for indirect association studies that depend on common SNPs serving as markers for neighboring disease-risk loci. The correlation of nearby alleles within the genome is described as linkage disequilibrium; these regions of association represent chromosomal regions or "haplotypes" descended from ancestral chromosomes. Haplotypes tend to be smaller in older populations (as a result of greater time for recombination) and larger in more recent population isolates. Because of the nonrandom distribution of SNPs throughout the genome, mapping haplotypes in different populations will likely prove important. Otherwise, a panel of SNPs might be quite redundant for large linkage disequilibrium blocks and may entirely miss regions with such blocks that are smaller. A haplotype map in a given population would, in theory, permit full genomic coverage by SNPs at maximal efficiency.

A fourth development is the continuing production of ever more sophisticated computational and statistical approaches to make sense of the vast amounts of sequence data that are being generated almost daily. Without powerful computational tools it would have been impossible to order the sequences already available; but, if the flood of new sequence data is to be manageable, even more effective mathematical approaches will be needed to solve the genetics of complex disorders. The final development of note is the concurrent sequencing of the genomes of other species—of yeast, fruit flies, nematodes, and especially the mouse. Until now our understanding of the functions of most known genes has come largely from experiments involving natural or induced mutations in these organisms, from specific knockouts of genes by homologous recombination, and from the creation of transgenic animals. The functions of only a modest number of genes have been determined first in humans, generally following the discovery of a distinctive disease phenotype like those listed in McKusick's On-Line Mendelian Inheritance in Man (OMIM).

We readily acknowledge that it is difficult (and will probably remain difficult) to determine the way in which genes contribute to behavioral phenotypes, and for most cognitive disorders it may be impossible to create fully satisfactory animal models. But there are a number of steps that can be taken once a gene has been associated with a behavioral disorder. The first is to screen for the homologous genes in other organisms since these may provide important clues about the possible function of the gene in question. The second is to determine where the gene is expressed (in the brain and elsewhere), and the final step is to establish in what way the function of the protein product of the affected gene is disturbed. Recent experience with AD and the spinocerebellar ataxias has shown how valuable animal models can be in elucidating the pathogenesis of CNS disorders (see Price et al. 2000, Zoghbi & Orr 2000).

If, as we have suggested, in the short term the most significant of these developments for our understanding of the genetics of mental illnesses is the availability of the human genome sequence, this question arises: In what ways can the human genome sequence throw light on the etiology and pathogenesis of mental illness and how is it likely to facilitate diagnosis and treatment of the various disorders? At present our answers to these questions can be only provisional and to a large extent speculative, given that the genome sequence has only recently become available and is still incomplete. But, with these caveats, we anticipate that progress is likely to be made in the following directions.

First, the availability of the genome sequence will enable investigators to focus their attention on all the genes encompassed by the many genetic loci that have been associated with specific mental illnesses. Since at this stage we have no way of knowing how many of the loci will prove to be relevant, they should probably all be subjected to this type of scrutiny, although at first those with the most significant lod scores and those that have had some degree of corroboration will deserve the greatest attention. This will be no easy task since most loci span a region of 10 to 20 cM or more, and in some cases may contain 50 or more different genes. The identification of all the open reading frames (ORFs) within the region will not be difficult, but knowing what they encode will be problematic. The reason for this is that although at present we can usually identify sequences that encode exons (by signature intron-exon boundary sequences) and, in many instances, sequences that encode specific protein domains (like tyrosine kinase and PDZ domains), we know even approximate functions of fewer than a quarter of the 40,000 or so genes in the human genome. As was mentioned above, the genes whose functions we can predict with some confidence are generally homologues of genes previously analyzed in one or another model organism. It may be several years before the functions of most of the uniquely human genes and, importantly, their protein products, are established. Many of the as yet unknown genes will probably turn out to be members of known gene families, so it should be possible to predict, at least tentatively, their likely functions. Second, it will be helpful to know which of the genes in the various linkage regions are uniquely or predominantly expressed in the brain. For this, the increasing availability of microarrays that include as many as 20,000–40,000 different EST or gene sequences that can be used to hybridize to RNAs from brain tissue will be critical. In principle, this approach should effectively be able to rule out genes that are not normally expressed in either the developing or adult brain.

Having identified all the ORFs that are expressed in the brain and lie within disease-associated loci, the next step will be to look for deviations from the published sequences, either in the form of mutations or polymorphisms, using material from a group of affected individuals. The finding of a distinct allele within a linkage region in several affected individuals can be regarded as prima facie evidence for the involvement of that gene in the relevant disorder. The occurrence of a recognizable polymorphism may be more problematic until it can be shown that other (nonaffected) family members do not carry the observed polymorphism. Of course, the presence of a polymorphism in unaffected family members does not necessarily rule out the possibility that the polymorphism in question is involved in the disorder; it may well be that the existence of several different polymorphisms (as well as mutations) is required for the disease phenotype to manifest. Indeed, a feature of genetically complex disorders may be that no one gene variant is necessary or sufficient for the development of the phenotype. Furthermore, until we know a great deal more about the range of polymorphisms throughout the genome in several different human populations, the finding of isolated polymorphisms in patients affected with a disorder will be difficult to interpret.

Once a mutation or a significant polymorphism has been found within a given linkage region, two further steps will be required. First, it will be critical to establish the function of the product of the relevant gene and to know when and where it is expressed within the brain. Second, the search for other mutated or polymorphic genes in other linkage regions will have to be identified since all the available evidence suggests that in most (and possibly in all) mental illnesses several different genes are involved. These considerations make one appreciate the complexity of the genetics of mental illness and the enormity of the challenge their discovery poses for investigators. Because each step will have to be repeated for each identified locus (and for most disorders several other loci remain to be identified), it is impossible to predict how soon we will know fully the underlying genetic basis of the major mental illnesses.

Altered gene function in the nervous system poses two further problems. First, since all neurons function as components of specific neural circuits rather than as separate elements, any disturbance of gene expression that interferes with the cells' ability to respond to synaptic inputs, to conduct impulses, or to act upon other neurons can disrupt the operation of the entire circuit. In the case of mutations that interfere with any one of a number of events in neural development, the circuit itself may fail to assemble appropriately and in consequence be unable to function normally. Second, since most neuronal activity involves a succession of intracellular events, mutations in any of several different genes may give rise to the same functional phenotype. To cite just one example, the response of a neuron to synaptic activation may involve not only the appropriate neurotransmitter receptor but also several intracellular signaling events. During the past decade we have learned from the genetic analysis of patients suffering from *retinitis pigmentosa* that the disorder can result from mutations not only in the gene for rhodopsin but in any of the other genes involved in the phototransduction cascade. Because most neural signaling is at least as complex as this, some mental illnesses (and many neurological disorders) may similarly result from multiple genetic disruptions within intracellular circuits as well as from multiple mutations (and/or polymorphisms) in genes involving different neuronal elements that comprise a system-wide neural circuit. Similarly, not all mutations may produce a significant phenotype by themselves because of compensatory changes within the network as a whole.

One further complication should be mentioned. This is the possibility that some of the mutations responsible for a mental illness occur, not in the coding region of a gene, but in one of its regulatory regions. Such mutations may not be too difficult to identify if they lie within or close to the promoter region of the gene, but if they involve more distant enhancers or suppressors (that can be located 100 kilobases or more from the coding regions and in most cases have not yet been characterized) their identification could be extremely difficult, even when the entire genome sequence is available.

While this may sound depressingly pessimistic, we believe that there are reasons for optimism. For example, despite the difficulties inherent in linkage analysis that Lander, Botstein, and others have pointed out, there are striking instances of the discovery of susceptibility loci in other complex disorders. The recent report of one such discovery (of a major locus at chromosome 10p13 for susceptibility to leprosy) is a case in point (see Siddiqui et al. 2001). And the discovery of a mutation in the *WNT2* gene in two families in which several members were affected with autism is an example of how following up on a previously identified linkage can lead to the discovery of a specific gene defect (see Wassink et al. 2001). There are reasons to be confident that similar discoveries will be made for other complex disorders and for other mental illnesses within the next four or five years. Furthermore, new methods of genome analysis, such as whole genome association studies, may become feasible.

An immediate benefit that derives from identifying specific mutations or distinctive polymorphisms is that it adds a degree of certainty and precision to diagnosis. As we mentioned in the introduction, there is often considerable uncertainty and subjectivity in the diagnosis of psychiatric illnesses. The same was true until recently for many other types of disorders in which diagnosis was uncertain. In this context the spinocerebellar ataxias are informative. Until it was established that each of the several different forms of the disorder is associated with a trinucleotide (CAG) repeat in a different gene, distinguishing between them was often difficult and controversial (see Zoghbi & Orr 2000). It is not too much to hope that the problem posed by the heterogeneity of BPD and schizophrenia and its related disorders will similarly be clarified once the relevant genes are known.

Beyond this there is good reason for thinking that once we know the affected genes (and what their disruption leads to in terms of the biology of the neurons involved and to the neural circuits in which they are engaged) the pharmaceutical and biotechnology industries will have a sound basis for developing new therapeutic agents aimed at specific molecular targets. Although many of the drugs that are in current use for the treatment of mental illness are very effective, it is salutary to remember that most were initially discovered serendipitously. Since then, newer drugs that cause fewer side effects have been introduced, but none has been specifically designed to act at the site of a defined molecular defect. Recent experience with rational drug design in other areas of medicine (like that for the recently introduced treatment of chronic myelogenous leukemia) gives one hope that when the specific genes are identified, highly selective, safe, and effective new therapies will become available for the treatment of the common mental illnesses in the not too distant future.

APPENDIX I

THE HISTORY OF THE HUMAN GENOME PROJECT

> Galileo... took us away from the idea that we were the center of the universe. The theory of evolution took us away from the conviction that we were a unique life form. And this work [on the human genome] will eventually tell us what makes our brains work and therefore our minds. It will tell us what we are.
>
> John Sulston

On June 26, 2000, an announcement was made jointly from the White House in Washington, D.C. and from 10 Downing Street in London that the sequencing of the human genome, biomedical research's largest and in many ways its most important project, had been completed. In a flurry of mutual congratulations, the principals—Francis Collins, representing the international Human Genome Project (HGP) and Craig Venter, president and CEO of the privately owned company Celera Genomics—each spoke of the magnitude of this accomplishment, and President Clinton was moved to say: "[This is] the most wondrous map ever produced by humankind," and "Today we are learning the language in which God created life." In reality, what was reported was the completion of a first or working draft of the genome. About 90% of the genome had been sequenced, but there were many gaps in the sequence and much annotation remained to be done before the full complement of genes could be identified.

In separate multi-authored papers that appeared in February 2001, the International Consortium's findings were reported in *Nature* (Lander et al. 2001), and on the same day, the work of Venter and colleagues was published in *Science* (Venter et al. 2001). For the first time, the scientific community could see the scope of the human genome and could savor the fruits of the past 15 years since the project was first proposed. Astonishment at the magnitude of the accomplishment was to some extent mitigated by the realization that the human genome comprised not 100,000 or 80,000 genes (as had been widely believed) but somewhere between 30,000 and 40,000. The realization that the "book of life" (to use Collins's phrase) and what Walter Gilbert had identified as "biology's holy grail" contained only 50% more genes than the one-millimeter-long nematode worm, only twice as many as the experimental geneticist's favorite organism—the fruit fly—and only five times as many as the lowly baker's yeast, came as something of a shock.[1] Our sense of superiority was restored only when commentators pointed out that because the messenger RNAs that are transcribed from genes can be spliced (that is, cut and recombined) in different ways, the human complement of 30,000 to 40,000 genes could give rise to several times that number of proteins.

Despite the initial hype that surrounded the announcement of the completion of the HGP, nothing could diminish the fact that this was not only a great scientific achievement but also one of the greatest importance for medicine. As someone remarked, "in the last analysis, all human illnesses are attributable, at least in part, to alterations in genes or to disturbances in gene function." So at the very least, having the complete array of human genes in hand, there is good reason to be confident that we have entered a new phase in biomedical research that will ultimately lead to a new era in medical practice. This is true even though at present we know something of the function of only about one third of the genes that have been identified and our understanding of their roles in development and in later life is still rather sketchy.

Public support for the HGP was largely predicated upon the promise that mapping and sequencing the genome would, in time, transform all branches of medicine and make a major contribution to world health. As James Watson expressed it in 1990:

> When finally interpreted, the genetic messages encoded within our DNA molecules will provide the ultimate answers to the chemical underpinnings of human existence. They will not only help us understand how we function as healthy human beings, but will also explain, at the chemical level, the role of genetic factors in a multitude of diseases, such as cancer, Alzheimer's disease, and schizophrenia, that diminish the individual lives of so many millions of people.

Others have written about the importance of the HGP for internal medicine, pediatrics, and other medical specialties (e.g., see Collins & McKusick 2001). Less attention has been paid to its potential impact on neuroscience and its cognate clinical disciplines of neurology and psychiatry. Elsewhere we have considered this in some detail (Cowan & Kandel 2001); here we need add only that it is our firm conviction that the application of the data and the methodological tools made available by the HGP will not only provide a more secure foundation for the diagnosis of neurological and psychiatric disorders, but will also greatly facilitate the development of new, more rational therapies. Coupled with our growing awareness of environmental risk factors, these data and tools may lead to preventative strategies that could ultimately lessen or even obviate the need for therapy.

Although the highlights of the HGP have been well covered by the media, the events that led up to the public announcement of the "completion" of the genome sequence in June 2000 and to the publication of the available data from both the public and private sources in February 2001 are not generally known. The following account of the history of the HGP from its inception in the mid-1980s to its near completion earlier this year is intended to fill this gap.[2]

The first public suggestion that the human genome might be mapped and sequenced was made in 1986 by Charles DeLisi, the director of the Department of

Energy's Office of Health and Environmental Research. The Energy Department had for many years been concerned with the genetic consequences of the atomic bombs dropped on Hiroshima and Nagasaki and, more generally, on radiation as a major cause of mutations and birth defects. Because of its long experience with genetic studies, DeLisi felt that the Energy Department was well positioned to launch a large-scale effort on the human genome. As a first step towards this, he organized a workshop in Santa Fe that brought together a number of well-known geneticists to explore the possibility. While the workshop participants seemed to agree that DeLisi's was a worthwhile goal, many were concerned that the necessary sequencing technology was not available and so urged that initially the effort should be directed toward the creation of physical and genetic maps of the genome. Still, others questioned whether the Department of Energy was the appropriate agency for such a major biological initiative.

In May 1985, several months before DeLisi's Sante Fe workshop, Robert Sinsheimer, chancellor of the University of California, Santa Cruz, had invited a group of scientific advisors to discuss the feasibility of setting up a DNA sequencing institute, which he hoped would establish his university as a major force in biomedical research. Although the proposed institute did not materialize, the meeting at Santa Cruz is still remembered as the occasion when Gilbert (who had shared the Nobel Prize with Frederick Sanger for the development of techniques for sequencing DNA) referred to the human genome sequence as "the grail of human genetics" and expressed the view that it would be "an incomparable tool for the investigation of every aspect of human function" (Cook-Deegan 1994).

One of the most important events in the early history of the HGP was a commentary in the March 7, 1986 issue of *Science* magazine by Renato Dulbecco, recipient of the Nobel Prize for his pioneering work on tumor viruses and at the time president of the Salk Institute. Although his own interests lay primarily in understanding how tumor viruses cause cancer in animals, Dulbecco called for a national effort to sequence the entire human genome, not only because it would have a major impact on cancer research by uncovering the genes that, when mutated, could lead to tumor formation, but also because it would be of the greatest importance for our understanding of all diseases and of normal human development. As he stated:

> [This] would be comparable to ... the effort that led to the conquest of space, and it should be carried out with the same spirit. Even more appealing would be to make it an international undertaking, because the sequence of the human DNA is the reality of our species, and everything that happens in the world depends on those sequences.

Stimulated by Dulbecco's commentary and no doubt worried that one of the potentially most significant projects in biology might be taken over by the Energy Department rather than the National Institutes of Health (NIH), which had always been the major supporter of biomedical research, Watson, director of the Cold Spring Harbor Laboratory, convened a meeting at the laboratory's conference center to discuss the feasibility of such a "big science" project. At this meeting

Gilbert first clearly defined the scope of the project: Given that the human genome comprises some three billion base pairs and that at a conservative estimate, the cost of sequencing might be about $1 per base, the overall project, might cost as much as $3 billion. Since at the time this represented a sizable proportion of the NIH's budget—even allowing that the effort would be spread over many years—considerable concern was expressed that to launch a project of this magnitude would seriously impact the ongoing activities of the NIH and diminish the resources available for individual investigator-instigated research. In response to these concerns Gilbert announced that he would found a new biotechnology company, The Genome Corporation, which would employ a large number of scientists and technicians to carry out the necessary sequencing. The company would make money by "selling" sequence data to the pharmaceutical industry and other organizations. In the event, the stock market crash of 1987 made it impossible for him to raise the necessary capital, and the project died on the vine.[3]

At about the same time, the Howard Hughes Medical Institute (HHMI), the nation's largest private philanthropy, saw an opportunity to move the genome effort forward by funding a number of ongoing activities that were in need of support.[4] At the suggestion of one of the institute's external advisors, Charles Scriver, the HHMI trustees approved support for the database of identified human genes, which had been started in the late 1970s by Frank Ruddle at Yale, for Victor McKusick's On-line Mendelian Inheritance In Man (OMIM, the catalog of all known human genetic disorders maintained at the Johns Hopkins School of Medicine), and for the Centre d'Etude du Polymorphisme Humaine, founded by the Nobelist, Jean Dausset. Later the institute put up a substantial amount to enable the international Human Genome Organization to get off the ground and provided the organization with a suite of offices in Bethesda to serve as its U.S. headquarters. It also developed laboratories at the University of Utah for one of its investigators, Ray White, who had initiated a program to analyze restriction fragment length polymorphisms among some of the extended Mormon families. Later, when federal and other sources of funding became available, HHMI support for these various activities (excepting White's research) was gradually phased out.

Another issue that was debated at the time (and for several years after) was whether the planned genome effort should aim at sequencing the entire human genome or focus, initially at least, on the expressed genes. Sidney Brenner, one of the more astute participants whose early contributions to the field of molecular biology (including the discovery of messenger RNA and, with Francis Crick, deciphering the genetic code) argued strongly for the latter approach. As he pointed out, if the expressed genes account for 5% or less of the genome, rapid progress could be made by concentrating on this relatively small part, while leaving until much later those regions that he and others facetiously referred to as "junk DNA."[5]

Later in 1987, a special panel of the U.S. National Research Council, chaired by Bruce Alberts (soon to be president of the National Academy of Sciences), having considered this issue and its medical and financial implications, recommended that an international effort be launched to map and sequence the entire human genome. It was further recommended that the effort begin by mapping the genome (and at the same time the genomes of certain simpler organisms that are widely used in genetic research, including yeast, mice, fruit flies, nematode worms, and some bacteria like *E. coli*) and that sequencing be delayed until the technology had improved and the cost significantly reduced. Using Gilbert's estimates, the panel further recommended that the U.S. government fund this effort at an annual rate of about $200 million for 15 years. The panel left open the question whether this effort should be spearheaded by the Department of Energy, which had already submitted a request to Congress for an additional $12 million for this purpose.

In presentations before Congress, James Wyngaarden, then director of NIH, argued persuasively that the U.S. genome effort should find its natural locus in the NIH and suggested that an initial amount of $50 million would be required to create a new Office of Human Genome Research. There was a general feeling, most forcibly articulated by Watson, that if such an office were to be established it should be led by an active scientist rather than an administrator. To most people in the scientific community there was only one person for the position: Watson himself. Accordingly, in May 1988, Watson was offered and accepted the directorship of what in October 1989 became the National Center for Human Genome Research (NCHGR) with a budget for the first year of $60 million.

Watson brought several distinct strengths to the position. It is no exaggeration to say that were it not for his firm grasp of what needed to be done, how best to get it done, and what the potential problems were—scientific, political, and social—the project might never have succeeded. One example of his political and social sensitivity was his early decision to set aside 3% of the center's budget for what he termed the "ethical, legal and social implications" of the project (Watson 1990).

Not everyone was convinced that Watson's plan to begin the effort by extensively mapping the genome before attempting sequencing was the right way to proceed. In retrospect, the insistence on this approach may have slowed the completion of the project. It was, however, strongly endorsed by many biomedical researchers who were primarily interested in the discovery of disease-related genes, including Collins, who later was appointed director of the NCHGR following Watson's resignation in April 1992 over a disagreement with Bernadine Healey, then director of NIH, who favored the patenting of partial cDNA sequences, known as ESTs.

While many of those involved in what soon become known as the International Genome Initiative were largely preoccupied with various mapping strategies, John (now Sir John) Sulston at Cambridge and Bob Waterston in St. Louis quietly proceeded to sequence the genome of the nematode worm, *C. elegans*, which Brenner had introduced as a useful model system for developmental biology in the early 1960s. Without waiting for the long-promised improvements in sequencing technology and simply by turning their laboratories into large-scale, round-the-clock, sequencing facilities, they had (with generous support from the Wellcome Trust and from the NCHGR, respectively) convincingly demonstrated that substantial amounts of highly accurate sequence data could be generated daily. About the same time, Venter, who had left the NIH because in his judgment it had neither adequately supported nor sufficiently recognized the significance of his work on ESTs, founded a nonprofit organization. The Institute of Genome Research (TIGR). In addition to producing many thousands of additional ESTs, Venter and his colleagues at TIGR succeeded in sequencing the genome of an important human pathogen, *H. influenza*. They were ably assisted in this effort by Hamilton Smith, who in 1978 had shared the Nobel Prize for the introduction and use of restriction enzymes that enabled one to "cut" DNA at sequence specific sites. Smith suggested that sequencing could be greatly accelerated if the genome were simply cut into smaller fragments with suitable restriction enzymes, and each fragment independently sequenced. The trick then was to correctly align the fragments with the aid of an appropriate computer program that identified sequence overlap at the ends of adjoining fragments.

Prompted by their success in sequencing the genome of *H. influenzae* and by the progress made by Sulston & Waterston, Venter began to consider the possibility of sequencing large genomes, including ultimately the human genome, using Smith's "shotgun" approach. At about the same time, Waterston & Sulston (1998) pointed out in a letter to *Science* that using conventional sequencing methods and the kinds of management methods developed in their laboratories, the human genome sequence could be completed well ahead of the originally proposed date if several laboratories in the United Kingdom, United States, and elsewhere were to commit themselves to the effort. Although initially their proposal seemed to fall on deaf ears, events in the private sector were soon to change the attitudes of the leaders of the international HGP. For some time it had been evident that a number of privately owned companies were generating considerable amounts of genomic data to which most scientists could either not obtain access or could do so only under very burdensome intellectual property constraints. This contrasted strikingly with the open access under which the publicly supported efforts operated. The rapid accumulation of thousands of human ESTs by private companies like Human Genome Sciences on the East Coast and Incyte on the West Coast, coupled with their well-advertised attempts to patent these partial gene sequences, promised to seriously undermine the publicly supported effort. In an attempt to counteract this, Waterston's group in St. Louis, with generous funding from Merck, generated hundreds of thousands of human ESTs that were immediately deposited in GenBank, the publicly accessible database. Later, with the help of the HHMI, they also generated over 400,000 mouse ESTs that were similarly deposited in GenBank.

The development in the late 1990s of a new and substantially better automated DNA sequencer provided Venter with the critical tool to launch his most ambitious project yet, the sequencing of the human genome. The ABI PRISM 3700 sequencer (which sold for $300,000) was based on a capillary flow system using fluorescently labeled dyes to distinguish the various bases. Used in conjunction with a robotic arm for handling the samples, the new sequencer greatly increased the rate of sequencing and at the same time reduced the cost involved by about 90%. Working optimally, it was capable of sequencing about a million bases per day. The developer of the PRISM 3700, Mike Hunkapiller, approached Venter and his colleague Mark Adams about the possibility of a joint venture that would not only underwrite the use of the shotgun approach to sequence the human genome,[6] but would also control and sell the data generated. Out of these discussions came the company Celera Genomics (*celera* is from the Latin for swift—the same root from which we have "accelerate"). Celera's business plan was to carry out large-scale sequencing (not only of the human genome, but also the genomes of other organisms, including select bacteria and the mouse), to patent certain genes of interest, and to sell to pharmaceutical and biotechnology companies access to Celera's database and computerized screening facilities.

When Venter disclosed his plans, including his prediction that Celera would sequence the human genome at least two years earlier than the completion date that had been planned for the publicly funded genome effort, he set off a storm of protest and criticism. While some claimed that the shotgun approach could not possibly succeed or, at best, would result in a sequence with innumerable gaps, most of the criticism was leveled at the inappropriateness of commercializing what rightly belonged to the entire human family. And it seemed especially egregious to some that Celera could draw freely on the publicly funded data that were being posted daily on the internet, while other scientists could not gain access to the company's data.

The immediate effect of Venter's announcement was to light a fire under the leaders of the publicly funded effort—including Harold Varmus, the director of the NIH, Collins, director of what is now the National Human Genome Research Institute, and Richard Morgan (who headed the Wellcome Trust's genome effort). No one stated publicly that Venter's proposals were the occasion for a sudden change in the pace and direction of the public HGP, but whatever the case, the results were dramatic. In Britain, the Wellcome Trust pledged a substantial increase in its support for the Sanger Center, the principal sequencing facility in the United Kingdom, directed by Sulston. And in the United States, Collins announced that: (*a*) a considerably higher proportion of the NIH's genome effort would be devoted to sequencing; (*b*) selected sequencing centers would receive increased funding for additional equipment and personnel; and (*c*) the immediate goal of these changes would be to produce a "first draft" of the human genome within approximately two years. The race was on. That there was a race to get the human genome sequenced (or at least to produce a first draft of the sequence) was repeatedly denied, however, by both Venter and the leaders of the public HGP.[7]

Meanwhile, Venter was anxious to try out his shotgun method and his expanded sequencing facilities. He accordingly approached Gerry Rubin, who for some years had been leading the NIH's effort to sequence the *Drosophila* genome, about a shotgun attempt to complete the fruit fly sequence within about 6 months instead of the 12 to 18 months that Rubin had originally anticipated. Despite some opposition from Watson and other leaders of the HGP, Rubin agreed to Venter's request once he had been assured that all the sequence data would be made public and that there were no strings attached to Venter's offer. The outcome of this public/private collaboration was striking. By the summer of 1999, within just four months, virtually the full sequence of the *Drosophila* genome (consisting of about 120 million bases) was completed and the genes it contains identified. This success surprised and almost silenced many of Venter's critics, but some continued to insist that the shotgun approach would not be successful when applied to the human genome, with its 3 billion base pairs and its high content of repeat sequences, which would make assembly extremely difficult. Undeterred, Venter and his colleagues proceeded with their effort to sequence the human genome, while at the same time taking full advantage of the publicly available data from the international HGP.

The public versus private approaches continued to provoke controversy—at times openly and not without vitriol. Suffice it to say, by June 2000 a truce had been called; the nonrace had ended in a tie, and preparations were made for a joint announcement that a "working draft" of the human genome had been completed. In some respects the announcement was anticlimactic—at least for most biologists. The greatest immediate surprise was that the human genome contained less than half the number of genes everyone had expected. Even the most recent textbooks and reviews repeated the earliest estimates of around 100,000 genes. This figure had been based on Gilbert's simple calculation of the number of bases present in the genome (three billion—itself an estimate derived from the total amount of DNA in human cells) and certain assumptions about the average size of a human gene (thought to be about 30,000 bases in length). Apart from this, the sequence appeared at first glance to be a virtually endless succession of As, Ts, Cs, and Gs (from the initial letters of the four bases adenine, thymine, cytosine, and guanine), much as everyone had expected following Watson & Crick's discovery of the structure of DNA in 1953. It had also been anticipated that genes would represent only a small fraction of the genome. We now could see that at most, they account for about 1.5% of the genome. What about the other 98.5%, which had been branded "junk DNA?" Here there was another surprise. About half of the noncoding DNA appears to consist of innumerable, fairly short sequences that during the past 500 million years or more have parasitized vertebrate genomes and have contrived to be replicated and passed on from generation to generation, jumping around from place to place within the genome. If any DNA warrants the name "selfish DNA" (to use a term that the English biologist Richard Dawkins introduced some years ago), these sequences are prime candidates. As far as anyone knows they contribute nothing useful: They appear to be concerned only with their own survival.

There are, in addition to these "nonhuman" sequences, long repetitive stretches in which the same groups of bases recur. Their function is also unknown, but their presence has made sequencing unnecessarily complicated. Lastly, the noncoding DNA also includes many important regulatory sequences that control the level of expression of the genes to which they are linked. Some of these regulatory regions are close to the gene itself or are even within the intervening regions, or introns, that separate the sequences, known as exons, which are finally translated into proteins. Other regulatory regions can be far removed from the gene, often by as much as 100,000 or more bases. Identifying these regulatory regions is one of the important issues that remains to be worked out.

There are several other matters that have still to be addressed. The first is that the overall sequence has to be completed. Both the public and private sequence contain gaps, some purely by chance, others because certain regions, like those near the ends of chromosomes and close to centromeres, have proved to be difficult to sequence. Then the public and private sequences need to be compared and whatever differences exist between them need to be resolved. In saying this we are not implying that either sequence is inadequate or unreliable. In fact, where they have been compared they show considerable concordance and it is estimated (by repeating each sequence several times) that their accuracy is at least 99.99%. Only when the sequence is truly complete and our gene detection algorithms improved will we know the exact number of human genes. At present, we know that there are at least 30,000 but probably no more than 40,000. Compare this imprecision with the precise number of genes found in the nematode worm (19,099).

Even more important than determining the actual number of genes is the elucidation of their functions in health and disease. At best, we know something about the functions of 20%–30% of the genes; some of the known genes were identified from the study of clinical disorders but even more are known from studies of their homologues in mice, flies, worms, and yeast. It is not surprising and singularly fortunate for investigators that so many genes have been conserved during the course of evolution. For many people it was surprising to learn that 40% of the genes in the nematode worm and 50% of those in the fruit fly have clearly recognizable counterparts in the human genome, despite the many millions of years that separate their divergent evolution. It is particularly noteworthy that the sequence of the mouse genome is now also available, not only because mice share many of the same genes often in the same order on their chromosomes as humans, but also because mice lend themselves so well to the kinds of experimental genetic manipulation that are critical for determining gene function.

It is reasonable to ask, "In what way will knowing the sequence of the human genome and the identification of all the encoded genes be of help in the diagnosis and treatment of human disorders?" The simple answer is that the completion of the human genome sequence marks the beginning of a new era that promises to completely transform the entire field of medicine, improving diagnosis, providing unprecedented tools for studying pathophysiology, and ultimately providing a rich array of new and more effective therapies. For those interested in neurological disorders and mental illness, the initial benefit of this new knowledge is that it will permit more certain diagnosis of the many disorders for which we know there is a significant genetic component but have not yet been able to identify the actual genes involved.

The search for the genes involved in mental illnesses will continue to depend on the identification of the relevant loci within the genome. This process will be greatly facilitated by another important development that has occurred in human genome studies over the past three or four years. This is the creation of a large public database documenting the locations of almost two million SNPs, the most common type of variation in the genome. More than 95% of the SNPs in the public domain have been located in the past 18 months, and at the time of writing (in July 2001), they are present at a density of roughly one SNP every 1.3 kilobases. Because additional SNPs are being discovered almost weekly (Celera claims to already have 2.8 million in its database), this additional tool promises to be of ever increasing importance in human genetics. In addition to learning which variants contribute to some of the more complex human disorders, these methods should also permit us to determine why individuals respond differently to various drugs. Beyond all this will be the continuing effort to elucidate the manifold ways in which the environment and our personal life experiences interact with our genes to make us what we are and either make us susceptible to, or protect us from, certain disorders. The completion of the human genome sequence is an important first step toward this goal; not an end in itself, but a major landmark. As T. S. Elliot so aptly expressed it in his poem "Little Gidding:"

... to make an end is to make a beginning.

The end is where we start from.

NOTES TO THE HISTORY OF THE HUMAN GENOME PROJECT

1. It is important to point out that the numbers of genes in different organisms is different from the relative sizes of their genomes. The human genome is about 30 times as large as that of the fruit fly or the nematode worm and roughly 200 times as large as that of baker's yeast. The much greater size of the human genome is largely attributable to the extensive tracts of noncoding (and often repetitive) DNA that it contains. The greater size of human genes is due mainly to the greater size and number of introns that they contain.

2. There are several good accounts of the early history of the HGP and the work that led up to it. Among the most useful are those of Watson (1990), Wills (1991), Cook-Deegan (1994), and Davies (2000).

3. Although Gilbert's plan to create a genomic company was aborted, it was resurrected several years later when Venter launched Celera Genomics.

4. In February 1986, two weeks before DeLisi's Santa Fe meeting, Don Frederickson, then President of HHMI, had invited several leading geneticists to a meeting at its headquarters in Coconut Grove, Florida, to discuss how best to handle the growing body of information about genetic markers, and new DNA probes, etc., and to consider if the time was ripe for a large-scale attack on the human genome.

5. As we now know, expressed genes account for only about 1.5% of the genome.

6. A whole-genome shotgun approach to sequencing the human genome appears to have been suggested first by James Weber & Eugene Myers in the spring of 1997 (Weber & Myers 1997).

7. "[Dr Venter and I] are intending to be partners in every possible way . . . this is not a race." Collins in testimony before the House Subcommittee on Energy and the Environment, June 17, 1998.

APPENDIX II

A number of basic terms and concepts are central to understanding the genetics involved in studying the inheritance of complex traits. For convenience many of the terms used in this review are summarized here.

Phenotype refers to the overt character or trait. For example, in BPD the phenotype is characterized by periods of depression with episodes of manic behavior marked by grandiosity, decreased need for sleep, and excessive involvement in pleasurable but often irrational behavior. *Genotype*, on the other hand, refers to the underlying gene or genes (at either a specific chromosomal locus or at a combination of loci) that ultimately determine phenotypic traits.

In *mendelian inheritance* the phenotype is due to a single genetic locus. The relevant gene is, therefore, passed on to future generations in a predictable manner. If only one copy of the gene is transmitted from the father (paternal inheritance) or the mother (maternal inheritance), the offspring are *heterozygous*; if the same gene is inherited from both parents, the offspring are *homozygous*. If the phenotype is evident in an individual who is heterozygous for the relevant gene, the pattern of inheritance is *dominant*. If a trait is only expressed when the subject is homozygous, the gene is *recessive*. It is often relatively easy to locate the gene responsible for a monogenic disorder because affected individuals pass the relevant trait on to their children in a clear, well-defined pattern of inheritance that follows Mendel's laws. *Complex genetic disorders*, on the other hand, pose a variety of challenges. (*a*) The disorder may be *oligogenic*, that is, caused by mutations in a small number of different genes. (*b*) *Genetic* or *locus heterogeneity* refers to the situation in which polymorphisms or mutations in one or more genes can lead to the same phenotype. (*c*) If both environmental and genetic factors are responsible for the disorder its etiology is said to be *multifactorial*. (*d*) *Penetrance* can be problematic (even in mendelian disorders). Simply defined, penetrance refers to the probability that a person with the appropriate genotype will manifest the character (or trait). That is to say, the gene may be always (and fully) manifest in the offspring of an affected individual, or it may vary widely in the degree to which it is expressed. Discriminating between a nonfully penetrant mendelian disorder and a complex genetic illness is often difficult.

Compared to many other illnesses, the identification of genes associated with psychiatric disorders is made more difficult by the variability of the *disease phenotype*. Most psychiatric disorders have a core of symptoms and signs that define the disorder and enable a diagnosis to be made. But beyond this core, there can be a spectrum of abnormalities in different patients; some of these provide the basis for the formal subclassifications of disorders like BPD and schizophrenia in the DSM-IV. Such distinctions cannot always be made in clinical practice, and differences in diagnosis often lead to a very heterogeneous sample population that further complicates genetic analysis.

Traditional approaches to studying the genetics of psychiatric diseases include genetic epidemiology (family studies), twin, and adoption studies. *Genetic epidemiology* studies assess the occurrence of a disorder within a family or within a specific population, looking for the segregation, or aggregation of the disorder in individual families or groups of families. For instance, as was pointed out in the text, genetic epidemiology studies have shown that both BPD and schizophrenia aggregate in families, and the risk for illness increases with familial relatedness; thus, a child or a sibling of an affected individual has a greater risk of acquiring the disease than a more distant relative, such as a cousin.

Twin studies compare the concordance rate (i.e., the likelihood that if one twin suffers from a disorder the second twin will develop the illness) of monozygotic (MZ) twins with that of dizygotic (DZ) twins. While MZ twins have the same genes, DZ twins share only 50% of their genes, and in this sense do not differ from other nontwin siblings. When the concordance rate for a disorder among MZ twins is appreciably greater than for DZ twins, it is strongly suggestive that the underlying cause of the illness has a genetic component because it is reasonable to assume that siblings living together are subject to the same environmental influences. *Adoption studies* ascertain the concordance rates between an affected adoptee and his/her biological relatives as compared with members of his/her adoptive families. A higher concordance rate between affected adoptees and their biological relatives again suggests that genetic factors (rather than environmental) play a greater role in the etiology of the disorder. The study of MZ twins adopted into different families is especially valuable in distinguishing genetic from environmental influences.

Methodological and statistical factors are important in most other genetic analyses of complex disorders. We comment here only on the differences between *parametric* and *nonparametric linkage analyses*. Parametric linkage analyses are powerful for scanning the genome and locating mendelian-inherited disease genes, but they are less useful when applied to complex disorders. The *logarithm of the odds* (usually *the lod score*) assesses the likelihood that a known genetic marker cosegregates with a disease gene to a greater extent than would be expected to occur by chance. In assessing the lod score, one needs to know the precise genetic model used, the mode of inheritance assumed (whether dominant or recessive), the gene frequencies, and the probable penetrance of each genotype. As we have pointed out, most psychiatric disorders are nonmendelian, but rare families have been identified that include affected individuals whose inheritance follows a mendelian pattern, usually with reduced penetrance. In such cases, parametric tests can be performed by analyzing only the affected family members (whereas normally, parametric linkage studies take into account all members of an affected individual's family).

Nonparametric linkage analyses allow researchers to use a model-free method. In such studies, unaffected individuals are essentially ignored, while an attempt is made to identify alleles or chromosomal markers that the affected individuals have in common. *Affected sib pair* (ASP) or *affected pedigree member* (APM) analyses are two examples of model-free nonparametric methods. ASP analyses are compared at specific marker locations and chromosomal regions to determine if the linkage exceeds the random ratio of 1:2:1. Unfortunately, sib pair analysis does not permit one to identify markers in a complex disorder that are as close to a disease locus as are markers that one identifies in a mendelian disorder using parametric analysis. There are two reasons for this. First, in complex disorders, it is highly unlikely that all sib pairs share a specific susceptibility locus; and second, sib pairs often share many loci purely by chance, and some loci may be adjacent to a susceptibility locus.

Currently, many psychiatric geneticists use a statistical program called GENE-HUNTER (which is based on an earlier program MAPMAKER/SIBS) (Kruglyak et al. 1996). GENEHUNTER allows one to include any number of loci in a multipoint analysis. Given a specific genetic model, an adequate pedigree, and other information, the program computes a parametric lod score, which can then be expressed as a *nonparametric lod* (NPL) score. (A lod score associated with ASP analysis is a form of NPL score.) According to Lander & Kruglyak (1995), an idealized ASP full-genome scan analysis can lead to the following types of linkage: (*a*) suggestive; (*b*) significant; (*c*) highly significant; or (*d*) confirmed. A *significant linkage* is expected to occur with a 5% probability, or 0.05 times in a genome scan. The lod score associated with a significant linkage is 3.6; the associated *P* value is 2.2×10^{-5}. A *suggestive linkage* has a lod score of 2.2 and a *P* value of 7.4×10^{-4}. (Note that in this analysis the lod score is concerned with the ratio of two probabilities, whereas the *P* values refer to a single absolute probability.) As Lander & Kruglyak (1995) define it, a *maximum lod score* (MLS; i.e., the lod score maximized over a set of parameters) is "the log-likelihood ratio of the data under the hypothesis that the allele sharing proportion has the observed value $\pi(x)$ as compared to the hypothesis that there is no excess sharing.... A *P* value reflects the pointwise chance of observing a deviation as high as $\pi(x)$ under independent assortment."

The *Annual Review of Neuroscience* is online at http://neuro.annualreviews.org

LITERATURE CITED

Aita VM, Liu J, Knowles JA, Terwilliger JD, Baltazar R, et al. 1999. A comprehensive linkage analysis of chromosome 21q22 supports prior evidence for a putative bipolar affective disorder locus. *Am. J. Hum. Genet.* 64:210–17

Amir RE, Zoghbi HY. 2000. Rett syndrome: methyl-CpG-binding protein 2 mutations and phenotype-genotype correlations. *Am. J. Med. Genet.* 97:147–52

Antonarakis SE, Blouin JL, Curran M. 1996. Linkage and sib-pair analysis reveal a potential schizophrenia susceptibility gene on chromosome 13q32. *Am. J. Hum. Genet.* 59:A210

Antonarakis SE, Blouin JL, Pulver AE, Wolyniec P, Lasseter VK, et al. 1995. Schizophrenia susceptibility and chromosome 6p24-22. *Nat. Genet.* 11:235–36

Ashley-Koch A, Wolpert CM, Menold MM, Zaeem L, Basu S, et al. 1999. Genetic studies of autistic disorder and chromosome 7. *Genomics* 61:227–36

Bailey A, Le Couteur A, Gottesman I, Bolton P,

Simonoff E, et al. 1995. Autism as a strongly genetic disorder: evidence from a British twin study. *Psychol. Med.* 25:63–77

Barden N, Morissette J. 1999. Chromosome 13 workshop report. *Am. J. Med. Genet.* 88:260–62

Baron M. 1977. Linkage between an X-chromosome marker (deutan color blindness) and bipolar affective illness. Occurrence in the family of a lithium carbonate-responsive schizo-affective pro band. *Arch. Gen. Psychiatry* 34:721–25

Baron M. 2001. Genetics of schizophrenia and the new millennium: progress and pitfalls. *Am. J. Hum. Genet.* 68:299–312

Baron M, Freimer NF, Risch N, Lerer B, Alexander JR, et al. 1993. Diminished support for linkage between manic depressive illness and X-chromosome markers in three Israeli pedigrees. *Nat. Genet.* 3:49–55

Baron M, Risch N, Hamburger R, Mandel B, Kushner S, et al. 1987. Genetic linkage between X-chromosome markers and bipolar affective illness. *Nature* 326:289–92

Bass MP, Menold MM, Wolpert CM, Donnelly SL, Ravan SA, et al. 1999. Genetic studies in autistic disorder and chromosome 15. *Neurogenetics* 2:219–26

Berrettini WH, Ferraro TN, Goldin LR, Weeks DE, Detera-Wadleigh S, et al. 1994. Chromosome 18 DNA markers and manic-depressive illness: evidence for a susceptibility gene. *Proc. Natl. Acad. Sci. USA* 91:5918–21

Berrettini WH, Goldin LR, Gelernter J, Gejman PV, Gershon ES, Detera-Wadleigh S. 1990. X-chromosome markers and manic-depressive illness. Rejection of linkage to Xq28 in nine bipolar pedigrees. *Arch. Gen. Psychiatry* 47:366–73

Bird TD, Wijsman EM, Nochlin D, Leehey M, Sumi SM, et al. 1997. Chromosome 17 and hereditary dementia: linkage studies in three non-Alzheimer families and kindreds with late-onset FAD. *Neurology* 48:949–54

Blackwood DH, He L, Morris SW, McLean A, Whitton C, et al. 1996. A locus for bipolar affective disorder on chromosome 4p. *Nat. Genet.* 12:427–30

Blazer DG, Kessler RC, McGonagle KA, Swartz MS. 1994. The prevalence and distribution of major depression in a national community sample: The National Comorbidity Survey. *Am. J. Psychiatry* 151:979–86

Blouin JL, Dombroski BA, Nath SK, Lasseter VK, Wolyniec PS, et al. 1998. Schizophrenia susceptibility loci on chromosomes 13q32 and 8p21. *Nat. Genet.* 20:70–73

Bolton P, Macdonald H, Pickles A, Rios P, Goode S, et al. 1994. A case-control family history study of autism. *J. Child Psychol. Psychiatry* 35:877–900

Brzustowicz LM, Hodgkinson KA, Chow EW, Honer WG, Bassett AS. 2000. Location of a major susceptibility locus for familial schizophrenia on chromosome 1q21-q22. *Science* 288:678–82

Brzustowicz LM, Honer WG, Chow EW, Little D, Hogan J, et al. 1999. Linkage of familial schizophrenia to chromosome 13q32. *Am. J. Hum. Genet.* 65:1096–103

Busatto GF, Costa DC, Ell PJ, Pilowsky LS, David AS, Kerwin RW. 1994. Regional cerebral blood flow (rCBF) in schizophrenia during verbal memory activation: a 99mTc-HMPAO single photon emission tomography (SPET) study. *Psychol. Med.* 24:463–72

Callicott JH, Mattay VS, Bertolino A, Finn K, Coppola R, et al. 1999. Physiological characteristics of capacity constraints in working memory as revealed by functional MRI. *Cereb. Cortex* 9:20–26

Cannon TD, Kaprio J, Lonnqvist J, Huttunen M, Koskenvuo M. 1998. The genetic epidemiology of schizophrenia in a Finnish twin cohort. A population-based modeling study. *Arch. Gen. Psychiatry* 55:67–74

Cao Q, Martinez M, Zhang J, Sanders AR, Badner JA, et al. 1997. Suggestive evidence for a schizophrenia susceptibility locus on chromosome 6q and a confirmation in an independent series of pedigrees. *Genomics* 43:1–8

Cardno AG, Gottesman II. 2000. Twin studies of schizophrenia: from bow-and-arrow concordances to Star Wars Mx and functional genomics. *Am. J. Med. Genet.* 97:12–17

Cardno AG, Holmans PA, Rees MI, Jones LA,

McCarthy GM, et al. 2001. A genome-wide linkage study of age at onset in schizophrenia. *Am. J. Med. Genet.* 105:439–45

Choi DW. 2000. Hearing before the Appropriations Subcommittee of the House Committee on Labor, Health and Human Services, 106th Cong, 2nd sess. http://www.sfn.org/news

Collaborative Linkage Study of Autism, Barrett S, Beck JC, Bernier R, Bisson E, et al. 1999. An autosomal genomic screen for autism. *Am. J. Med. Genet.* 88:609–15

Collins FS, McKusick VA. 2001. Implications of the Human Genome Project for medical science. *JAMA* 285:540–44

Cook-Deegan R. 1994. *The Gene Wars*. New York: Norton

Cook EH Jr, Courchesne RY, Cox NJ, Lord C, Gonen D, et al. 1998. Linkage-disequilibrium mapping of autistic disorder, with 15q11-13 markers. *Am. J. Hum. Genet.* 62:1077–83

Coon H, Holik J, Hoff M, Reimherr F, Wender P, et al. 1994a. Analysis of chromosome 22 markers in nine schizophrenia pedigrees. *Am. J. Med. Genet.* 54:72–79

Coon H, Jensen S, Holik J, Hoff M, Myles-Worsley M, et al. 1994b. Genomic scan for genes predisposing to schizophrenia. *Am. J. Med. Genet.* 54:59–71

Cowan WM, Kandel ER. 2001. Prospects for neurology and psychiatry. *JAMA* 285:594–600

Dann J, DeLisi LE, Devoto M, Laval S, Nancarrow DJ, et al. 1997. A linkage study of schizophrenia to markers within Xp11 near the MAOB gene. *Psychiatry Res.* 70:131–43

Davies K. 2001. *Cracking the Genome*. New York: Free Press

Del Zompo M, Bocchetta A, Goldin LR, Corsini GU. 1984. Linkage between X-chromosome markers and manic-depressive illness. Two Sardinian pedigrees. *Acta Psychiatr. Scand.* 70:282–87

DeLisi LE, Devoto M, Lofthouse R, Poulter M, Smith A, et al. 1994a. Search for linkage to schizophrenia on the X and Y chromosomes. *Am. J. Med. Genet.* 54:113–21

DeLisi LE, Friedrich U, Wahlstrom J, Boccio-Smith A, Forsman A, et al. 1994b. Schizophrenia and sex chromosome anomalies. *Schizophr. Bull.* 20:495–505

Detera-Wadleigh SD, Badner JA, Berrettini WH, Yoshikawa T, Goldin LR, et al. 1999. A high-density genome scan detects evidence for a bipolar-disorder susceptibility locus on 13q32 and other potential loci on 1q32 and 18p11.2. *Proc. Natl. Acad. Sci. USA* 96:5604–9

Detera-Wadleigh SD, Badner JA, Goldin LR, Berrettini WH, Sanders AR, et al. 1996. Affected-sib-pair analyses reveal support of prior evidence for a susceptibility locus for bipolar disorder, on 21q. *Am. J. Hum. Genet.* 58:1279–85

Detera-Wadleigh SD, Badner JA, Yoshikawa T, Sanders AR, Goldin LR, et al. 1997. Initial genome scan of the NIMH genetics initiative bipolar pedigrees: chromosomes 4, 7, 9, 18, 19, 20, and 21q. *Am. J. Med. Genet.* 74:254–62

Detera-Wadleigh SD, Hsieh WT, Berrettini WH, Goldin LR, Rollins DY, et al. 1994. Genetic linkage mapping for a susceptibility locus to bipolar illness: chromosomes 2, 3, 4, 7, 9, 10p, 11p, 22, and Xpter. *Am. J. Med. Genet.* 54:206–18

Edenberg HJ, Foroud T, Conneally PM, Sorbel JJ, Carr K, et al. 1997. Initial genomic scan of the NIMH genetics initiative bipolar pedigrees: chromosomes 3, 5, 15, 16, 17, and 22. *Am. J. Med. Genet.* 74:238–46

Egan MF, Goldberg TE, Kolachana BS, Callicott JH, Mazzanti CM, et al. 2001. Effect of COMT Val108/158 Met genotype on frontal lobe function and risk for schizophrenia. *Proc. Natl. Acad. Sci. USA* 98:6917–22

Egeland JA, Gerhard DS, Pauls DL, Sussex JN, Kidd KK, et al. 1987. Bipolar affective disorders linked to DNA markers on chromosome 11. *Nature* 325:783–87

Evans KL, Muir WJ, Blackwood DH, Porteous DJ. 2001. Nuts and bolts of psychiatric genetics: building on the human genome project. *Trends Genet.* 17:35–40

Ewald H, Degn B, Mors O, Kruse TA. 1998a. Significant linkage between bipolar affective disorder and chromosome 12q24. *Psychiatr. Genet.* 8:131–40

Ewald H, Degn B, Mors O, Kruse TA. 1998b. Support for the possible locus on chromosome 4p16 for bipolar affective disorder. *Mol. Psychiatry* 3:442–48

Ewald H, Mors O, Flint T, Koed K, Eiberg H, Kruse TA. 1995. A possible locus for manic depressive illness on chromosome 16p13. *Psychiatr. Genet.* 5:71–81

Faraone SV, Matise T, Svrakic D, Pepple J, Malaspina D, et al. 1998. Genome scan of European-American schizophrenia pedigrees: results of the NIMH Genetics Initiative and Millennium Consortium. *Am. J. Med. Genet.* 81:290–95

Fava M, Kendler KS. 2000. Major depressive disorder. *Neuron* 28:335–41

Fisher SE, Vargha-Khadem F, Watkins KE, Monaco AP, Pembrey ME. 1998. Localisation of a gene implicated in a severe speech and language disorder. *Nat. Genet.* 18:168–70

Folstein S, Rutter M. 1977. Infantile autism: a genetic study of 21 twin pairs. *J. Child Psychol. Psychiatry* 18:297–321

Freimer NB, Reus VI, Escamilla MA, McInnes LA, Spesny M, et al. 1996. Genetic mapping using haplotype, association and linkage methods suggests a locus for severe bipolar disorder (BPI) at 18q22-q23. *Nat. Genet.* 12:436–41

Friddle C, Koskela R, Ranade K, Hebert J, Cargill M, et al. 2000. Full-genome scan for linkage in 50 families segregating the bipolar affective disease phenotype. *Am. J. Hum. Genet.* 66:205–15

Fuster JM. 2000. Prefrontal neurons in networks of executive memory. *Brain Res. Bull.* 52:331–36

Gejman PV, Detera-Wadleigh S, Martinez MM, Berrettini WH, Goldin LR, et al. 1990. Manic depressive illness not linked to factor IX region in an independent series of pedigrees. *Genomics* 8:648–55

Goldman-Rakic PS. 1994. Working memory dysfunction in schizophrenia. *J. Neuropsychiatry Clin. Neurosci.* 6:348–57

Gottesman II. 1991. *Schizophrenia Genesis: The Origins of Madness.* New York: Freeman

Hallmayer J, Hebert JM, Spiker D, Lotspeich L, McMahon WM, et al. 1996. Autism and the X chromosome. Multipoint sib-pair analysis. *Arch. Gen. Psychiatry* 53:985–89

Harrison PJ. 1999. The neuropathology of schizophrenia. A critical review of the data and their interpretation. *Brain* 122(Pt 4): 593–624

Hauser ER, Boehnke M, Guo SW, Risch N. 1996. Affected-sib-pair interval mapping and exclusion for complex genetic traits: sampling considerations. *Genet. Epidemiol.* 13:117–37

Hovatta I, Varilo T, Suvisaari J, Terwilliger JD, Ollikainen V, et al. 1999. A genomewide screen for schizophrenia genes in an isolated Finnish subpopulation, suggesting multiple susceptibility loci. *Am. J. Hum. Genet.* 65:1114–24

Hyman SE. 1999. The neurobiology of mental disorders. In *The Harvard Guide to Psychiatry*, ed. AM Nicholi, pp. 134–54. Cambridge, MA: Harvard Univ. Press

Hyman SE. 2000. Mental illness: genetically complex disorders of neural circuitry and neural communication. *Neuron* 28:321–23

International Molecular Genetic Study of Autism Consortium. 1998. A full genome screen for autism with evidence for linkage to a region on chromosome 7q. *Hum. Mol. Genet.* 7:571–78

International Molecular Genetic Study of Autism Consortium. 2001. A genome wide screen for autism: strong evidence for linkage to chromosomes 2q, 7q, and 16p. *Am. J. Hum. Genet.* 69:570–81

Jensen J, Coon H, Hoff M, Rosenthal J, Reimherr F, et al. 1998. Search for a schizophrenia susceptibility gene on chromosome 13. *Psychiatr. Genet.* 8:239–43

Kallman FJ. 1938. *The Genetics of Schizophrenia*. Locust Valley, NY: J. J. Augustin

Kalsi G, Chen CH, Smyth C. 1996. Genetic analysis in an Icelandic/British sample fails to exclude the putative chromosome 13q14.1-q32 schizophrenia susceptibility locus. *Am. J. Hum. Genet.* 59:A388

Karayiorgou M, Gogos JA, Galke BL, Wolyniec PS, Nestadt G, et al. 1998. Identification of sequence variants and analysis of the role of the catechol-O-methyl-transferase gene in schizophrenia susceptibility. *Biol. Psychiatry* 43:425–31

Kelsoe JR, Ginns EI, Egeland JA, Gerhard DS, Goldstein AM, et al. 1989. Re-evaluation of the linkage relationship between chromosome 11p loci and the gene for bipolar affective disorder in the Old Order Amish. *Nature* 342:238–43

Kelsoe JR, Spence MA, Loetscher E, Foguet M, Sadovnick AD, et al. 2001. A genome survey indicates a possible susceptibility locus for bipolar disorder on chromosome 22. *Proc. Natl. Acad. Sci. USA* 98:585–90

Kendler KS, Gruenberg AM. 1984. An independent analysis of the Danish Adoption Study of Schizophrenia. VI. The relationship between psychiatric disorders as defined by DSM-III in the relatives and adoptees. *Arch. Gen. Psychiatry* 41:555–64

Kendler KS, MacLean CJ, O'Neill FA, Burke J, Murphy B, et al. 1996. Evidence for a schizophrenia vulnerability locus on chromosome 8p in the Irish Study of High-Density Schizophrenia Families. *Am. J. Psychiatry* 153:1534–40

Kendler KS, Tsuang MT, Hays P. 1987. Age at onset in schizophrenia. A familial perspective. *Arch. Gen. Psychiatry* 44:881–90

Kety SS, Rosenthal D, Wender PH, Schulsinger F, Jacobsen B. 1976. Mental illness in the biological and adoptive families of adopted individuals who have become schizophrenic. *Behav. Genet.* 6:219–25

Knowles JA, Rao PA, Cox-Matise T, Loth JE, de Jesus GM, et al. 1998. No evidence for significant linkage between bipolar affective disorder and chromosome 18 pericentromeric markers in a large series of multiplex extended pedigrees. *Am. J. Hum. Genet.* 62:916–24

Kruglyak L, Daly MJ, Reeve-Daly MP, Lander ES. 1996. Parametric and nonparametric linkage analysis: a unified multipoint approach. *Am. J. Hum. Genet.* 58:1347–63

Lachman HM, Kelsoe JR, Remick RA, Sadovnick AD, Rapaport MH, et al. 1997. Linkage studies suggest a possible locus for bipolar disorder near the velo-cardio-facial syndrome region on chromosome 22. *Am. J. Med. Genet.* 74:121–28

Lamb JA, Moore J, Bailey A, Monaco AP. 2000. Autism: recent molecular genetic advances. *Hum. Mol. Genet.* 9:861–68

Lander E, Kruglyak L. 1995. Genetic dissection of complex traits: guidelines for interpreting and reporting linkage results. *Nat. Genet.* 11:241–47

Lander ES, Linton LM, Birren B, Nusbaum C, Zody MC, et al. 2001. Initial sequencing and analysis of the human genome. *Nature* 409:860–921

Lasseter VK, Pulver AE, Wolyniec PS, Nestadt G, Meyers D, et al. 1995. Follow-up report of potential linkage for schizophrenia on chromosome 22q: part 3. *Am. J. Med. Genet.* 60:172–73

Leonard S, Gault J, Moore T, Hopkins J, Robinson M, et al. 1998. Further investigation of a chromosome 15 locus in schizophrenia: analysis of affected sibpairs from the NIMH Genetics Initiative. *Am. J. Med. Genet.* 81:308–12

Levinson DF, Mahtani MM, Nancarrow DJ, Brown DM, Kruglyak L, et al. 1998. Genome scan of schizophrenia. *Am. J. Psychiatry* 155:741–50

Lijam N, Paylor R, McDonald MP, Crawley JN, Deng CX, et al. 1997. Social interaction and sensorimotor gating abnormalities in mice lacking Dvl1. *Cell* 90:895–905

Lin MW, Curtis D, Williams N, Arranz M, Nanko S, et al. 1995. Suggestive evidence for linkage of schizophrenia to markers on chromosome 13q14.1-q32. *Psychiatr. Genet.* 5:117–26

Lin JP, Bale SJ. 1997. Parental transmission

and D18S37 allele sharing in bipolar affective disorder. *Genet. Epidemiol.* 14:665–68

Lin MW, Sham P, Hwu HG, Collier D, Murray R, Powell JF. 1997. Suggestive evidence for linkage of schizophrenia to markers on chromosome 13 in Caucasian but not Oriental populations. *Hum. Genet.* 99:417–20

Liu J, Nyholt DR, Magnussen P, Parano E, Pavone P, et al. 2001. A genomewide screen for autism susceptibility loci. *Am. J. Hum. Genet.* 69:327–40

Martinez M, Goldin LR, Cao Q, Zhang J, Sanders AR, et al. 1999. Follow-up study on a susceptibility locus for schizophrenia on chromosome 6q. *Am. J. Med. Genet.* 88:337–43

McInnes LA, Escamilla MA, Service SK, Reus VI, Leon P, et al. 1996. A complete genome screen for genes predisposing to severe bipolar disorder in two Costa Rican pedigrees. *Proc. Natl. Acad. Sci. USA* 93:13060–65

McMahon FJ, Hopkins PJ, Xu J, McInnis MG, Shaw S, et al. 1997. Linkage of bipolar affective disorder to chromosome 18 markers in a new pedigree series. *Am. J. Hum. Genet.* 61:1397–404

Mendlewicz J, Linkowski P, Guroff JJ, Van Praag HM. 1979. Color blindness linkage to bipolar manic-depressive illness. New evidence. *Arch. Gen. Psychiatry* 36:1442–47

Mendlewicz J, Linkowski P, Wilmotte J. 1980. Linkage between glucose-6-phosphate dehydrogenase deficiency and manic-depressive psychosis. *Br. J. Psychiatry* 137:337–42

Mendlewicz J, Simon P, Sevy S, Charon F, Brocas H, et al. 1987. Polymorphic DNA marker on X chromosome and manic depression. *Lancet* 1:1230–32

Michaud CM, Murray CJ, Bloom BR. 2001. Burden of disease-implications for future research. *JAMA* 285:535–39

Millar JK, Wilson-Annan JC, Anderson S, Christie S, Taylor MS, et al. 2000. Disruption of two novel genes by a translocation cosegregating with schizophrenia. *Hum. Mol. Genet.* 9:1415–23

Mirnics K, Middleton FA, Marquez A, Lewis DA, Levitt P. 2000. Molecular characterization of schizophrenia viewed by microarray analysis of gene expression in prefrontal cortex. *Neuron* 28:53–67

Moises HW, Yang L, Kristbjarnarson H, Wiese C, Byerley W, et al. 1995. An international two-stage genome-wide search for schizophrenia susceptibility genes. *Nat. Genet.* 11:321–24

Morissette J, Villeneuve A, Bordeleau L, Rochette D, Laberge C, et al. 1999. Genome-wide search for linkage of bipolar affective disorders in a very large pedigree derived from a homogeneous population in Quebec points to a locus of major effect on chromosome 12q23-q24. *Am. J. Med. Genet.* 88:567–87

Murphy KC, Jones LA, Owen MJ. 1999. High rates of schizophrenia in adults with velocardio-facial syndrome. *Arch. Gen. Psychiatry* 56:940–45

Murray CJL, Lopez AD. 1996. *The Global Burden of Disease.* Cambridge, MA: Harvard Univ. Press

Nothen MM, Cichon S, Rohleder H, Hemmer S, Franzek E, et al. 1999. Evaluation of linkage of bipolar affective disorder to chromosome 18 in a sample of 57 German families. *Mol. Psychiatry* 4:76–84

Nurnberger JI Jr, DePaulo JR, Gershon ES, Reich T, Blehar MC, et al. 1997. Genomic survey of bipolar illness in the NIMH genetics initiative pedigrees: a preliminary report. *Am. J. Med. Genet.* 74:227–37

Ostrander EA, Stanford JL. 2000. Genetics of prostate cancer: too many loci, too few genes. *Am. J. Hum. Genet.* 67:1367–75

Pekkarinen P, Terwilliger J, Bredbacka PE, Lonnqvist J, Peltonen L. 1995. Evidence of a predisposing locus to bipolar disorder on Xq24-q27.1 in an extended Finnish pedigree. *Genome Res.* 5:105–15

Philippe A, Martinez M, Guilloud-Bataille M, Gillberg C, Rastam M, et al. 1999. Genome-wide scan for autism susceptibility genes. Paris Autism Res. Int. Sibpair Study. *Hum. Mol. Genet.* 8:805–12

Pickles A, Bolton P, Macdonald H, Bailey A, Le

Couteur A, et al. 1995. Latent-class analysis of recurrence risks for complex phenotypes with selection and measurement error: a twin and family history study of autism. *Am. J. Hum. Genet.* 57:717–26

Polymeropoulos MH, Coon H, Byerley W, Gershon ES, Goldin L, et al. 1994. Search for a schizophrenia susceptibility locus on human chromosome 22. *Am. J. Med. Genet.* 54:93–99

Price DL, Wong PC, Markowska AL, Lee MK, Thinakaren G, et al. 2000. The value of transgenic models for the study of neurodegenerative diseases. *Ann. NY Acad. Sci.* 920:179–91

Pulver AE, Karayiorgou M, Wolyniec PS, Lasseter VK, Kasch L, et al. 1994. Sequential strategy to identify a susceptibility gene for schizophrenia: report of potential linkage on chromosome 22q12-q13.1: part 1. *Am. J. Med. Genet.* 54:36–43

Pulver AE, Mulle J, Nestadt G, Swartz KL, Blouin JL, et al. 2000. Genetic heterogeneity in schizophrenia: stratification of genome scan data using co-segregating related phenotypes. *Mol. Psychiatry* 5:650–53

Repetto GM, White LM, Bader PJ, Johnson D, Knoll JH. 1998. Interstitial duplications of chromosome region 15q11q13: clinical and molecular characterization. *Am. J. Med. Genet.* 79:82–89

Riley BP, McGuffin P. 2000. Linkage and associated studies of schizophrenia. *Am. J. Med. Genet.* 97:23–44

Risch N, Botstein D. 1996. A manic depressive history. *Nat. Genet.* 12:351–53

Risch N, Spiker D, Lotspeich L, Nouri N, Hinds D, et al. 1999. A genomic screen of autism: evidence for a multilocus etiology. *Am. J. Hum. Genet.* 65:493–507

Robins E, Guze SB. 1970. Establishment of diagnostic validity in psychiatric illness: its application to schizophrenia. *Am. J. Psychiatry* 126:983–87

Robins LN, Regier DA. 1991. *Psychiatric Disorders in America: The Epidemiologic Catchment Area Study.* New York: Free Press

Sanders AR, Detera-Wadleigh SD, Gershon ES. 1999. Molecular genetics of mood disorders. In *Neurobiology of Mental Illness*, pp. 299–316. New York: Oxford Univ. Press

Sasaki T, Billett E, Petronis A, Ying D, Parsons T, et al. 1996. Psychosis and genes with trinucleotide repeat polymorphism. *Hum. Genet.* 97:244–46

Schwab SG, Albus M, Hallmayer J, Honig S, Borrmann M, et al. 1995. Evaluation of a susceptibility gene for schizophrenia on chromosome 6p by multipoint affected sib-pair linkage analysis. *Nat. Genet.* 11:325–27

Schwab SG, Eckstein GN, Hallmayer J, Lerer B, Albus M, et al. 1997. Evidence suggestive of a locus on chromosome 5q31 contributing to susceptibility for schizophrenia in German and Israeli families by multipoint affected sib-pair linkage analysis. *Mol. Psychiatry* 2:156–60

Schwab SG, Hallmayer J, Albus M, Lerer B, Eckstein GN, et al. 2000. A genome-wide autosomal screen for schizophrenia susceptibility loci in 71 families with affected siblings: support for loci on chromosome 10p and 6. *Mol. Psychiatry* 5:638–49

Schwab SG, Hallmayer J, Albus M, Lerer B, Hanses C, et al. 1998a. Further evidence for a susceptibility locus on chromosome 10p14-p11 in 72 families with schizophrenia by nonparametric linkage analysis. *Am. J. Med. Genet.* 81:302–7

Schwab SG, Hallmayer J, Lerer B, Albus M, Borrmann M, et al. 1998b. Support for a chromosome 18p locus conferring susceptibility to functional psychoses in families with schizophrenia, by association and linkage analysis. *Am. J. Hum. Genet.* 63:1139–52

Shahbazian MD, Zoghbi HY. 2001. Molecular genetics of Rett syndrome and clinical spectrum of MECP2 mutations. *Curr. Opin. Neurol.* 14:171–76

Shaw SH, Kelly M, Smith AB, Shields G, Hopkins PJ, et al. 1998. A genome-wide search for schizophrenia susceptibility genes. *Am. J. Med. Genet.* 81:364–76

Siddiqui MR, Meisner S, Tosh K, Balakrishnan K, Ghei S, et al. 2001. A major susceptibility locus for leprosy in India maps to chromosome 10p13. *Nat. Genet.* 27:439–41

Silverman JM, Greenberg DA, Altstiel LD, Siever LJ, Mohs RC, et al. 1996. Evidence of a locus for schizophrenia and related disorders on the short arm of chromosome 5 in a large pedigree. *Am. J. Med. Genet.* 67:162–71

Sobell JL, Lind TJ, Hebrink DD, Heston LL, Sommer SS. 1997. Screening the monoamine oxidase B gene in 100 male patients with schizophrenia: a cluster of polymorphisms in African-Americans but lack of functionally significant sequence changes. *Am. J. Med. Genet.* 74:44–49

Spurlock G, Williams J, McGuffin P, Aschauer HN, Lenzinger E, et al. 1998. European multicentre association study of schizophrenia: a study of the DRD2 Ser311Cys and DRD3 Ser9Gly polymorphisms. *Am. J. Med. Genet.* 81:24–28

Stine OC, McMahon FJ, Chen L, Xu J, Meyers DA, et al. 1997. Initial genome screen for bipolar disorder in the NIMH genetics initiative pedigrees: chromosomes 2, 11, 13, 14, and X. *Am. J. Med. Genet.* 74:263–69

Stine OC, Xu J, Koskela R, McMahon FJ, Gschwend M, et al. 1995. Evidence for linkage of bipolar disorder to chromosome 18 with a parent-of-origin effect. *Am. J. Hum. Genet.* 57:1384–94

Stober G, Saar K, Ruschendorf F, Meyer J, Nurnberg G, et al. 2000. Splitting schizophrenia: periodic catatonia-susceptibility locus on chromosome 15q15. *Am. J. Hum. Genet.* 67:1201–7

Straub RE, Lehner T, Luo Y, Loth JE, Shao W, et al. 1994. A possible vulnerability locus for bipolar affective disorder on chromosome 21q22.3. *Nat. Genet.* 8:291–96

Straub RE, MacLean CJ, Martin RB, Ma Y, Myakishev MV, et al. 1998. A schizophrenia locus may be located in region 10p15-p11. *Am. J. Med. Genet.* 81:296–301

Straub RE, MacLean CJ, O'Neill FA, Burke J, Murphy B, et al. 1995. A potential vulnerability locus for schizophrenia on chromosome 6p24-22: evidence for genetic heterogeneity. *Nat. Genet.* 11:287–93

Straub RE, MacLean CJ, O'Neill FA, Walsh D, Kendler KS. 1997. Support for a possible schizophrenia vulnerability locus in region 5q22-31 in Irish families. *Mol. Psychiatry* 2:148–55

Straub RE, Speer MC, Luo Y, Rojas K, Overhauser J, et al. 1993. A microsatellite genetic linkage map of human chromosome 18. *Genomics* 15:48–56

Sullivan PF, Neale MC, Kendler KS. 2000. Genetic epidemiology of major depression: review and meta-analysis. *Am. J. Psychiatry* 157:1552–62

Tsuang MT, Faraone SV. 1990. *The Genetics of Mood Disorders.* Baltimore, MD: Johns Hopkins Univ. Press

US Dep. Health Hum. Serv. 1999. *Mental Health: A Report of the Surgeon General.* Rockville, MD: US Dep. Health Hum. Serv., NIH

Veenstra-Vander Weele J, Gonen D, Leventhal BL, Cook EH Jr. 1999. Mutation screening of the UBE3A/E6-AP gene in autistic disorder. *Mol. Psychiatry* 4:64–67

Venter JC, Adams MD, Myers EW, Li PW, Mural RJ, et al. 2001. The sequence of the human genome. *Science* 291:1304–51

Vincent JB, Herbrick JA, Gurling HM, Bolton PF, Roberts W, Scherer SW. 2000. Identification of a novel gene on chromosome 7q31 that is interrupted by a translocation breakpoint in an autistic individual. *Am. J. Hum. Genet.* 67:510–14

Wassink TH, Piven J, Vieland VJ, Huang J, Swiderski RE, et al. 2001. Evidence supporting WNT2 as an autism susceptibility gene. *Am. J. Med. Genet.* 105:406–13

Waterston R, Sulston JE. 1998. The Human Genome Project: reaching the finish line. *Science* 282:53–54

Watson JD. 1990. The human genome project: past, present, and future. *Science* 248:44–49

Weber JL, Myers EW. 1997. Human whole-genome shotgun sequencing. *Genome Res.* 7:401–9

Weinberger DR, Egan MF, Bertolino A, Callicott JH, Mattay VS, et al. 2001. Prefrontal neurons and the genetics of schizophrenia. *Biol. Psychiatry* 50:825–44

Williams J, McGuffin P, Nothen M, Owen MJ. 1997. Meta-analysis of association between the 5-HT2a receptor T102C polymorphism and schizophrenia. Eur. Multicent. Assoc. Study Schizophr. (EMASS) Group. *Lancet* 349:1221

Williams J, Spurlock G, McGuffin P, Mallet J, Nothen MM, et al. 1996. Association between schizophrenia and T102C polymorphism of the 5-hydroxytryptamine type 2a-receptor gene. Eur. Multicent. Assoc. Study Schizophr. (EMASS) Group. *Lancet* 347:1294–96

Wills C. 1991. *The Science Behind the Human Genome Project*. New York: Basic Books

Winokur G, Clayton PJ, Reich T. 1969. *Manic-Depressive Illness*, pp 112–25. St. Louis: CV Mosby

Zoghbi HY, Orr HT. 2000. Glutamine repeats and neurodegeneration. *Annu. Rev. Neurosci.* 23:217–47

AUDITORY SYSTEM DEVELOPMENT: Primary Auditory Neurons and Their Targets

Edwin W. Rubel[1] and Bernd Fritzsch[2]

[1]*Virginia Merrill Bloedel Hearing Research Center, Department of Otolaryngology/Head and Neck Surgery, University of Washington, Seattle, Washington 98195-7923; email: rubel@u.washington.edu*
[2]*Department of Biomedical Science, Creighton University, Omaha, Nebraska 68178; email: fritzsch@creighton.edu*

Key Words cochlear ganglion, ear, cochlear nuclei, embryology

■ **Abstract** The neurons of the cochlear ganglion transmit acoustic information between the inner ear and the brain. These placodally derived neurons must produce a topographically precise pattern of connections in both the inner ear and the brain. In this review, we consider the current state of knowledge concerning the development of these neurons, their peripheral and central connections, and their influences on peripheral and central target cells. Relatively little is known about the cellular and molecular regulation of migration or the establishment of precise topographic connection to the hair cells or cochlear nucleus (CN) neurons. Studies of mice with neurotrophin deletions are beginning to yield increasing understanding of variations in ganglion cell survival and resulting innervation patterns, however. Finally, existing evidence suggests that while ganglion cells have little influence on the differentiation of their hair cell targets, quite the opposite is true in the brain. Ganglion cell innervation and synaptic activity are essential for normal development of neurons in the cochlear nucleus.

INTRODUCTION

The acoustico-vestibular ganglion neurons are derived from the otic placode. Early in development, a group of premitotic and postmitotic neuronal precursors migrate across the basal lamina delimiting the otic vesicle and acquire a position between the developing inner ear and the closely apposed tissue of the rhombencephalon. These neuroblasts form the primary neurons of the auditory and vestibular pathways, the cochlear and vestibular ganglia, linking the inner ear and the central nervous system (CNS). Their development and the integrity of their descendants in the mature animal are essential for processing of acoustic and vestibular information. Moreover, maintaining the integrity of the cochlear ganglion cells following congenital or postnatal hearing loss is critically important when we consider current therapies such as cochlear implants, or future therapies such as hair cell regeneration. In addition, studies of the interactions of cochlear ganglion cells with their central and peripheral targets may provide useful principles for understanding cell-cell interactions in the developing nervous system. We limit this review to studies of development of the cochlear ganglion neurons in birds and mammals and the interactions of these neurons with their central and peripheral partners. For more comprehensive reviews of inner ear development see several recent reviews: Fritzsch et al. 1998, Pujol et al. 1998, Fekete 1999, Frago et al. 2000. Those interested in functional development and CNS auditory pathway development should consult Rubel et al. 1998 and Friauf & Lohmann 1999.

ADULT ORGANIZATION

The mammalian cochlea consists of two types of receptor cells (the outer and inner hair cells) and is innervated by at least two distinct ganglion neurons, the Type I and Type II spiral sensory (cochlear ganglion) neurons. The two types of hair cells form a total of four rows (three rows of outer and one row of inner hair cells; Figure 1). In the adult, inner hair cells are innervated by Type I cochlear ganglion neurons, and outer hair cells are innervated by Type II sensory neurons. Innervation varies between species and along the apical-to-basal direction (Ryugo 1992). About three times more Type I afferents appear to converge on each inner hair cell of the base than on the apex. In contrast, synaptic contacts between Type II cochlear ganglion neurons and outer hair cells is about two times higher in the apex than in the base. Each Type I ganglion neuron contacts only a single inner hair cell, but each Type II cochlear ganglion neuron contacts about 30–60 outer hair cells. Thus, 15–20 Type I cochlear ganglion neurons provide parallel channels from a single inner hair cell to the CNS. In contrast, the small number of Type II cochlear ganglion neurons integrate information from many outer hair cells.

DEVELOPMENT OF COCHLEAR GANGLION NEURONS AND INNERVATION OF INNER EAR

Development of the Inner Ear

The inner ear begins as a placodal thickening that is induced in the ectoderm by the nearby hindbrain and the underlying mesoderm (Fritzsch et al. 1998, Fekete 1999, Baker & Bronner-Fraser 2001). This placode undergoes invagination and compartmentalization to form the inner ear. Further development of the ear can be subdivided into a number of parallel and sequential processes with largely unknown interrelationships. Very roughly and in chronological order, these processes include placodal induction, placodal invagination, otocyst morphogenesis, sensory neuron (cochlear ganglion neuron) and hair cell proliferation, cochlear elongation, and, peculiar to mammals, coiling (Rubel 1978, Morsli et al. 1998, Hutson et al. 1999, Cantos et al. 2000). The three-dimensional structure that emerges is uniquely

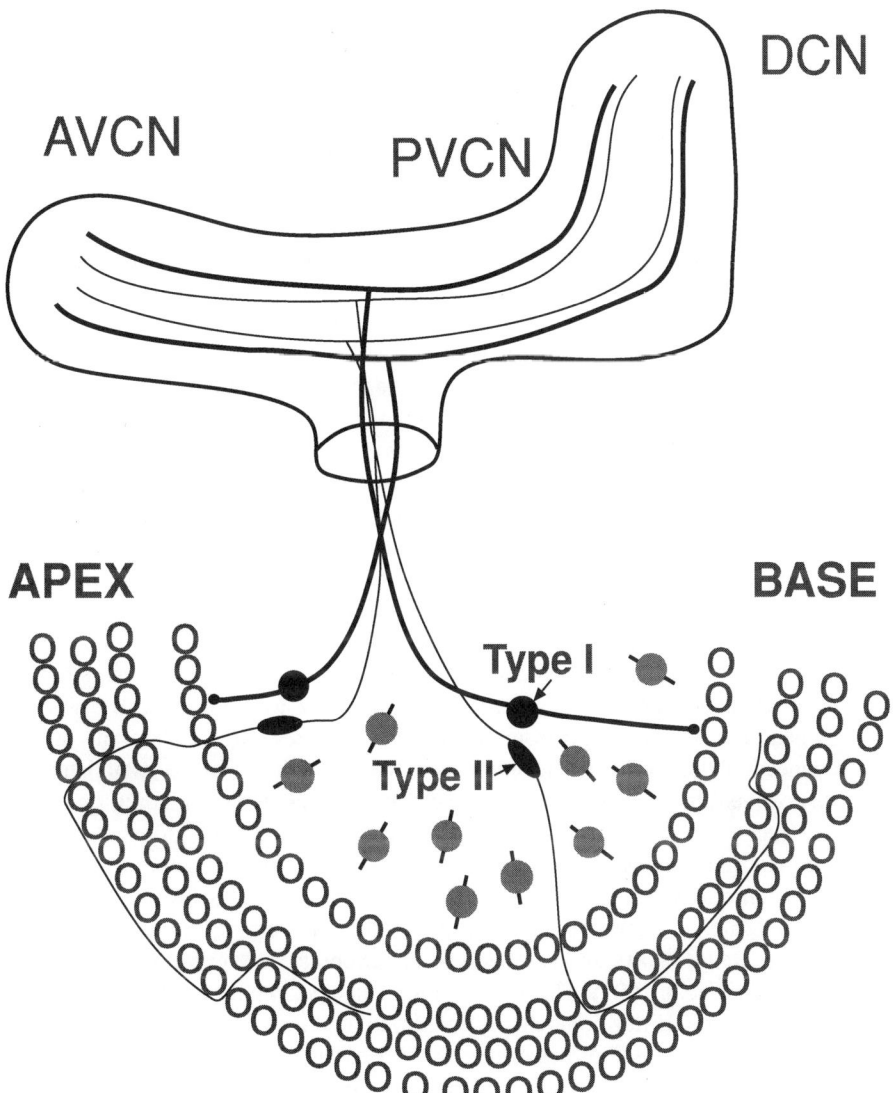

Figure 1 Basic organization of hair cells, cochlear ganglion neurons, and cochlear nucleus (CN) subdivisions in mammals. Ten to twenty Type I neurons converge on each inner hair cell. In contrast, innervation of outer hair cells Type II neurons is highly divergent and less topographic. Innervation varies between species and along the apical-to-basal dimension but, in general, there are at least four to five times more Type I neurons in the basal turn, and two times more contacts are formed between Type II neurons and apical outer hair cells. Central axons of both Type I and Type II neurons branch upon entering the brain to form a tonotopic projection in each subregion of the CN complex.

suited to extract specific components of the mechanical stimuli that reach the ear. Morphological maturation of the otocyst is accompanied by the formation of ganglion neurons and distinct sensory patches of hair cells and supporting cells. These sensory patches are positioned so as to enable them to convert the specific directional and frequency components of mechanical stimuli into electric signals that are transmitted by the ganglion neurons of the ear to the brain. In mammals and birds, this results in a cochlea designed to extract the spectral properties of sound by way of variations in structure and in cellular functional attributes along its longitudinal axis. The cochlea is also designed to provide a digital representation in the pattern of ganglion cell unitary discharges sent to the brain.

The ear is connected to the brain via two different kinds of axons. The afferent axons of otocyst-derived sensory neurons (vestibular and cochlear ganglion neurons) connect the cochlea and vestibular organs with the auditory and vestibular nuclei in the brainstem. The auditory components of this connection are organized such that an orderly topography of frequency-selective responses to sound along the cochlea is retained in the form of a frequency-specific map, the tonotopic organization of cochlear ganglion cells, and their connections with the brain. In addition, the ear is innervated by efferent fibers (Warr 1992, Berlin 1999).

The expression of many transcription factors and their effect on ear morphogenesis has been elucidated within recent years (Fekete 1999, Cantos et al. 2000, Represa et al. 2000, Karis et al. 2001). However, how the region of ganglion cell precursors is determined and delineated in the developing otocyst is not yet clear. Understanding inner ear ganglion neuron development requires resolution of ganglion neuron origin from the otocyst and knowledge of how the orderly connection between these ganglion neurons and the cochlea is established. In this context, recent data on insect development show that pathfinding properties of insect-ciliated sensory neurons are determined by both the global and the local patterning processes and are mediated by transcription factors that underly the specific sensory-organ developmental program (Ghysen & Dambly-Chaudiere 2000). Many of the genes related to these global patterning programs in insects (Sato & Saigo 2000) are also involved in vertebrate development (Cantos et al. 2000, Karis et al. 2001). This conservation extends from molecular homology to several apparently developmentally conserved functions of these transcription factors (Adam et al. 1998, Fritzsch et al. 2000, Hassan & Bellen 2000, Eddison et al. 2001). Thorough understanding of the avian and mammalian inner ear innervation development requires much greater understanding in general of the molecular programs of ear development.

Origin and Timing of Cochlear Ganglion Neuron Proliferation

Delamination of postmitotic ganglion neurons appears to happen at the level of the placode and during invagination and cochlear elongation in mammals and birds (Rubel 1978, Carney & Silver 1983, Adam et al. 1998). Various in vitro and in vivo manipulations suggest that all ganglion neurons derive from the anteroventral quadrant of the otic vesicle (Noden & van de Water 1986). In mice, cells were found to delaminate from the rostrolateral wall (Carney & Silver 1983), whereas data from chicken suggest a somewhat more medial origin (Hemond & Morest 1991b, Adam et al. 1998). From the literature, however, it is usually not clear whether early delaminating neurons will contribute to the auditory or vestibular neuronal group. It is also unclear through what means the ganglion neurons find their way through the basement membrane surrounding the otocyst (Legan & Richardson 1997). Ganglion neurons do not derive from the same sites of the otic vesicle they later innervate (Noden & van de Water 1986). However, this conclusion has not been verified using selective tracing and remains open to alternative interpretations. Observations in mice suggest that cochlear ganglion neurons emigrate from the anlagen of the cochlear epithelium (Altman & Bayer 1982) and project back to the same region of the cochlea along the route of delamination and migration (Carney & Silver 1983). But again, this suggestion has yet to be verified using modern labeling methods.

Some early suggestions on the origin of ganglion cells have been substantiated recently using the expression of neurotrophin marker genes. Basal- and middle-turn cochlear ganglion neurons transiently express the neurotrophin 3 (NT-3) at the time of or soon after delamination from the developing sensory epithelium, and the brain-derived neurotrophic factor (BDNF) is expressed in the developing apical turn neurons (Fariñas et al. 2001). In addition, these delaminating neurons all express the neuronal developmental marker (NeuroD) (Ma et al. 1998, Kim et al. 2001), a differentiation-regulating transcription factor. Delaminating ganglion neurons can also be identified based on the early expression of the LIM-gene islet 1 and other early neuronal markers (Adam et al. 1998). Most diagnostic may be the early expression of the zinc finger gene GATA-3 in cochlear ganglion neurons (Rivolta & Holley 1999). Karis et al. (2001) suggest that GATA-3 is exclusively expressed in delaminating cochlear ganglion neurons, providing additional evidence that these neurons do not derive from neural crest (which is GATA-3 negative) and indicating that they are molecularly distinct from the nearby vestibular ganglion neurons at this early stage of development. Other genes, such as fibroblast growth factor (FGF)3, FGF10, and BF1, also appear to be exclusively expressed in delaminating inner ear–ganglion neurons (Hatini et al. 1999, Pirvola et al. 2000). Conversely, all Schwann cells around the inner ear–ganglion neurons are derived from neural crest (Noden & van de Water 1986). In summary, the in vivo evidence in mice suggests multiple sites of delamination of ganglion neurons. It remains unclear whether delaminating ganglion neurons project precisely back to their place of origin in an orderly fashion, thereby establishing the tonotopic organization of cochlear ganglion cells.

The evidence from chicken and mice suggests that many of these delaminating cells are neuroblasts that may undergo further divisions before they differentiate and express neuron-specific markers (D'Amico-Martel 1982, Adam et al. 1998, Fariñas et al. 2001). However, judging from the expression of NeuroD, some cells may become postmitotic inside the otocyst (Ma et al. 1998). Recent experimental tracing studies have shown in mice an occasional differentiated cochlear ganglion neuron that projects to the brain while the cell body remains in the otocyst (Bruce et al. 1997). It therefore remains unclear how many cochlear ganglion neurons become postmitotic in the otocyst wall compared to the number that are derived from the delaminated neuroblasts.

In mice, Ruben (1967) showed opposite spatial-temporal gradients of hair cell and cochlear ganglion neuron proliferation (Figure 2). Auditory ganglion neurons become postmitotic around embryonic day E11.5 to E15.5 (with peak at E13.5) in a basal-to-apical gradient. In contrast, hair cells become postmitotic between E11.5 and E15.5 (peak E13.5) in an apical-to-basal gradient. These data suggest that the earliest maturing cochlear ganglion neurons in the basal turn project to the latest forming hair cells, provided there is no radical change in the developmental timetable. Specifically, at about E12.5, processes from the earliest differentiating basal turn ganglion neurons (Tello 1931) reach an area of the cochlea that has no postmitotic inner hair cells until about E13.5.

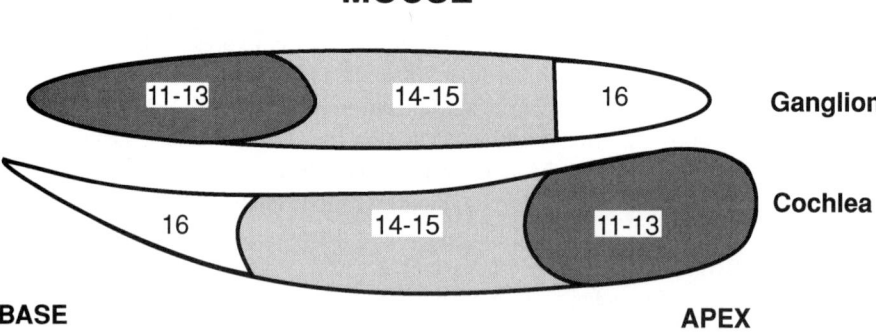

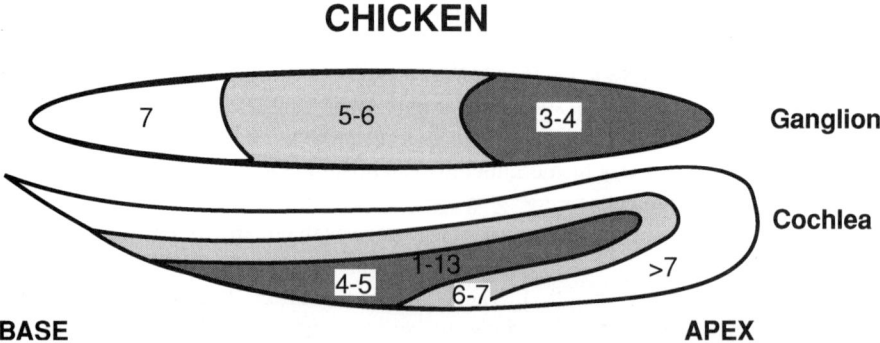

Figure 2 The "birthdates" of cochlear ganglion neurons and cochlear hair cells are shown for mice (*top*) and chickens (*bottom*) as revealed with ^{3}H-thymidine.

In chickens, proliferation of cochlear ganglion neurons is between E4 and E7 in an apical-to-basal gradient (D'Amico-Martel 1982). Proliferation of hair cells in the cochlea is between E5 and E8 (Katayama & Corwin 1989). Thus some early fiber outgrowth of ganglion neurons (day 3–5; Whitehead & Morest 1985) may happen before hair cells are postmitotic, as may happen with the basal turn in mice. The pattern of hair cell proliferation shows some gradients but not any opposite to those of the ganglion cells. In addition, unlike the mouse there is a largely nonoverlapping proliferation period of ganglion neurons and hair cells in the chicken. Figure 2 describes these patterns graphically. Although they define important developmental events and windows for cellular interactions, it remains unclear whether the topography and timing of proliferation of ganglion neurons relate to the topography of connections with the cochlea.

Proneuronal Genes and Development of Cochlear Ganglion Neurons

In mice, formation of ganglion neuron precursors that express the proneuronal gene neurogenin-1 (ngn-1) is an early event in ear development (Ma et al. 1998, 2000). Ngn-1 is expressed in delaminating ganglion neuron precursors from the anteroventral quadrant of the E9 mouse otocyst. However, later expression of ngn-1 in the growing cochlea has not been investigated. Ngn-1 is rapidly followed by the expression of NeuroD in the otocyst and in delaminating ganglion neurons as well as the genes for delta and notch that are involved in further selection of the proneuronal cells (Ma et al. 1998, Adam et al. 1998, Liu et al. 2000, Kim et al. 2001). Null mutants of ngn-1 show absence of ganglion neurons at any stage in development (Ma et al. 1998, 2000). In addition to the loss of all ganglion neurons, ngn-1-null mutants show a variety of defects in the sensory epithelia, including loss of hair cells; the saccule is most severely affected (Ma et al. 2000). In other systems such as the neural crest, sensory neuron precursor cells die in the ngn-1-null mutants (Ma et al. 1999); logical extension suggests that ganglion neuron precursor cells also die in the ear of ngn-1-null mutants, but this has not been determined. Interestingly, the zebrafish mutant mindbomb shows an exuberant formation of inner–ear ganglion neurons and hair cells, likely owing to disregulation of the delta/notch system (Eddison et al. 2001). However, more data on ngn-1 and Math1 expression are needed in these mutants to fully understand if the delta/notch disregulation affects the expression of these genes.

Are there homologies between insect sensory organs and the cochlear ganglion cells in mammals and birds (Hassan & Bellen 2000)? The insect homolog of neurogenins, tap, is not expressed in mechanosensory organs, but tap is expressed in a small subset of gustatory sensory neurons (Goulding et al. 2000). Likewise, the single known ancestral chordate homolog of neurogenins, Amphineurogenin (Holland et al. 2000), is not expressed in the peripheral nervous system of the lancelet Amphioxus, a chordate without an ear. It appears therefore that the involvement of ngn-1 in the inner-ear ganglion neuron formation is a phylogenetic novelty that comes about by co-opting a novel bHLH gene, ngn-1, into the otherwise conserved developmental program of a mechanosensor (Fritzsch et al. 2000). Interestingly, this co-option coincides with the formation of a unique set of cells in vertebrate ear development. These cells are the ganglion neurons that contact the mechano-electric transducers, the hair cells. The unique dependence of inner–ear ganglion neurons on ngn-1 (Ma et al. 1998, 2000), a gene not expressed in any insect mechanosensory organ (Goulding et al. 2000), renders less likely the idea of homology between the insect bristle mechanosensory neurons and the vertebrate inner-ear ganglion neurons (Adam et al. 1998, Eddison et al. 2001). Clearly, the entire problem of how ganglion cell dendrites find their hair cell targets does not exist in insects, in which the dendrites contain the mechano-electric transducers. Thus, in this case, mechanisms cannot be conserved across phyla.

Establishing Polarity

During further differentiation, cochlear ganglion neurons must establish polarity and produce one process growing toward the brain (an axon) and another process growing toward the cochlea (a dendrite). The growing peripheral processes must become sorted at their targets according to at least two properties: i) They eventually innervate only inner or outer hair cells (Figure 1). Type I ganglion cells project to inner hair cells via the radial fiber bundle. Type II ganglion cells innervate outer hair cells and form the outer spiral fibers. ii) Cochlear ganglion neurons establish a precise longitudinal cochleotopic projection with the hair cells that is maintained in the connections with the cochlear nuclei. This cochleotopic projection is the anatomical substrate for frequency-specific sound processing, one of the basic principles of hearing (Dallos et al. 1996).

In birds there does not appear to be a strict segregation of ganglion cell types and hence of differential innervation to hair cell types. Nevertheless, there is a clear gradient of the density of afferent innervation across the superior-to-inferior dimension of the cochlea (Takasaka & Smith 1971, Whitehead & Morest 1985, Fisher 1992). How this gradient emerges during development and how it might relate to molecular events are completely unknown. Most interesting is the fact that some hair cells do not appear to receive any afferent innervation (Fischer 1992), an observation that is hard to reconcile with a sensory function for these hair cells, unless they are electrically coupled to other cochlear elements such as supporting or hyaline cells.

Available neuroanatomical data suggest that the polarity of delaminating ganglion neurons may already be established inside the otic wall. In fact, it is conceivable that polarity is retained during the asymmetric divisions that give rise to these cells, as is suggested for neurons (Trimmer 1999). If that is the case, the leading process of these delaminating cells could be viewed as the axon precursor, whereas the trailing process would be the designated dendrite. Divergent views also exist (Hemond & Morest 1991b).

Anatomical studies of early cochlear ganglion neuron morphology suggest that all ganglion neurons are bipolar, with one process directed toward the brain and another toward the cochlea (Retzius 1893, Perkins & Morest 1975, Ginzberg & Morest 1983, Whitehead & Morest 1985, Hemond & Morest 1991a). As noted above, some cells may establish an axonal projection to the brainstem before they actually delaminate (Bruce et al. 1997), but clearly they are a minority. Most ganglion neurons extend processes into the developing sensory epithelia before they have reached their adult position (Carney & Silver 1983) and prior to formation of a central process. In chickens, available evidence suggests a uniform phase of fiber outgrowth toward the cochlea, followed by a period of invasion into the sensory epithelium (Whitehead & Morest 1985). Invasion of the sensory epithelium by processes appears to start at the basal end of the sensory epithelium in both birds and mammals (Tello 1931). In contrast to the progressive basal-to-apical innervation pattern seen in mice, the apex and then the middle part of the cochlea appear to follow innervation of the base in chickens (Hemond & Morest 1991a).

Establishing the precise time delays between mitosis, delamination, establishing polarity, and initial process formation is necessary in order to evaluate the molecular cues with which the ganglion neurons establish polarity and topographically correct connections with the cochlea. Should the suggestions of Carney & Silver (1983) prove correct, pathfinding of ganglion neuron projection into the sensory epithelium would be a simple matter of the delaminating neuroblast process following already established "highways." More recent evidence showing origin of delaminating ganglion neurons from various areas of the otocyst is compatible with the idea that ganglion neurons project back to their areas of origin (Fariñas et al. 2001). However, a rigorous experimental evaluation of these ideas is needed.

Development of Types and Topography of Afferent Innervation of the Cochlea

The development of individually labeled axons and small groups of afferent axons to hair cells in the cochlea has been studied in a variety of species (Retzius 1893, Perkins & Morest 1975, Ginzberg & Morest 1983, Echteler 1992, Sobkowicz 1992, Bruce et al. 1997, Pujol et al. 1998). In adult and neonatal mammals, ganglion neurons project via the radial fiber bundles to the cochlea. How neurons select a given radial bundle during embryonic development is unclear. Ultimately, this decision may determine what frequency region of the cochlea a given ganglion neuron axon enters in the sensory epithelium. As beautifully demonstrated in the Golgi preparations of Lorente de Nó (1981), a fiber may either enter a bundle ajacent to its ganglion neuron, or a bundle several radial bundles away. In fact, classic work by Retzius (1893) suggests that single developing cochlear ganglion neurons may form multiple branches, each entering a different radial bundle in neonatal rodents. To our knowledge, this has not been confirmed by modern cell labeling methods. One view of development suggests that afferent fibers may form a dense plexus that will be reduced upon reaching the cochlea by differential pruning (Sobkowicz 1992). If this view can be substantiated by experimental tracing studies, it implies that pruning of afferent fibers is used to produce a precise topographic distribution of ganglion cell to hair cell connections in embryos, prior to the ontogeny of frequency-specific sound activation of ganglion cells that occurs postnatally in most mammals (Rübsamen & Lippe 1998). The mechanism(s) by which this pruning would be governed to achieve the appropriate topography of connections remains unclear.

Irrespective of issues concerning the longitudinal topography of afferent innervation of the cochlea, there is a major issue concerning the radial innervation. At a given longitudinal position along the cochlea, how do Type I and Type II ganglion cell innervation become segregated to inner and outer hair cells, respectively? One view is that the determination of a ganglion cell to the Type I or Type II phenotype is not made until after its peripheral connections are established (peripheral instruction hypothesis). Another view is that cell type is determined upon completing the final mitotic division, well before innervation of hair cells (cell-autonomous instruction hypothesis). Studies on neonatal kittens, gerbils, and mice demonstrate that early in development some afferents branch more extensively than in adults and are sometimes difficult to assign to either of the two recognized cochlear ganglion neuron types. It is possible that each fiber branches to both inner and outer hair cells (Perkins & Morest 1975, Echteler 1992, Sobkowicz 1992, Simmons 1994, Bruce et al. 1997, Wiechers et al. 1999). Some older studies using Golgi techniques claimed that there are three types of terminals in neonatal animals (Ryugo 1992). In neonatal cats and rats, so-called giant fibers project to a group of inner hair cells rather than to a single inner hair cell (Perkins & Morest 1975). Moreover, some investigators claim that three types of spiral neurons are present in some pathological material (Rosbe et al. 1996), and some physiological data also suggest that there are three distinct types of cochlear ganglion neurons (Liberman & Oliver 1984). Likewise, others have identified a third, possibly transient, type of cochlear ganglion neurons that contribute fibers to the inner spiral bundle for a short distance (Sobkowicz 1992).

The most detailed experimental studies of this issue suggest that cochlear ganglion neuron processes arrive simultaneously at a given radial position across the cochlea and form arbors to both inner and outer hair cells (Simmons et al. 1991, Pujol et al. 1998). Ultrastructural data suggest that development of afferent terminals is, in fact, rather simultaneous for both inner and outer hair cells with a very short delay to the outer hair cells (Pujol et al. 1998). Immunocytochemical data using neurofilament immunocytochemistry suggest widespread distribution of fibers in the developing cochlea (Chen & Segil 1999), which is also noted in recent axon tracing studies (Bruce et al. 2000). Kainate sensitivity experiments suggest that individual ganglion cells innervate both inner and outer hair cells early in development (Pujol et al. 1998). Following this period of exuberant innervation in the radial dimension, arbors mature over several days and appear to become segregated to individual hair cell types by differential pruning (Echteler 1992).

Some studies assigned a specific type of cochlear ganglion neuron to each fiber investigated and suggest that each fiber type has additional side branches during early development, which are eventually pruned in later neonatal stages (Perkins & Morest 1975, Ginzberg & Morest 1984, Simmons 1994, Wiechers et al. 1999). For example, those afferents projecting to inner hair cells first arborize among several adjacent inner hair cells and later become restricted to a single inner hair cell (Simmons et al. 1991). Some of these radial afferents to the inner hair cells may develop temporary side branches that extend for several cells along the inner spiral bundle (Ginzberg & Morest 1984) and to outer hair cells (Simmons 1994, Wiechers et al. 1999).

The basal-to-apical gradient of ganglion neuron proliferation may be retained as a basal-to-apical gradient of ganglion neuron differentiation in mammals (Rubel 1978). In contrast, differentiation appears to be more uniform along the chicken cochlea (Hemond & Morest 1991b), with no discernable relationship to the known proliferation gradient of ganglion neurons. Recent immunocytochemical data suggest correlations between the timing of expression of specific molecules and segregation processes related to innervation of the cochlea (Whitlon et al. 1999b). Tenascin-C, n-CAM, and other adhesion molecules have been found in the developing cochlea during this general time period (Legan & Richardson 1997). Other data correlate fiber growth with laminin and fibronectin expression (Hemond & Morest 1991a). However, no experimental evidence for causal relationships exists. Furthermore, no molecular signals that may be responsible for early pathfinding of cochlear ganglion neuron processes toward the cochlea have been experimentally identified.

In summary, all cochlear ganglion neurons are identified by the unique markers BF1 (Hatini et al. 1999), GATA-3 (Karis et al. 2001), and FGF10 (Pirvola et al. 2000). Speculations about differential origin of the two types of mammalian cochlear ganglion neurons (Pujol et al. 1998) are not supported by these data. It remains unclear whether the phenotype (Type I or II) is determined prior to or only after innervation of hair cells and if the innervation pattern plays a role in this process. It is conceivable that there is excess branching and pruning during early development to correct for innervation pattern "errors" in the radial or longitudinal dimensions, but the evidence for any major reorganization, particularly along the longitudinal axis, is weak at best. Immunocytochemical data are thus far of little help as they can distinguish the types of cochlear ganglion neurons only after their peripheral innervation has acquired its distinct morphology in neonates (Hafidi & Romand 1989, Pujol et al. 1998, Hafidi 1999a). Combined studies using neuroanatomical tracing or immunocytochemistry in conjunction with birthdating by ^{3}H-thymidine or bromodeoxyuridine (BrdU) are needed to establish whether or not birthdates along the length of the cochlear ganglion translate into ganglion neuron phenotype. In this regard, it is interesting that possibly more Type II ganglion neurons are present among the latest-forming cochlear ganglion neurons of the apex (Ruben 1967, Ryugo 1992).

In Vitro Studies on Fiber Growth Mechanisms

In general, the process of axonal guidance appears to depend on interactions between the growth cones, extracellular matrix, and target epithelium (Hong et al. 2000). These interactions are controlled by adhesive and chemotropic influences that mediate attraction, repulsion, and distinct checkpoints for turns. Of the many neuronal guidance molecules identified in the last few years (e.g., semaphorins, neuropilins, netrins, ephrins) and various adhesion molecules, few have been well characterized in the developing inner ear. Studies of the distributions of ephrins and eph receptors have yielded interesting distribution patterns, but targeted mutations of EphB2 reveal only some delay in the onset of hair cell innervation (Bianchi & Liu 1999, Cowan et al. 2000). An emerging picture of a temporospatial map of adhesive molecules in the developing mouse cochlea (Legan & Richardson 1997, Whitlon et al. 1999a,b) appears quite promising, but a causal relation has yet to be established between any of these molecules and innervation patterns.

Guidance by neurotrophins has been studied in great detail in chicken tissue cultures. In fact, BDNF was the first factor to be linked to ganglion cell survival and neuritogenesis in the developing chicken inner ear (Lindsay et al. 1985, Robinson et al. 1996). FGFs, known to be expressed in developing ganglia (Pirvola et al. 2000), seem to enhance expression of the BDNF-specific receptor trkB (Brumwell et al. 2000), and thereby promote neuritogenesis. In addition, evidence exists for an as yet uncharacterized neurotropic factor (or factors) that attracts the growing afferent neurites to the developing sensory epithelia (Hemond & Morest 1992, Bianchi & Cohan 1993). These experiments also established that neither nerve growth factor (NGF) nor BDNF plays this role (Bianchi & Cohan 1993).

Some information on the role of neurotrophins in ganglion cell development also exists from in vitro experiments on mouse tissue (van de Water et al. 1992), but the roles of murine neurotrophic factors are even less well characterized than those in chickens. Early in vitro work suggested that NGF is released in the murine inner ear and plays an important role in neuritogenesis but not in survival of cochlear ganglion cells (van de Water et al. 1992). However, detailed examinations using in situ hybridization for neurotrophins and their high-affinity receptors could not detect any expression of NGF in the developing rodent ear (Pirvola et al. 1994) and only very transient expression of the NGF-specific receptor, trkA, in delaminating ganglion neuron precursors (Fritzsch et al. 1999). It is unlikely that this transient expression of trkA could mediate the reported neuritogenesis-promoting activity. Like they may do in the chicken, FGFs may play an important role in neuritogenesis and migration of cochlear ganglion neurons in the mouse (Hossain & Morest 2000).

Cultures of mouse cochlea develop afferent arbors comparable to same-aged littermates. This suggests that afferent fibers may be capable of innervating inner hair cells located as far as 600 μm away (Sobkowicz 1992). In contrast to data from chickens, organ culture experiments in mice in which hair cells have been eliminated demonstrate that cochlear ganglion neurons nonetheless grow normally toward their target (Sobkowicz 1992), suggesting that pathfinding mechanisms for ganglion neuron afferents are present within the developing spiral limbic tissue. These data argue against a simple sorting mechanism via attraction toward hair cells. Moreover, studies of ganglion cells from the Bronx-Waltzer mutant, which loses inner hair cells, show that axons grow abundantly to the outer hair cells. Conversely, destroying outer hair cells with gentamycin results in extensive looping of what has been interpreted to be Type II fibers around inner hair cells (Sobkowicz 1992).

Together, these in vitro studies suggest that some fiber outgrowth from ganglion neurons is independent of both a tropic and trophic signal from the hair cells, but probably requires permissive (or even instructive) substrates for navigation toward specific cochlear targets. This conclusion is supported by the apparent difference in timing of hair cell and sensory neuron proliferation, especially in chickens (D'Amico-Martel 1982, Katayama & Corwin 1989). These in vitro data also suggest that some properties of Type I and Type II spiral neurons are intrinsic to the neurons and only expressed during establishment of innervation with the cochlea hair cells rather than being induced by the cochlea.

Pathfinding in Mutant Mammals

The availability of mutant mice with severely altered hair cell development and/or afferent projection patterns to the cochlea allows exploration of some of these issues for the first time in vivo. Two of these mutations belong to the POU family of genes (Brn 3a and 3c, also known as Brn 3.0 and Brn 3.2). Initial data in Brn 3c–null mutants suggested absence of hair cells and loss of all innervation in neonates (Erkman et al. 1996, Xiang et al. 1997). However, closer examination revealed the presence of undifferentiated hair cells (Xiang et al. 1998) as well as a rather normal supply of axons at birth (B. Fritzsch, unpublished observations). These data suggest that either undifferentiated hair cells can provide the proper cues for normal axon outgrowth or that hair cells are not necessary for appropriate axon outgrowth. Preliminary data in Math1-null mutants, which never develop any hair cells (Bermingham et al. 1999), also support the notion that some pathfinding by ganglion neurons is not mediated by hair cells.

Early evidence suggested that Brn 3a has a direct effect on migration and survival of cochlear ganglion cells (McEvilly et al. 1996, Ryan 1997). Closer examinations have confirmed defects in projection and survival of inner-ear ganglion neurons but suggest that Brn 3a regulates a single neurotrophin receptor, trkC (Huang et al. 1999), and that the innervation defects in the cochlea can be largely attributed to the absence of this receptor (Huang et al. 2001). Additional innervation deficits that cannot be correlated with neurotrophin-related defects exist in the vestibular system and suggest direct involvement of Brn 3a in pathfinding of some but not all inner-ear ganglion neurons (Huang et al. 2001). The neuronal differentiation gene NeuroD also affects ganglion neuron migration, survival, and pathfinding (Kim et al. 2001). Again, only the neurotrophin receptors have been identified as immediate downstream genes thus far.

The Role of Neurotrophins and Neurotrophin Receptors in Ganglion Neuron Survival

The family of neurotrophin ligands and their receptors consists of four mammalian ligands: NGF, BDNF, NT-3, and neurotrophin 4/5 (NT4/5). The three high-affinity receptors are tyrosine kinase (trk)A, trkB, and trkC. Each neurotrophin forms a homodimer that causes homodimerization of the specific receptor for appropriate intracellular signaling (NGF with trkA; BDNF and NT4/5 with trkB; NT-3 with trkC). In addition, the inner ear contains the low-affinity receptor, p75 (von Bartheld et al. 1991). In general, neurotrophins are thought to provide molecular signals that mediate survival of neurons (Reichardt & Fariñas 1999).

BDNF was the first neurotrophin associated with inner-ear ganglion neuron development (Lindsay et al. 1985). In situ hybridization showed that BDNF and NT-3 are synthesized in the sensory epithelium of the otic vesicle and their high-affinity receptors, trkB and trkC, are synthesized in cochlear ganglion neurons of mammals (Pirvola et al. 1992, 1994, Schecterson & Bothwell 1994, Wheeler et al. 1994) and birds (Pirvola et al. 1997, Cochran et al. 1999). Targeted deletions of the appropriate gene alone or in combination have shown that the ligands, BDNF and NT-3, and their cognate receptors, trkB and trkC, are essential for survival. For example, in neonatal mice with targeted deletions of both the BDNF and NT-3 genes, there is a complete loss of all ganglion neurons (Ernfors et al. 1995, Liebl et al. 1997). Likewise, neonatal mice in which both the trkB and trkC receptor have been deleted lose all ganglion neurons (Fritzsch et al. 1995, Minichiello et al. 1995, Schimmang et al. 1997).

Variations in the patterns of innervation caused by each single receptor or ligand deletion initially suggested a simple solution to many of the questions posed above regarding innervation of the cochlea. Work by Ernfors et al. (1995) suggests that BDNF supports innervation of outer hair cells, whereas NT-3 supports the innervation of inner hair cells. The loss of about 85% of cochlear ganglion neurons in NT-3-null mutants (Fariñas et al. 1994, Ernfors et al. 1995) and of about 15% of cochlear ganglion neurons in BDNF-null mutants (Jones et al. 1994, Ernfors et al. 1995) relate closely to the proportion of Type I and Type II cochlear ganglion neurons, around 92% and 8%, respectively (Romand & Romand 1987). Apparently, confirming this conclusion were data claiming a complete loss of afferent innervation of the inner hair cells in mice with deletion of the trkC gene and of afferent innervation to outer hair cells in trkB-null mice (Schimmang et al. 1995).

Unfortunately, further analysis revealed that the segregation of Type I and Type II afferents on the basis of the specific neurotrophins is not that simple. Detailed studies using in situ hybridization for neurotrophins and their receptors (Pirvola et al. 1992, 1994; Wheeler et al. 1994), immunocytochemical investigations of neurotrophin receptor distribution (Fariñas et al. 2001), and analyses of neurotrophins using the sensitive lacZ reporter technique (Fariñas et al. 2001) all suggest complete overlap of the two neurotrophin receptors, trkB and trkC, in all ganglion cochlear neurons during the relevant phases in embryonic development. Likewise, whereas NT-3 becomes concentrated into inner hair cells in neonates (Ernfors et al. 1995, Fritzsch et al. 1999), neither BDNF nor NT-3 shows segregation into inner and outer hair cells, respectively, during embryonic development (Fariñas et al. 2001). TEM analysis showed afferent synapses on inner hair cells in trkC-null mutants, and afferent and efferent synapses were found on basal-turn outer hair cells in trkB-null mutants (Fritzsch et al. 1997a). This was recently confirmed for mice with BDNF-null mutations using immunocytochemistry (Wiechers et al. 1999).

A different but equally interesting pattern of changes in ganglion cells and innervation following manipulations of neurotrophin genes has recently emerged. BDNF and trkB-null mutant mice show the most pronounced effect in the apex, whereas the effects of either trkC or NT-3-null mutations are most severe at the cochlear base (Fritzsch et al. 1995, 1999). In fact, in NT-3-null mutants all cochlear ganglion neurons in the basal turn are absent at birth. Labeling of afferents using DiI as a tracer reveals that afferent axons from more apical locations innervate groups of inner hair cells while most outer hair cells of the basal turn remain uninnervated (Fritzsch et al. 1997b). These data could not establish whether these afferents are derived from Type I or Type II ganglion cells.

In summary, the attraction of Type I and Type II afferents to inner and outer hair cells, respectively, is not simply related to the expression of specific neurotrophins. Rather, it appears that a given neurotrophin always mediates the reduction or loss of Type II afferent innervation to outer hair cells. This reduction is most severe at the base in the case of an NT-3 deletion and at the apex when BDNF is deleted. The effects of both neurotrophins are not related simply to a radial influence but appear to be related to a longitudinal gradient that is transformed into a radial effect at each end of the cochlea.

Nevertheless, there appears to be some connection between neurotrophins and the specificity of hair cell innervation in the cochlea. Type II outer spiral fibers always enter the outer hair cell region relatively apically and turn toward the base to innervate a series of outer hair cells. In NT-3-null mutants this organization is disrupted and axons turn in both directions (Fritzsch et al. 1997b). It is possible that a longitudinal-temporal (or spatial-temporal) gradient of neurotrophin expression (Fariñas et al. 2001) conveys this peculiar feature of outer spiral fibers. The mechanism could be opposing attractions of BDNF and NT-3 on axon growth (Song & Poo 1999). In fact, the various disruptions of pathway selection found in mutant mice missing a single neurotrophin gene suggest that at least two neurotrophins are needed to establish the appropriate conditions for innervation and for the trajectory of Type II cochlear ganglion neuron axons (Figure 1).

Data from transgenic mice in which NT-3 has been replaced by BDNF fully support the idea that the cellular specificity between cochlear ganglion neurons and hair cells is not mediated by specific neurotrophins. As expected by the well-established uniform distribution of trkB and trkC in spiral neurons, expression of BDNF instead of NT-3 leads to a complete rescue of the NT-3 phenotype in embryos (Fariñas et al. 2001). Expression of BDNF in place of NT-3 can rescue the BDNF-null phenotype in neonates, thereby establishing that the temporal dynamics of expression of these neurotrophins is critically important (Coppola et al. 2001).

As an aside, it is interesting to note that the apparently novel evolution of inner-ear ganglion neurons (Fritzsch et al. 2000) is correlated with another evolutionary novelty, neurotrophin-mediated cell survival (Hallböök 1999). The most prominent and evolutionary-conserved receptor expressed in the ear is trkB, which also may be the ancestral trk receptor gene (Hallböök 1999). It appears that among land vertebrates NT-3 has evolved into the main supporting neurotrophin for the cochlear ganglion neurons of the mammalian ear. In contrast, only limited expression of NT-3 has been found in chickens (Pirvola et al. 1997). Importantly, NT-3 dependent neurons innervate the unique high-frequency, basal, part of the mouse cochlea, which represents a novel addition of ganglion neurons with numerous features not shared with nonmammalian vertebrates.

In the foregoing descriptions we have been considering neurotrophin expression during the early period of differentiation and process growth by cochlear ganglion neurons. These interactions occur at approximately E14–18 in mice and E5–8 in chicks. Later in development there is a period of programmed cell death of cochlear ganglion cells that may be mediated by neurotrophins as well. In neonatal rats approximately 22% of cochlear ganglion neurons die between postnatal day (P)0 and P6 (Rueda et al. 1987) and 25%–33% of chicken cochlear ganglion cells die at E8-14 (Ard & Morest 1984). Distinctive patterns of changes of neurotrophin expression have been described in neonatal mice and gerbils both in vitro and in vivo (Mou et al. 1997, 1998, Wiechers et al. 1999) that may be related to neuronal survival (Hegarty et al. 1997) and may tie into the extensive fiber reorganization noticed during that period in mammals (Pujol et al. 1998). Neurotrophins can also rescue mature cochlear ganglion neurons following various insults (Agerman et al. 1999).

Synaptogenesis Between Hair Cells and Afferent Fibers

Both light- and electron-microscopic data suggest that processes of cochlear ganglion neurons reach the hair cells very early in development (Retzius 1893, Hafidi & Romand 1989, Pujol et al. 1998). Afferent fibers of ganglion neurons have been found so early in the vicinity of both outer and inner hair cells that speculations have recurred many times suggesting a role for afferents in hair cell maturation (Rubel 1978, Schimmang et al. 1995, Pujol et al. 1998). However, a variety of studies have shown that chick otocysts can develop morphologically normal hair cells when transplanted in the absence of the cochlear ganglion neurons (Waddington 1937, Corwin & Cotanche 1989, Swanson et al. 1990). Waterman (1938), for example, described rather normal development of hair cells in transplanted rabbit ears. Furthermore, autonomous morphological differentiation of hair cells is observed in organ culture of otocysts from chicks (Fell 1928; Friedmann 1956; Orr 1981, 1986; Sokolowski et al. 1993; Ard et al. 1985) and mammals (van de Water 1983, van de Water et al. 1992, Sobkowicz 1992) that were stripped of ganglion neurons. Data on various neurotrophin mutants as well as on ngn-1-null mutants are consistent with this conclusion (Silos-Santiago et al. 1997, Ma et al. 2000). Recent physiological studies suggest that auditory and vestibular hair cells can acquire their normal specialized physiological properties in the absence of innervation (He & Dallos 1997, Rüsch et al. 1998). An influence of innervation on long-term maintenance of hair cells has been suggested (Walsh et al. 1998) and awaits further stringent testing. Recent data suggest that hair cells can survive in long-term denervated ears of NeuroD-null mutants in vivo (Kim et al. 2001).

Clearly, a major, as yet unexplored issue is how the over 10 Type I afferent endings per inner hair cell manage to converge and compete for synaptic space on a single inner hair cell. One idea is that Type I afferents arrive earlier and occupy the available space on the inner hair cells, thus leaving synaptic space only on the outer hair cells for the Type II spiral afferents (Echteler 1992). More recently, synaptic reorganization in neonates was studied in some detail (Knipper et al. 1995, Wiechers et al. 1999). These studies do not address the issue of how inner hair cells receive multiple Type I afferents, since these are already in place at birth (Echteler 1992, Simmons 1994, Bruce et al. 1997, Fritzsch et al. 1997b).

Little is known about the embryonic growth and development of afferents before they reach the inner hair-cell region or about the mechanisms through which many Type I afferents converge onto a single inner hair cell. This lack of knowledge is due to a variety of factors. Most studies on this topic have used stains that label all axons (Sobkowicz 1992), thereby being unable to distinguish between afferents and efferents (Lorente de Nó 1981). Others have focused on only a restricted area of the cochlea such as the apex (Echteler 1992) or exclusively on postnatal ages (Simmons 1994, Wiechers et al. 1999).

Detailed ultrastructural data suggest that cochlear ganglion neurons contact both inner and outer hair cells almost at the same time and develop classic afferent synapses prior to the formation of hair cell–specific apical specializations (Pujol et al. 1998). In addition, both pioneering work (Pujol et al. 1998) and recent DiI data suggest that a second type of ending, the efferent fibers of neurons that segregate from facial branchial motoneurons (Fritzsch & Nichols 1993, Bruce et al. 1997, Karis et al. 2001), reach the developing hair cells at about the same time. This suggests that establishing functional contacts is mediated by factors unrelated to the electrosensory properties of hair cells. Adult synapses appear after a period during which afferents show numerous filopodia extending in the nearby greater epithelial ridge and numerous presynaptic bodies are found in hair cells.

There is a large body of literature dealing with the predominantly postnatal segregation of afferent and efferent contacts to outer and inner hair cells. These discussions revolve around the possible role played by the changing topography of afferent and efferent innervation to outer and inner hair cells (Pujol et al. 1998, Wiechers et al. 1999, Liberman et al. 2000, Bruce et al. 2000). A clear picture has not yet emerged and the reader is referred to the various positions held by the above cited authors for detail. Two issues are important to keep in mind: The cochlea shows a progressive basal-to-apical gradient of development and early contacts may not be readily identifiable as belonging to a specific fiber type. We summarize here what has been confirmed by several studies and likely will not be modified by future, more sophisticated analyses.

Data in other developing systems suggest that synapse formation and some electrochemical transmission start within hours or a few days after initial contacts are established. These early electrical events are often rhythmic and may not be related to any external stimulus (Sanes & Walsh 1998). This suggests that the onset of synaptic transmission has to happen in mice, for example, several days prior to birth. In fact, recent studies suggest that the neurotransmitter receptor necessary for the function of efferent fibers ending on hair cells is expressed prenatally in mice, around the time the efferent fibers have first been seen near cochlear hair cells (Bruce et al. 1997, 2000; Zuo et al. 1999). The temporal pattern of expression of this receptor follows a base-to-apex and inner hair cell to outer hair cell progression (Simmons & Morley 1998, Zuo et al. 1999). This upregulation of expression starts at E16 in the base of the mouse cochlea and reaches the apex by P4. This pattern of expression suggests that maturation of efferent synaptic transmission, as evidenced by the expression of the main postsynaptic receptor, extends over at least eight days of development. Consequently, future analyses describing efferent development should specify in detail the area of the cochlea examined.

The details of afferent transmitter and receptor development are not as well worked out as those of the efferent transmission. Nevertheless, recent studies of various glutamate receptors during development have shown a dynamic change of their expression (Wiechers et al. 1999) and composition (Luo et al. 1995a), and some synaptic vesicle release proteins have been analyzed (Knipper et al. 1995). Minimally, we need more information on the development of synaptic transmission from hair cells to afferent axons and the relationship of these events to changes in the topography of connections. Such studies should be feasible with the discoveries of specific proteins related to glutamatergic transmission (Jahn & Südhof 1999, Takamori et al. 2000, Fukuda et al. 2000, Verhage et al. 2000) and methods for labeling individual axons.

DEVELOPMENT OF CENTRAL PROJECTIONS AND TROPHIC REGULATION OF CNS TARGETS

The development of central projections of eighth-nerve ganglion cells has been studied in a variety of ways, ranging from descriptive studies using classical silver staining or the Ramon y Cajal/Golgi methods to more contemporary methods using cell-specific markers or axonal tracing. In this section, we summarize recent descriptive and mechanistic studies on the development of the eighth-nerve projection to the brainstem in avian and mammalian species. It is useful by way of organization to consider the ontogenetic series of events that take place in the ganglion cells and their surrounding environment.

After the immature neuron has delaminated from the developing otocyst and undergone its final mitotic division, it forms a centrally directed process that traverses the basal lamina surrounding the lateral aspect of the rhombencephalon and enters the brain parenchyma. This protoplasmic process, the eighth-nerve axon, bifurcates one or more times to send branches into the presumptive CN subdivisions (Lorente de Nó 1981, Fekete et al. 1984). In mammals, most eighth-nerve axons are thought to provide afferents to all three major CN subdivisions, the anteroventral cochlear nucleus (AVCN), the posteroventral cochlear nucleus (PVCN), and the dorsal cochlear nucleus (DCN). In birds, a branch is sent to each of two subnuclei, n. magnocellularis (NM) and n. angularis (NA). During development of these projections, the axons must arrange themselves in precisely the same order as their peripheral targets in the cochlea. In other words, the frequency/place organization of the sensory epithelium that is mapped onto the population of ganglion cells must be exactly recreated in the organization of projections into each division of the cochlear nucleus. This mapping of the receptor surface onto the cells in each division of the CN establishes the precise tonotopic organization seen physiologically and anatomically in the mature animal.

At the same time or shortly after entering the CN, the eighth-nerve axons form different highly stereotyped synaptic specializations that are unique to each target region. The morphology and physiology of the eighth-nerve synapse onto postsynaptic cells in the AVCN become markedly different from those expressed by a collateral of the same axon in the DCN or PVCN. The contacts and synaptic activity transmitted by the cochlear nerve axons can have dramatic influences on the development and maintenance of the target cells in the subnuclei of the cochlear nuclear complex.

In the remainder of this review, we consider each of these topics in order: i) eighth-nerve growth into the brain parenchyma; ii) development of synaptic contacts between eighth-nerve axons and target cells in the cochlear nucleus; iii) emergence of topographic (tonotopic) organization in the cochlear nucleus; and iv) trophic interactions between the eighth-nerve and CN cells. In each topic area, we consider the current descriptive information and the level of understanding of cellular and molecular mechanisms, and we offer suggestions for future investigations.

Axon Development

As early cochlear ganglion axons grow through peripheral connective tissue rich in fibronectin and laminin (Hemond & Morest 1991a,b), they are thought to fasciculate with axons of the vestibular nerve, which have already penetrated the developing rhombencephalon. In the chick, the final mitosis of cochleo-vestibular ganglion cells occurs between E2 and E7, with cochlear cells developing later than vestibular cells (D'Amico-Martel 1982). While cell division is still occurring (i.e., by E3 or H & H Stage 19), some axons enter the medulla (Windle & Austin 1936, Hemond & Morest 1991a). The cochlear processes are probably delayed relative to the vestibular processes by about one day. By Stage 25–26 (E5), many cochlear axons have penetrated the brain parenchyma (Knowlton 1967, Book & Morest 1990). However, it is not clear from the literature exactly when eighth-nerve axons become intercalated among the developing NM and NA neurons (Knowlton 1967, Rubel et al. 1976).

This same close association between the birthdate of ganglion cells and central axon formation in birds is present in mammals. For example, in the mouse, most cochlear ganglion cells are born at around E13.5 (Ruben 1967), but cochlear axons enter the brain by E13–14 (Willard 1993, 1995). While the precise timing of ganglion cell birth dates has not been studied in many species, the age at which eighth-nerve axons enter the brain has been examined in a variety of mammals, including pig (Shaner 1934), rat (Angulo et al. 1990), human (Moore et al. 1997, Ulatowska-Blaszyk & Bruska 1999), and the marsupial *Monodelphis domestica* (Willard 1993).

The growth of axons into the brain occurs well before the onset of hearing as defined by physiological responses to acoustic stimuli. In the E16 rat, axons from ganglion cells originating at the basal turn of the cochlea invade both the AVCN and the PVCN (Angulo et al. 1990). Over the next two to three days, axons from the middle and apical turns enter the nuclear subdivisions. This is approximately two weeks before the rat hears airborne sounds. Similar data are available for the ferret (Moore 1991), opossum (Willard & Martin 1986, Willard 1990), and hamster (Schweitzer & Cant 1984). In the human embryo, cochlear nerve fibers invade the VCN by 16 weeks of gestation, whereas physiological and behavioral responses to sound are not apparent until about 26 weeks (Moore et al. 1997). These studies suggest that innervation forms independently of auditory input.

When the axons enter the brain, collaterals form to innervate the subdivisions of the cochlear nucleus. These collaterals must grow, and must stop growing when they encounter the appropriate target. The cellular and molecular mechanisms underlying the growth and fasciculation, targeting, -branching and cessation of growth of these axons are almost completely unknown. In spite of the emerging wealth of information on axonal pathfinding cues in other systems (Flanagan & Vanderhaeghen 1998, Mueller 1999, Brose & Tessier-Lavigne 2000, Raper 2000), the molecules that alter the pathway selection of auditory nerve axons in the brain have not been identified. Similarly, the cellular interactions that induce growing eighth-nerve axons to bifurcate once or twice upon entering the brainstem and to stop upon entering their targets are also unknown. In fact, in most species, it is not agreed upon whether auditory nerve axons grow into their final position and provide the attractive signals for postmitotic neuronal precursors to coalesce around them (e.g., see Morest 1969) or the neuronal precursors of the brainstem auditory nuclei begin their migration and "attract" the growing eighth-nerve axonal process. Willard (1990) argues that the auditory nerve grows into the brainstem prior to the migration of auditory neurons. Migrating postmitotic neurons may then be attracted to these axons and cease migrating. Some support for this view is found in both experimental and descriptive studies of the developing chick brainstem. Parks (1979) showed that NA cells migrated into ectopic positions in the brainstem following early otocyst removal at E2.5, which eliminated development of the cochleo-vestibular ganglion cell. In the mouse, neuroblasts forming the main targets of the cochlear nerve leave mitosis on days 10–14 (Taber Pierce 1967, Martin & Rickets 1981). These dates completely coincide with the generation of ganglion cells (Ruben 1967). Therefore, migration of most CN neurons is likely to be occurring at about the same time as most of the eighth-nerve axons are arriving. Without experimental manipulations, it is difficult to understand the interactions between migrating neuroblasts that form the CN and the growing cochlear nerve axons. Some progress has been made toward identifying both intracellular and secreted molecules that may be important for these interactions. For example, Represa and colleagues (San Jose et al. 1997) have begun examining cytoskeletal changes in the growing axons of chicks. In this same species, Morest and colleagues are examining the timing and spatial pattern of FGF-2 and its receptors in relation to ganglion cell and brainstem development (e.g., Brumwell et al. 2000). Finally, the developmental patterns of expression of neurotrophins and their receptors in the mammalian and chick auditory brainstem are being examined (Hafidi et al. 1996, Hafidi 1999b, Cochran et al. 1999).

It is now possible to experimentally address the key issues discussed above. We are in need of studies combining careful descriptive, developmental methods with experimental manipulations that eliminate either the eighth-nerve axons (Ma et al. 1998, 2000) or the hindbrain regions that form the anlagen of the auditory nuclei (e.g., Studer et al. 1998, Cramer et al. 2000a). For example, the target cells within the developing brainstem could be removed to test the role of targets in specifying axonal branching patterns.

Development of Contacts Between Ganglion Cell Axons and Cochlear Nucleus Neurons

As axons of the cochlear nerve arrive in the brainstem, they interact with the cell bodies and processes of postmitotic neuroblasts in several important ways. For example, they form synaptic connections to establish the information-processing network of the auditory pathways. One fascinating property of this process is that the different collaterals of the auditory nerve form synaptic connections with very different morphologies. In the AVCN of mammals and NM of birds, the predominant synaptic morphology is a calyx surrounding much of the cell body, known as the end bulb of Held (Lorente de Nó 1981). This presynaptic ending is highly stereotyped and provides a phase-locked, powerful excitatory connection, known to be important for temporal processing. In other regions of the cochlear nuclear complex, more common boutonal synapses are made. Considerable work has been done on the developmental dynamics of end bulb development in the AVCN and NM. A comparison of the developmental changes seen in the chick and mouse is shown in Figure 3.

Jhaveri & Morest (1982a,b) show rather elegantly that postsynaptic NM neurons initially have extensively ramifying dendritic processes among which the

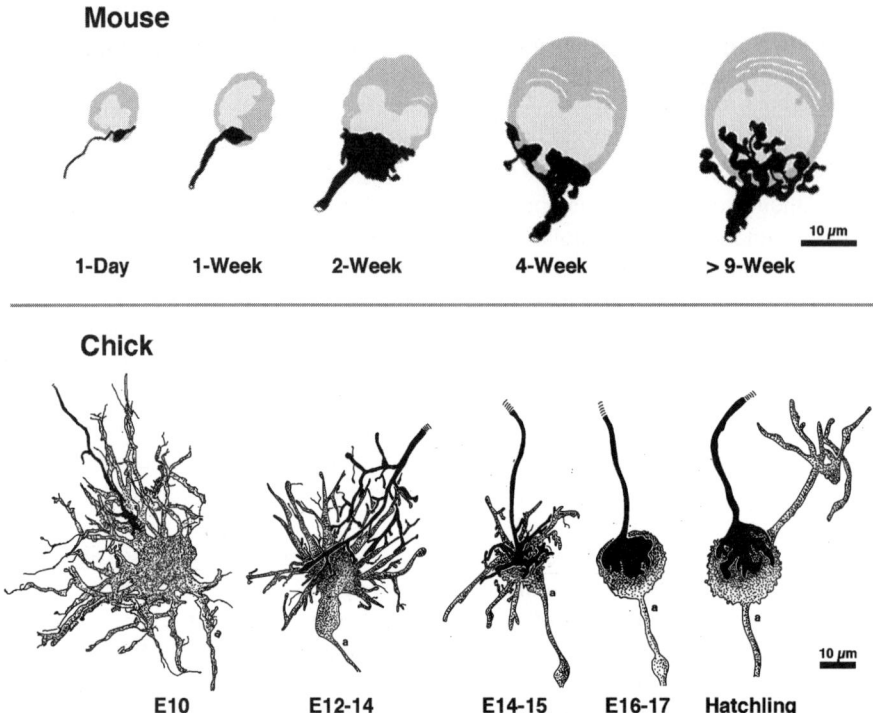

Figure 3 Development of end bulb of Held in mouse AVCN (*top*) and chick NM (*bottom*). In each series the structure of the postsynaptic cell and afferent axon is shown before the onset of auditory function (*left* drawing in each sequence), during the early stages of synaptic and auditory function (*middle* 3 panels), and when hearing is relatively mature (*right*-most drawings). Note that while the end bulbs look very similar by maturity, the proposed sequences of development appear quite different. Some caution in this conclusion is warranted, however, as different methods were used to evaluate end bulb development in the two species. Mouse drawings from HRP filled axons by Limb & Ryugo (2000); chick drawings from neurons stained by the Golgi-Kopsch method by Jhaveri & Morest (1982a).

ingrowing auditory nerve axons branch and form at least transient synaptic connections. Then, coincident with the early stages of auditory function, the dendritic arbors become resorbed (see also Parks & Jackson 1984, Young & Rubel 1986), and 2–3 end bulbs form on the cell body of each NM neuron. Formation of the end bulb may be due to coalescence of many terminal arbors or the dramatic expansion of a few of the initial presynaptic structures. Dendritic resorption and end bulb formation begin at the rostromedial (high-frequency) area of NM and progress caudolaterally along the tonotopic axis of the nucleus (Parks & Jackson 1984, Young & Rubel 1986). The divergence of eighth-nerve axons to neighboring cells in NM, and the convergence of axons onto single NM neurons have been examined in the chick (Jackson & Parks 1982). The large, complex dendritic arbors of NM neurons at E9–12 (Young & Rubel 1986) make it difficult to draw firm conclusions about divergence of presynaptic arbors at the age synaptic connections are forming. But clearly, there is a modest decrease in preterminal axonal branching between E14 and E17. In addition, physiological analyses have shown a small decrease (from 4 to 2.4) in the number of unitary EPSPs excited by stimulation of eighth-nerve inputs over the same age range (Jackson & Parks 1982). While these results are often cited as supporting the idea of widespread exuberance of axonal connections in the developing nervous system, there is little evidence supporting such an interpretation. The decrease in convergence is quite limited, and there is no evidence that the one to two supernumerary axons come from different cochlear regions. In the next section, we provide evidence that, from the outset, connections appear to form precisely in the brainstem auditory pathways.

Developmental studies of the end bulbs of Held have also been carried out in the mouse, rat, cat and barn owl (Mattox et al. 1982, Neises et al. 1982, Ryugo & Fekete 1982, Carr & Boudreau 1996, Limb & Ryugo 2000). While early development has not been studied in detail, the abundance of synaptic profiles in the neuropil and on somatic processes in newborn rats and barn owls suggests that the pattern is quite similar to that described in the chick. On the other hand, the developmental pattern has been described rather completely in the cat (Ryugo & Fekete 1982) and mouse (Limb & Ryugo 2000, Figure 3). These papers describe a series of changes in end bulb morphology, from a simple spoon-shaped ending to an elaborate series of filopodia engulfing the somata of AVCN neurons. Interestingly, at all ages described, the ending is elaborated on the somata of the developing AVCN neuron, and the neuron itself is rather adendritic (see Figure 3).

The stereotyped structure of the end bulb of Held provides a unique opportunity to consider the relative contributions of the axon collateral versus the target cell in specifying this presynaptic phenotype. Since other collaterals of the same axons—those terminating in DCN and PVCN—possess boutonal type endings, it seems logical to speculate that the form is specified by the target. Parks et al. (1990) addressed this question experimentally by taking advantage of the earlier discovery that NM neurons make ectopic projections to contralateral NM when the contralateral otocyst is removed. NM neurons normally make boutonal synapses onto n. laminaris (NL), the third-order neurons in the avian auditory system. At the light microscopic level, the ectopically projecting NM-to-NM axons form boutons, which suggests that the cell of origin, not the target cell, specifies synaptic morphology. However, some ultrastructural features resemble the eighth-nerve synapse on NM neurons. Thus, it appears from the studies that both axons and target cells determine synaptic morphology.

A second issue is whether eighth-nerve action potential generation and synaptic activity influence the development of contacts between the nerve and CN neurons. In the chick, electrophysiological studies have shown that NM neurons are responsive to eighth-nerve stimulation at day 10–11 of embryogenesis (Jackson et al.

1982, Pettigrew et al. 1988). Responses to sound are seen in brainstem recordings by E12–13 (Saunders et al. 1973). Therefore, it is possible that the resorption of dendritic arbors seen in NM neurons and/or the changes in end bulb morphology are dependent on afferent activity. While end bulb morphology has not been studied carefully in the chick, neither the time course nor the tonotopic gradient of dendritic changes appears influenced by the presence or activity of eighth-nerve axons (Parks & Jackson 1984). In the kitten, however, there is considerable evidence that the presence and activity of the eighth-nerve influence the complexity and size of the end bulbs and their ultrastructural characteristics (Saada et al. 1996; Ryugo et al. 1997, 1998; Niparko 1999; Redd et al. 2000).

The presence of spatial-temporal gradients in the relationship between eighth-nerve axons and the developing CN has been observed in a variety of studies (Rubel et al. 1976, Jackson et al. 1982, Schweitzer & Cant 1984, Kubke et al. 1999). For example, Schweitzer & Cant found that fibers from the basal portion of the hamster cochlea are the first to enter the DCN, followed by axons from the middle and apical turns, respectively. How such gradients among the axons or the postsynaptic cells in the CN are established remains a mystery awaiting molecular discovery. However, they do appear to be independent of sensory input from the ear (Parks & Jackson 1984).

Finally, it is important to mention that during the time period when connections are forming between cochlear nerve axons and CN neurons, both elements are likely to be changing in a large variety of cellular and molecular respects, including transmitter and modulator expression and release kinetics, neurotransmitter receptor pharmacology (e.g., Zhou & Parks 1992, Code & McDaniel 1998, Kubke & Carr 1998, Lawrence & Trussell 2000, Parks 2000, Zirpel et al. 2000a), ion channel characteristics (Perney et al. 1992, Garcia-Diaz 1999), and other synaptic specializations (e.g., Lurie et al. 1997, Hack et al. 2000). The intracellular and intercellular molecular pathways influencing such changes await further research.

Development of Topographic (Tonotopic) Connections

In the visual, somatosensory, and auditory pathways of most organisms, there is a highly stereotyped, topographic relationship between the receptor surface and the collections of neurons in nuclei or specific brain areas at each level of the ascending sensory pathways. These maps of the receptive surfaces of the organism are defined anatomically by preservation of neighbor relationship projections to each brain region. Physiologically, they are demonstrated by an orderly array of receptive fields seen in postsynaptic responses as one moves an electrode in small increments through a sensory area of the brain. Such maps represent physical space in the visual and somatosensory systems. In the auditory systems of birds and mammals, the maps provide a representation of a quite different stimulus/response attribute: the "best frequency" or "characteristic frequency" of the neuronal response to acoustic stimulation. This mapping property is a function of the remarkably precise coding of frequency along the cochlea (von Békésy 1960, Rhode 1978, Dallos 1992) and the precise topography of connections between cochlear ganglion cells and hair cells along the sensory epithelium, discussed above.

When considering the development of topographic (tonotopic) organization of ganglion cell projections to the cochlear nucleus, three issues need to be addressed. First, does the map emerge from relatively indiscriminate connections, or is there a degree of precision as soon as the projection is evident? If some precision is evident from the onset of function, does the "grain" of the map change during further development? Second, a popular belief is that rough characteristics initially form and that these are refined during use. What role, if any, does auditory experience or neuronal activity independent of sound-driven activity have on the development or maintenance of this topography? Finally, and most important, what are the cellular signals responsible for the establishment and maintenance of the tonotopic map?

In the developing auditory system of birds and mammals, available evidence suggests that the topography of connections between the cochlea, the ganglion cells, and the cochlear nuclei develops quite precisely, well before acoustic information is processed by these cells. For results relevant to this issue, see anatomical studies in the rat (Angulo et al. 1990, Friauf 1992, Friauf & Kandler 1993), mouse (Fritzsch et al. 1997b), opossum (Willard 1993), hamster (Schweitzer & Cant 1984, Schweitzer & Cecil 1992), and cat (Snyder & Leake 1997). No single study has labeled neighboring cells in the spiral ganglion and examined the relative alignment of terminal fields in the CN or done a similar analysis by retrograde transport (e.g., see Agmon et al. 1995). Demonstrations that terminal arbors in the CN are initially small and precisely oriented provide indirect evidence for a great deal of initial precision; terminal arbors grow as the nucleus expands in volume (Schweitzer & Cecil 1992). Furthermore, well before hearing onset in opossum and cat, small injections of HRP into the spiral ganglion label discrete bands of terminals in the CN, and the size of these bands does not change with age (Willard 1993, Snyder & Leake 1997). While it is impossible to state that the precision, or "grain", of the map does not change with experience, there is no compelling evidence for such a viewpoint at this time.

Physiological studies that have addressed the development of tonotopic organization at the level of the CN or other brainstem nuclei lead to similar conclusions. Physiological mapping studies invariably find a precise tonotopic organization early during development (Lippe & Rubel 1985, Sanes et al. 1989, Sterbing et al. 1994, Lippe 1995). Similarly, studies using pure-tone acoustic stimuli to modulate metabolic markers (c-FOS, 2-DG) have found discrete bands of label in the CN as early as stimuli elicit a metabolic response (Ryan & Woolf 1988, Friauf 1992, Friauf & Kandler 1993). It appears fashionable to propose that the early topographic organization is somewhat crude or rough (meaning less well ordered, we presume) and that it is "fine tuned" by auditory experience (e.g., see Friauf & Lohmann 1999). However, little evidence exists for any role of auditory experience toward shaping the tonotopic organization of connections between the cochlea and the cochlear nuclei. In both birds and mammals, this organization appears before one can readily record responses to acoustic stimuli. There appear to be no gross "mistakes" in the orderly arrangement of connections, and the overall growth of the brain regions can account for the changes that are seen in the degree of specificity of axonal connections. Although the precision of the early eighth-nerve to CN projections has not been studied in detail, the pattern has been studied at the next synaptic level. Young & Rubel (1986) examined the topography of the ipsilateral projection between NM and NL, and Sanes & Rubel (1988) studied the development of bilateral connections to the lateral superior olive in the gerbil. Young & Rubel used single cell reconstructions to show that by E9, which is well before an auditory response can be found, the ipsilateral projections from NM to NL are as precise as they will ever be. In fact, subsequent development causes a loss of one dimension of specificity. Sanes & Rubel showed that at the age responses to sound can first be recorded in the lateral superior olive (P14–15), the matching of excitatory and inhibitory frequency tuning is virtually perfect. These results suggest again that the tonotopy at the level of the CN must already be mature.

Having established that the tonotopic organization of projections from the cochlear ganglion to the CN emerges prior to responsiveness to external acoustic stimulation, it becomes important to ascertain whether activity that is independent of acoustic stimulation (spontaneous activity) plays an important role in the establishment and maintenance of appropriate connections. In this case, we are considering spontaneous activity as action potential generation in the eighth nerve or CN that is not driven by acoustic stimuli, but does not preclude hair cell origin. As noted above, synaptic connections with the CN are formed and appear to be precisely ordered before the onset of peripheral responses to sound in chicks and mammals (Jackson et al. 1982, Kandler & Friauf 1995, Snyder & Leake 1997). On the other hand, Leake et al. (2001) report a modest decrease in the relative size (corrected for overall CN growth) of eighth-nerve axonal projections in the CN from small groups of labeled spiral ganglion neurons between birth and P6 in kittens. They hypothesize that spontaneous activity is involved in these changes. Spontaneous activity can be recorded soon after synaptic connections are seen physiologically or anatomically in chicks (Lippe 1994), wallabies (Gummer & Mark 1994), kittens (Walsh & McGee 1988), and gerbils (W. R. Lippe, personal communication). At this time, however, there are no convincing data suggesting that the spontaneous activity plays a role in the establishment of topographic connections. Lippe (1994) has described rhythmic activity that is of cochlear origin and shows a gradient in its developmental properties along the tonotopic axis. However, at E14, the age when this gradient is seen, the tonotopically organized projection from the ganglion cells to NM is already well established (E. W. Rubel, unpublished observations).

Virtually nothing is known about the molecules that determine the tonotopic axis of the cochlear nuclei or guide the establishment of connections in an orderly way along this axis. It is clear, however, that both the presynaptic axons and the postsynaptic target cells must express some sort of signaling molecules that specify the tonotopic axis. Two interesting experiments support this conclusion. First, the resorption of dendrites in the chick NM takes place along a rostromedial to caudolateral spatial "gradient" that matches the tonotopic organization (Rubel & Parks 1975). Remarkably, the dendritic resorption, its time course, and its spatial organization appear independent of presynaptic input from the cochlear nerve (Parks & Jackson 1984). Second, abnormal connections to NM will form a normal orderly array along the tonotopic axis. This was shown by mapping the ectopic connection that forms between the two NMs when a unilateral otocyst removal is performed very early in development (Jackson & Parks 1988). Lippe et al. (1992) recorded from NM neurons while stimulating the contralateral ear in animals in which this projection was induced. Normally, NM axons innervate only NL neurons on the ipsilateral and contralateral sides of the brain (Young & Rubel 1983). When these axons are induced to innervate the contralateral NM, they produce a tonotopic organization indistinguishable from the normal ipsilateral eighth-nerve input. This finding suggests again that the tonotopic axis is somehow encoded by the NM neurons and can be communicated to ectopic auditory afferents as well as its normal ipsilateral afferents from the eighth nerve.

While there is little known about the molecules or cellular interactions participating in the establishment of the tonotopic organization of the CN in birds or mammals, developmental gradients in the ingrowth of eighth-nerve fibers and of CN properties appear to correspond to the tonotopic axis (Rubel et al. 1976, Rubel 1978, Jackson et al. 1982, Schweitzer & Cant 1984, Willard 1993, Kubke et al. 1999). Timing alone is unlikely to provide the signal (Holt 1984; Holt & Harris 1993, 1998), but these gradients may provide clues to discover candidate molecules. Several growth factors and receptors have been examined in the ganglion cells and CN. Some of those growth factors and receptors appear to be expressed at approximately the time that connections are being established or that auditory function matures (e.g., see Luo et al. 1995b, Riedel et al. 1995). However, gradients of expression that match the tonotopic axis at the time topographic connections are forming have not been reported. Understanding gradients of molecules along topographic axes is an important and timely problem in developmental neurobiology, in general, and the auditory pathways may be particularly advantageous for experimentally examining it. Eighth-nerve ganglion and cochlear nuclei are derived from entirely separated epithelial compartments that can be separately manipulated. Further, there is a single, functionally defined, axis of orientation.

To adequately address the molecular identities responsible for the establishment of topography in the auditory pathways, two areas of research are initially needed. First, we need detailed analyses of the timing of the development of topographic connections at a single cell level in a few "model" species. Second, detailed analyses of the spatial and temporal distribution of candidate molecules that have provided important new information in other systems (e.g., Eph receptors and ephrins) are likely to prove important (e.g., see O'Leary & Wilkinson 1999, Wilkinson 2000). For example, recent studies of the developmental distribution of trkB and EphA4 show remarkable and provocative patterns of expression that are likely to be important for determining the laminar specificity of connections between NM and NL (Cochran et al. 1999, Cramer et al. 2000b). Further study of these classes of molecules may be helpful for understanding the development of tonotopy in the cochlear nuclei.

Influence of Cochlear Nerve on Development of Cochlear Nucleus

In this final section, we see that the trophic relationships between the cochlear nerve and its central targets, the cochlear nucleus, are fundamentally different from the peripheral interactions with hair cells. Whereas hair cell development appears largely independent of innervation by ganglion cells, the cells of the cochlear nuclei are dramatically influenced by manipulations of the developing inner ear and ganglion cells.

The classic study by Levi-Montalcini (1949) provided one set of fundamental observations underlying our approach to this problem. Levi-Montalcini removed the otocyst, the origin of the sensory cells and ganglion cells of the inner ear, at 2–2 1/2 days of development in chick embryos. This manipulation deprived the embryos of normal input to the developing cochlear and vestibular nuclei of the brainstem. By studying the brainstem in silver-stained sections at various developmental time points, she discovered that the cochlear nuclei (NM and NA) develop normally until approximately E11. After this time, however, the overall volume and the number of neurons in both nuclei decrease dramatically. These observations were later replicated and extended in Rubel's lab. Parks (1979) carefully followed the progression of events after otocyst removal and found that both NA and NM displayed normal nuclear volume, cell size, and neuron number until E11, after which they rapidly deteriorated. Jackson et al. (1982) then determined that E11 was the first age at which postsynaptic action potentials in NM could be evoked by eighth-nerve stimulation. This pair of results has two important implications. The first is that most developmental events take place independently of excitatory afferent activity, even though the eighth-nerve fibers are in the vicinity of the cells of the CN early in development. Proliferation, early migration and the establishment of afferent and efferent topographic connections all occur before functional afferent synaptic connections are made. The second implication is that, at the time normal synaptic input occurs, the postsynaptic neurons suddenly become metabolically dependent on the establishment of functional synapses. Without afferent stimulation, there is cell death, atrophy of the remaining neurons, abnormal migration, and a variety of other abnormalities.

The dependence of the postsynaptic neuron on presynaptic input does not seem to be permanent in most species and most sensory systems. For example, if we consider the trophic role of eighth-nerve on CN cells exclusively, it terminates somewhere between six weeks and one year of age in the chicken (Born & Rubel 1985), at about 14 days after birth (P14) in the mouse (Mostafapour et al. 2000), at about P9 in the gerbil (Hashisaki & Rubel 1989, Tierney et al. 1997), and between P5 and P24 in the ferret (Moore 1990). This differential sensitivity of the postsynaptic neurons to presynaptic manipulations is usually referred to as a critical period or sensitive period. In addition to cell death, a large variety of metabolic and structural changes have been examined in neurons and glial cells after cochlear manipulations at different ages in birds and mammals. (See earlier reviews by Rubel 1978, Rubel & Parks 1988, Rubel et al. 1990, Moore 1992, Parks 1997, Zirpel et al. 1997, and Friauf & Lohmann 1999 for much of this information.) In the remainder of this review, we consider such changes only as they relate to the following questions: (*a*) What is the signal from the presynaptic neuron that maintains the integrity of the postsynaptic cell? (*b*) What is the cascade of cellular events in the postsynaptic cell that leads to cell death or cell survival following cochlear removal? (*c*) What are the biological mechanisms underlying the critical period during which peripheral input is essential for normal development? (*d*) What is the nature of the variability in cell survival following early deafferentation; why do some cells live and others die?

SIGNALS The first question to address is the nature of the signals transmitted from the cochlear nerve to CN neurons and glia that influence their survival, structure, and metabolism. An extensive literature, beginning with the landmark papers of Wiesel & Hubel (1963, 1965), suggested that patterned acoustic information may be of critical importance. Webster and colleagues (Webster & Webster 1977, 1979; Webster 1983a,b,c, 1988a) and Coleman (Coleman & O'Connor 1979, Coleman et al. 1982) suggested that neonatal acoustic deprivation in mice and rats produced by a conductive hearing impairment (ear plug, closing ear canal, or disarticulation of middle ear bones) causes reduced neuronal size (atrophy) and reduced neuropil volume in the cochlear nucleus. However, in several other species a chronic conductive hearing loss did not cause atrophy of CN neurons, including chick (Tucci & Rubel 1985), ferret (Moore et al. 1989), gerbil (E. W. Rubel, unpublished observations), or rhesus monkey (Doyle & Webster 1991). Several explanations for this apparent discrepancy have been proposed. The most parsimonious explanation at this time is based on studies comparing both spontaneous eighth-nerve activity and cell size changes following purely conductive vs. sensorineural hearing loss. Tucci et al. (1987) showed that a purely conductive hearing loss does not disrupt high levels of spontaneous activity in the auditory nerve, and this activity is sufficient to preserve normal neuronal numbers and morphology in the chick NM. However, inner ear manipulations that produce a sensorineural hearing loss always reduce or eliminate spontaneous eighth-nerve activity and result in rapid changes in neuronal size. This explanation is supported by studies of experimentally induced sensorineural hearing loss using pharmacological inhibition of eighth-nerve spikes or aminoglycosides, as well as by studies of animals with congenital hair cell loss (Webster 1985; Born & Rubel 1988; Pasic & Rubel 1989, 1991; Lippe 1991; Sie & Rubel 1992; Dodson et al. 1994; Saada et al. 1996; Saunders et al. 1998). It seems entirely possible, in light of our current knowledge, that the conductive manipulations performed by Webster & Coleman resulted in secondary sensorineural damage to the basal part of the rodent cochlea, especially when produced in young animals. Electrophysiological data support this interpretation (Clopton 1980, Evans et al. 1983, Money et al. 1995).

The studies cited above clearly show that the integrity of the auditory nerve is essential for normal development of CN neurons. Patterned activity appears not to be an essential signal at this level of the auditory pathways. A long series of studies in chicks and gerbils have attempted to determine the signal or signals that are essential for preserving normal development of CN neurons. The first approach was to ask if eliminating eighth-nerve activity without damaging the sensory or neural cells would produce the same postsynaptic changes in the CN as total destruction of the cochlea. This was accomplished by infusion of the sodium channel blocker, tetrodotoxin, into the inner ear. Complete blockade of eighth-nerve action potentials, in fact, produced rapid changes in NM neurons and AVCN neurons that were indistinguishable from those resulting from complete destruction of the cochlea (Born & Rubel 1988, Pasic & Rubel 1989, Sie & Rubel 1992, Garden et al. 1994). These results strongly suggest that the voltage-dependent release of glutamate or a molecule coreleased with glutamate is essential for normal maintenance of CN neurons in young animals. Further support for this conclusion comes from a series of studies on rodents and chicks showing that neuronal atrophy and decreased protein synthesis induced by eighth-nerve action potential blockade or sensorineural hearing loss can be reversed by restoration of presynaptic activity (Born & Rubel 1988, Webster 1988b, Pasic & Rubel 1991, Lippe 1991, Saunders et al. 1998). In addition, a number of investigators have attempted to use cochlear implants to reverse atrophy of CN cells in cats deafened as neonates or adults. The results are contradictory at this time (Ni et al. 1993, Lustig et al. 1994, Kawano et al. 1997).

Activity in the presynaptic elements during a critical period is essential for maintaining cellular integrity and neuronal morphology in NM and AVCN. Is synaptic stimulation necessary? One may recall that the same question was addressed in the neuromuscular system many years ago (Drachman & Witzke 1972, Lomo & Rosenthal 1972). To address this question, Hyson & Rubel (1989, 1995) asked if the morphological changes seen in NM neurons could be prevented by electrical stimulation of the eighth nerve (orthodromic stimulation), and if so, could they be equally well prevented by antidromic stimulation of the NM neurons? The results of in vitro orthodromic and antidromic stimulation experiments demonstrated that the early events following deafferentation and activity deprivation, decreased protein synthesis and ribosomal integrity, could be prevented by orthodromic stimulation. However, antidromic stimulation actually exacerbated these degenerative events. More recent experiments have shown that propidium iodide incorporation, a common measure of dying cells, is also prevented by orthodromic stimulation (Zirpel et al. 1998). Finally, blocking neurotransmitter release from the eighth-nerve fibers by bathing the preparation in low Ca^{2+} or blocking metabotropic glutamate receptors reversed the positive effects of orthodromic stimulation (see Rubel et al. 1990, Zirpel et al. 1997). Taken together, these results provide strong evidence that the trophic influences of the eighth nerve on its target neurons in the CN are mediated by voltage-dependent release of glutamate or of a molecule coreleased with glutamate and that the influences require activation of one or more glutamate receptors on NM neurons. Conversely, deprivation of glutamate release or receptor activation in young animals activates a cascade of events culminating in cell death or atrophy of the postsynaptic neurons. In vitro experiments comparing the effects of antidromic and orthodromic stimulation have not been replicated in the mammalian AVCN. However, the effects of deprivation and of pharmacological blockade of the eighth nerve on deprivation-induced postsynaptic changes in NM and AVCN are strikingly similar. These findings, coupled with the clear homology between NM and AVCN, strongly suggest that similar conclusions can be made for the identity of the signals regulating trophic influences on CN neurons in mammals.

POSTSYNAPTIC EVENTS The immediate and long-term changes in the CN following deafferentation or deprivation have been examined primarily in chicks, rodents, and cats, with differing goals. Most of the studies on cats and guinea pigs have focused on the long-term phenotype of the CN neurons and on whether some or all of the effects of deprivation can be reversed by stimulation through cochlear prostheses. This clinically oriented goal has important implications for interventions in young children suffering serious and profound hearing loss. The second goal is trying to understand the sequelae of events following alterations in afferent activity and determining their causal relationships. This approach can add new information and concepts toward understanding the role of activity in nervous system development and the plasticity of the developing nervous system.

Before it was appreciated that deprivation of eighth-nerve activity produced the same sequence of initial events in the postsynaptic CN neurons as did deafferentation, several investigators removed the cochlea (usually including the ganglion cell bodies) in animals of varying ages and examined the CN weeks or months later (Levi-Montalcini 1949; Powell & Erulkar 1962; Parks 1979; Trune 1982a,b; Nordeen et al. 1983). Large reductions in neuron size, neuropil volume (including dendritic size), nuclear volume, neuron number, and concomitant increases in neuron packing density were seen when cochlea removal was performed in young or embryonic birds and mammals. In general, the changes seen in mature animals were less severe and did not include deafferentation-induced cell death of CN neurons. Changes comparable to those seen in young birds and mammals were also described in frog auditory nuclei after otocyst removal (Fritzsch 1990).

A series of papers on the chick CN beginning in 1985 led to new ways of thinking about the cascade of cellular events that may lead to these long-term changes. Born & Rubel (1985) carefully examined the time course and age dependence of the morphological changes in NM neurons following cochlea removal. Remarkably, cell death and cell atrophy after cochlear removal occur extremely rapidly, within two days in young chickens. Dramatic cytoplasmic changes in Nissl staining are evident at 12–24 h. Furthermore, there is no difference in outcome between removing the cochlea alone versus removing the cochlea and the ganglion cells, thereby directly severing the eighth-nerve central process. These rapid changes in the responses of the postsynaptic neurons as well as the morphological details described by Born & Rubel suggested that deafferentation evokes an apoptotic-like process in NM neurons. This interpretation has been strengthened by studies showing that protein synthesis, RNA synthesis, ribosome integrity, and ribosomal RNA content all decrease within 30 min to a few hours after eliminating eighth-nerve activity or removing the cochlea (Steward & Rubel 1985; Rubel et al. 1991; Garden et al. 1994, 1995a,b; Hartlage-Rübsamen & Rubel 1996). These early events are distinctly biphasic. During the initial 3–4 h after the onset of deprivation, there appears to be a generalized decrease in synthetic activity, with only minor changes in cytoplasmic ultrastructure. This is reflected in quantitative measures as an overall, unimodal, shift in the distributions of labeling densities. By about 6 h after deafening, depending on the specific parameter under investigation, a clearly bimodal distribution of NM cells emerges. Approximately 70% of the neurons show partial recovery of protein synthesis and RNA synthesis and no obvious structural alterations in cytoplasmic ribosomes. The remaining 30% of NM neurons shows no synthetic activity (by our measures), a complete loss of polyribosomes in their cytoplasm and loss of staining for ribosomal RNA (see Rubel et al. 1991; Garden et al. 1994, 1995a,b). This latter group represents the neurons that die over the next two days; while ~70% of neurons that show less severe changes atrophy, but survive. The effects of deprivation on CN cells are more rapid than expected and that the ultimate fate of the deprived neurons is predictable quite early in the process, by about 6 h after the beginning of deprivation.

A variety of other rapid and long-term changes in presynaptic and postsynaptic elements of auditory neurons in chicks and mammals have been observed after elimination or reduced eighth-nerve activity. These include the expected decrease in glucose uptake in young and adult animals (Lippe et al. 1980, Born et al. 1991, Tucci et al. 1999), dramatic and rapid changes in calcium-binding proteins in mature guinea pig and rat (Winsky & Jacobowitz 1995, Caicedo et al. 1997, Forster & Illing 2000; but see Parks et al. 1997), and changes in cFOS protein and mRNA expression (Gleich & Strutz 1997, Luo et al. 1999; see also Zhang et al. 1996). On the other hand, some proteins such as GAP-43 transiently increase expression (Illing et al. 1997), which could be related to some spreading of inhibitory connections after deafferentation (Benson et al. 1997; but see Code et al. 1990). One of the most dramatic and rapid changes in NM neurons that has been observed after activity deprivation in vivo is in the density of antibody staining to the cytoskeletal proteins, tubulin, actin, and MAP-2 (Kelley et al. 1997). Within 3 h after deafening, immunoreactivity of NM neurons to antibodies to these three proteins is dramatically reduced, without a concomitant decrease in mRNA. It was hypothesized that the cytoskeletal proteins change configuration to allow the cells to change shape. Within four days, antigenicity in the surviving NM neuron begins to recover. Finally, Durham and colleagues found a biphasic response of Kreb's cycle enzymes and mitochondria density in chick NM neurons (Durham & Rubel 1985; Hyde & Durham 1990, 1994a,b; Durham et al. 1993). During the first 24–36 h there are increases in enzyme activity and the density of mitochondria, which are followed by a smaller but sustained decrease. These results are discussed further at the conclusion of this section.

The rapid time course and the patterns of structural and ultrastructural changes in CN neurons suggest that the activity-dependent trophic interactions rely on a rather simple interaction, such as activation of a receptor tyrosine kinase and/or maintenance of normal intracellular signaling pathways. The distributions of some trk receptors during development have recently been described in both birds and mammals (Hafidi et al. 1996, Cochran et al. 1999), and ligands for these receptors are present in the mammal CN (Hafidi 1999b). Other families of growth factors are also being examined (e.g., Riedel et al. 1995). However, a role for any of these receptor/ligand pairs has not been tested.

The role of afferent activity on the homeostasis of intracellular calcium $[Ca^{2+}]_i$ and the importance of $[Ca^{2+}]_i$ for trophic regulation of NM neurons have been studied extensively during the past few years (Zirpel et al. 1995, 1997, 1998, 2000a,b; Zirpel & Rubel 1996). Intuitively, it might be expected that deprivation of presynaptic activity would lead to a decrease in $[Ca^{2+}]_i$ in postsynaptic neurons. Surprisingly, just the opposite is true of NM neurons. Elimination of eighth-nerve activity leads to a rapid, threefold increase in $[Ca^{2+}]_i$ in NM neurons, which is reversed entirely by electrical stimulation of the nerve or activation of metabotropic glutamate receptors (mGluRs). When the eighth nerve is stimulated in the presence of mGluR antagonists, $[Ca^{2+}]_i$ increases dramatically. Furthermore, activation of mGluRs is required for maintenance of ribosomal RNA (Hyson 1998). A direct link between elevated $[Ca^{2+}]_i$ and increased cell death has been established by Zirpel et al. (1998). Finally, Zirpel et al. (2000a,b) provides convincing evidence that activation of Group 1 mGluRs is necessary for maintaining normal $[Ca^{2+}]_i$ in NM neurons, and that in the absence of mGluR activation, influx of Ca^{2+} through AMPA receptors is involved in creating the hypercalcemic condition.

In order to understand the relationship between mGluR activation and $[Ca^{2+}]_i$ homeostasis in NM neurons, it is important to remember that most eighth-nerve axons, NM neurons, and AVCN neurons have extremely high levels of ongoing "spontaneous" activity, even in silence (Dallos & Harris 1978, Liberman 1978, Tucci et al. 1987, Warchol & Dallos 1990, Born et al. 1991). In addition, Ca^{2+}-permeable AMPA receptors appear to be required for the faithful processing of temporally precise, high-frequency information (Trussell 1998, Parks 2000). This combination seems to place auditory brainstem neurons at high risk for calcium cytotoxicity or a calcium-activated apoptotic-like cascade. Perhaps to adapt to this challenge, auditory neurons are rich in calcium-binding proteins (Takahashi et al. 1987, Braun 1990, Kubke et al. 1999, Hack et al. 2000) and mitochondria, and appear to have specialized (or highly expressed) intracellular pathways by which Group 1 mGluRs inhibit cytoplasmic buildup of Ca^{2+}. A series of studies using ratiometric Ca^{2+} imaging suggests that Ca^{2+} permeability and intracellular Ca^{2+} release are dramatically regulated by mGluR activation (Lachica et al. 1995a,b; Kato et al. 1996, Kato & Rubel 1999). These studies need confirmation by direct measurements of Ca^{2+} conductance, but they suggest a novel set of pathways whereby transmitter release can independently regulate activity and $[Ca^{2+}]_i$ in the postsynaptic neuron. The results to date, therefore, suggest a working model: Glutamate release is necessary to activate one or more mGluRs, which in turn, prevent large increases in $[Ca^{2+}]_i$ by a variety of mechanisms, including, but not limited to, inhibition of Ca^{2+} permeability of AMPA channels, high-voltage activated Ca^{2+} channels, and Ca^{2+} release from intracellular stores. Activity deprivation then releases this inhibition, which allows a cascade of events beginning with a rise in $[Ca^{2+}]_i$ that is subsequently similar or identical to excitotoxicity (Mattson et al. 2000). While many more experiments are needed to fill in the details of the intracellular events that lead to cell death or cell phenotype changes, support for this model is emerging (Wilson & Durham 1995, Solum et al. 1997, Caicedo et al. 1998, Zirpel & Parks 2001).

CRITICAL PERIOD   As noted above, some of the transneuronal structural and metabolic interactions between the eighth nerve and CN neurons occur throughout life, whereas others appear limited to a specific period of development. As seen in other developing sensory systems, there appears to be a critical period for trophic regulation of CN neurons by their presynaptic partners. Trune (1982a) showed extensive cell death in mouse CN after neonatal deafferentation but did not test adults. Nordeen et al. (1983), Born & Rubel (1985), Hashisaki & Rubel (1989), and Moore (1990) provide convincing evidence for differential effects of cochlea removal on CN neuronal survival and atrophy in neonatal and adult chicks, gerbils, and ferrets. Young animals were much more susceptible than adults. However, not until a recent report by Tierney et al. (1997) was it appreciated how sharp the window of this critical period could be. Tierney and colleagues report that between P7 and P9 there is an abrupt change in the survival of gerbil CN neurons following deafferentation. Cochlea removals before seven days of age result in 45%–88% cell death in the CN; at nine days of age or older, this same manipulation results in no reliable cell death.

The remarkably rapid changes in susceptibility of CN neurons to deprivation-induced cell death suggest that some simple molecular switch is controlling susceptibility to afferent deprivation. To address this possibility, a series of studies examining the critical period for trophic regulation in mice has been initiated. The first studies described the temporal boundaries of the critical period and time course of cell death following deafening (Mostafapour et al. 2000). In addition, experiments with bcl-2-null and bcl-2 overexpression mice have shown dramatic modulation of this critical period. CN neurons in adult mice lacking the bcl-2 gene appear to be equivalently susceptible to deafening as wild-type neonatal mice. Conversely, overexpression of bcl-2 prevents all transneuronal cell death in neonatal mice (Mostafapour & Rubel 2001). These results should not be over-interpreted. It is not clear if bcl-2 modulation is due to a direct role of bcl-2 (or to related gene family members) in determining the critical period or if this protein is playing a role downstream of such a molecule. In any case, these results may provide a beginning toward understanding the biological basis of this critical period.

A LIFE OR DEATH DECISION   One of the most intriguing and medically important questions is understanding why, after afferent deprivation in young animals, some postsynaptic neurons live and others die. The proportion of CN neurons that die varies dramatically with species as well as with age. For example, cochlea removals in three-day-old gerbils result in almost 90% neuron loss within two weeks, but the same manipulation at P7 results in only 50% loss (Tierney et al. 1997). A similar decrease in susceptibility is seen in the mouse during the first 10 postnatal days (Mostafapour et al. 2000). In chicks, however, only about 30% cell loss is seen at the most vulnerable times (Born & Rubel 1985).

What determines which neurons survive deprivation or deafferentation and which die? Two major possibilities emerge. The most favored hypothesis is that there is a bimodal population of neurons with an intrinsic difference in susceptibility to deafferentation. It is possible that particular differences in receptor phenotypes, for example, cause two groups of neurons to respond fundamentally differently to deafferentation. While this explanation is particularly attractive for nuclear regions with mixed cell types, such as the mammalian CN following deafening, there is, in fact, little supporting evidence (see Tierney et al. 1997). Furthermore, in the avian NM there appears to be only a single neuron type throughout most of the nucleus, and repeated attempts to discover two or more distinct populations on the basis of structure or protein expression have failed (Rubel & Parks 1988, Kubke et al. 1999).

A second hypothesis for explaining the differences in neuronal fate after deafferentation was first explicitly proposed by Garden et al. (1994) and Hyde & Durham (1994a). It was hypothesized that the neuronal populations are not bimodal with respect to susceptibility to afferent deprivation-induced cell death. Instead, it is possible that the deprivation condition elicits two competing intracellular responses. The first response is activation of an apoptotic-like pathway and the second is activation of a survival pathway. This model further suggested that activation of the survival pathways is delayed by a few hours compared to the apoptotic-like pathway. The resulting amount of cell death would then be a function of the relative effectiveness of these competing pathways and survival or death of individual cells would be stochastically determined during the period of susceptibility.

There are several lines of evidence supporting this second hypothesis. First, from the initial deafferentation experiments, it was recognized that there is no consistent spatial pattern of cell death in the CN and that there is high variability in the absolute amount of cell death during the period of susceptibility (Born & Rubel 1985, Moore 1990). Second, many of the early degradative events following the onset of afferent deprivation are uniform across the population of NM neurons. These events include decreases in protein synthesis, RNA synthesis, ribosomal antigenicity, and cytoskeletal protein antigenicity (Steward & Rubel 1985; Born & Rubel 1988; Garden et al. 1994, 1995a; Kelley et al. 1997). There are, of course, variations across the population of NM neurons in these responses to deprivation, but there is no hint of a population of neurons that does not respond at all. Third, since 1985 it has been recognized that oxidative enzyme activity actually shows a biphasic response following afferent deprivation. Beginning at about 6 h and continuing for 24–30 h, there is a dramatic increase in enzyme activity; this is followed by a long-lasting decrease as has been described in other sensory regions following deprivation (Durham & Rubel 1985, Hyde & Durham 1990, Durham et al. 1993). Concomitant with the increase in oxidative enzyme activity is an increase in the density of mitochondria in the cytoplasm of NM neurons (Hyde & Durham 1994b). These lines of evidence, in addition to growing bodies of literature showing mitochondrial influences on Ca^{2+} homeostasis and cell survival (Mostafapour et al. 1997, Nicholls & Budd 2000), suggest that the survival mechanisms in deafferented NM neurons involve the mitochondria response seen 6–24 h following deprivation.

Five different experiments have now examined the role of mitochondrial protein synthesis on deafferentation-induced changes in ribosomes (Garden et al. 1994, 1995b; Hartlage-Rübsamen & Rubel 1996), zinc translocation (Wilson & Durham 1995), and cell death (Hyde & Durham 1994a). These studies indicate that decreasing or preventing mitochondrial protein synthesis for the first 12–24 h following the initiation of deprivation dramatically increases the early degradative changes and the number of NM neurons that subsequently die. For example, NM neuron death five days after cochlea removal increases from 30% to 60%–80% in chloremphenicol-treated animals (Hyde & Durham 1994a). Inhibition of cytoplasmic protein synthesis with cycloheximide, on the other hand, has no effect on the response of NM neurons to afferent deprivation (Garden et al. 1994).

ACKNOWLEDGMENTS

The authors would like to thank Karina Cramer, Jennifer Stone, Sarah Woolley, David Molea, David Ryugo, Julie Harris, and Kirk Beisel for valuable contributions to this manuscript and Laurie Johnson for expert manuscript preparation. The authors' laboratories are supported by NIH Grants DC00395, DC03829, DC02854, and DC00215.

The *Annual Review of Neuroscience* is online at http://neuro.annualreviews.org

LITERATURE CITED

Adam J, Myat A, LeRoux I, Eddison M, Henrique D, et al. 1998. Cell fate choices and the expression of notch, delta and serrate homologues in the chick inner ear: parallel with *Drosophila* sense-organ development. *Development* 125:4645–54

Agerman K, Canlon B, Duan M, Ernfors P. 1999. Neurotrophins, NMDA receptors, and nitric oxide in development and protection of the auditory system. *Ann. NY Acad. Sci.* 884:131–42

Agmon A, Yang LT, Jones EG, O'Dowd DK. 1995. Topological precision in the thalamic projection to neonatal mouse barrel cortex. *J. Neurosci.* 15:549–61

Altman J, Bayer S. 1982. Development of the cranial nerve ganglia and related nuclei in the rat. *Adv. Anat. Embryol. Cell Biol.* 74:1–90

Angulo A, Merchán JA, Merchán MA. 1990. Morphology of the rat cochlear primary afferents during prenatal development: a Cajal's reduced silver and rapid Golgi study. *J. Anat.* 168:241–55

Ard MD, Morest DK. 1984. Cell death during development of the cochlear and vestibular ganglia in the chick. *Int. J. Dev. Neurosci.* 2:535–47

Ard MD, Morest DK, Hauger SH. 1985. Trophic interaction between the cochleovestibular ganglion of chick embryo and its synaptic targets in culture. *Neuroscience* 16:151–70

Baker CV, Bronner-Fraser M. 2001. Vertebrate cranial placodes I. Embryonic induction. *Dev. Biol.* 232:1–61

Benson CG, Gross JS, Suneja SK, Potashner SJ. 1997. Synaptophysin immunoreactivity in the cochlear nucleus after unilateral cochlear or ossicular removal. *Synapse* 25:243–57

Berlin CI, ed. 1999. *The Efferent Auditory System.* San Diego, CA: Singular

Bermingham NA, Hassan BA, Price SD, Vollrath MA, Ben-Arie N, et al. 1999. Math1, an essential gene for the generation of inner ear hair cells. *Science* 284:1837–40

Bianchi LM, Cohan CS. 1993. Effects of the neurotrophins and CNTF on developing statoacoustic neurons: comparison with an otocyst-derived factor. *Dev. Biol.* 159:353–65

Bianchi LM, Liu H. 1999. Comparison of Ephrin-A ligand and EphA receptor distribution in the developing inner ear. *Anat. Rec.* 254:127–34

Book KJ, Morest DK. 1990. Migration of neuroblasts by perikaryal translocation: role of cellular elongation and axonal outgrowth in the acoustic nuclei of the chick embryo medulla. *J. Comp. Neurol.* 297:55–76

Born DE, Durham D, Rubel EW. 1991. Afferent influences on brainstem auditory nuclei of the chick: nucleus magnocellularis neuronal activity following cochlea removal. *Brain Res.* 557:37–47

Born DE, Rubel EW. 1985. Afferent influences on brain stem auditory nuclei of the chicken: neuron number and size following cochlea removal. *J. Comp. Neurol.* 231:435–45

Born DE, Rubel EW. 1988. Afferent influences on brain stem auditory nuclei of the chicken: presynaptic action potentials regulate protein synthesis in nucleus magnocellularis neurons. *J. Neurosci.* 8:901–19

Braun K. 1990. Calcium-binding proteins in avian and mammalian central nervous system: localization, development and possible functions. *Prog. Histochem. Cytochem.* 21:1–64

Brose K, Tessier-Lavigne M. 2000. Slit proteins: key regulators of axon guidance, axonal branching, and cell migration. *Curr. Opin. Neurobiol.* 10:95–102

Bruce LL, Christensen MA, Warr WB. 2000. Postnatal development of efferent synapses in the rat cochlea. *J. Comp. Neurol.* 423:532–48

Bruce LL, Kingsley J, Nichols DH, Fritzsch B. 1997. The development of vestibulocochlear efferents and cochlear afferents in mice. *Int. J. Devel. Neuro.* 15:671–92

Brumwell CL, Hossain WA, Morest DK, Bernd P. 2000. Role for basic fibroblast growth factor (FGF-2) in tyrosine kinase (TrkB) expression in the early development and innervation of the auditory receptor: in vitro and in situ studies. *Exp. Neurol.* 162:121–45

Caicedo A, d'Aldin C, Eybalin M, Puel JL. 1997. Temporary sensory deprivation changes calcium-binding proteins levels in the auditory brainstem. *J. Comp. Neurol.* 378:1–15

Caicedo A, Kungel M, Pujol R, Friauf E. 1998. Glutamate-induced Co^{2+} uptake in rat auditory brainstem neurons reveals developmental changes in Ca^{2+} permeability of glutamate receptors. *Eur. J. Neurosci.* 10:941–54

Cantos R, Cole LK, Everett L, Green E, Simeone A, Wu DK. 2000. Patterning of the mammalian cochlea. *Proc. Natl. Acad. Sci. USA* 97:11707–13

Carney PR, Silver J. 1983. Studies on cell migration and axon guidance in the developing distal auditory system of the mouse. *J. Comp. Neurol.* 215:359–69

Carr CE, Boudreau RE. 1996. Development

of the time coding pathways in the auditory brainstem of the barn owl. *J. Comp. Neurol.* 373:467–83

Chen P, Segil N. 1999. p27 kip links cell proliferation to morphogenesis in the developing organ of Corti. *Development* 126:1581–90

Clopton BM. 1980. Neurophysiology of auditory deprivation. In *Birth Defects: Original Article Series Vol XVI:4, Morphogenesis and Malformations of the Ear*, ed. RJ Garlin, pp. 271–88. New York: Alan R. Liss

Cochran SL, Stone JS, Bermingham-McDonogh O, Akers SR, Lefcort F, Rubel EW. 1999. Ontogenetic expression of Trk neurotrophin receptors in the chick auditory system. *J. Comp. Neurol.* 413:271–88

Code RA, Durham D, Rubel EW. 1990. Effects of cochlea removal on GABAergic terminals in nucleus magnocellularis of the chicken. *J. Comp. Neurol.* 301:634–54

Code RA, McDaniel AE. 1998. Development of dynorphin-like immunoreactive auditory nerve terminals in the chick. *Dev. Brain Res.* 106:165–72

Cohen GM, Cotanche DA. 1992. Development of the sensory receptors and their innervation in the chick cochlea. See Romand 1992, pp. 101–38

Coleman J, Blatchley BJ, Williams JE. 1982. Development of the dorsal and ventral cochlear nuclei in rat and effects of acoustic deprivation. *Dev. Brain Res.* 4:119–23

Coleman JR, O'Connor P. 1979. Effects of monaural and binaural sound deprivation on cell development in the anteroventral cochlear nucleus of rats. *Exp. Neurol.* 64:553–66

Coppola V, Kucera J, Palko ME, Martinez-De Velasco J, Lyons WE, et al. 2001. Dissection of NT3 functions in vivo by gene replacement strategy. *Development* 128:4315–27

Corwin JT, Cotanche DA. 1989. Development of location-specific hair cell stereocilia in denervated embryonic ears. *J. Comp Neurol.* 288:529–37

Cowan CA, Yokoyama N, Bianchi LM, Henkemeyer M, Fritzsch B. 2000. EphB2 guides axons at the midline and is necessary for normal vestibular function. *Neuron* 26:417–30

Cramer KS, Fraser SE, Rubel EW. 2000a. Embryonic origins of auditory brain-stem nuclei in the chick hindbrain. *Dev. Bio.* 224:138–51

Cramer KS, Rosenberger MH, Frost DM, Cochran SL, Pasquale EB, Rubel EW. 2000b. Developmental regulation of EphA4 expression in the chick auditory brainstem. *J. Comp. Neurol.* 426:270–78

Dallos P. 1992. The active cochlea. *J. Neurosci.* 12:4575–85

Dallos P, Harris D. 1978. Properties of auditory nerve responses in absence of outer hair cells. *J. Neurophysiol.* 41:365–82

Dallos P, Popper AN, Fay RR. 1996. *The Cochlea.* New York: Springer. 551 pp.

D'Amico-Martel A. 1982. Temporal patterns of neurogenesis in avian cranial sensory and autonomic ganglia. *Am. J. Anat.* 163:351–72

Dodson HC, Bannister LH, Douek EE. 1994. Effects of unilateral deafening on the cochlear nucleus of the guinea pig at different ages. *Brain Res.* 80:261–67

Doyle WJ, Webster DB. 1991. Neonatal conductive hearing loss does not compromise auditory function and structure in rhesus monkeys. *Hear. Res.* 54:145–51

Drachman DB, Witzke F. 1972. Trophic regulation of acetylcholine sensitivity of muscle: effect of electrical stimulation. *Science* 176:514–16

Durham D, Matschinsky FM, Rubel EW. 1993. Altered malate dehydrogenase activity in nucleus magnocellularis of the chicken following cochlea removal. *Hear. Res.* 70:151–59

Durham D, Rubel EW. 1985. Afferent influences on brain stem auditory nuclei of the chicken: changes in succinate dehydrogenase activity following cochlea removal. *J. Comp. Neurol.* 231:446–56

Echteler SM. 1992. Developmental segregation in the afferent projections to mammalian auditory hair cells. *Proc. Natl. Acad. Sci. USA* 89:6324–27

Eddison M, LeRoux I, Lewis J. 2001. Notch signaling in the development of the inner ear: lessons from *Drosophila. Proc. Natl. Acad. Sci. USA.* In press

Erkman L, McEvilly RJ, Luo L, Ryan AK, Hooshmand F, et al. 1996. Role of transcription factors Brn 3.1 and Brn 3.2 in auditory and visual system development. *Nature* 381:603–6

Ernfors P, van de Water T, Loring J, Jaenisch R. 1995. Complementary roles of BDNF and NT-3 in vestibular and auditory development. *Neuron* 14:1153–64

Evans WJ, Webster DB, Cullen JK Jr. 1983. Auditory brainstem responses in neonatally sound deprived CBA/J mice. *Hear. Res.* 10:269–77

Fariñas I, Jones KR, Backus C, Wang X-Y, Reichardt LF. 1994. Severe sensory and sympathetic deficits in mice lacking neurotrophin-3. *Nature* 369:658–61

Fariñas I, Jones KR, Tessarollo L, Vigers AJ, Huang E, et al. 2001. Spatial shaping of cochlear innervation by temporally-regulated neurotrophin expression. *J. Neurosci.* 21:6170–80

Fekete DM. 1999. Development of the vertebrate ear: insights from knockouts and mutants. *Trends Neurosci.* 22:263–69

Fekete DM, Rouiller EM, Liberman MC, Ryugo DK. 1984. The central projections of intracellularly labeled auditory nerve fibers in adult cats. *J. Comp. Neurol.* 229:432–50

Fell HB. 1928. The development in vitro of the isolated otocyst of the embryonic fowl. *Arch. Exp. Zellforsch.* 7:69–81

Fischer FP. 1992. Quantitative analysis of the innervation of the chicken basilar papilla. *Hear. Res.* 61:167–78

Flanagan JG, Vanderhaeghen P. 1998. The ephrins and the eph receptors in neural development. *Annu. Rev. Neurosci.* 21:309–45

Forster CR, Illing RB. 2000. Plasticity of the auditory brainstem: cochleotomy-induced changes of calbindin-D28k expression in the rat. *J. Comp. Neurol.* 416:173–87

Frago LM, Camerero G, Canon S, Paneda C, Sanz C, et al. 2000. Role of diffusible and transcription factors in inner ear development: implications in regeneration. *Histol. Histopathol.* 15:657–66

Friauf E. 1992. Tonotopic order in the adult and developing auditory system of the rat as shown by c-fos immunocytochemistry. *Eur. J. Neurosci.* 4:798–812

Friauf E, Kandler K. 1993. Cell birth, formation of efferent connections, and establishment of tonotopic order in the rat cochlear nucleus. See Merchán et al. 1993, pp. 19–28

Friauf E, Lohmann C. 1999. Development of auditory brainstem circuitry: Activity-dependent and activity-independent processes. *Cell Tissue Res.* 297:187–95

Friedmann I. 1956. In vitro culture of the isolated otocyst of the embryonic fowl. *Ann. Otol.* 65:98–107

Fritzsch B. 1990. Experimental reorganization in the alar plate of the clawed toad, *Xenopus laevis*. I. Quantitative and qualitative effects of embryonic otocyst extirpation. *Dev. Brain Res.* 51:113–22

Fritzsch B, Barald KF, Lomax MI. 1998. Early embryology of the vertebrate ear. See Rubel et al. 1998, pp. 80–145

Fritzsch B, Beisel KW, Bermingham NA. 2000. Developmental evolutionary biology of the vertebrate ear: conserving mechanoelectric transduction and developmental pathways in diverging morphologies. *Neuroreport* 11:R35–44

Fritzsch B, Fariñas I, Reichardt LF. 1997b. Lack of NT-3 causes losses of both classes of spiral ganglion neurons in the cochlea in a region specific fashion. *J. Neurosci.* 17:6213–25

Fritzsch B, Nichols DH. 1993. DiI reveals a prenatal arrival of efferent neurons at developing ears of mice. *Hear. Res.* 65:51–60

Fritzsch B, Pirvola U, Ylikoski J. 1999. Making and breaking the innervation of the ear: neurotrophic support during ear development and its clinical implications. *Cell Tissue Res.* 295:369–82

Fritzsch B, Silos-Santiago I, Bianchi L, Fariñas I. 1997a. Neurotrophins, neurotrophin receptors and the maintenance of the afferent inner ear innervation. *Sem. Cell Dev. Biol.* 8:277–84

Fritzsch B, Silos-Santiago I, Smeyne R, Fagan AM, Barbacid M. 1995. Reduction and loss of inner ear innervation in trkB and trkC receptor knockout mice: a whole mount DiI and scanning electron microscopic analysis. *Audit. Neurosci.* 1:401–17

Fukuda R, McNew JA, Weber T, Parlati F, Engel T, et al. 2000. Functional architecture of an intracellular membrane t-SNARE. *Nature* 407:198–202

Garden GA, Canady KS, Lurie DI, Bothwell M, Rubel EW. 1994. A biphasic change in ribosomal conformation during transneuronal degeneration is altered by inhibition of mitochondrial, but not cytoplasmic protein synthesis. *J. Neurosci.* 14:1994–2008

Garden GA, DeWulf VR, Rubel EW. 1995a. Afferent influences on brain stem auditory nuclei of the chicken: regulation of transcriptional activity following cochlea removal. *J. Comp. Neurol.* 359:412–23

Garden GA, Hartlage-Rübsamen M, Rubel EW. 1995b. Protein masking of a ribosomal RNA epitope is an early event in afferent deprivation induced neuronal death. *Mol. Cell. Neurosci.* 6:293–310

Garcia-Diaz JF. 1999. Development of a fast transient potassium current in chick cochlear ganglion neurons. *Hear. Res.* 135:124–34

Ghysen A, Dambly-Chaudiere C. 2000. A genetic programme for neuronal connectivity. *Trends Genet.* 16:221–22

Ginzberg RD, Morest K. 1983. A study of cochlear innervation in the young cat with the Golgi method. *Hear. Res.* 10:227–46

Ginzberg RD, Morest K. 1984. Fine structure of cochlear innervation in the cat. *Hear. Res.* 14:109–27

Gleich O, Strutz J. 1997. Age-dependent effects of the onset of a conductive hearing loss on the volume of the cochlear nucleus subdivisions and the expression of c-fos in the mongolian gerbil (*Meriones unguiculatus*). *Audiol. Neurootol.* 2:113–27

Goulding SE, White NM, Jarman AP. 2000. Cato encodes a basic helix-loop-helix transcription factor implicated in the correct differentiation of *Drosophila* sense organs. *Dev. Biol.* 221:120–31

Gummer AW, Mark RE. 1994. Patterned neural activity in brain stem auditory areas of a prehearing mammal, the tammar wallaby (*Macropus eugenii*). *Neuroreport* 5:685–88

Hack NJ, Wride MC, Charters KM, Kater SB, Parks TN. 2000. Developmental changes in the subcellular localization of calretinin. *J. Neurosci.* 20:RC67 http://www.jneurosci.org

Hafidi A. 1999a. Peripherin-like immunoreactivity in type II spiral ganglion cell body and projections. *Brain Res.* 805:181–90

Hafidi A. 1999b. Distribution of BDNF, NT-3 and NT-4 in the developing auditory brainstem. *Int. J. Dev. Neurosci.* 17:285–94

Hafidi A, Moore T, Sanes DH. 1996. Regional distribution of neurotrophin receptors in the developing auditory brainstem. *J. Comp. Neurol.* 367:454–64

Hafidi A, Romand R. 1989. First appearance of type II neurons during ontogenesis in the spiral ganglion of the rat. An immunocytochemical study. *Dev. Brain Res.* 48:143–49

Hallböök F. 1999. Evolution of the vertebrate neurotrophin and trk receptor gene families. *Curr. Opin. Neurobiol.* 9:616–21

Hartlage-Rübsamen M, Rubel EW. 1996. Influence of mitochondrial protein synthesis inhibition on deafferentation-induced ultrastructural changes in nucleus magnocellularis of developing chicks. *J. Comp. Neurol.* 371:448–60

Hashisaki GT, Rubel EW. 1989. Effects of unilateral cochlea removal on anteroventral cochlear nucleus neurons in developing gerbils. *J. Comp. Neurol.* 283:465–73

Hassan BA, Bellen HJ. 2000. Doing the Math: is the mouse a good model for fly development? *Genes Dev.* 14:1852–65

Hatini V, Ye X, Balas G, Lai E. 1999. Dynamics of placodal lineage development revealed by targeted transgene expression. *Dev. Dyn.* 215:332–43

He DZ, Dallos P. 1997. Expression of potassium channels in gerbil outer hair cells during development does not require neural induction. *Dev. Brain. Res.* 103:95–97

Hegarty JL, Kay AR, Green SH. 1997. Trophic support of cultured spiral ganglion neurons by depolarization exceeds and is additive with that by neurotrophins or cAMP and requires elevation of $[Ca^{2+}]_i$ within a set range. *J. Neurosci.* 17:1959–70

Hemond SG, Morest DK. 1991a. Formation of the cochlea in the chicken embryo: sequence of innervation and localization of basal lamina-associated molecules. *Dev. Br. Res.* 61:87–96

Hemond SG, Morest DK. 1991b. Ganglion formation from the otic placode and the otic crest in the chick embryo: mitosis, migration and the basal lamina. *Anat. Embrol.* 184:1–13

Hemond SG, Morest DK. 1992. Trophic effects of otic epithelium on cochleo-vestibular ganglion fiber growth in vitro. *Anat. Rec.* 232:273–84

Holland LZ, Schubert M, Holland ND, Neuman T. 2000. Evolutionary conservation of the presumptive neural plate markers AmphiSox 1/2/3 and AmphiNeurogenin in the invertebrate chordate amphioxus. *Dev. Biol.* 226:18–33

Holt CE. 1984. Does timing of axon outgrowth influence initial retinotectal topography in *Xenopus*? *J. Neurosci.* 4:1130–52

Holt CE, Harris WA. 1993. Position, guidance, and mapping in the developing visual system. *J. Neurobiol.* 24:1400–22

Holt CE, Harris WA. 1998. Target selection: invasion, mapping and cell choice. *Curr. Opin. Neurobiol.* 8:98–105

Hong K, Nishiyama M, Henley J, Tessier-Lavigne M, Poo M. 2000. Calcium signalling in the guidance of nerve growth by netrin-1. *Nature* 403:93–98

Hossain WA, Morest DK. 2000. Fibroblast growth factors (FGF-1, FGF-2) promote migration and neurite growth of mouse cochlear ganglion cells in vitro: immunohistochemistry and antibody perturbation. *Neurosci. Res.* 62:40–55

Huang EJ, Zang K, Schmidt A, Saulys A, Xiang M, Reichardt LF. 1999. POU domain factor Brn-3a controls the differentiation and survival of trigeminal neurons by regulating Trk receptor expression. *Development* 126:2869–82

Huang HJ, Liu W, Fritzsch B, Bianchi LM, Reichardt LF, Xiang M. 2001. Brn-3a is a transcriptional regulator of soma size, target field innervation, and axon pathfinding of inner ear sensory neurons. *Development* 128:2421–2430

Hutson MR, Lewis JE, Nguyen-Luu D, Lindberg KH, Barald KF. 1999. Expression of Pax2 and patterning of the chick inner ear. *J. Neurocytol.* 28:795–807

Hyde GE, Durham D. 1990. Cytochrome oxidase response to cochlea removal in chicken auditory brainstem neurons. *J. Comp. Neurol.* 297:329–39

Hyde GE, Durham D. 1994a. Increased deafferentation-induced cell death in chick brainstem auditory neurons following blockade of mitochondrial protein synthesis with chloramphenicol. *J. Neurosci.* 14:291–300

Hyde GE, Durham D. 1994b. Rapid increase in mitochondrial volume in nucleus magnocellularis neurons following cochlea removal. *J. Comp. Neurol.* 339:27–48

Hyson RL. 1998. Activation of metabotropic glutamate receptors is necessary for trans-neuronal regulation of ribosomes in chick auditory neurons. *Brain Res.* 809:214–20

Hyson RL, Rubel EW. 1989. Transneuronal Regulation of Protein Synthesis in the Brain Stem Auditory System of the Chick Requires Synaptic Activation. *J. Neurosci.* 9:2835–45

Hyson RL, Rubel EW. 1995. Activity-Dependent Regulation of a Ribosomal RNA Epitope in the Chick Cochlear Nucleus. *Brain Res.* 672:196–204

Illing RB, Horvath M, Laszig R. 1997. Plasticity of the auditory brainstem: effects of cochlear ablation on GAP-43 immunoreactivity in the rat. *J. Comp. Neurol.* 382:116–38

Jackson H, Hackett JT, Rubel EW. 1982. Organization and development of brain stem auditory nuclei in the chick: ontogeny of postsynaptic responses. *J. Comp. Neurol.* 210:80–86

Jackson H, Parks TN. 1982. Functional synapse elimination in the developing avian cochlear nucleus with simultaneous reduction in cochlear nerve axon branching. *J. Neurosci.* 2:1736–43

Jackson H, Parks TN. 1988. Induction of aberrant functional afferents to the chick cochlear nucleus. *J. Comp. Neurol.* 271:106–14

Jahn R, Südhof TC. 1999. Membrane fusion and exocytosis. *Annu. Rev. Biochem.* 68:863–911

Jhaveri S, Morest DK. 1982a. Sequential alterations of neuronal architecture in nucleus magnocellularis of the developing chicken: a Golgi study. *Neuroscience* 7:837–53

Jhaveri S, Morest DK. 1982b. Sequential alterations of neuronal architecture in nucleus magnocellularis of the developing chicken: an electron microscope study. *Neuroscience* 7:855–70

Jones KR, Fariñas I, Backus C, Reichardt LF. 1994. Targeted disruption of the brain-derived neurotrophic factor gene perturbs brain and sensory but not motor neuron development. *Cell* 76:989–100

Kandler K, Friauf E. 1995. Develoment of glycinergic and glutamatergic synaptic transmission in the auditory brainstem of perinatal rats. *J. Neurosci.* 15:6890–904

Karis A, Pata I, van Doornick JH, Grosveld F, De Zeeuw CI, et al. 2001. Transcription factor GATA-3 alters pathway selection of olivocochlear neurons and affects morphogenesis of the ear. *J. Comp. Neurol.* 429:615–30

Katayama A, Corwin JT. 1989. Cell production in the chicken cochlea. *J. Comp. Neurol.* 281:129–35

Kato BM, Lachica EA, Rubel EW. 1996. Glutamate modulates intracellular Ca^{2+} stores in brainstem auditory neurons. *J. Neurophysiol.* 76:646–50

Kato BM, Rubel EW. 1999. Glutamate regulates IP3-Type and CICR stores in the avian cochlear nucleus. *J. Neurophysiol.* 81:1587–96

Kawano A, Seldon HL, Clark GM, Hakuhisa Efumasaka S. 1997. Effects of chronic electrical stimulation on cochlear nuclear neuron size in deaf kittens. *Acta Otolaryngol.* 52:33–35

Kelley MS, Lurie DI, Rubel EW. 1997. Rapid regulation of cytoskeletal proteins and their mRNAs following afferent deprivation in the avian cochlear nucleus. *J. Comp. Neurol.* 389:469–83

Kim W-Y, Fritzsch B, Serls A, Bakel LA, Huang EJ, et al. 2001. NeuroD-null mice are deaf due to a severe loss of the inner ear sensory neurons during development. *Development* 128:417–426

Knipper M, Zimmermann U, Rohbock K, Köpschall I, Zenner H-P. 1995. Synaptophysin and Gap-43 proteins in efferent fibers of the inner ear during postnatal development. *Dev. Brain Res.* 89:73–86

Knowlton VY. 1967. Correlation of the development of membranous and bony labyrinths, acoustic ganglia, nerves, and brain centers of the chick embryo. *J. Morphology* 121:179–208

Kubke MF, Carr CE. 1998. Development of AMPA-selective glutamate receptors in the auditory brainstem of the barn owl. *Microsc. Res. Tech.* 41:176–86

Kubke MF, Gauger B, Basu L, Wagner H, Carr CE. 1999. Development of calretinin immunoreactivity in the brainstem auditory nuclei of the barn owl (*Tyto alba*). *J. Comp. Neurol.* 415:189–203

Lachica EA, Rübsamen R, Zirpel L, Rubel EW. 1995a. Glutamatergic inhibition of voltage operated calcium channels in the avian cochlear nucleus. *J. Neurosci.* 15:1724–34

Lachica EA, Zirpel L, Rubel EW. 1995b. Intracellular mechanisms involved in the afferent regulation of neurons in the avian cochlear nucleus. In *Auditory Systems Plasticity and Regeneration*, ed. RJ Salvi, D Henderson, V Coletti, F Fiorino. New York: Thieme Med.

Lawrence JJ, Trussell LO. 2000. Long-term specification of AMPA receptor properties after synapse formation. *J. Neurosci.* 20:4864–70

Leake PA, Snyder RL, Hradek GT. 2001. Postnatal refinement of auditory nerve projections to the cochlear nucleus in cats. *J. Comp. Neurol.* In press

Legan PK, Richardson GP. 1997. Extracellular

matrix and cell adhesion molecules in the developing ear. *Sem. Cell Dev. Biol.* 8:217–24

Levi-Montalcini R. 1949. The development of the acousticovestibular centers in the chick embryo in the absence of the afferent root fibers and of descending fiber tracts. *J. Comp. Neurol.* 91:209–41

Liberman MC. 1978. Auditory-nerve responses from cats raised in a low-noise chamber. *J. Acoust. Soc. Am.* 63:442–55

Liberman MC, O'Grady DF, Dodds LW, McGee J, Walsh AJ. 2000. Afferent innervation of outer and inner hair cells is normal in neonatally de-efferented cats. *J. Comp. Neurol.* 423:123–39

Liberman MC, Oliver ME. 1984. Morphometry of intracellularly labeled neurons of the auditory nerve: correlations with functional properties. *J. Comp. Neurol.* 223:163–76

Liebl DJ, Tessarollo L, Palko ME, Parada LF. 1997. Absence of sensory neurons before target innervation in brain-derived neurotrophic factor-, neurotrophin3-, and trkC-deficient embryonic mice. *J. Neurosci.* 17:9113–27

Limb CJ, Ryugo DK. 2000. Development of primary axosomatic endings in the anteroventral cochlear nucleus of mice. *JARO* 1:103–19

Lindsay RM, Barde YA, Davies AM, Rohrer H. 1985. Differences and similarities in the neurotrophic growth factor requirements of sensory neurons derived from neural crest and neural placode. *J. Cell Sci. Suppl.* 3:115–29

Lippe WR. 1991. Reduction and recovery of neuronal size in the cochlear nucleus of the chicken following aminoglycoside intoxication. *Hear. Res.* 51:193–202

Lippe WR. 1994. Rhythmic spontaneous activity in the developing avian auditory system. *J. Neurosci.* 14:1486–95

Lippe WR. 1995. Relationship between frequency of spontaneous bursting and tonotopic position in the developing avian auditory system. *Brain Res. Dev. Brain Res.* 703:205–13

Lippe WR, Fuhrmann DS, Yang W, Rubel EW. 1992. Aberrant projection induced by otocyst removal maintains normal tonotopic organization in the chick cochlear nucleus. *J. Neurosci.* 12:962–69

Lippe WR, Rubel EW. 1985. Ontogeny of tonotopic organization of brain stem auditory brain stem auditory nuclei in the chicken: implications for development of the place principle. *J. Comp. Neurol.* 238:371–81

Lippe WR, Steward O, Rubel EW. 1980. The effect of unilateral basilar papilla removal upon nuclei laminaris and magnocellularis of the chick examined with [^{3}H]2-deoxy-D-glucose autoradiography. *Brain Res.* 196:45–58

Liu M, Pereira FA, Price SD, Chu MJ, Shope C, et al. 2000. Essential role of BETA2/NeuroD1 in development of the vestibular and auditory systems. *Genes Dev.* 14:2839–54

Lomo T, Rosenthal J. 1972. Control of Ach sensitivity by muscle activity in the rat. *J. Physiol.* 221:493–513

Lorente de Nó R. 1981. *The Primary Acoustic Nuclei*. New York: Raven

Luo L, Brumm D, Ryan AF. 1995a. Distribution of non-NMDA Ggutamate receptor mRNAs in the developing rat cochlea. *J. Comp. Neurol.* 361:372–82

Luo L, Moore JK, Baird A, Ryan AF. 1995b. Expression of acidic FGF mRNA in rat auditory brainstem during postnatal maturation. *Dev. Brain Res.* 86:24–34

Luo L, Ryan AF, Saint Marie RL. 1999. Cochlear ablation alters acoustically induced c-fos mRNA expression in the adult rat auditory brainstem. *J. Comp. Neurol.* 404:271–83

Lurie DI, Pasic TR, Hockfield SJ, Rubel EW. 1997. Development of CAT-301 immunoreactivity in auditory brainstem nuclei of the gerbil. *J. Comp. Neurol.* 380:319–34

Lustig LR, Leake PA, Snyder RL, Rebscher SJ. 1994. Changes in the cat cochlear nucleus following neonatal deafening and chronic intracochlear electrical stimulation. *Hear. Res.* 74:29–37

Ma Q, Anderson DJ, Fritzsch B. 2000. Neurogenin1 null mutant ears develop fewer, morphologically normal hair cells in smaller sensory epithelia devoid of innervation. *J. Assoc. Res. Otolaryngol.* 1:129–43

Ma Q, Chen Z, del Barco Barrantes I, de la Pompa JL, Anderson DJ. 1998. Neurogenin1 is essential for the determination of neuronal precursors for proximal cranial sensory ganglia. *Neuron* 20:469–82

Ma Q, Fode C, Guillemot F, Anderson DJ. 1999. Neurogenin1 and neurogenin2 control two distinct waves of neurogenesis in developing dorsal root ganglia. *Genes Dev.* 13:1717–28

Martin MR, Rickets C. 1981. Histogenesis of the cochlear nucleus of the mouse. *J. Comp. Neurol.* 197:169–84

Mattox DE, Neises GR, Gulley RL. 1982. A freeze-fracture study of the maturation of synapses in the anteroventral cochlear nucleus of the developing rat. *Anat. Rec.* 204:281–87

Mattson MP, LaFerla FM, Chan SL, Leissring MA, Shepel PN, Geiger JD. 2000. Calcium signaling in the ER: its role in neuronal plasticity and neurodegenerative disorders. *Trends Neurosci.* 23:222–29

McEvilly RJ, Erkman L, Luo L, Sawchenko PE, Ryan AF, Rosenfeld MG. 1996. Requirement for Brn-3.0 in differentiation and survival of sensory and motor neurons. *Nature* 384:574–77

Merchán MA, Juiz JM, Godfrey DA, Mugnaini E, eds. 1993. *The Mammalian Cochlear Nuclei: Organization and Function*. New York: Plenum

Minichiello L, Piehl F, Vazquez E, Schimmang T, Hökfelt T, et al. 1995. Differential effects of combined trk receptor mutations on dorsal root ganglion and inner ear. *Development* 121:4067–75

Money MK, Pippin GW, Weaver KE, Kirsch JP, Webster DB. 1995. Auditory brainstem responses of CBA/J mice with neonatal conductive hearing losses and treatment with GM1 ganglioside. *Hear. Res.* 87:104–13

Moore DR. 1990. Auditory brainstem of the ferret: early cessation of development sensitivity of neurons in the cochlear nucleus to removal of the cochlea. *J. Comp. Neurol.* 302:810–23

Moore DR. 1991. Development and plasticity of the ferret auditory system. In *Neurobiology of Hearing: The Central Auditory System*, ed. RA Altschuler, RP Bobbin, BM Clopton, DW Hoffman, pp. 461–75. New York: Raven

Moore DR. 1992. Developmental plasticity of the brainstem and midbrain auditory nuclei. See Romand 1992, pp. 297–320

Moore JK, Guan YL, Shi SR. 1997. Axogenesis in the human fetal auditory system, demonstrated by neurofilament immunohistochemistry. *Anat. Embrol.* 195:15–30

Moore DR, Hutchings ME, King AJ, Kowalchuk NE. 1989. Auditory brain stem of the ferret: some effects of rearing with a unilateral ear plug on the cochlea, cochlear nucleus, and projections to the inferior colliculus. *J. Neurosci.* 9:1213–22

Morest DK. 1969. The differentiation of cerebral dendrites: a study of the post-migratory neuroblast in the medial nucleus of the trapezoid body. *Z. Anat. Entwicklung.* 128:271–89

Morsli H, Choo D, Ryan A, Johnson R, Wu DK. 1998. Development of the mouse inner ear and origin of its sensory organs. *J. Neurosci.* 18:3327–35

Mostafapour SP, Cochran SL, del Puerto NM, Rubel EW. 2000. Patterns of cell death in mouse AVCN neurons after unilateral cochlea removal. *J. Comp. Neurol.* 426:561–72

Mostafapour SP, Lachica EA, Rubel EW. 1997. Mitochondrial regulation of calcium in the avian cochlear nucleus. *J. Neurophysiol.* 78:1928–34

Mostafapour SP, Rubel EW. 2001. Bcl-2 overexpression saves neonatal cochlear nucleus neurons from transneuronal degeneration. Presented at Assoc. Res. Otolaryngol. 24th, St. Petersburg Beach, FL

Mou K, Adamson CL, Davis RL. 1998. Time-dependence and cell-type specificity of synergistic neurotrophin actions on spiral ganglion neurons. *J. Comp. Neurol.* 402:123–39

Mou K, Hunsberger CL, Cleary JM, Davis RL. 1997. Synergistic effects of BDNF and NT-3 on postnatal sipral ganglion neurons. *J. Comp. Neurol.* 386:529–39

Mueller BK. 1999. Growth cone guidance: first steps towards a deeper understanding. *Annu. Rev. Neurosci.* 22:351–88

Neises GR, Mattox DE, Gulley RL. 1982. The maturation of the end bulb of Held in the rat anteroventral cochlear nucleus. *Anat. Rec.* 204:271–79

Ni D, Seldon HL, Shepherd RK, Clark GM. 1993. Effect of chronic electrical stimulation on cochlear nucleus neuron size in normal hearing kittens. *Acta Otolaryngol.* 113:489–97

Nicholls DG, Budd SL. 2000. Mitochondria and neuronal survival. *Physiol. Rev.* 80:315–60

Niparko JK. 1999. Activity influences on neuronal connectivity within the auditory pathway. *Laryngoscope* 109:1721–30

Noden DM, van de Water TR. 1986. The developing ear: tissue origins and interactions. In *The Biology of Change in Otolaryngology*, ed. RJ Ruben, TR van de Water, EW Rubel, pp. 15–46. Amsterdam: Elsevier,

Nordeen KW, Killackey HP, Kitzes LM. 1983. Ascending projections to the inferior colliculus following unilateral cochlear ablation in the neonatal gerbil, *Meriones unguiculatus*. *J. Comp. Neurol.* 214:144–53

O'Leary DD, Wilkinson DG. 1999. Eph receptors and ephrins in neural development. *Curr. Opin. Neurobiol.* 9:65–73

Orr MF. 1981. Anatomical development of the embryonic chick otocyst in organ culture. *Anat. Rec.* 199:188A

Orr MF. 1986. Development of acoustic ganglia in tissue cultures of embryonic chick otocysts. *Exp. Cell Res.* 40:68–77

Parks TN. 1979. Afferent influences on the development of the brain stem auditory nuclei of the chicken: otocyst ablation. *J. Comp. Neurol.* 183:665–77

Parks TN. 1997. Effects of early deafness on development of brain stem auditory neurons. *Ann. Otol. Rhinol. Laryngol. Suppl.* 168:37–43

Parks TN. 2000. The AMPA receptors of auditory neurons. *Hear. Res.* 147:77–91

Parks TN, Code RA, Taylor DA, Solum DA, Strauss KI, et al. 1997. Calretinin expression in the chick brainstem auditory nuclei develops and is maintained independently of cochlear nerve input. *J. Comp. Neurol.* 383:112–21

Parks TN, Jackson H. 1984. A developmental gradient of dendritic loss in the avian cochlear nucleus occurring independently of primary afferents. *J. Comp. Neurol.* 227:459–66

Parks TN, Taylor DA, Jackson H. 1990. Adaptations of synaptic form in an aberrant projection to the avian cochlear nucleus. *J. Neurosci.* 3:975–84

Pasic TR, Rubel EW. 1989. Rapid changes in cochlear nucleus cell size following blockade of auditory nerve electrical activity in gerbils. *J. Comp. Neurol.* 283:474–80

Pasic TR, Rubel EW. 1991. Cochlear nucleus cell size is regulated by auditory nerve electrical activity. *J. Oto-HNS.* 104:6–13

Perkins RE, Morest DK. 1975. A study of cochlear innervation patterns in cats and rats with the Golgi method and Nomarski optics. *J. Comp. Neurol.* 163:129–58

Perney TM, Marshall J, Martin KA, Hockfield S, Kaczmarek LK. 1992. Expression of the mRNAs for the Kv3.1 potassium channel gene in the adult and developing rat brain. *J. Neurophysiol.* 68:756–66

Pettigrew AG, Ansselin AD, Bramley JR. 1988. Development of functional innervation in the second and third order auditory nuclei of the chick. *Development* 104:575–88

Pirvola U, Arumae U, Moshnyakov M, Palgi J. Saarma M, Ylikoski J. 1994. Coordinated expression and function of neurotrophins and their receptors in the rat inner ear during target innervation. *Hear. Res.* 75:131–44

Pirvola U, Hallbook F, Xing-Qun L, Virkkala J, Saarma M, Ylikoski J. 1997. Expression of neurotrophins and Trk receptors in the developing, adult, and regenerating avian cochlea. *J. Neurobiol.* 33:1019–33

Pirvola U, Spencer-Dene B, Xing-Qun L, Kettunen P, Thesleff I, et al. 2000. FGFR-2(IIIb) signalling is essential for inner ear morphogenesis. *J. Neurosci.* 20:6125–34

Pirvola U, Ylikoski J, Palgi J, Lehtonen E, Arumae U, Saarma M. 1992. Brain-derived neurotrophic factor and neurotrophin 3 mRNAs in the peripheral target fields of developing inner ear ganglia. *Proc. Natl. Acad. Sci. USA* 89:9915–19

Powell TPS, Erulkar SD. 1962. Transneuronal cell degeneration in the auditory relay nuclei of the cat. *J. Anat.* 96:249–68

Pujol R, Lavigne-Rebillard M, Lenoir M. 1998. Development of sensory and neural structures in the mammalian cochlea. See Rubel et al. 1998, pp. 146–92

Raper JA. 2000. Semaphorins and their receptors in vertebrates and invertebrates. *Curr. Opin. Neurobiol.* 10:88–94

Redd EE, Pongstaporn T, Ryugo DK. 2000. The effects of congenital deafness on auditory nerve synapses and globular bushy cells in cats. *Hear. Res.* 147:160–74

Reichardt LF, Fariñas I. 1999. Early actions of neurotrophic factors. In *Neurotrophins and the Neural Crest*, ed. M Sieber-Blum, pp. 1–27. Boca Raton: CRC

Represa J, Frenz D, van de Water TR. 2000. Genetic patterning of embryonic inner ear development. *Acta Otolaryngol.* 120:5–10

Retzius G. 1893. Zur Entwicklung der Zellen des Ganglion Spirale Acustici und zur Endigungsweise des Gehörnerven bei den Säugethieren. *Biol. Untersuch.* 4:52–57

Rhode WS. 1978. Some observations on cochlear mechanics. *J. Acoust. Soc. Am.* 64:158–76

Riedel B, Friauf E, Grothe C, Unsicker K. 1995. Fibroblast growth factor-2-like immunoreactivity in auditory brainstem nuclei of the developing and adult rat: correlation with onset and loss of hearing. *J. Comp. Neurol.* 354:353–60

Rivolta MN, Holley MC. 1999. GATA-3 is downregulated during hair cell differentiation in the mouse cochlea. *J. Neurocytol.* 27:637–47

Robinson M, Adu J, Davies AM. 1996. Timing and regulation of trkB and BDNF mRNA expression in placode-derived sensory neurons and their targets. *Eur. J. Neurosci.* 8:2399–406

Romand R, ed. 1992. *Development of Auditory and Vestibular Systems II*. Amsterdam: Elsevier

Romand MR, Romand R. 1987. The ultrastructure of spiral ganglion cells in the mouse. *Acta Otolaryngol.* 104:29–39

Rosbe KW, Burgess BJ, Glynn RJ, Nadol JB. 1996. Morphologic evidence for three cell types in the human spiral ganglion. *Hear. Res.* 93:120–7

Rubel EW. 1978. Ontogeny of structure and function in the vertebrate auditory system. In *Handbook of Sensory Physiology Vol. IX, Development of Sensory Systems*, ed. M Jacobsen, pp. 135–37. New York: Springer-Verlag

Rubel EW, Hyson RL, Durham D. 1990. Afferent regulation of neurons in the brain stem auditory system. *J. Neurobiol.* 21:169–96

Rubel EW, Oesterele EC, Weisleder P. 1991. Hair cell regeneration in the avian inner ear. *Ciba Foundation Symposium, Regeneration of Vertebrate Sensory Receptor Cells*, pp. 77–96. New York: John Wiley

Rubel EW, Parks TN. 1975. Organization and development of brain stem auditory nuclei of the chicken: tonotopic organization of projections from n. magnocellularis to n. laminaris. *J. Comp. Neurol.* 164:435–48

Rubel EW, Parks TN. 1988. Organization and development of the avian brain-stem auditory system. In *Auditory Function: Neurobiological Bases of Hearing*, ed. GM Edelman, WE Gall, WM Cowans, pp. 3–92. New York: John Wiley

Rubel EW, Popper AN, Fay RR, eds. 1998. *Development of the Auditory System*. New York: Springer

Rubel EW, Smith DJ, Miller LC. 1976. Organization and development of brain stem auditory nuclei of the chicken: ontogeny of n.

magnocellularis and n. laminaris. *J. Comp. Neurol.* 166:469–90

Ruben RJ. 1967. Development of the inner ear of the mouse: a radioautographic study of terminal mitoses. *Acta Otolaryngol.* 220:1–44

Rübsamen R, Lippe WR. 1998. The development of cochlear function. See Rubel et al. 1998, pp. 193–270

Rueda J, de la Sen C, Juiz JM, Merchán JA. 1987. Neuronal loss in the spiral ganglion of young rats. *Acta Otolaryngol.* 104:417–21

Rüsch A, Lysakowski A, Eatock RA. 1998. Postnatal development of type I and type II hair cells in the mouse utricle: acquisition of voltage-gated conductances and differentiated morphology. *J. Neurosci.* 18:7487–501

Ryan AF. 1997. Transcription factors and the control of inner ear development. *Sem. Cell Dev. Biol.* 8:249–56

Ryan AF, Woolf NK. 1988. Development of tonotopic representation in the Mongolian gerbil: a 2-deoxyglucose study. *Brain Res.* 469:61–70

Ryugo DK. 1992. The auditory nerve, peripheral innervation, cell body morphology, and central projections. In *The Mammalian Auditory Pathway: Neuroanatomy, Springer Handbook of Auditory Research, Volume 1*, ed. DB Webster, AN Popper, RR Fay, pp. 23–65. New York: Springer-Verlag

Ryugo DK, Fekete DM. 1982. Morphology of primary axosomatic endings in the anteroventral cochlear nucleus of the cat: a study of the endbulbs of Held. *J. Comp. Neurol.* 210:239–57

Ryugo DK, Pongstaporn T, Huchton DM, Niparko JK. 1997. Ultrastructural analysis of primary endings in deaf white cats: morphologic alterations in endbulbs of Held. *J. Comp. Neurol.* 385:230–44

Ryugo DK, Rosenbaum BT, Kim PJ, Niparko JK, Saada AA. 1998. Single unit recordings in the auditory nerve of congenitally deaf white cats: morphological correlates in the cochlea and cochlear nucleus. *J. Comp. Neurol.* 397:532–48

Saada AA, Niparko JK, Ryugo DK. 1996. Morphological changes in the cochlear nucleus of congenitally deaf white cats. *Brain Res.* 736:315–28

San Jose I, Vasquez E, Garcia-Atares N, Huerta JJ, Vega JA, Represa J. 1997. Differential expression of microtubule associated protein MAP-2 in developing cochleovestibular neurons and its modulation by neurotrophin-3. *Int. J. Dev. Biol.* 41:509–19

Sanes DH, Merickel M, Rubel EW. 1989. Evidence for an alteration of the tonotopic map in the gerbil cochlea during development. *J. Comp. Neurol.* 279:436–44

Sanes DH, Rubel EW. 1988. The ontogeny of inhibition and excitation in the gerbil lateral superior olive. *J. Neurosci.* 8:682–700

Sanes DH, Walsh EJ. 1998. The development of central auditory processing. See Rubel et al. 1998, pp. 271–314

Sato M, Saigo K. 2000. Involvement of pannier and u-shaped in regulation of decapentaplegic-dependent wingless expression in developing *Drosophila* notum. *Mech. Dev.* 93:127–38

Saunders JC, Adler JH, Cohen YE, Smullen S, Kazahaya K. 1998. Morphometric changes in the chick nucleus magnocellularis following acoustic overstimulation. *J. Comp. Neurol.* 390:412–26

Saunders JC, Coles RB, Gates GR. 1973. The development of auditory evoked responses on the cochlea and cochlear nuclei of the chick. *Brain Res.* 63:59–74

Schecterson LC, Bothwell M. 1994. Neurotrophin and neurotrophin mRNA expression in developing inner ear. *Hear. Res.* 73:92–100

Schimmang T, Alvarez-Bolado G, Minichiello L, Vazquez E, Giraldez F, et al. 1997. Survival of inner ear sensory neurons in trk mutants. *Mech. Dev.* 64:77–85

Schimmang T, Minichiello L, Vazquez E, San Jose I, Giraldez F, et al, 1995. Developing inner ear sensory neurons require TrkB and TrkC receptors for innervation of their peripheral targets. *Development* 121:3381–91

Schweitzer L, Cant NB. 1984. Development of the cochlear innervation of the dorsal cochlear nucleus of the hamster. *J. Comp. Neurol.* 225:228–43

Schweitzer L, Cecil T. 1992. Morphology of HRP-labelled cochlear nerve axons in the dorsal cochlear nucleus of the developing hamster. *Hear. Res.* 60:34–44

Shaner RF. 1934. The development of the nuclei and tracts related to the acoustic nerve in the pig. *J. Comp. Neurol.* 60:5–19

Sie KCY, Rubel EW. 1992. Rapid changes in protein synthesis and cell dize in the cochlear nucleus following eighth nerve activity blockade and cochlea ablation. *J. Comp. Neurol.* 320:501–8

Silos-Santiago I, Fagan AM, Garber M, Fritzsch B, Barbacid M. 1997. Severe sensory deficits but normal CNS development in newborn mice lacking TrkB and TrkC tyrosine protein kinase receptors. *Eur. J. Neurosci.* 9:2045–56

Simmons DD. 1994. A transient afferent innervation of outer hair cells in the postnatal cochlea. *Neuroreport* 5:1309–12

Simmons DD, Manson-Gieseke L, Hendrix TW, Morris K, Williams SJ. 1991. Postnatal maturation of spiral ganglion neurons: a horseradish peroxidase study. *Hear. Res.* 55:81–91

Simmons DD, Morley BJ. 1998. Differential expression of the 9 nicotinic acetylcholine receptor subunit in neonatal and adult cochlear hair cells. *Molec. Brain Res.* 56:287–92

Snyder RL, Leake PA. 1997. Topography of spiral ganglion projections to cochlear nucleus during postnatal development in cats. *J. Comp. Neurol.* 384:293–311

Sobkowicz HM. 1992. The development of innervation in the organ of Corti. See Romand 1992, pp. 59–100

Sokolowski BHA, Stahl LM, Fuchs PA. 1993. Morphological and physiological development of vestibular hair cells in the organcultured otocyst of the chick. *Dev. Biol.* 155:134–46

Solum D, Hughes D, Major MS, Parks TN. 1997. Prevention of normally occurring and deafferentation-induced neuronal death in chick brainstem auditory neurons by periodic blockade of AMPA/kainate receptors. *J. Neurosci.* 17:4744–52

Song HJ, Poo MM. 1999. Signal transduction underlying growth cone guidance by diffusible factors. *Curr. Opin. Neurobiol.* 9:355–63

Sterbing SJ, Schmidt U, Rübsamen R. 1994. The postnatal development of frequency-place code and tuning characteristics in the auditory midbrain of the phyllostomid bat, *Carollia perspicillata*. *Hear. Res.* 76:133–46

Steward O, Rubel EW. 1985. Afferent influences on brain stem auditory nuclei of the chicken: cessation of amino acid incorporation as an antecedent to age-dependent transneuronal degeneration. *J. Comp. Neurol.* 231:385–95

Studer M, Gavalas A, Marshall H, Ariza-McNaughton L, Riji F, et al. 1998. Genetic interactions between *Hoxa1* and *Hoxb1* reveal new roles in regulation of early hindbrain patterning. *Development (Great Britain)* 125:1025–36

Swanson GJ, Howard M, Lewis J. 1990. Epithelial autonomy in the development of the inner ear of a bird embryo. *Dev. Biol.* 137:243–57

Taber Pierce E. 1967. Histogenesis of the dorsal and ventral cochlear nuclei in the mouse. An autobiographic study. *J. Comp. Neurol.* 131:27–54

Takahashi TT, Carr CE, Brecha N, Konishi M. 1987. Calcium binding protein-like immunoreactivity labels the terminal field of nucleus laminaris of the barn owl. *J. Neurosci.* 7:1843–56

Takamori S, Rhee JS, Rosenmund C, Jahn R. 2000. Identification of a vesicular glutamate transporter that defines a glutamatergic phenotype in neurons. *Nature* 407:189–94

Takasaka T, Smith CA. 1971. The structure and innervation of the pigeon's basilar papilla. *Ultrastr. Res.* 35:20–65

Tello JF. 1931. Le reticule des cellules ciliees du labyrinth chez la souris et son independance des terminaisons nerveuses de la huitieme paire. *Trav. Lab Rech. Biol.* 27:151–86

Tierney TS, Russell FA, Moore DR. 1997. Susceptibility of developing cochlear nucleus neurons to deafferentation-induced death abruptly ends just before the onset of hearing. *J. Comp Neurol.* 378:295–306

Trimmer JS. 1999. Sorting out receptor trafficking. *Neuron* 22:411–17

Trune DR. 1982a. Influence of neonatal cochlear removal on the development of mouse cochlear nucleus: I. Number, size and density of its neurons. *J. Comp. Neurol.* 209:409–24

Trune DR. 1982b. Influence of neonatal cochlear removal on the development of mouse cochlear nucleus: II. Dendritic morphometry of its neurons. *J. Comp. Neurol.* 209:425–34

Trussell LO. 1998. Control of time course of glutamatergic synaptic currents. *Prog. Brain Res.* 116:59–69

Tucci DL, Born DE, Rubel EW. 1987. Changes in spontaneous activity and CNS morphology associated with conductive and sensorineural hearing loss in chickens. *Ann. Otol. Rhinol. Laryngol.* 96:343–50

Tucci DL, Cant NB, Durham D. 1999. Conductive hearing loss results in a decrease in central auditory system activity in the young gerbil. *Laryngoscope* 109:1359–71

Tucci DL, Rubel EW. 1985. Afferent influences on brain stem auditory nuclei of the chicken: effects of conductive and sensorineural hearing loss on n. magnocellularis. *J. Comp. Neurol.* 238:371–81

Ulatowska-Blaszyk K, Bruska M. 1999. The cochlear ganglion in human embryos of developmental stages 18 and 19. *Folia Morphol. (Warsz.)* 58:29–35

van de Water TR. 1983. Embryogenesis of the inner ear: "in vitro studies". In *Development of Auditory and Vestibular Systems*, ed. R Romand, pp. 337–74. New York: Academic

van de Water TR, Frenz DA, Giraldez F, Represa J, Lefebvre PP, et al. 1992. Growth factors and development of the stato-acoustic system. See Romand 1992, pp. 1–32

Verhage M, Maia AS, Plomp JJ, Brussaard AB, Heeroma JH, et al. 2000. Synaptic assembly of the brain in the absence of neurotransmitter secretion. *Science* 287:864–69

von Bartheld CS, Patterson SL, Heuer JG, Wheeler EF, Bothwell M, Rubel EW. 1991. Expression of nerve growth factor (NGF) receptors in the developing inner ear of chick and rat. *Development* 113:455–70

von Békésy G. 1960. *Experiments in Hearing.* New York: Am. Instit. Phys.

Waddington CH. 1937. The determination of the auditory placode in the chick. *J. Exp. Biol.* 14: 232–39

Walsh EJ, McGee J. 1988. Rhythmic discharge properties of caudal cochlear nucleus neurons during postnatal development in cats. *Hear. Res.* 36:233–47

Walsh E, McGee J, McFadden S, Liberman M. 1998. Longterm effects of sectioning the olivocochlear bundle in neonatal cats. *J. Neurosci.* 18:3859–69

Warchol ME, Dallos P. 1990. Neural coding in the chick cochlear nucleus. *J. Comp. Physiol.* 166:721–34

Warr WB. 1992. Organization of olivocochlear efferent systems in mammals. In: *The Anatomy of the Mammalian Auditory Pathways*, ed. RR Fay, AN Popper, DB Webster, pp. 410–48. New York: Springer

Waterman AJ. 1938. The development of the inner ear rudiment of the rabbit embryo in a foreign environment. *Amer. J. Anat.* 63:161–217

Webster DB. 1983a. Auditory neuronal sizes after a unilateral conductive hearing loss. *Exp. Neurol.* 79:130–40

Webster DB. 1983b. Late onset of auditory deprivation does not affect brainstem auditory neuron soma size. *Hear. Res.* 12:145–47

Webster DB. 1983c. A critical period during postnatal auditory development of mice. *Int. J. Pediatr. Otorhinolaryngol.* 6:107–18

Webster DB. 1985. The spiral ganglion and cochlear nuclei of deafness mice. *Hear. Res.* 18:19–27

Webster DB. 1988a. Conductive hearing loss affects the growth of the cochlear nuclei over an extended period of time. *Hear. Res.* 32:185–92

Webster DB. 1988b. Sound amplification

negates central effects of a neonatal conductive hearing loss. *Hear. Res.* 32:193–95

Webster DB, Webster M. 1977. Neonatal sound deprivation affects brain stem auditory nuclei. *Arch. Otolaryngol.* 103:392–96

Webster DB, Webster M. 1979. Effects of neonatal conductive hearing loss on brain stem auditory nuclei. *Ann. Otol. Rhinol. Laryngol.* 88:684–88

Wheeler EF, Bothwell M, Schecterson LC, von Bartheld CS. 1994. Expression of BDNF and NT-3 mRNA in hair cells of the organ of Corti: quantitative analysis in developing rats. *Hear. Res.* 73:46–56

Whitehead MC, Morest DK. 1985. The development of innervation patterns in the avian cochlea. *Neuroscience* 14:255–76

Whitlon DS, Zhang X, Kusakabe M. 1999a. Tenascin-C in the cochlea of the developing mouse. *J. Comp. Neurol.* 406:361–74

Whitlon DS, Zhang X, Pecelunas K, Greiner MA. 1999b. A temporospatial map of adhesive molecules in the organ of corti of the mouse cochlea. *J. Neurocytol.* 28:955–68

Wiechers B, Gestwa G, Mack A, Carroll P, Zenner H-P, Knipper M. 1999. A changing pattern of brain-derived neurotrophic factor expression correlates with rearrangement of fibers during cochlear development of rats and mice. *J. Neurosci.* 19:3033–42

Wiesel TN, Hubel DH. 1963. Single-cell responses in striate cortex of kittens deprived of vision in one eye. *J. Neurophys.* 26:1003–17

Wiesel TN, Hubel DH. 1965. Comparison of the effects of unilateral and bilateral eye closure on cortical unit responses in kittens. *J. Neurophysiol.* 6:1029–40

Wilkinson DG. 2000. Eph receptors and ephrins: regulators of guidance and assembly. *Int. Rev. Cytol.* 196:177–244

Willard F. 1990. Analysis of the development of the human auditory system. *Sem. Hearing* 11:107–23

Willard FH. 1993. Postnatal development of auditory nerve projections to the cochlear nucleus in *Monodelphis domestica*. See Merchán et al. 1993, pp. 29–42

Willard FH. 1995. Development of the mammalian auditory hindbrain. In *Advances in Neural Science*, Vol 2, ed. S Malhotra, pp. 205–34. Greenwich, CT: JAI

Willard FH, Martin GE. 1986. The development and migration of large multipolar neurons into the cochlear nucleus of the North American opossum. *J. Comp. Neurol.* 248:119–32

Wilson CJ, Durham D. 1995. Changes in the distribution of zinc in chick cochlear nucleus during deafferentation-induced neuronal cell death. *Assoc. Res. Otolaryngol.* (Abstr.) 18:36

Windle WF, Austin MF. 1936. Neurofibrillar development in the central nervous system of chick embryos up to 5 days' incubation. *J. Comp. Neurol.* 63:431–63

Winsky L, Jacobowitz DM. 1995. Effects of unilateral cochlea ablation on the distribution of calretinin mRNA and immunoreactivity in the guinea pig ventral cochlear nucleus. *J. Comp. Neurol.* 354:564–82

Xiang M, Gan L, Li D, Chen ZY, Zhou L, et al. 1997. Essential role of POU-domain factor Brn-3c in auditory and vestibular hair cell development. *Proc. Natl. Acad. Sci. USA* 94:9445–50

Xiang M, Gao W-Q, Hasson T, Shin JJ. 1998. Requirement for Brn-3c in maturation and survival, but not fate determination of inner ear hair cells. *Development* 125:3935–46

Young SR, Rubel EW. 1983. Frequency specific projections of individual neurons in chick brain stem auditory nuclei. *J. Neurosci.* 7:1373–78

Young SR, Rubel EW. 1986. Embryogenesis of arborization pattern and topography of individual axons in n. laminaris of the chicken brain stem. *J. Comp. Neurol.* 254:425–59

Zhang JS, Haenggeli CA, Tempini A, Vischer MW, Moret V, Rouiller EM. 1996. Electrically induced fos-like immunoreactivity in the auditory pathway of the rat: effects of survival time, duration, and intensity of stimulation. *Brain Res. Bull.* 39:75–82

Zhou N, Parks TN. 1992. Developmental changes in the effects of drugs acting at NMDA or non-NMDA receptors on synaptic transmission in the chick cochlear nucleus (nuc. magnocelluaris). *Brain Res. Dev. Brain Res.* 67:145–52

Zirpel L, Janowiak MA, Taylor DA, Parks TN. 2000a. Developmental changes in metabotropic glutamate receptor-mediated calcium homeostasis. *J. Comp. Neurol.* 421:95–106

Zirpel L, Janowiak MA, Veltri CA, Parks TN. 2000b. AMPA receptor-mediated, calcium-dependent CREB phosphorylation in a subpopulation of auditory neurons surviving activity deprivation. *J. Neurosci.* 20:6267–75

Zirpel L, Lachica EA, Rubel EW. 1995. Activation of a metabotropic glutamate receptor increases intracellular calcium concentrations in neurons of the avian cochlear nucleus. *J. Neurosci.* 15:214–22

Zirpel L, Lachica EA, Rubel EW. 1997. Afferent regulation of cochlear nucleus neurons. In *Neurotransmission and Hearing Loss: Basic Science, Diagnosis, and Management.*

Ch 3, ed. CI Berlin. pp. 47–76. San Diego: Singular

Zirpel L, Lippe WR, Rubel EW. 1998. Activity-dependent regulation of intracellular calcium in avian cochlear nucleus neurons: roles of protein kinases A and C and relation to cell death. *J. Neourophysiol.* 79:2288–302

Zirpel L, Parks TN. 2001. Zinc inhibition of group I mGluR-mediated calcium homeostasis in auditory neurons. *J. Assoc. Res. Otolaryngol.* 2:180–87

Zirpel L, Rubel EW. 1996. Eighth nerve activity regulates the intracellular calcium concentration of cochlear nucleus neurons in the embryonic chick via a metabotopic glutamate receptor. *J. Neurophysiol.* 76:4127–39

Zuo J, Treadaway J, Buckner TW, Fritzsch B. 1999. Visualization of $\alpha 9$ acetylcholine receptor expression in hair cells of transgenic mice containing a modified bacterial artificial chromosome. *Proc. Natl. Acad. Sci. USA* 96:14100–5

AMPA RECEPTOR TRAFFICKING AND SYNAPTIC PLASTICITY

Roberto Malinow[1] and Robert C. Malenka[2]

[1]*Cold Spring Harbor Laboratory, Cold Spring Harbor, New York 11724;*
email: malinow@cshl.org
[2]*Nancy Pritzker Laboratory, Department of Psychiatry and Behavioral Sciences,*
Stanford University School of Medicine, Palo Alto, California 94304;
email: malenka@stanford.edu

Key Words excitatory, transmission, memory, LTP, LTD

■ **Abstract** Activity-dependent changes in synaptic function are believed to underlie the formation of memories. Two prominent examples are long-term potentiation (LTP) and long-term depression (LTD), whose mechanisms have been the subject of considerable scrutiny over the past few decades. Here we review the growing literature that supports a critical role for AMPA receptor trafficking in LTP and LTD, focusing on the roles proposed for specific AMPA receptor subunits and their interacting proteins. While much work remains to understand the molecular basis for synaptic plasticity, recent results on AMPA receptor trafficking provide a clear conceptual framework for future studies.

INTRODUCTION

It is widely believed that a long-lasting change in synaptic function is the cellular basis of learning and memory (Alkon & Nelson 1990, Eccles 1964, Hebb 1949, Kandel 1997). The most thoroughly characterized examples of such synaptic plasticity in the mammalian nervous system are long-term potentiation (LTP) and long-term depression (LTD). A remarkable feature of LTP and LTD is that a short period of synaptic activity (either high- or low-frequency stimulation) can trigger persistent changes of synaptic transmission lasting at least several hours and often longer. This single property initially led investigators to suggest that these forms of plasticity are the cellular correlate of learning (Bliss & Gardner-Medwin 1973, Bliss & Lomo 1973). Work over the past 25 years that has elucidated many properties of LTP and LTD reinforces this view as well as suggests their involvement in various other physiological as well as pathological processes (Martin et al. 2000, Zoghbi et al. 2000).

Much effort in the field has been directed toward understanding the detailed molecular mechanisms that account for the change in synaptic efficacy. For many years, a most basic question remained intractable: Is the change in synaptic strength during these forms of plasticity primarily due to a pre- or postsynaptic modification? Numerous experiments using a variety of approaches were directed toward answering this question. Surprisingly, they often yielded conflicting conclusions (Kullmann & Siegelbaum 1995). Although many studies suggested primarily postsynaptic modifications (Davies et al. 1989, Kauer et al. 1988, Manabe et al. 1992, Muller et al. 1988), a consistent finding was a change in synaptic failures after LTP (Isaac et al. 1996, Kullmann & Nicoll 1992, Malinow & Tsien 1990, Stevens & Wang 1994). Because synaptic failures were assumed to be due to failure to release transmitter (a presynaptic property), these results were in apparent contradiction. A resolution arrived with the identification of postsynaptically "silent synapses" and the demonstration that they could be converted to active synapses by a postsynaptic modification (Durand et al. 1996, Isaac et al. 1995, Liao et al. 1995). Synapses are postsynaptically silent if they show an NMDA but no AMPA receptor response. Thus, at resting potentials NMDA receptors (NMDARs) are minimally opened, and transmitter release at such a synapse is recorded as a failure. The wholesale appearance of an AMPA response at such synapses during LTP, with no change in the NMDA response, strongly supports a postsynaptic modification consisting of a functional recruitment of AMPA receptors (AMPARs). One potential mechanism envisioned was the rapid delivery of AMPARs from nonsynaptic sites to the synapse, via a mechanism analogous to the exocytosis of presynaptic vesicles during transmitter release.

Two early studies provided support for postsynaptic exocytosis playing a role in synaptic plasticity. One study in hippocampal slices showed that loading postsynaptic cells with toxins that specifically perturb membrane fusion could block LTP (Lledo et al. 1998). Thus, postsynaptic events such as exocytosis were implicated. A separate study in dissociated cultured neurons identified a form of dendritic exocytosis that was mediated by activation of CaMKII (Maletic-Savatic et al. 1998), an enzyme believed to play a critical role in LTP (Lisman et al. 1997). Thus, dendritic exocytosis was further linked to synaptic plasticity. These studies along with the demonstration of the role of silent synapses in LTP provided strong motivation for the development of cellular and molecular techniques that could monitor and perturb trafficking of AMPARs to and away from synapses.

MOLECULAR INTERACTIONS OF AMPA RECEPTORS

AMPA receptors (AMPARs) are hetero-oligomeric proteins made of the subunits GluR1 to GluR4 (also known as GluRA-D) (Hollmann & Heinemann 1994, Wisden & Seeburg 1993). Each receptor complex is thought to contain four subunits (Rosenmund et al. 1998). In the adult hippocampus two species of AMPAR appear to predominate: receptors made of GluR1 and GluR2 or those composed of GluR3 and GluR2 (Wenthold et al. 1996). Immature hippocampus, as well as other mature brain regions, express GluR4, which also complexes with GluR2 to form a receptor (Zhu et al. 2000). Although the extracellular and transmembrane regions of AMPAR subunits are very similar, their intracellular cytoplasmic tails are distinct (Figure 1). GluR1, GluR4, and an alternative splice form of GluR2 (GluR2L) have longer cytoplasmic tails and are homologous. In contrast, the predominant splice form of GluR2, GluR3, and an alternative splice form of GluR4 that is primarily expressed in cerebellum (GluR4c) have shorter, homologous cytoplasmic tails. Through their C-terminal tails, each subunit interacts with specific cytoplasmic proteins (Figure 1). Most of these AMPAR-interacting proteins thus far identified have single or multiple PDZ domains, which are well-characterized protein-protein interaction motifs that often interact with the extreme C-terminal tails of target proteins (Sheng & Sala 2001). GluR1 forms a group I PDZ ligand, while GluR2, GluR3, and GluR4c form group II PDZ ligands. GluR4 and GluR2L have variant C-terminal tails, and if they interact with classical PDZ-domain proteins is unclear. In a variety of cell types, proteins containing PDZ domains have been implicated in playing important roles in the targeting and clustering of membrane proteins to specific subcellular domains (Sheng & Sala 2001).

GluR1 interacts with the PDZ-domain regions of SAP97 (Leonard et al. 1998) and RIL [reversion-induced LIM gene (Schulz et al. 2001)]. SAP97 is closely related to a family of proteins (SAP90/PSD95, chapsyn110/PSD93, and SAP102) that interact with NMDAR subunits. RIL, on the other hand, may link AMPA receptors to actin. GluR2 and GluR3 interact with glutamate receptor-interacting protein (GRIP) (Dong et al. 1997, 1999) and AMPA receptor binding protein (ABP)/GRIP2 (Dong et al. 1999, Srivastava et al. 1998), proteins with six or seven PDZ domains. GluR2 and GluR3 as well as GluR4c also interact with PICK1 (protein interacting with C-kinase) (Dev et al. 1999, Xia et al. 1999), which contains a single PDZ domain that interacts with both PKCα and GluR2. Other group II PDZ-domain-containing proteins that interact with GluR2, GluR3, and GluR4c have recently been identified and include rDLG6 (Inagaki et al. 1999) and afadin (Rogers et al. 2001). No binding partners have yet been reported for GluR4 and GluR2L.

Some additional proteins interact with the cytoplasmic tails of AMPAR subunits at regions that are not at the exact C terminus. GluR1 interacts with band 4.1N and is linked through it to actin (Shen et al. 2000). The interaction occurs at a region on GluR1 that is homologous with all other subunits, and thus band 4.1N may interact with other AMPAR subunits as well. A surprising finding is that the cytoplasmic tail of GluR2, in addition to interacting with PDZ proteins, also binds to NSF (NEM-sensitive-factor) (Nishimune et al. 1998, Osten et al. 1998, Song et al. 1998), an ATPase known to play an essential role in the membrane fusion processes that underlie intracellular protein trafficking and presynaptic vesicle exocytosis (Rothman 1994). This was particularly surprising given that GluR2 has no sequence homology with the extensively characterized SNARE proteins that previously were thought to be the unique targets for NSF action. Another key component of membrane fusion machinery, α and β SNAPS (soluble NSF attachment proteins) could be co-immunoprecipitated with AMPARs containing GluR2 (Osten et al. 1998), although the molecular regions mediating the presumed interaction between AMPARs and SNAPS remain to be determined.

Because these AMPAR-interacting proteins either contain PDZ domains, are proteins implicated in membrane fusion, or interact with the actin cytoskeleton,

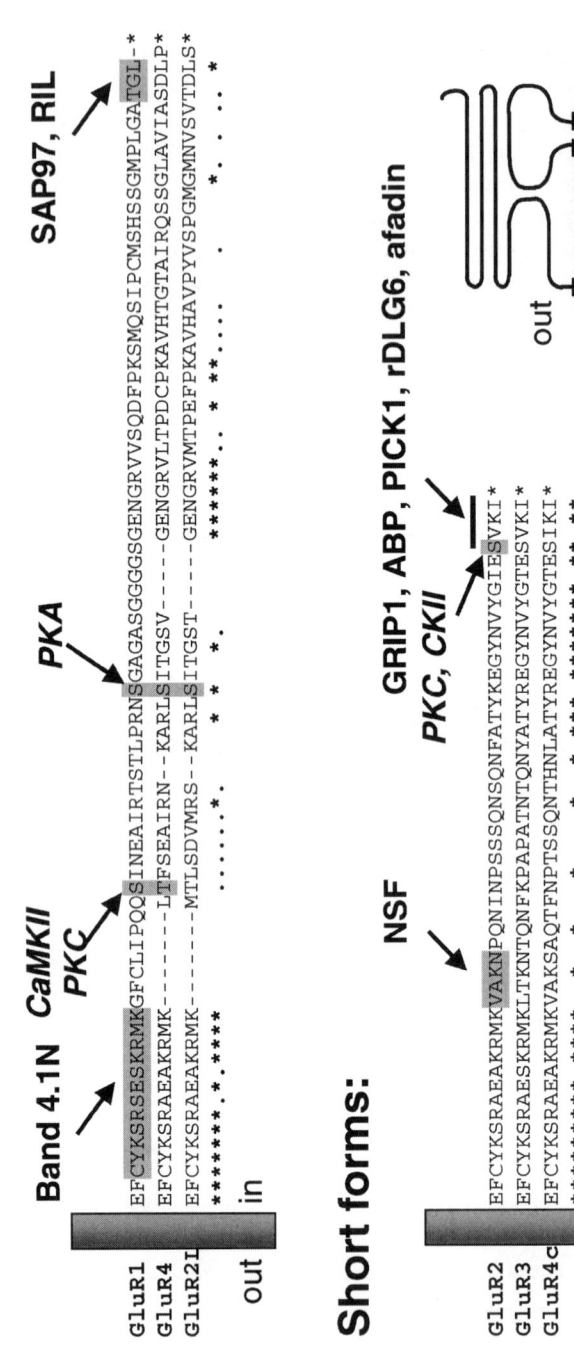

Figure 1 Cytoplasmic carboxyl tail regions of AMPARs, phosphorylation sites, and interacting proteins. Long and short forms of AMPARs are aligned and homologies are indicated (*, identity; .., similarity). Interacting proteins and approximate sites of interaction indicated (...). Protein kinases (italics) and sites of phosphorylation indicated (...). See text for details.

they have been suggested to play important roles in controlling the trafficking of AMPARs and/or their stabilization at synapses. The proposed specific functions of each of these proteins in controlling AMPAR behavior are discussed in greater detail in the following sections.

AMPA RECEPTOR ENDOCYTOSIS AND LONG-TERM DEPRESSION

Activity-Dependent Loss of Synaptic AMPA Receptors

The first experiments to test directly the idea that activity could influence the number of AMPARs at individual synapses involved chronic (several days) pharmacological manipulations of network activity in dissociated cultured neurons. Increasing network activity by blocking inhibitory synaptic transmission with GABA-A receptor antagonists caused a significant decrease in the proportion of synapses containing epitope-tagged AMPARs with no detectable effect on the level of NMDAR expression at synapses (Lissin et al. 1998). Consistent with this decrease in the content of AMPARs, prolonged increases in the activity of spinal cord and cortical cultures caused a decrease in the amplitude of miniature AMPAR-mediated synaptic currents (O'Brien et al. 1998, Turrigiano et al. 1998). Conversely, applying AMPAR antagonists for hours to days caused an increase in the surface expression of AMPARs at synapses (Liao et al. 1999, O'Brien et al. 1998) and a decrease in the proportion of anatomically defined silent synapses (Liao et al. 1999).

Although these studies demonstrated that prolonged manipulations of activity affected the synaptic distribution of AMPARs, the question remained whether a much more rapid movement of AMPARs might occur. Such rapid movement is mandatory if changes in AMPAR content at synapses contribute to synaptic plasticity such as LTD. One of the first pieces of evidence for a rapid redistribution of AMPARs came from the demonstration that brief applications of glutamate to hippocampal cultures cause a significant loss of AMPARs, detected immunohistochemically, from synaptic sites (Lissin et al. 1999). Consistent with the idea that NMDARs are relatively more stable in the synaptic plasma membrane (Allison et al. 1998), NMDARs appeared to be unaffected. It is now clear that a similar rapid loss of synaptic AMPARs can be triggered by the activation of several different receptors including AMPARs, NMDARs, metabotropic glutamate receptors, and insulin receptors (Beattie et al. 2000, Carroll et al. 1999a, Ehlers 2000, Man et al. 2000, Snyder et al. 2001).

Endocytosis of AMPA Receptors During LTD

Immunocytochemical techniques that distinguish AMPARs in the plasma membrane from AMPARs in intracellular pools have directly demonstrated that the loss of AMPARs due to pharmacological activation of glutamate or insulin receptors is indeed due to internalization of receptors originally on the cell surface. Several lines of evidence indicate that the endocytosis of AMPARs is analogous to the stimulated endocytosis of G protein–coupled receptors in that AMPAR endocytosis occurs via clathrin-coated pits and requires the activity of dynamin. First, inhibition of clathrin-mediated endocytosis by high concentrations of sucrose, the expression of a dominant negative form of dynamin, or peptide-mediated disruption of the dynamin-amphiphysin complex each blocks the triggered endocytosis of AMPARs (Carroll et al. 1999a, Man et al. 2000, Wang & Linden 2000). Second, following internalization, AMPARs exhibit an increased colocalization and interaction with the clathrin adaptor protein AP2 (Carroll et al. 1999a, Man et al. 2000).

All of the experiments reviewed thus far used pharmacological manipulations to induce AMPAR endocytosis. Thus, the critical question remained whether AMPAR endocytosis actually contributed to LTD. The first experimental support for this idea came from the use of immunocytochemical techniques to examine the distribution of AMPARs following the generation of NMDAR-dependent LTD in hippocampal cultures (Carroll et al. 1999b). LTD caused a decrease in the proportion of synapses containing detectable surface AMPARs while having no effect on the distribution of synaptic NMDARs. Generation of LTD in the hippocampus in vivo subsequently was found to cause a decrease in the number of AMPARs in synaptoneurosomes, providing further evidence for the role of AMPAR endocytosis (Heynen et al. 2000). That the loss of synaptic AMPARs during LTD involves their clathrin-mediated endocytosis is further supported by experiments in which LTD was blocked by loading CA1 pyramidal neurons or cerebellar Purkinje cells with a peptide that disrupts dynamin function (Luscher et al. 1999, Wang & Linden 2000). Importantly, these results were the first demonstration that two forms of LTD that previously were thought to be mechanistically distinct, cerebellar LTD and NMDAR-dependent LTD in the hippocampus, appear to share a common mechanism of expression. Inhibition of endocytosis also blocked the actions of insulin, which can cause a depression of synaptic currents that occludes LTD (Lin et al. 2000, Man et al. 2000).

Intracellular Signaling Pathways That Trigger AMPA Receptor Endocytosis

Given that LTD involves the endocytosis of AMPARs, a critical question is what intracellular signaling pathways trigger this increase in AMPAR trafficking. For NMDAR-dependent LTD a predominant hypothesis is that activation of a calcium-dependent protein phosphatase cascade involving calcineurin and protein phosphatase 1 (PP1) is required for its triggering (Lisman 1989, Mulkey et al. 1994, Mulkey et al. 1993). Thus, a number of laboratories have investigated the role of calcium and protein phosphatase activity and have found that both are involved in the triggering of AMPAR endocytosis. Specifically, the internalization of AMPARs caused by activation of NMDARs was blocked by removing extracellular calcium or application of the membrane permeable calcium chelator BAPTA-AM as well as by specific inhibitors of calcineurin (Beattie et al. 2000, Ehlers 2000). Similarly, the AMPAR endocytosis triggered by application of AMPA or insulin was blocked by calcineurin inhibitors (Beattie et al. 2000, Lin et al. 2000). The mechanisms by which calcineurin facilitates AMPAR endocytosis are unknown. One attractive hypothesis, based on work on the mechanisms mediating the activity-dependent increase in the endocytosis of presynaptic vesicles, is that calcineurin facilitates endocytosis via its association with dynamin/amphiphysin and the consequent dephosphorylation of components of the endocytic machinery (Beattie et al. 2000, Lai et al. 1999, Slepnev et al. 1998).

The role of PP1 in AMPAR endocytosis is less clear. Pharmacological inhibition of PP1 has been reported to block the endocytosis of AMPARs triggered by NMDA application (Ehlers 2000) but also has been found to have the opposite effect: enhancement of AMPAR endocytosis (Beattie et al. 2000, Lin et al. 2000). Differences in the techniques used to detect internalized AMPARs may have identified different subpopulations of AMPARs in these studies and therefore may have contributed to these differing results. Study of LTD in cerebellar Purkinje cells provides further complexity to the signaling pathways triggering AMPAR endocytosis. Cerebellar LTD requires activation of PKC (Linden 1994), which is required to stimulate the internalization of AMPARs in these cells (Matsuda et al. 2000, Xia et al. 2000). In fact, activation of PKC with a phorbol ester can also drive AMPAR internalization in cultured cortical neurons (Chung et al. 2000). When all of the studies on the signaling pathways involved in AMPAR endocytosis are considered together, it appears that the regulation of AMPAR endocytosis may be cell-type specific. This likely is because the detailed subunit composition of AMPARs differs among cell types and as a consequence so do the protein-protein interactions involving AMPARs that regulate endocytosis.

In addition to the regulated endocytosis of AMPARs that plays a critical role in LTD, it is clear that synaptic AMPARs also undergo a constitutive endocytosis that contributes to the basal cycling of AMPARs. Such cycling has been observed using electrophysiological (Kim & Lisman 2001, Luscher et al. 1999, Lüthi et al. 1999, Nishimune et al. 1998, Noel et al. 1999, Shi et al. 2001), biochemical (Ehlers 2000), and immunocytochemical (Lin et al. 2000, Zhou et al. 2001) techniques; the role of specific AMPAR subunits in this cycling has been addressed (see below). Another form of activity-dependent endocytosis that does not require calcium influx or calcineurin can be stimulated by ligand binding to AMPARs, either by competitive antagonists or by AMPA in the absence of receptor activation (Ehlers 2000, Lin et al. 2000). These forms of AMPAR endocytosis are further distinguished from regulated AMPAR endocytosis by differences in the effects of mutations in the carboxyl terminus of GluR2 expressed in HEK293 cells (Lin et al. 2000).

Role of AMPA Receptor Interacting Proteins

NSF-GluR2 INTERACTIONS The first protein that interacts with an AMPAR subunit to receive attention for its possible role in AMPAR trafficking was NSF, which as mentioned above plays a key role in membrane fusion events such as synaptic vesicle exocytosis (Rothman 1994). Loading CA1 pyramidal cells with peptides that disrupt the interaction of NSF with GluR2 causes a fairly rapid decrease in the size of synaptic currents, which suggests a loss of synaptic AMPARs (Kim & Lisman 2001, Lüscher et al. 1999, Lüthi et al. 1999, Nishimune et al. 1998, Noel et al. 1999, Shi et al. 2001). Depression by such peptides does not occur in animals lacking GluR2, which indicates that the depressive effect is specific for interactions mediated by GluR2 (Shi et al. 2001). More prolonged expression of the peptide caused a decrease in the expression of surface AMPARs identified by immunocytochemistry (Lüscher et al. 1999, Noel et al. 1999), a decrease in the responses of neurons to local application of AMPA (Nishimune et al. 1998), and an ~50% decrease in evoked synaptic transmission (Shi et al. 2001). These results suggest that the synaptic expression of some AMPARs may require an NSF-GluR2 interaction.

This interaction may also be important for controlling the synaptic expression of the population of AMPARs that play a role in synaptic plasticity. Peptide-mediated disruption of the NSF-GluR2 interaction was observed to occlude LTD in hippocampal CA1 pyramidal cells while saturation of LTD occluded any further reduction in synaptic currents due to the peptide (Lüthi et al. 1999, Noel et al. 1999). LTP also was impaired by manipulations that disrupt the function of NSF in postsynaptic cells (Lledo et al. 1998), although these effects were likely not due to a specific disruption of the NSF-GluR2 complex because LTP can still be elicited in knockout mice lacking GluR2 (Jia et al. 1996, Mainen et al. 1998) and because overexpression of the cytoplasmic tail of GluR2 does not block LTP (Shi et al. 2001).

While the results using peptide-mediated disruption of the NSF-GluR2 interaction have supported an important role for this interaction in AMPAR trafficking, the interpretation of these results is dependent on the specificity of the peptide's actions, in particular that the peptides have no effect on the function of NSF in the membrane fusion events required for protein trafficking through the secretory pathway. That the peptide does not have a clear effect in knockout mice lacking GluR2 supports the specificity of its actions (Shi et al. 2001). A complementary approach that obviates this concern is to examine the behavior of mutant forms of GluR2 that do not bind NSF. When such a mutant form of GluR2 is expressed in hippocampal slice cultures, it is not present in the synaptic plasma membrane, whereas wild-type GluR2 can be delivered to synapses without difficulty (Shi et al. 2001). This result is consistent with the experiments using peptides and suggests that the NSF-GluR2 interaction is required either for the delivery of AMPARs to the plasma membrane at synapses or to stabilize AMPARs at the synapses, making them resistant to endocytosis. In contrast, however, a similar construct can be detected at the synaptic surface in hippocampal cultured neurons (Braithwaite & Malenka 2001). In response to AMPA or NMDA application, this construct exhibits enhanced internalization, a result consistent with the suggestion that the NSF-GluR2 interaction plays a role in the stabilization of surface AMPARs. Clearly more work needs to be done on the role of NSF in AMPAR trafficking as well as the role of α- and β-SNAPs that are also closely associated with GluR2 (Osten et al. 1998).

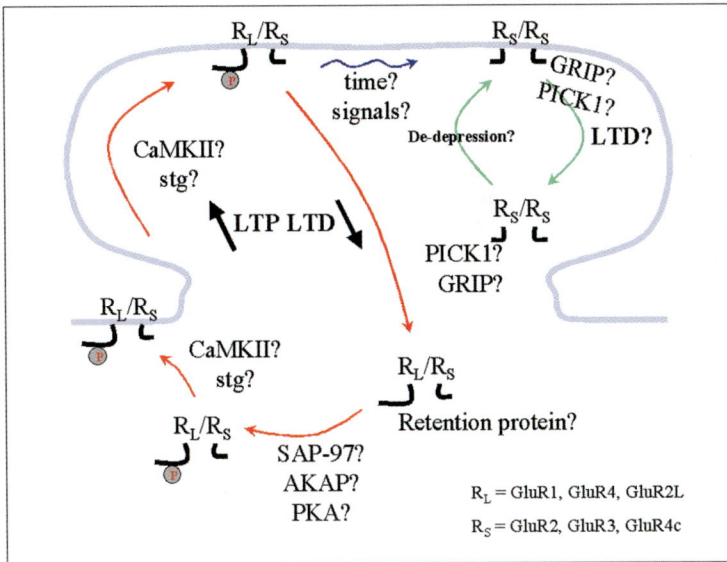

Figure 2 Different steps in trafficking of AMPARs. Note participation in regulated (*red*) or continuous (*green*) trafficking depends on AMPAR composition. Potential molecules or processes controlling specific steps are shown. Phosphorylation at Ser845 by PKA is indicated. See text for details.

PDZ PROTEINS-GluR2 INTERACTIONS The roles of GRIP, ABP/GRIP2, and PICK1 in AMPAR endocytosis and trafficking have also been addressed using techniques similar to those described above. Loading CA1 pyramidal cells with a peptide that disrupts the GluR2/3-GRIP/ABP interaction caused an increase in synaptic currents in a subset of cells and prevented the generation of LTD (Daw et al. 2000). These results are consistent with the hypothesis that the binding of GRIP/ABP to GluR2/3 stabilizes AMPARs in an intracellular pool and prevents their insertion into the synaptic plasma membrane. Consistent with this idea, following the generation of LTD (which should increase the pool of intracellular AMPARs due to their endocytosis), the peptide increased synaptic strength in a much higher proportion of cells (Daw et al. 2000). Experiments in which a mutant form of GluR2 that does not bind to GRIP/ABP was expressed in hippocampal neurons have also been performed. In one study (Osten et al. 2000), such a mutant GluR2 was targeted appropriately to the synaptic plasma membrane, but its accumulation at synapses was significantly reduced when compared to wild-type GluR2. These findings were consistent with the idea that the association of GluR2 with GRIP and/or ABP is essential for maintaining AMPARs at synapses, perhaps by limiting their endocytosis. Similar results were obtained when a mutant GluR2 that could not bind GRIP/ABP or PICK1 was expressed in hippocampal slice cultures, in that there was no detectable surface expression at synapses of the mutant GluR2, whereas wild-type GluR2 could be readily detected (Shi et al. 2001). Another study, however, found no effect of mutating the GRIP/ABP and PICK1 binding site on GluR2 on its targeting to synapses in hippocampal cultures, although this mutant did exhibit a smaller degree of regulated endocytosis (Braithwaite & Malenka 2001), perhaps because it was not retained in an intracellular pool following its internalization. Taken together, these experiments that attempt to define the role of the interactions of PDZ proteins with GluR2/3 are confusing and do not allow definitive conclusions to be reached. It is conceivable that GRIP/ABP subserves several functions in the delivery, stabilization, and endocytosis of synaptic AMPARs. For example, GRIP/ABP appears to be found both at the membrane and in the cytosol of neurons and thus might stabilize AMPARs in both locations.

Although GRIP/ABP and PICK1 bind to the same sites on GluR2 and GluR3, their interactions, at least with GluR2, can be regulated independently. Phosphorylation of serine 880 in the PDZ-binding domain of GluR2 greatly decreases the affinity of GluR2 for GRIP/ABP but not for PICK1 (Chung et al. 2000, Matsuda et al. 1999). This phosphorylation is enhanced by activation of PKC leading to a decreased association of GRIP with GluR2 in vitro, in HEK293 cells, and in Purkinje cells (Chung et al. 2000, Matsuda et al. 1999). In cultured neurons, activation of PKC with phorbol esters also causes a redistribution of PICK1 and PKCα to synaptic sites (Chung et al. 2000, Perez et al. 2001) and a decrease in the level of GluR2 in the synaptic plasma membrane (Perez et al. 2001), presumably due to an increase in its internalization (Chung et al. 2000, Matsuda et al. 2000). Thus the phosphorylation state of serine 880 on GluR2, via its differential effects on the binding of GluR2 to GRIP/ABP and PICK1, may be important for influencing the subcellular localization of AMPARs and may play a role in some forms of LTD.

Work on the role of GluR2/3-PDZ protein interactions in hippocampal and cerebellar LTD suggests that the functions of GRIP/ABP and PICK1 may differ among cell types. As mentioned above, a peptide that inhibits GRIP/ABP and PICK1 interaction with GluR2 impairs LTD in CA1 pyramidal cells and causes an increase in synaptic strength in most cells following the prior induction of LTD (Daw et al. 2000). In contrast, a peptide designed to inhibit specifically only the PICK1-GluR2/3 interaction has no effect on either LTD or basal synaptic strength. The investigators also found that the increase in synaptic currents due to the peptide was prevented by PKC inhibitors, which suggests that PKC activity is required for the recycling of internalized AMPARs back to the plasma membrane (Daw et al. 2000). Taken together these results are consistent with the hypothesis that disrupting the binding of GRIP/ABP to GluR2 impairs the retention of AMPARs in an intracellular pool and allows them to return to the synaptic plasma membrane, thereby preventing the maintenance of LTD. Furthermore, the return of AMPARs to the plasma membrane requires PKC activity.

In contrast to NMDAR-dependent LTD in the hippocampus, cerebellar LTD studied in cultured Purkinje cells was impaired by several different manipulations aimed at specifically disrupting the PICK1-GluR2/3 interaction: loading cells with peptides designed to disrupt this interaction or antibodies directed against the PDZ domain of PICK1 as well as expression of mutant PICK1-GST fusion proteins (Xia et al. 2000). Furthermore, application of a phorbol ester, which elicits LTD in cultured Purkinje cells, causes phosphorylation of serine 880 on GluR2 and internalization of GluR2-containing surface AMPARs (Matsuda et al. 2000). Although these results do not rule out a role of GRIP/APB in cerebellar LTD, they are most consistent with a role for PICK1 binding to GluR2/3 in priming AMPARs for endocytosis or stabilizing endocytosed receptors in intracellular pools. Consistent with the first of these alternatives, overexpression of PICK1 in cultured hippocampal neurons was found to reduce the level of surface expression of GluR2 (Perez et al. 2001). Indeed, in contrast to the conclusions of Daw et al. (2000), a similar role for the PICK1-GluR2/3 interaction in hippocampal LTD has recently been suggested based on the findings that induction of LTD in hippocampal slices increases phosphorylation of serine 880 in GluR2 and a peptide that specifically inhibits the PICK1-GluR2/3 interaction enhances basal synaptic strength and impairs LTD in CA1 pyramidal cells (Kim et al. 2001).

Clearly, we currently do not have a thorough understanding of the functions of the cytosolic proteins that interact with AMPAR subunits; however, it is equally clear that these interactions do play important roles in controlling AMPAR content at synapses and thereby synaptic strength. Some of the confusion on their role in LTD may be because there are cell-type specific differences in the role these proteins play. In addition, each protein may play multiple roles in the delivery, stabilization, and/or removal of synaptic AMPARs, and thus, the consequences of perturbing the interactions of an individual protein with AMPARs may vary depending on the specific method used to impair protein function and the assays used to monitor the consequences of this manipulation. For example, optical and electrophysiological methods may have different sensitivities for the detection of synaptic AMPARs. It is also possible that AMPAR trafficking in different experimental preparations (e.g., dissociated cultured neurons versus brain slices) may differ. Perhaps regulatory mechanisms are differentially expressed as a consequence of different levels of signals (synaptic or growth factors, etc.) impinging on neurons in these different preparations.

AMPA RECEPTOR DELIVERY TO SYNAPSES AND LTP

Subcellular Steady-State Distribution of AMPA Receptors

A number of studies over the past few years have tested the notion that silent synapses lack AMPARs and that AMPARs are rapidly delivered to synapses during LTP. An important requirement for this model is that there be a pool of nonsynaptic AMPARs near synapses available for delivery. Several studies have used microscopic techniques to examine distribution of glutamate receptors at and near synapses in rat brains (Baude et al. 1995, Kharazia et al. 1996, Martin et al. 1993, Molnar et al. 1993, Nusser et al. 1998, Petralia et al. 1999, Petralia & Wenthold 1992, Takumi et al. 1999). While the concentration of AMPARs is normally higher at synapses, these studies generally find ample amounts of nonsynaptic AMPARs on both surfaces and intracellular regions of dendrites. Indeed, given the much larger space occupied by nonsynaptic regions, nonsynaptic AMPARs appear to outnumber synaptic AMPARs by quite a large margin (Shi et al. 1999). The distance between these nonsynaptic receptors and synaptic regions is a few microns, a distance that could be traversed in seconds by membrane-trafficking processes. Importantly, recent studies using postembedding immunogold techniques (Nusser et al. 1998, Petralia et al. 1999, Takumi et al. 1999) found that a sizable fraction of synapses in CA1 hippocampus lacks or has very few AMPARs, while most synapses have NMDARs. The fraction of synapses lacking AMPARs is greater earlier in development, consistent with the electrophysiological observations that silent synapses are more prevalent at these ages (Durand et al. 1996, Isaac et al. 1997, Liao & Malinow 1996, Rumpel et al. 1998, Wu et al. 1996). While some studies in dissociated cultured neurons support these views (Gomperts et al. 2000, Liao et al. 1999), others do not (Renger et al. 2001), possibly due to different culture conditions.

Optical Detection of Recombinant AMPA Receptor Trafficking During LTP

To monitor AMPAR trafficking in living tissue, one study generated and acutely expressed GFP-tagged GluR1 receptors in organotypic hippocampal slices (Shi et al. 1999). While slices of tissue provide a more challenging experimental preparation to examine receptor trafficking, this tissue was used, rather than dissociated neurons, since there had been little success in generating LTP using standard electrophysiological protocols in dissociated neurons. These recombinant GluR1-GFP receptors are functional, and their cellular distribution can be monitored with two-photon microscopy. Upon expression, these receptors distribute diffusely throughout the dendritic tree. Interestingly, they remain in the dendritic shaft regions, with little encroachment into dendritic spines, which are the sites of excitatory contacts. This restriction from synapses is in contrast with what is found in dissociated cultured neurons in which expression of recombinant GluR1 concentrates at synapses (Lissin et al. 1998, Shi et al. 1999). In slices, little movement of GluR1-GFP was detected in the absence of stimulation. However, high-frequency synaptic activation generated LTP-induced movement of GFP-tagged receptors to the surface of dendritic shafts as well as to dendritic spines. These movements of GFP-tagged receptors were detected over the course of about 15–30 min and were prevented by blockade of NMDARs. The tagged receptors remained in at least some spines for at least 50 min. This study concluded that GuR1-containing receptors are maintained in reserve at the dendritic shaft and can be delivered to synapses during LTP.

A number of studies have made findings that strengthen these conclusions. Adult knockout mice lacking GluR1 cannot generate LTP, indicating that this subunit plays a critical role (Zamanillo et al. 1999). In a follow-up study, GluR1-GFP was genetically inserted into these GluR1 knockout mice and GFP fluorescence was detected in dendritic spines (Mack et al. 2001). This distribution differs from what is observed when GluR1-GFP is acutely expressed in hippocampal slices before LTP but resembles the distribution after LTP. These observations are consistent with the view that an LTP-like process drives the genetically expressed GluR1-GFP into synapses when the animals are alive. This study also found that LTP was rescued by expression of only $\sim$10% of the normal amount of GluR1. This further supports the view that normally there is an overabundance of GluR1 available for generating LTP.

Electrophysiological Tagging to Monitor Synaptic Delivery of Recombinant AMPA Receptors

While optical studies provide important information regarding receptor distribution, the location of a receptor (even with electron microscopic resolution) cannot unambiguously reveal its contribution to synaptic transmission. One approach to address this issue used electrophysiologically tagged recombinant AMPARs. Such receptors differed in their rectification from endogenous receptors. Rectification is an intrinsic biophysical property of a receptor that can be detected as the ratio of the response observed at -60 mV to that at $+40$ mV. Most endogenous AMPARs contain the GluR2 subunit and can pass current equally well in both inward and outward directions. In contrast, AMPARs lacking GluR2 (or containing GluR2 that is genetically modified) exhibit profound inward rectification such that they can pass minimal current in the outward direction when the cell is depolarized to $+40$ mV. Thus, incorporation of recombinant AMPARs into synapses and their contribution to real synaptic transmission can be monitored functionally. With this assay for AMPAR delivery, it has been possible to show that LTP and overexpression of active CaMKII induce delivery of GluR1-containing receptors into synapses (Hayashi et al. 2000). An interaction between GluR1 and a PDZ-domain protein is necessary for LTP or CaMKII to drive synaptic delivery of GluR1, as point mutations in the PDZ-binding region of GluR1 prevent its synaptic delivery. Neither the identity of the GluR1-interacting PDZ-domain protein(s) responsible for LTP, nor the subcellular site where this (these) critical interaction(s) occurs is known.

One important finding observed in mice lacking GluR1 is that LTP is neither absent in all brain regions [e.g., LTP in dentate gyrus is present (Zamanillo et al. 1999)] nor in all ages [e.g., LTP in CA1 is present in juvenile animals (Mack et al. 2001)]. This suggests that AMPAR subunits other than GluR1 may play critical roles in activity-dependent synaptic plasticity. Indeed, the CA1 hippocampal region in immature animals, as well as the dentate gyrus in older animals, contain GluR4, a subunit with considerable homology to GluR1. Studies using electrophysiological assays to monitor the synaptic delivery of recombinant GluR4 indicate that this subunit mediates activity-dependent AMPAR delivery in immature hippocampus (Zhu et al. 2000). Interestingly, this delivery of recombinant GluR4 to synapses required NMDAR activity (i.e., delivery was blocked by APV) but not CaMKII activity.

As expression of GluR4 in hippocampus decreases to nearly undetectable levels by postnatal day 10, the LTP observed in CA1 hippocampus of juvenile (∼postnatal day 28) animals that lack GluR1 may be mediated by other AMPAR subunits. It is possible that this role is played by GluR2L, the alternative splice form of GluR2 with a cytoplasmic tail that resembles GluR1 and GluR4 (Hollmann & Heinemann 1994, Wisden & Seeburg 1993) (Figure 1). Indeed, recent results indicate activity-driven synaptic delivery of recombinant GluR2L (J. Zhu, J. Esteban, R. Malinow, unpublished observations). Results from experiments that used peptides loaded directly into postsynaptic cells suggest that interactions between GluR2 and PDZ-domain proteins may also be necessary to execute some other forms of synaptic potentiation in hippocampus (Daw et al. 2000) as well as in spinal cord (Li et al. 1999).

Synaptic Delivery of Endogenous Receptors

While the studies described above monitored synaptic delivery of recombinant AMPARs, other studies have tested if such a process occurs for endogenous receptors. In one study (Heynen et al. 2000) the amount of GluR1 and GluR2 in synaptoneurosomes (a fraction of brain extracts enriched for synaptic membranes) was increased following generation of LTP in the hippocampus in vivo. This increase was blocked by APV, and no change was detected in the level of NR1, the major NMDAR subunit. This study is consistent with a previous in vivo study that found increased AMPAR binding in hippocampal regions following LTP induction (Maren et al. 1993). At the mossy fiber synapse onto granule cells in cerebellum, tetanic stimulation induces replacement of calcium-permeable AMPARs with calcium-impermeable AMPARs (Liu & Cull-Candy 2000). This protocol does not enhance transmission but is likely to represent a protective mechanism. One study in cultured hippocampal slices detected an increase of AMPARs in synaptic membranes following pharmacological induction of LTP (Broutman & Baudry 2001). This was blocked by brefeldin A, implicating a process involving protein trafficking between ER/Golgi and surface membrane. A number of in vitro studies on dissociated cultured neurons have also examined delivery of endogenous AMPARs to synapses. In one study, an LTP of miniature excitatory postsynaptic currents could be triggered following brief pharmacological stimulation of NMDA receptors (Lu et al. 2001). This LTP was accompanied by a rapid insertion of endogenous AMPARs that could be blocked by interfering with SNARE-dependent exocytosis. Similarly, Liao et al. show in dissociated cultured neurons chronically kept in APV that washing out APV allows spontaneous activity to stimulate NMDARs, and this leads to the rapid delivery of native AMPARs into synapses previously without AMPARs (e.g., silent synapses) (Liao et al. 2001). Such a stimulus can also recruit CaMKII to synapses (Liao et al. 2001, Shen & Meyer 1999), where it may bind to NMDA receptors (Bayer et al. 2001, Leonard et al. 1999, Strack & Colbran 1998, Strack et al. 2000). The relation between CaMKII translocation to synapses and AMPAR delivery to synapses has not been established, although an intriguing model proposes a necessary link between these processes (Lisman & Zhabotinsky 2001). An actin-dependent process underlying the delivery of AMPARs during LTP is supported by experiments showing that agents that perturb actin function block LTP (Kim & Lisman 1999, Krucker et al. 2000).

Role of AMPA Receptor Phosphorylation in Synaptic Delivery

There has been considerable evidence indicating that protein kinases play critical roles in the generation of LTP (Bliss & Collingridge 1993, Madison et al. 1991, Malenka & Nicoll 1999). Some kinases [e.g., CaMKII; (Lisman et al. 1997)] are thought to mediate directly the signals leading to LTP, while others [e.g., PKA; (Blitzer et al. 1995)] may "gate" (i.e., modulate) its generation. The targets of these kinases responsible for mediating or gating LTP have been the source of considerable investigation. During LTP the CaMKII-phosphorylation site on GluR1, Ser831, is phosphorylated (Barria et al. 1997a, Barria et al. 1997b, Mammen et al. 1997). Such phosphorylation can increase conductance through GluR1 receptors (Derkach et al. 1999), and AMPARS show increased conductance during LTP (Benke et al. 1998). Thus, it was of considerable interest to determine if phosphorylation of Ser831 is required for synaptic delivery of GluR1-containing receptors. However, mutations on GluR1-Ser831 that prevent its phosphorylation by CaMKII do not prevent delivery of the receptor to synapses by active CaMKII (Hayashi et al. 2000) or by LTP (S-H. Shi & R. Malinow, unpublished observations). Thus, CaMKII must be acting on a different target to effect synaptic delivery of GluR1. Recent studies indicate that CaMKII can control a synaptic rasGAP (Chen et al. 1998, Kim et al. 1998) and thereby increase ras activity. Ras activity appears to be necessary to generate LTP and is the downstream effector of CaMKII that drives synaptic delivery of AMPARs (J. Zhu, L. Van Aelst, & R. Malinow unpublished observations). This conforms with results indicating a critical role for MAP kinase, a downstream effector for ras, in LTP (English & Sweatt 1996, 1997).

Interestingly, mutations at Ser845, the PKA phosphorylation site of GluR1 (Roche et al. 1996), do prevent delivery of GluR1 to synapses by active CaMKII or LTP (Shi & Malinow 2001). Phosphorylation at this site of GluR1 also accompanies surface reinsertion of receptors (Ehlers 2000) and LTP induction after prior LTD (Lee et al. 2000). Phosphorylation at this site by exogenous application of drugs that raise cAMP does not induce delivery of recombinant GluR1 (Shi & Malinow 2001). Thus, PKA phosphorylation of GluR1 is necessary, but not sufficient, for its synaptic delivery, i.e., phosphorylation of Ser845 acts as a gate. Of note, the PKA-scaffolding molecule, AKAP (a kinase anchoring protein), binds to SAP97 and thereby effectively brings PKA to GluR1 (Colledge et al. 2000). Thus, it is possible that the PDZ mutation on GluR1 blocks its synaptic delivery, at least in part, because it prevents PKA phosphorylation at Ser845. Of note, SAP97 associates with GluR1 primarily in intracellular sites (Sans et al. 2001), consistent with its playing a role in making GluR1 competent for synaptic delivery (Figure 2, see color insert).

Recent studies indicate that activity-driven phosphorylation of GluR4 by PKA is necessary and sufficient for delivery of these recombinant AMPARs to synapses during early development (Esteban & Malinow 2001). Such phosphorylation relieves a retention interaction that, in the absence of synaptic activity, maintains GluR4-containing receptors away from the synapse. Thus, a mechanism (PKA phosphorylation of AMPARs) that mediates plasticity early in development (with GluR4) becomes a gate for plasticity (with GluR1) later in development. Increasing requirements over development may be one way that plasticity becomes more specific and also recalcitrant with age.

GENERAL TRAFFICKING MECHANISMS

A key question has been if plasticity acts by directly modulating a process that is responsible for turning over receptors at synapses (e.g., increasing rate of delivery or decreasing rate of removal) or if there are distinct processes responsible for plasticity and receptor turnover. One recent study (Shi et al. 2001) examined this question and argues for distinct AMPARs responsible for LTP and receptor turnover (Figure 2, see color insert). AMPARs composed of GluR1 and GluR2 (or any receptor with a long cytoplasmic tail along with GluR2) participate in regulated delivery. In the absence of electrical activity, these receptors are restricted from accessing synapses. LTP (for GluR1-containing receptors) or spontaneous activity (for GluR4-containing receptors) drives these receptors (along with associated scaffolding) into synapses. The long cytoplasmic tails, and not the short cytoplasmic tails, of GluR1/GluR2 heteromers are critical for this activity-dependent synaptic delivery. Receptors composed of GluR2 and GluR3 continuously replace synaptic GluR2/GluR3 receptors in a manner that maintains transmission constant. How can this model explain long-term changes in synaptic receptor number following plasticity that enhances transmission? At some point after their synaptic delivery, receptors containing GluR1 or GluR4 become replaceable by GluR2/GluR3 receptors. The scaffolding associated with GluR1 or GluR4 [called "slot" complexes (Shi et al. 2001)] must somehow control this. One study provides evidence for replacement of synaptic GluR4-containing receptors by GluR2/GluR3 receptors (Zhu et al. 2000). This occurs over the course of days after the activity-driven delivery of GluR4-containing receptors.

A recent independent study in cultured dissociated neurons used expression of recombinant AMPAR subunits that contained a thrombin cleavable epitope tag in their extracellular domains to monitor surface delivery of receptors. This clever study supports this view that plasticity and constitutive receptor turnover are controlled by different AMPAR species (Passafaro et al. 2001). Furthermore, this study provided important spatial and temporal information regarding synaptic delivery of these different AMPAR complexes. Receptors containing GluR2/GluR3 are continuously delivered to the spine surface with a time constant of $\sim$15 min and unaffected by agents that perturb plasticity. On the other hand, the rate of surface delivery of GluR1-containing receptors is greatly enhanced by stimulation (in this dissociated cultured system, by either glycine, NMDA, or insulin). Their surface appearance takes place in dendritic (extrasynaptic) regions, with subsequent movement into spine regions.

The movement of AMPARs from cytosol to dendritic surface and subsequently to synaptic regions, described above for receptors containing GluR1, had previously been proposed for GluR4-containing receptors by a study examining the protein stargazin (STG) (Chen et al. 2000). Animals without the transmembrane protein STG lack AMPAR responses in granule cells of the cerebellum (Hashimoto et al. 1999). AMPAR responses to both synaptic and exogenously applied transmitter are absent, although intracellular AMPARs levels are normal. Acute expression of STG rescues responses to synaptic and exogenous transmitter application. Interestingly, expression of STG with a mutation at its cytoplamic tail rescues responses to exogenous application, but not synaptic responses. This suggests that interactions controlled by STG cytoplasmic tail are important for movement of receptors from extrasynaptic (surface) regions to synaptic regions. The relation between STG and regulated or continuous receptor delivery described above has not been established.

In the context of the model that proposes two AMPAR species (Figure 2, see color insert), one participating in activity-dependent delivery, the other in continuous replacement, it is not clear what receptor types are removed by LTD. Both GluR1/GluR2 and GluR2/GluR3 contain GluR2 carboxyl tails that appear to be critical for LTD (see above). The occlusion of continuous receptor cycling following LTD (Lüthi et al. 1999) suggests that LTD acts by removal of GluR2/GluR3 receptors. However, Shi (2000) finds that overexpression of GluR3 carboxyl tails depresses transmission (presumably by interfering with continuous GluR2/GluR3 delivery) but does not prevent LTD. Overexpression of GluR1 carboxyl tails neither depresses transmission nor blocks LTD. This seems to argue that LTD removes GluR1/GluR2 heteromers by interactions requiring GluR2. Agents that perturb GluR2 interactions may independently block continuous receptor cycling (of GluR2/GluR3) and activity-dependent removal of GluR1/GluR2, receptors that normally do not participate in cycling.

OPEN QUESTIONS

Lynch & Baudry (1984) proposed almost two decades ago that LTP is due to an increase in the number of synaptic glutamate receptors. However, the idea did not gain universal favor and a vigorous exchange over the ensuing decades debated the pre- and postsynaptic contributions to the expression of LTP. Thus, the general acceptance of postsynaptic silent synapses and AMPAR trafficking as playing important roles in synaptic plasticity represent significant advances in the field. They provide a clear conceptual framework that should continue to facilitate studies aimed at determining which molecules play critical roles in LTP and LTD and exactly what role they play. Historically, many molecules have been suggested to be necessary for LTP since their genetic removal or pharmacological inhibition impairs LTP (Sanes & Lichtman 1999). However, to state that a molecule or a process is necessary for LTP (or LTD) is no longer particularly meaningful without specifying exactly what role it plays as a direct mediator of these synaptic phenomena or as a modulator of the molecular events known to be necessary for their triggering or expression.

From our vantage point, a number of significant questions remain open. 1. What are all of the critical targets downstream of NMDAR activation that are responsible for LTP and LTD? 2. How do these downstream targets communicate with the machinery that controls AMPAR trafficking? 3. Does AMPAR trafficking play the same important role in NMDAR-independent forms of LTP and LTD, of which there are several? 4. Which proteins, known to associate with AMPARs, control regulated and continuous trafficking? 5. What other proteins interact with and control AMPAR trafficking? 6. How is a transient delivery/removal of AMPARs transformed to a persistent increase/decrease in the number of synaptic receptors? 7. How is this persistence maintained in the face of protein turnover? 8. What are the differences in AMPAR trafficking between different cell types? Different brain regions? 9. What different circuit functions do different trafficking properties subserve? 10. How do environmental and behavioral stimuli influence AMPAR trafficking? 11. How does AMPAR trafficking impact behavior? 12. Do dysfunctions in AMPAR trafficking contribute to neurological or neuropsychiatric disorders? Given the large number of creative and hardworking scientists actively working on many of these questions, it is our hope and belief that many of these will be answered before we contribute another review on AMPAR trafficking and synaptic plasticity.

The *Annual Review of Neuroscience* is online at http://neuro.annualreviews.org

LITERATURE CITED

Alkon DL, Nelson TJ. 1990. Specificity of molecular changes in neurons involved in memory storage. *FASEB J.* 4:1567–76

Allison DW, Gelfand VI, Spector I, Craig AM. 1998. Role of actin in anchoring postsynaptic receptors in cultured hippocampal neurons: differential attachment of NMDA versus AMPA receptors. *J. Neurosci.* 18:2423–36

Barria A, Derkach V, Soderling T. 1997a. Identification of the Ca2+/calmodulin-dependent protein kinase II regulatory phosphorylation site in the alpha-amino-3-hydroxyl-5-methyl-4-isoxazole-propionate-type glutamate receptor. *J. Biol. Chem.* 272:32,727–30

Barria A, Muller D, Derkach V, Griffith LC, Soderling TR. 1997b. Regulatory phosphorylation of AMPA-type glutamate receptors by CaM-KII during long-term potentiation. *Science* 276:2042–45

Baude A, Nusser Z, Molnar E, McIlhinney RAJ, Somogyi P. 1995. High-resolution immunogold localization of AMPA-type glutamate receptor subunits at synaptic and nonsynaptic sites in rat hippocampus. *Neuroscience* 69:1031–55

Bayer K-U, De Konnick P, Leonard AS, Hell JW, Schulman H. 2001. Interaction with the NMDA receptor locks CaMKII in an active conformation. *Nature* 411:801–5

Beattie EC, Carroll RC, Yu X, Morishita W, Yasuda H, et al. 2000. Regulation of AMPA receptor endocytosis by a signaling mechanism shared with LTD. *Nat. Neurosci.* 3:1291–1300

Benke TA, Luthi A, Isaac JT, Collingridge GL. 1998. Modulation of AMPA receptor unitary conductance by synaptic activity. *Nature* 393:793–97

Bliss TV, Collingridge GL. 1993. A synaptic model of memory: long-term potentiation in the hippocampus. *Nature* 361:31–39

Bliss TV, Gardner-Medwin AR. 1973. Long-lasting potentiation of synaptic transmission in the dentate area of the unanaesthetized rabbit following stimulation of the perforant path. *J. Physiol.* 232:357–74

Bliss TV, Lomo T. 1973. Long-lasting potentiation of synaptic transmission in the dentate area of the anaesthetized rabbit following stimulation of the perforant path. *J. Physiol.* 232:331–56

Blitzer RD, Wong T, Nouranifar R, Iyengar R, Landau EM. 1995. Postsynaptic cAMP pathway gates early LTP in hippocampal CA1 region. *Neuron* 15:1403–14

Braithwaite SP, Malenka RC. 2001. *Differential roles for NSF and PDZ domain proteins in the endocytosis of AMPA receptors*. Presented at 27th Soc. Neurosci. Mtg., San Diego

Broutman G, Baudry M. 2001. Involvement of the secretory pathway for AMPA receptors in NMDA-induced potentiation in hippocampus. *J. Neurosci.* 21:27–34

Carroll RC, Beattie EC, Xia H, Lüscher C, Altschuler Y, et al. 1999a. Dynamin-dependent endocytosis of ionotropic glutamate receptors. *Proc. Natl. Acad. Sci. USA* 96:14,112–17

Carroll RC, Lissin DV, von Zastrow M, Nicoll RA, Malenka RC. 1999b. Rapid redistribution of glutamate receptors contributes to long-term depression in hippocampal cultures. *Nat. Neurosci.* 2:454–60

Chen HJ, Rojas-Soto M, Oguni A, Kennedy MB. 1998. A synaptic Ras-GTPase activating protein (p135 SynGAP) inhibited by CaM kinase II. *Neuron* 20:895–904

Chen L, Chetkovich DM, Petralia RS, Sweeney NT, Kawasaki Y, et al. 2000. Stargazing regulates synaptic targeting of AMPA receptors by two distinct mechanisms. *Nature* 408:936–43

Chung HJ, Scannevin RH, Zhang X, Huganir RL. 2000. Phosphorylation of the AMPR receptor subunit GluR2 differentially regulates its interaction with PDZ domain-containing proteins. *J. Neurosci.* 20:7258–67

Colledge M, Dean RA, Scott GK, Langeberg LK, Huganir RL, Scott JD. 2000. Targeting of PKA to glutamate receptors through a MAGUK-AKAP complex. *Neuron* 27:107–19

Davies SN, Lester RA, Reymann KG, Collingridge GL. 1989. Temporally distinct pre- and post-synaptic mechanisms maintain long-term potentiation. *Nature* 338:500–3

Daw MI, Chittajallu R, Bortolotto ZA, Dev KK, Duprat F, et al. 2000. PDZ proteins interacting with C-terminal GluR2/3 are involved in a PKC-dependent regulation of AMPA receptors at hippocampal synapses. *Neuron* 28:873–86

Derkach V, Barria A, Soderling TR. 1999. Ca2+/calmodulin-kinase II enhances channel conductance of alpha-amino-3-hydroxy-5-methyl-4-isoxazolepropionate type glutamate receptors. *Proc. Natl. Acad. Sci. USA* 96:3269–74

Dev KK, Nishimune A, Henley JM, Nakanishi S. 1999. The protein kinase C alpha binding protein PICK1 interacts with short but not long form alternative splice variants of AMPA receptor subunits. *Neuropharmacology* 38:635–44

Dong H, O'Brien RJ, Fung ET, Lanahan AA, Worley PF, Huganir RL. 1997. GRIP: a synaptic PDZ domain-containing protein that interacts with AMPA receptors. *Nature* 386:279–84

Dong H, Zhang P, Song I, Petralia RS, Liao D, Huganir RL. 1999. Characterization of the glutamate receptor-interacting proteins GRIP1 and GRIP2. *J. Neurosci.* 19:6930–41

Durand GM, Kovalchuk Y, Konnerth A. 1996. Long-term potentiation and functional synapse induction in developing hippocampus. *Nature* 381:71–75

Eccles JC. 1964. *The Physiology of Synapses.* New York: Academic Press. xi, 316 pp.

Ehlers MD. 2000. Reinsertion or degradation of AMPA receptors determined by activity-dependent endocytic sorting. *Neuron* 28:511–25

English JD, Sweatt JD. 1996. Activation of p42 mitogen-activated protein kinase in hippocampal long term potentiation. *J. Biol. Chem.* 271:24,329–32

English JD, Sweatt JD. 1997. A requirement for the mitogen-activated protein kinase cascade in hippocampal long term potentiation. *J. Biol. Chem.* 272:19,103–6

Esteban JA, Malinow R. 2001. *A molecular mechanism for the regulated synaptic delivery of GluR4-containing AMPA receptors.* Presented at 27th Soc. Neurosci. Mtg., San Diego

Gomperts SN, Carroll R, Malenka RC, Nicoll RA. 2000. Distinct roles for ionotropic and metabotropic glutamate receptors in the maturation of excitatory synapses. *J. Neurosci.* 20:2229–37

Hashimoto K, Fukaya M, Qiao X, Sakimura K, Watanabe M, Kano M. 1999. Impairment of AMPA receptor function in cerebellar granule cells of ataxic mutant mouse stargazer. *J. Neurosci.* 19:6027–36.

Hayashi Y, Shi S-H, Esteban JA, Piccini A, Poncer JC, Malinow R. 2000. Driving AMPA receptors into synapses by LTP and CaMKII: requirement for GluR1 and PDZ domain interaction. *Science* 287:2262–67

Hebb D. 1949. *The Organization of Behavior.* New York: Wiley

Heynen AJ, Quinlan EM, Bae DC, Bear MF. 2000. Bidirectional, activity-dependent regulation of glutamate receptors in the adult hippocampus in vivo. *Neuron* 28:527–36

Hollmann M, Heinemann S. 1994. Cloned glutamate receptors. *Annu. Rev. Neurosci.* 17:31–108

Inagaki H, Maeda S, Lin KH, Shimizu N, Saito T. 1999. rDLG6: a novel homolog of Drosophila DLG expressed in rat brain. *Biochem. Biophys. Res. Commun.* 265:462–68.

Isaac JT, Crair MC, Nicoll RA, Malenka RC. 1997. Silent synapses during development of thalamocortical inputs. *Neuron* 18:269–80

Isaac JT, Hjelmstad GO, Nicoll RA, Malenka RC. 1996. Long-term potentiation at single fiber inputs to hippocampal CA1 pyramidal cells. *Proc. Natl. Acad. Sci. USA* 93:8710–15

Isaac JT, Nicoll RA, Malenka RC. 1995. Evidence for silent synapses: implications for the expression of LTP. *Neuron* 15:427–34

Jia Z, Agopyan N, Miu P, Xiong Z, Henderson J, et al. 1996. Enhanced LTP in mice deficient in the AMPA receptor GluR2. *Neuron* 17:945–56

Kandel ER. 1997. Genes, synapses, and long-term memory. *J. Cell. Physiol.* 173:124–25

Kauer JA, Malenka RC, Nicoll RA. 1988. A persistent postsynaptic modification mediates long-term potentiation in the hippocampus. *Neuron* 1:911–17

Kharazia VN, Wenthold RJ, Weinberg RJ. 1996. GluR1-immunopositive interneurons in rat neocortex. *J. Comp. Neurol.* 368:399–412

Kim CH, Lisman JE. 1999. A role of actin filament in synaptic transmission and long-term potentiation. *J. Neurosci.* 19:4314–24

Kim CH, Lisman JE. 2001. A labile component of AMPA receptor-mediated synaptic transmission is dependent on microtubule motors, actin, and N-ethylmaleimide-sensitive factor. *J. Neurosci.* 21:4188–94

Kim CH, Chung HJ, Lee HK, Huganir RL. 2001. Interaction of the AMPA receptor subunit GluR2/3 with PDZ domains regulates hippocampal long-term depression. *Proc. Natl. Acad. Sci. USA* 98:11725–30

Kim JH, Liao D, Lau LF, Huganir RL. 1998. SynGAP: a synaptic RasGAP that associates with the PSD-95/SAP90 protein family. *Neuron* 20:683–91

Krucker T, Siggins GR, Halpain S. 2000. Dynamic actin filaments are required for stable long-term potentiation (LTP) in area CA1 of the hippocampus. *Proc. Natl. Acad. Sci. USA* 97:6856–61

Kullmann DM, Nicoll RA. 1992. Long-term potentiation is associated with increases in quantal content and quantal amplitude. *Nature* 357:240–44

Kullmann DM, Siegelbaum SA. 1995. The site of expression of NMDA receptor-dependent LTP: new fuel for an old fire. *Neuron* 15:997–1002

Lai MM, Hong JJ, Ruggiero AM, Burnett PE, Slepnev VI, et al. 1999. The calcineurin-dynamin 1 complex as a calcium sensor for synaptic vesicle endocytosis. *J. Biol. Chem.* 274:25963–66

Lee H-K, Barbarosie M, Kameyama K, Bear MF, Huganir RL. 2000. Regulation of distinct AMPA receptor phosphorylation sites during bidirectional synaptic plasticity. *Nature* 405:955–59

Leonard AS, Davare MA, Horne MC, Garner CC, Hell JW. 1998. SAP97 is associated with the alpha-amino-3-hydroxy-5-methylisoxazole-4-propionic acid receptor GluR1 subunit. *J. Biol. Chem.* 273:19,518–24

Leonard AS, Lim IA, Hemsworth DE, Horne MC, Hell JW. 1999. Calcium/calmodulin-dependent protein kinase II is associated with the N-methyl-D-aspartate receptor. *Proc. Natl. Acad. Sci. USA* 96:3239–44

Li P, Kerchner GA, Sala C, Wei F, Huettner JE, et al. 1999. AMPA receptor-PDZ interactions in facilitation of spinal sensory synapses. *Nat. Neurosci.* 2:972–77

Liao D, Hessler NA, Malinow R. 1995. Activation of postsynaptically silent synapses during pairing-induced LTP in CA1 region of hippocampal slice. *Nature* 375:400–4

Liao D, Malinow R. 1996. Deficiency in induction but not expression of LTP in hippocampal slices from young rats. *Learn. Mem.* 3:138–49

Liao D, Scannevin RH, Huganir R. 2001. Activation of silent synapses by rapid activity-dependent synaptic recruitment of AMPA receptors. *J. Neurosci.* 21:6008–17

Liao D, Zhang X, O'Brien R, Ehlers MD, Huganir RL. 1999. Regulation of morphological postsynaptic silent synapses in developing hippocampal neurons. *Nat. Neurosci.* 2:37–43

Lin JW, Ju W, Foster K, Lee SH, Ahmadian G, et al. 2000. Distinct molecular mechanisms and divergent endocytotic pathways of AMPA receptor internalization. *Nat. Neurosci.* 3:1282–90

Linden DJ. 1994. Long-term synaptic

depression in the mammalian brain. *Neuron* 12:457–72

Lisman J. 1989. A mechanism for the Hebb and the anti-Hebb processes underlying learning and memory. *Proc. Natl. Acad. Sci. USA* 86:9574–78

Lisman J, Malenka RC, Nicoll RA, Malinow R. 1997. Learning mechanisms: the case for CaM-KII. *Science* 276:2001–2

Lisman JE, Zhabotinsky AM. 2001. A model of synaptic memory: a CaMKII/PP1 switch that potentiates transmission by organizing an AMPA receptor anchoring assembly. *Neuron* 31:191–201

Lissin DV, Carroll RC, Nicoll RA, Malenka RC, von Zastrow M. 1999. Rapid, activation-induced redistribution of ionotropic glutamate receptors in cultured hippocampal neurons. *J. Neurosci.* 19:1263–72

Lissin DV, Gomperts SN, Carroll RC, Christine CW, Kalman D, et al. 1998. Activity differentially regulates the surface expression of synaptic AMPA and NMDA glutamate receptors. *Proc. Natl. Acad. Sci. USA* 95:7097–102

Liu SQ, Cull-Candy SG. 2000. Synaptic activity at calcium-permeable AMPA receptors induces a switch in receptor subtype. *Nature* 405:454–58

Lledo PM, Zhang X, Sudhof TC, Malenka RC, Nicoll RA. 1998. Postsynaptic membrane fusion and long-term potentiation. *Science* 279:399–403

Lu W, Man H, Ju W, Trimble WS, MacDonald JF, Wang YT. 2001. Activation of synaptic NMDA receptors induces membrane insertion of new AMPA receptors and LTP in cultured hippocampal neurons. *Neuron* 29:243–54

Lüscher C, Xia H, Beattie EC, Carroll RC, von Zastrow M, et al. 1999. Role of AMPA receptor cycling in synaptic transmission and plasticity. *Neuron* 24:649–58

Lüthi A, Chittajallu R, Duprat F, Palmer MJ, Benke TA, et al. 1999. Hippocampal LTD expression involves a pool of AMPARs regulated by the NSF-GluR2 interaction. *Neuron* 24:389–99

Lynch G, Baudry M. 1984. The biochemistry of memory: a new and specific hypothesis. *Science* 224:1057–63

Mack V, Burnashev N, Kaiser KM, Rozov A, Jensen V, et al. 2001. Conditional restoration of hippocampal synaptic potentiation in Glur-A-deficient mice. *Science* 292:2501–4

Madison DV, Malenka RC, Nicoll RA. 1991. Mechanisms underlying long-term potentiation of synaptic transmission. *Annu. Rev. Neurosci.* 14:379–97

Mainen ZF, Jia Z, Roder J, Malinow R. 1998. Use-dependent AMPA receptor block in mice lacking GluR2 suggests postsynaptic site for LTP expression. *Nat. Neurosci.* 1:579–86

Malenka RC, Nicoll RA. 1999. Long-term potentiation—a decade of progress? *Science* 285:1870–74

Maletic-Savatic M, Koothan T, Malinow R. 1998. Calcium-evoked dendritic exocytosis in cultured hippocampal neurons. Part II: mediation by calcium/calmodulin-dependent protein kinase II. *J. Neurosci.* 18:6814–21

Malinow R, Tsien RW. 1990. Presynaptic enhancement shown by whole-cell recordings of long-term potentiation in hippocampal slices. *Nature* 346:177–80

Mammen AL, Kameyama K, Roche KW, Huganir RL. 1997. Phosphorylation of the alpha-amino-3-hydroxy-5-methylisoxazole4-propionic acid receptor GluR1 subunit by calcium/calmodulin-dependent kinase II. *J. Biol. Chem.* 272:32,528–33

Man H-Y, Lin JW, Ju WH, Ahmadian G, Liu L, et al. 2000. Regulation of AMPA receptor–mediated synaptic transmission by clathrin-dependent receptor internalization. *Neuron* 25:649–62

Manabe T, Renner P, Nicoll RA. 1992. Postsynaptic contribution to long-term potentiation revealed by the analysis of miniature synaptic currents. *Nature* 355:50–55

Maren S, Tocco G, Standley S, Baudry M, Thompson RF. 1993. Postsynaptic factors in the expression of long-term potentiation (LTP): increased glutamate receptor binding following LTP induction in vivo. *PNAS (USA)* 90:9654–58

Martin LJ, Blackstone CD, Levey AI, Huganir RL, Price DL. 1993. AMPA glutamate receptor subunits are differentially distributed in rat brain. *Neuroscience* 53:327–58

Martin SJ, Grimwood PD, Morris RG. 2000. Synaptic plasticity and memory: an evaluation of the hypothesis. *Annu. Rev. Neurosci.* 23:649–711

Matsuda S, Launey T, Mikawa S, Hirai H. 2000. Disruption of AMPA receptor GluR2 clusters following long-term depression induction in cerebellar Purkinje neurons. *EMBO J.* 19:2765–74

Matsuda S, Mikawa S, Hirai H. 1999. Phosphorylation of serine-880 in GluR2 by protein kinase C prevents its C terminus from binding with glutamate receptor-interacting protein. *J. Neurochem.* 73:1765–68

Molnar E, Baude A, Richmond SA, Patel PB, Somogyi P, McIlhinney RAJ. 1993. Biochemical and immunocytochemical characterization of antipeptide antibodies to a cloned GluR1 glutamate receptor subunit: cellular and subcellular distribution in the rat forebrain. *Neuroscience* 53:307–26

Mulkey RM, Endo S, Shenolikar S, Malenka RC. 1994. Involvement of a calcineurin/inhibitor-1 phosphatase cascade in hippocampal long-term depression. *Nature* 369:486–88

Mulkey RM, Herron CE, Malenka RC. 1993. An essential role for protein phosphatases in hippocampal long-term depression. *Science* 261:1051–55

Muller D, Joly M, Lynch G. 1988. Contributions of quisqualate and NMDA receptors to the induction and expression of LTP. *Science* 242:1694–97

Nishimune A, Isaac JT, Molnar E, Noel J, Nash SR, et al. 1998. NSF binding to GluR2 regulates synaptic transmission. *Neuron* 21:87–97

Noel J, Ralph GS, Pickard L, Williams J, Molnar E, et al. 1999. Surface expression of AMPA receptors in hippocampal neurons is regulated by an NSF-dependent mechanism. *Neuron* 23:365–76

Nusser Z, Lujan R, Laube G, Roberts JD, Molnar E, Somogyi P. 1998. Cell type and pathway dependence of synaptic AMPA receptor number and variability in the hippocampus. *Neuron* 21:545–59

O'Brien RJ, Kamboj S, Ehlers MD, Rosen KR, Fischbach GD, Huganir RL. 1998. Activity-dependent modulation of synaptic AMPA receptor accumulation. *Neuron* 21:1067–78

Osten P, Khatri L, Perez JL, Kohr G, Giese G, et al. 2000. Mutagenesis reveals a role for ABP/GRIP binding to GluR2 in synaptic surface accumulation of the AMPA receptor. *Neuron* 27:313–25

Osten P, Srivastava S, Inman GJ, Vilim FS, Khatri L, et al. 1998. The AMPA receptor GluR2 C terminus can mediate a reversible, ATP-dependent interaction with NSF and alpha- and beta-SNAPs. *Neuron* 21:99–110

Passafaro M, Piech V, Sheng M. 2001. Subunit-specific temporal and spatial patterns of AMPA receptor exocytosis in hippocampal neurons. *Nat. Neurosci.* 4:917–26

Perez JL, Khatri L, Chang C, Srivastava S, Osten P, Ziff EB. 2001. PICK1 targets activated protein kinase C alpha to AMPA receptor clusters in spines of hippocampal neurons and reduces surface levels of the AMPA-type glutamate receptor subunit 2. *J. Neurosci.* 21:5417–28

Petralia RS, Esteban JA, Wang YX, Partridge JG, Zhao HM, et al. 1999. Selective acquisition of AMPA receptors over postnatal development suggests a molecular basis for silent synapses. *Nat. Neurosci.* 2:31–36

Petralia RS, Wenthold RJ. 1992. Light and electron immunocytochemical localization of AMPA-selective glutamate receptors in the rat brain. *J. Comp. Neurol.* 318:329–54

Renger JJ, Egles C, Liu G. 2001. A developmental switch in neurotransmitter flux enhances synaptic efficacy by affecting AMPA receptor activation. *Neuron* 29:469–84

Roche KW, O'Brien RJ, Mammen AL, Bernhardt J, Huganir RL. 1996. Characterization of multiple phosphorylation sites on the

AMPA receptor GluR1 subunit. *Neuron* 16: 1179–88

Rogers CA, Maron C, Schulteis C, Allen W-R, Heinemann S-F. 2001. *Afadin, a link between AMPA receptors and the actin cytoskeleton.* Presented at 27th Soc. Neurosci. Mtg., San Diego

Rosenmund C, Stern-Bach Y, Stevens CF. 1998. The tetrameric structure of a glutamate receptor channel. *Science* 280:1596–99

Rothman JE. 1994. Mechanisms of intracellular protein transport. *Nature* 372:55–63

Rumpel S, Hatt H, Gottmann K. 1998. Silent synapses in the developing rat visual cortex: evidence for postsynaptic expression of synaptic plasticity. *J. Neurosci.* 18:8863–74

Sanes JR, Lichtman JW. 1999. Can molecules explain long-term potentiation? *Nat. Neurosci.* 2:597–604

Sans N, Racca C, Petralia RS, Wang YX, McCallum J, Wenthold RJ. 2001. Synapse-associated protein 97 selectively associates with a subset of AMPA receptors early in their biosynthetic pathway. *J. Neurosci.* 21: 7506–16

Schulz W, Nakagawa T, Kim J-H, Sheng M, Seeburg PH, Osten P. 2001. *Novel interaction of the GluR-A AMPA receptor subuint with the PDZ-LIM domain protein RIL.* Presented at 27th Soc. Neurosci. Mtg., San Diego

Shen K, Meyer T. 1999. Dynamic control of CaMKII translocation and localization in hippocampal neurons by NMDA receptor stimulation. *Science* 284:162–66

Shen L, Liang F, Walensky LD, Huganir RL. 2000. Regulation of AMPA receptor GluR1 subunit surface expression by a 4.1N-linked actin cytoskeletal association. *J. Neurosci.* 20:7932–40

Sheng M, Sala C. 2001. PDZ domains and the organization of supramolecular complexes. *Annu. Rev. Neurosci.* 24:1–29

Shi S-H. 2000. *Molecular mechanisms for synaptic regulaion of AMPA receptor.* PhD thesis. State Univ. New York, Stony Brook

Shi S-H, Hayashi Y, Esteban JA, Malinow R. 2001. Subunit-specific rules governing AMPA receptor trafficking to synapses in hippocampal pyramidal neurons. *Cell* 105: 331–43

Shi S-H, Hayashi Y, Petralia RS, Zaman SH, Wenthold RJ, et al. 1999. Rapid spine delivery and redistribution of AMPA receptors after synaptic NMDA receptor activation. *Science* 284:1811–16

Shi S-H, Malinow R. 2001. *Synaptic trafficking of AMPA-Rs containing GluR1 is gated by PKA phosphorylation at Ser845.* Presented at 27th Soc. Neurosci. Mtg., San Diego

Slepnev VI, Ochoa GC, Butler MH, Grabs D, Camilli PD. 1998. Role of phosphorylation in regulation of the assembly of endocytic coat complexes. *Science* 281:821–24

Snyder EM, Philpot BD, Huber KM, Dong X, Fallon JR, Bear MF. 2001. Internalization of ionotropic glutamate receptors in response to mGluR activation. *Nat. Neurosci.* 4:1079–85

Song I, Kamboj S, Xia J, Dong H, Liao D, Huganir RL. 1998. Interaction of the N-ethylmaleimide-sensitive factor with AMPA receptors. *Neuron* 21:393–400

Srivastava S, Osten P, Vilim FS, Khatri L, Inman G, et al. 1998. Novel anchorage of GluR2/3 to the postsynaptic density by the AMPA receptor-binding protein ABP. *Neuron* 21:581–91

Stevens CF, Wang Y. 1994. Changes in reliability of synaptic function as a mechanism for plasticity. *Nature* 371:704–7

Strack S, Colbran RJ. 1998. Autophosphorylation-dependent targeting of calcium/calmodulin-dependent protein kinase II by the NR2B subunit of the N-methyl-D-aspartate receptor. *J. Biol. Chem.* 273:20,689–92

Strack S, McNeill RB, Colbran RJ. 2000. Mechanism and regulation of calcium/calmodulin-dependent protein kinase II targeting to the NR2B subunit of the N-methyl-D-aspartate receptor. *J. Biol. Chem.* 275:23,798–806

Takumi Y, Ramírez-León V, Laake P, Rinvik E, Ottersen OP. 1999. Different modes of expression of AMPA and NMDA receptors in hippocampal synapses. *Nat. Neurosci.* 2:618–24

Turrigiano GG, Leslie KR, Desai NS, Rutherford LC, Nelson SB. 1998. Activity-dependent scaling of quantal amplitude in neocortical neurons. *Nature* 391:892–96

Wang YT, Linden DJ. 2000. Expression of cerebellar long-term depression requires postsynaptic clathrin-mediated endocytosis. *Neuron* 25:635–47

Wenthold RJ, Petralia RS, Blahos J II, Niedzielski AS. 1996. Evidence for multiple AMPA receptor complexes in hippocampal CA1/CA2 neurons. *J. Neurosci.* 16:1982–89

Wisden W, Seeburg PH. 1993. Mammalian ionotropic glutamate receptors. *Curr. Op. Neurobiol.* 3:291–98

Wu G, Malinow R, Cline HT. 1996. Maturation of a central glutamatergic synapse. *Science* 274:972–76

Xia J, Chung HJ, Wihler C, Huganir RL, Linden DJ. 2000. Cerebellar long-term depression requires PKC-regulated interactions between GluR2/3 and PDZ domain-containing proteins. *Neuron* 28:499–510

Xia J, Zhang X, Staudinger J, Huganir RL. 1999. Clustering of AMPA receptors by the synaptic PDZ domain-containing protein PICK1. *Neuron* 22:179–87

Zamanillo D, Sprengel R, Hvalby O, Jensen V, Burnashev N, et al. 1999. Importance of AMPA receptors for hippocampal synaptic plasticity but not for spatial learning. *Science* 284:1805–11

Zhou Q, Xiao M, Nicoll RA. 2001. Contribution of cytoskeleton to the internalization of AMPA receptors. *PNAS (USA)* 98:1261–66

Zhu JJ, Esteban JA, Hayashi Y, Malinow R. 2000. Synaptic potentiation during early development: delivery of GluR4-containing AMPA receptors by spontaneous activity. *Nat. Neurosci.* 3:1098–1106

Zoghbi HY, Gage FH, Choi DW. 2000. Neurobiology of disease. *Curr. Op. Neurobiol.* 10:655–60

MOLECULAR CONTROL OF CORTICAL DENDRITE DEVELOPMENT

Kristin L. Whitford, Paul Dijkhuizen, Franck Polleux,* and Anirvan Ghosh

Department of Neuroscience, Johns Hopkins University School of Medicine, Baltimore, Maryland 21205; *INSERM U371, 69675 BRON Cedex, France; email: kwhit@jhmi.edu; pdijkhui@jhmi.edu; polleux@lyon151.inserm.fr; aghosh@jhmi.edu

Key Words dendrites, spines, semaphorins, neurotrophins, Notch, Rho GTPases

■ **Abstract** Dendritic morphology has a profound impact on neuronal information processing. The overall extent and orientation of dendrites determines the kinds of input a neuron receives. Fine dendritic appendages called spines act as subcellular compartments devoted to processing synaptic information, and the dendritic branching pattern determines the efficacy with which synaptic information is transmitted to the soma. The acquisition of a mature dendritic morphology depends on the coordinated action of a number of different extracellular factors. Here we discuss this evidence in the context of dendritic development in the cerebral cortex. Soon after migrating to the cortical plate, neurons extend an apical dendrite directed toward the pial surface. The oriented growth of the apical dendrite is regulated by Sema3A, which acts as a dendritic chemoattractant. Subsequent dendritic development involves signaling by neurotrophic factors and Notch, which regulate dendritic growth and branching. During postnatal development the formation and stabilization of dendritic spines are regulated in part by patterns of synaptic activity. These observations suggest that extracellular signals play an important role in regulating every aspect of dendritic development and thereby exert a critical influence on cortical connectivity.

INTRODUCTION

The structural and molecular differences that characterize axons and dendrites form the basis of information flow in neuronal circuits. The presence of voltage-gated sodium channels allows axons to propagate action potentials to synaptic terminals, which leads to the release of neurotransmitter into the synaptic cleft. The postsynaptic elements at excitatory synapses in the central nervous system (CNS) are often dendritic spines. Dendritic spines contain neurotransmitter receptors, which mediate ligand-dependent influx of ions into the postsynaptic neuron. Although it was long considered that dendrites then passively transmit this information to the soma, recent evidence suggests that dendrites act as dynamic integrators of synaptic input (Hausser et al. 2000). The transmission of the synaptic signal to the soma is greatly affected by the branching pattern of the dendritic tree, and the striking variations in dendritic morphology in the CNS therefore have an enormous consequence for neuronal information processing (Hausser et al. 2000).

In this review we focus on our current understanding of the mechanisms that regulate the principal features of dendritic morphology in the developing cortex. Most cortical neurons are pyramidal in morphology and are characterized by a prominent apical dendrite that allows these neurons to integrate information from superficial cortical layers. Perhaps the most distinguishing feature of dendrites is a highly branched morphology, with decreasing diameter at successive branch points, which gives it the appearance of a tree (dendron in Greek means tree). The specific branching pattern and the distribution of ion channels on dendrites influence how synaptic signals decay as they propagate toward the soma. The dendrites in most cortical neurons are studded with small protrusions called spines, which are the major sites of excitatory synaptic transmission. We begin this review with a consideration of how neurons acquire a polarized morphology and how the growth of the apical dendrite toward the pial surface is regulated. We then move to a discussion of the influence of extracellular signals in regulating dendritic branching patterns. Finally we will consider the mechanisms that regulate the formation of dendritic spines and close with a discussion of intracellular signaling pathways that mediate dendritic growth and remodeling. The main generalization that can be drawn from these studies is that a large variety of extracellular signals, which includes both proteins and neurotransmitters, plays a critical role in specifying the mature dendritic morphology. There has been a great deal of recent progress in identifying these extracellular signals, and the challenge ahead is to understand how the action of these signals is coordinated so that they induce the acquisition of stereotyped neuronal morphologies specialized to process particular kinds of information.

MORPHOLOGICAL FEATURES OF CORTICAL DENDRITES

The mammalian neocortex can be divided into six layers, which can be distinguished based on neuronal morphology and density. Much of our knowledge of dendritic architecture in the cortex is based on anatomical studies of the visual cortex (Gilbert 1983). Each cortical layer serves a specific function, illustrated by their afferent inputs and efferent projections. The superficial layers 2, 3, and 4 are mainly responsible for intracortical projections. On the other hand, the deeper layers 5 and 6 contain neurons that project subcortically. Cells in layer 5 project to the superior colliculus, pons, and spinal cord, while layer 6 neurons project to the claustrum and thalamus.

There is a well-defined relationship between the morphology of cortical neurons and their function. The main excitatory neuronal subtypes are the pyramidal cells and the spiny stellate cells (Figure 1, see color insert). The other main cell type is the smooth stellate cell, which is thought to be inhibitory (Houser et al. 1983, Prieto et al. 1994). Pyramidal neurons are the dominant cell type in the neocortex. They typically have an apical dendrite that branches out in an apical tuft that terminates in layer 1. In addition, pyramidal neurons contain several highly branched basal dendrites that emanate from the cell body. These neurons are mainly present in layers 2, 3, 5, and 6. Within a layer, pyramidal neurons can be morphologically classified based on their projection areas. In layer 6 for instance, neurons projecting to the lateral geniculate nucleus have an apical dendrite that terminates in layer 3, with side branches in layer 4 and 5, while claustrum-projecting neurons have an apical that extends to layer 1, with side branches only in layer 5 (Katz 1987). In contrast to pyramidal neurons, spiny stellate cells are exclusively found in layer 4. They usually have many spiny dendrites of similar length radiating from the cell body. Inhibitory smooth stellate cells have varying nonpyramidal morphologies and are present in all layers of the cortex.

During early development, just after they have reached the cortical plate, all excitatory cortical neurons share a common morphological phenotype. They contain a single apical dendrite that branches within layer 1 (Marin-Padilla 1992, Miller 1981). With time, basal dendrites appear and oblique side branches emerge from the apical shaft. Spines start to appear as the arborization of the apical and basal dendrites becomes more complex. When pyramidal neurons reach their mature morphology, they have a highly complex dendritic arbor and are covered with spines. Interestingly, early spiny stellate neurons in layer 4 also start out with a pyramidal morphology. These neurons, however, acquire a stellate morphology by retracting their apical dendrite at an early postnatal age (Vercelli et al. 1992). A similar phenomenon has also been described for certain layer 5 pyramidal neurons. Callosally projecting pyramidal neurons were found to specifically retract their apical dendrite to obtain their characteristically short pyramidal morphology (Koester & O'Leary 1992). This is in contrast to corticotectal-projecting layer 5 pyramidal neurons, which maintain their apical dendrite to layer 1. These results emphasize that early dendritic development of cortical neurons is mediated in part by an intrinsic growth program, but specific refinement occurs later on to generate class-specific dendritic morphologies.

NEURONAL POLARITY AND THE REGULATION OF APICAL DENDRITE ORIENTATION BY SEMAPHORIN SIGNALING

Most cortical neurons are generated from precursors proliferating in the germinal zones lining the ventricle (Figure 2A, see color insert). Recent evidence suggests that radial glial cells are the main neural precursors in the developing cortex (Malatesta et al. 2000, Miyata et al. 2001, Noctor et al. 2001). These dividing precursors display a polarized localization of signaling molecules such as Notch, Numb, Numblike, and β-catenin (Chenn & McConnell 1995, Chenn et al. 1998, Zhong et al. 1997). Therefore, one model to explain the emergence of neuronal polarity postulates that postmitotic neurons leaving the ventricular zone are polarized due to asymmetric targeting of signaling molecules.

Dendritic differentiation, as determined by expression of dendrite-specific genes such as MAP-2, does not begin until the cells have completed their migration.

Following migration, pyramidal neurons extend an axon toward the ventricle and an apical dendrite toward the pial surface (Figure 2B, see color insert). Is this directed growth of axons and dendrites simply a consequence of the intrinsic polarity of neurons evident during radial migration, or are there extrinsic cues that direct the growth of axon and dendrites in appropriate directions? To test the role of the local cortical environment in directing the growth of nascent axons and dendrites, an in vitro assay was developed in which dissociated neurons from a donor cortex were plated onto cortical slices and cultured organotypically (Polleux et al. 1998). Only two to three hours after plating, the vast majority of neurons extended an axon directed toward the ventricle (Figure 2C, see color insert). This demonstrates the existence of extracellular cues that are sufficient to induce the directed outgrowth of the axon toward the ventricle (Polleux et al. 1998). This ventrally oriented outgrowth is due to the chemorepulsive action of a diffusible semaphorin (Sema3A) expressed in a gradient with highest levels near the marginal zone.

Cortical neurons in the slice overlay assay do not begin to extend a well-differentiated dendrite until a day or two after the axon has emerged. Strikingly, the dendrites of neurons plated on cortical slices behave just like the endogenous pyramidal neurons and extend an apical dendrite toward the pial surface. Cell biological experiments indicate that the oriented growth of dendrites toward the pia is regulated independently from the axons and is due to the action of a chemoattractant present near the marginal zone. Remarkably, the dendritic chemoattractant is Sema3A, the factor that repels cortical axons (Polleux et al. 2000). These experiments indicate that the differential response of axons and dendrites to the same chemotropic cue leads to the specification of the basic pyramidal morphology that characterizes most cortical projection neurons (Figure 2D, see color insert).

What are the mechanisms that might lead to the generation of opposite responses in two compartments of the same neuron? The work of Mu-Ming Poo and his colleagues had previously shown that elevation of the intracellular level of cGMP is sufficient to convert Sema3A from a chemorepellant to a chemoattractant in Xenopus spinal cord axons (Song et al. 1998). The ability of cGMP to switch chemotropic responses appears to be involved in the differential response of axons and dendrites to Sema3A. The enzyme that regulates cGMP production, soluble guanylate cyclase (sGC), is localized asymmetrically in immature cortical neurons and is preferentially targeted to the emerging apical dendrite (Figure 2E, see color insert) (Polleux et al. 2000). Pharmacological inhibition of sGC activity or one of its downstream targets, cGMP-dependent protein kinase, abolishes the ability of Sema3A to attract apical dendrites (Polleux et al. 2000). Thus the basis of the differential response of axons and dendrites to Sema3A appears to be asymmetric targeting of a signaling molecule to the emerging dendrite.

These findings show that the asymmetric targeting of signaling molecules can serve as a mechanism for generating a polarized neuronal response to an extracellular signal, as had been shown previously in amoeba and leukocytes (Parent & Devreotes 1999) and in budding yeast (Pruyne & Bretscher 2000). During chemotaxis, for example, chemoattractant receptors and G proteins are not clustered at the leading edge of the chemotaxing cells (Servant et al. 1999, Xiao et al. 1997). Instead specific intracellular effectors such as cytosolic regulator of adenylate cyclase (CRAC) are rapidly redistributed to the leading edge of chemotaxing cells through specific targeting to the membrane via pleckstrin homology domains (Parent et al. 1998). Interestingly, this is similar to the cortical neuron response to Sema3A, during which the semaphorin receptor Neuropilin-1 is present both on the axon and the developing apical dendrite (Polleux et al. 2000), but a polarized response is generated due to asymmetric localization of the downstream effector molecule sGC.

Until recently, most of our knowledge concerning the emergence of neuronal polarity was based on studies using the classic cell culture system developed by Banker and colleagues (Bradke & Dotti 2000, Craig & Banker 1994). This assay takes advantage of the fact that dissociated hippocampal neurons represent a morphologically homogenous population of pyramidal neurons and rapidly become polarized in vitro following a reproducible and well-characterized program (Dotti et al. 1988). Four stages have been distinguished: At stage 1 the cells are round following dissociation and immediately after plating; at stage 2, shortly after plating, hippocampal neurons extend 4–5 neurites; and within 24 h after plating, one of the neurites starts to elongate to become the axon (stage 3). At this stage if the axon is cut, then other neurites can still be converted into axons instead of becoming dendrites (Dotti & Banker 1987). At stage 4, dendrites acquire their morphological and structural features and lose their plasticity to form a new axon when it is sectioned. This model has been used extensively to explore the factors controlling both the emergence of neuronal polarity and the polarized sorting of axonal and dendritic cues (Bradke & Dotti 2000).

Using this assay, Dotti and collaborators have proposed a model in which a stochastic process marks one neurite to become the axon. The chosen neurite then inhibits other neurites from growing and expressing axon-specific features (Bradke & Dotti 1997, 2000). The same group has shown that local perfusion of the actin-depolymerizing agent, cytochalasin D, onto a randomly chosen growth cone of a stage 2 neuron induces it to grow as an axon, which suggests that actin depolymerization is sufficient for axon specification (Bradke & Dotti 1999). Observations from the slice overlay assay described above do not fully support this model of axon specification through neurite selection. Time-lapse imaging experiments indicate that only one process initially emerges from cells plated onto slices, and that process goes on to become an axon (F. Polleux & A. Ghosh, unpublished observation). Instead the emergence of the axon at a specific pole of the cell and a later emergence of the apical dendrite appear to be regulated by extracellular cues present in the local environment of cortical neurons (Polleux et al. 1998).

REGULATION OF DENDRITIC GROWTH AND BRANCHING BY NEUROTROPHIC FACTORS

Studies over the last few years provide compelling evidence that neurotrophic factors play an important role in regulating dendritic growth and branching in cortical neurons. Many of these studies have focused on the role of neurotrophins, which consist of nerve growth factor (NGF), brain-derived neurotrophic factor (BDNF), neurotrophin-3 (NT-3), and NT-4 (Huang & Reichardt 2001). These factors exert their effects through the Trk family of tyrosine kinase receptors. NGF binds to TrkA, BDNF and NT-4 bind to TrkB, and NT-3 preferentially binds to TrkC. Experiments in which the effects of neurotrophins on dendritic growth control have been examined in slice cultures indicate that in general, neurotrophins increase the dendritic complexity of pyramidal neurons by increasing total dendritic length, the number of branchpoints, and/or the number of primary dendrites (Baker et al. 1998, McAllister et al. 1995, Niblock et al. 2000). The response is rapid and an increase in dendritic complexity is readily apparent within 24 h of neurotrophin exposure. There is a clear specificity in the short-term response of pyramidal neurons of different cortical layers to each of the neurotrophins. For instance, NT-3 strongly increases dendritic complexity in layer 4 neurons but has no apparent effect on layer 5 neurons. In addition, basal dendrites in specific layers respond most strongly to single neurotrophins, whereas apical dendritic growth is increased by a wider array of neurotrophins. Live imaging of layer 2/3 neurons expressing BDNF shows a high level of dendrite dynamics. Both dendritic branches and spines are rapidly lost and gained in BDNF transfected neurons (Baker et al. 1998, McAllister et al. 1995, Niblock et al. 2000). BDNF overexpression favors addition of primary dendrites and proximal branches at the expense of more distal segments. Similarly, overexpression of TrkB in layer 6 pyramidal neurons results in a predominance of short proximal basal dendrites (Yacoubian & Lo 2000).

Recently, osteogenic protein-1 (OP-1) (aka BMP7), which is a member of the transforming growth factor-beta superfamily, was shown to increase total dendritic growth and branching from dissociated embryonic cortical neurons (Le Roux et al. 1999). Furthermore, insulin-like growth factor-1 (IGF-1) has been shown to affect dendrite growth and branching of postnatal layer 2 cortical neurons (Niblock et al. 2000). In contrast to neurotrophins, IGF affected both basal and apical dendritic growth and remodeling, illustrating that the final dendritic complexity of pyramidal neurons is likely to be influenced by the action of multiple neurotrophic factors.

How do neurotrophic factors mediate the morphological changes linked with dendritic remodeling? The observed short-term dynamics indicate a rapid modulation of cytoskeletal elements by neurotrophic factor signaling. Of the major signaling pathways activated by Trk receptors and most other tyrosine kinase receptors, the MAP kinase and PI-3Kinase pathways have been implicated in neurite formation in both neuronal cell lines and primary neurons (Posern et al. 2000, Wu et al. 2001). It is likely that these signaling pathways influence neuronal morphology by regulating the activity of the Rho family GTPases, which mediate actin cytoskeleton dynamics and are known to induce rapid dendritic remodeling. Experiments in neuronal cell lines show that NGF can activate the small GTPase Rac1 in a PI-3kinase–dependent manner, and this activation is necessary for neurite elaboration (Kita et al. 1998, Posern et al. 2000, Yasui et al. 2001). It will be of interest to determine if this pathway is involved in neurotrophin-induced dendritic morphogenesis of cortical neurons.

Part of the neurotrophic factor effect on dendritic morphogenesis may also include the control of expression of structural proteins because long-term exposure to neurotrophins leads to net dendritic growth. It was recently reported that BDNF can upregulate local protein synthesis in dendrites within hours (Aakalu et al. 2001). In addition, specific mRNAs for several cytoskeletal proteins are present in dendrites (Kuhl & Skehel 1998). This raises the interesting possibility that local synthesis of structural components is involved in neurotrophic factor control of dendritic growth.

REGULATION OF DENDRITIC GROWTH AND BRANCHING BY NOTCH SIGNALING

The diversity of signals that can influence dendritic morphology is underscored by a series of recent studies on the role of mammalian Notch proteins in regulating dendritic growth and branching. Originally identified in Drosophila, Notch is a type I cell–surface protein, approximately 300 kDa in size, which functions as a receptor (Artavanis-Tsakonas et al. 1995, 1999; Weinmaster 1997). Proteolytic processing of full-length Notch generates two fragments that associate at the plasma membrane to form a receptor complex. The mechanism of Notch receptor activation involves cleavage and nuclear translocation of the intracellular domain of the receptor (Weinmaster 2000). The intracellular domain of Notch enters the nucleus and binds the transcription factor Suppressor of Hairless [Su(H)], activating gene transcription. Mammalian homologs of Notch (Notch1–4), the Notch ligands Delta (Delta1–3) and Serrate (Jagged1, Jagged2), the transcription factors Su(H) (CBF1/RBP-Jk), and E(Spl) (Hes1-5) have been isolated (Weinmaster 2000). Several of these genes are expressed in the developing brain and spinal cord and are likely to control various aspects of neural development (Bettenhausen et al. 1995, de la Pompa et al. 1997, Del Amo et al. 1992, Dunwoodie et al. 1997, Furukawa et al. 1992, Lardelli et al. 1994, Lindsell et al. 1996, Nakamura et al. 2000, Ohtsuka et al. 1999, Reaume et al. 1992).

The possibility that Notch might play a role in regulating dendritic patterning was suggested by immunocytochemical localization studies that showed that mammalian Notch1 is expressed by both dividing cells in the ventricular zone and postmitotic neurons in the cortical plate (Redmond et al. 2000, Sestan et al. 1999). Several observations suggest that Notch signaling mediates contact-dependent inhibition of neurite outgrowth. For example, in postmitotic neurons there is an inverse correlation between Notch1 expression and total neurite length, and overexpression of a constitutively active Notch1 construct leads to a reduction in the total neurite length (Sestan et al. 1999). Cocultures of cortical neurons with Delta- or Jagged-expressing cell lines, or addition of soluble ligands leads to a decrease in total neurite length, suggesting that Delta or Jagged are the relevant Notch1 ligands (Sestan et al. 1999). Also, overexpression of Numb and Numblike, intracellular modulators that inhibit Notch activation via Su(H)/CBF1, leads to an increase in total neurite length (Sestan et al. 1999). Berezovska and coworkers (1999) have also found that expression of constitutively active Notch1 in hippocampal neurons leads to an inhibition of neurite outgrowth. A study examining Notch function in neuroblastoma cells came to a similar conclusion regarding the effects of Notch signaling on neurite length (Franklin et al. 1999). Together these observations indicate that Notch signaling has an inhibitory effect on process outgrowth.

Recent experiments (Redmond et al. 2000) indicate that, in addition to restricting length, Notch signaling in cortical neurons has a major influence on dendritic branching. The effects of Notch1 signaling on dendrite morphology were examined by measuring several parameters of dendrite complexity, including process and branch point number, dendrite length, and branching index. Inhibition of Notch1 signaling by overexpression of a dominant negative Notch1 construct or with antisense oligonucleotide treatment leads to a decrease in dendritic branching in neurons. In addition, overexpression of a constitutively active Notch1 construct decreases average dendrite length but increases the branching index, resulting in an overall increase in dendritic complexity. Taken together these experiments reveal a positive role for Notch in dendrite branching and a negative role in dendrite and total neurite length.

It is not yet known whether Notch regulation of dendritic development involves the same effectors that regulate cell fate decisions, or if a different set of effector proteins mediates dendritic development. Some of the Notch effects on dendrites may be mediated by molecules previously implicated in cell fate decisions because Hes-1, Neurogenin-1, and MASH-1 are all involved in neuronal differentiation and neurite outgrowth (Castella et al. 1999, Olson et al. 1998, Yavari et al. 1998). An alternate mechanism for the effects of Notch signaling on process outgrowth has recently been proposed by Giniger (Giniger 1998). In a series of genetic studies in *Drosophila*, Notch and the tyrosine kinase Abl were shown to interact synergistically in producing axonal defects. In addition, both Notch and Abl are present in the axon, and Notch can biochemically interact with the Abl-interacting protein Disabled. Since Abl and Disabled are thought to control the axonal cytoskeleton, they could provide a link between Notch activation and cytoskelton changes, thus mediating the effects of Notch signaling on axonal morphology. It will be interesting to determine if Abl and Disabled are present in dendrites and whether they play a role in mediating the effects of Notch on dendritic patterning.

REGULATION OF DENDRITIC SPINE DEVELOPMENT

Normal Development of Dendritic Spines: Golgi Studies

The final step in the acquisition of a mature dendritic morphology is the development of spines. Historically, spines have been categorized on the basis of morphology as thin spines (with thin necks and bulbous heads), mushroom spines (with broader heads), and stubby spines (with no necks). These small dendritic protrusions harbor the vast majority of excitatory synapses and contain receptors and other proteins necessary for synaptic transmission (Kennedy 2000). The function of spines may be to modulate the synaptic response in a compartment separate from the dendritic shaft.

The formation of spines appears to be a developmentally regulated program, which may be influenced by activity. Miller (1981) performed a systematic Golgi survey of dendritic development in early postnatal rat visual cortex. He examined the density of spines on the apical dendrites of both layer 2/3 and layer 5 pyramidal cells, which mature similarly despite the three-day difference in their birthdates. During the first week after birth, there are very few spines present along the apical shaft. However, at the end of the first week, the number of spines increases dramatically, especially between postnatal day 6 (P6) and P9. A second increase in spine density occurs between P12 and 15. The appearance of spines at P6–9 and P12–15 correlates with the arrival of geniculate axons in the cortical plate and eye opening, respectively, which suggests that activity may regulate some aspects of spine development.

Spine density continues to increase through the first postnatal month, after which it declines slightly. Initially, spines are distributed relatively evenly along the apical dendrite, excluding only the area immediately adjacent to the soma. However, by P15, spine density increases with distance from the cell body. This trend is observed along the first 125 μm, after which the density remains fairly stable or declines slightly.

Studies in other animal models and cortical areas have reached similar conclusions about the dynamics of spine development yet reveal important morphological differences between neurons in different layers and areas. For example, work done in primates allows more precise laminar localization of Golgi-impregnated neurons. In the visual cortex of *Macaca nemestrina* monkeys (pig-tailed macaques), Boothe et al. (1979) studied the development of pyramidal neurons in layers 3B and 6 and spiny stellate neurons in layers 4Cα and 4Cβ. During the first eight weeks of life, corresponding to the critical period for development of normal binocular vision, the density of spines increases greatly for both populations of pyramidal neurons, after which it declines. Like the rat neurons described by Miller (1981), these pyramidal neurons show a higher density of spines in the proximal apical dendrite. In contrast, the layer 4 spiny neurons have a constant density of spines throughout the length of their dendrites. In addition, the layer 4 cells demonstrated a much more gradual rise in the density of spines during the critical period. These differences between the two populations of cells presumably reflect differences in the nature of their synaptic input and response to activity during the critical period.

Petit et al. (1988) carried out a similar study, but in the rat sensorimotor cortex. These investigators found an increase in spine density until P30 on the apical dendrites of layer 5 pyramidal neurons but no period of accelerated spine formation. This may reflect an intrinsic difference between the wiring of visual and somatosensory cortices. Interestingly, when they examined the terminal dendritic branches of the apical dendrites, they found that the spine density continued to increase until adulthood.

Most studies on dendritic spines have focused on one region of the cortex at a time, such as visual or somatosensory cortex. Jacobs et al. (2001) arranged eight Brodmann's areas into a functional hierarchy based on the complexity of neural processing occurring in that area. For example, primary motor cortex was considered to be a region of low integration, while the supplementary motor cortex was designated a region of high integration. Next, they compared the dendritic complexity and spine density of Golgi-stained human tissue from the different Brodmann's areas. Although there were significant interindividual and interarea differences, areas of high integration consistently had longer basal dendrites and larger numbers of dendritic spines. This study indicates that an increased density of spines is associated with increased complexity of synaptic integration.

Normal Development of Dendritic Spines: Time-Lapse Studies

How do dendritic spines form? Early studies combined Golgi staining with electron microscopy (EM), revealing the detailed ultrastructure of dendrites and spines once a neuron had been identified. Miller & Peters (1981) studied the development of layer 5 pyramidal neurons in the rat visual cortex with such an approach. During the first postnatal week, before a significant number of spines had formed, dendrites had a large number of filopodia-like processes and stubby spines. The filopodia were long and thin, lacking a bulbous head. Each appeared to be directly apposed to an axon, and while no synapses were present, both the axonal and filopodial membranes were thickened, suggesting that a synapse was forming. Stubby spines, on the other hand, were associated with synapses, but smaller and symmetric ones. However, more protuberant stubby spines formed the asymmetric junctions associated with excitatory synapses, which suggested that stubby spines might grow into taller spines as the synapse matured. As the number of mature spines increased, the number of filopodia and stubby spines decreased, which suggested that this transition is linked to the development of spines.

Time-lapse experiments in which DiI-labeled hippocampal neurons in slices were imaged (Dailey & Smith 1996) revealed that dendritic filopodia are highly dynamic projections, rapidly changing length and shape but lasting only short periods of time. These filopodia would either disappear or become more stable protospines or spines. This suggested a model in which highly protrusive filopodia explore the local environment, make contact with an axon, and guide it back to the dendrite to form a spine. However, it was not clear from these time-lapse studies whether the filopodia directly transition into spines or whether a synapse must first form directly on the shaft. An EM study (Fiala et al. 1998) found that 70% of the synapses in the CA1 region of the hippocampus were present on the dendritic shaft or at the base of a filopodia, which suggests that filopodia are involved in inducing shaft synapses.

A recent paper (Marrs et al. 2001) appears to help resolve this issue of spine development. Particle-mediated gene delivery was used to transfect early postnatal hippocampal slices with a cDNA construct encoding PSD-95 [a component of the postsynaptic density (PSD)] tagged with enhanced green fluorescent protein (GFP). This protein localized to PSDs, which could then be visualized. PSDs were found to be highly dynamic and were able to appear, move, and disappear in a matter of minutes. These PSD95-GFP clusters largely co-localized with synapsin-I, a marker of presynaptic terminals, and were found in mature spines. Transient filopodia were observed; often they would regress spontaneously, but in those cases where a PSD95-GFP cluster developed, the structure would stabilize into a protospine or spine. This suggests that the cluster formed because of a synaptic contact and that this contact was responsible for transforming the filopodia into a spine. Additionally, some spines were formed directly by extension from the shaft, which suggests that shaft synapses can directly form spines.

Dendritic Spines and Activity-Dependent Plasticity

The imaging studies described above were performed in the hippocampus rather than in the neocortex, but presumably the mechanisms of spine formation are the same. Spines and excitatory synapses in the hippocampus have been extensively studied in order to understand how the structure of the brain may change in response to experience. The hippocampus is known to be essential for learning and memory, and synapses in the hippocampus can change the strength of their activation in response to activity, a phenomenon known as long-term potentiation (LTP).

Dendritic spines in the hippocampus undergo morphological changes in response to activity. This area has been extensively reviewed recently (Yuste & Bonhoeffer 2001), but to summarize, there is evidence that synaptic activity results in the enlargement of spines and the shortening of spine necks. These changes reflect the number of presynaptic-docked vesicles and postsynaptic receptors and may affect the efficiency with which an excitatory postsynaptic potential is transmitted to the dendritic shaft or that calcium is extruded. Additionally, increases in activity result in the formation of new spines.

What is the evidence for activity-dependent changes in cortical spines? The majority of reports have been Golgi-staining studies following various deprivation paradigms. For example, Valverde (1967) raised mice in the dark for the first three postnatal weeks and then examined the apical dendrites of layer 5 pyramidal cells where the dendrites traverse through layer 4. He found a significant reduction in the density of spines in the visual cortex. Spine density in the temporal region was unaffected, which leads to the conclusion that the decrease in spine density was due to the reduction in visual input.

Riccio & Matthews (1985) further examined this phenomenon by injecting tetrodotoxin into one eye to abolish action potentials, thereby completely silencing the input from the injected eye for three weeks. Spine densities in layer 5 pyramidal cells in sham-injected animals and internal controls (corresponding to the uninjected eye) are very similar. However, the tetrodotoxin treatment results in a 26% reduction in spines, Changes in spine density, however, have not been seen in every study. Vees et al. (1998) studied the somatosensory cortex of rats following whisker plucking between one and two months of age. The spines of spiny stellate neurons in layer 4 were examined following serial EM reconstruction. Although there were no differences in spine density, spines contralateral to the deprived side tended to have decreased volume and surface area of the spine head and increased length of the spine neck.

In a recent technological advance, it has become possible to study the effect of experience-dependent plasticity in vivo (Lendvai et al. 2000). The barrel cortices of P8–18 postnatal rats were infected with Sindbis virus containing the gene for eGFP, and labeled layer 2/3 neurons were imaged in vivo with time-lapse two-photon laser scanning microscopy. Dendritic protrusions were captured every 10 min and quantified as the length of individual protrusions as a function of time. Their imaging demonstrated that both filopodia and spines are highly motile in vivo. Since filopodial motility had previously been linked to changes in synapse formation, Lendvai et al. deprived the barrel cortex of activity by trimming the rat's whiskers one to three days before imaging. During a brief period, P11–13, spine motility was reduced by the deprivation. This time period corresponds to a period of intense synaptogenesis, when the rats first start using their whiskers to explore their environment. However, deprivation did not affect spine density, length, or morphology, in contrast to some of the other anatomical studies.

Mechanisms of Spine Motility

What is responsible for the rapid motility of dendritic spines and filopodia? A study in the hippocampus (Fischer et al. 1998) demonstrated that the motility is actin based. Cultured hippocampal cells were transfected with GFP-tagged actin, which accumulates in the actin-rich spines. Time-lapse imaging of the GFP fluorescence demonstrated that spines can be very motile, changing their shape on a timescale of seconds or less. This phenomenon is actin dependent because drugs interfering with actin polymerization rapidly abolish the spine motility.

These findings were extended to cortical neurons in slice culture by Dunaevsky et al. (2001) using particle-mediated transfection of GFP in cultured mouse brain slices. Time-lapse imaging revealed these spines to be highly motile. Futhermore, this motility was developmentally regulated. When slices cultured at the day of birth were examined at one week, 74% of spines were motile, compared with 50% of the spines in slices that had been cultured for three weeks. Similar findings were reported by Lendvai et al., who found that protrusive spine motility in the somatosensory cortex decreases with age.

There is also evidence for the regulation of spine formation and spine motility by the neurotrophin BDNF. In addition to affecting dendritic morphology as described earlier, particle-mediated gene transfer of BDNF into ferret slices affected spines (Horch et al. 1999). Over a 16-h period, there was a 2.5-fold reduction in spine density. Of the spines that remained, a higher percentage had an increased turnover rate. This is an exciting finding because it provides a mechanism by which local activity-dependent activation of BDNF could cause local destabilization of spines, thus facilitating synaptic remodeling. While a number of perturbations have now been shown to influence spine formation or spine motility, the consequences of these treatments on synaptic function are not known. Combining morphological analysis of spines with single-cell electrophysiological recordings should provide important insight into the relationship between dendritic morphology and synaptic function.

Dendritic Spines and Mental Retardation

The proper development of dendritic spines is essential for normal cognitive development. Many abnormalities associated with mental retardation (MR) have been identified. For example, Purpura (1975) recognized that many individuals with nonsyndromic MR (MR not associated with a known genetic defect) have dendritic spines that are abnormally sparse, long, and thin, reminscent of the filopodia seen during early spine development. Improper spine (and synapse) formation may therefore be the anatomical cause of some forms of MR.

Some genetic disorders associated with MR also demonstrate defects in dendritic spines, including Down syndrome, Rett syndrome, and Fragile-X (FraX) syndrome [reviewed extensively by (Kaufmann & Moser 2000)]. Mouse models of MR syndromes may help elucidate some of the mechanisms of normal and abnormal development of dendritic spines. For example, Comery et al. (1997) studied the dendritic spines of mice with a targeted deletion in FMR1, the FraX gene, at four months of age. Dendritic spines in knockout mice were longer and spine density was higher than in wild-type controls, consistent with a role for the FMR1 protein in spine maturation and pruning.

More recent work (Nimchinsky et al. 2001) has extended the analysis of the FMR1 mice to earlier developmental stages. Layer 5 barrel cortex neurons were imaged with two-photon microscopy following infection with eGFP-Sindbis virus. At one week after birth, dendritic spines were longer and more dense in knockout animals. However, these differences decreased or disappeared after the second postnatal week. This study suggests that FMR1 is required at two different stages of spine formation or function because of this spine normalization after the first week.

It is clear that similar studies in the future will provide great insight into the molecular mechanisms of dendritic spine formation. In addition to mouse models of MR, the roles of other genes could be studied after their introduction via viral vectors to determine what effects their products might have on spine formation, stability, or motility.

LOOKING INSIDE DENDRITES: mRNAs, PROTEINS, AND CONTROL OF THE DENDRITIC CYTOSKELETAL

Dendritic Targeting of mRNAs and Proteins

Ultrastructurally, dendrites contain almost all of the organelles present in the soma, especially mRNAs and free or membrane-bound ribosomes, which suggests an efficient translational activity (Craig & Banker 1994). Beyond the axonal hillock, only few ribosomes and mRNAs can be found (Craig & Banker 1994, but see also Bassell et al. 1998). This property of local protein synthesis is believed to be important in mediating long-term adaptive responses in neurons, but the mechanisms by which mRNAs are targeted to dendrites and how their translation within dendrites is regulated are not well understood.

The specialized function of dendrites requires selective trafficking of proteins to the dendritic compartment. Over the past two decades, the mechanisms leading to the polarized sorting of membrane proteins have been extensively studied in the epithelial Madin-Darby canine kidney (MDCK) cells, a useful experimental model in which specific proteins are selectively sorted either to the apical or basolateral domains. Dotti & Simons (1990) have proposed that neurons and epithelial cells might use similar cellular mechanisms to generate a polarized distribution of membrane proteins. This was based on a study showing that hippocampal neurons infected with vesicular stomatitis virus or influenza virus display a targeted localization of some viral proteins such as G proteins to the somatodendritic domains. These proteins are normally found in the basolateral domain of MDCK cells. On the other hand, influenza hemagglutinin, which is targeted to the apical domain of MDCK cells, was targeted to the axon of hippocampal neurons. Subsequent studies have shown that mutation of sequences required for basolateral localization in MDCK cells also disrupt dendritic targeting, which suggests the involvement of common sorting and targeting mechanisms (Jareb & Banker 1998). However, the generality of this analogy is not clear since for some proteins, such as transferrin receptor, dendritic targeting is mediated by a signal distinct from that mediating basolateral targeting (Haass et al. 1995, Tienari et al. 1996, Wozniak & Limbird 1998).

Recently, an elegant study performed by Stowell & Craig (1999) has provided evidence that specific sequences in the C-terminal of the cytoplasmic tail of metabotropic glutamate receptors (mGluR) are sufficient for their appropriate targeting to the somatodendritic or axonal domain. The authors took advantage of the fact that mGluR2 is targeted to dendrites and excluded from axons, whereas mGluR7 is targeted both to axons and dendrites. Using viral-mediated expression of chimeric or deleted proteins in hippocampal neurons, they showed that axonal exclusion of mGluR2 versus axon-specific targeting of mGluR7 is mediated by a 60–amino acid-long C-terminal cytoplasmic region. Addition of the mGluR7 C-terminal sequence to mGluR2 or to an unrelated somatodendritic protein was sufficient to induce axonal targeting. The mechanism by which this sequence regulates protein targeting is not yet known.

Rho GTPases and the Regulation of Actin Dynamics

Actin and microtubules are the cytoskeletal components that provide the structural basis for dendrites. Microtubules form the core of dendritic shafts, while actin is located at the rim and at the tips of dendrites. Actin components of the neuronal cytoskeleton drive exploratory activity, while microtubules stabilize newly formed processes. Regulation and restructuring of both actin and microtubule components therefore forms the basis for dendritic growth and remodeling.

The Rho family of small GTPases has lately received a lot of attention for its role in mediating changes in dendritic shape. Rho family GTPases are regulators of actin dynamics and act as molecular switches. They cycle between an active GTP-bound state and an inactive GDP-bound state. In their GTP-bound state they are able to bind and activate downstream effector proteins. The transition from inactive to active state is mediated by guanosine nucleotide exchange factors (GEFs), and the return to the inactive GDP-bound state is catalyzed by GTPase activating proteins (GAPs). The Rho family of GTPases consists of ten members, and the best-studied members are RhoA, Rac1, and Cdc42. Recent studies have demonstrated the central role for these proteins in mediating dendrite growth and remodeling (Lee et al. 2000, Li et al. 2000, Nakayama et al. 2000, Ruchhoeft et al. 1999, Threadgill et al. 1997).

Experiments in vertebrates and invertebrates indicate that RhoA influences the growth of dendritic arbors. Expression of a constitutively active form of RhoA in fly mushroom body neurons (Lee et al. 2000), Xenopus and chick retinal ganglion cells (Ruchhoeft et al. 1999, Wong et al. 2000), Xenopus tectal neurons (Li et al. 2000), and hippocampal neurons (Nakayama et al. 2000) generally leads to a dramatic decrease in dendritic growth. Active RhoA not only prevents the formation of new dendrites but also seems to induce retraction of existing branches. Conversely, blocking RhoA function in these systems promotes growth of dendritic segments. This is nicely demonstrated in *Drosophila*, in which selective removal of RhoA in individual fly mushroom body neurons leads to overextension of dendrites into areas normally not occupied by these neurons (Lee et al. 2000). However, in most other systems, blocking RhoA function results in only a mild phenotypic defect. Perhaps the RhoA pathway is normally inactive to allow dendrite extension and is only activated locally when dendrite arbor growth should be restricted.

Rac1, and to a lesser extent Cdc42, appear to control dendritic branching and remodeling. The most striking phenotype of Rac1 activation in several systems is the selective increase in dendrite branch additions and retractions (Li et al. 2000, Wong et al. 2000). This restructuring induced by Rac1 is rapid, and while Rac1 has been reported to increase branching complexity, the overall dendritic morphology is not greatly affected (Li et al. 2000, Nakayama et al. 2000, Ruchhoeft et al. 1999, Threadgill et al. 1997, Wong et al. 2000).

Rho proteins also exert considerable influence on dendritic spines. Transfection of Rac1 and RhoA constructs into cultured mouse cortical slices perturbs spine formation (Tashiro et al. 2000). Constitutively active Rac1 promotes spine formation, while constitutively active RhoA reduces the number of spines and the length of spine necks. Similar observations have been made in Purkinje cells and hippocampal neurons (Luo et al. 1997, Nakayama et al. 2000, Tashiro et al. 2000). Spines are actin-based structures, which explains the strong influence of Rac1 on spine morphology.

The regulators of the RhoGTPases, GEFs, and GAPs are also likely to be important for spine formation. Consistent with that possibility, Kalirin-7, a RhoGEF with exchange activity for Rac1, was recently found to be targeted to postsynaptic densities in spines through its association with PSD-95 (Penzes et al. 2001). In cultured cortical neurons, transfection of Kalirin-7 resulted in the production of spine-like protrusions of various morphology, reminiscent of Rac activation. Targeting of the GEF to the PSD may be a mechanism that controls spine formation and morphology.

The effects of RhoA on dendritic morphology appear to be mediated by the Rho-associated kinase (ROCK). Blocking ROCK activation prevents RhoA-induced dendritic simplification of hippocampal neurons, whereas expression of activated ROCK mimics the effect of RhoA (Nakayama et al. 2000). ROCK has been shown to activate actomyosin-based contractility and to suppress microtubule assembly in neuroblastoma cells (Hirose et al. 1998), indicating a possible mechanism by which ROCK mediates dendritic retraction. Less is known about downstream effectors mediating the effects of Rac1 and Cdc42 on dendrite remodeling. Given the similarity of the morphological changes induced by activated Rac1 and Cdc42, it is likely that they signal through a common effector protein. A well-known effector that can be activated by both Rac1 and Cdc42 is the serine/threonine kinase PAK1. PAK1 activation has been shown to induce neurite formation in PC12 cells (Daniels et al. 1998). Future experiments should clarify the role of PAK1 and other downstream effectors on Rho family GTPase-induced morphological changes.

Considering the large body of work that shows the role for Rho family GTPases in dendritic patterning, it is surprising how little is known about the mechanisms by which extracellular signals regulate Rho GTPase activity. According to estimates based on the human genome sequence, there are about fifty GEFs, the activators of Rho family GTPases, and more than fifty GAPs, which inactivate Rho family GTPases. It is likely that many of these are expressed in neurons. This high number suggests that multiple factors influence Rho GTPase activity with a significant level of spatial and temporal specificity. Experiments in Xenopus tectal neurons indicate that modulation of dendritic development by NMDA receptor activation is influenced by RhoA (Li et al. 2000) and provides one of the few examples for which Rho GTPase function has been linked to dendritic remodeling induced by an extracellular signal. It is likely that dendritic remodeling induced by other extracellular factors will also be mediated at least in part by Rho family GTPases, and exploring the link between specific extracellular signals and the activation of particular GEFs and GAPs will likely be an active area of investigation in the coming years.

Microtubules and Dendritic Stability

While actin plays a critical role in regulating cytoskeletal dynamics, microtubules provide structural integrity to dendrites. Stabilization of newly formed dendritic branches and consolidation of the dendritic tree therefore requires microtubule invasion. Microtubules are oriented strictly with their plus-end distal to the cell body throughout the axon (Baas et al. 1989, Heidemann et al. 1981). In contrast, microtubules in the proximal and middle regions of dendrites are nonuniformly oriented (Baas et al. 1988, 1989). Given that the polarity of microtubules is relevant to both the dynamics and transport properties, these distinct patterns could provide

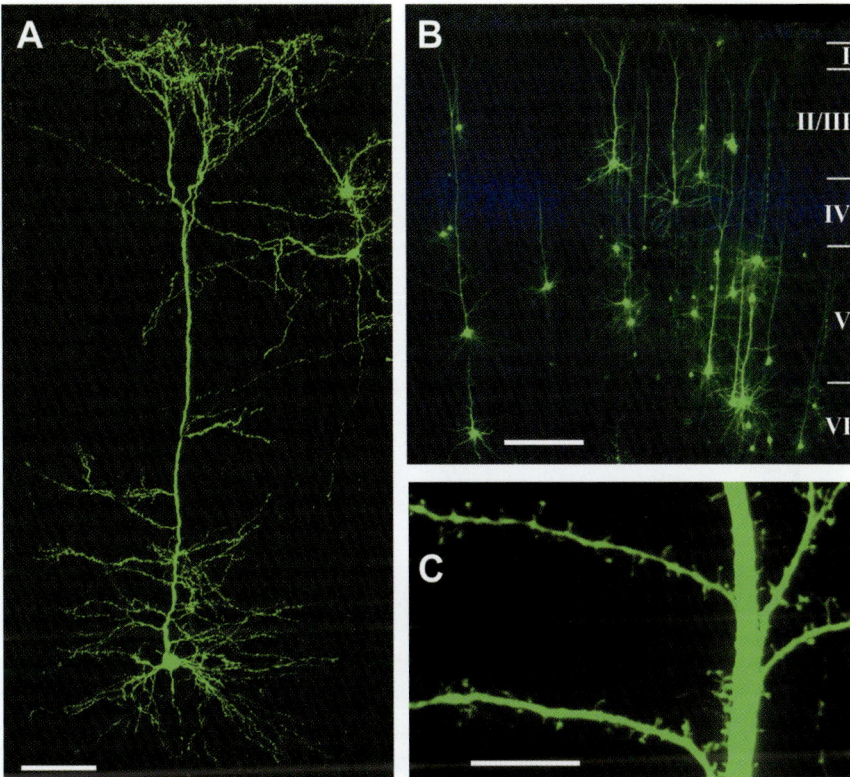

Figure 1 Morphological features of cortical dendrites as revealed by biolistic transfection of GFP into rat cortical slices. (*A*) Example of a layer 5 pyramidal neuron at P10. The salient features of pyramidal neuron morphology include an apical dendrite that extends toward the pial surface and terminates in an apical dendritic tuft, numerous side branches that emerge from the apical dendrite, and several basal dendrites that emerge from the soma. Scale bar, 100 µm. (*B*) Low magnification image of cortical neurons transfected at P12. Note that the majority of neurons have a pyramidal morphology and that the extent of the apical dendrites varies between deep-layer neurons. Scale bar, 250 µm. (*C*) High magnification image of apical dendrite and side branches in a cortical neuron transfected at P18. Note that the dendrites are studded with dendritic filopodia and spines, which are sites of excitatory synaptic transmission. Scale bar, 10 µm. (Images courtesy of Paul Dijkhuizen, Daniele Peters, and Vivian Fenstermaker.)

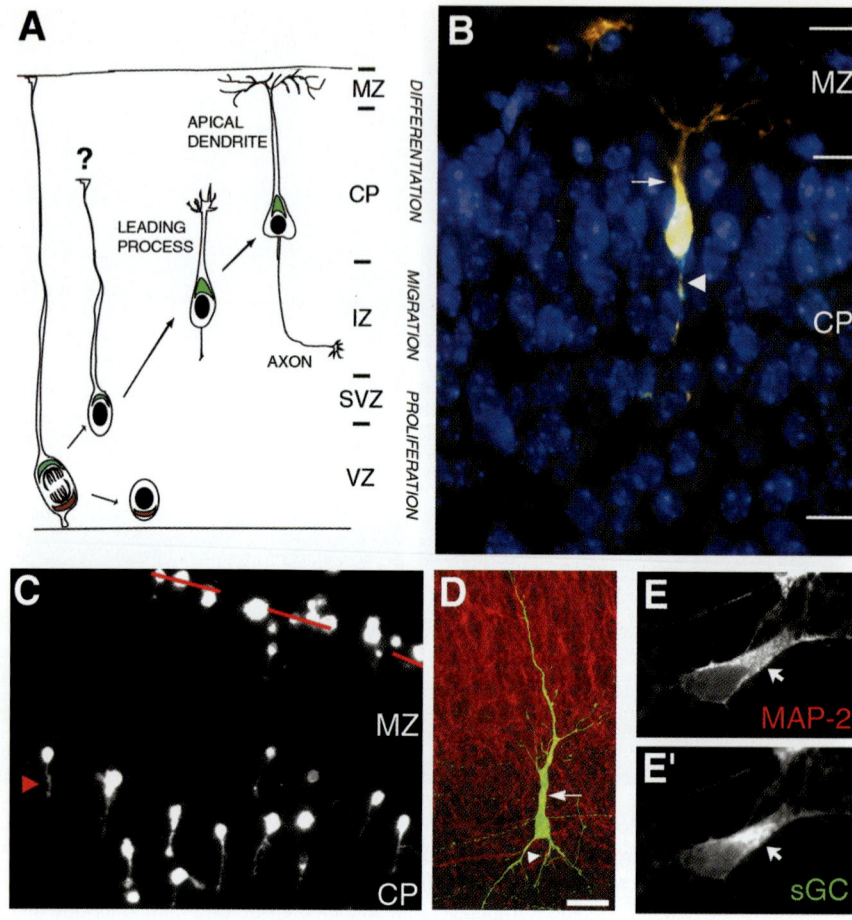

Figure 2 (opposite) Emergence of axonal and dendritic polarity of cortical neurons in vivo and in vitro. (*A*) Summary of the different steps in cortical neuron differentiation. Recent results suggest that radial glial cells are the main neuronal precursors in the ventricular zone (VZ; *left hand side*). These precursors display asymmetric localization of signaling components such as Notch (*green crescent*) and Numb (*brown crescent*). The precursors divide asymmetrically to give rise to two daughter cells that inherit distinct amounts of the cues located at the two poles of the cell. Once in the cortical plate (CP), neurons start differentiating by extending an axon toward the intermediate zone (IZ) and an apical dendrite toward the pial surface. (*B*) Morphology of immature pyramidal neurons in the cortex. This cell was labeled retrogradely by the fluorescent carbocyanine (DiI, *yellow*) injected in the intermediate zone at E18 in the mouse. This labeling reveals the developing apical dendrite growing dorsally (*arrow*) and the axon growing ventrally (*arrowhead*). The cytoarchitecture of the cortex is revealed by a nuclear counterstaining with bisbenzimide (*blue*). (*C*) In vitro study of initial steps of neuronal polarization using the slice overlay assay (Polleux et al. 1998). Three hours after plating, 80% of E18 dissociated cortical neurons differentiating in the CP emit only one neurite (the future axon) growing ventrally toward the IZ (*arrowhead*), thereby mimicking the in vivo situation. This demonstrates the presence of extracellular cues in the CP that are able to direct axon growth in cortical neurons. The *red dotted line* indicates the pial surface. (*D*) In vitro study of factors controlling the polarized outgrowth of apical dendrites in the cortex using the slice overlay assay (Polleux et al. 2000). Four days after plating on cortical slices, GFP-expressing neurons (*green*) growing in the CP have a well-differentiated apical dendrite (*arrow*) directed toward the pial surface and an axon (*arrow*) directed toward the IZ. The red counterstaining is obtained using MAP2 immunofluorescence. (*E-E'*) Asymmetric localization of soluble guanylate cyclase (sGC) in cortical pyramidal neurons. E18 rat cortical neurons were cultured in vitro for two days and double immunofluorescent staining was performed against MAP2 (*E*; a dendritic marker) and sGC (*E'*). Value of the scale bar indicated in *D*: 15 microns for *A*; 20 microns for *C*; 30 microns for *D*; 10 microns for *E* and *E'*.

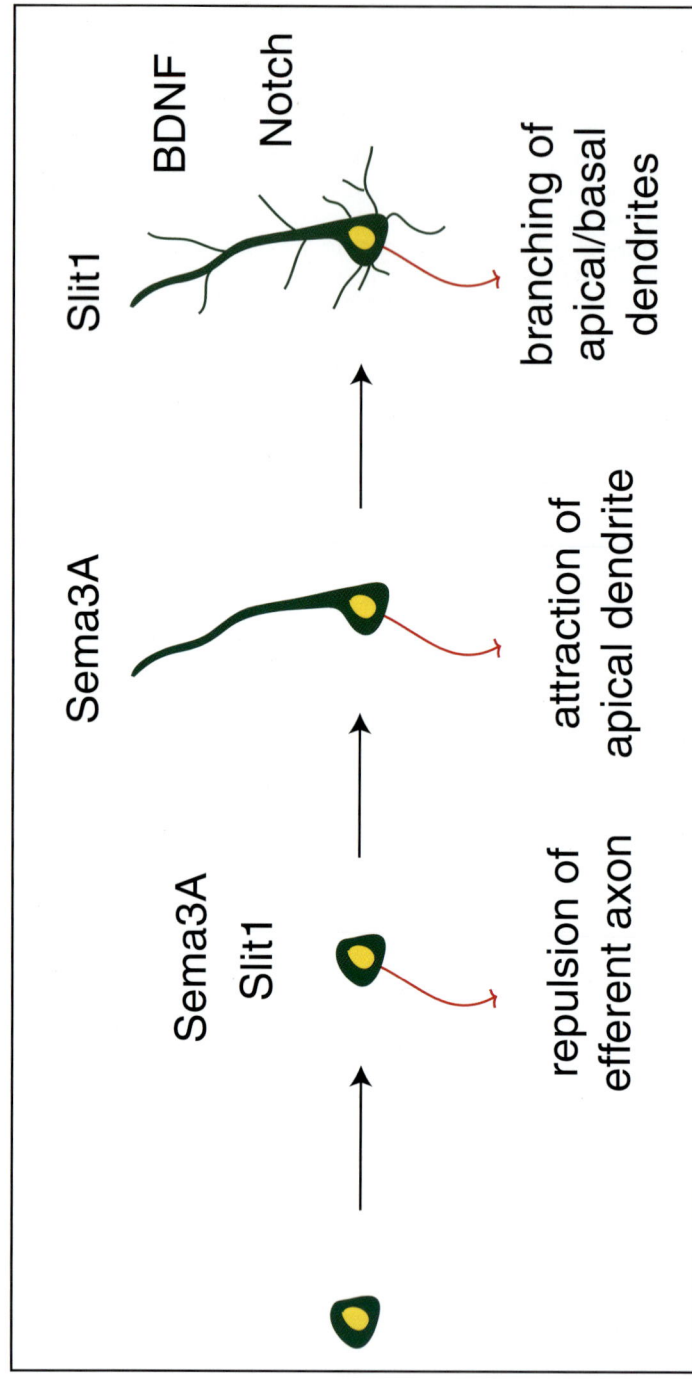

Figure 3 A model of how sequential action of extracellular factors might specify cortical neuron morphology. A newly postmitotic neuron arrives at the cortical plate, where it encounters a gradient of Sema3A (Polleux et al. 1998), which directs the growth of the axon toward the white matter. The same gradient of Sema3A attracts the apical dendrite of the neuron toward the pial surface (Polleux et al. 2000). Other factors, such as BDNF and Notch, control the subsequent growth and branching of dendrites.

the basis for some of the functional and morphological differences characterizing axons and dendrites (Black & Baas 1989).

Emerging dendritic branches are first invaded by plus-end-distal microtubules, followed by the gradual addition of minus-end-directed microtubules. Active transport of microtubules may be essential for establishing the microtubule array in developing dendrites (Baas 1999). Several reports have shown that proteins involved in microtubule transport are essential for dendritic development. For instance, mutations in the minus-end-directed motor protein dynein or its associated protein Lis1 inhibit dendritic growth and branching in neurons of the *Drosophila* mushroom body (Liu et al. 2000). Interestingly, Lis1 haploinsufficiency in humans causes lissencephaly or "smooth brain." Although this defect is caused mainly by defects in cortical neuronal migration, reduced function of Lis1 in mice also affects the morphology of early postmitotic cortical neurons (Cahana et al. 2001). Another microtubule motor protein involved in dendritogenesis is CHO1/MKLP1, which is believed to transport minus-end-distal microtubules into dendrites. Antisense depletion of this motor protein in early sympathetic cultures prevents formation of dendrites (Sharp et al. 1997) and alters the microtubule composition of dendrites in older cultures (Yu et al. 2000).

Little is known about how extracellular signals can influence microtubule dynamics or how interactions between the actin and microtubule cytoskeleton are orchestrated. The importance of coordinated interaction of actin and microtubule structures in dendritogenesis is illustrated by the gene *Kakapo*. The *Kakapo* gene, and its vertebrate homolog *MACF*, encode a cross-linking factor that contains a domain that binds to actin and a domain that binds to and stabilizes microtubules (Leung et al. 1999). *Drosophila* CNS neurons carrying a mutation in the *Kakapo* gene display a reduction in dendritic branching and sprouting (Gao et al. 1999, Prokop et al. 1998), showing that interactions between the actin and microtubule cytoskeleton are essential for dendritic arbor development. Exploring the link between extracellular control of the actin cytoskeleton and the subsequent mobilization of microtubules should provide us with a greater understanding of the molecular mechanisms that specify neuronal morpohology.

CONCLUSIONS

Research on dendritic development and dendritic function has progressed rapidly in the past few years. There is now compelling evidence that dendritic development is a highly dynamic process and that extracellular signals play an important role in specifying the final dendritic morphology of neurons (Figure 3, see color insert). The orientation of dendritic growth appears to be regulated by gradients of local environmental cues, and the findings with semaphorins indicate that a common set of extracellular signals regulate axonal and dendritic development. Interestingly, guidance cues have distinct effects on axonal and dendritic development, and this differential response is mediated in part by asymmetric distribution of intracellular signaling molecules to the two compartments. The growth and branching of dendrites is regulated by a number of factors including trophic factors, Notch, and neuronal activity. An area of particularly rapid advance has been the study of dendritic filopodia and spines. Observations on dendritic filopodial and spine dynamics suggest that highly motile filopodia play an active role in searching out synaptic contacts and that synaptic activity can influence the final morphology of dendritic spines. What is wonderful is that all of this new knowledge about dendritic growth control has come at a time when investigations of the consequences of dendritic morphology on neuronal information processing are in bloom. The emerging view is that dendrites are not just passive conveyors of synaptic input but that they process the incoming information based on the physical and electrophysiological characteristics of the dendritic tree. The ability to manipulate dendritic morphology by molecular perturbation provides a golden opportunity to address long-standing questions regarding the relationship between dendritic form and function.

ACKNOWLEDGMENTS

Our research on dendritic development is supported by grants from NINDS, the Pew Foundation, the Merck Foundation, and the March of Dimes Birth Defects Foundation.

The *Annual Review of Neuroscience* is online at http://neuro.annualreviews.org

LITERATURE CITED

Aakalu G, Smith WB, Nguyen N, Jiang C, Schuman EM. 2001. Dynamic visualization of local protein synthesis in hippocampal neurons. *Neuron* 30:489–502

Artavanis-Tsakonas S, Matsuno K, Fortini ME. 1995. Notch signaling. *Science* 268:225–32

Artavanis-Tsakonas S, Rand MD, Lake RJ. 1999. Notch signaling: cell fate control and signal integration in development. *Science* 284:770–76

Baas PW. 1999. Microtubules and neuronal polarity: lessons from mitosis. *Neuron* 22:23–31

Baas PW, Black MM, Banker GA. 1989. Changes in microtubule polarity orientation during the development of hippocampal neurons in culture. *J. Cell. Biol.* 109:3085–94

Baas PW, Deitch JS, Black MM, Banker GA. 1988. Polarity orientation of microtubules in hippocampal neurons: uniformity in the axon and nonuniformity in the dendrite. *Proc. Natl. Acad. Sci. USA* 85:8335–39

Baker RE, Dijkhuizen PA, Van Pelt J, Verhaagen J. 1998. Growth of pyramidal, but not non-pyramidal, dendrites in long-term organotypic explants of neonatal rat neocortex chronically exposed to neurotrophin-3. *Eur. J. Neurosci.* 10:1037–44

Bassell GJ, Zhang H, Byrd AL, Femino AM, Singer RH, et al. 1998. Sorting of beta-actin mRNA and protein to neurites and growth cones in culture. *J. Neurosci.* 18:251–65

Berezovska O, McLean P, Knowles R, Frosh M, Lu FM, et al. 1999. Notch1 inhibits neurite outgrowth in postmitotic primary neurons. *Neuroscience* 93:433–39

Bettenhausen B, Hrabe de Angelis M, Simon D, Guenet JL, Gossler A. 1995. Transient and restricted expression during mouse embryogenesis of Dl11, a murine gene closely related to Drosophila Delta. *Development* 121:2407–18

Black MM, Baas PW. 1989. The basis of polarity in neurons. *Trends Neurosci.* 12:211–14

Boothe RG, Greenough WT, Lund JS, Wrege K. 1979. A quantitative investigation of spine and dendrite development of neurons in visual cortex (area 17) of Macaca nemestrina monkeys. *J. Comp. Neurol.* 186:473–89

Bradke F, Dotti CG. 1997. Neuronal polarity: vectorial cytoplasmic flow precedes axon formation. *Neuron* 19:1175–86

Bradke F, Dotti CG. 1999. The role of local actin instability in axon formation. *Science* 283:1931–34

Bradke F, Dotti CG. 2000. Establishment of neuronal polarity: lessons from cultured hippocampal neurons. *Curr. Opin. Neurobiol.* 10:574–81

Cahana A, Escamez T, Nowakowski RS, Hayes NL, Giacobini M, et al. 2001. Targeted mutagenesis of Lis1 disrupts cortical development and LIS1 homodimerization. *Proc. Natl. Acad. Sci. USA* 98:6429–34

Castella P, Wagner JA, Caudy M. 1999. Regulation of hippocampal neuronal differentiation by the basic helix-loop-helix transcription factors HES-1 and MASH-1. *J. Neurosci. Res.* 56:229–40

Chenn A, McConnell SK. 1995. Cleavage orientation and the asymmetric inheritance of Notch1 immunoreactivity in mammalian neurogenesis. *Cell* 82:631–41

Chenn A, Zhang YA, Chang BT, McConnell SK. 1998. Intrinsic polarity of mammalian neuroepithelial cells. *Mol. Cell. Neurosci.* 11:183–93

Comery TA, Harris JB, Willems PJ, Oostra BA, Irwin SA, et al. 1997. Abnormal dendritic spines in fragile X knockout mice: maturation and pruning deficits. *Proc. Natl. Acad. Sci. USA* 94:5401–4

Craig AM, Banker G. 1994. Neuronal polarity. *Annu. Rev. Neurosci.* 17:267–310

Dailey ME, Smith SJ. 1996. The dynamics of dendritic structure in developing hippocampal slices. *J. Neurosci.* 16:2983–94

Daniels RH, Hall PS, Bokoch GM. 1998. Membrane targeting of p21-activated kinase 1 (PAK1) induces neurite outgrowth from PC12 cells. *EMBO J.* 17:754–64

de la Pompa JL, Wakeham A, Correia KM, Samper E, Brown S, et al. 1997. Conservation of the Notch signalling pathway in mammalian neurogenesis. *Development* 124:1139–48

Del Amo FF, Smith DE, Swiatek PJ, Gendron-Maguire M, Greenspan RJ, et al. 1992. Expression pattern of Motch, a mouse homolog of Drosophila Notch, suggests an important role in early postimplantation mouse development. *Development* 115:737–44

Dotti CG, Banker GA. 1987. Experimentally induced alteration in the polarity of developing neurons. *Nature* 330:254–56

Dotti CG, Simons K. 1990. Polarized sorting of viral glycoproteins to the axon and dendrites of hippocampal neurons in culture. *Cell* 62:63–72

Dotti CG, Sullivan CA, Banker GA. 1988. The establishment of polarity by hippocampal neurons in culture. *J. Neurosci.* 8:1454–68

Dunaevsky A, Blazeski R, Yuste R, Mason C. 2001. Spine motility with synaptic contact. *Nat. Neurosci.* 4:685–86

Dunwoodie SL, Henrique D, Harrison SM, Beddington RS. 1997. Mouse Dll3: a novel divergent *Delta* gene which may complement the function of other Delta homologues during early pattern formation in the mouse embryo. *Development* 124:3065–76

Fiala JC, Feinberg M, Popov V, Harris KM. 1998. Synaptogenesis via dendritic filopodia in developing hippocampal area CA1. *J. Neurosci.* 18:8900–11

Fischer M, Kaech S, Knutti D, Matus A. 1998. Rapid actin-based plasticity in dendritic spines. *Neuron* 20:847–54

Franklin JL, Berechid BE, Cutting FB, Presente A, Chambers CB, et al. 1999. Autonomous and non-autonomous regulation of mammalian neurite development by Notch1 and Delta1. *Curr. Biol.* 9:1448–57

Furukawa T, Maruyama S, Kawaichi M, Honjo T. 1992. The Drosophila homolog of the immunoglobulin recombination signal-binding protein regulates peripheral nervous system development. *Cell* 69:1191–97

Gao FB, Brenman JE, Jan LY, Jan YN. 1999. Genes regulating dendritic outgrowth, branching, and routing in Drosophila. *Genes Dev.* 13:2549–61

Gilbert CD. 1983. Microcircuitry of the visual cortex. *Annu. Rev. Neurosci.* 6:217–47

Giniger E. 1998. A role for Abl in Notch signaling. *Neuron* 20:667–81

Haass C, Koo EH, Capell A, Teplow DB, Selkoe DJ. 1995. Polarized sorting of beta-amyloid precursor protein and its proteolytic products in MDCK cells is regulated by two independent signals. *J. Cell. Biol.* 128:537–47

Hausser M, Spruston N, Stuart GJ. 2000. Diversity and dynamics of dendritic signaling. *Science* 290:739–44

Heidemann SR, Landers JM, Hamborg MA. 1981. Polarity orientation of axonal microtubules. *J. Cell. Biol.* 91:661–65

Hirose M, Ishizaki T, Watanabe N, Uehata M, Kranenburg O, et al. 1998. Molecular dissection of the Rho-associated protein kinase (p160ROCK)-regulated neurite remodeling in neuroblastoma N1E-115 cells. *J. Cell. Biol.* 141:1625–36

Horch HW, Kruttgen A, Portbury SD, Katz LC. 1999. Destabilization of cortical dendrites and spines by BDNF. *Neuron* 23:353–64

Houser CR, Hendry SH, Jones EG, Vaughn JE. 1983. Morphological diversity of immunocytochemically identified GABA neurons in the monkey sensory-motor cortex. *J. Neurocytol.* 12:617–38

Huang EJ, Reichardt LF. 2001. Neurotrophins: roles in neuronal development and function. *Annu. Rev. Neurosci.* 24:677–736

Jacobs B, Schall M, Prather M, Kapler E, Driscoll L, et al. 2001. Regional dendritic and spine variation in human cerebral cortex: a quantitative Golgi study. *Cereb. Cortex* 11:558–71

Jareb M, Banker G. 1998. The polarized sorting of membrane proteins expressed in cultured hippocampal neurons using viral vectors. *Neuron* 20:855–67

Katz LC. 1987. Local circuitry of identified projection neurons in cat visual cortex brain slices. *J. Neurosci.* 7:1223–49

Kaufmann WE, Moser HW. 2000. Dendritic anomalies in disorders associated with mental retardation. *Cereb. Cortex* 10:981–91

Kennedy MB. 2000. Signal-processing machines at the postsynaptic density. *Science* 290:750–54

Kita Y, Kimura KD, Kobayashi M, Ihara S, Kaibuchi K, et al. 1998. Microinjection of activated phosphatidylinositol-3 kinase induces process outgrowth in rat PC12 cells through the Rac-JNK signal transduction pathway. *J. Cell Sci.* 111:907–15

Koester SE, O'Leary DD. 1992. Functional classes of cortical projection neurons develop dendritic distinctions by class-specific sculpting of an early common pattern. *J. Neurosci.* 12:1382–93

Kuhl D, Skehel P. 1998. Dendritic localization of mRNAs. *Curr. Opin. Neurobiol.* 8:600–6

Lardelli M, Dahlstrand J, Lendahl U. 1994. The novel Notch homologue mouse Notch 3 lacks specific epidermal growth factor-repeats and is expressed in proliferating neuroepithelium. *Mech. Dev.* 46:123–36

Le Roux P, Behar S, Higgins D, Charette M. 1999. OP-1 enhances dendritic growth from cerebral cortical neurons in vitro. *Exp. Neurol.* 160:151–63

Lee T, Winter C, Marticke SS, Lee A, Luo L. 2000. Essential roles of Drosophila RhoA in the regulation of neuroblast proliferation and dendritic but not axonal morphogenesis. *Neuron* 25:307–16

Lendvai B, Stern EA, Chen B, Svoboda K. 2000. Experience-dependent plasticity of dendritic spines in the developing rat barrel cortex in vivo. *Nature* 404:876–81

Leung CL, Sun D, Zheng M, Knowles DR, Liem RK. 1999. Microtubule actin crosslinking factor (MACF): a hybrid of dystonin and dystrophin that can interact with the actin and microtubule cytoskeletons. *J. Cell. Biol.* 147:1275–86

Li Z, Van Aelst L, Cline HT. 2000. Rho GTPases regulate distinct aspects of dendritic arbor growth in Xenopus central neurons in vivo. *Nat. Neurosci.* 3:217–25

Lindsell CE, Boulter J, diSibio G, Gossler A, Weinmaster G. 1996. Expression patterns of

Jagged, *Delta1*, *Notch1*, *Notch2*, and *Notch3* genes identify ligand-receptor pairs that may function in neural development. *Mol. Cell. Neurosci.* 8:14–27

Liu Z, Steward R, Luo L. 2000. Drosophila Lis1 is required for neuroblast proliferation, dendritic elaboration and axonal transport. *Nat. Cell Biol.* 2:776–83

Luo L, Jan LY, Jan YN. 1997. Rho family GTP-binding proteins in growth cone signalling. *Curr. Opin. Neurobiol.* 7:81–86

Malatesta P, Hartfuss E, Gotz M. 2000. Isolation of radial glial cells by fluorescent-activated cell sorting reveals a neuronal lineage. *Development* 127:5253–63

Marin-Padilla M. 1992. Ontogenesis of the pyramidal cell of the mammalian neocortex and developmental cytoarchitectonics: a unifying theory. *J. Comp. Neurol.* 321:223–40

Marrs GS, Green SH, Dailey ME. 2001. Rapid formation and remodeling of postsynaptic densities in developing dendrites. *Nat. Neurosci.* 4:1006–13

McAllister AK, Lo DC, Katz LC. 1995. Neurotrophins regulate dendritic growth in developing visual cortex. *Neuron* 15:791–803

Miller M. 1981. Maturation of rat visual cortex. I. A quantitative study of Golgi-impregnated pyramidal neurons. *J. Neurocytol.* 10:859–78

Miller M, Peters A. 1981. Maturation of rat visual cortex. II. A combined Golgi-electron microscope study of pyramidal neurons. *J. Comp. Neurol.* 203:555–73

Miyata T, Kawaguchi A, Okano H, Ogawa M. 2001. Asymmetric inheritance of radial glial fibers by cortical neurons. *Neuron* 31:727–41

Nakamura Y, Sakakibara S, Miyata T, Ogawa M, Shimazaki T, et al. 2000. The bHLH gene *hes1* as a repressor of the neuronal commitment of CNS stem cells. *J. Neurosci.* 20:283–93

Nakayama AY, Harms MB, Luo L. 2000. Small GTPases Rac and Rho in the maintenance of dendritic spines and branches in hippocampal pyramidal neurons. *J. Neurosci.* 20:5329–38

Niblock MM, Brunso-Bechtold JK, Riddle DR. 2000. Insulin-like growth factor I stimulates dendritic growth in primary somatosensory cortex. *J. Neurosci.* 20:4165–76

Nimchinsky EA, Oberlander AM, Svoboda K. 2001. Abnormal development of dendritic spines in FMR1 knock-out mice. *J. Neurosci.* 21:5139–46

Noctor SC, Flint AC, Weissman TA, Dammerman RS, Kriegstein AR. 2001. Neurons derived from radial glial cells establish radial units in neocortex. *Nature* 409:714–20

Ohtsuka T, Ishibashi M, Gradwohl G, Nakanishi S, Guillemot F, Kageyama R. 1999. Hes1 and Hes5 as notch effectors in mammalian neuronal differentiation. *EMBO J.* 18:2196–207

Olson EC, Schinder AF, Dantzker JL, Marcus EA, Spitzer NC, Harris WA. 1998. Properties of ectopic neurons induced by Xenopus neurogenin1 misexpression. *Mol. Cell. Neurosci.* 12:281–99

Parent CA, Blacklock BJ, Froehlich WM, Murphy DB, Devreotes PN. 1998. G protein signaling events are activated at the leading edge of chemotactic cells. *Cell* 95:81–91

Parent CA, Devreotes PN. 1999. A cell's sense of direction *Science* 284:765–70

Penzes P, Johnson RC, Sattler R, Zhang X, Huganir RL, et al. 2001. The neuronal Rho-GEF Kalirin-7 interacts with PDZ domain-containing proteins and regulates dendritic morphogenesis. *Neuron* 29:229–42

Petit TL, LeBoutillier JC, Gregorio A, Libstug H. 1988. The pattern of dendritic development in the cerebral cortex of the rat. *Brain Res.* 469:209–19

Polleux F, Giger RJ, Ginty DD, Kolodkin AL, Ghosh A. 1998. Patterning of cortical efferent projections by semaphorin-neuropilin interactions. *Science* 282:1904–6

Polleux F, Morrow T, Ghosh A. 2000. Semaphorin 3A is a chemoattractant for cortical apical dendrites. *Nature* 404:567–73

Posern G, Rapp UR, Feller SM. 2000. The Crk signaling pathway contributes to the bombesin-induced activation of the small GTPase Rap1 in Swiss 3T3 cells. *Oncogene* 19:6361–68

Prieto JJ, Peterson BA, Winer JA. 1994.

Morphology and spatial distribution of GABAergic neurons in cat primary auditory cortex (AI). *J. Comp. Neurol.* 344:349–82

Prokop A, Uhler J, Roote J, Bate M. 1998. The kakapo mutation affects terminal arborization and central dendritic sprouting of Drosophila motorneurons. *J. Cell. Biol.* 143:1283–94

Pruyne D, Bretscher A. 2000. Polarization of cell growth in yeast. *J. Cell Sci.* 113:571–85

Purpura DP. 1975. Dendritic differentiation in human cerebral cortex: normal and aberrant developmental patterns. *Adv. Neurol.* 12:91–134

Reaume AG, Conlon RA, Zirngibl R, Yamaguchi TP, Rossant J. 1992. Expression analysis of a Notch homologue in the mouse embryo. *Dev. Biol.* 154:377–87

Redmond L, Oh SR, Hicks C, Weinmaster G, Ghosh A. 2000. Nuclear Notch1 signaling and the regulation of dendritic development. *Nat. Neurosci.* 3:30–40

Riccio RV, Matthews MA. 1985. Effects of intraocular tetrodotoxin on dendritic spines in the developing rat visual cortex: a Golgi analysis. *Brain Res.* 351:173–82

Ruchhoeft ML, Ohnuma S, McNeill L, Holt CE, Harris WA. 1999. The neuronal architecture of Xenopus retinal ganglion cells is sculpted by rho-family GTPases in vivo. *J. Neurosci.* 19:8454–63

Servant G, Weiner OD, Neptune ER, Sedat JW, Bourne HR. 1999. Dynamics of a chemoattractant receptor in living neutrophils during chemotaxis. *Mol. Biol. Cell* 10:1163–78

Sestan N, Artavanis-Tsakonas S, Rakic P. 1999. Contact-dependent inhibition of cortical neurite growth mediated by notch signaling. *Science* 286:741–46

Sharp DJ, Yu W, Ferhat L, Kuriyama R, Rueger DC, Baas PW. 1997. Identification of a microtubule-associated motor protein essential for dendritic differentiation. *J. Cell. Biol.* 138:833–43

Song H, Ming G, He Z, Lehmann M, McKerracher L, et al. 1998. Conversion of neuronal growth cone responses from repulsion to attraction by cyclic nucleotides. *Science* 281:1515–18

Stowell JN, Craig AM. 1999. Axon/dendrite targeting of metabotropic glutamate receptors by their cytoplasmic carboxy-terminal domains. *Neuron* 22:525–36

Tashiro A, Minden A, Yuste R. 2000. Regulation of dendritic spine morphology by the rho family of small GTPases: antagonistic roles of Rac and Rho. *Cereb. Cortex* 10:927–38

Threadgill R, Bobb K, Ghosh A. 1997. Regulation of dendritic growth and remodeling by Rho, Rac, and Cdc42. *Neuron* 19:625–34

Tienari PJ, De Strooper B, Ikonen E, Simons M, Weidemann A, et al. 1996. The beta-amyloid domain is essential for axonal sorting of amyloid precursor protein. *EMBO J.* 15:5218–29

Valverde F. 1967. Apical dendritic spines of the visual cortex and light deprivation in the mouse. *Exp. Brain Res.* 3:337–52

Vees AM, Micheva KD, Beaulieu C, Descarries L. 1998. Increased number and size of dendritic spines in ipsilateral barrel field cortex following unilateral whisker trimming in postnatal rat. *J. Comp. Neurol.* 400:110–24

Vercelli A, Assal F, Innocenti GM. 1992. Emergence of callosally projecting neurons with stellate morphology in the visual cortex of the kitten. *Exp. Brain Res.* 90:346–58

Weinmaster G. 1997. The ins and outs of notch signaling. *Mol. Cell. Neurosci.* 9:91–102

Weinmaster G. 2000. Notch signal transduction: a real rip and more. *Curr. Opin. Genet. Dev.* 10:363–69

Wong WT, Faulkner-Jones BE, Sanes JR, Wong RO. 2000. Rapid dendritic remodeling in the developing retina: dependence on neurotransmission and reciprocal regulation by Rac and Rho. *J. Neurosci.* 20:5024–36

Wozniak M, Limbird LE. 1998. Trafficking itineraries of G protein–coupled receptors in epithelial cells do not predict receptor localization in neurons. *Brain Res.* 780:311–22

Wu GY, Deisseroth K, Tsien RW. 2001. Spaced stimuli stabilize MAPK pathway activation and its effects on dendritic morphology. *Nat. Neurosci.* 4:151–58

Xiao Z, Zhang N, Murphy DB, Devreotes PN. 1997. Dynamic distribution of chemoattractant receptors in living cells during chemotaxis and persistent stimulation. *J. Cell. Biol.* 139:365–74

Yacoubian TA, Lo DC. 2000. Truncated and full-length TrkB receptors regulate distinct modes of dendritic growth. *Nat. Neurosci.* 3:342–49

Yasui H, Katoh H, Yamaguchi Y, Aoki J, Fujita H, et al. 2001. Differential responses to nerve growth factor and epidermal growth factor in neurite outgrowth of PC12 cells are determined by Rac1 activation systems. *J. Biol. Chem.* 276:15298–305

Yavari R, Adida C, Bray-Ward P, Brines M, Xu T. 1998. Human metalloprotease-disintegrin Kuzbanian regulates sympathoadrenal cell fate in development and neoplasia. *Hum. Mol. Genet.* 7:1161–67

Yu W, Cook C, Sauter C, Kuriyama R, Kaplan PL, Baas PW. 2000. Depletion of a microtubule-associated motor protein induces the loss of dendritic identity. *J. Neurosci.* 20: 5782–91

Yuste R, Bonhoeffer T. 2001. Morphological changes in dendritic spines associated with long-term synaptic plasticity. *Annu. Rev. Neurosci.* 24:1071–89

Zhong W, Jiang MM, Weinmaster G, Jan LY, Jan YN. 1997. Differential expression of mammalian Numb, Numblike and Notch1 suggests distinct roles during mouse cortical neurogenesis. *Development* 124:1887–97

FUNCTIONAL MRI OF LANGUAGE: New Approaches to Understanding the Cortical Organization of Semantic Processing

Susan Bookheimer
Brain Mapping Center, UCLA School of Medicine, Los Angeles, California 90095; email: sbook@ucla.edu

Key Words fMRI, functional neuroimaging, inferior frontal gyrus, speech, comprehension, semantics

■ **Abstract** Until recently, our understanding of how language is organized in the brain depended on analysis of behavioral deficits in patients with fortuitously placed lesions. The availability of functional magnetic resonance imaging (fMRI) for in vivo analysis of the normal brain has revolutionized the study of language. This review discusses three lines of fMRI research into how the semantic system is organized in the adult brain. These are (*a*) the role of the left inferior frontal lobe in semantic processing and dissociations from other frontal lobe language functions, (*b*) the organization of categories of objects and concepts in the temporal lobe, and (*c*) the role of the right hemisphere in comprehending contextual and figurative meaning. Together, these lines of research broaden our understanding of how the brain stores, retrieves, and makes sense of semantic information, and they challenge some commonly held notions of functional modularity in the language system.

INTRODUCTION

Over 150 years of research into the organization of language in the brain is based on a lesion deficit approach, which deduces the functional significance of a brain area through observation of deficit following either temporary or permanent brain lesions. These methods reveal brain areas that, when disrupted, produce a complete or near-complete breakdown in the patient's ability to perform a task, and by deduction, lead to models of the underlying functional role of the affected brain region. This fundamental philosophy has led neuroscientists toward what we may term a large-module conceptualization of functional organization, an approach in which rather widespread territories of cortex are deemed responsible for broad categories of function. A large-module philosophy is particularly prevalent within the realm of language research; texts and reviews describe the language system as composed primarily of two broad-domain regions: Broca's area in inferior frontal cortex and Wernicke's area in the posterior superior temporal region. Consider, for instance, the commonly stated functions attributed to Broca's area [pars opercularis of the inferior frontal gyrus (IFG) according to Broca, but encompassing a broader territory within the IFG in most neuropsychological reviews]. Lesions to this area produce a wide range of deficits known collectively as Broca's aphasia, including those involving articulation, sequential production of speech, sentence production, syntax, naming, and comprehension of some complex syntactic structures. The corresponding argument—that Broca's area must therefore be responsible for executing these functions—appears unlikely given their scope; it is difficult to conceive of a set of neural computations complex enough to account for them all simultaneously. Large systems models have given rise to similarly phrased cognitive descriptions of brain systems, such as the semantic system, seeming in the same way to encompass extensive networks of meaning in a large centralized locale.

Small numbers of case reports over the years have suggested holes in this general philosophy, most notably the findings of some patients with category-specific memory or naming deficits, or the rare cases with specific deficits in certain syntactic constructions. Yet the small numbers and uncertain or inconsistent localization of these cases have made it difficult to generate comprehensive models of language organization that respect both the behavioral data and the available anatomical data. More recent studies using structural MRI to identify brain regions common to aphasia subtypes such as Broca's have found a great deal of variability on location of the critical lesion. In identifying MRI lesions with specific aspects of Broca's aphasia, Alexander et al. (1990) found different anatomical structures around the IFG correlated with different aphasic symptoms within the Broca's syndrome. Damage to the underlying white matter of all regions appears necessary to produce the complete syndrome. Similarly, resting metabolic studies using positron emission tomography (PET) indicate that brain regions far distal to the lesion site are affected in patients with diagnosed aphasia (Metter 1991). Taken together, the data have indicated increasingly that large modules do not adequately describe the organization of complex brain functions such as language, although few opportunities for testing alternative models had been available before the advent of functional neuroimaging.

Functional brain imaging, particularly activation PET and functional magnetic resonance imaging (fMRI), rely on a very different fundamental approach to understanding brain organization. These techniques reveal brain areas involved in, though not necessarily essential to, the ongoing performance of a task. Recent years have shown a tremendous increase in the number of imaging studies of language in the normal brain. From this corpus or research, several long-held notions of language organization are openly challenged, and new findings add substantial detail to our understanding of how language is organized and retrieved. In particular, it is apparent that large-module theories are clearly incorrect; rather, the language system is organized into a large number of relatively small but tightly clustered and interconnected modules with unique contributions to language processing. There is increasing evidence that language regions in the brain—even classic Broca's area—are not specific to language, but rather involve more reductionist processes that give rise to language as well as nonlinguistic functions. Finally, functional imaging has indicated far greater involvement of the right hemisphere in some aspects of language processing than previously appreciated.

This review focuses on three areas of fMRI research in language organization: subdivisions within the IFG, emphasizing the role of the IFG in semantic processing; the nature of category-specific organization of semantic information in the temporal lobe; and contributions of the right hemisphere to language comprehension. The number of recent fMRI publications on language far exceeds what can be reviewed here; therefore, this review focuses on a few representative papers in each area. Advances in fMRI studies of language are by no means limited to these areas; in particular, there is now a large body of work on reading that has strongly supported the dual-route theory in reading and substantial, novel findings relevant to auditory comprehension in the posterior temporal parietal regions. A recent review by Price (2000) emphasizes both of these aspects of language research. In addition, progress in our understanding of semantic memory and how the brain accesses store verbal memories is described in recent reviews by Gabrieli et al. (1998) and Buckner & Wheeler (2001).

Functional Magnetic Resonance Imaging: Some Basic Principles

Activation PET and fMRI share fundamental similarities in that they measure blood flow changes during the performance of a cognitive task, in comparison with another task or condition. Blood flow serves as an indirect marker of neural activity, although the two are tightly coupled under most conditions. One key limitation of blood flow as a marker for neural activity is that the brain's vascular response is sluggish, beginning about 2 sec after onset of neural activity and peaking only after about 5–7 sec (Savoy et al. 1994, Cohen 1997). Thus, unlike unit recordings or evoked potentials, the temporal resolution of functional imaging under the most ideal conditions is quite slow compared to actual neural activity. fMRI carries the additional disadvantage of having no reliable means to quantify neural activity in an absolute sense. Unlike PET, which measures blood flow directly, fMRI measures blood flow indirectly by detecting susceptibility changes associated with the relative concentration of oxy- and deoxy-hemoglobin on the venous side of the capillary bed (see Cohen & Bookheimer 1994 for an introduction to fMRI). Measuring blood flow using fMRI requires at least two experimental conditions. Differences observed in the MRI signal between two cognitive states are therefore relative, and consequently, results from activation imaging experiments depend on skill with which one designs both the experimental and control task. Even in cases in which the experimental and control conditions appear well matched based on solid psychological grounds, subtle differences in task difficulty, response styles, and strategies can easily affect the magnitude, spatial extent, and even the location of brain regions in imaging experiments (Raichle et al. 1994).

Approaches to account for these problems include taking simultaneous behavioral measures, using multiple baseline conditions, and employing paradigm designs that minimize potential confounds. As none of these methods works perfectly, one must approach interpretations of fMRI results with a great deal of skepticism.

To describe increases in MRI signal intensity during task vs. control comparisons, I use the terms activation or increased brain activity, along with decreased brain activity in cases in which the MR signal intensity is lower in the experimental compared with the control task. Because MRI signal intensity changes are comparative by their nature, we cannot determine whether they represent an actual increase in blood flow for the experimental task or a decrease in the control or some combination. Thus, the term activation implies only relative changes in MRI signal intensity. Although this review emphasizes fMRI studies, I have included a review of important PET papers where fMRI data are lacking.

Recent Advances in Paradigm Design

Nearly all of the early PET activation studies prior to the advent of fMRI employed a hierarchical subtraction model for isolating cognitive operations. Petersen and colleagues' (1988) paper on single-word processing typifies this approach: A resting baseline is first attained and then subtracted from a sensory control, in this case, passive presentation of either visual or auditory words. This task is subtracted in turn from an output condition, reading or repeating the words, which is then subtracted from an association condition. The model makes several key assumptions, most importantly that brain activity in lower levels of the hierarchy remains constant across hierarchical levels and that passive presentation invokes primarily sensory regions, to name only a few. More recently, several new designs and analysis approaches that rely less on assumptions of hierarchical organization have begun to dominate the imaging literature. These include common baseline designs, in which all experimental tasks are compared to a single, simple baseline (Bookheimer et al. 1995); parametric designs (i.e., Price et al. 1992), in which the level or load of the dimension of interest is varied; and selective attention designs (e.g., Corbetta et al. 1990, Dapretto & Bookheimer 1999), in which subjects see identical or nearly identical stimuli but selectively attend to one or another feature within the stimulus set.

A new model possible only for fMRI research is the single-trial or event-related (ER) design (Savoy et al. 1994). The blocked designs listed above require that subjects enter a steady state in which they perform multiple trials of a particular task, usually in blocks of half a minute or more. The ER design presents one stimulus at a time, allowing the blood flow response to rise and fall for that particular item before presenting a second stimulus. Items of different categories (experimental vs. control items, e.g.) are presented randomly, making it impossible for subjects to develop an effective strategy for only one stimulus type. This design has additional advantages in that the blood flow response may ebb over time in a longer block, and thus the magnitude of the blood flow response may be preserved in single-trial studies (though this has not been demonstrated formally). Also, ER designs distribute fatigue effects evenly across trials, whereas in blocked designs subjects may tire over the course of the experiment, affecting late-occurring conditions in particular. The original ER designs allowed time for the blood flow response to reach baseline (about 12–16 sec after each stimulus) before proceeding to the next, which greatly lengthens the task. In these designs the control state is rest, which most agree is not an adequate control. Newer approaches present stimuli more rapidly and perform direct contrasts of height epochs across stimulus types. This has the advantage of allowing far more trials in an imaging session at the cost of making additional assumptions about the linearity of the blood flow response when trials of the same type occur in succession, assumptions that also have not been validated. In practice, the ER designs are most important in studies in which strategies invoked by blocked paradigms would mar interpretation of the study results. For more detailed discussions of ER procedures and effective utilization of this approach, see D'Esposito et al. 1999 and Buckner et al. 2000. Whereas most of the fMRI studies of language continue to employ blocked designs, ER approaches have become increasingly popular and may be essential in cases where strategic effects of blocks could obscure effects due to different stimulus types.

The Universal Language of Brain Mapping

Nearly all fMRI and PET experiments report results in Talairach coordinate space (Talairach & Tournoux 1988). This system is based on the postmortem analysis of one person's brain and is published as an atlas, though newer atlases containing the average of several hundred brains are now accessible through this system. In the Talairach system, all locations within the three-dimensional space of the brain are represented as a number from left to right [x-dimension: -65 mm left hemisphere (LH) to $+65$ mm right hemisphere (RH)]; from anterior to posterior (y-dimension: $+70$ mm anterior to -90 posterior); and from inferior to superior (z-dimension: -40 mm inferior to $+65$ mm superior). In this system all brains are normalized to fit a template that is centered in z along the line connecting the anterior and posterior commissures, $x = 0$ is at the midline and $y = 0$ at the anterior commissure. In this way, a three-number coordinate defines the spatial location of any point in the brain, usually representing either the highest peak of activation in a region or the geometric center of a three-dimensional blob of activity. The coordinate system makes it possible for imaging investigators to compare the locations of brain activity across centers, imaging modalities, and subjects. They can compare the locations with reasonable certainty and in a common language that circumvents the need to agree on anatomical boundaries and naming conventions. All tables in this paper refer to the Talairach coordinate system; see Talairach & Tournoux 1988 for a more detailed description.

THE ROLE OF THE INFERIOR FRONTAL GYRUS IN SEMANTIC PROCESSING

Since Broca's original report of a patient with a motor speech disturbance, the IFG, especially Brodmann's area 44, has been attributed the role of producing language. Most models of Broca's area function refer to the role of the IFG in expressive speech. A strong challenge to this simple idea emphasizes the variability in location of Broca's area across individuals. In particular, Ojemann's cortical stimulation mapping studies (Ojemann et al. 1989) indicate striking variability in frontal temporal and parietal lobes for brain regions that disrupt object naming. Furthermore, the effects of language disruption in anterior vs. posterior language areas do not follow the classic posterior-comprehension, anterior-expression dissociation. Rather, particular within the frontal lobe, Ojemann found areas specialized for semantic processing and phonology as well as articulation. Luria (1966) noted that Broca's area patients made comprehension errors in syntactically complex sentences such as passive constructions. For instance, Broca's aphasics had difficulty answering the question "A lion was fatally attacked by a tiger. Which animal died?" but no problem with the active construction "The tiger fatally attacked the lion." Clearly, comprehension was intact at the word level, but meaning at the sentence level was lost under conditions in which function words or knowledge of the syntactic structure were essential for comprehension. Several anatomical correlation studies found little evidence for area 44's exclusive association with Broca's aphasia. Alexander et al. (1990), Dronkers (1996), and others have found evidence for functional heterogeneity in IFG for different deficits among Broca's aphasics including those in articulation, syntax, and naming. Rather than indicating a high degree of variability in lesion location, the data tend to show that multiple regions are involved in expressive language and that only disruption to all of them produces the catastrophic breakdown of language as revealed in aphasics. As Broca's aphasia patients have a variety of seemingly diverse impairments, it is likely that these skills have different neural representations.

In the past five years, an increasing number of fMRI studies have identified small brain regions within the IFG that respond to specific aspects of language. From the earliest language activation studies using PET, functional imaging studies consistently have demonstrated increased blood flow during tasks in which subjects did not make an overt verbal response (Gabrieli et al. 1998). In one such study using PET, we compared Broca's area activity in both silent and oral object naming and word reading (Bookheimer et al. 1995). Both silent and oral tasks produced Broca's area activity; indeed, in the case of reading, there was greater IFG activity for silent than for oral reading, a pattern also seen in posterior temporal cortex. This suggests that greater semantic processing in silent vs. oral reading produces greater IFG activity. As speaking produces head motion artifacts that contaminate fMRI pictures, covert verbal responses in fMRI studies of language are the norm. Wildgruber et al. (1996) used a covert speech paradigm to exam lateral differences in motor cortex during speech, finding ample activation in motor cortex even though subjects did not actually speak. Studies comparing overt and covert speech have found differences represented primarily in magnitude of fMRI activation rather than in location, with the exception of motor areas (Palmer et al. 2001). Several investigations of sentence comprehension wherein covert pronunciation was unlikely have also shown IFG activity. One possible explanation is that strongly connected brain regions (i.e., Wernicke's and Broca's areas) jointly activate each other during language processing even if the neural activity is not critical to the process. However, increasing evidence suggests that, in addition to neural activity that might occur as a result of functional connectivity, activity in the IFG in the absence of speech production reflects functional activity in those regions specific for other aspects of language processing (Gabrieli et al. 1998). Recent fMRI research has identified at least three separate regions of functional specialization within the IFG separate from those involved principally in motor speech. These are syntax (reflecting both the production and comprehension of syntactic information), semantics, and phonology.

In their classic PET experiment, Petersen et al. (1988) identified a region of activity in the anterior, inferior portion of the IFG that was selectively engaged when subjects generated a semantic association to a presented noun. While at the time this conclusion generated considerable controversy, numerous studies using a variety of imaging methods and paradigm designs have consistently replicated this finding and elaborated on the role of the anterior IFG in semantic processing. This region lies in the junction between the pars triangularis and pars orbitalis of the left IFG in what is likely Brodmann's area 47. It appears to represent a unique brain region involved not in decoding meaning of individual words but in processing semantic relationships between words or phrases, or in retrieving semantic information. Evidence supporting this general idea has emerged from experimental designs that differ substantially in their specific task demands, input modality, and type of stimuli employed. This literature is now too extensive to review completely, but examples of recent fMRI and PET studies of this brain region using several different theoretical and experimental approaches follow below.

Priming Effects in the IFG

In the cognitive psychology literature, semantic priming paradigms have proven effective in identifying benefits in performance when subjects respond to a stimulus that follows a semantically related prime. In functional imaging research, the effect of priming in the brain is demonstrated by a decrease in the amount of brain activation for a stimulus that has been primed either by repetition or by following a semantically related stimulus. Demb et al. (1995) demonstrated semantic priming effects—a decrease in blood flow for repeated words during a semantic decision task—in the IFG, encompassing Brodmann's areas 45, 46, and 47. They varied the depth by which subjects encoded a word list. Increased semantic encoding (making a concrete vs. abstract judgment) was compared with making a case judgment or alphabetic order judgment on the typed words. The latter tasks differed substantially from the experimental task in difficulty level with one more and one less difficult, which makes it possible to separate out difficulty effects from those most critical to the experimental task. Even though this study aimed principally at understanding the effects of deep vs. shallow encoding on memory, the results are relevant to language processing as well. Demb et al. (1995) found that deep semantic encoding produced increased fMRI activation in the lateral inferior prefrontal cortex; the exact location was not stated but was roughly centered around the anterior IFG. This finding was not due to increased task difficulty, as the more difficult, nonsemantic encoding task produced no significant blood flow changes in this region. Furthermore, when repeated items that were semantically encoded in a second task were presented, the same regions showed decreased activation. The authors have argued that this region may serve as a central executive for retrieving semantic information.

Wagner et al. (2000) used a similar repetition priming paradigm to differentiate priming effects in the anterior vs. posterior IFG. They varied task demands so that a list of words were processed at different levels of depth, i.e., either semantically or perceptually. In the perceptual or nonsemantic conditions, subjects determined whether letters in the target words were presented in upper- or lowercase. In semantic-processing conditions, they determined whether the words were concrete or abstract nouns. During the scans, subjects saw either novel or repeated words from the word lists and had to make either perceptual or semantic judgments. The stimuli were crossed such that items initially processed semantically could be repeated under either semantic or perceptual task instructions, and likewise for items initially encoded perceptually.

Although both the anterior and posterior portions of the IFG showed increased MRI activity during initial processing, Wagner et al. (2000) found an interaction between regions within the IFG and task-specific priming effects. In the posterior IFG, they found priming in both within-task and across-task repetitions, while in the anterior IFG, priming was specific for the task performed [i.e., items were primed only in the semantic task instructions when they had been initially encoded at a deep (semantic) level].

This study indicates that the priming effects in the IFG can not be explained as solely due to the identicality of the individual stimuli, but must also involve a semantic analysis of those stimuli, which strongly implicates a primary role of the IFG in semantic processing. In a separate study, Wagner et al. (1997) examined whether different stimulus content (words vs. pictures) produced priming in the same area within IFG. It was found that anterior IFG region showed reduced activation for items that had been presented previously, regardless of whether they were presented as pictures or words.

Buckner and colleagues (2000) examined repetition priming effects across a series of experiments varying in their task demands and sensory input modality. They had subjects perform a visual word stem–completion task in which they saw several letters and had to generate a complete word, for example, 'bas—' (basket). The same stems were shown over four repeated trials, and fMRI analysis identified those brain regions showing repetition priming effects, that is, reduced brain activation for repeated trials. They also had subjects generate a verb from a visually presented noun to identify those brain regions that were not task specific. In other words, if the priming effects are truly conceptual as opposed to perceptual in nature, both sets of task demands should reveal the same priming-related regional brain changes. Two regions in the IFG show repetition effects across tasks: one in superior, posterior IFG and another the anterior, inferior IFG. Buckner validated these results using auditory input in a word stem–completion paradigm. Here, subjects heard initial phonemes of possible words (e.g., pur) and had to generate a complete word (perfect). Comparing new word stems with those that were repeated, nearly identical regions in the IFG were found (see Table 1). The latter regions correspond well with the data of Demb et al. (1995) and others, whereas the former is more closely associated with phonological or articulatory processing (see below). Because the repetition priming effects were both item specific (unprimed items did not show priming) and were observed across task demands and input mode, the data provide support for the notion that these priming effects are conceptual in nature. Since repetition priming involves priming not only of the semantic content of the stimuli but also access to the motor plans for producing a response (either overtly or covertly), as well as the selection of a response (Thompson-Schill et al. 1997), the regions identified could reflect any of these components. Different experimental approaches, however, have tended to support the semantic retrieval hypothesis.

In the above studies, semantic information processing was deduced by observing a decrease in activation for repeated items. Other studies using paradigms not dependent on repetition priming have implicated the anterior IFG in semantic processing; these studies produce increases in functional activity. For instance, Poldrack et al. (1999) had subjects perform a semantic decision task and a phonological decision task in comparison with a perceptual control to differentiate areas within the IFG responsible for semantics and phonology, respectively. Task instructions involved either making a case judgment (perceptual task), counting syllables (phonological task), or judging whether the words were concrete or abstract (semantic task). This investigation revealed increased IFG activity in the anterior portion selectively during semantic processing, while other IFG regions showed less task specificity. Because the stimuli were matched across conditions, defining the critical variable in producing IFG increases must have been the semantic analysis required to make a decision. Gabrieli et al. (1998) note that the same region producing signal decreases in the priming tasks showed increases in the judgment task, which suggests that this region responds dynamically during semantic processing tasks.

Selective Attention to Meaning

Dapretto & Bookheimer (1999) modified a selective attention paradigm to differentiate syntactic and semantic aspects of sentence processing in a task in which

TABLE 1 Talairach coordinates of inferior frontal gyrus activations

First author	Task	Syntax	Semantics	Phonology
Moro 2001	Sentence judgment—acceptability	−28 34 8		
Friederici 2000	Function words	−45 12 6		
Dapretto 1999	Semantic vs. syntactic Sentence judgment	−40 30 14 −52 10 28	−46 30 −6	
Stromswold 1994	Plausibility of syntactically complex sentences	−46 10 4		
Caplan 1998	Replication of Stromswold	−42 18 24		
Kang 1999	Detecting syntactic anomalies—noun-verb combinations	−50 15 12 −45 25 4		
Wagner 2000	Repetition priming Semantic judgment		−43 34 12	
Wagner 2001	Low association words—Semantic comparison		−45 27 −12 −51 21 −3	
Petersen 1988	Word generation		−33 32 −6[a] −38 25 8[a]	
Buckner 2000	Word stem priming—visual Word stem priming—auditory		−43 34 3 −34 31 3	−43 9 34 −43 6 25
Wagner 2001	Semantic priming: Weak associations vs. > strong		−45 27 −12	
Thompson-Schill 1997	Classification—hi vs. low selection Generation—hi vs. low selection		−49 8 30 −38 15 30	
Thompson-Schill 1999	Generation—new vs. repeated		−44 15 22	
Muller 2001	Selection of a tone pattern		−42 27 −9	−48 0 33
Poldrack 1999	Phonological vs. case match Semantic vs. case match		−47 20 −3	−47 28 16
Demonet 1992	Phoneme monitoring vs. tones			−50 18 20
Demonet 1994	Sequential/ambiguous phoneme detection			−42 6 28
Zatorre 1996	Phonetic monitoring Phonetic discrimination			−44 8 27 −35 20 21
Burton 2000	Phonological segmentation			−47 15 29
Zatorre 1992[b]	Phonetic discrimination vs. speech Vs. pitch discrimination			−48 3 24 −56 6 29

[a]Transformed to Talairach & Tournoux 1988 coordinates.
[b]1996 reanalysis.

selection demands were minimal and held constant across conditions. In this study, subjects heard pairs of nearly identical sentences in which either a single word or the word order were varied, and they had to judge whether the meaning of the sentences remained the same or differed after this change. Although both tasks involve a semantic analysis of the sentences, additional semantic processing at the single-word level was required in the semantic condition in order to perform the task. Subjects were given the same set of instructions for both tasks and were not informed of the experimental manipulation, making it less likely that they used different strategies for performing the tasks. Both semantic and syntactic conditions activated a large network of regions in language areas principally in the left hemisphere. However, the semantic manipulation produced additional activity in anterior IFG, and the syntactic task produced a more posterior, superior area of selective MR activity. Because the selection demands across tasks were identical, as was task difficulty, the data are most consistent with the IFG's role in some aspect of semantic processing. Figure 1 (see color insert) illustrates the foci of the activations from Dapretto & Bookheimer (1999).

Controversies Over the Semantic Processing Hypothesis

While most have held that the anterior IFG is integral to some aspect of semantic processing, Thompson-Schill and colleagues (1997, 1999) have argued that the IFG performs the more general task of selective task-relevant stimulus attributes from amid a field of competing responses. To test this model, Thompson-Schill and colleagues (1997) performed an fMRI experiment in which subjects made a semantic decision in the face of smaller or larger competing demands. They presented a cue followed by two targets, and the subjects had to choose the one that was semantically related in a low-selection condition, vs. similar on a single-semantic dimension or feature in a high selection condition. Thompson-Schill et al. argue that the former task does not require selection because the comparisons made on the basis of global similarity do not require selection. If selection is the key variable influencing anterior IFG activity, then this condition should not elicit IFG activity, and they indeed found none. They also argue that a semantic account of IFG would predict more IFG activity when there are more semantic targets and that the region responds merely to the selection process. In a second experiment (Thompson-Schill et al. 1999), subjects performed a word generation task in which they generated a verb in response to a target noun. Words were primed with relevant or irrelevant information. If IFG depends on selection demands, the latter condition should show more activity in the primed condition relative to unprimed because the irrelevant information should increase the selection demands. They found significant priming in the same condition and a small increase in activity for primed vs. unprimed in the different condition, consistent with the selection hypothesis.

Intense debate on this issue continues, questioning whether semantic processing is a necessary component of anterior IFG activity. Wagner and colleagues (2000) have noted that the region identified by Thompson-Schill et al. (1999) as important to selection falls significantly posterior to that indicated during semantic priming by Wagner et al. (1997), Poldrack et al. (1999), and others. In a second study, Wagner et al. (2001) argued that the anterior IFG is involved in controlled semantic retrieval, a process that could be prone to interference by competing alternatives but that is not dependent on selection. They varied the associative strength and the number of targets in a semantic decision task, in which subjects had to determine whether a word was globally related to a target, given a choice of either two or four words. If the semantic retrieval hypothesis is correct, one could predict that the IFG activation is modulated by both associative strength and number of targets. In contrast, if the selection hypothesis is correct, one would expect that neither factor should increase IFG activity. Wagner et al. (2001) found that weak associations, which should require more strenuous controlled semantic retrieval, activated the anterior IFG most strongly. The posterior IFG was active across all conditions, consistent with the predictions of Thompson-Schill and colleagues (1997, 1999). This posterior IFG region overlapped with that found in studies of phonological processing (see Table 1). Overall, Wagner's data help explain the disparate findings across studies relevant to the selection hypothesis and provide further support for the role of anterior IFG in semantic processing: specifically controlled semantic retrieval. Table 1 shows the coordinates in Thompson-Schill's studies (1997, 1999) in comparison to those of other groups; this is depicted graphically in Figure 2 (see color insert). Whereas the other studies reviewed show a tight cluster of activation foci in anterior, inferior IFG, the areas Thompson-Schill identifies as representing selection processes are located substantially more posterior and superior, supporting the spatial dichotomy suggested by Wagner and colleagues (2001).

Dissociations Among IFG Regions: Syntax

Several lines of evidence indicate that the anterior IFG region active during semantic processing is clearly different in both spatial location and function from other areas in the IFG responsible for other aspects of language processing (Fiez 1997). The most commonly dissociated processes are syntax and phonology.

Although rare, there have been several reports of patients with selective impairments in syntax or in some aspects of syntactic production or comprehension (Berndt & Caramazza 1980). fMRI investigations of syntax have largely supported the notion of a specialized region in IFG for processing syntactic aspects of sentence comprehension. PET studies by both Caplan et al. (1998) and Just et al. (1996) found increased activity in the IFG during additional resource allocation to syntactic complexity, though neither study differentiated among subregions within the IFG. Stromswold (1994) compared different manipulations of syntactic structure in a PET task. They compared right-branching sentences (e.g., The child spilled the juice that stained the rug) to the more difficult center-embedded structures (The juice that the child spilled stained the rug), finding increased activity in Brodmann's area 44 for the more complex constructions.

Caplan et al. (1998) varied the syntactic complexity of sentences using PET and the same stimuli as Stromswold (1994). The focus of activity in the IFG was close but did not precisely replicate their prior results; it is more in line with other studies of syntax. In a second experiment, they varied the number of propositions in sentences ("The magician performed the student that included the joke" vs. "The magician performed the stunt and the joke"). In this experiment, differences were found only in temporal lobe regions, not in IFG. Caplan et al. (1998) argue that in the latter experiment, the increased memory load is associated with the products of sentence comprehension, whereas in the former experiment, the load is with the "determination of the sentence's meaning." If so, this suggests that the frontal lobe contribution to syntactic comprehension may depend on an interaction with syntactic and semantic processes toward the determination of meaning. Such a view is supported by several other imaging studies of syntax.

For instance, Dapretto & Bookheimer (1999) contrasted syntactic with semantic aspects of sentence processing in an auditory sentence judgment task (see above) and had subjects make a semantic judgment of the form of the sentence. Subjects had to determine whether two sentences conveyed the same meaning. The sentences contained identical words but differed in word order. In each case, the novel word order combinations were grammatically correct and plausible, so that subjects had to rely solely on a comparison of the effect of word order on meaning to generate a correct response. The actual sentences and the syntactic complexity were matched across syntactic and semantic conditions so that activations depended not on the stimuli themselves but on the process of comparing the syntactic forms. They found an area in the superior portion of BA 45 that showed enhanced activation for the syntactic condition alone. The results from this study are illustrated in Figure 1 (see color insert).

In the Dapretto and Bookheimer (1999) study, subjects focused their attention on the effect of syntactic structure on meaning; that is, there was by design a strong relationship between syntactic variations and emergent meaning. One may question whether the brain regions reflect the attention to syntax alone, or the integration of syntactic and semantic information. One way to distinguish among these alternatives is to present syntactic information in the absence of semantic information. Friederici et al. (2000) used "Jabberwocky" sentences in comparison to normal sentences, word strings containing only content words, and nonwords. Jabberwocky sentences contain function words in appropriate placements between nonsense words taking the place of meaningful referents such as nouns and verbs ("Twas brillig, and the slithy toves..."). Subjects made a decision based on both semantic and syntactic grounds (subjects determined if there was a legal syntactic structure, or if there was a content word present). Regardless of whether sentences contained real or nonwords, those with normal placement of function words (normal sentences and Jabberwocky sentences) showed temporal lobe activation (Wernicke's area), and Jabberwocky sentences produced additional activity in area 44. Their results suggest this area is specific not simply for syntax, but for increased selective attention to syntactic structure. In contrast, the temporal lobes, from Heschl's gyrus to the planum polare, show a relative increase for syntactic vs. nonsyntactic sentences types though all conditions produced significant activation. This suggests that the frontal lobe may play a more executive role in syntactic processing such as controlled retrieval of syntactic information, responding dynamically as the task demands require.

Furthering the notion that syntactic and semantic modules in the frontal lobe are mutually interactive, Keller et al. (2001) varied the cognitive load on syntactic and lexical processing. Using written sentences as stimuli, they varied the frequency of key words as well as the syntactic complexity of the sentences. Presumably, brain regions general to task difficulty should increase activity under both conditions, while those related solely to lexical semantics vs. syntactic processing should show task-specific regional activation. Because they used a region of interest approach and did not differentiate among regions within the IFG, and did not report results in Talairach coordinates, it is not possible to directly contrast their result with other studies dissociating among these processes. However, in the broadly defined IFG, they found an interaction between syntactic and semantic complexity, indicating increased engagement of IFG when both the syntactic and semantic demands of the task were greater. They argue that, even within a modularity framework, one must consider such modules to be highly interactive.

The above studies all used sentences as stimuli, presented either visually or auditorally. In several of these tasks, the more complex syntactic structures that produced IFG activation also were more difficult and involved greater working memory demands. Thus, the IFG activity associated with syntactic processing may reflect not the syntactic components of the task but rather working memory load or general difficulty, factors that may be independent from syntactic processing. Kang et al. (1999) used two-word, noun-verb combinations in the context of an event-related paradigm to differentiate brain regions associated with semantic vs. syntactic processes. Subjects saw these pairs printed and simply had to read them. Both semantic and syntactic anomalies were contained within the lists (e.g., "Grew heard" vs. "Ate suitcases") but most pairs were logical ("Wore glasses," "Broke rules"). Images associated with syntactic and semantic anomalies were averaged separately. Both stimulus types produced activity in the center of area 44, whereas syntactic anomalies produced additional activity in two other regions in the IFG (see Table 1). Whereas one of these regions is close to that most associated with semantic processing as described above, the second closely approximates regions found by other investigators for syntactic comprehension (see Table 1). The study did not require subjects to effortfully process the syntactic structure of the stimuli, and the anomalies occurred at low frequency. This suggests that at least some aspects of syntactic processing may be independent of executive functions including working memory, effortful attention, or retrieval of stimulus-specific information. However, the co-activation of anterior IFG suggests an interaction with processes relevant to semantic processing during the detection of syntactic anomalies.

Dissociations Among IFG Regions: Phonology

In addition to a general deficit in speech production, Broca's aphasia patients have specific problems in accessing, sequencing, and monitoring phonemes. Direct cortical stimulation of area 44 in patients undergoing surgical removal of the epileptic focus disrupts phoneme monitoring even when patients were not required to articulate (Ojemann & Mateer 1979). In an early PET study by Demonet et al. (1992) using PET, subjects performed a phoneme sequencing task in which they had to attend to the order of phonemes within nonwords in comparison to a tone control task. A second task asked subjects to monitor words for content (determine whether an adjective-noun combination was positive or negative in connotation). The area of activity selective for phoneme (vs. tone) comprehension lies in the superior portion of BA 44/45, while the word meaning task produced activity predominantly within the temporal lobes. A second study by Demonet et al. (1994) investigated the relative effects of perceptual ambiguity and sequence processing in a set of phoneme monitoring tasks. The tasks were designed to differentiate between two models of frontal lobe involvement in phonology: One holds that the IFG activity reflects rehearsal in working memory, corresponding to the articulatory loop proposed by Baddely (1992), while the other focuses on phonological processes directly, which may include phoneme sequencing or discrimination. Using nonwords as stimuli, they asked subjects to detect the presence of a letter occurring either at the beginning or within a word, either in isolation or if preceded by a second letter. Ambiguous stimuli were embedded in a group of consonants. They found increased blood flow in posterior IFG only in the condition in which subjects made the sequential judgment for embedded (ambiguous) phonemes. Simpler detection tasks produced temporal lobe activation only, as did sequencing tasks alone. They suggest that this part of the IFG likely performs sensorimotor encoding of auditory phonetic input, consistent with the rehearsal account.

Also using PET, Zatorre et al. (1992) identified a region in the posterior IFG selectively engaged when subjects made a phonological vs. a pitch discrimination on auditorally presented syllables. More recently, Zatorre et al. (1996) added an additional pair of tasks to differentiate frontal regions that may be involved primarily with working memory from those involved solely in phonetic analysis. Phonetic discrimination—in this case, determining whether real words ended in the same letter—produced activity in BA 44/45. Phoneme monitoring (detecting a target letter within a word) produced activity in posterior IFG. In comparison, lexical judgment did not show increases in this region, suggesting that the area 44 findings are specific to the phonetic processing demands of the task.

Other PET investigators reported similar results in Brodmann's area 44 on phonological tasks, including Paulesu et al. (1993) in a task of phonological recoding and rehearsal. Although it supports the critical role of the IFG in phonological processing, the PET technology may lack sufficient spatial resolution to test the hypothesis of a unique center for phonological processing in the IFG. Other fMRI studies have demonstrated increased IFG during phonological processing, but with a region of interest approach that could not differentiate posterior IFG from the central IFG regions seen in most language tasks (e.g., Pugh et al. 1996). Poeppel (1996) argued on the basis of earlier PET work that the results in Broca's area for phonological tasks did not replicate well. However, more recent work with higher resolution PET (Zatorre et al. 1996) and with fMRI has generally supported the role of posterior IFG in phonological processing.

Several paradigms sensitive to phonological processing but using approaches differing substantially from the earlier works by Zatorre (1992), Demonet et al. (1992), and others have demonstrated posterior IFG activity. For instance, Friederici and colleagues (2000) report increased posterior IFG activity in sentence processing tasks that involved either Jabberwocky sentences, sentences composed of content but no function words, and pseudoword lists. The left hemisphere junction of the inferior frontal sulcus and inferior precentral sulcus showed increased activity for all conditions compared to normal sentences. This result could merely represent a general area for increased attention to any verbal stimuli but may reflect an increased need for attention to the word sequences or to articulate them covertly, which is consistent with the view that this area forms a part of the articulatory loop in a working memory circuit (Baddely 1992).

Although activations around posterior IFG appear frequently across laboratories and phonological processing tasks, some studies show activation in other portions of the IFG. In Zatorre et al.'s (1996) very similar tasks of phoneme discrimination and phoneme identification, for instance, the focus of activity in the former task was in the center of the IFG, whereas the centroid for discrimination was found in posterior IFG. Activation during phonological processing could interact with other task demands in the language system, producing new or overlapping activations in IFG or other brain regions. In the priming literature, for instance, lexical decision tasks can produce both semantic and phonological priming (Neely et al. 1989, Berent et al. 2001), although the priming effects may be independent (Cronk 2001). Poldrack et al. (1999) compared semantic with phonological processing by requiring subjects to count syllables in real words or pseudowords presented visually, and also to judge the case of the same words. They found IFG activity during the phonological tasks, most notably in the posterior, superior IFG during the pseudoword syllable counting condition. However, the same region was significantly more active in the semantic task. Because all tasks involved reading printed words, it is likely that phonological processing took place in all conditions, which would tend to minimize differences in direct task-task comparisons. Nonetheless, the results suggest that semantic processing automatically engages regions within the IFG responsible for phonological processing, at least during reading.

Activation of Left Inferior Gyrus in Nonlanguage Tasks

The most widely accepted theories posit separate modules for semantic, syntactic, and phonological processing, but how we interpret the basic functions of these regions is still a matter of considerable debate. In particular, arguments have centered

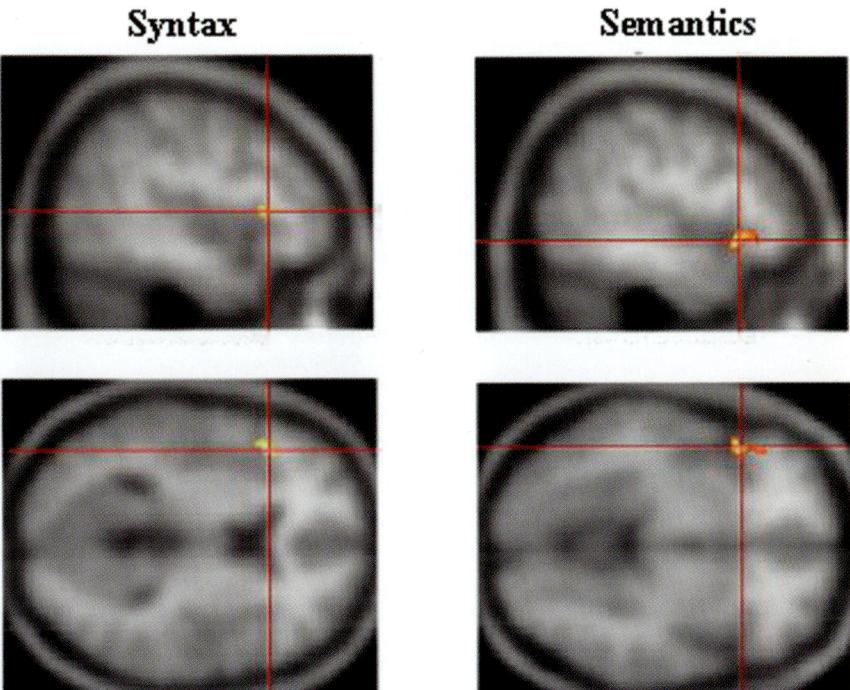

Figure 1 Activation in the inferior frontal gyrus (IFG) for semantic and syntactic aspects of sentence processing, from Dapretto & Bookheimer (1999). For both tasks, subjects determined whether two sentences conveyed the same meaning. In the semantic condition, a single word was changed; in half of the trials, the word was a synonym and in half the word was unrelated (e.g., "The boy went to the store." "The boy went to the market" vs. "The boy went to the school"). In the syntactic condition, the words remained constant, but the word order was changed to make a syntactically plausible sentence. In half the cases, the meaning remained the same and in half the meaning differed ("The city is west of the lake." "The lake is west of the city" vs. "West of the city is the lake"). Syntactic complexity was matched across conditions. Results here show direct comparisons of semantics vs. syntax (revealing the anterior, inferior IFG–pars orbitalis) and syntax vs. semantics (showing the middle IFG area 45–pars triangularis).

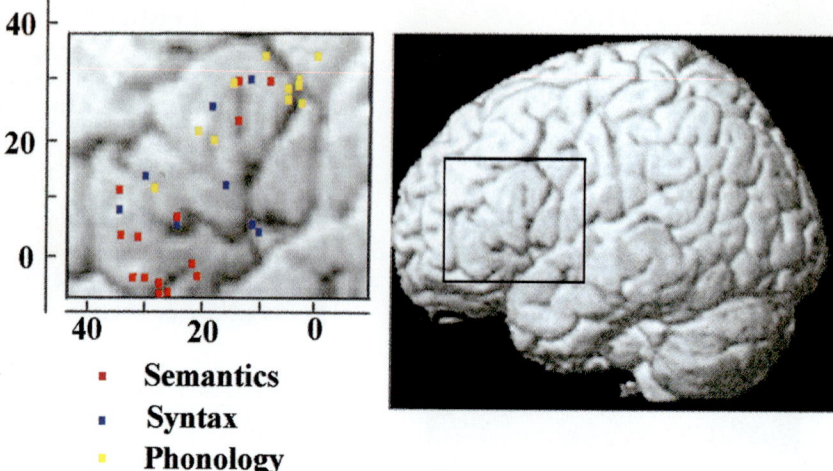

Figure 2 Summary of inferior frontal gyrus (IFG) activations across studies. Table 1 shows the list of studies and centers of activation for semantic, syntactic, and phonological processing experiments showing significant activity in the left IFG. Semantic areas (shown in *red*) cluster around the anterior, inferior IFG (pars orbitalis); phonological regions center around the posterior superior IFG at the border of Brodmann's areas 44 and 6; syntax regions fall in the center near middle IFG in pars triangularis, area 44/45.

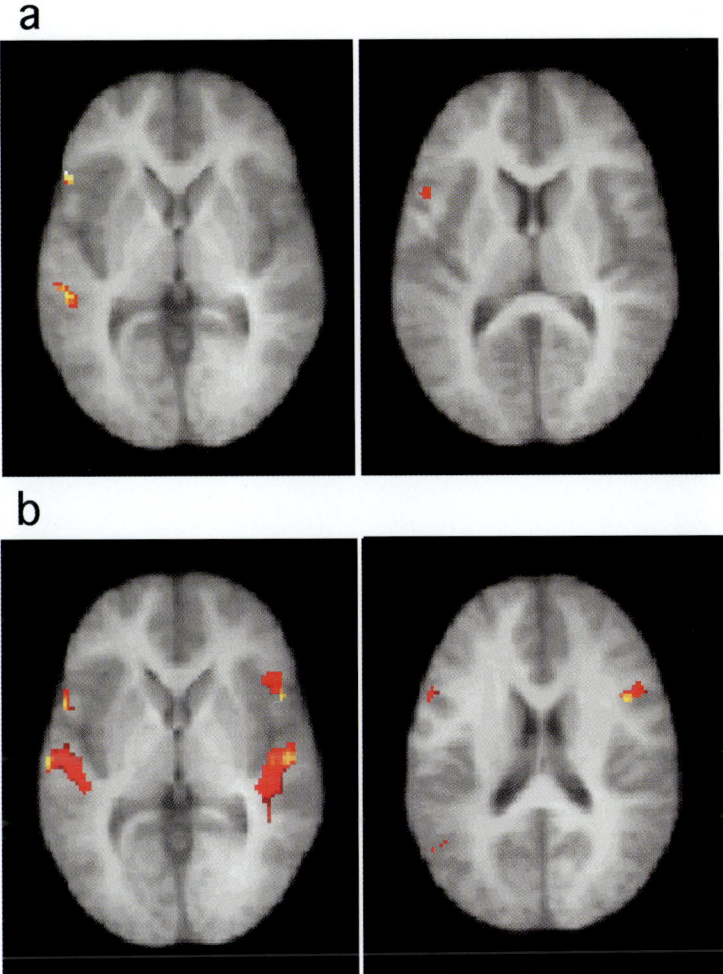

Figure 3 FMRI activation during logical reasoning and topic maintenance from Caplan & Dapretto (2001). (*a*) shows areas of increased MR signal for logic compared with topic maintenance; (*b*) shows the reverse comparison.

on how specific these subdivisions are to language processing vs. whether some more general processes may underlie language as well as nonlinguistic functions. Typically, in studying the properties of traditional language cortices, investigators employ tasks involving some component of language exclusively. A few studies have reported increased fMRI activation in Broca's Area for tasks that do not contain a language component. Such studies offer a unique perspective on the underlying properties of neurons in this region that give rise to language, though not necessarily exclusively, and cast doubt on the concept of the language-specific processing module.

Muller et al. (2001) examined the role of Broca's area in making tone discriminations. Specifically, subjects had to attend to 600 Hz tones that either remained stable, rose, or fell by 50% in a very brief duration (200 msec). Subjects listened to these tones and pressed a button whenever they heard the rising tone. In the comparison condition, subjects heard bursts of white noise and pressed a button whenever they heard the sound. Thus, the conditions differed in several ways; the quality of the sound, the duration over which the sound occurred, the nature of the task including monitoring a rapid temporal change, and selecting from among competing alternatives. Consequently, any of these features could have produced the results, although in no case could language processing account for any task-related activity. In the inferior frontal lobe, they found significant activation in the anterior, inferior portion of the IFG (areas 45/47), corresponding closely to the anterior IFG region demonstrated by Poldrack et al. 1999, Wagner et al. 2000, and others. These data may argue in support of the Thompson-Schill et al. (1997) model of anterior IFG function, which reflects selection from among competing responses as opposed to semantic processing. Muller et al. 2001 also reported a very large area of activity in the superior posterior portion of the IFG, a region more closely associated with phonological processing and verbal working memory.

Iacoboni and colleagues (1999) have performed a series of experiments on motor imitation that have produced increased fMRI responses in traditional Broca's area. Imitation is thought to represent a core, prerequisite skill for developing language; autistic children have notoriously poor imitative skills, but their ability to imitate oromotor movements appears to predict language acquisition. In this task subjects were asked to perform a simple finger movement either in response to a visual cue or an imitation of the same response; all other task parameters were held constant, including visual input. They found two brain regions that responded to imitated movements, specifically, the right parietal lobe and pars opercularis of the left IFG. This research argues for a neural basis of imitation that corresponds to similar neural responses seen in nonhuman primates. In addition, they suggest that motor imitation may underlie aspects of language acquisition. Binkofski et al. (2000) performed a detailed cytoarchitectonic analysis in combination with fMRI on a similar task in which imagery of motor movements was compared with imagery of observed hand movements. They found left hemisphere activation in area 44 when subjects imagined observing a movement, and right opercular activation when subjects imagined a moving target (a light point moving on a screen). Studying the activity in these areas in spatially normalized histological sections, they found that these regions corresponded to cytoarchitectonic area 44; this area was clearly differentiated from nearby area 6 involved in motor planning. Thus, the authors conclude that area 44 of the left hemisphere plays a key role in the imagery of motion, which is thought to represent an efferent copy of a planned motor action, supporting the notion that human area 44 corresponds to the mirror neurons described in F5 of the monkey (Rizzolatti & Arbib 1998). Together, the data suggest that Broca's area retains function that is not directly related to language processing, but that neurons here have response characteristics that may give rise to imitation of complex motor behaviors including language. However, in a recent meta-analysis of studies involving imagery and observation of motor movements in comparison to speech, Grezes & Decety (2001) argue that this portion of Broca's area is only rarely activated during motor imagery; more commonly, activity is associated with silent speech processing. Indeed, that Grezes & Decety (2001) found activation in this area during observation of movement and not during observation made in order to imitate tends to contradict this hypothesis.

Summary of the IFG Hypothesis

In summary, recent fMRI research strongly supports the notion that there are separate subsystems within the IFG responsible for different aspects of language processing. In particular, there is strong support for the role of anterior IFG (Brodmann's area 47) in some aspects of semantic processing. By most accounts, the region is not modality specific, nor is there any evidence for specificity of content. Rather, this region appears important for executive aspects of semantic processing that involve semantic working memory, directing semantic search, or drawing comparisons between semantic concepts in working memory. While the argument that this region plays a general role in selection from among alternatives has not held up well, it remains quite possible that this region performs a more general function. It perhaps involves making comparisons or judgments among information held in working memory that underlies this aspect of semantic processing as well as other nonlinguistic processes. Other regions within the IFG appear to be specialized for syntactic and phonological processing as well. Figure 2 (see color insert) presents a summary of the studies reported here that directly contrast one of these processes with another (a more complete list of studies comparing phonological and semantic processing regions can be found in Poldrack et al. 2001). Of note, many of the regions shown to have greater activation for one process still demonstrate significant increases in the comparison tasks. In some studies (e.g., Keller et al. 2001), interaction effects between these distinctive processes suggest that regions within the IFG are highly connected, influencing both brain activity and processing efficiently for other language skills. Together, such a network of unique but highly interactive, compact modules should give rise to the tremendously complex language processing of which humans are capable. Perhaps a tight spatial and functional coupling of these small modules enables rapid, efficient processing of language, but such an arrangement may also make us vulnerable to the catastrophic loss of broad-based language skills that comprise Broca's aphasia. Thus, we may characterize Broca's aphasia not as resulting from a lesion in a single Broca's area, but rather as a lesion affecting a family of Broca's areas, each contributing uniquely to one deficit but creating a whole syndrome that is far greater than the sum of its parts.

CONTENT-SPECIFIC ORGANIZATION OF SEMANTICS

Reports of patients with category-specific deficits in naming (Hart & Gordon 1992, Warrington & Shallice 1984), in addition to a large literature on semantic priming in both normal control subjects and patients with aphasia (Blumstein et al. 1982), suggest that the brain stores semantic information along categorical lines. While the frontal lobe semantic region appears to be both modality- and content-independent, there is increasing evidence to suggest that, in the temporal lobe, semantic content is highly organized and spatially segregated.

Living vs. Nonliving Categories

The most common categorical dissociation in the lesion literature separates living from nonliving entities. Several reports using functional imaging provide anatomical confirmation of this dissociation in the brain. The most common finding is increased occipital activation for animals compared with tools (Martin et al. 1996, Moore & Price 1999, Mummery et al. 1998, Perani et al. 1999). For instance, Perani et al. (1999) had subjects make a same-different judgment of objects that were either animals or tools while undergoing PET. In one experiment, they compared activation for pictures of objects. For direct task-task comparisons, animals showed specific activation in the fusiform and occipital gyrus, areas associated generally with object processing. Tools, in contrast, showed specific activation in the left hemisphere near principal language areas, namely BA 45/46 and 21/20. In a second experiment, subjects performed the same task in response to printed words. Areas replicated across experiments were the left fusiform gyrus for animals and the left middle temporal lobe (area 21) for tools.

A critical question in understanding the basis for category-specific semantic organization addresses the basis upon which these dissociations are made. One could, for instance, argue that they are generated by visual similarity (for instance, natural objects tend to have no straight lines or sharp edges, whereas manmade objects do). Several investigators have probed this issue. Mummery and colleagues (1998) contrasted living things and artifacts (tools) in two tasks: In one case they determined if the objects depicted by two words shared a perceptual characteristic (the same color); in another, they judged whether two words were associated (usually found in the same locations). A control task had subjects count the syllables in each word. Relatively large differences were found in contrasting the operation performed on the stimuli (location vs. color), whereas smaller activations represented category-specific stimuli. In particular, no significant effects for living things were found that spanned task demands. This suggests that the retrieval task performed in accessing categorical information is at least as important as the category itself in defining focal brain activation.

Chao et al. (1999) tested this hypothesis by comparing categorical stimuli (animals, tools, houses, and faces) when presented as either pictures or words. Across tasks, several regions showed specific activation for animals and tools under different task conditions (viewing pictures, naming objects, matching pictures, or reading the names of objects). Reading produced the least replicable results, whereas other conditions showed relative consistency across tasks. Animals tended to show increased activity in bilateral medial and inferior occipital cortex, lateral fusiform gyrus, and superior temporal sulcus (near Wernicke's area). Tools appeared to activate the medial fusiform and the middle temporal gyrus in a region most often associated with visual motion perception. Both the lateral fusiform and superior temporal sulcus regions were also significantly activated for faces, which suggests a more general response to living things, whereas the medial fusiform showed a preference for nonliving things (though animals showed smaller levels of activation here as well). Within these categories, however, there were finer distinctions in the precise location of activation peaks that suggested a more refined separation of specific categories of objects. Similar dissociations have been demonstrated using these and other categories of objects (Moore & Price 1999, Okada et al. 2000, Smith et al. 2001, Spitzer et al. 1995).

Do category-specific findings in imaging necessarily imply unique modules for each entity? Ishai et al. (1999), using fMRI, found specific activations for several categories of objects, including chairs and houses. Within subjects, activation of separate regions for each category was very reliable. However, when examining the hemodynamic response across all stimulus types, they noted increased brain activity for other categories as well. Their model argues that, rather than representing distinct, independent modules for a single category, the neural representations of categorical information overlap, forming a continuous representation of information across the cortex, potentially represented as attributes or shared features.

A critical question in interpreting category-specific effects concerns whether such dissociations reflect organization based on critical features of those objects or categories (such as visual features, associations with visual motion, etc.) or semantic knowledge about those objects or categories (for a complete review of theoretical accounts of category specificity, see Caramazza et al. 1990). Likely, object knowledge reflects an amalgam of information widely distributed in the brain and includes feature information, as well as associative information. However, there is increasing evidence that, at least in the temporal lobe, the organization of semantic categories is not random, but may tie directly to critical features of the categories, for instance, as tools are associated with reaching and nonbiological movement (Martin et al. 1996). For instance, Chao & Martin (2000) had subjects view or name tools in comparison to animals, faces, and houses, imaging regions in the frontal and parietal lobes associated with reaching and grasping in monkeys (e.g., Rizzolatti & Arbib 1998). Both of these areas (intraparietal sulcus and ventral premotor cortex) show specific activation for tools compared with all other categories, suggesting that one aspect of object knowledge includes associations with sensorimotor correlates of their use.

Using PET, Moore & Price (1999) addressed the question of whether category-specific effects were due to features of physical similarity between objects. This was done in two ways: First, they compared nonobjects (realistic-looking drawings that were not actual objects) to real objects; second, they varied the amount of available perceptual information by comparing black-and-white vs. colored objects as well as object complexity. This comprehensive study examined two categories of natural objects (animals and fruit) and two categories of manmade objects (vehicles and tools). Further, they used two separate tasks (naming and word-picture matching) to evaluate the effect of task demands on category-specific organization.

Right hemisphere occipital and temporal occipital/fusiform cortex show a preference for more visually complex objects in all categories. As in other studies, the left posterior temporal region showed increased activation for tools, but this was also seen for nonobjects. The authors suggest that, at least in the right occipital and fusiform cortices, reports in the literature implicating these areas in category-specific representations may have measured only an aspect of visual complexity (objects with multiple components like animals and complex tools).

The left posterior temporal region appeared to show category specificity for manmade vs. natural objects, but the finding that nonobjects activate the same area casts doubt on whether the region can be regarded as semantically based. The left anterior temporal cortex specificity for natural objects was found only for black-and-white objects, although the authors argue this is consistent with the semantic category-specific hypotheses. Overall, the data are consistent with the notion that natural objects are organized on the basis of shared perceptual information, while manmade objects are organized along the lines of functional information.

The authors also report, but do not interpret, one result that may be inconsistent with a clear dichotomy in these categories. They found increased activity in the posterior sensory cortex at the junction of the intraparietal cortex for fruit and for simple tools. This brain region is strongly associated with reaching for objects (see Binkofski et al. 1998, Johnson et al. 1996). Both fruit and simple tools are easily graspable objects, which might have caused co-activation here. One would not necessarily argue for the presence of a new category, reachable objects, nor do the data support a single structural/functional dichotomy for natural vs. manmade objects, respectively. Rather, both recognized objects may automatically activate shared associations with similar motor movements as well as having other distinctive features (perceptual and functional) on which they differ. This interpretation suggests that there is no single, unifying feature that allows for a dissociation between these categories, nor for a single anatomical locus for category-specific representations. Rather, categorical differences may be expressed in multiple, nonoverlapping areas in the brain. Such a model is consistent with a diffusely organized semantic system based on multiple features, both perceptual and functional and likely involving other bases of similarity as well.

Body Parts

Ample clinical data implicate the parietal lobes in spatial representation, including representations of one's own body and manipulation of numbers. Le Clec'H et al. (2000) used a combination of blocked and event-related designs to identify brain regions active for these categories that were independent of sensory input modality. In the numerals condition, they asked subjects to compare a target number to a specified amount (larger, smaller, or equal); in the body parts condition, they asked subjects whether a target body part was higher, lower, or at the same level as the shoulders. For both conditions, stimuli were presented either verbally or visually. Although each task produced activation in numerous brain regions, those that responded to both input modalities were more specific: the right inferior parietal region for numbers, the left parietal lobe and left premotor cortex for body parts. The latter network is essentially the same as that identified in monkeys for reaching and maintaining plans for motor actions (Iacoboni et al. 1999). It is also close to that identified by Martin et al. (1996) for tools. This provides further evidence that categorical organization of semantic information may have emerged from nonlinguistic representations of sensory and motor behavior.

Word-Specific Effects: Concreteness

Several fMRI studies have investigated the cortical representations of different categories of words, including function (e.g., nouns vs. verbs), regularity (for instance, in past-tense production), and concreteness among nouns. By some theoretical models, concrete nouns can be represented both visually and auditorally, whereas abstract nouns can be represented only auditorally. Early accounts by Paivio (1986) and others suggested a left-hemisphere verbal–right hemisphere visual distinction would predict lateralized effects for both word types, though neuroanatomical data on word processing do not support a laterality model (Price et al. 1994, Bookheimer et al. 1995). Keihl et al. (1999) tested this model directly by using fMRI while subjects performed a lexical decision task; in the task nonwords were interspersed with concrete or abstract printed words. Both concrete and abstract words activated an extensive network of right and left hemisphere regions; in direct comparisons, however, only abstract words produced additional activation in the right temporal lobe, in a region corresponding to tip of the superior temporal gyrus. This region is anterior to primary auditory cortex and was the only region showing significant activity in direct task-task comparisons. One reason for the small differences between conditions may have been the choice of a lexical decision task; because this task involves no substantial semantic processing, differences between word types may have been missed. Mellet et al. (1998) had subjects listen to concrete vs. abstract nouns and, in the case of concrete nouns, generate a mental image. As in the Keihl study (1999), this task showed increased right temporal lobe activity for abstract words, but also showed significant activity for concrete words throughout the inferior temporal/fusiform region, primarily in the right temporal lobe. These regions are the same as those involved in object recognition and naming (Bookheimer et al. 1995), and apparently reflect the visual imagery engaged by subjects. In addition, left hemisphere activation during concrete as opposed to abstract words may reflect the extent to which the words and their associations are verbalizable. No coherent models for right hemisphere specialization for abstract words have been brought forth, but such words may demand context for their comprehension. A right hemisphere role for linguistic context has been reported in the literature and is discussed at length in the next section. However, an event-related fMRI study by Jessen and colleagues (2000) did not replicate the concrete/abstract dissociation reported above. In their task, subjects read words and were told to remember them for later use, but did not have to judge their concreteness or perform any other task. In this case, concrete words produced increased activity in the left superior frontal sulcus and in the left inferior parietal cortex, while abstract words produced increased activity relative to concrete in the right lateral occipital lobe and left IFG. Note that no brain areas overlapped across studies.

Summary

While there are some inconsistencies in the literature regarding the loci of each of the categories of stimuli discussed, overall there appears to be agreement that, particularly within the individual, there are separate peak foci of activity in several brain regions for categories of objects and concepts. Regions of focal activity reflect several different dimensions of object naming and knowledge: visual features, associations with object uses, and associations with semantically related objects. Within each of these broad categories may lie many associated features, thus making complete object representations broadly distributed in the brain. How to conceptualize organization by content remains an extremely contentious topic in the literature (see Thompson-Schill et al. 1999 and Caramazza 2000 for a very lively discussion of this issue). One general principle, however, appears to emerge consistently regardless of which category is under study: Whether the task that subjects perform requires naming, generating an associate, or matching stimuli on perceptual or conceptual grounds, the brain regions identified do not appear randomly distributed, but rather spatially proximal to brain regions with strong sensory or motor associations with the conceptual category. Objects that are manipulable activate brain regions associated with reaching and grasping; objects that move show activation close to visual motion centers, and objects that must be discriminated from many exemplars of similar objects (like faces) activate visual form recognition areas. Martin & Chao (2001) argue for a gradient of detail in which posterior brain regions process information more generically, whereas increasing uniqueness (a specific face or unique object like the White House) is processed in anterior temporal lobe regions and suggest that anterior temporal lobe integrates some aspects of visual information (see also Damasio et al. 1996). The available functional neuroimaging data do not clearly support one model of the functional architecture of the semantic system, but the technique holds promise for adding the data necessary for generating more plausible and comprehensive models than are available today.

Assuming that different aspects of sensory, conceptual, and associative semantic information have separate and diffuse organization in the brain, how do we then integrate such knowledge in the service of language? Martin & Chao (2001) argue that a good candidate model for this integration could be subserved by the left anterior IFG as discussed in detail above. This region is likely involved in the executive control of semantic information processing, including retrieving, integrating, comparing, and possibly selecting the diverse pieces of semantic information in the brain. Currently there is no clear evidence supporting or contradicting this model. Direct tests of this model could utilize functional connectivity approaches, which measure the correlated activity during brain activation across different brain regions (see Horwitz et al. 1999 for an excellent description of functional connectivity methods).

RIGHT HEMISPHERE CONTRIBUTIONS TO LANGUAGE COMPREHENSION

Language research, from the perspectives of both lesion-deficit and activation imaging approaches, has focused primarily on small units: speech sounds and words. Much of communication, however, relies on processing language at a level that goes well beyond information expressed by single words or in literal interpretations of words and sentences. Recent fMRI research has explored contributions to understanding language that emphasize metaphor, connotative meaning, prosody, and processes relevant to comprehending language at a level above that of literal meaning of words of sentences. Increasingly, this line of research has revealed a critical role for the right hemisphere in language that has received only a little attention in the lesion literature.

Figurative Language and Metaphor

The first major contribution in this field was made using (PET) by Bottini et al. (1994), who explored the role of the right hemisphere in processing metaphors. Metaphors depart from semantic rules in that one may be able to decode the sounds and apply meaning to each word (and correctly parse them syntactically) and yet not comprehend the meaning of the phrase or sentence. The correct meaning is implied through association and comparison of similarities between different experiences that are not stated explicitly; indeed, relying on interpretation of literal meaning destroys the true, connotative meaning of the metaphor. Our ability to reject literal meaning and accept an unspoken, connotative meaning requires both a traditional linguistic analysis and a contextual analysis. To differentiate these processes, Bottini et al. had subjects listen to sentences and to random word strings, making either a plausibility judgment or a lexical decision (in the control task). Metaphoric, plausible sentences consisted of metaphors that either made sense (e.g., The investors were squirrels collecting nuts) or were implausible (The investors were trams). Plausibility, then, was determined by whether the metaphor was meaningful though not necessarily familiar. Literal sentences were also judged plausible or implausible, but this was based on the literal meaning and logical relations between specific words or phrases (e.g., "The boy used stones as paperweights" vs. "Tim used feathers as paperweights"). In both cases, a correct answer requires accurate knowledge about the words and concepts, but only the former required an understanding of connotative meaning and rejection of literal meaning.

The comparison of interest—making plausibility judgments of metaphoric vs. literal sentences—revealed a striking dissociation between right and left hemisphere activation, respectively. Metaphoric sentence judgments produced relatively greater activation in several areas in the right hemisphere, most notably in right IFG, right pre-motor cortex, and right posterior temporal cortex. The latter regions are roughly right hemisphere analogs of Broca's and Wernicke's areas.

Several very different processes may take place in making judgments about metaphors; the authors suggest that reference to long-term, episodic memories may be specifically required in judging metaphors, whereas no such reference is required in judging literal meaning. Frontal lobe activation may reflect the additional need to search long-term memories for information on which to compare meanings, or the activation may be involved in generating visual imagery to facilitate decision making. In general, though, metaphoric analysis may require emphasis on holistic aspects of language processing, whereas judging literal meaning requires only the sequential, one-to-one mapping of words with known literal meanings and not necessarily knowledge of the context.

An unfortunate problem with this study is that the metaphors were far more difficult for subjects, who performed significantly less accurately than in the literal meaning conditions. Unmatched level difficulty can produce changes in the magnitude, spatial extent, and location of brain activity in ways that may not relate closely to the process under study, and it is possible that the results reported by Bottini et al. (1994) reflect this alone.

Using PET, Nichelli et al. (1995) examined how subjects derived the figurative meaning of passages by asking them to monitor the moral of a story, in comparison with a semantic or syntactic detail of the story. Subjects listened to Aesop's fables and made a judgment about either the figurative or literal meaning of the fable. Although left hemisphere language regions were engaged in all cases, making a judgment regarding the moral of the story produced relative activity increases in right frontal and temporal regions. Because subjects always heard the same or similar stimuli and made a judgment in all cases, right hemisphere activity must have been due specifically to the process of drawing a figurative inference from the passage. Several underlying processes could be involved, however: These include making an inference from limited or less apparent information, or attending to the context of the passage as opposed to individual elements of a passage. Several other studies have examined linguistic context and reasoning independently; these are discussed below.

Linguistic Context

Probably the most widely reported single paradigm in the language-imaging literature is the word or verb generation paradigm (e.g., Petersen et al. 1988, Cuenod et al. 1995, McCarthy et al. 1993, Rueckert et al. 1994). This paradigm has subjects generate a word or a list of words in response to a category cue or to a single-word cue (e.g., dog-bark), and has demonstrated activation in left frontal cortex consistently. Kircher et al. (2001) modified this basic design to have subjects generate a single word in response to a sentence cue, e.g., "These days the weather is rather—," such that subjects use the context of the whole sentence to generate a response. In comparison to control tasks in which subjects either chose one of two displayed responses or simply read a complete sentence, this generation task produced unique activity in the right temporal lobe, in cortex roughly homologous to Wernicke's area. In contrast, areas of the brain associated with production (typically the MFG in a generation task) were limited to the left hemisphere. The data suggest that subjects use the left hemisphere language system to access a verbal associate but used a right hemisphere system to analyze the spoken information. Since the right temporal region has not been reported in single-word generation studies, the RTL activation likely represents processing of the content of the sentence as a whole.

In the absence of a task requirement to actively synthesize contextual information or produce a response, increased right hemisphere activation appears limited to the temporal lobe. St. George et al. (1999) imaged subjects listening to a verbal passage with or without a title that oriented the reader to the context. In this case, the passage consisted of a set of instructions and tips on riding a horse. With the title provided, the paragraph makes perfect sense; without the title the paragraph is nearly incomprehensible. In this study, participants read paragraphs with and without titles, but made no decision or response. Titled paragraphs showed activity in standard language and reading areas: inferior frontal cortex, basal temporal region; and posterior temporal cortex, with a strong left lateralization. In contrast, reading untitled paragraphs showed overall greater activity in the right hemisphere, and was significantly greater than titled passages in the right temporal lobe. Since the actual content heard was identical across conditions (counterbalanced across subjects), right temporal lobe activation was likely due to the increased effort in attempting to place seemingly unrelated sentences in context to create a coherent presentation of the text. Thus, it is not simply sentence length comprehension that drives the right temporal lobe activation, but presumably the process involved in trying to make a passage coherent.

Is the critical variable in elucidating the RH contribution to semantic processing in sentences the duration over which the semantic information occurs? If so, shorter phrases should not elicit RH activity. Kang et al. (1999) looked at simple noun-verb combinations in an event-related fMRI design in which subjects detected anomalies based on either semantics (ate-suitcases vs. broke-rules) or syntax (wore-glasses vs. grew-heard). In this case, the temporal durations were brief and the context minimal. While both conditions showed left hemisphere activity in traditional BA (44), the semantic task showed additional activation in anterior, inferior pars triangularis of the right hemisphere. The data suggest that reliance on semantic context, rather than temporal duration, underlie the right hemisphere contributions to language comprehension.

Reasoning and Logic

Several studies demonstrate that deriving meaning from text relies on right hemisphere structures, but drawing logical inferences from text is an exception. Goel et al. (2000) tested the hypothesis that deductive reasoning relies on syntactic aspects of language—that knowledge of the structural properties of words and phrases underlies logical deduction—making it an inherently linguistic process. Using event-related fMRI, they compared congruent and incongruent phrases with and without semantic content (e.g., All pets are poodles; all poodles are vicious; thus, all pets are vicious) vs. the similarly framed arguments without semantic content (e.g., All B are C). They compared these phrases with matched sets in which all sentences were unrelated, or in which only the first two were related; subjects had to judge the validity of the conclusions. Both types of stimuli engaged primarily left hemisphere regions associated with language processing (BA 44/45 and 21/22) though the stimuli with less semantic content also engaged bilateral parietal and premotor cortex (BA6). However, in right hemisphere analogs of language regions, no conditions showed preferential activation. While this paper argues for parallel systems in drawing logical conclusions depending on whether subjects relied on content vs. spatial processing, it also provides a contrast to other studies of text comprehension that show primarily right hemisphere activation. In the Goel et al. (2000) task, logical inference relied exclusively on the serial position of individual statements. In contrast, in studies by Bottini et al. (1994), St. George et al. (1999), and others, subjects could derive meaning from passages as a whole, regardless of the strict serial order of sentences or elements within sentences. This suggests that inference per se does not determine lateralization but may depend instead on whether the context as a whole must be preserved to reach a conclusion.

In a recent study, Caplan & Dapretto (2001) examined different aspects of reasoning separately to evaluate right vs. left hemisphere contributions to logical inference. They predicted that whereas the left hemisphere was important in assessing logic in discourse, the right hemisphere participated in the implicit assessment of topic maintenance. To test this, they had subjects make judgments about whether pairs of sentences made sense. In each case, the sentences were presented as questions with responses. "Do you believe in angels?" "Yes, I have my own special angel" (on topic response); "Yeah, I like to go to camp" (off topic response).

To test logic they asked: "Do you like having fun?" "Yes because it makes me happy" (logical); "No, because it makes me happy" (illogical). The subjects thus heard the same types of stimuli and in every case made a judgment response, and they were unaware of the experimental manipulation. In direct comparisons of topic maintenance with logic reasoning, the logic task produced significantly greater activity in the left hemisphere language areas (BA 44/45 and 22). In contrast, the topic maintenance condition produced a right hemisphere bias in both homologous regions in addition to other cortical areas (dorsal prefrontal cortex, angular gyrus, and supplementary motor area). Their data suggest that, in making sense in a conversation, both hemispheres participate but do so uniquely; the right hemisphere appear to play a specific role in integrating semantic information into the context as a whole, while the left hemisphere may implicitly sense the sequential logic of a conversation. Figure 3 (see color insert) illustrates the areas of increased activation for topic maintenance vs. logic conditions in the Caplan & Dapretto (2001) study.

Cohesion and Repair

Several imaging studies have examined the brain's response to linguistic anomalies—incorrect groups of words or phrases. How the brain responds to these anomalies differs depending on whether the anomalies are based in syntax or semantics; and how the task demands that subjects respond to these anomalies. In the Kang et al. (1999) study, in which subjects passively heard but did not make an overt or covert response to semantic anomalies in word pairs (e.g., heard-shirts), only right frontal lobe activity for semantic anomalies was found. However, the investigators did not report on brain regions showing an early hemodynamic rise, which is more characteristic of temporal lobe responses, so it is possible that such activations may have been overlooked. Further, when subjects make no behavioral response, it is difficult to know how well they are attending to the stimuli and what cognitive operations are in effect.

In discourse and text, we tend to join sentences together with cohesive ties that help to build on the coherence of the passages as a whole. Such ties may reflect causality of agents and actions, facilitating comprehension. Ferstl & von Cramon (2001) used fMRI to examine brain activity while subjects judged whether pairs of sentences were related to one another pragmatically. Four types of sentence pairs were presented: (*a*) coherent and cohesive sentences both maintained a logical relationship and contained cohesive ties (Mary's exam was about to start. Therefore, her palms were sweaty); (*b*) incoherent and cohesive ties (Mary's exam was about to start. Her friends remembered her birthday); (*c*) similarly constructed without cohesive ties (e.g., Mary's exam was about to start. The palms were sweaty); and (*d*) nonword sentences served as a control. Sentence pairs in general activated traditional anterior and posterior language areas in the left hemisphere. In addition, cohesive ties in the context of incoherent sentence pairs produced left frontal lobe activation, particularly in areas 44 and 44/6, but no regions were increased for incohesive or incoherent pairs relative to cohesive and coherent pairs. These regional activations were likely due to increase effort in resolving the apparent conflict between the logic of the sentences and the cohesive ties. Unlike similar studies that use auditory input, the Ferstl & von Cramon (2001) study used printed sentences in both activation and control tasks. The role of the right hemisphere in judging coherence may not be apparent in tasks using printed text alone, given the strong left hemisphere bias for reading.

"Repair" refers to the act of fixing anomalies in speech. A study by Meyer et al. (2000) had subjects repair anomalies in sentences. They presented subjects sentences auditorally that were either grammatically correct or contained an error in one of several ways, such as case disagreement or word order violation. Both groups of subjects had to determine whether a sentence was grammatically correct; one group of subjects had the additional job of silently repairing the incorrect sentences. In judging whether or not the sentences were correct, activation was reported bilaterally in anterior, middle, and posterior temporal portions of the superior temporal gyrus (STG); however, when subjects had to repair those anomalies, significant increases were seen in the right middle temporal gyrus and the right frontal lobe (44/45). In contrast, the posterior STG showed bilateral activation for all conditions. Possibly, greater right hemisphere activity for repair vs. detection of anomalies, may reflect the greater demands placed on processing the context or global intent of language.

Prosody

Information about intent, connotative meaning, and some aspects of the form of speech can be expressed though intonation and emphasis in discourse that is irrespective of the semantic content of the words themselves. Prosody in speech comprehension encompasses a range of features, including intonations relevant to emotion, importance (e.g., stress or accents), and linguistic forms (questions vs. imperatives) that are associated with right hemisphere function.

Emotional prosody refers to changes in stress and intonation in either sentences or words that convey information about the speaker's emotional state, such as anger, surprise, sadness, or happiness. Buchanan et al. (2000) compared brain activity in normal subjects when listening to a set of four rhyming words differing in the initial phoneme, which were presented in each of the above four emotional intonations. In two verbal conditions subjects monitored words for the appearance of one of the initial phonemes; in the prosody condition, subjects monitored words for one of two emotional states. Thus, this design follows the selective attention approach in which subjects always receive the same sensory input and perform the same type of decision and output, but the design differs in the processes that give rise to the subjects' response. Looking exclusively at both prosody conditions independent of conveyed emotion, they found that activity in the right IFG and right inferior parietal regions was greater than that in the verbal tasks. In comparison, attention to the initial phoneme of the same stimuli produced left hemisphere activity in the anterior IFG (47) and in posterior temporal cortex. In comparison to a resting baseline, both verbal and prosody conditions produced activation in bilateral temporal cortex; in the posterior portions of auditory cortex, all conditions showed greater right than left temporal lobe activity. Possibly, prosodic information is processed automatically and preferentially in the right temporal lobe and is less affected by selective attention than frontal lobe regions. Interestingly, the location of the right temporal lobe region is very close to that reported by Burton et al. (2000) for tone discrimination vs. speech and close to the region for integrative meaning in metaphors and discourse (Bottini et al. 1994, St. George et al. 1999). An earlier study using PET to identify brain areas involved in detecting emotional prosody (George et al. 1996) in sentences also reveals a right prefrontal activation similar to the Buchanan et al. (2000) study. The right prefrontal cortex has a well-documented role in processing emotion generally; for instance, imaging studies of emotional face processing also reveal right frontal lobe activity (Hariri et al. 1999), and emotional deficits among RH damages patients span not only comprehension and expression of affective prosody but also interpretation and expression of facial affect (Montreys & Borod 1998).

Friederici (2001) also reports on a study contrasting normal sentences, Jabberwocky sentences, and sentences filtered so that subjects could detect the intonations, but not the actual content, of spoken sentences. Subjects had to determine whether the sentences had an active or passive construction. In this case, the attention to prosodic features was based on linguistic, as opposed to affective, prosody. The investigators found bilateral activity in both frontal and temporal regions, with additional right frontal activity during the prosodic condition in the right pars opercularis. In this study the experimental conditions likely differ in their difficulty (behavioral data not reported), and it is thus difficult to determine what brain changes, if any, are due solely to the general effect of task difficulty and increased effort. Nonetheless the data are consistent with the notion that increased attention to the syntactic content of sentences produces increases deep in the frontal operculum. Unfortunately, the authors do not provide coordinates of the region of interest centers, making comparison across studies difficult.

In examining the loci of right hemisphere activations across studies, there is striking consistency in the center of mass. Table 2 presents the centers of activation in the frontal and temporal lobes for studies reported here. In the frontal lobe, the right IFG activations occur primarily in Brodmann's area 45, and are homologous to Broca's area activity seen in traditional language tasks; in temporal lobe, the activations appear analogous to the posterior superior temporal sulcus area specific to auditory word processing (Wise et al. 2001) experiment four: $-64, -34, 2$. The right IFG appears particularly important in tasks in which subjects are required to make a decision or judgment about information, as opposed to passive comprehension, although this is not universal.

TABLE 2 Talairach coordinates of right hemisphere activation in fMRI studies of language and logic

Author	Task	Frontal	Temporal
	Reasoning		
Goel et al. 2000	Deductive reasoning—language control	−52 14 20	−50 −42 2
Caplan & Dapretto 2001	Logic vs. topic maintenance judgment	−50 14 12 −54 24 4	−54 −34 4
	Figurative Language		
St. George et al. 1999	Discourse/integration Comprehension/passive		r. posterior temporal
Bottini et al. 1994	Metaphors vs. sentences Comprehension/judgment	40 28 8	56 −38 0
	Linguistic context		
Kang et al. 1999	Semantic vs. syntactic anomalies passive	40 23 4	
Kircher et al. 2001	Generate word to complete a sentence; reading control		58 −8 9 58 −36 9
Caplan & Dapretto 2001	Topic maintenance vs. logic judgment	44 20 28 40 24 −6 44 12 18	52 −14 2 46 −30 −2
	Cohesion and repair		
Meyer 2000	Repair of incoherent sentences	48 11 10	51 −19 9
Ferstl 2001	Cohesive ties/ coherent sentences Judgment	−46 20 19 −46 28 5	
	Prosody		
Buchanan et al. 2000	Prosody vs. verbal attention Prosodic stimuli vs. rest (COM)[a]	44, 20, 16	49 −42 12[a]
	Tones		
Demonet et al. 1994	Tone vs. phonemes	46 4 28	48 −42 20 50 −62 0
Burton et al. 2000	Tone discrimination vs. speech		62 −42 15

[a]Calculated center of mass in the defined ROI.

Summary

Across studies, we can identify three large clusters of right hemisphere activity during a wide range of tasks that measure aspects of language relevant to figurative, contextual, or connotative meaning. In posterior temporal lobe, these regions cluster most strongly around cortex that is roughly contralateral to Wernicke's area, while an additional cluster more anterior in the STG is closer to primary auditory cortex. The posterior temporal region overlaps with that involved in making judgments about tones, which clearly requires no language interpretation. Thus, we can presume that the nature of the right temporal lobe neurons gives rise to a variety of processes that can be used in the service of language as well as other processes. Many of the tasks showing RH activation require subjects to integrate information over time in order to reach a correct conclusion. This general principle can apply to language tasks from the paragraph level as in Bottini et al. (1994) to the word level as in Buchanan et al. (2000), but it may also apply in the case of nonlanguage tasks including tone-discrimination tasks. In contrast, sentence level tasks involving processes like sequential logic or cohesion may not require integration over the whole set of information but rather require one to extract individual, relevant elements of information from the whole to make a decision; such tasks tend to show left hemisphere activity. In other ways, right hemisphere activity during language tasks appears analogous to the homologous left hemisphere regions: Tasks that involve analyzing sensory input generally produce more posterior activity, and those that require making an active response or judgment, or generating a solution give rise to frontal lobe activation. Taken as a whole, the fMRI data add anatomical detail to the growing acceptance of the crucial role the RH plays in language comprehension.

GENERAL CONCLUSIONS

This review has emphasized three primary results from fMRI research into language. First, the broadly defined terms Broca's area and Wernicke's area do not correspond with the reality of how language processes are organized in the brain. Rather than demonstrating that large brain regions (like the IFG) are responsible for several different functions, the data suggest that within a large brain area are small, compact zones with relatively narrow functions, but which may interact to a high degree with one another. While this review discusses only three of these functions within the IFG, other regions in this area also have specialized functions relevant to language, and a similar organization in Wernicke's area is also likely (Wise et al. 2001).

Second, the concept of either a single system for semantic information or an organizational structure formed along strictly categorical lines, cannot be supported by the imaging data. Rather, information about natural categories and about specific—environmentally and perhaps culturally specific—categories is spread diffusely in sensory and cortical association areas. However, the organization is not random but rather reflects associations with visual or other sensory features, associations with use or actions, and associations with linguistic attributes. Likely, these diverse and diffuse representations are bound together by an executive system in the frontal lobe that is relatively specific for semantic information processing.

Third, the right hemisphere makes a substantial contribution to many aspects of language comprehension, though not at the single-word level. Whereas the right hemisphere appears to lack both the one-to-one mapping of information with words and the sequential analysis in discourse that the left hemisphere performs with ease, the right hemisphere appears to make unique contributions to keeping track of the topic, drawing inferences from text and in conversation, and integrating prosodic information into a complete representation of meaning and intent, aspects of language that are critical to social communication.

Finally, a few tidbits of new imaging data suggest that each of the regions critical for language processing may in fact have some more general underlying function, bringing us closer to understanding language in basic neural terms.

While many imaging studies of language have aimed to confirm theories already suggested by lesion-deficit research, fMRI has added anatomical precision and a level of complexity unavailable to lesion-based methods. The principal new findings emerging from fMRI investigations of semantic organization in the brain highlight the complexity of this organization, reflecting not only a high degree of specialization for specific aspects of language, but also a high degree of interactivity and interdependence. While this review has focused primarily on semantics, it is striking to note how broadly regions contributing to semantic processing are distributed in the brain. This is clearly at odds with the standard neurological models on language comprehension, which continue to teach that comprehension is completed in Wernicke's area, as well as with the cognitive psychological models positing a central semantic store.

As brain imaging data are nearly always presented in terms of focal centers of task-related activity, it is commonly assumed that all brain imaging research assumes a strict interpretation of the principles of modularity as detailed by Fodor (1983). fMRI research easily lends itself to that interpretation, but researchers are increasingly utilizing new techniques that go beyond this account. In particular, the use of functional connectivity techniques in image processing reveal patterns of mutual engagement of different brain areas; new analysis tools allow for greater use of interactions among the data, and paradigm designs that rely less on assumptions of hierarchical organization of cognitive processes have become the norm. New approaches such as these will undoubtedly accelerate advances in language research. Also the wide availability of MRI scanners and the noninvasiveness of this technique guarantees that our understanding of how language is organized in the brain will continue to burgeon in the coming years.

The *Annual Review of Neuroscience* is online at http://neuro.annualreviews.org

LITERATURE CITED

Alexander MP, Naeser MA, Palumbo C. 1990. Broca's area aphasias: aphasia after lesions including the frontal operculum. *Neurology* 40:353–62

Baddely A. 1992. Working memory. *Science* 255:556–59

Berent I, Bouissa R, Tuller B. 2001. The effect of shared structure and content on reading nonwords: evidence for a CV skeleton. *J. Exp. Psychol. Learn. Mem. Cogn.* 27:1042–57

Berndt RS, Caramazza A. 1980. Semantic operations deficits in sentence comprehension. *Psychol. Res.* 41:169–77

Binkofski F, Amunts K, Stephan K, Posse S, Schormann T, et al. 2000. Broca's region subserves imagery of motion: a combined cytoarchitectonic and fMRI study. *Hum. Brain Mapp.* 11:273–85

Binkofski F, Dohle C, Posse S, Stephan K, Hefter H, et al. 1998. Human anterior intraparietal area subserves prehension: a combined lesion-functional MRI activation study. *Neurology* 50:1253–59

Blumstein S, Milberg W, Shrier R. 1982. Semantic processes in aphasia: evidence from an auditory lexical decision task. *Brain Lang.* 17:301–16

Bookheimer S, Zeffiro T, Blaxton T, Gaillard W, Theodore W. 1995. Regional cerebral blood flow changes during object naming and word reading. *Hum. Brain Mapp.* 3:93–106

Bottini G, Corcoran R, Sterzi R, Paulesu E, Schenone P, et al. 1994. The role of the right hemisphere in the interpretation of figurative aspects of language. A positron emission tomography activation study. *Brain* 117:1241–53

Buchanan T, Lutz K, Mirzazade S, Specht K, Shah N, et al. 2000. Recognition of emotional prosody and verbal components of spoken language: an fMRI study. *Cogn. Brain Res.* 9:227–38

Buckner RL, Koutstaal W, Schacter DL, Rosen BR. 2000. Functional MRI evidence for a role of frontal and inferior temporal cortex in amodal components of priming. *Brain* 123(Part 3):620–40

Buckner RL, Wheeler ME. 2001. The cognitive neuroscience of remembering. *Nat. Rev. Neurosci.* 2:624–34

Burton M, Small S, Blumstein S. 2000. The role of segmentation in phonological processing: an fMRI investigation. *J. Cogn. Neurosci.* 12:679–90

Caplan D, Alpert N, Waters G. 1998. Effects of syntactic structure and propositional number on patterns of regional cerebral blood flow. *J. Cogn. Neurosci.* 10:541–52

Caplan R, Dapretto M. 2001. Making sense during conversation: an fMRI study. *Neuroreport.* In press

Caramazza A. 2000. Minding the facts: a comment on Thompson-Schill et al.'s "A neural basis for category and modality specificity of semantic knowledge." *Neuropsychologia* 38:944–49

Caramazza A, Hillis A, Rapp B, Romani C. 1990. The multiple semantics hypothesis: multiple confusions? *Cogn. Neuropsychol.* 7:161–89

Chao LL, Haxby JV, Martin A. 1999. Attribute-based neural substrates in temporal cortex for perceiving and knowing about objects. *Nat. Neurosci.* 2:913–19

Chao LL, Martin A. 2000. Representation of manipulable man-made objects in the dorsal stream. *Neuroimage* 12:478–84

Cohen MS. 1997. Parametric analysis of fMRI data using linear systems methods. *Neuroimage* 6:93–103

Cohen M, Bookheimer S. 1994. Localization of brain function using magnetic resonance imaging. *TINS* 17:268–77

Corbetta M, Miezin FM, Dobmeyer S, Shulman GL, Petersen SE. 1990. Attentional modulation of neural processing of shape, color, and velocity in humans. *Science* 248:1556–59

Cronk BC. 2001. Phonological, semantic, and repetition priming with homophones. *J. Psycholinguist. Res.* 30:365–78

Cuenod CA, Bookheimer SY, Hertz-Pannier L, Frank JA, Zeffiro TA, et al. 1995. Functional MRI during word generation using conventional equipment: a potential tool for language localization in clinical environment. *Neurology* 34:1821–27

D'Esposito M, Zarahn E, Aguirre GK. 1999. Event-related functional MRI: implications for cognitive psychology. *Psychol. Bull.* 125:155–64

Damasio H, Grabowski T, Tranel D, Hichwa R, Damasio A. 1996. A neural basis for lexical retrieval. *Nature* 380:645–51

Dapretto M, Bookheimer SY. 1999. Form and content: dissociating syntax and semantics in sentence comprehension. *Neuron* 24:427–32

Demb J, Desmond J, Wagner A, Vaidya C, Glover G, Gabrieli J. 1995. Semantic encoding and retrieval in the left inferior prefrontal cortex: a functional MRI study of task difficulty and process specificity. *J. Neurosci.* 15:5870–78

Demonet JF, Chollet F, Ramsay S, Cardebat D, Nespoulous J-L, et al. 1992. The anatomy of phonological and semantic processing in normal subjects. *Brain* 115:1753–68

Demonet JF, Price C, Wise R, Frackowiak RS. 1994. A PET study of cognitive strategies in normal subjects during language tasks. Influence of phonetic ambiguity and sequence processing on phoneme monitoring. *Brain* 117:671–82

Dronkers N. 1996. A new brain region for coordinating speech articulation. *Nature* 384:159–61

Ferstl E, von Cramon D. 2001. The role of coherence and cohesion in text comprehension: an event-related fMRI study. *Cogn. Brain Res.* 11:325–40

Fiez J. 1997. Phonology, semantics, and the role of the left interior prefrontal cortex. *Hum. Brain Mapp.* 5:79–83

Fodor JA. 1983. *The Modularity of Mind*. Cambridge: MIT Press

Friederici A, Optiz B, von Cramon D. 2000. Segregating semantic and syntactic aspects of processing in the human brain: an fMRI investigation of different word types. *Cereb. Cortex* 10:698–705

Friederici AD. 2001. Syntactic, prosodic, and semantic processes in the brain: evidence from event-related neuroimaging. *J. Psycholinguist. Res.* 30:237–50

Gabrieli J, Poldrack R, Desmond J. 1998. The role of left prefrontal cortex in language and memory. *Proc. Natl. Acad. Sci. USA* 95:906–13

George MS, Parekh PI, Rosinsky N, Ketter TA, Kimbrell TA, et al. 1996. Understanding emotional prosody activates right hemisphere regions. *Arch. Neurol.* 53:665–70

Goel V, Buchel C, Frith C, Dolan R. 2000. Dissociation of mechanisms underlying syllogistic reasoning. *Neuroimage* 12:504–14

Grezes J, Decety J. 2001. Functional anatomy of execution, mental simulation, observation, and verb generation of actions: a meta-analysis. *Hum. Brain Mapp.* 12:1–19

Hariri A, Bookheimer S, Mazziotta J. 1999. A neural network for modulating the emotional response to faces. *Neuroreport* 11:43–48

Hart J, Gordon B. 1992. Neural subsystems for object knowledge. *Nature* 359:60–64

Horwitz B, Tagamets MA, McIntosh AR. 1999. Neural modeling, functional brain imaging, and cognition. *Trends Cogn. Sci.* 3:91–98

Iacoboni M, Woods R, Brass M, Bekkering H, Mazziotta J, Rizzolatte G. 1999. Cortical mechanisms of human imitations. *Science* 286:2526–28

Ishai A, Ungerleider LG, Martin A, Schouten JL, Haxby JV. 1999. Distributed representation of objects in the human ventral visual pathway. *Proc. Natl. Acad. Sci. USA* 96:9379–84

Jessen F, Heun R, Erb M, Granath D, Klose U, et al. 2000. The concreteness effect: evidence for dual coding and context availability. *Brain Lang.* 74:103–12

Johnson P, Ferraina S, Bianchi L, Caminiti R. 1996. Cortical networks for visual reaching: physiological and anatomical organization of frontal and parietal lobe arm regions. *Cereb. Cortex* 6:102–99

Just MA, Carpenter PA, Keller TA, Eddy WF, Thulborn KR. 1996. Brain activation modulated by sentence comprehension. *Science* 274:114–16

Kang A, Constable R, Gore J, Avrutin S. 1999. An event-related fMRI study of implicit phrase level syntactic and semantic processing. *Neuroimage* 10:555–61

Keller T, Carpenter P, Just MA. 2001. The neural bases of sentence comprehension: an fMRI examination of syntactic and lexical processing. *Cereb. Cortex* 11:223–37

Keihl K, Liddle P, Smith A, Mendrek A, Forster B, Hare R. 1999. Neural pathways involved in the processing of concrete and abstract words. *Hum. Brain Mapp.* 7:225–33

Kircher T, Brammer M, Tous Andreu N, Williams S, McGuire P. 2001. Engagement of right temporal cortex during processing of linguistic context. *Neuropsychologia* 39:798–809

Le Clec'H G, Dehaene S, Cohen L, Mehler J, Dupoux E, et al. 2000. Distinct coritcalk areas for names of numbers and body parts independent of language and input modality. *Neuroimage* 12:381–91

Luria A. 1966. *The Higher Cortical Function in Man*. New York: Basic Books

Martin A, Chao LL. 2001. Semantic memory and the brain: structure and processes. *Curr. Opin. Neurobiol.* 11:194–201

Martin A, Wiggs CL, Ungerleider LG, Haxby JV. 1996. Neural correlates of category-specific knowledge. *Nature* 379:649–52

McCarthy G, Blamire AM, Rothman DL, Gruetter R, Shulman RG. 1993. Echo-planar magnetic resonance imaging studies of frontal cortex activation during word generation in humans. *Proc. Natl. Acad. Sci. USA* 90: 4952–56

Mellet E, Tzourio N, Denis M, Mazoyer B. 1998. Cortical anatomy of mental imagery of concrete nouns based on their dictionary definition. *Neuroreport* 9:803–8

Metter E. 1991. Brain-behavior relationships in aphasia studied by positron emission tomography. *Ann. NY Acad. Sci.* 620:153–64

Meyer M, Friederici AD, von Cramon DY. 2000. Neurocognition of auditory sentence comprehension: event related fMRI reveals sensitivity to syntactic violations and task demands. *Cogn. Brain Res.* 9:19–33

Montreys CR, Borod JC. 1998. A preliminary evaluation of emotional experience and expression following unilateral brain damage. *Int. J. Neurosci.* 96:269–83

Moore CJ, Price CJ. 1999. A functional neuroimaging study of the variables that generate category-specific object processing differences. *Brain* 122:943–62

Moro A, Tettamanti M, Perani D, Donati C, Cappa SF, Fazio F. 2001. Syntax and the brain: disentangling grammar by selective anomalies. *Neuroimage* 13:110–18

Muller R, Kleinhaus N, Courchesne E. 2001. Broca's area and the distribution of frequency transitions: an fMRI study. *Brain Lang.* 76:70–76

Mummery CJ, Patterson K, Hodges JR, Price CJ. 1998. Functional neuroanatomy of the semantic system: divisible by what? *J. Cogn. Neurosci.* 10:766–77

Neely J, Keefe D, Ross K. 1989. Semantic priming in the lexical decision task: roles of prospective prime generated expectancies and prospective semantic matching. *J. Exp. Psych: Learn. Mem. Cog.* 15:1003–19

Nichelli P, Grafman J, Pietrini P, Clark K, Lee KY, Miletich R. 1995. Where the brain appreciates the moral of a story. *Neuroreport* 6:2309–13

Ojemann G, Mateer C. 1979. Human language cortex: localization of memory, syntax, and sequential motor-phoneme identification systems. *Science* 205:1401–3

Ojemann G, Ojemann J, Lettich E, Berger M. 1989. Cortical language localization in left, dominant hemisphere. An electrical stimulation mapping investigation in 117 patients. *J. Neurosurg.* 71:316–26

Okada T, Tanaka S, Nakai T, Nishizawa S, Inui T, et al. 2000. Naming of animals and tools: a functional magnetic resonance imaging study of categorical differences in the human brain areas commonly used for naming visually presented objects. *Neurosci. Lett.* 15:33–36

Paivio A. 1986. *Mental Representation: A Dual Coding Approach*. New York: Oxford Univ. Press

Palmer ED, Rosen HJ, Ojemann JG, Buckner RL, Kelley WM, Petersen SE. 2001. An event-related fMRI study of overt and covert word stem completion. *Neuroimage* 14:182–93

Paulesu P, Frith CD, Bench CJ, Bottini G, Grasby G, Frackowiak SJ. 1993. Functional anatomy of working memory: the articulatory loop. *J. Cereb. Blood Flow Metab.* 13: 551

Perani D, Schnur T, Tettamanti M, Gorno-Tempini M, Cappa SF, Fazio F. 1999. Word and picture matching: a PET study of

semantic category effects. *Neuropsychologia* 37:293–306
Petersen SE, Fox PT, Posner MI, Mintun M, Raichle ME. 1988. Positron emission tomographic studies of the cortical anatomy of single-word processing. *Nature* 331:585–89
Poeppel D. 1996. A critical review of PET studies of phonological processing. *Brain Lang.* 55:317–51
Poldrack R, Wagner A, Prull M, Desmond J, Glover G, Gabrieli JD. 1999. Functional specialization for semantic and phonological processing in the left interior prefrontal cortex. *Neuroimage* 10:15–35
Poldrack RA, Temple E, Protopapas A, Nagarajan S, Tallal P, et al. 2001. Relations between the neural bases of dynamic auditory processing and phonological processing: evidence from fMRI. *J. Cogn. Neurosci.* 13:687–97
Price C. 2000. The anatomy of language: contribution from functional neuroimaging. *J. Anat.* 197:335–59
Price C, Wise R, Ramsay S, Friston K, Howard D, et al. 1992. Regional response differences within the human auditory cortex when listening to words. *Neurosci. Lett.* 146:179–82
Price C, Wise R, Watson J, Patterson K, Howard D, Frackowiak R. 1994. Brain activity during reading. *Brain* 117:1255–69
Pugh K, Shaywitz B, Shaywitz S, Constable R, Skudlarski P, et al. 1996. Cerebral organization of component processes in reading. *Brain* 119:1221–38
Raichle ME, Fiez JA, Videen TO, MacLeod AK, Pardo JV, et al. 1994. Practice-related changes in human brain functional anatomy during nonmotor learning. *Cereb. Cortex* 4:8–26
Rizzolatti G, Arbib MA. 1998. Language within our grasp. *Trends Neurosci.* 21:188–94
Rueckert L, Appollonio I, Grafman J, Jezzard P, Johnson R Jr, et al. 1994. Magnetic resonance imaging functional activation of left frontal cortex during covert word production. *J. Neuroimag.* 4:67–70
Savoy R, O'Craven K, Weisskoff R, Davis T, Baker J, Rosen B. 1994. *Exploring the temporal boundaries of fMRI: measuring responses to very brief visual stimuli.* Presented at SfN, 24th, Miami Beach
Smith C, Anderson A, Kryscio R, Schmitt F, Kindy M, et al. 2001. Differences in functional magnetic resonance imaging activation by category in a visual confrontation naming task. *J. Neuroimag.* 11:165–70
Spitzer M, Kwong K, Kennedy W, Rosen B, Belliveau J. 1995. Category-specific activaion in fMRI during picture naming. *Neuroreport* 6:2109–12
St. George M, Kutas M, Martinez A, Sereno MI. 1999. Semantic integration in reading: engagement of the right hemisphere during discourse processing. *Brain* 122:1317–25
Stromswold K. 1994. The cognitive and neural bases of language acquisition. *Cogn. Neurosci.*
Talairach J, Tournoux P. 1988. *Co-Planar Stereotaxic Atlas of the Human Brain.* New York: Thieme Med.
Thompson-Schill S, Aguirre G, D'Esposito M, Farah M. 1999. A neural basis for category and modality specificity of semantic knowledge. *Neuropsychologia* 37:671–76
Thompson-Schill S, D'Esposito M, Aguirre JK, Farah MJ. 1997. Role of the left inferior prefrontal cortex in retrieval of semantic knowledge: a reevaluation. *Proc. Natl. Acad. Sci. USA* 94:14792–97
Wagner A, Desmond J, Demb J, Glover G, Gabrieli JD. 1997. Semantic repetition priming for verbal and pictorial knowledge. *J. Cogn. Neurosci.* 9:714–26
Wagner A, Koustaal W, Maril A, Schacter D, Buckner R. 2000. Task-specific repetition priming in left inferior prefrontal cortex. *Cereb. Cortex* 10:1176–84
Wagner AD, Pare-Blagoev EJ, Clark J, Poldrack RA. 2001. Recovering meaning: left prefrontal cortex guides controlled semantic retrieval. *Neuron* 31:329–38
Warrington EK, Shallice T. 1984. Category specific semantic impairments. *Brain* 107:829–54
Wildgruber D, Ackermann H, Klose U, Kardatzki, Grodd W. 1996. Functional lateralization of speech production at primary

motor cortex: a fMRI study. *Neuroreport* 7:2791–95

Wise R, Scott S, Blamk S, Mummery C, Murphy K, Warburton E. 2001. Separate neural subsystems within 'Wernicke's area.' *Brain* 124:83–95

Zatorre RJ, Evans AC, Meyer E, Gjedde A. 1992. Lateralization of phonetic and pitch discrimination in speech processing. *Science* 256:846–49

Zatorre RJ, Meyer E, Gjedde A, Evans AC. 1996. PET studies of phonetic processing of speech: review, replication, and reanalysis. *Cereb. Cortex* 6:21–30

INTENTIONAL MAPS IN POSTERIOR PARIETAL CORTEX

Richard A. Andersen and Christopher A. Buneo
Division of Biology, California Institute of Technology, Mail Code 216-76, Pasadena, California 91125; email: andersen@vis.caltech.edu; chris@vis.caltech.edu

Key Words eye movements, arm movements, optic flow, spatial representations, neural prosthetics

■ **Abstract** The posterior parietal cortex (PPC), historically believed to be a sensory structure, is now viewed as an area important for sensory-motor integration. Among its functions is the forming of intentions, that is, high-level cognitive plans for movement. There is a map of intentions within the PPC, with different subregions dedicated to the planning of eye movements, reaching movements, and grasping movements. These areas appear to be specialized for the multisensory integration and coordinate transformations required to convert sensory input to motor output. In several subregions of the PPC, these operations are facilitated by the use of a common distributed space representation that is independent of both sensory input and motor output. Attention and learning effects are also evident in the PPC. However, these effects may be general to cortex and operate in the PPC in the context of sensory-motor transformations.

INTRODUCTION

The posterior parietal cortex (PPC) has traditionally been viewed as a sensory "association" area, associating different modalities and having higher-level sensory functions such as spatial attention and spatial awareness. In this review, we highlight a new view of the PPC that is emerging. It is proposed that the PPC, rather than serving a purely sensory or motor role, subserves higher-level cognitive functions related to action. Among these higher cognitive functions is the formation of intentions, or early plans for movement. These intentions are anatomically segregated within the PPC, with regions specialized for the planning of saccades, reaches, and grasps. Moreover, these intentions are highly abstract and are evident in the discharge of single neurons even when a specific intention is not carried out.

The different intention-related regions of the PPC appear to participate in operations critical to the earliest stages of movement planning: multisensory integration and coordinate transformations. These functions are facilitated by employing a rather unique, distributed code. The response fields of neurons in at least two regions are in retinal coordinates, independent of both the sensory modality used to cue target locations (i.e., audition vs. vision) and the action that will ultimately be performed (i.e., reaches vs. saccades). However, these retinal fields are also gain modulated by eye, head, and limb positions. As a result, groups of parietal cells do not generally represent space in a single, defined spatial reference frame. Rather, they code locations in a distributed manner, which can be read out by other groups of neurons in a variety of reference frames.

We describe a potential medical application that utilizes the finding that the PPC encodes movement intentions. The intention-related activity in the PPC can, in principle, be used to operate a neural prosthesis for paralyzed patients. Such a neural prosthesis would consist of recording the activity of PPC neurons, interpreting the movement intentions of the subject with computer algorithms, and using these predictions of the subject's intentions to operate external devices such as a robot limb or a computer. We describe preliminary investigations in healthy monkeys that estimate the number of parietal cells needed to operate such a prosthesis (Meeker et al. 2001, Shenoy et al. 1999b). We also describe a recent finding that monkeys can use this intended movement activity to position a cursor on a computer screen just by thinking about a reach movement, without actually generating a reach (D. Meeker, S. Cao, J. W. Burdick & R. A. Andersen, unpublished observations). This result was obtained without extensive training and strongly suggests that we are in fact tapping into the highly abstract neural signals that represent the earliest plans for movement.

THE PPC SUBSERVES COGNITIVE FUNCTIONS RELATED TO ACTION

Many of the deficits observed following lesions of the PPC are consistent with the area playing a high-level, cognitive role in sensory-motor integration. Patients with PPC lesions do not have primary sensory or motor deficits. However, when they attempt to connect these functions, for instance during sensory guided movements, then defects become apparent. Patients with PPC lesions often suffer from optic ataxia; that is, difficulty in estimating the location of stimuli in 3D space, as indicated by pronounced errors in reaching movements (Balint 1909, Rondot et al. 1977). Patients with PPC lesions can also suffer from one or more of the apraxias, a class of deficits characterized by the inability to plan movements (Geshwind & Damasio 1985). These can range from a complete inability to follow verbal commands for simple movements, to difficulty in performing sequences of movements. Patients with parietal lobe damage also have difficulty correctly shaping their hands as they prepare to grasp objects, which again points to a disconnection between the visual sensory apparatus that registers the shape of objects and the motor systems that shape the configuration of the hand (Goodale & Milner 1992, Perenin & Vighetto 1988).

Neglect is another deficit commonly attributed to lesions of the PPC, although there is currently some debate about whether it is damage to the PPC or to the nearby superior temporal gyrus that is the source of this defect (Critchley 1953, Karnath et al. 2001). The hallmark of neglect is the lack of awareness within the personal and extrapersonal space contralateral to the lesioned hemisphere, with the most profound deficits seen with right hemisphere lesions in right-handed humans.

These clinical results are extremely informative and useful and have helped guide much of the neurophysiological investigation of the PPC. However, to understand the neural mechanisms and circuits within the PPC that are involved in sensory-motor integration requires that the investigator, rather than relying on the happenstance of medical defects, be able to control the parameters of the experiments. Moreover, refined techniques need to be applied. In the case of humans, this has generally taken the form of fMRI studies, and in the case of monkeys, electrophysiological recording and anatomical studies. The monkey has proven to be a good model for the study of the PPC, since sophisticated motor behaviors such as hand-eye coordination are similar in the two species of primates, and there is extensive evidence to suggest that the PPC in both species performs similar functions (Connolly et al. 2000, DeSouza et al. 2000, Rushworth et al. 2001b). Evidence from these studies provides additional support for the concept that the PPC is neither strictly sensory nor motor but rather is involved in high-level cognitive functions related to action (Mountcastle et al. 1975, Andersen 1987, Goodale & Milner 1992). These functions include early-movement planning, particularly the coordinate transformations required for sensory-guided movement. The activity of PPC may also be influenced by spatial attention and learning. However, these functions are general to cortex, and in the PPC appear to operate in the more specific context of sensory-motor operations.

INTENTION

Intention is an early plan for a movement. It specifies the goal of a movement and the type of movement. For instance, "I wish to pick up the coffee cup" specifies both the goal and type of movement. An intention is high level and abstract. For instance, we can have intentions without actually acting upon them. Moreover, a neural correlate of intention does not necessarily contain information about the details of a movement, for instance the joint angles, torques, and muscle activations required to make a movement. As discussed below, intentions are initially coded in visual coordinates in at least some of the cortical areas within the PPC. This encoding is consistent with a more cognitive representation of intentions, specifying the goals of movements rather than the exact muscle activations required to execute the movement.

An intention is also a broad category of cortical functions, which include decision making (Gold & Shadlen 2001) and "motor attention" (Rushworth et al. 2001a). For instance, decision making can be considered a competition between potential movement intentions (Platt & Glimcher 1999). It may also be the case that the earliest intentions sit atop a sequence of increasingly more specific movement plans. In the example above, the earliest intention may reflect the desire to grasp the cup, with further specifications including which limb (right or left), the trajectory of the movement to avoid obstacles, the coordination of eye and hand movements, the speed of the movement, etc. Only further research will be able to resolve which parameters of a movement are coded at which stages in the sensory-motor pathway.

Distinguishing Intention from Attention

The issue of intention versus attention has been most prominent in the study of the PPC, which is perhaps not surprising considering this area is at the interface between sensory and motor systems. Mountcastle and colleagues (1975) first noted neural activity in the PPC related to the behaviors of monkeys. Robinson and colleagues (1978) later argued that these effects could be due to sensory stimulation and attention during movement. In experiments designed to tease apart sensory and movement components of activity, Andersen et al. (1987) found both, which is consistent with a role for this area in sensory-motor transformations.

One common method of separating sensory from motor components is the so-called memory task (Hikosaka & Wurtz 1983) in which an animal is cued as to the location for a movement by a briefly flashed stimulus but must withhold the response until a go signal. Typically, PPC neurons show bursts of activity to the cue and the movement, indicating both sensory- and motor-related activity. However, during the memory period the cells in many parietal areas have persistent activity, even in the dark (Gnadt & Andersen 1988, Snyder et al. 1997). This persistent activity by and large does not represent the sensory memory of the target. This can be demonstrated using tasks in which animals memorize the locations of two stimuli and subsequently make movements to both locations. For eye and arm movements the persistent activity in the delay period for nearly all neurons in the PPC is only present for the next planned movement (Batista & Andersen 2001, Mazzoni et al. 1996a), even though the animals must hold in memory two cued locations. This result indicates that the sensory memory of the target locations is either contained in a very small subset of neurons within the PPC, or in areas outside the PPC, perhaps in the frontal lobe (Tian et al. 2000).

The results of the double movement tasks rule out the coding of a sensory memory in the delay period activity. However, this activity could reflect either the direction of a movement plan or the direction of attention. Experimentally it has been very difficult to distinguish movement planning or preparation from spatial attention. Most studies of attention in monkeys use experimental paradigms that require animals to make eye or limb movements as part of the experimental design, or have the potential artifact of the animal covertly planning these movements. This fact is reason for concern in studies of the dorsal, sensory-motor pathway since there is extensive overlap of circuitry concerned with attention and eye movements, as demonstrated by fMRI experiments in humans (Corbetta et al. 1998). The finding that the locus of spatial attention can affect the metrics of saccades electrically evoked from the superior colliculus (SC) further argues for a very tight coupling of spatial attention and eye movements (Kustov & Robinson 1996). These and other results have led Rizzolatti and colleagues to argue for a motor theory of spatial attention (1994). They propose that spatial attention is an early form of motor preparation, at least for eye movements.

Some investigators have used antisaccade and antireach tasks to separate sensory from movement processing. In these paradigms, animals are trained to make movements in the opposite direction from flashed visual targets. For the case of reaches, activity in the medial intraparietal area (MIP) has been reported to code mostly the direction of the movement, and not the location of the stimulus (Eskandar & Assad 1999, Kalaska 1996). Gottlieb & Goldberg (1999) have reported that the reverse is true in the lateral intraparietal area (LIP) for eye movements, i.e., that cells respond to the stimulus and not the direction of planned movement. However, a recent report by Zhang & Barash (2000) indicates that, after a brief transient linked to the stimulus, most cells in LIP code the direction of the planned eye movement. Moreover, a smaller class of cells encode both the location of the stimulus and the movement plan, which suggests that LIP is involved in the intermediate stages of the sensory-motor transformations required for the antisaccade task. Overall, these antisaccade and antireach results reinforce the idea that PPC cells have both sensory- and movement-related responses, and occupy an intermediate stage in the sensory-motor transformation process.

We recently conducted an experiment specifically designed to separate the effects of spatial attention from those of intention (Snyder et al. 1997). In this experiment, animals attended to a flashed target and planned a movement to it during a delay period, but in one case they were instructed to plan a saccade and in the other a reach (see Figure 1*a*, see color insert). The only difference in the task during the memory period was the movement the animals were planning to make. We reasoned that if PPC activity reflected a sensory memory or attention, it should be the same in the two conditions, but if it reflected the movement plan it should be different.

Figure 1 shows two intention-specific neurons, one from area LIP (*b*) and one from an area we refer to as the parietal reach region (PRR) (*c*). In this task the monkey plans an eye or an arm movement to the same location in space. The activity of the LIP neuron illustrated in panel (*b*) shows a transient response to the onset of the briefly flashed target. This is followed by activity during the delay period if the animal is planning an eye movement (left histogram), but not if he is planning an arm movement to the same location (right histogram). In contrast, the cell in panel (*c*) shows no activity above baseline in the delay period when the animal is planning an eye movement, but strong activity when he is planning an arm movement. Such results were typical in the PPC: In general, we found that during eye movement planning area LIP was much more active, and during limb movement planning PRR was more active. PRR included MIP, 7a, and the dorsal aspect of the parieto-occipital (PO) area, though MIP was found to have the highest concentration of reach-related neurons. The results from both LIP and PRR argue strongly for a role of the PPC in movement planning.

A subsequent experiment showed that activity in the PPC is also related to the shifting of movement plans, when spatial attention is held constant (Snyder et al. 1998a). Cells with a particular movement preference (reach or saccade) showed greater activity if a plan was changed from the nonpreferred to the preferred movement (for the same target location), compared to simply reaffirming the preferred plan. This result is reminiscent of proposals that the PPC plays a role in shifting attention (Steinmetz & Constantinidis 1995), but in this case it is the intended movement that shifts, and not the spatial locus of attention.

Default Plans

The experiments by Snyder et al. (1997) were not the first to attempt to separate attention from intended movement activity. Bushnell and colleagues (1981) trained animals to either reach or saccade to a target while recording from PPC neurons. They reasoned that if the PPC was involved in attention then they should see the same level of activity regardless of the motor output, and this was what they reported. However, they recorded from only nine cells, and inspection of their Figure 1 suggests that the animal may have looked to the stimulus after the reach. Thus the animal may have been planning an eye movement as well as an arm movement during the task.

This potential problem of covert planning of eye movements is a general problem for experiments examining attention to targets placed away from the fixation point. While the formation of covert plans is unlikely to be critical for studies of attention in the ventral visual pathway, which current evidence suggests is largely involved in visual recognition, it is certainly a problem when studying the dorsal pathway, which is involved in movement planning. The issue of covert planning was directly addressed in Snyder et al. (1997). In the population of cells from which we recorded, 68% were significantly modulated in the delay period by one movement plan (reach or saccade) but not the other. Interestingly, even during the cue period 44% showed this specificity. We reasoned that the remaining cells showing significant activity for both movement plans might reflect covert plans for movement, since it is very natural to look to where you reach. To control for this possibility, we had the animals also perform a "dissociation" task in which they simultaneously planned an eye and an arm movement in different directions, with one movement into the response field and the other outside.

Figure 2 shows an example of a neuron that had activity for both eye and arm movements in the single-movement task. In the top row of histograms, the

Figure 2 A posterior parietal cortex neuron whose motor specificity was revealed by a dissociation task. In saccade (*top*) and reach (*middle*) tasks, delay period activity was greater before movements in the preferred direction (*left*) compared to the null direction (*right*). Thus, in single-movement tasks, this neuron appeared to code remembered target location independent of movement intent. However, firing was vigorous in the delay period preceding a reach in the preferred direction when this reach was combined with a saccade in the null direction (*bottom left*), but firing was nearly absent before a saccade in the preferred direction combined with a null reach. Thus, when both a reach and a saccade were planned, delay-period activity reflected the intended reach and not the intended saccade. Panel formats similar to Figure 1 (see color insert) except that every other action potential is shown in the rasters. (Modified from Snyder et al. 1997.)

monkey performed only saccades, making eye movements into the response field (left histogram) or in the opposite direction (right histogram). In the middle row, the animal made reaches instead of saccades, and a similar level of activity is seen when the animal reaches into the response field. The bottom row of histograms shows activity from the same neuron while the animal was performing the dissociation

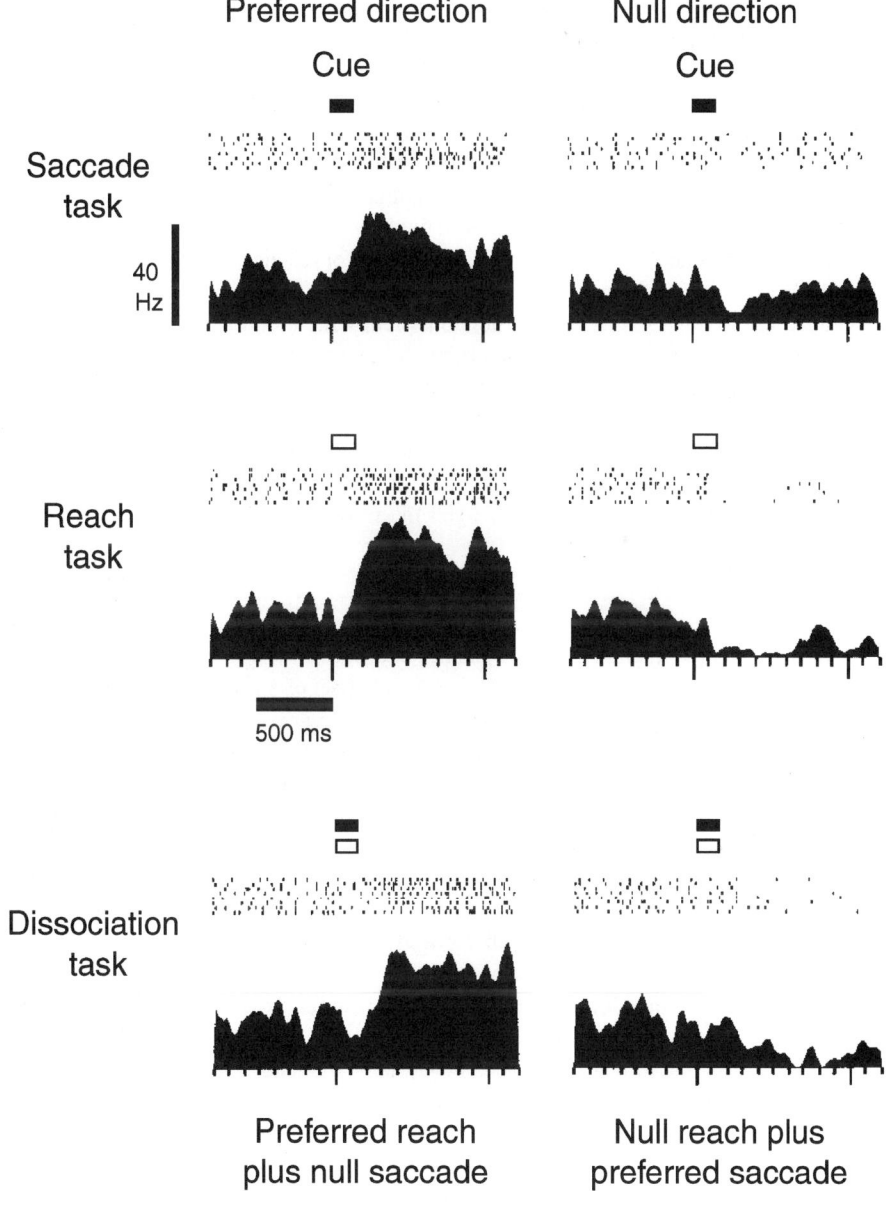

task. In the histogram on the left, the animal was simultaneously planning an arm movement into the response field and an eye movement out of the response field. In this case the cell was very active. In the histogram on the right, the animal was planning an eye movement into the response field, but an arm movement out of the field. Although this is the same eye movement plan that evoked activity in the single-movement case, now there is no activity or even a slight suppression. The pattern of activity of this neuron can be explained if the animal is forming a covert plan in the single-movement cases, in this example a covert arm movement plan. Of the neurons in the population, 62% that were not specific for single movements were specific in the dissociation task, bringing to 84% the number of cells that showed movement planning specificity in the delay period. Interestingly, more cells also revealed specificity for the cue response in the dissociation task, with a total of 45% being specific for reaches and 62% for saccades.

Covert planning may explain activity that is seen in go/no-go tasks. In these tasks a stimulus appears in the response field and the animal is later cued whether to make a movement to it or not. Activity in area 5 for reaches (Kalaska & Crammond 1995), and in LIP for saccades (Pare & Wurtz 1997) continues when the animal is cued not to move. This result is not consistent with attention or intention activity, since the target is no longer important to the animal's behavior (Pare & Wurtz 1997). However, it is consistent with a covert or default plan, which remains if no new movement plans are being formed. Evidence for this alternative explanation comes from experiments in which the plan is cancelled, and a new movement plan is put in place (Bracewell et al. 1996). An example is shown in Figure 3*a* and 3*b*, in which the monkey changed the movement plan three times before a

---

Figure 3 Responses of PPC neurons when movement plans are changed. (*a*) Activity of an LIP neuron when saccades are planned into, out of, and then into the receptive field. The long horizontal bar indicates the onset and offset of a visual fixation point. Short horizontal bars represent the timings of three successive target presentations. Spike rasters show every action potential recorded in a given trial. Spike-density histograms were constructed using 50 ms binwidths. Thin horizontal lines below each histogram represent the animal's vertical eye position. Importantly, no saccades were initiated until after the fixation point was extinguished; thus changes in activity during the time of fixation correspond to changes in the animal's plans or intentions. (Modified from Bracewell et al. 1996). (*b*) Activity of the same neuron when saccades are planned out of, into, and then out of the cell's receptive field. (Modified from Bracewell et al. 1996). (*c*) Activity of a neuron in the parietal reach region when the type of movement plan, but not its direction, is changed. Activity resulting from an instruction to plan a reach (R1) was abolished when a second flash changed the plan to a saccade (S2). An initial instruction to plan a saccade elicited only a transient response (S1) but when the plan changed to a reach activity increased (R2). Each ribbon represents the mean response of 8–12 trials +/−1 SE. Data were smoothed with a 121-point digital low-pass filter, transition band 20–32 Hz. Dashed rectangles indicate the timings target flashes. (Modified from Snyder et al. 1998a.)

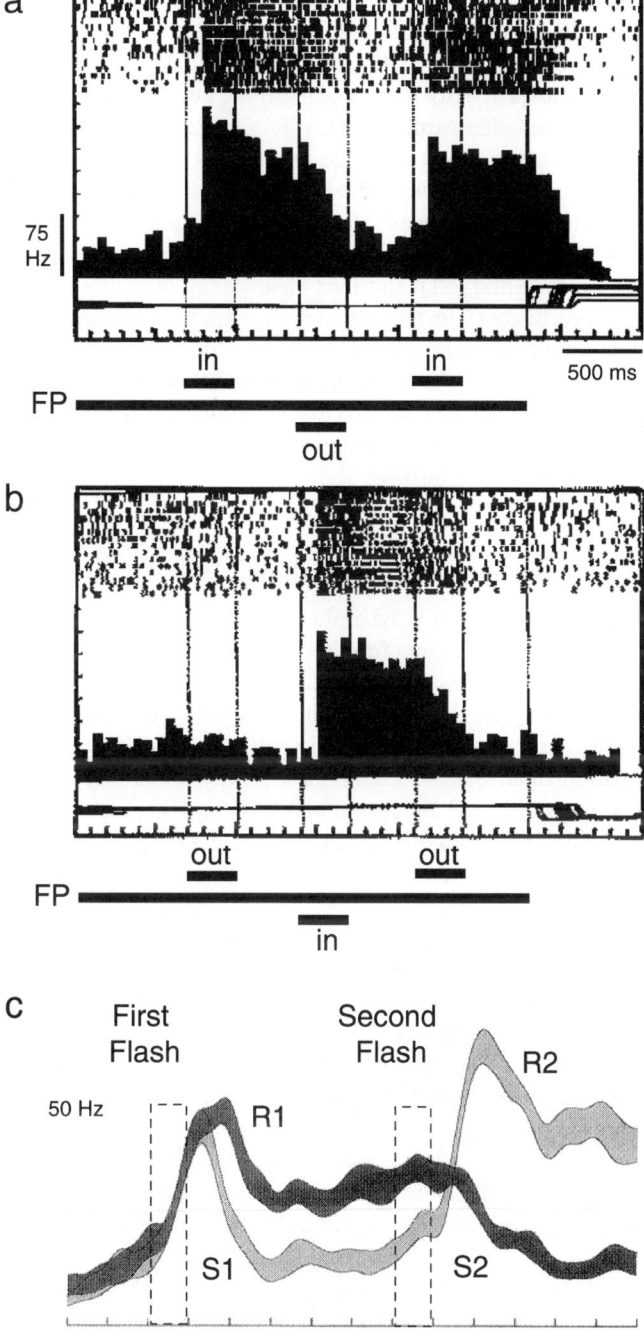

saccade, alternating between planning into, out of, and into the response field (*a*), or out, in, and out (*b*). A similar result is found even when the type of movement plan is changed, but not the direction (Snyder et al. 1998a). In Figure 3*c*, the dark histogram shows activity for a reach-specific neuron when the first cue presented in the response field instructs a reach (R1), and the second stimulus appears at the same location instructing a change in plan to a saccade (S2). Note that although no movement is made during the time shown in the histogram, the activity turns off when the animal changes to the nonpreferred plan for a movement to the same location. The lighter histogram shows activity for the same cell when the monkey plans a saccade first (S1), and then changes the plan to a reach (R2). Again the activity is consistent with the cell's activity expressing the intent of the animal, with baseline activity after the cue transient when the animal is planning a saccade, and high activity during the delay when he changes his plan to a reach. Taken together, the data from various labs suggest that default plans are formed in parietal areas to stimuli of behavioral significance in the case of no alternative plans, but are erased if alternative plans are formed.

Dynamic Evolution of Intention-Related Activity

Several studies point to a dynamic evolution in the relation of PPC activity to task requirements, changing from sensory to cognitive to motor as the demands of the task change. For instance, we recently examined the activity of PRR neurons when monkeys plan reaches to auditory versus visual targets in a memory-reach task. We found that at cue onset activity for visually cued trials carried more information about spatial location than activity for auditory cued trials. However, as the trials progressed and the animal was preparing a movement, the amount of spatial location information increased for the auditory cued trials so that by the time of the reach movement, it was not significantly different from the visually cued trials (Y. C. Cohen & R. A. Andersen, unpublished observations).

In another study, we trained animals to make saccades to a specific location cued on an object, but after the cue and before the saccade the object was rotated. Early in the task area LIP cells carried information about the location of the cue and the orientation of the object, both pieces of information being important for solving the task. However, near the time of the eye movement many of these same neurons predominately coded just the direction of the intended movement (Breznen et al. 1999).

Platt & Glimcher (1999) showed in a delayed eye movement task that the early activity of LIP neurons varied as a function of the expected probability that a stimulus was a target for a saccade, as well as the amount of reward previously associated with the target. However, during later periods of the trial the cells coded only the direction of the planned eye movement. A similar evolution has been shown in LIP and dorsal prefrontal cortex in eye movement tasks instructed by motion signals. The strength of the motion signal is an important determinant of activity in the beginning of the trial, but at the end of the trial the activity codes the decision or movement plan of the animal (Leon & Shadlen 1999, Shadlen & Newsome 1996). These studies emphasize the fact that the circuits involved in sensory-motor transformations are distributed in nature, involving parietal, frontal, and prefrontal areas (Chaffee & Goldman-Rakic 1998). Moreover, activity in these circuits can evolve dynamically to reflect sensory, cognitive, and movement components of behavior.

Intentional Maps

The above studies point to a map of intentions within the PPC (Figure 4). Area LIP is more specialized for saccade planning, and area MIP for reaching. Work by other investigators implicates areas 5, PO, 7m, and PEc as additional reaching-related regions within the posterior parietal cortex (Battaglia-Mayer et al. 2000, Ferraina et al. 2001, Ferraina et al. 1997, Kalaska 1996). Recent studies by Sakata and colleagues (1995, 1997) point to the anterior intraparietal area (AIP) as specialized for grasping. Cells in this area respond to the shapes of objects and the configuration of the hand for grasping the objects. Reversible inactivations of AIP produce deficits in shaping the hand prior to grasping in monkeys. This deficit is reminiscent of problems in shaping the hands prior to grasping found in humans with parietal lobe damage (Perenin & Vighetto 1988). The medial superior temporal area (MST)

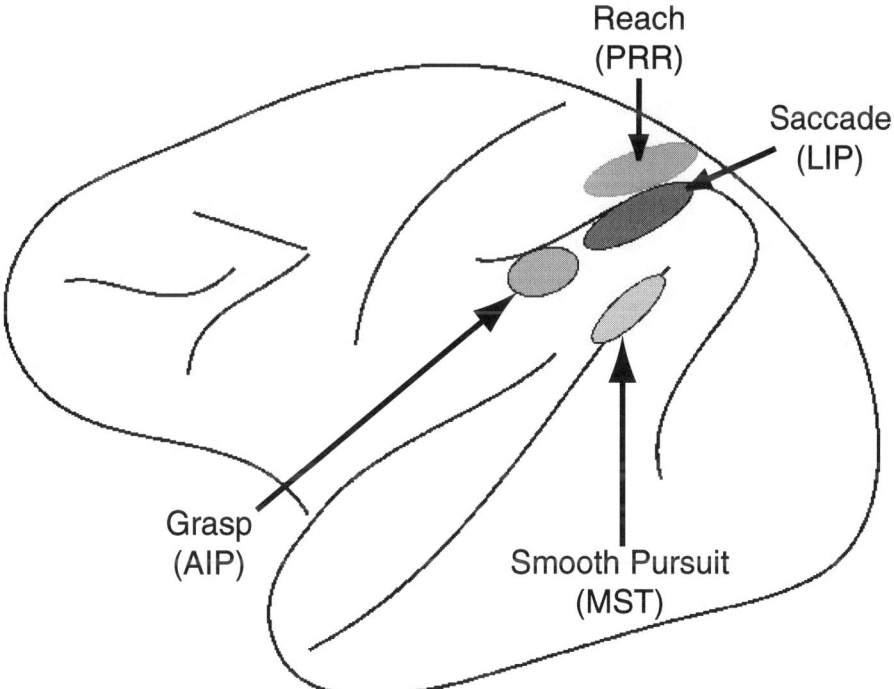

Figure 4 Anatomical map of intentions in the PPC.

appears to play a specialized role in smooth-pursuit eye movements. Cells in this area are active for pursuit, even during brief periods when the pursuit target is extinguished (Newsome et al. 1988). Inactivations of this area produce pursuit deficits that are not a result of sensory deficits (Dursteler & Wurtz 1988).

Experiments using fMRI in humans are consistent with the monkey results. Rushworth and colleagues (2001b) found that peripheral attention tasks activated the lateral bank of the intraparietal sulcus, whereas planning manual movements activated the medial bank. They concluded that their results were consistent with the monkey studies, with the medial bank specialized for manual movements and the lateral bank for attention and eye movements. A similar result has recently been reported by Connolly et al. (2000) using event-related fMRI and an eye and hand movement task similar to the one employed by Snyder et al. (1997). An area specialized for grasping has also been identified in the anterior aspect of the intraparietal sulcus in humans (Binkofski et al. 1998). This area may be homologous to monkey AIP.

Multisensory Integration and Coordinate Transformations

Producing a movement in response to a sensory stimulus requires that a host of problems be solved. From the sensory side, different sensory modalities are coded in different reference frames, vision in retinal or eye-centered coordinates, sound in head-centered coordinates, and touch in body-centered coordinates. These different coordinate frames need to be resolved in some way, since a particular movement might need to be directed to a visual, auditory, or somatosensory stimulus or any combination of these. From the motor side, the locations of these stimuli must ultimately be transformed into the natural coordinates of the muscles in order to make movements.

Lesions to the PPC can produce optic ataxia, where patients mislocalize stimuli to which they are reaching. These mislocalizations are apparent in all three dimensions. Since these effects are often present with no primary sensory or motor defects, they suggest that the PPC is important for multisensory integration as well as for the coordinate transformations required for making sensory-guided movements. In this section, we discuss electrophysiological experiments supporting this view. In particular, we provide evidence that spatial locations are represented in a common coordinate frame in at least some parts of the PPC, independent of sensory input or motor output.

AREA LIP: SACCADE PLANNING IN EYE-CENTERED COORDINATES We can easily make eye movements to visual or auditory targets. If area LIP is involved in making eye movements, then cells in this area should respond when an animal is planning an eye movement, regardless of the sensory modality of the stimulus. Recently we have found this to be the case (Grunewald et al. 1999, Linden et al. 1999, Mazzoni et al. 1996b). However, this observation raises the question, do these two modalities share a common reference frame, and if so, what is it? Cells in the intermediate layers of the SC use a common, eye-position-dependent reference frame for representing saccades to visual, auditory, or somatosensory stimuli (Groh & Sparks 1996a, Groh & Sparks 1996b, Jay & Sparks 1987a, Jay & Sparks 1987b). This is not surprising given that the SC is near the final motor output stage for saccades and that motor error is expressed in eye-centered coordinates. However, area LIP is intermediate between sensory and motor areas; thus it is not immediately apparent what reference frame should be used to represent visual and auditory targets in this region.

In experiments in which monkeys made saccades to auditory targets, we found that a majority of the neurons coded these targets in eye-centered coordinates, although some also coded auditory targets in head-centered coordinates, or in a reference frame intermediate between the eye and head reference frames (Stricanne et al. 1996). Moreover, many of the response fields of LIP neurons were gain modulated by eye position. These data suggest that area LIP may be one of the sites involved in the transformation of auditory signals from head- to eye-centered coordinates. Recent experiments examining cells in the temporo-parietal cortex (TpT) (Wu & Andersen 2001), an auditory association area that projects into the PPC (Pandya & Kuypers 1969), and the inferior colliculus (Groh et al. 2001), indicate that cells antecedent to LIP code auditory locations in head-centered coordinates, with many neurons also gain modulated by eye position. These results support a model in which head-centered auditory signals are gain modulated by eye position and are then read out at subsequent levels in eye-centered coordinates (Xing & Andersen 2000b).

PRR: REACH PLANNING IN EYE-CENTERED COORDINATES   The LIP results suggest that this area encodes sensory stimuli as motor error for saccades. If this is the case, then one might predict that PRR would code sensory stimuli as motor error as well, i.e., in limb coordinates. We tested this prediction by training monkeys to reach to targets from two different initial arm positions while fixating their gaze in two different directions. As illustrated for one PRR neuron in Figure 5 (see color insert), the response field did not vary with changes in limb position (a,b), but shifted with gaze direction (c,d). This result indicates that PRR codes limb movements in eye-centered coordinates. This result, as well as those obtained in LIP, indicates that the PPC is capable of encoding intended movements in eye-centered coordinates independent of the type of movement to be made, i.e., saccades (LIP) and reaches (PRR) (see Figure 6).

The finding that area LIP encodes intended saccades in eye-centered coordinates for both visual and auditory stimuli, as well as the finding that area PRR encodes reaches in eye-centered coordinates, led us to an unusual prediction: that PRR would code reaches to auditory stimuli in eye-centered coordinates. This prediction is based on the assumption that the PPC may use a common reference frame for movement planning, independent of sensory input or motor output. Such a result would be quite surprising since sounds, which are initially coded in head-centered coordinates, could simply be converted to body- and then limb-centered coordinates—a transformation to eye-centered coordinates is not required. In a study in which monkeys planned reaches to sounds in complete darkness, we found that this prediction was correct. Under these conditions, many cells in PRR encoded the intended movement in eye-centered coordinates (Cohen & Andersen 2000).

EYE-CENTERED CODING IN OTHER AREAS   Recent stimulation studies suggest that the SC, rather than coding desired gaze displacement or gaze direction in space, encodes the desired gaze direction in retinal coordinates (Klier et al. 2001). Electrophysiological studies of the SC have provided evidence for an eye-centered coding of limb movements in this structure as well (Stuphorn et al. 2000). The ventral premotor cortex also appears to contain neurons that code the location of reach targets in eye-centered coordinates (Mushiake et al. 1997), though these cells may coexist with others having more arm-centered properties (Graziano et al. 1994, 1997). These results, as well as those obtained in the PPC, support the existence of a distributed network devoted to eye-hand coordination that uses a common eye-centered reference frame for representing the spatial aspects of eye and arm movements (Figure 6).

COMPENSATION FOR EYE MOVEMENTS   If saccade and reach plans are coded in eye-centered coordinates, then problems can arise in situations where a movement plan is formed and an intervening saccade is made before the movement is executed. The problem is particularly acute in cases of movements planned to remembered locations in the dark. Mays & Sparks (1980) found that under these circumstances, activity shifts within the eye movement map of the SC to compensate for the intervening saccade and to still code the correct motor vector. Gnadt & Andersen (1988) reported a similar result in area LIP. Duhamel et al. (1992) extended these results by showing that it was not necessary to make an eye movement for this updating to take place. They interpreted this updated activity as sensory and a mechanism for maintaining perceptual stability across eye movements. The results of Snyder et al. (1997, 1998a) provide an alternative explanation: that this activity reflects a default plan for an eye movement. Whether this shift also accounts for perceptual stability remains a possibility and requires additional investigation.

Accounting for eye movements is also a problem when reach plans are coded in eye-centered coordinates. Imagine that an animal plans an arm movement in eye-centered coordinates to the remembered location of a stimulus in the dark and then makes an intervening eye movement before the arm movement takes place. If the reach plan is not adjusted to take into account the retinal position of the stimulus after the eye movement, then areas downstream of PRR will use the previous retinal position of the stimulus to calculate the motor error. This will result in an error corresponding to the size and direction of the intervening saccade.

We directly tested the effect of intervening saccades on intended reach activity in PRR. Figure 5e–g (see color insert) shows the design of the experiment and results from one cell. When the flash occurred outside the response field there was no response (e), and when it fell within the response field there was robust planning activity (f). Note that the histograms demonstrate planning activity; the actual arm movement occurs at a time later than that shown. In (g), the task began with the same configuration of eye, hand, and stimulus as in (e). However, after the stimulus was extinguished, the animal was instructed to make a saccade to a new location on the board, bringing the response field over the location on the board where the animal was planning the reach. The activity shifted in PRR such that the cell was now active, coding the correct location of the planned reach in eye coordinates even though the reach cue never appeared in the response field. Thus the cell compensated for the saccade to maintain the correct coding of the reach target in eye coordinates. All 34 PRR cells tested with this paradigm showed such compensation for saccades. A remapping of reach plans in eye coordinates has been demonstrated in psychophysical experiments in humans (Henriques et al. 1998), consistent with this physiological finding.

Another possible example of this type of compensation for eye movements, which must still be experimentally verified, is the compensation for smooth-pursuit eye movements that must occur for self-motion perception. During forward locomotion, self-motion perception is estimated from the focus of expansion of the visual field. However, when subjects make smooth eye movements during forward locomotion, as would occur when tracking an object on the ground, these eye movements introduce an additional, laminar motion on the retinas that disrupts the focus of expansion, generally shifting it in the direction of the eye movement. Cells in the dorsal subdivision of the medial superior temporal area (MSTd) are thought to play a role in self-motion perception because they are sensitive to optic flow stimuli and because they are tuned to the spatial location of the focus of expansion (Duffy & Wurtz 1995). In experiments from our laboratory, we found that these focus tuning curves shift to compensate for smooth-pursuit gaze movements. This compensation appears to depend on both efference copies of commands to move the eye or head and the visual information in the optic flow pattern (Bradley et al. 1996, Shenoy et al. 1999a). To guide locomotion, this signal would eventually need to be coded in body- or world-centered coordinates; however, it is currently not known in what reference frame MSTd neurons code focus-position signals. It would be consistent with the data from LIP and PRR if the MSTd cells compensated for the eye movements to maintain the correct heading direction in eye-centered coordinates.

GAIN FIELDS   The common representation of space in the PPC is embodied not only in eye-centered response fields, but also in the gain modulation of these fields by body-position signals. These gain field effects are found throughout the PPC and include modulation of retinotopic fields by eye-, head-, body-, and limb-position signals (Andersen et al. 1993). Computational studies have shown that these gain effects can be the mechanism for transforming between coordinate frames (Salinas & Abbott 1995, Zipser & Andersen 1988). Moreover, groups of neurons with retinal response fields, modulated by various body part–position signals, can conceivably be read out in multiple frames of reference (Pouget &

Snyder 2000, Xing & Andersen 2000b) as would be needed to direct movements of the eyes, head, or hand. Thus the representation of space in the PPC is distributed and is comprised of eye-centered response fields with gain modulation.

A potential problem with this type of representation is the "curse of dimensionality." For example, if it takes 10 cells to tile each dimension in visual space, and 10 for each dimension of eye position, head position, etc., the number of cells required to represent all possible combinations of such signals quickly exceeds the number of neurons in the brain. One method the PPC employs to avoid this combinatorial explosion is to code only a limited number of variables in each of its subdivisions (Snyder et al. 1998b). We have found that area LIP and area 7a both carry information about head position that is used to gain modulate the cells in this area. In area LIP this information is derived from neck proprioceptive signals indicating the orientation of the head on the body, whereas in area 7a the information is derived from vestibular signals and indicates the orientation of the head in the world. Thus, activity in LIP can be read out in body-centered coordinates, while activity in 7a can be read out in world-centered coordinates. One possible reason for this paucity of dimensionality is that area LIP may be concerned primarily with representing space for gaze shifts and eye-head coordination, whereas area 7a may be more related to representing space for navigation. In other words, these areas and possibly other cortical areas may only represent as many dimensions as are needed for the particular functions they perform. Knowledge of those dimensions may provide clues to the function of a particular area.

GAIN FIELDS AND REMAPPING   As mentioned above, eye movements are compensated for within PPC representations by shifting activity within eye-centered maps. Such a remapping of activity is necessary if coordinate transformations are to be accurately achieved using a gain mechanism. For instance, if the eyes move, the new location of a stimulus or planned movement must be adjusted in eye coordinates to correctly read out the head-centered location of the target. An important question is how this remapping is achieved. It could be accomplished using an eye displacement signal, or it could be accomplished using an eye position signal. Both eye displacement and eye position–related signals are found in the PPC (Mountcastle et al. 1975).

The experimental protocol typically used for examining remapping is the "double saccade" paradigm, in which an animal remembers two sequentially flashed targets and makes eye movements to the remembered locations of the targets in the order of their appearance. Activity in LIP appears for the next impending movement and disappears for the previous movement (Gnadt & Andersen 1988, Mazzoni et al. 1996a). More importantly, the activity for the second saccade specifies the direction and amplitude of the planned saccade, not the location on the retina in which the second flash occurred prior to the first eye movement. This compensation requires taking into account either the eye displacement for the first saccade or the new eye position after the first saccade. Patients with PPC lesions performing double saccades can make the first eye movement into the unhealthy visual field, but are not able to generate an accurate second saccade (Heide et al. 1995). Although it has been argued that this proves that an eye-displacement signal mediates remapping, eye displacement and eye position are in fact confounded in this task. The deficit was seen when the displacement of the eyes was in the direction of the unhealthy field, but also when the eye position after the first movement was in the unhealthy field.

In other experiments, area LIP was reversibly inactivated in monkeys in experiments designed to directly examine whether eye-displacement or eye-position signals are used for remapping in double-saccade experiments (Li & Andersen 2001). Both initial eye position and the direction of eye movements were varied in individual trials in order to tease apart eye-position and eye-displacement contributions. It was found that the largest deficits were seen when the animal made the first eye movement into the unhealthy visual field, largely independent of the direction of the eye movement. This result suggests that eye-position signals play a large role in the compensation for intervening saccades.

A recent computational study illustrates that dynamic neural networks can be trained to perform the double saccade task using eye-position signals (Xing & Andersen 2000a). These networks show activity similar to that recorded from LIP when monkeys perform the same task. These include eye-centered response fields that are gain modulated by eye position, and activity that shifts within an eye-centered map of visual space to correct for intervening saccades. Thus, the gain field mechanism can account for dynamic compensation for intervening eye movements in eye-centered coordinates.

GAIN FIELDS: OTHER USES   Since their discovery in areas of the PPC, gain effects have been identified throughout the brain. This suggests that multiplicative and additive interactions between different inputs to neurons may reflect a general method of neural computation. Although the role of gain fields in coordinate transformations has been highlighted in this review, gain fields appear to play a role in many other functions, including attention, navigation, decision making, and object recognition. Some examples are discussed briefly below (see also Salinas & Thier 2000).

The direction of attention can modulate the activity of V4 neurons (McAdams & Maunsell 2000, Reynolds et al. 2000), and this effect has been proposed to play a role in the binding of features in objects (Salinas & Abbott 1997). In addition, although smooth pursuit shifts the focus tuning of many MSTd neurons, as mentioned above, other MSTd neurons do not shift their focus tuning but are gain modulated by the pursuit signal (Bradley et al. 1996, Shenoy et al. 1999a). This gain modulation is consistent with an intermediate step toward the production of shifting focus-tuning curves, and thus may play a role in the perception of self-motion for navigation. Monkeys and humans have been shown to choose between two targets for a reach depending on eye position, essentially choosing targets that tend to center the reach with respect to the head (Scherberger et al. 1999). Eye-position gain effects have been shown in PRR and may bias the decision of animals to choose targets based on eye position (Scherberger & Andersen 2001).

We have recently trained monkeys to make object-based saccades by cueing a location on an object, extinguishing the object, and then presenting the object again at a different orientation. In this task, the animals must saccade to the previously cued location on the object to obtain their reward (Breznen et al. 1999). We find that area LIP does not code the cued location in an explicit object-centered reference frame in this task, even though it requires the animal to code the target in such a reference frame. Rather, cells in area LIP carry information about the cued location, movement vector, and orientation of the object, all in retinal coordinates. Some cells show a gain modulation of the cue or movement vector activity by object orientation. This result is surprising given the finding that lesions to the PPC in humans often produce deficits in object-centered coordinates (Arguin & Bub 1993, Driver & Mattingley 1998). However, computational studies show that it is not necessary to use cells with object-centered response fields to solve object-based tasks; rather, distributed coding using retinal response fields for target locations and object orientations, and gain modulations between the two, is sufficient (Pouget & Sejnowski 1997). This distributed representation can be used to form response fields in object-centered coordinates, as has been reported in the supplementary eye fields (SEF) (Olson 2001). However, the SEF results may also be explained as a result of gain modulation of retinal response fields by object position (Pouget & Sejnowski 1997), and more thorough mapping of the response fields will be required to distinguish between the two possibilities.

A COMMON DISTRIBUTED CODE FOR INTENDED MOVEMENTS IN AREAS LIP AND PRR   The above results, summarized in Figure 6, suggest that LIP and PRR use a common space representation in which response fields are represented in eye-centered coordinates. This representation exists independent of whether the targets are visual or auditory. Likewise, this representation is used regardless of whether the output is to move the limb or make an eye movement. This general scheme generated a nonintuitive, but correct, prediction that auditory targets for reach would be coded in eye-centered coordinates in PRR. Currently, we do not know if somatosensory stimuli, such as proprioceptive signals coding the position of the hand, are coded in eye coordinates in PRR and LIP. This would be an interesting question for future experimentation, and if true, would provide further evidence for the generality of this model. In both LIP and PRR, the eye-centered response fields are gain modulated by eye-, head-, and limb-position signals. This gain modulation may provide the mechanism for converting stimuli in various reference frames into eye-centered coordinates. Likewise, these gain modulations may allow other areas to read out signals from LIP and PRR in different coordinate frames, including eye-, head-, body-, and limb-centered coordinates.

Why use a common coordinate frame for PRR and LIP? One possibility is to facilitate hand-eye coordination. Presumably the orchestration of these movements would be facilitated if they used a common reference frame (Battaglia-Mayer et al. 2000). A second reason may be that vision is the most accurate spatial sense in

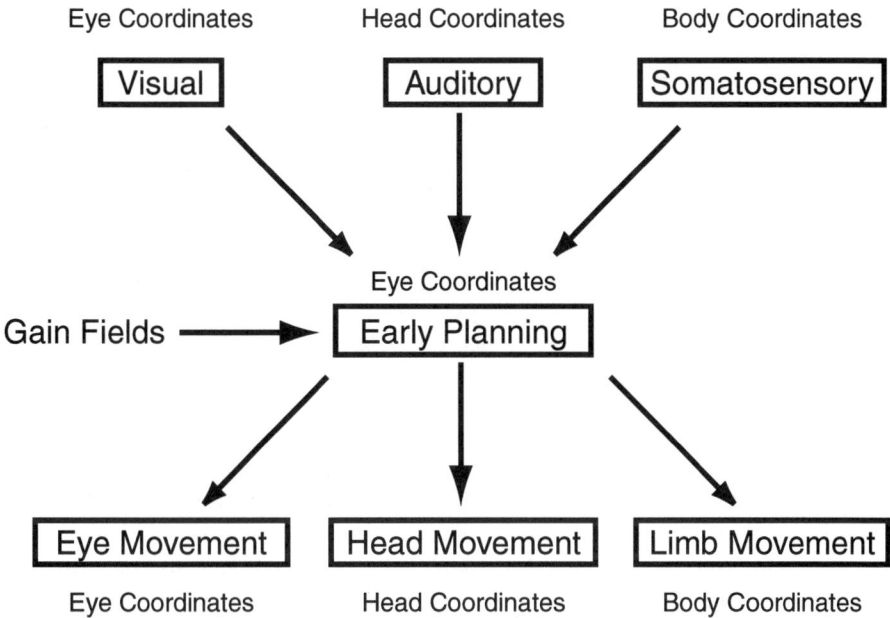

Figure 6 Theory of multisensory integration and coordinate transformations subserved by the PPC.

primates. This dominance of vision may explain certain illusions, such as the ventriloquist effect, in which the spatial locations of sounds are referred to seen objects.

COORDINATE TRANSFORMATIONS FOR REACH—DIRECT TRANSFORMATION To make visually guided reaching movements, the location of the target for the reach must be converted from eye- to limb-centered coordinates. There have been two general schemes for this transformation in the literature. One we refer to as the sequential method, shown in Figure 7a (see color insert) (Flanders et al. 1992; McIntyre et al. 1997, 1998). In this scheme, visual signals in retinal coordinates are combined with eye-position signals to represent targets in head-centered coordinates. Next, head position is combined with the representation of target location in head-centered coordinates to form a representation of the target in body-centered coordinates. Finally, the current location of the limb (in body-centered coordinates) is subtracted from the location of the target (in body-centered coordinates) to generate the motor vector, in limb-centered coordinates. There are two drawbacks to this method. One is that it requires a number of stages and separate computations, which would likely require a large number of neurons and cortical areas. The second is that, although there are some reports of cells in the PPC coding targets in head-centered coordinates (Duhamel et al. 1997, Galletti et al. 1993), the vast majority of PPC cells code visual targets in eye-centered coordinates and not head- or body-centered coordinates.

A second scheme is referred to as the combinatorial method. As shown in Figure 7*b* (see color insert), retinal target location, eye, head, and limb position are all combined at once, and the target location in limb-centered coordinates is then read out from this representation (Battaglia-Mayer et al. 2000). As mentioned above, a drawback to this approach is the "curse of dimensionality."

A third scheme we refer to as the direct method. Figure 7*c* (see color insert) shows that this approach subtracts the current position of the hand (in eye coordinates) from the position of the target (in eye coordinates) to directly generate the motor vector in limb coordinates. An advantage of this approach over the sequential method is that it requires fewer computational stages. In addition the computation is restricted to only dimensions in eye coordinates and does not suffer the curse of dimensionality of the combinatorial approach.

Recently, we have provided evidence for the direct transformation scheme. Single cells in area 5, a somatomotor cortical area within the PPC, have been found to code target locations simultaneously in eye- and limb-centered coordinates (Buneo et al. 2002). This result suggests that the PPC transforms target locations directly between these two frames of reference. Moreover, cells in PRR code the target location in eye-centered coordinates, but the initial hand position introduces a gain on this response that is also eye centered. These two findings, taken together, suggest that a simple gain field mechanism underlies the transformation from eye- to limb-centered coordinates. A convergence of input from cells in PRR onto area 5 neurons can perform this transformation directly (see Figure 7*d*) without having to resort to intermediate coordinate frames or a large combination of retina-, eye-, head-, and limb-position signals.

Psychophysical evidence supporting a sequential scheme has been provided by Flanders et al. (1992), Henriques et al. (1998), and McIntyre et al. (1997, 1998). These results, as well as our own physiological studies supporting an alternative direct scheme, may reflect an underlying context dependence in the coordinate transformations that subserves visually guided reaching (Carrozzo et al. 1999). For example, direct transformations may be the preferred scheme when both target location and the current hand position are simultaneously visible, even for a brief instant. In contrast, a sequential scheme may be used when visual information about the current position of the hand is unavailable.

MOVEMENT DECISIONS

Experiments in LIP by Platt & Glimcher (1999) and by Shadlen and colleagues (Kim & Shadlen 1999, Shadlen & Newsome 1996) have found activity related to the decision of a monkey to make eye movements. Both the prior probability and amount of reward influence the effectiveness of visual stimuli in LIP, consistent with a role for this area in decision making. As monkeys accumulate sensory information to make a movement plan, activity increases for neurons in LIP and the prefrontal cortex (Kim & Shadlen 1999, Leon & Shadlen 1999, Shadlen & Newsome 1996). These results are consistent with these areas weighting decision variables for the purpose of planning eye movements (Gold & Shadlen 2001). The fact that these effects appear in multiple brain areas suggests that decision making is a distributed function that includes the PPC.

ATTENTION

The PPC has been classically thought to play a central, perhaps controlling role in attention. Strong evidence for this idea is the finding of neglect, an inability to attend to the contralateral visual field, after PPC lesions in humans (Critchley 1953). However, many of the processes involved in visual-motor transformations, for example the shaping of the hand for grasping, appear to operate unconsciously (Goodale & Milner 1992). In fact, lesions to the PPC in monkeys produce visual-motor deficits and not neglect (Faugier-Grimaud et al. 1978, Lamotte & Acuna 1978). Rather, it has been reported that lesions to the superior temporal gyrus produce neglect similar to that found in humans (Watson et al. 1994). Interestingly, a recent report by Karnath and colleagues (2001) suggests that the superior temporal gyrus damage may also be the source of neglect seen in humans. Thus, the locus of cortical lesions that produce neglect is still an open question that will likely be resolved with further research.

Although there have been several studies reporting attentional effects on neural activity in the PPC, these experiments have been performed in conjunction with eye or limb movements, or in peripheral attention paradigms where animals are likely to form covert plans for eye movements. As yet no experiments have been performed similar to those of Snyder et al. (1997, 1998a). In those experiments, intention was isolated from attentional effects; similar experimental designs are needed to isolate attentional effects from intentional effects.

It has been argued that the lower degree of activation of LIP neurons when monkeys reach rather than saccade to targets is due to less attention being required for reaching (Colby & Goldberg 1999). This reasoning would predict less activity in PRR as well, but in fact the reverse is true. Figure 8 shows the population activity from one monkey for recordings obtained in LIP and PRR. When this monkey planned, a saccade activity was high in LIP and low in PRR. On the other hand, when reaches were planned the reverse was true. This figure also shows that when covert planning was controlled in dual-movement trials the separation for saccades and reaches was even greater. This double dissociation between saccades and reaches for LIP and PRR shows that the effects are due to planning, and not a general reduction in attention when the animal reaches.

In a recent study, Powell & Goldberg (2000) flashed stimuli around the time of eye movements and found responses for LIP neurons even while the animals were planning eye movements outside of the cells' response fields. They argued that this demonstrated that LIP is more involved in registering the salience of visual stimuli than in planning eye movements. This interpretation is at odds with that of Mazzoni et al. (1996a), who showed that, when monkeys were planning eye movements outside of the response field of LIP neurons, a flash in the center of their response field produced only a very brief transient before the activity was suppressed. A closer inspection of the data in Powell & Goldberg shows

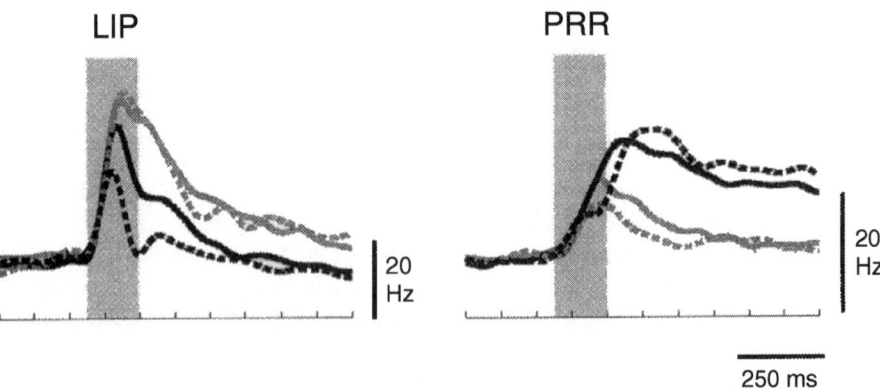

Figure 8 Population response from one monkey for areas LIP (*left*) and PRR (*right*). Cells had significant activity during the delay period of either the reach or saccade task of Snyder et al. (1997). Solid gray traces represent the average activity of the population of cells for saccades into the response field. Solid black traces represent activity for reaches into the response field. Dashed lines represent activity in the dual-movement task, with gray traces representing saccades into the response field and black traces representing reaches. Histograms were smoothed with a 181-point digital low-pass filter with a −3 dB point at 9 Hz. (From Snyder et al. 2000.)

a similar effect, with activity dying out prior to and during the eye movement. A similar sensory transient can be seen when a reach target is flashed within an LIP response field, which quickly dies away in this case hundreds of msec before the reach movement (see Figure 1*b*, right histogram, see color insert). Interestingly, units in simulated networks programmed to hold a movement plan for one location while a distractor stimulus is briefly flashed at another location [conditions similar to the experiment of Powell & Goldberg (2000)] also exhibit input transients (Xing & Andersen 2000a). However, lateral inhibitory connections in these networks quickly suppressed the activity due to the distractor, similar to the suppression seen in LIP experiments.

In conclusion, it is likely that pure attentional effects will be found in the PPC. However, because this area is specialized for sensory-motor integration, there is the additional challenge of designing paradigms that rule out movement planning as a source of activity. This concern is less problematic in areas of the visual cortex that are involved in recognition, where there is ample evidence of attentional effects. Attentional effects in PPC may be general, since attention-related activity has been reported throughout the cerebral cortex. Thus attention effects in PPC would be related to planning movements, much like attention effects in inferotemporal cortex would be related to visual recognition. Interestingly, recent studies suggest PPC and prefrontal structures may regulate spatial aspects of attention in the ventral, recognition pathway (Kastner & Ungerleider 2000). Whether PPC is specialized for attention and is the controller for attention throughout cortex is an important question.

LEARNING AND ADAPTATION

Like attention, learning is a distributed function and like atttentional effects, learning effects in the PPC tend to be most apparent in the context of sensory-motor operations, for example prism adaptation. As first demonstrated by Held & Hein (1963), when human subjects reach to visual targets while wearing displacing prisms, they initially miss-reach in the direction of target displacement but gradually recover and reach correctly if provided with appropriate feedback about their errors. Using positron emission tomography (PET) to monitor changes in cerebral blood flow, Clower and colleagues (1996) showed that the prism adaptation process results in selective activation of the PPC contralateral to the reaching arm, when confounding sensory, motor, and cognitive effects are ruled out. Similarly, Rosetti and colleagues (1998) found that hemispatial neglect resulting from damage to the right hemisphere can be at least partially ameliorated by first having affected patients make reaching movements in the presence of a prismatic shift, then removing the prisms. They interpreted these effects as resulting from the stimulation of neural structures responsible for sensorimotor transformations, including the PPC as well as the cerebellum. A recent electrophysiological study employing a prism adaptation paradigm suggests that the ventral premotor cortex plays a role in this process as well (Kurata & Hoshi 1999).

An example of the effects of learning in the PPC was revealed in a recent electrophysiological study of LIP (Grunewald et al. 1999). The responses of LIP neurons to auditory stimuli in a passive fixation task were examined before and after animals were trained to make saccades to auditory targets. Before such training, the number of cells responding to auditory stimuli in LIP was statistically insignificant. After training, however, 12% showed significant responses to auditory stimuli. This indicates that at least some LIP neurons become active for auditory stimuli only after an animal has learned that these stimuli are important for oculomotor behavior. As with the learning effects discussed above, effects of this nature have been reported in other areas of cortex, e.g., area 3a for tactile discrimination (Recanzone et al. 1992), premotor cortex for arbitrary associations (Mitz et al. 1991), and the frontal eye fields (FEF) for visual search training (Bichot et al. 1996), highlighting the distributed nature of learning in cortex.

READING OUT INTENTIONS—THE PPC AND NEURAL PROSTHETICS

The experimental results reviewed above indicate that activity related to an animal's intentions to make reaches or saccades is strong and robust in the PPC. These experiments are typical of neurophysiological experiments in demonstrating correlations between behaviors or perception and neural activity. It would be powerful to be able to test these proposed linkages more directly. One method to achieve a more direct demonstration of intention coding is to perform experiments that close the loop. This type of experiment is demonstrated in Figure 9.

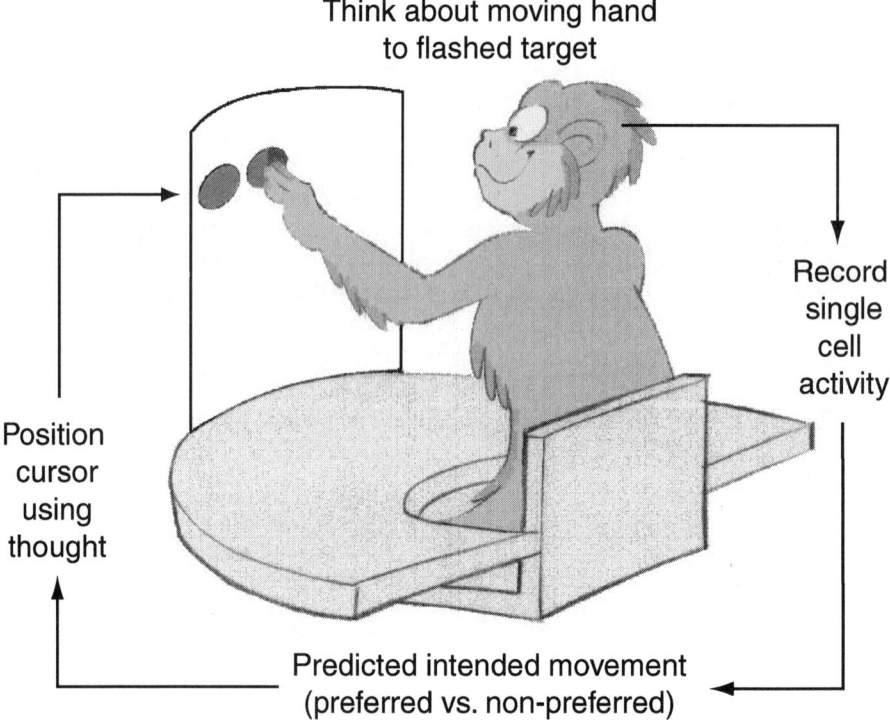

Figure 9 Schematic of an experiment in "closing the loop." In this experiment, the activity of a single-PRR neuron is first isolated and a database of the neuron's responses for reaches in the preferred and nonpreferred direction is constructed. In the relevant trials, the animal fixates and touches a central cursor on a display screen and a target is then presented either in the preferred or nonpreferred direction of the cell. Based on the response of the neuron for this one trial, and the statistics of the previously constructed database, we then predict whether this animal is intending to reach in either the preferred or nonpreferred direction. Importantly, the animal never actually makes the intended movement on these trials. A cursor moves to the location predicted from the cell activity and the animal is rewarded if the prediction corresponds to the cued location.

Rather than simply recording the monkeys' reach intentions from PRR and then rewarding them for making a reach, we record their intentions and use them to move a cursor to the location on the display screen where we predict, based on the neural activity, that they are intending to reach. This cursor provides feedback to the animals, indicating where the reach is predicted to end up, based purely on the animals' thoughts, i.e., without the animals making any actual reaches.

In preliminary studies we have used PRR activity to predict one out of two possible reach directions in real time (D. Meeker, S. Cao, J. W. Burdick & R. A. Andersen, unpublished observations). In other words, the animals perform a task in which they locate a cursor on the computer screen by using intended movement activity without making an actual reach. In other studies, offline analysis of single-cell recordings from PRR indicates that reliably predicting one out of eight directions could require only a small number of simultaneously recorded neurons, perhaps in the range of 10 to 15 (Meeker et al. 2001, Shenoy et al. 1999b).

The high-level planning activity observed in PRR could be used in the control of a neural prosthetic for paralyzed patients. Patients with paralysis due to peripheral neuropathies, trauma, and stroke can often still think about making movements but cannot execute them. The idea of a cortical prosthetic is to record these intentions to move, interpret the intentions using real-time decode algorithms running on computers, and then convert these decoded intentions to control signals that can operate external devices. These external devices could include stimulators imbedded in the patient's muscles that would allow the patient to move his/her own body, a robot limb, or a computer interface for communication.

Research on neural prosthetics is a burgeoning and young field. Several groups are working toward using motor cortex for such a prosthesis (Chapin et al. 1999, Isaacs et al. 2000, Wessberg et al. 2000, Kennedy et al. 2000, Maynard et al. 1999), which makes sense because it is the area of cortex closest to the motor output. There may also be advantages for using higher, cognitive areas of the sensory-motor system such as PRR for the control of prosthetics. Because this is a field that is still at its infancy, it is not clear if one area is optimal for prosthetic control. Moreover, because different areas of the sensory-motor pathway no doubt provide different, useful information, it may turn out to be most optimal to develop multiarea prosthetics that can read out and decode these different signals. Below are listed some potential attributes of PRR for prosthetic control.

1.) Motor cortex is known to undergo degradation as a result of paralysis. For instance with spinal cord lesions, cortico-spinal neurons are destroyed and the somatosensory reafferent signals to this area of the cortex are also lost. However, PRR may undergo less degradation after paralysis as it is more closely tied to the visual system.

2.) Learning is an important aspect of neural prosthetic success. For instance, cochlear prosthetics do not produce natural auditory sensations with electrical stimulation, but patients learn to interpret these stimulus-induced percepts, for instance in understanding speech. Implanted arrays of electrodes in humans will likely only sample a part of the workspace of the subject. Although other parts of space not well sampled can be inferred from the activity of cells in sampled regions, the resolution cannot be as good as would result from a more even sampling. Thus, neural plasticity would be an important advantage for a cortical prosthetic and, as mentioned above, there is evidence that the PPC does play a role in adjusting the registration of sensory-motor representations for accurate behaviors.

3.) To successfully close the loop, patients require feedback to the cortex regarding the success of the movement. Whereas these feedback signals are lost with paralysis in motor areas, they are largely intact in PRR, as the reafference to this region is largely visual. Moreover, the evidence presented above, that the PPC performs early coordinate transformations required for reaching in retinal coordinates, suggests that vision can be used by the patients for correcting motor error computations.

4.) The fact that the intended movement signal is a high-level, cognitive signal may have advantages. For instance, if PRR is coding intentions in abstract or general terms, it may require fewer neurons to control devices that are dissimilar to the human limb. In addition, the fact that this planning activity can be sustained for long periods of time may help in the decoding of the movement plan by providing prolonged, stable signals related to the intentions of the patient.

5.) We have recently found that, during the delays when monkeys are planning a movement, there are broadband gamma oscillations in the local field potentials (LFPs) in both LIP and PRR. The LFPs in both areas are tuned to the direction of planned movements and change their strength with behavioral state, becoming much larger in amplitude when the animal plans a movement and rapidly decreasing during the movement execution. Although it is somewhat difficult to record single-cell activity over long periods of time with chronically implanted electrodes, LFPs are stable and relatively easy to record with chronic electrodes, since they reflect the activity of cortical columns rather than single cells. Thus, using LFPs in LIP and PRR may be an important breakthrough for obtaining long-term recordings with existing technology.

CONCLUSION

The PPC is important for sensory-motor integration, particularly the forming of intentions or high-level cognitive plans for movement. There is a map of intentions within the PPC, with subregions for saccades, reaching, and grasping. These regions appear to be specialized for multisensory integration and coordinate transformations. Within each subregion is a map of the working space. In PRR and LIP these maps are in eye-centered coordinates, regardless of sensory input or movement plan. Within these maps, activity is gain modulated by eye-, head-, and limb-position signals. This distributed, abstract representation is consistent with intentions in this area being high level and cognitive.

These ideas perhaps raise more questions than they answer. For instance, what other features of intended movements are coded in the activity of PPC neurons? Are the trajectories, distance of reaching, or the dynamics of the planned movement coded in PRR activity? Is the transition from a high-level, cognitive intention to the motor output of the cortex one of multiple stages and continual refinement of the plan, or is the intention to move converted to an executable plan in one step of convergence onto motor cortical areas? Are there additional cortical areas in PPC for intending other types of movement, such as leg movements and head movements?

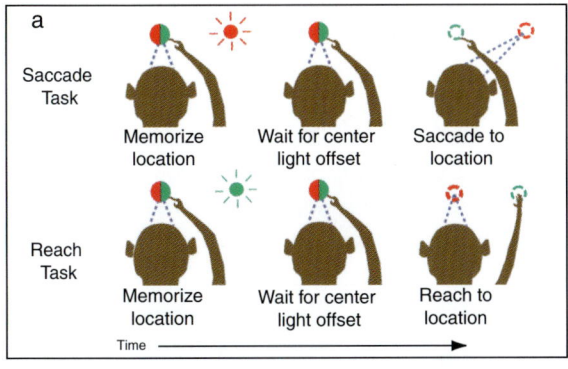
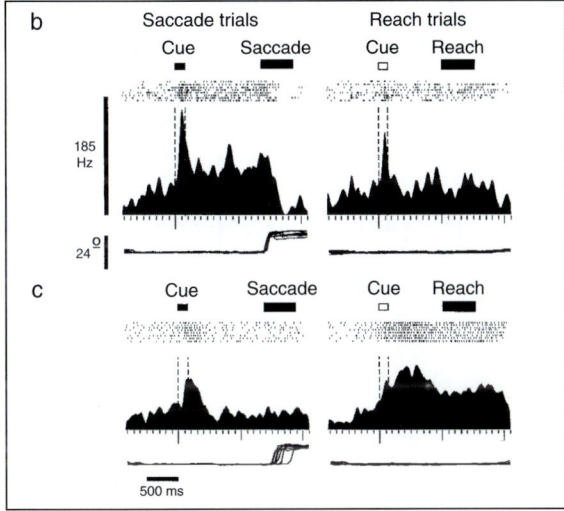

Figure 1 (*a*) Tasks used in Snyder et al. (1997) to separate the effects of spatial attention from those of intention. Animals made either a saccade (*top*) or a reach (*bottom*) to the remembered location of a flashed visual target (red flash: saccade, green flash: reach). Movements were made in complete darkness, after a delay period. (*b*) A lateral intraparietal area cell showing elevated delay period activity before a saccade (*left*) but not before a reach (*right*). *Vertical dashed lines* and *short horizontal bars* indicate the timing of target ("Cue") presentation (red flashes: filled bars, green flashes: open bars) and *long horizontal bars* indicate the timing of the motor response ("Saccade" or "Reach"). Each panel shows eight rasters of tick marks corresponding to every third action potential recorded during each of eight trials. Below each set of rasters is a spike density histogram representing the average rate of action potential firing over all trials (generated by convolution with a triangular kernel) that is aligned on cue presentation. *Thin horizontal lines* below each histogram represent the animal's vertical eye position on each trial. During the delay interval (150–600 msec after target extinction) firing depended specifically on motor intent. For illustration purposes, data for this cell were collected using a fixed delay interval. (*c*) A PRR cell showing reach rather than saccade specificity during the delay interval. (Modified from Snyder et al. 1997)

Different initial hand positions
Different eye positions
a
b
c
d
120
Hz
300
ms
H
E

e
f
g
40
Hz
36º
H
E

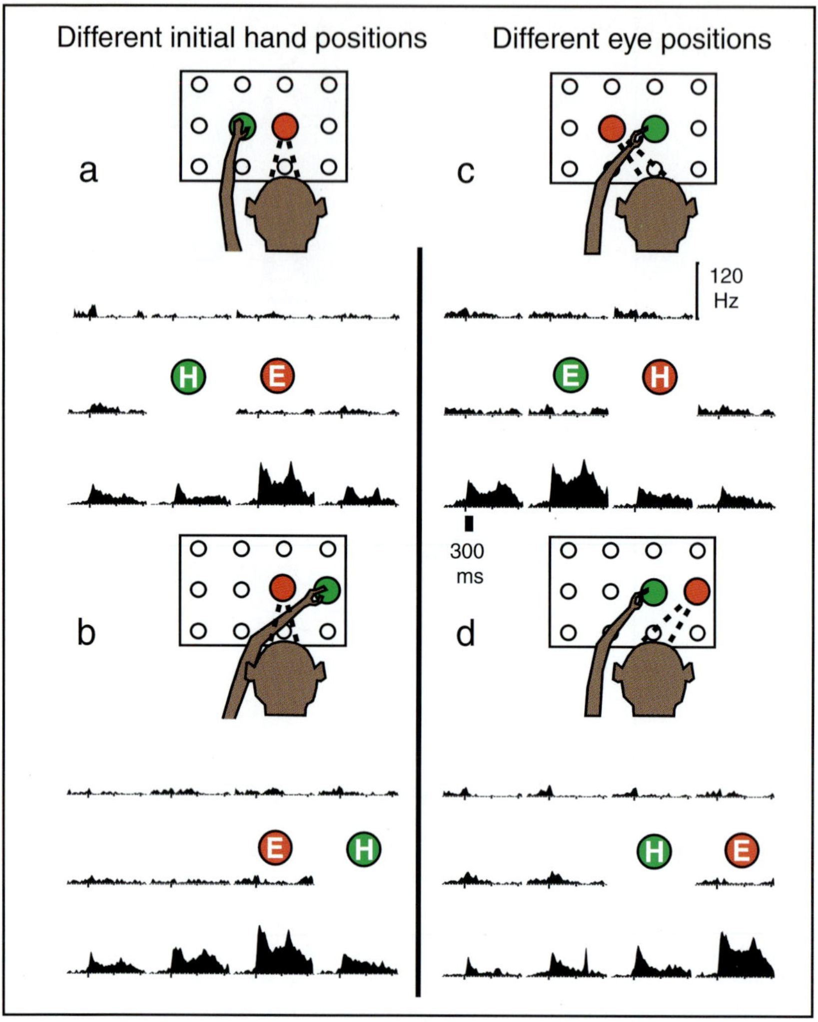

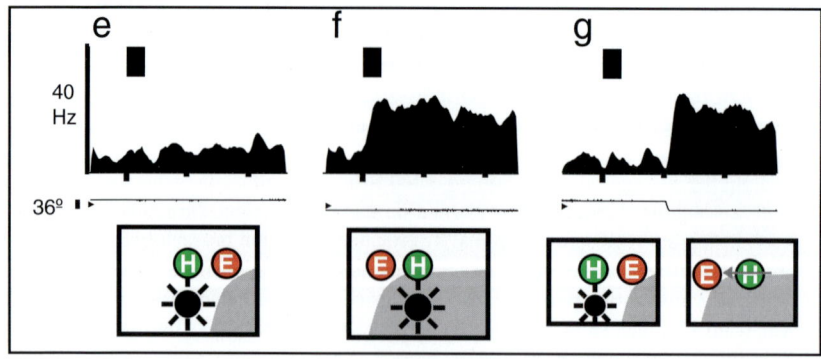

Figure 5 (*a–d*) A PRR neuron that codes target locations in eye-centered coordinates. Icons depict the four possible behavioral conditions at the beginning of a trial; initial hand position and the point of visual fixation are represented by *green* and *red circles*, respectively. *Open circles* represent target locations on a vertically oriented board of push buttons. Below each icon, spike density histograms (aligned at cue onset and smoothed as in Figure 1) are plotted at positions corresponding to the target locations on the board (11 locations in *a, b,* and *d*; 10 locations in *c*). *Short horizontal bar* below the histograms in (*c*) represents the timing of the cue. The response field of this neuron did not vary with changes in limb position (*a, b*) but shifted with gaze direction (*c, d*). (Modified from Batista et al. 1999). (*e–g*) Activity of a PRR neuron in an intervening saccade experiment. Each spike density histogram shows the response of the cell for the experimental conditions illustrated in the corresponding icon (*below*). *Shaded region* represents the spatial extent of the cell's response field. (*e*) Activity when the target is presented outside of the response field. (*f*) Response when the target is in the response field. (*g*) Activity when an eye movement carries the reach goal into the neuron's response field. This cell compensated for the saccade to maintain the correct coding of the reach target in eye coordinates. The H and E in the icons below the histograms indicate the position of the hand and eye on the reach board, the *black target* indicates the location of a flashed reached target, and the *shaded area* indicates the spatial extent of the response field of a PRR neuron. The *squares* above the histograms show the time of the flashed target, and the *traces* below the histograms are the recorded eye positions. (Modified from Batista et al. 1999).

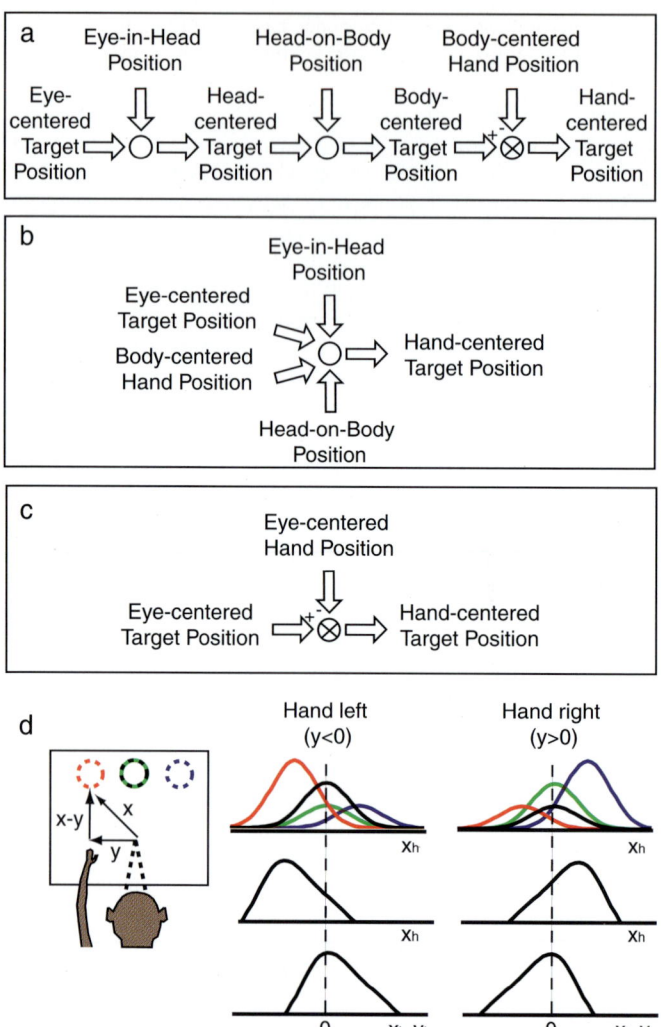

Figure 7 (*a–c*) Schemes for transforming target position from eye-centered to hand-centered coordinates. (*a*) Sequential method. (*b*) Combinatorial method. (*c*) Direct method. (*d*) Illustration showing how convergent input from neurons encoding target position (x) and initial hand position (y) in eye coordinates can drive downstream responses that encode target position in hand-centered coordinates (x-y), as in the direct method. *Top row of curves* represents the responses of four idealized neurons at two initial hand positions. *Middle* and *bottom rows* show the responses of a downstream neuron, derived as a weighted sum of the responses in the top row minus a constant (Salinas & Abbott 1995, Salinas & Thier 2000). When initial hand position varies, the peak response of the downstream neuron shifts in eye-centered coordinates (*middle row*), but remains fixed in hand-centered coordinates (*bottom row*). For simplicity, only the horizontal components of responses are shown.

There are a number of outstanding questions regarding the finding of a common coordinate frame for spatial location in LIP and PRR. For instance, do other parietal areas, such as the grasp area, also code in retinal coordinates with gain modulation by eye- and other body part–position signals? Are somatosensory signals also represented in eye coordinates in these areas? Does this common reference frame also support intended head movements? Are there tasks in which the intentions to move do not use this common, distributed reference frame? Experiments designed to answer these questions will help to determine how generally the concepts of cognitive intentions and common reference frames can be applied to the PPC.

ACKNOWLEDGMENTS

We wish to acknowledge the generous support of the James G. Boswell Foundation, the National Institutes of Health (NIH), the Defense Advanced Research Projects Agency (DARPA), Sloan-Swartz Center for Theoretical Neurobiology, the Office of Naval Research (ONR), and the Christopher Reeves Foundation. We also thank Dr. Paul Glimcher for helpful comments.

The *Annual Review of Neuroscience* is online at http://neuro.annualreviews.org

LITERATURE CITED

Andersen RA. 1987. *The Role of the Inferior Parietal Lobule in Spatial Perception and Visual-Motor Integration.* Bethesda, MD: Am. Physiol. Soc., pp. 483–518

Andersen RA, Essick GK, Siegel RM. 1987. Neurons of area 7a activated by both visual stimuli and oculomotor behavior. *Exp. Brain Res.* 67:316–22

Andersen RA, Snyder LH, Li CS, Stricanne B. 1993. Coordinate transformations in the representation of spatial information. *Curr. Opin. Neurobiol.* 3:171–76

Arguin M, Bub DN. 1993. Evidence for an independent stimulus-centered spatial reference frame from a case of visual hemineglect. *Cortex* 29:349–57

Balint R. 1909. Seelenlahmung des "Schauens," optische Ataxie, raumliche Storung der Aufmerksamkeit. *Monatsschr. Psychiatr. Neurol.* 25:51–81

Batista AP, Buneo CA, Snyder LH, Andersen RA. 1999. Reach plans in eye-centered coordinates. *Science* 285:257–60

Batista AP, Andersen RA. 2001. The parietal reach region codes the next planned movement in a sequential reach task. *J. Neurophysiol.* 85:539–44

Battaglia-Mayer A, Ferraina S, Mitsuda T, Marconi B, Genovesio A, et al. 2000. Early coding of reaching in the parietooccipital cortex. *J. Neurophysiol.* 83:2374–91

Bichot NP, Schall JD, Thompson KG. 1996. Visual feature selectivity in frontal eye fields induced by experience in mature macaques. *Nature* 381:697–99

Binkofski F, Dohle C, Posse S, Stephan KM, Hefter H, et al. 1998. Human anterior intraparietal area subserves prehension: a combined lesion and functional MRI activation study. *Neurology* 50:1253–59

Bracewell RM, Mazzoni P, Barash S, Andersen RA. 1996. Motor intention activity in the macaque's lateral intraparietal area. II. Changes of motor plan. *J. Neurophysiol.* 76:1457–64

Bradley DC, Maxwell M, Andersen RA, Banks MS, Shenoy KV. 1996. (Neural) mechanisms of heading perception in primate visual cortex. *Science* 273:1544–47

Breznen B, Sabes PN, Andersen RA. 1999. Parietal coding of object-based saccades: reference frames. *Soc. Neurosci. (Abstr.)* 25:1547

Buneo CA, Jarvis MR, Batista AP, Andersen RA. 2002. Direct visuomotor transformations for reaching. *Nature*. In press

Bushnell MC, Goldberg ME, Robinson DL. 1981. Behavioral enhancement of visual responses in monkey cerebral cortex. I. Modulation in posterior parietal cortex related to selective visual attention. *J. Neurophysiol.* 46:755–74

Carrozzo M, McIntyre J, Zago M, Lacquaniti F. 1999. Viewer-centered and body-centered frames of reference in direct visuomotor transformations. *Exp. Brain Res.* 129:201–10

Chaffee MV, Goldman-Rakic PS. 1998. Matching patterns of activity in primate prefrontal area 8a and parietal area 7ip during a spatial working memory task. *J. Neurophysiol.* 79:2919–40

Chapin JK, Moxon KA, Markowitz RS, Nicolelis MA. 1999. Real-time control of a robot arm using simultaneously recorded neurons in the motor cortex. *Nat. Neurosci.* 2:664–70

Clower DM, Hoffman JM, Votaw JR, Faber TL, Woods RP, Alexander GE. 1996. Role of posterior parietal cortex in the recalibration of visually guided reaching. *Nature* 383:618–21

Cohen YE, Andersen RA. 2000. Reaches to sounds encoded in an eye-centered reference frame. *Neuron* 27:647–52

Colby CL, Goldberg ME. 1999. Space and attention in parietal cortex. *Annu. Rev. Neurosci.* 22:319–49

Connolly JD, Menon RS, Goodale MA. 2000. Human frontoparietal areas active during a pointing but not a saccade delay. *Soc. Neurosci. (Abstr.)* 26:1329

Corbetta M, Akbudak E, Conturo TE, Snyder AZ, Ollinger JM, et al. 1998. A common network of functional areas for attention and eye movements. *Neuron* 21:761–73

Critchley M. 1953. *The Parietal Lobes.* London: Arnold

DeSouza JF, Dukelow SP, Gati JS, Menon RS, Andersen RA, Vilis T. 2000. Eye position signal modulates a human parietal pointing region during memory-guided movements. *J. Neurosci.* 20:5835–40

Driver J, Mattingley JB. 1998. Parietal neglect and visual awareness. *Nat. Neurosci.* 1:17–22

Duffy CJ, Wurtz RH. 1995. Response of monkey MSTd neurons to optic flow stimuli with shifted centers of motion. *J. Neurosci.* 15:5192–208

Duhamel JR, Colby CL, Goldberg ME. 1992. The updating of the representation of visual space in parietal cortex by intended eye movements. *Science* 255:90–92

Duhamel JR, Bremmer F, Ben Hamed S, Grof W. 1997. Spatial invariance of visual receptive fields in parietal cortex neurons. *Nature* 389:845–48

Dursteler MR, Wurtz RH. 1988. Pursuit and optokinetic deficits following chemical lesions of cortical areas Mt and Mst. *J. Neurophysiol.* 60:940–65

Eskandar EN, Assad JA. 1999. Dissociation of visual, motor and predictive signals in parietal cortex during visual guidance. *Nat. Neurosci.* 2:88–93

Faugier-Grimaud S, Frenois C, Stein DG. 1978. Effects of posterior parietal lesions on visually guided behavior in monkeys. *Neuropsychologia* 16:151–68

Ferraina S, Battaglia-Mayer A, Genovesio A, Marconi B, Onorati P, Caminiti R. 2001. Early coding of visuomanual coordination during reaching in parietal area PEc. *J. Neurophysiol.* 85:462–67

Ferraina S, Garasto MR, Battaglia-Mayer A, Ferraresi P, Johnson PB, et al. 1997. Visual control of hand-reaching movement: activity in parietal area 7m. *Eur. J. Neurosci.* 9:1090–95

Flanders M, Helms-Tillery SI, Soechting JF.

1992. Early stages in a sensorimotor transformation. *Behav. Brain Sci.* 15:309–62

Galletti C, Battaglini PP, Fattori P. 1993. Parietal neurons encoding spatial locations in craniotopic coordinates. *Exp. Brain Res.* 96: 221–29

Geshwind N, Damasio AR. 1985. Apraxia. In *Handbook of Clinical Neurology*, ed. PJ Vinken, GW Bruyn, HL Klawans, pp. 423–32. Amsterdam: Elsevier

Gnadt JW, Andersen RA. 1988. Memory related motor planning activity in posterior parietal cortex of macaque. *Exp. Brain Res.* 70:216–20

Gold JI, Shadlen MN. 2001. Neural computations that underlie decisions about sensory stimuli. *Trends Cogn. Sci.* 5:10–16

Goodale MA, Milner AD. 1992. Separate visual pathways for perception and action. *Trends Neurosci.* 15:20–25

Gottlieb J, Goldberg ME. 1999. Activity of neurons in the lateral intraparietal area of the monkey during an antisaccade task. *Nat. Neurosci.* 2:906–12

Graziano MS, Hu XT, Gross CG. 1997. Visuospatial properties of ventral premotor cortex. *J. Neurophysiol.* 77:2268–92

Graziano MS, Yap GS, Gross CG. 1994. Coding of visual space by premotor neurons. *Science* 266:1054–57

Groh JM, Sparks DL. 1996a. Saccades to somatosensory targets. 2. Motor convergence in primate superior colliculus. *J. Neurophysiol.* 75:428–38

Groh JM, Sparks DL. 1996b. Saccades to somatosensory targets. 3. Eye-position-dependent somatosensory activity in primate superior colliculus. *J. Neurophysiol.* 75:439–53

Groh JM, Trause AS, Underhill AM, Clark KR, Inati S. 2001. Eye position influences auditory responses in primate inferior colliculus. *Neuron* 29:509–18

Grunewald A, Linden JF, Andersen RA. 1999. Responses to auditory stimuli in macaque lateral intraparietal area. I. Effects of training. *J. Neurophysiol.* 82:330–42

Heide W, Blankenburg M, Zimmermann E, Kompf D. 1995. Cortical control of double-step saccades—implications for spatial orientation. *Ann. Neurol.* 38:739–48

Held R, Hein A. 1963. Movement produced stimulation in the development of visually guided behavior. *J. Comp. Physiol. Psychol.* 53:236–41

Henriques DY, Klier EM, Smith MA, Lowy D, Crawford JD. 1998. Gaze-centered remapping of remembered visual space in an open-loop pointing task. *J. Neurosci.* 18:1583–94

Hikosaka O, Wurtz RH. 1983. Visual and oculomotor functions of monkey substantia nigra pars reticulata. III. Memory-contingent visual and saccade responses. *J. Neurophysiol.* 49:1268–84

Isaacs RE, Weber DJ, Schwartz AB. 2000. Work toward real-time control of a cortical neural prothesis. *IEEE Trans. Rehabil. Eng.* 8:196–98

Jay MF, Sparks DL. 1987a. Sensorimotor integration in the primate superior colliculus. 1. Motor convergence. *J. Neurophysiol.* 57:22–34

Jay MF, Sparks DL. 1987b. Sensorimotor integration in the primate superior colliculus. 2. Coordinates of auditory signals. *J. Neurophysiol.* 57:35–55

Kalaska JF. 1996. Parietal cortex area 5 and visuomotor behavior. *Can. J. Physiol. Pharmacol.* 74:483–98

Kalaska JF, Crammond DJ. 1995. Deciding not to GO: neural correlates of response selection in a GO/NOGO task in primate premotor and parietal cortex. *Cerebr. Cortex* 5:410–28

Karnath H, Ferber S, Himmelbach M. 2001. Spatial awareness is a function of the temporal not the posterior parietal lobe. *Nature* 411:950–53

Kastner S, Ungerleider LG. 2000. Mechanisms of visual attention in the human cortex. *Annu. Rev. Neurosci.* 23:315–41

Kennedy PR, Bakay RAE, Moore MM, Adams K, Goldwaithe J. 2000. Direct control of a computer from the human central nervous system. *IEEE Trans. Rehabil. Eng.* 8:198–202

Kim JN, Shadlen MN. 1999. Neural correlates of a decision in the dorsolateral prefrontal cortex of the macaque. *Nat. Neurosci.* 2:176–85

Klier EM, Wang H, Crawford JD. 2001. The superior colliculus encodes gaze commands in retinal coordinates. *Nat. Neurosci.* 4:627–32

Kurata K, Hoshi E. 1999. Reacquisition deficits in prism adaptation after muscimol microinjection into the ventral premotor cortex of monkeys. *J. Neurophysiol.* 81:1927–38

Kustov AA, Robinson DL. 1996. Shared neural control of attentional shifts and eye movements. *Nature* 384:74–77

Lamotte RH, Acuna C. 1978. Defects in accuracy of reaching after removal of posterior parietal cortex in monkeys. *Brain Res.* 139:309–26

Leon MI, Shadlen MN. 1999. Effect of expected reward magnitude on the response of neurons in the dorsolateral prefrontal cortex of the macaque. *Neuron* 24:415–25

Li CS, Andersen RA. 2001. Inactivation of macaque lateral intraparietal area delays initiation of the second saccade predominantly from contralesional eye positions in a double-saccade task. *Exp. Brain Res.* 137:45–57

Linden JF, Grunewald A, Andersen RA. 1999. Responses to auditory stimuli in macaque lateral intraparietal area. II. Behavioral modulation. *J. Neurophysiol.* 82:343–58

Mays LE, Sparks DL. 1980. Saccades are spatially, not retinocentrically, coded. *Science* 208:1163–65

Maynard EM, Hatsopoulos NG, Ojakangas CL, Acuna BD, Sanes JN, et al. 1999. Neuronal interactions improve cortical population coding of movement direction. *J. Neurosci.* 19:8083–93

Mazzoni P, Bracewell RM, Barash S, Andersen RA. 1996a. Motor intention activity in the macaque's lateral intraparietal area. I. Dissociation of motor plan from sensory memory. *J. Neurophysiol.* 76:1439–56

Mazzoni P, Bracewell RM, Barash S, Andersen RA. 1996b. Spatially tuned auditory responses in area LIP of macaques performing delayed memory saccades to acoustic targets. *J. Neurophysiol.* 75:1233–41

McAdams CJ, Maunsell JHR. 2000. Attention to both space and feature modulates neuronal responses in macaque area V4. *J. Neurophysiol.* 83:1751–55

McIntyre J, Stratta F, Lacquaniti F. 1997. Viewer-centered frame of reference for pointing to memorized targets in three-dimensional space. *J. Neurophysiol.* 78:1601–18

McIntyre J, Stratta F, Lacquaniti F. 1998. Short-term memory for reaching to visual targets: psychophysical evidence for body-centered reference frames. *J. Neurosci.* 18:8423–35

Meeker D, Shenoy KV, Cao S, Pesaran B, Scherberger H, et al. 2001. Cognitive control signals for prosthetic systems. *Soc. Neurosci. (Abstr.)* 27:63.6

Mitz AR, Godschalk M, Wise SP. 1991. Learning-dependent neuronal-activity in the premotor cortex-activity during the acquisition of conditional motor associations. *J. Neurosci.* 11:1855–72

Mountcastle VB, Lynch JC, Georgopoulos A, Sakata H, Acuna C. 1975. Posterior parietal association cortex of the monkey: command functions for operations within extrapersonal space. *J. Neurophysiol.* 38:871–908

Mushiake H, Tanatsugu Y, Tanji J. 1997. Neuronal activity in the ventral part of premotor cortex during target-reach movement is modulated by direction of gaze. *J. Neurophysiol.* 78:567–71

Newsome WT, Wurtz RH, Komatsu H. 1988. Relation of cortical areas MT and MST to pursuit eye movements. II. Differentiation of retinal from extraretinal inputs. *J. Neurophysiol.* 60:604–20

Olson CR. 2001. Object-based vision and attention in primates. *Curr. Opin. Neurobiol.* 11:171–79

Pandya DN, Kuypers HG. 1969. Cortico-cortical connections in the rhesus monkey. *Brain Res.* 13:13–36

Pare M, Wurtz RH. 1997. Monkey posterior parietal cortex neurons antidromically activated from superior colliculus. *J. Neurophysiol.* 78:3493–97

Perenin MT, Vighetto A. 1988. Optic ataxia: a specific disruption in visuomotor mechanisms. I. Different aspects of the deficit in reaching for objects. *Brain* 111:643–74

Platt ML, Glimcher PW. 1999. Neural correlates of decision variables in parietal cortex. *Nature* 400:233–38

Pouget A, Sejnowski TJ. 1997. A new view of hemineglect based on the response properties of parietal neurones. *Philos. Trans. R. Soc. Lond. B. Biol. Sci.* 352:1449–59

Pouget A, Snyder LH. 2000. Computational approaches to sensorimotor transformations. *Nat. Neurosci.* 3:1193–98

Powell KD, Goldberg ME. 2000. Response of neurons in the lateral intraparietal area to a distractor flashed during the delay period of a memory-guided saccade. *J. Neurophysiol.* 84:301–10

Recanzone GH, Merzenich MM, Jenkins WM. 1992. Frequency discrimination training engaging a restricted skin surface results in an emergence of a cutaneous response zone in cortical area 3a. *J. Neurophysiol.* 67:1057–70

Reynolds JH, Pasternak T, Desimone R. 2000. Attention increases sensitivity of V4 neurons. *Neuron* 26:703–14

Rizzolatti G, Riggio L, Sheliga B. 1994. Space and selective attention. In *Attention and Performance*, ed. C Umilta, M Moscovitch, pp. 231–65. Cambridge, MA: MIT Press

Robinson DL, Goldberg ME, Stanton GB. 1978. Parietal association cortex in the primate: sensory mechanisms and behavioral modulations. *J. Neurophysiol.* 41:910–32

Rondot P, Recondo J, de Ribadeau Dumas J. 1977. Visuomotor ataxia. *Brain* 100:355–76

Rossetti Y, Rode G, Pisella L, Farne A, Li L, et al. 1998. Prism adaptation to a rightward optical deviation rehabilitates left hemispatial neglect. *Nature* 395:166–69

Rushworth MFS, Ellison A, Walsh V. 2001a. Complementary localization and lateralization or orienting and motor attention. *Nat. Neurosci.* 4:656–61

Rushworth MFS, Paus T, Sipila PK. 2001b. Attention systems and the organization of the human parietal cortex. *J. Neurosci.* 21:5262–71

Sakata H, Taira M, Murata A, Mine S. 1995. Neural mechanisms of visual guidance of hand action in the parietal cortex of the monkey. *Cereb. Cortex* 5:429–38

Sakata H, Taira M, Kusunoki M, Murata A, Tanaka Y. 1997. The TINS lecture. The parietal association cortex in depth perception and visual control of hand action. *Trends Neurosci.* 20:350–57

Salinas E, Abbott LF. 1995. Transfer of coded information from sensory to motor networks. *J. Neurosci.* 15:6461–74

Salinas E, Abbott LF. 1997. Invariant visual responses from attentional gain fields. *J. Neurophysiol.* 77:3267–72

Salinas E, Thier P. 2000. Gain modulation: a major computational principle of the central nervous system. *Neuron* 27:15–21

Scherberger H, Andersen RA. 2001. Neural activity in the posterior parietal cortex during decision processes for generating visually-guided eye and arm movements in the monkey. *Soc. Neurosci. (Abstr.)* 27:237.7

Scherberger H, Goodale MA, Andersen RA. 1999. Reaching and saccadic target selection both follow a head- rather than trunk-centered reference frame during visual double simultaneous stimulation in the monkey. *Soc. Neurosci. (Abstr.)* 25:2189

Shadlen MN, Newsome WT. 1996. Motion perception: seeing and deciding. *Proc. Natl. Acad. Sci. USA* 93:628–33

Shenoy KV, Bradley DC, Andersen RA. 1999a. Influence of gaze rotation on the visual response of primate MSTd neurons. *J. Neurophysiol.* 81:2764–86

Shenoy KV, Kureshi SA, Meeker D, Gillikan BL, Dubowitz DJ, et al. 1999b. Toward prosthetic systems controlled by parietal cortex. *Soc. Neurosci. (Abstr.)* 25:383

Snyder LH, Batista AP, Andersen RA. 1997. Coding of intention in the posterior parietal cortex. *Nature* 386:167–70

Snyder LH, Batista AP, Andersen RA. 1998a. Change in motor plan, without a change in the spatial locus of attention, modulates activity in posterior parietal cortex. *J. Neurophysiol.* 79:2814–19

Snyder LH, Grieve KL, Brotchie P, Andersen RA. 1998b. Separate body- and world-referenced representations of visual space in parietal cortex. *Nature* 394:887–91

Snyder LH, Batista AP, Andersen RA. 2002. Intention-related activity in the posterior parietal cortex: a review. *Vision Res.* 40:1433–41

Steinmetz MA, Constantinidis C. 1995. Neurophysiological evidence for a role of posterior parietal cortex in redirecting visual attention. *Cereb. Cortex* 5:448–56

Stricanne B, Andersen RA, Mazzoni P. 1996. Eye-centered, head-centered, and intermediate coding of remembered sound locations in area LIP. *J. Neurophysiol.* 76:2071–76

Stuphorn V, Bauswein E, Hoffman K-P. 2000. Neurons in the primate superior colliculus coding for arm movements in gaze-related coordinates. *J. Neurophysiol.* 83:1283–99

Tian J, Schlag J, Schlag-Rey M. 2000. Testing quasi-visual neurons in the monkey's frontal eye field with the triple-step paradigm. *Exp. Brain Res.* 130:433–40

Watson RT, Valenstein E, Day A, Heilman KM. 1994. Posterior neocortical systems subserving awareness and neglect—neglect associated with superior temporal sulcus but not area-7 lesions. *Arch. Neurol.* 51:1014–21

Wessberg J, Stambaugh CR, Kralik JD, Beck PD, Laubach M, et al. 2000. Real-time prediction of hand trajectory by ensembles of cortical neurons in primates. *Nature* 408:361–65

Wu S, Andersen RA. 2001. The representation of auditory space in temporo-parietal cortex. *Soc. Neurosci. (Abstr.)* 27:166.15

Xing J, Andersen RA. 2000a. Memory activity of LIP neurons for sequential eye movements simulated with neural networks. *J. Neurophysiol.* 84:651–65

Xing J, Andersen RA. 2000b. Models of the posterior parietal cortex which perform multimodal integration and represent space in several coordinate frames. *J. Cogn. Neurosci.* 12:601–14

Zhang M, Barash S. 2000. Neuronal switching of sensorimotor transformations for antisaccades. *Nature* 408:971–75

Zipser D, Andersen RA. 1988. A back-propagation programmed network that simulates response properties of a subset of posterior parietal neurons. *Nature* 331:679–84

BEYOND PHRENOLOGY: What Can Neuroimaging Tell Us About Distributed Circuitry?

Karl Friston

The Wellcome Department of Cognitive Neurology, University College London, Queen Square, London, WC1N 3BG United Kingdom; email: k.friston@fil.ion.ucl.ac.uk

Key Words neuroimaging, predictive coding, generative model, information theory, effective connectivity

■ **Abstract** Unsupervised models of how the brain identifies and categorizes the causes of its sensory input can be divided into two classes: those that minimize the mutual information (i.e., redundancy) among evoked responses and those that minimize the prediction error. Although these models have the same goal, the way that goal is attained, and the functional architectures required, are fundamentally different. This review describes the differences, in the functional anatomy of sensory cortical hierarchies, implied by the two models. We then consider how neuroimaging can be used to disambiguate between them. The key distinction reduces to whether backward connections are employed by the brain to generate a prediction of sensory inputs. To ascertain whether backward influences are evident empirically requires a characterization of functional integration among brain systems. This review summarizes the approaches to measuring functional integration in terms of effective connectivity and proceeds to address the question posed by the theoretical considerations. In short, it will be shown that the conjoint manipulation of bottom-up and top-down inputs to an area can be used to test for interactions between them, in elaborating cortical responses. The conclusion, from these sorts of neuroimaging studies, points to the prevalence of top-down influences and the plausibility of generative models of sensory brain function.

INTRODUCTION

Functional neuroimaging, or human brain mapping, has enjoyed an enormous amount of success in systems and cognitive neuroscience over the past decade. Much of this success rests on being able to identify functionally specialized areas. Implicit in the term mapping is cartography, which some have referred to as neo-phrenology. Functional cartography, by itself, is clearly not going to reveal the principles that underlie the brain's functional architectures. However, it is an important prelude. This review is about using neuroimaging to answer questions about organizational principles, in terms of distributed and coupled interactions among specialized brain systems. The question, chosen to illustrate this sort of application, concerns the respective roles of forward and backward connections among cortical areas and how this coupling mediates perceptual synthesis and categorization. The first half of this review establishes the potential importance of backward connections using ideas from theoretical neurobiology and machine learning. The ensuing predictions are then addressed using empirical examples from neuroimaging.

The article starts by reviewing two fundamental principles of brain organization, namely functional specialization and functional integration, and how they rest on the anatomy and physiology of cortico-cortical connections in the brain. The second section deals with the nature of representations from a theoretical or computational perspective. This section contrasts information theoretic approaches and those predicated on predictive coding. This section concludes that predictive coding architectures are more plausible because they lend themselves to a Bayesian formulation, in which constraints from higher levels of a cortical hierarchy provide contextual guidance to lower levels of processing. This confers a context-sensitivity on evoked responses.

Empirical evidence from electrophysiological studies of animals and functional neuroimaging studies of human subjects is presented in the third and fourth sections to illustrate the context-sensitive nature of functional specialization and how its expression depends on functional integration among remote cortical areas. The third section (on generative models and the brain) looks at extra-classical effects in electrophysiology, in terms of the predictions afforded by generative models of brain function. The theme of context-sensitive evoked responses is pursued at a cortical level in human functional neuroimaging studies in the subsequent section (on functional architectures and brain imaging). The critical focus of this section is evidence for the interaction of bottom-up and top-down influences in determining regional brain responses. These interactions can be considered signatures of a predictive coding strategy.

FUNCTIONAL SPECIALIZATION AND INTEGRATION

The brain appears to adhere to two fundamental principles of functional organization, functional integration and functional specialization, where the integration within and among specialized areas is mediated by effective connectivity. The distinction relates to that between "localizationism" and "[dis]connectionism," which dominated thinking about cortical function in the nineteenth century. Since the early anatomic theories of Gall, the identification of a particular brain region with a specific function has become a central theme in neuroscience. However, functional localization per se was not easy to demonstrate: For example, a meeting that took place on August 4, 1881 addressed the difficulties of attributing function to a cortical area, given the dependence of cerebral activity on underlying connections (Phillips et al. 1984). This meeting was entitled "Localization of Function in the Cortex Cerebri." Goltz, although accepting the results of electrical stimulation in dog and monkey cortex, considered that the excitation method was inconclusive in that the behaviors elicited might have originated in related pathways, or current could have spread to distant centers. In short the excitation method could not be used to infer functional localization because localizationism discounted interactions, or functional integration among different brain areas. It was proposed that lesion studies could supplement excitation experiments. Ironically, it was observations on patients with brain lesions some years later (see Absher & Benson 1993) that led to the concept of disconnection syndromes and the refutation of localizationism as a complete or sufficient explanation of cortical organization. Functional localization implies that a function can be localized in a cortical area, whereas specialization suggests that a cortical area is specialized for some aspects of perceptual or motor processing, in which this specialization can be anatomically segregated within the cortex. The cortical infrastructure supporting a single function may then involve many specialized areas whose union is mediated by the functional integration among them. Functional specialization and integration are not exclusive; they are complementary. Functional specialization is only meaningful in the context of functional integration and vice versa.

Functional Specialization

The functional role, played by any component (e.g., cortical area, subarea, neuronal population, or neuron) of the brain, is defined largely by its connections. Certain patterns of cortical projections are so common that they could amount to rules of cortical connectivity. "These rules revolve around one, apparently, overriding strategy that the cerebral cortex uses—that of functional segregation" (Zeki 1990). Functional segregation demands that cells with common functional properties be grouped together. This architectural constraint in turn necessitates both convergence and divergence of cortical connections. Extrinsic connections between cortical regions are not continuous but occur in patches or clusters. This patchiness has, in some instances, a clear relationship to functional segregation. For example, the secondary visual area V2 has a distinctive cytochrome oxidase architecture, consisting of thick stripes, thin stripes, and interstripes. When recordings are made in V2, directionally selective (but not wavelength or color selective) cells are found exclusively in the thick stripes. Retrograde (i.e., backward) labeling of cells in V5 is limited to these thick stripes. All the available physiological evidence suggests that V5 is a functionally homogeneous area that is specialized for visual motion. Evidence of this nature supports the notion that patchy connectivity is the anatomical infrastructure that underpins functional segregation and specialization.

THE ANATOMY AND PHYSIOLOGY OF CORTICO-CORTICAL CONNECTIONS If specialization depends on connectivity, then important principles underpinning specialization should be embodied in the neuroanatomy and physiology of extrinsic connections. Extrinsic connections couple different cortical areas, whereas intrinsic connections are confined to the cortical sheet. There are certain features of cortico-cortical connections that provide strong clues about their functional role. In brief, there appears to be a hierarchical organization that rests on the distinction

TABLE 1 Some key characteristics of extrinsic cortico-cortical connections in the brain

Hierarchical organization
- The organization of the visual cortices can be considered as a hierarchy (Felleman & Van Essen 1991).
- The notion of a hierarchy depends on a distinction between forward and backward extrinsic connections.
- This distinction rests on different laminar specificity (Rockland & Pandya 1979, Salin & Bullier 1995).
- Backward connections are more numerous and transcend more levels.
- Backward connections are more divergent than forward connections (Zeki & Shipp 1988).

Forward connections	**Backward connections**
Sparse axonal bifurcations	Abundant axonal bifurcation
Topographically organized	Diffuse topography
Originate in supragranular layers	Originate in bilaminar/infragranular layers
Terminate largely in layer VI	Terminate predominantly in supragranular layers
Postsynaptic effects through fast AMPA (1.3–2.4 ms decay) and GABA$_A$ (6 ms decay) receptors	Modulatory afferents activate slow (50 ms decay) voltage-sensitive NMDA receptors

between forward and backward connections. The designation of a connection as forward or backward depends primarily on its cortical layers of origin and termination. Some characteristics of cortico-cortical connections are summarized in Table 1. In brief, the anatomy and physiology of cortico-cortical connections suggest that forward connections are driving and commit cells to a prespecified response given the appropriate pattern of inputs. Backward connections, on the other hand, are less topographic and are in a position to modulate the responses of lower areas to driving inputs from either higher or lower areas (see Table 1). Reversible inactivation (e.g., Sandell & Schiller 1982, Girard & Bullier 1989) and functional neuroimaging (e.g., Büchel & Friston 1997) studies are consistent with this distinction. The notion that forward connections are concerned with the promulgation and segregation of sensory information is consistent with their (*a*) sparse axonal bifurcation, (*b*) patchy axonal terminations, and (*c*) topographic projections. In contradistinction modulatory, backward connections are generally considered to have a role in mediating contextual effects and in the coordination of processing channels. This is consistent with their (*a*) frequent bifurcation, (*b*) diffuse axonal terminations, and (*c*) nontopographic projections (Salin & Bullier 1995, Crick & Koch 1998).

In summary, backward connections are abundant and are in a position to exert powerful effects on evoked responses, in lower levels, that define the specialization of any area or neuronal population. The idea pursued in this review is that specialization depends on backward connections and, due to the greater divergence of the latter, can embody contextual effects. Appreciating this is important for understanding the role of functional integration in dynamically reshaping the specialization of brain areas that mediate perceptual synthesis and adaptive behavioral responses.

Functional Integration and Effective Connectivity

Electrophysiology and imaging neuroscience have firmly established functional specialization as a principle of brain organization in man. The functional integration of specialized areas has proven more difficult to assess. Functional integration refers to the interactions among specialized neuronal populations and how these interactions depend on the sensorimotor or cognitive context. Functional integration is usually assessed by examining the correlations among activity in different brain areas or trying to explain the activity in one area in relation to activities elsewhere. "Functional connectivity" is defined as correlations between remote neurophysiological events. However, correlations can arise in a variety of ways. For example in multi-unit electrode recordings they can result from stimulus-locked transients evoked by a common input, or they can reflect stimulus-induced oscillations mediated by synaptic connections (Gerstein & Perkel 1969). Integration within a distributed system is usually better understood in terms of effective connectivity. Effective connectivity refers explicitly to the influence that one neuronal system exerts over another, either at a synaptic (i.e., synaptic efficacy) or population level. It has been proposed that "the [electrophysiological] notion of effective connectivity should be understood as the experiment- and time-dependent, simplest possible circuit diagram that would replicate the observed timing relationships between the recorded neurons" (Aertsen & Preißl 1991). This speaks to two important points: (*a*) Effective connectivity is dynamic, i.e., activity- and time-dependent and (*b*) it depends on a model of the interactions. The models employed in functional neuroimaging can be divided into those based on regression models (Friston 1995) and those based on structural equation models (McIntosh & Gonzalez-Lima 1994). A more important distinction is whether these models are linear or nonlinear. Recent characterizations of effective connectivity have focused on nonlinear models that accommodate the modulatory or nonlinear effects described above. The most general model one could envisage is provided by nonlinear system identification through the use of Volterra series (see Box 1). This model has been used to address nonlinear coupling among brain areas induced by attention (Friston & Büchel 2000). We will use this example in the functional architectures assessed with brain imaging section.

Box 1 Dynamical Systems, Volterra Kernels, and Effective Connectivity

Input-State-Output Systems and Volterra Series
Neuronal systems are inherently nonlinear and lend themselves to modeling by nonlinear dynamical systems. However, due to the complexity of biological systems it is difficult to find analytic equations that describe them adequately.

Even if these equations were known the state variables are often not observable. An alternative approach to identification is to adopt a very general model (Wray & Green 1994) and focus on the inputs and outputs. Consider the single input–single output system:

$$\dot{x}(t) = f(x(t), u(t))$$

$$y(t) = \lambda(x(t)).$$

The Fliess fundamental formula (Fliess et al. 1983) describes the causal relationship between the outputs and the recent history of the inputs. This relationship can be expressed as a Volterra series, which expresses the output $y(t)$ as a nonlinear convolution of the inputs $u(t)$, critically without reference to the state variables $x(t)$. This series is simply a functional Taylor expansion of $y(t)$.

$$y(t) = F(u(t-\sigma)) = \kappa_0 + \sum_{i=1}^{\infty} \int_0^t \cdots \int_0^t \kappa_i(\sigma_1, \ldots \sigma_i) u(t-\sigma_1) \ldots u(t-\sigma_i) d\sigma_1 \ldots d\sigma_i$$

$$\kappa_i(\sigma_1, \ldots \sigma_i) = \frac{\partial^i y(t)}{\partial u(t-\sigma_1) \ldots \partial u(t-\sigma_i)},$$

where $\kappa_i(\sigma_1, \ldots \sigma_i)$ is the ith-order kernel. Volterra series have been described as a "power series with memory" and are generally thought of as a high-order or nonlinear convolution of the inputs to provide an output. See Bendat (1990) for a fuller discussion. When the inputs and outputs are measured neuronal activity, the Volterra kernels have a special interpretation.

Volterra Kernels and Effective Connectivity

Volterra kernels are essential in characterizing the effective connectivity or influences that one neuronal system exerts over another because they represent the causal characteristics of the system in question. Neurobiologically they have a simple and compelling interpretation—they are synonymous with effective connectivity:

$$\kappa_1(\sigma_1) = \frac{\partial y(t)}{\partial u(t-\sigma_1)}, \quad \kappa_2(\sigma_1, \sigma_2) = \frac{\partial^2 y(t)}{\partial u(t-\sigma_1) \partial u(t-\sigma_2)}.$$

It is evident that the first-order kernel embodies the response evoked by a change in input at $t - \sigma_1$. In other words it is a time-dependant measure of *driving* efficacy. Similarly, the second-order kernel reflects the *modulatory* influence of the input at $t - \sigma_1$ on the evoked response at $t - \sigma_2$, and so on for higher orders.

> The important thing about this formulation of effective connectivity is that it can be defined and estimated using just the inputs and responses of a system (e.g., cortical area). In other words, effective connectivity does not refer to the (hidden) state variables (e.g., depolarization of every cell membrane, the electrochemical status of every cell compartment, or the configuration of every membrane channel) that actually mediate the input-output transformation. This is important because these state variables are often unmeasurable, particularly in functional neuroimaging.

THEORETICAL AND COMPUTATIONAL PERSPECTIVES

This section compares and contrasts two prevalent computational approaches to perceptual categorization and synthesis: information theoretic and predictive coding frameworks. This section restricts itself to sensory processing in cortical hierarchies. This precludes a discussion of other important ideas [e.g., reinforcement learning (Sutton & Barto 1990, Friston et al. 1994), neuronal selection (Edelman 1993), and dynamical systems theory (Freeman & Barrie (1994)].

The relationship between modeled and real neuronal architectures is central to basic and cognitive neuroscience. This section addresses this relationship, in terms of representations. We start with an overview of representations so that the distinctions among various approaches can be seen clearly. An important focus of this section is the interactions among "causes" of sensory input. These interactions posit the problem of contextual invariance, which has severe implications for supervised (i.e., connectionist) models of cognitive architectures. In brief the problem of contextual invariance points to the adoption of unsupervised models in which interactions among causes of a percept are modeled explicitly. Within the class of unsupervised models we compare classical information theoretic approaches and predictive coding. These schemes allow the emergence of natural representations that can accommodate contextual invariance but do so in a very different way. The question then reduces to how this difference would be expressed in terms of measurable brain responses and effective connectivity. This issue is taken up in subsequent sections.

The Nature of Representations

What is a representation? Here a representation is taken to be a neuronal event that represents some cause in the sensorium. It can be defined operationally as the neuronal transient evoked by the cause being represented. In very general terms, let us frame the problem of representing real world causes $u(t)$ in terms of the system of equations

$$\dot{y}(t) = f(y(t), u(t)), \qquad 1.$$

where u is a vector describing the expression of causes in the environment (e.g., the presence of a particular object, direction of radiant light, etc.), and y represents sensory inputs. $\dot{y}(t)$ denotes the rate of change of y at time t. The function f can be highly nonlinear and allows for both the current state of the sensory inputs and their causes to interact when inducing changes in the activity of sensory units. The sensory input can be shown to be a function of, and only of, the causes and their recent history.

$$y(t) = F(u(t - \sigma))$$

$$= \sum_{i=1}^{\infty} \int_0^t \cdots \int_0^t \frac{\partial^i y(t)}{\partial u(t - \sigma_1) \ldots \partial u(t - \sigma_i)} u(t - \sigma_1) \ldots u(t - \sigma_i) \, d\sigma_1 \ldots d\sigma_i.$$

2.

Equation 2 is simply a functional Taylor expansion to cover dynamical systems of the sort implied by Equation 1. This expansion is called a Volterra series and can be thought of as a nonlinear convolution of the causes to give the inputs (see Box 1). Convolution is like smoothing, in this instance smoothing over time. The importance of this formulation is that it highlights (*a*) the dynamical aspects of sensory input and (*b*) the role of interactions among the causes of the sensory input. For example the second-order terms with $i = 2$ in Equation 2 represent pairwise interactions among u, possibly at different points in time. Interactions can be viewed as contextual effects, for which the expression of a particular cause is highly dependent on the context induced by another. For example, the extraction of motion from the visual field depends on there being sufficient luminance or wavelength contrast. Another ubiquitous example, from early visual processing, is the occlusion of one object by another. In the absence of interactions we would see a linear superposition of both objects, but the visual input caused by the nonlinear mixing of these two objects renders one occluded by the other. At a more cognitive level the cause associated with the word hammer will depend on the semantic context (that determines whether the word is a verb or a noun). These contextual effects are profound and must be discounted before the representations of the underlying causes can be considered veridical. The problem the brain has to contend with is how to find a function of the input $y(t)$ that represents the underlying causes. To do this, the brain must effectively undo the interactions to reveal contextually invariant causes. In other words the brain must perform some form of nonlinear unmixing of causes and context without ever knowing either. Furthermore, because of the convolution implied by Equation 2, it must deconvolve the inputs to obtain these causes. In estimation theory this general problem is sometimes called blind deconvolution because the estimation is blind to the underlying causes that are convolved to give the observed variables.

Most models of perceptual categorization can be understood as trying to effect a blind deconvolution of sensory inputs to reveal the causes. Consider a formally similar system of equations to Equation 1 that represent the dynamics of the brain:

$$\dot{x}(t) = f_\theta(x(t), y(t))$$
$$v(t) = l_\theta(x(t)) \qquad 3.$$

and by analogy with Equation 2 $v(t) = F_\theta(y(t - \sigma))$.   4.

Here x represents the activity of neuronal units (i.e., neurons or populations of neurons) in the brain. A subset of units can be selected and passed through a nonlinear function to give some explicit or implicit representation v. The parameters θ of the functions in Equation 3 embody the series of dynamical transformations that the sensory input is subject to and can be thought of specifying the connection strengths and biases of a neuronal network model or effective connectivity. The problem of extracting causes from the input reduces to finding the right parameters such that the activity of the representational units v have some clearly defined relationship to the causes u. More formally one wants to find the parameters that maximize the mutual information or dependence between the dynamics of the representations and their causes. Models of neuronal computation try to solve this problem in the hope that the ensuing parameters can be interpreted in relation to real neuronal parameters. The greater the biological validity of the constraints under which these solutions are obtained, the more plausible the relationship becomes.

In what follows we will consider models based on information theory and those based on predictive coding. Each subsection below provides the background for the approach and then describes it using the formalism above. Figure 1 provides a graphical overview of the two schemes.

Information Theoretic Approaches

There have been many compelling developments in theoretical neurobiology that have used information theory (e.g., Barlow 1961, Optican & Richmond 1987, Linsker 1988, Oja 1989, Foldiak 1990, Tovee et al. 1993, Tononi et al. 1994). Many appeal to the principle of maximum information transfer (e.g., Linsker 1988, Atick & Redlich 1990, Bell & Sejnowski 1995). This principle has proven extremely powerful in predicting some of the basic receptive field properties of cells involved in early visual processing (e.g., Atick & Redlich 1990, Olshausen & Field 1996). This principle represents a formal statement of the common sense notion that neuronal dynamics in sensory systems should reflect, efficiently, what is going on in the environment (Barlow 1961). In the present context, the principle of maximum information transfer (infomax; Linsker 1988) suggests that a model's parameters should be configured to maximize the mutual information between the representations that they engender and the causes of sensory input. This maximization is usually considered in the light of some sensible constraints, e.g., the presence of noise in the sensory input (Atick & Redlich 1990) or dimension reduction (Oja 1988).

For any given causes we want to maximize the mutual information between $u(t)$ and the neuronal responses $v(t)$. Intuitively mutual information is like the

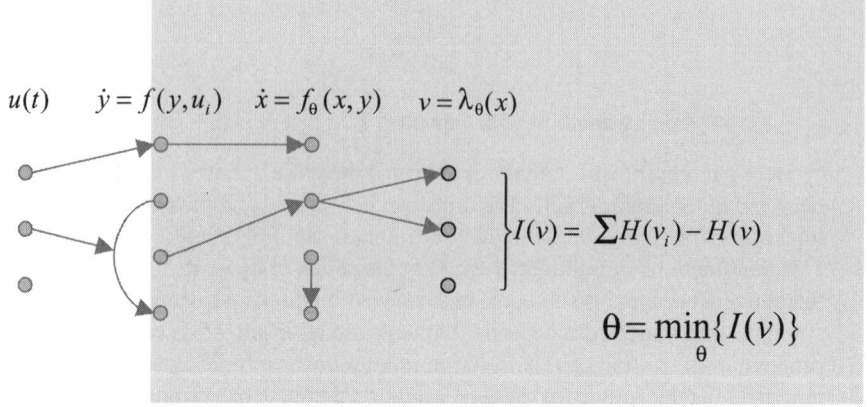

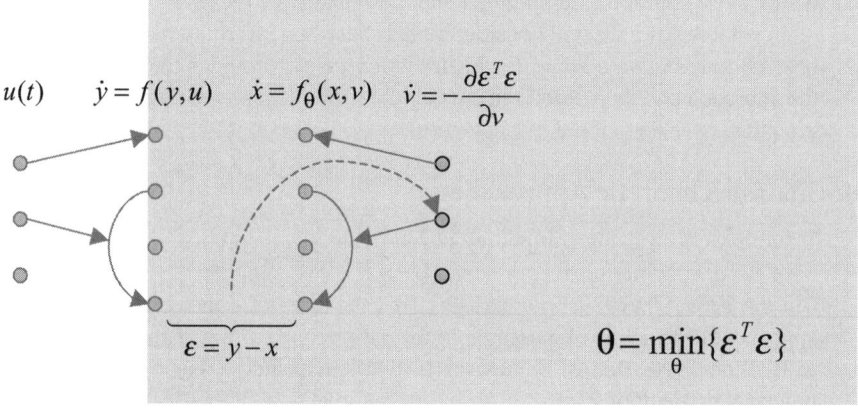

Figure 1 Schematic illustrating the architectures implied by information theory–based approaches and predictive coding. The circles represent nodes in a network and the arrows represent a few of the connections. See the main text for an explanation of the equations and designation of the variables each set of nodes represents. The light grey boxes encompass connections and nodes within the model. The strengths of connections within this area are determined by the free parameters of the model θ. Nonlinear effects are implied when one arrow connects with another. Nonlinearities can be construed as the modulation of responsiveness to one input by another (see Box 1 for a more formal account or interpretation). The broken arrow in the lower panel denotes connections that convey an error signal to the higher level from the input level.

covariance or correlation between two variables but extended to cover multivariate observations. In a similar way entropy can be regarded as the uncertainty or variability of an observation (cf. variance of a univariate observation). The mutual information is given by

$$I(u, v) = H(u) + H(v) - H(u, v)$$
$$= H(v) - H(v|u), \qquad 5.$$

where $H(v|u)$ is the conditional entropy or uncertainty in the representations, given the causes. For a deterministic system there is no such uncertainty and this term can be discounted (see Bell & Sejnowski 1995). More generally

$$\frac{\partial}{\partial \theta} I(u, v) = \frac{\partial}{\partial \theta} H(v). \qquad 6.$$

It follows that maximizing the mutual information between the outputs and the causes is the same as maximizing the entropy of the responses. The infomax principle (maximum information transfer) is closely related to the idea of efficient coding. Generally speaking, redundancy minimization and efficient coding are all variations on the same theme and can be considered as the infomax principle operating under some appropriate constraints. The key thing that distinguishes among the various information theoretic schema is the nature of the constraints under which entropy is maximized. These constraints render the infomax a viable approach to recovering the original causes of data, especially if one can enforce the outputs to comply with the same constraints as the causes. One useful way of looking at constraints is in terms of efficiency.

EFFICIENCY, REDUNDANCY, AND INFORMATION   Efficiency can be considered as the complement of redundancy (Barlow 1961); the less redundant, the more efficient a system will be. Redundancy is reflected in the dependencies or mutual information *among* the outputs (cf. Gawne & Richmond 1993).

$$I(v) = \sum H(v_i) - H(v). \qquad 7.$$

Here $H(v_i)$ is the entropy of the ith unit. Equation 7 implies that redundancy is the difference between the joint entropy and the sum of the entropies of the individual units (componential entropies). Intuitively this expression makes sense if one considers that the variability in activity of any one unit corresponds to its entropy. Therefore, an efficient system represents its inputs with the minimum changes in firing. Another way of thinking about Equation 7 is to note that maximizing efficiency is equivalent to minimizing the mutual information among the outputs. This is the basis of approaches that seek to decorrelate or orthogonalize the outputs.

To minimize redundancy one can either minimize the entropy of the output units or maximize their joint entropy. Olshausen & Field (1996) present a very nice analysis based on sparse coding. Sparse coding minimizes the redundancy by minimizing componential entropies. This minimization is implicit in sparse coding because a neuron that fires very sparsely will generally not be firing. We can, therefore, be relatively certain about its (quiescent) state, conferring low entropy upon it.

Approaches that seek to maximize the joint entropy of the outputs include principle component analysis (PCA), learning algorithms (that sample the subspace of the inputs that have the highest entropy) (e.g., Foldiak 1990, Friston et al. 1993), and independent component analysis (ICA). ICA finds nonlinear functions of the inputs that maximize the joint entropy (Common 1994, Bell & Sejnowski 1995). In PCA the componential entropies are constrained by setting the sum of squared connection strengths to be one. In ICA they are maintained at low levels by the application of a sigmoid squashing function to the outputs.

IMPLEMENTATION   In terms of the above formulation, information theoretic approaches can be construed as finding the parameters that maximize the efficiency or minimize the redundancy

$$\theta = \min_{\theta} I(v). \qquad 8.$$

Compared to supervised schemes this has the fundamental advantage that the algorithm is unsupervised (the causes do not enter into Equation 8). Furthermore, the inputs can, in principle be generated dynamically and can interact. However, for simple variants of this information theoretic approach (e.g., ICA) only linear mixtures of independent causes can be recovered (up to some permutation and scaling). For example a typical model adopted by PCA for Gaussian and ICA for non-Gaussian causes is

$$y(t) = F(u(t - \sigma)) = Wu(t), \qquad 9.$$

where W is linear mixing matrix. This example highlights the operational shortcomings of information theoretic approaches that are based on feedforward architectures: Not only does the model of real world mixing of causes in Equation 9 preclude any dynamics, it also ignores interactions among causes. Nonlinear variants of ICA and PCA do exist (e.g., Karhunen & Joutsensalo 1994, Dong & McAvoy 1996) and typically employ a "bottleneck" architecture that forces the inputs through a small number of nodes. The output from these nodes then diverges to predict the original inputs (see predictive coding below). However, these architectures are better regarded as generative models in the sense that the nonlinear transformations, from the bottleneck nodes to the output layer, recapitulate the nonlinear mixing of the original causes and constitute a generative model. Generative models are presented in the next subsection.

Finally ICA, like parallel and distributed processing models, assumes the existence of an operator that can deconvolve (a nonlinear function of) the causes out of the inputs. For very simple mixtures of causes this may be tenable. However, generally nonlinear mixing applied by the real world renders the existence of this deconvolution very questionable. In the inverse solution literature these problems are known as ill-posed or underdetermined. The solution is to render the problem tractable using constraints on the solution. This means that information theoretic approaches that try to solve the unmixing or inverse problem rely heavily on constraints (e.g., efficiency, sparse coding, etc.). In the alternative approach, considered here, we discuss predictive coding models that moderate this constraint-dependency and suggest a more natural form for representations.

Generative Models and Predictive Coding

Over the past years generative models have supervened over other modeling approaches to brain function and represent one of the most promising avenues, offered by computational neuroscience, to understanding neuronal dynamics in relation to perceptual categorization. In generative models the dynamics of units in a network are trying to predict the inputs. The representational aspects of any unit emerge spontaneously as the capacity to predict improves with learning. There is no a priori "labeling" of the units or any supervision in terms of what a correct response should be (cf. connectionism). The only correct response is one in which the implicit internal model of the causes and their nonlinear mixing is sufficient to predict the input with minimal error. There are many forms of generative models that range from conventional statistical models (e.g., factor and cluster analysis) and those motivated by Bayesian learning (e.g., Dayan et al. 1995, Hinton et al. 1995) to biologically plausible models of visual processing (e.g., Rao & Ballard 1998). Indeed many of the algorithms discussed under the heading of information theory can be formulated as generative models. The goal of generative models is "to learn representations that are economical to describe but allow the input to be reconstructed accurately" (Hinton et al. 1995). These models emphasize the role of backward connections in mediating the prediction, at lower or input levels, based on the activity of units in higher levels.

IMPLEMENTATION Predictive, or more generally, generative, models turn the inverse problem on its head. Instead of trying to find functions of the inputs that predict their causes, they find functions of estimated causes that predict the inputs. As in approaches based on information theory, the causes do not enter into the learning rules and they are therefore unsupervised. Furthermore, they do not require the convolution of causes, engendering the inputs to be invertible. This is because generative models instantiate the forward solution, not the inverse solution. Here the forward solution is the nonlinear mixing of causes that by definition must exist. The estimation of the causes still rests on constraints, but these are now framed in terms of the generative model and have a much more direct relationship to casual processes in the real world. The ensuing mirror symmetry of the architecture is illustrated in Figure 1. Notice that the connections within the model are now going backward. In the predictive coding scheme the outputs now become the inputs such that

$$\dot{x}(t) = f_\theta(x(t), v(t)) \Rightarrow$$
$$x(t) = F_\theta(v(t - \sigma)), \qquad\qquad 10.$$

cf. Equation 3. The parameters now change so as to minimize some function G of the prediction error at the input level

$$\theta = \min_{\theta} G(\varepsilon)$$
$$\varepsilon = y - F_{\theta}(v). \qquad 11.$$

The top-down inputs $v(t)$ now drive the predictions $x(t)$ of the input and the parameters of the backward connections forming these predictions change so as to minimize the prediction error. But what drives the top-down inputs? The casual estimates or representations change in the same way as the other free parameters of the model. They change to minimize prediction error, usually through gradient descent

$$\dot{v} = -\frac{\partial G(\varepsilon)}{\partial v}. \qquad 12.$$

The error is conveyed from the input layer to the higher layer by forward connections that are rendered as a broken line in the lower panel of Figure 1. This component of the predictive coding scheme has a principled (Bayesian) motivation that is described in the next subsection. For the moment consider what would happen after training or learning and prediction error is largely eliminated. This implies that the prediction of the input becomes very precise $x(t) \rightarrow y(t)$, and consequently from Equation 2 and Equation 10

$$F_{\theta}(v(t - \sigma)) \rightarrow F(u(t - \sigma)). \qquad 13.$$

In other words the brain's nonlinear convolution of the estimated causes reproduces exactly the real convolution of the real causes. In short there is a veridical (or at least sufficient) representation of both the causes and the dynamical structure of their mixing through the connections or parameters of F_{θ}.

The dynamics of representational units or populations implied by Equation 12 represents the essential difference between this class of approaches and others. Only in predictive coding is the activity of the units driven explicitly to improve the representational capacity of the system. Predictive coding is a strategy that has some compelling (Bayesian) underpinnings (see below) and is not simply using a connectionist architecture in auto-associative mode or using error minimization to maximize mutual information transfer. It is a real time, dynamical scheme that embeds two concurrent processes. (*a*) The parameters of the model are changing so that the generative model emulates the real world mixing of causes, using their current estimates, and (*b*) the representations are converging to the best estimate of the causes extant at any time, using the generative model. Both the parameters and the state variables change in a mathematically identical way to minimize prediction error. The predictive coding scheme can easily accommodate dynamical and nonlinear mixing of causes in the real world. It does not require this mixing to be invertible, and it only requires the sensory inputs to be known. Before considering how the brain might perform predictive coding, we look at its motivation from another point of view.

PREDICTIVE CODING AND BAYESIAN INFERENCE One of the most important aspects of generative models is that they emphasize the role of the brain as an inferential machine (Dayan et al. 1995). From this perspective functional architectures exist, not to unmix the input to obtain the causes, but to make inferences about the causes and test the predictions against observed input. A compelling aspect of predictive coding schemes is that they lend themselves to a hierarchical extension that can be viewed in terms of Bayesian inference. In the simplest extension, let us suppose we had some expected values $\bar{u}$ of the causes, which were used to generate a prior prediction error $G(v - \bar{u})$ not at the level of the inputs but at the higher level of the causal representations v. The changes in v are now required to minimize the error at both levels so that

$$\dot{v} = -\frac{\partial}{\partial v}[G(y - F_\theta(v)) + G(v - \bar{u})]. \qquad 14.$$

The addition of this extra term renders the ensuing estimation of causes a Bayesian one in the following way. Bayesian inference allows one to posit the probability of the causes of some input or data given that data. This is in contradistinction to maximum likelihood estimates, which simply identify the causes that maximize the likelihood of the input. The difference rests on Bayes' rule, which states that the probability of the cause and input occurring together is the probability of the cause given the input times the probability of the input. This, in turn, is the same as the probability of the input given the causes times the prior probability of the causes

$$p(u, y) = p(u|y)p(y) = p(y|u)p(u)$$
$$\Rightarrow p(u|y) \propto p(y|u)p(u). \qquad 15.$$

The Bayesian, posterior, or conditional estimator of the causes is that which is most likely given the data.

$$\max_u p(u|y) = \max_u \{\ln p(y|u) + \ln p(u)\}. \qquad 16.$$

This is referred to as the maximum posterior or MAP estimator. The first term on the right is known as the log likelihood or likelihood potential and the second is the prior potential. If we take the Gibb's form for $p(y|u) = \exp(-\frac{1}{2}G\{y - F(u)\})$, then Equation 16 becomes

$$\max_u p(u|y) = \min_u \{G\{y - F(u)\} + G(u - \bar{u})\}. \qquad 17.$$

A gradient descent to find the MAP estimator would be

$$\dot{u} = -\frac{\partial}{\partial u}[G(y - F(u)) + G(u - \bar{u})]. \qquad 18.$$

This is formally identical to Equation 14, the dynamics of the representations. This suggests that if the connectivity has properly captured the dynamical structure of the real world, i.e., $F_\theta(u) \to F(u)$ then the activities of representational units or populations strive to encode the most probable causes given the input. In this Bayesian formulation the state of the brain changes, not to minimize error per se, but to attain an estimate of the causes that maximizes both the likelihood of the input given that estimate and the prior probability of the estimate being true.

This notion can be extended in a hierarchical fashion to any number of levels as depicted in Figure 2. In the forgoing we simply assumed some expected values for the causes. These expected values can, of course, be predictions from higher level causes. This extension models the world as a hierarchy of dynamical systems in which supraordinate causes induce, and moderate, changes in subordinate causes. For example, the presence of a particular object in the visual field induces changes in the incident light falling on a particular part of the retina. A more abstract example, which illustrates the brain's inferential capacities, is presented in Figure 3. On reading the first sentence "Jack and Jill went up the hill" we perceive the word

Hierarchical architecture

$$p(u_i, \ldots, u_i \mid y) = p(y \mid u_1) p(u_1 \mid u_2) \ldots p(u_n)$$
$$\dot{u}_i = f^i(u_i, u_{i+1})$$

$$\varepsilon_0 = y - x_0 \qquad \varepsilon_1 = v_1 - x_1 \qquad \varepsilon_2 = v_2 - x_2$$

$$\dot{x}_i = f_\theta^i(x_i, v_{i+1})$$
$$\dot{v}_i = -\frac{\partial}{\partial v_i}\left(\varepsilon_{i-1}^T \varepsilon_{i-1} + \varepsilon_i^T \varepsilon_i\right)$$

Figure 2 Schematic depicting a hierarchical extension to the predictive coding architecture, with the same format as Figure 1. Here hierarchical arrangements within the model serve to provide predictions or priors to representations in the level below. The open circles are the predictions and the filled circles are the representations of causes in the environment. These representations change to minimize both the discrepancy between their predicted value and the mismatch incurred by their own prediction of the representation in the level below. These two constraints correspond to the prior and likelihood potentials respectively (see main text).

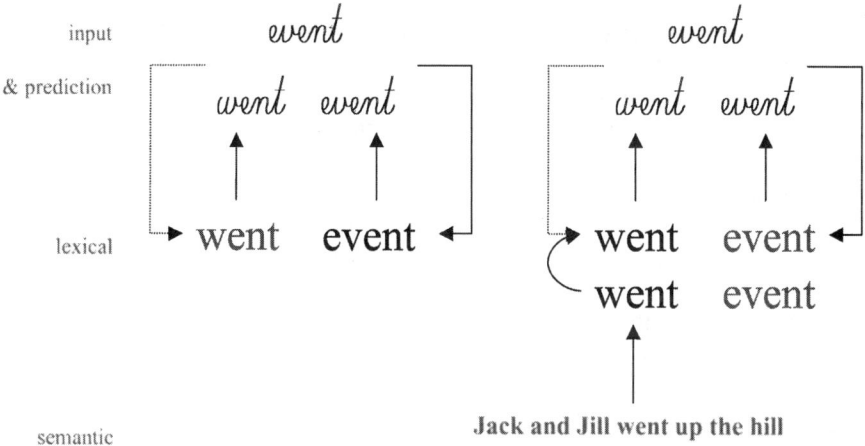

Figure 3 Schematic illustrating the role of priors in biasing toward one representation of an input or another. (*Left*) The word event is selected as the most like cause of the visual input. (*Right*) The word went is selected as the most likely word that is (*a*) a reasonable explanation for the sensory input and (*b*) conforms to prior expectations based on semantic context.

"event" as "went." However, in the absence of any hierarchical inference the best explanation for the pattern of visual stimulation incurred by the text is "event." This would correspond to the maximum likelihood estimate of the word and would be the most appropriate in the absence of prior information about the most likely word. However, within a hierarchical scheme, semantic context can provide top-down predictions, to which the posterior estimate is accountable. When this prior strongly biases in favor of "went," we tolerate a small error at a lower level of visual analysis to minimize the overall prediction error at both the visual and lexical level. This illustrates the role of higher level estimates in providing predictions or priors for subordinate levels. These priors offer contextual guidance toward the most likely cause of the input. Note that predictions at higher levels do not arise by magic. They are themselves subject to the same constraints; only the highest level (if there is one in the brain) is free to be directed solely by bottom-up influences.

The hierarchical structure of the real world literally comes to be "reflected" by the hierarchical architectures trying to minimize prediction error, not just at the level of sensory input but at all levels of the hierarchy (again notice the deliberate mirror symmetry in Figure 2). The nice thing about this architecture is that the dynamics of casual representations at the ith level v_i require only the error for the current level and the immediately preceding level. This follows from the Markov property of hierarchical systems, for which one only needs to know the immediately supraordinate causes to determine the evolution of causes at any level in question. The fact that only error from the current and lower level is required to drive the dynamics of v_i is important because it permits a biologically plausible implementation, where the connections driving the error minimization have only to run forward from one level to the next (see Box 2).

In summary the predictive coding approach lends itself naturally to a hierarchical Bayesian treatment, which considers the brain as an inferential device. This perspective arises because the dynamics of the units or populations are driven to minimize error at all levels and implicitly render themselves posterior estimates of the causes given the data. They can do this even with data generated by hierarchies of highly nonlinear dynamical systems. Unlike information theoretic approaches they do not require strong constraints to be built into the architecture; these constraints emerge spontaneously as priors from higher levels. The implicit Bayesian estimation can be formalized from a number of different perspectives. Rao & Ballard (1998) give an extremely nice example using the Kalman filter.

Box 2 Hierarchical Bayes in the Brain

> The biological plausibility of the scheme depicted in Figure 2 can be established very simply. Consider any level i in a cortical hierarchy containing units (neurons or neuronal populations), whose activity v_i is being predicted by equivalent units in the level above v_{i+1}. The prediction error being reflected in the activities of units is denoted by ε_i. Assuming the simplest generative model possible
>
> $$x_i = F_\theta^i(v_{i+1}) = \theta_i v_{i+1}$$
> $$\varepsilon_i = v_i - x_i,$$
>
> where θ_i are backwards connection strengths; we require units in the higher level v_{i+1} to maximize the probability of v_{i+1} given v_i. Assuming the errors have a Gaussian distribution with variance λ_i (i.e., $G_i(\varepsilon) = \varepsilon_i^T \varepsilon_i / \lambda_i$) we have
>
> $$p(v_{i+1}|v_i) \propto p(v_i|v_{i+1})p(v_{i+1}) \propto \exp\left(-\frac{1}{2}\left(\frac{\varepsilon_i^T \varepsilon_i}{\lambda_i} + \frac{\varepsilon_{i+1}^T \varepsilon_{i+1}}{\lambda_{i+1}}\right)\right).$$
>
> Both the dynamics of v_{i+1} and the connection strengths perform a gradient ascent on the log of this probability.

$$\dot{v}_{i+1} = \frac{\partial \ln p(v_{i+1}|v_i)}{\partial v_{i+1}} = \lambda_i^{-1}\theta_i^T \varepsilon_i - \lambda_{i+1}^{-1}\varepsilon_{i+1}$$

$$\dot{\theta}_i = \frac{\partial \ln p(v_{i+1}|v_i)}{\partial \theta_i} = \lambda_i^{-1}\varepsilon_i v_{i+1}^T$$

Despite the complicated nature of the hierarchical generative model and the abstract theorizing, three simple and biologically plausible things emerge: (a) The forwards and backwards connections are exactly the same, consistent with the reciprocity of anatomical connections; (b) changes in connection strengths reduce to simple Hebbian or associative plasticity; and (c) the dynamics of representational units v_{i+1} are subject to two (locally available) influences: a likelihood term mediated by forward afferents from the error units in the level below and a prior term conveyed by error units in the same level.

GENERATIVE MODELS AND THE BRAIN

The arguments in the preceding section clearly favor predictive coding over information theoretic frameworks as a more plausible account of functional brain architectures. However, it should be noted that the differences between them have been deliberately emphasized. For example, predictive coding and the implicit error minimization results in the maximization of information transfer. In other words, predictive coding conforms to the principle of maximum information transfer, but it does so in a very distinct way (see Olshausen & Field 1996 for a nice integration of predictive and sparse coding). Predictive coding is entirely consistent with the principle of maximum information. The infomax principle is a principle, whereas predictive coding represents a particular scheme that serves that principle. There are examples of infomax that do not employ predictive coding (e.g., transformations of stimulus energy in early visual processing; Atick & Redlich 1990) that may be specified genetically or epigenetically. However, predictive coding is likely to play a much more prominent role at higher levels of processing for the reasons detailed in the previous section.

Predictive coding, especially in its hierarchical formulation, also conforms to the same parallel and distributed processing principles that underpin connectionist schema (Rumelhart & McClelland 1986). The representation of any cause depends on the internally consistent representations of subordinate and supraordinate causes in lower and higher levels. These representations mutually induce and maintain themselves, across and within all levels of the sensory hierarchy, through dynamic and reentrant interactions (Edelman 1993). The same parallel and distributed processing phenomena (e.g., lateral interactions leading to competition among representations) may be observed. However, in predictive coding, these dynamics are driven explicitly by error minimization, whereas in connectionist simulations the activity is determined solely by the connection strengths that are established during training.

In addition to the theoretical bias toward generative models and predictive coding, the clear emphasis on backward and reentrant dynamics makes it a more natural framework for understanding neuronal infrastructures. Figure 1 shows the fundamental difference between infomax and generative schemes. In infomax schemes the connections are universally forward. In the predictive coding scheme the forward connections (broken line) drive the casual representations to minimize error, whereas backward connections (solid lines) use these representations to emulate mixing enacted by the real world. The nonlinear aspects of this mixing imply that backward connections are modulatory in predictive coding, whereas the nonlinear *un*mixing in infomax schemes is mediated by forward connections. The section on functional specialization and integration assembled some of the anatomical and physiological evidence that backward connections are prevalent in the real brain and can support nonlinear mixing through their modulatory characteristics. It is pleasing that purely theoretical considerations and neurobiological empiricism converge on the same architecture. Before turning to electrophysiological and functional neuroimaging evidence for backward connections, we consider the implications for classical views of receptive fields and the representational capacity of neurons.

Context, Causes, and Representations

The Bayesian perspective suggests something quite profound for the classical view of receptive fields. If neuronal responses encompass both a bottom-up likelihood term and top-down priors, then responses evoked by bottom-up input should change with the context established by prior expectations from higher levels of processing. In other words, when a neuron or population is predicted by top-down inputs, it will be much easier to drive than when it is not. Consider the example in Figure 3 again. Here a unit encoding the visual form of "went" responds when we read the first sentence at the top of this figure. When we read the second sentence "The last event was cancelled" it would not. If we recorded from this unit we might infer that our "went" unit was, in some circumstances, selective for the word event. Without an understanding of hierarchical inference and the semantic context the stimulus was presented in, this might be difficult to explain. In short, under a predictive coding scheme, the receptive fields of neurons should be context-sensitive. The remainder of this section deals with empirical evidence for these extra-classical receptive field effects.

Neuronal Responses and Representations

Classical models (e.g., classical receptive fields) assume that evoked responses will be invariably expressed in the same units or neuronal populations irrespective of the context. However, real neuronal responses are not invariant but depend on the context in which they are evoked. For example, visual cortical units have dynamic receptive fields that can change from moment to moment (cf. the nonclassical receptive field effects modeled in Rao & Ballard 1999). Another example is attentional modulation of evoked responses that can change the sensitivity of neurons to different perceptual attributes (e.g., Treue & Maunsell 1996). There are numerous examples of context-sensitive neuronal responses. Perhaps the simplest is short-term plasticity. Short-term plasticity refers to changes in connection strength, either potentiation or depression, following presynaptic inputs (e.g., Abbot et al. 1997). In brief, the underlying connection strengths, which define what that unit represents, are a strong function of the immediately preceding neuronal transient (i.e., preceding representation).

These sorts of effects are commonplace in the brain and are generally understood in terms of the dynamic modulation of receptive field properties by backward and lateral afferents. There is clear evidence that lateral connections in visual cortex are modulatory in nature (Hirsch & Gilbert 1991), speaking to an interaction between the functional segregation implicit in the columnar architecture of V1 and the neuronal dynamics in distal populations. These observations suggest that lateral and backward interactions may convey contextual information that shapes the responses of any neuron to its inputs (e.g., Kay & Phillips 1996, Phillips & Singer 1997) to confer on the brain the ability to make conditional inferences about sensory input. See also McIntosh (2000), who develops the idea from a cognitive neuroscience perspective "that a particular region in isolation may not act as a reliable index for a particular cognitive function. Instead, the neural context in which an area is active may define the cognitive function." His argument is predicated on careful characterizations of effective connectivity using neuroimaging.

AN EXAMPLE FROM ELECTROPHYSIOLOGY   In the next section we will illustrate the context-sensitive nature of cortical activations, and implicit specialization, in the infero-temporal (IT) lobe using neuroimaging. Here we consider the evidence for contextual representations in terms of single-cell responses, to visual stimuli, in the inferior temporal cortex of awake behaving monkeys. If the representation of a stimulus depends on establishing representations of subordinate and supraordinate causes at all levels of the visual hierarchy, then information about the high-order attributes of a stimulus must be conferred by top-down influences. Consequently, one might expect to see the emergence of selectivity, for high-level attributes, after the initial visually evoked response (it typically takes about 10 ms for volleys of spikes to be propagated from one cortical area to another and about a 100 ms to reach prefrontal areas). This is because the representations at higher levels must emerge before backward afferents can dynamically reshape the response profile of neurons in lower areas. This temporal delay, in the emergence of selectivity, is precisely what one sees empirically: Sugase et al. (1999) recorded neurons in macaque temporal cortex during the presentation of faces and objects. The faces were either human or monkey faces and were categorized in terms of identity (whose face it was) and expression (happy, angry, etc.). "Single neurones conveyed two different scales of facial information in their firing patterns, starting at different latencies. Global information, categorizing stimuli as monkey faces, human faces or shapes, was conveyed in the earliest part of the responses. Fine information about identity or expression was conveyed later," starting on average about 50 ms after face-selective responses. These observations demonstrate representations for facial identity or expression that emerge dynamically in a way that might rely on backward connections. These influences imbue neurons with a selectivity that is not intrinsic to the area but depends on interactions across levels of a processing hierarchy.

The preceding arguments have been based largely on electrophysiological responses. They can be extended to the population responses elicited in functional neuroimaging where functional specialization (cf. selectivity in unit recordings) is established by showing regionally specific responses to some sensorimotor attribute or cognitive component. At the level of cortical responses in neuroimaging, the dynamic and contextual nature of evoked responses means that regionally specific responses to a particular cognitive component may be expressed in one context but not another. In the next section we look at some empirical evidence from functional neuroimaging that confirms the idea that functional specialization is conferred in a context-sensitive fashion by backward connections from higher brain areas.

FUNCTIONAL ARCHITECTURES ASSESSED WITH BRAIN IMAGING

Information theory and predictive coding schema posit alternative architectures that the brain might adopt for perceptual synthesis. The former relies on forward connections, whereas the latter suggests that most of the brain's infrastructure would be used to predict the sensory input through a hierarchy of top-down projections. Clearly to adjudicate between these alternatives the existence of backward influences must be established. This is a slightly deeper problem for functional neuroimaging than might be envisaged. This is because making causal inferences about effective connectivity is not straightforward (see Pearl 2000). It might be thought that showing regional activity in one level was partially predicted by activity in a higher level would be sufficient to confirm the existence of backward influences. The problem is that this statistical dependency does not permit any causal inference. Statistical dependencies could easily arise in purely feedforward architecture because the higher level activity is predicated on activity in the lower level. One resolution of this problem is to perturb the higher level directly using transmagnetic stimulation or lesions. However, discounting these interventions, one is left with the difficult problem of inferring backward influences, based on measures that could be correlated because of forward connections. Although there are causal modeling techniques that can address this problem, we take a simpler approach and note that interactions between bottom-up and top-down influences cannot be explained by a feedforward architecture. This is because the top-down influences have no access to the bottom-up inputs. An interaction, in this context, can be construed as an effect of backward connections on the driving efficacy of forward connections. In other words, the response evoked by the same driving bottom-up inputs depends on the context established by top-down inputs.

In summary, a critical feature of the functional architectures implied by predictive coding is the expression of interactions between bottom-up and top-down influences from other brain regions at a unit, population, or cortical level. The remainder of this article focuses on the evidence for these interactions. From the point of view of functionally specialized responses these interactions manifest as context-sensitive or contextual specialization, where modality-, category-, or exemplar-specific responses, driven by bottom-up inputs are modulated by top-down influences induced by perceptual set. The first half of this section adopts this perspective. The second part of this section uses measurements of effective connectivity to establish interactions between bottom-up and top-down influences. All the examples presented below rely on attempts to establish interactions by trying to change sensory-evoked neuronal responses through putative manipulations of top-down influences. These involve eliciting changes in perceptual or cognitive (attentional) set.

Context-Sensitive Specialization

If the contextual nature of specialization is mediated by backward modulatory afferents then it should be possible to find cortical regions in which functionally specific responses, elicited by the same stimuli, are modulated by the activity in higher areas. The following example shows that this is indeed possible.

PSYCHOPHYSIOLOGICAL INTERACTIONS Psychophysiological interactions speak directly to the interactions between bottom-up and top-down influences, where one is modeled as an experimental factor and the other constitutes a measured brain response. An analysis of psychophysiological interactions tries to explain a regionally specific response in terms of an interaction between the presence of a sensorimotor or cognitive process and activity in another part of the brain (Friston et al. 1997). The supposition here is that the remote region is the source of backward or lateral modulatory afferents that confer functional specificity on the target region. For example, by combining information about activity in the posterior parietal cortex, mediating attentional or perceptual set pertaining to a particular stimulus attribute, can we identify regions that respond to that attribute when, and only when, activity in the parietal source is high? If such an interaction exists, then one might infer that the parietal area is modulating selective responses in the target area. The statistical model employed in testing for psychophysiological interactions is a simple regression model of effective connectivity that embodies nonlinear (second-order or modulatory effects). This class of model speaks directly to functional specialization of a nonlinear and contextual sort. Figure 4 illustrates a specific example (see Dolan et al. 1997 for details). Subjects were asked to view (degraded) faces and nonface (object) controls. The interaction between activity in the parietal region and the presence of faces was expressed most significantly in the right IT region. Changes in parietal activity were induced experimentally by pre-exposure of the (undegraded) stimuli before some scans but not others to

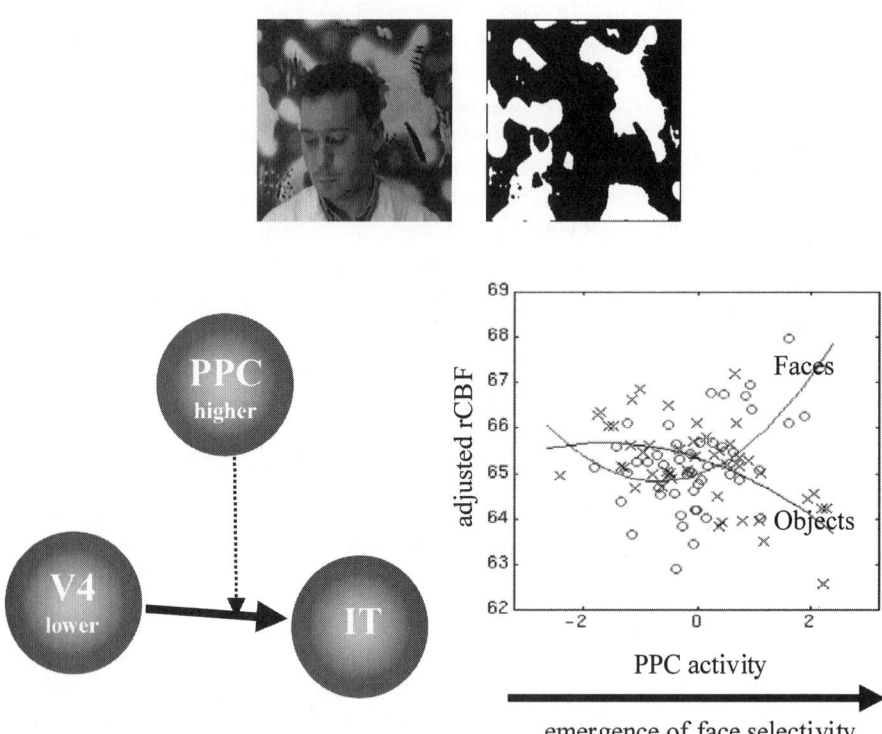

Figure 4 (*Top*) Examples of the stimuli presented to subjects. During the measurement of brain responses only degraded stimuli where shown (e.g., *the right-hand picture*). In half the scans the subject was given the underlying cause of these stimuli through presentation of the original picture (e.g., *left*) before scanning. This priming induced a profound difference in perceptual set for the primed, relative to nonprimed, stimuli. (*Right*) Activity observed in a right infero-temporal (IT) region, as a function of (mean corrected) PPC activity. This region showed the most significant interaction between the presence of faces in visually presented stimuli and activity in a reference location in the posterior (medial) parietal cortex (PPC). This analysis can be thought of as finding those areas that are subject to top-down modulation of face-specific responses by medial parietal activity. The crosses correspond to activity while viewing nonface stimuli and the circles to faces. The essence of this effect can be seen by noting that this region differentiates between faces and nonfaces when, and only when, medial parietal activity is high. The lines correspond to the best second-order polynomial fit. These data were acquired from six subjects using PET. Left: Schematic depicting the underlying conceptual model in which driving afferents from ventral form areas (here designated as V4) excite responses in IT regions subject to permissive modulation by PPC projections.

prime them. The data in the right panel of Figure 4 suggest that the IT region shows face-specific responses, relative to nonface objects, when, and only when, parietal activity is high. These results can be interpreted as a priming-dependent face-specific response, in IT regions that are mediated by interactions with medial parietal cortex. This is a clear example of contextual specialization that depends on top-down nonlinear effects.

Nonlinear Coupling Among Brain Areas

The previous example, demonstrating contextual specialization, is consistent with functional architectures implied by predictive coding. However, it does not provide definitive evidence for an interaction between top-down and bottom-up influences. In this subsection we look for direct evidence using functional imaging. This rests on being able to measure effective connectivity in a way that is sensitive to interactions among inputs. Linear models of effective connectivity assume that the multiple inputs to a brain region are linearly separable. This assumption precludes activity-dependent connections that are expressed in one context and not in another. The resolution of this problem lies in adopting nonlinear models, like the Volterra formulation, that include interactions among inputs (see Box 1 and the second section). These interactions can be construed as a context- or activity-dependent modulation of the influence that one region exerts over another. These nonlinearities can also be introduced into structural equation modeling using so-called moderator variables that represent the interaction between two regions when causing activity in a third (Büchel & Friston 1997). From a dynamical point of view, modulatory effects are modeled by second-order kernels. Within these models the influence of one region on another has two components: (*a*) the direct or driving influence of input from the first (e.g., hierarchically lower) region, irrespective of the activities elsewhere and (*b*) an activity-dependent, modulatory component that represents an interaction with inputs from the remaining (e.g., hierarchically higher) regions. These are mediated by the first- and second-order kernels respectively. The example provided in Figure 5 addresses the modulation of visual cortical responses by attentional mechanisms (e.g., Treue & Maunsell 1996) and the mediating role of activity-dependent changes in effective connectivity.

The bottom panel in Figure 5 shows a characterization of this modulatory effect in terms of the increase in V5 responses, to a simulated V2 input, when posterior parietal activity is zero (broken line) and when it is high (solid lines). In this study subjects were studied with fMRI under identical stimulus conditions (visual motion subtended by radially moving dots) while manipulating the attentional component of the task (detection of velocity changes). The brain regions and connections comprising the model are shown in the upper panel. The lower panel shows a characterization of the effects of V2 inputs on V5 and their modulation by posterior parietal cortex (PPC) using simulated inputs at different levels of PPC activity. It is evident that V2 has an activating effect on V5 and that PPC increases the responsiveness of V5 to these inputs. The insert shows all the voxels in V5 that

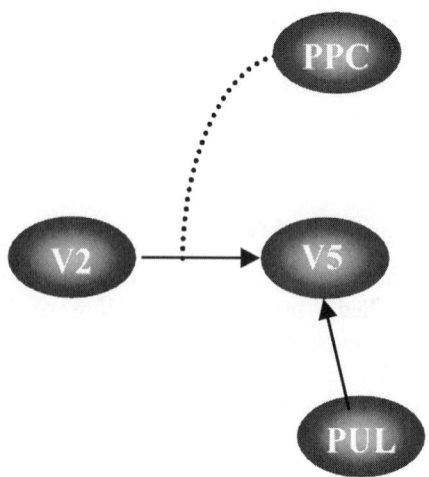

Changes in V5 responses to inputs from V2 with PPC activity

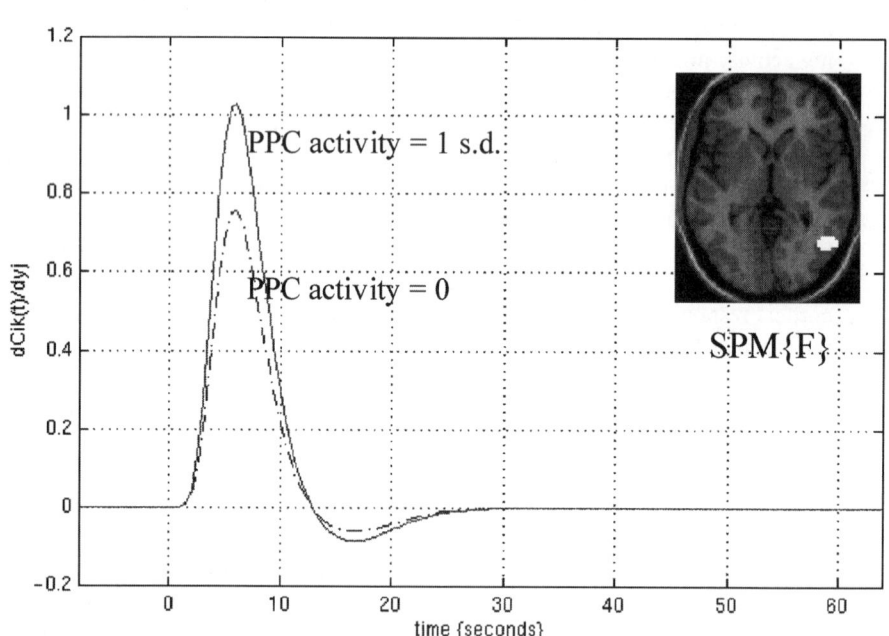

evidenced a modulatory effect ($p < 0.05$ uncorrected). These voxels were identified by thresholding statistical parametric maps of the F statistic (Friston et al. 1995) testing for the contribution of second-order kernels involving V2 and PPC while treating all other components as nuisance variables. The estimation of the Volterra kernels and statistical inference procedure is described in Friston & Büchel (2000).

This sort of result suggests that backward parietal inputs may be a sufficient explanation for the attentional modulation of visually evoked extrastriate responses. More importantly, they are consistent with the functional architecture implied by predictive coding. V5 cortical responses evidence an interaction between bottom-up input from early visual cortex and top-down influences from parietal cortex.

CONCLUSION

In conclusion, the representational capacity and inherent function of any neuron, neuronal population, or cortical area in the brain is dynamic and context sensitive. Functional integration, or interactions among brain systems, that employ driving (bottom-up) and backward (top-down) connections mediate this adaptive and contextual specialization. The arguments in this review were developed under generative models of brain function, where higher-level systems provide a prediction of the inputs to lower-level regions. Conflict between the two is resolved by changes in the higher-level representations, which are driven by the ensuing error in lower regions, until the mismatch is "cancelled." From this perspective the specialization of any region is determined both by bottom-up driving inputs and by top-down predictions. Specialization is therefore not an intrinsic property of any region but depends on both forward and backward connections with other areas. Because the latter have access to the context in which the inputs are generated, they are in a position to modulate the selectivity or specialization of lower areas.

Figure 5 (*Top*) Brain regions and connections comprising the model. (*Bottom*) Characterization of the effects of V2 inputs on V5 and their modulation by posterior parietal cortex (PPC). The broken lines represent estimates of V5 responses when PPC activity is zero, according to a second-order Volterra model of effective connectivity with inputs to V5 from V2, PPC, and the pulvinar (PUL). The solid curves represent the same response when PPC activity is one standard deviation of its variation over conditions. It is evident that V2 has an activating effect on V5 and that PPC increases the responsiveness of V5 to these inputs. The insert shows all the voxels in V5 that evidenced a modulatory effect ($p < 0.05$ uncorrected). These voxels were identified by thresholding a SPM (Friston et al. 1995) of the F statistic testing for the contribution of second-order kernels involving V2 and PPC (treating all other terms as nuisance variables). The data were obtained with fMRI under identical stimulus conditions (visual motion subtended by radially moving dots) while manipulating the attentional component of the task (detection of velocity changes).

The implications for classical models (e.g., classical receptive fields in electrophysiology, classical specialization in neuroimaging, and connectionism in cognitive models) are severe and suggest these models may provide incomplete accounts of real brain architectures. On the other hand, predictive coding, in the context of generative models, not only accounts for many extra-classical phenomena seen empirically but enforces a view of the brain as an inference machine, through its Bayesian motivation.

ACKNOWLEDGMENT

The Wellcome Trust funded this work.

The *Annual Review of Neuroscience* is online at http://neuro.annualreviews.org

LITERATURE CITED

Abbot LF, Varela JA, Sen K, Nelson SB. 1997. Synaptic depression and cortical gain control. *Science* 275:220–23

Absher JR, Benson DF. 1993. Disconnection syndromes: an overview of Geschwind's contributions. *Neurology* 43:862–67

Aertsen A, Preißl H. 1991. Dynamics of activity and connectivity in physiological neuronal networks. In *Non Linear Dynamics and Neuronal Networks*, ed. HG Schuster, pp. 281–302. New York: VCH

Atick JJ, Redlich AN. 1990. Towards a theory of early visual processing. *Neural Comput.* 2:308–20

Barlow HB. 1961. Possible principles underlying the transformation of sensory messages. In *Sensory Communication*, ed. WA Rosenblith. Cambridge: MIT

Bell AJ, Sejnowski TJ. 1995. An information maximisation approach to blind separation and blind de-convolution. *Neural Comput.* 7:1129–59

Bendat JS. 1990. *Nonlinear System Analysis and Identification from Random Data.* New York: Wiley

Büchel C, Friston KJ. 1997. Modulation of connectivity in visual pathways by attention: cortical interactions evaluated with structural equation modelling and fMRI. *Cereb. Cortex* 7:768–78

Common P. 1994. Independent component analysis, a new concept? *Signal Process.* 36:287–314

Crick F, Koch C. 1998. Constraints on cortical and thalamic projections: the no-strong-loops hypothesis. *Nature* 391:245–50

Dayan P, Hinton GE, Neal RM. 1995. The Helmholtz machine. *Neural Comput.* 7:889–904

Dolan RJ, Fink GR, Rolls E, Booth M, Holmes A, et al. 1997. How the brain learns to see objects and faces in an impoverished context. *Nature* 389:596–98

Dong D, McAvoy TJ. 1996. Nonlinear principal component analysis—based on principal curves and neural networks. *Comput. Chem. Eng.* 20:65–78

Edelman GM. 1993. Neural Darwinism: selection and reentrant signalling in higher brain function. *Neuron* 10:115–25

Felleman DJ, Van Essen DC. 1991. Distributed hierarchical processing in the primate cerebral cortex. *Cerebral Cortex* 1:1–47

Fliess M, Lamnabhi M, Lamnabhi-Lagarrigue F. 1983. An algebraic approach to nonlinear functional expansions. *IEEE Trans. Circuits Syst.* 30:554–70

Foldiak P. 1990. Forming sparse representations by local anti-Hebbian learning. *Biol. Cybern.* 64:165–70

Freeman W, Barrie J. 1994. Chaotic oscillations and the genesis of meaning in cerebral cortex. In *Temporal Coding in the Brain*, ed. G Buzsaki, R Llinas, W Singer, A Berthoz, T Christen, pp. 13–38. Berlin: Springer Verlag

Friston KJ. 1995. Functional and effective connectivity in neuroimaging: a synthesis. *Hum. Brain Mapp.* 2:56–78

Friston KJ, Büchel C. 2000. Attentional modulation of V5 in humans. *Proc. Natl. Acad. Sci. USA* 97:7591–96

Friston KJ, Büchel C, Fink GR, Morris J, Rolls E, Dolan RJ. 1997. Psychophysiological and modulatory interactions in neuroimaging. *NeuroImage* 6:218–29

Friston KJ, Frith CD, Frackowiak RSJ. 1993. Principal component analysis learning algorithms: a neurobiological analysis. *Proc. R. Soc. B* 254:47–54

Friston KJ, Holmes AP, Worsley KJ, Poline JB, Frith CD, Frackowiak RSJ. 1995. Statistical parametric maps in functional imaging: a general linear approach. *Hum. Brain Mapp.* 2:189–210

Friston KJ, Tononi G, Reeke GH, Sporns O, Edelman GE. 1994. Value-dependent selection in the brain: simulation in a synthetic neural model. *Neuroscience* 39:229–43

Gawne TJ, Richmond BJ. 1993. How independent are the messages carried by adjacent inferior temporal cortical neurons? *J. Neurosci.* 13:2758–71

Gerstein GL, Perkel DH. 1969. Simultaneously recorded trains of action potentials: analysis and functional interpretation. *Science* 164:828–30

Girard P, Bullier J. 1989. Visual activity in area V2 during reversible inactivation of area 17 in the macaque monkey. *J. Neurophysiol.* 62:1287–301

Hinton GE, Dayan P, Frey BJ, Neal RM. 1995. The "Wake-Sleep" algorithm for unsupervised neural networks. *Science* 268:1158–61

Hirsch JA, Gilbert CD. 1991. Synaptic physiology of horizontal connections in the cat's visual cortex. *J. Neurosci.* 11:1800–9

Karhunen J, Joutsensalo J. 1994. Representation and separation of signals using nonlinear PCA type learning. *Neural Netw.* 7:113–27

Kay J, Phillips WA. 1996. Activation functions, computational goals and learning rules for local processors with contextual guidance. *Neural Comput.* 9:895–910

Linsker R. 1988. Self-organisation in a perceptual network. *Computer* March:105–17

McIntosh AR. 2000. Towards a network theory of cognition. *Neural Netw.* 13:861–70

McIntosh AR, Gonzalez-Lima F. 1994. Structural equation modelling and its application to network analysis in functional brain imaging. *Hum. Brain Mapp.* 2:2–22

Oja E. 1989. Neural networks, principal components, and subspaces. *Int. J. Neural Syst.* 1:61–68

Olshausen BA, Field DJ. 1996. Emergence of simple-cell receptive field properties by learning a sparse code for natural images. *Nature* 381:607–9

Optican L, Richmond BJ. 1987. Temporal encoding of two-dimensional patterns by single units in primate inferior cortex. II. Information theoretic analysis. *J. Neurophysiol.* 57:132–46

Pearl J. 2000. *Causality, Models, Reasoning and Inference*. Cambridge, UK: Cambridge Univ. Press

Phillips CG, Zeki S, Barlow HB. 1984. Localisation of function in the cerebral cortex past present and future. *Brain* 107:327–61

Phillips WA, Singer W. 1997. In search of common foundations for cortical computation. *Behav. Brain Sci.* 20:57–83

Rao RPN, Ballard DH. 1999. Predictive coding in the visual cortex: a functional interpretation of some extra-classical receptive field effects. *Nat. Neurosci.* 2:79–87

Rockland KS, Pandya DN. 1979. Laminar origins and terminations of cortical connections of the occipital lobe in the rhesus monkey. *Brain Res.* 179:3–20

Rumelhart D, McClelland J. 1986. *Parallel Distributed Processing: Explorations in the Microstructure of Cognition*. Cambridge: MIT

Salin P-A, Bullier J. 1995. Corticocortical connections in the visual system: structure and function. *Psychol. Bull.* 75:107–54

Sandell JH, Schiller PH. 1982. Effect of cooling area 18 on striate cortex cells in the squirrel monkey. *J. Neurophysiol.* 48:38–48

Sugase Y, Yamane S, Ueno S, Kawano K. 1999. Global and fine information coded by single neurons in the temporal visual cortex. *Nature* 400:869–73

Sutton RS, Barto AG. 1990. Time derivative models of Pavlovian reinforcement. In *Learning and Computational Neuroscience: Foundations of Adaptive Networks*, ed. M Gabriel, J Moore, pp. 497–538. Cambridge: MIT

Tononi G, Sporns O, Edelman GM. 1994. A measure for brain complexity: relating functional segregation and integration in the nervous system. *Proc. Natl. Acad. Sci.* 91:5033–37

Tovee MJ, Rolls ET, Treves A, Bellis RP. 1993. Information encoding and the response of single neurons in the primate temporal visual cortex. *J. Neurophysiol.* 70:640–54

Treue S, Maunsell HR. 1996. Attentional modulation of visual motion processing in cortical areas MT and MST. *Nature* 382:539–41

Wray J, Green GGR. 1994. Calculation of the Volterra kernels of non-linear dynamic systems using an artificial neuronal network. *Biol. Cybern.* 71:187–95

Zeki S. 1990. The motion pathways of the visual cortex. In *Vision: Coding and Efficiency*, ed. C Blakemore, pp. 321–45. Cambridge, UK: Cambridge Univ. Press

Zeki S, Shipp S. 1988. The functional logic of cortical connections. *Nature* 335:311–17

TRANSCRIPTIONAL CODES AND THE CONTROL OF NEURONAL IDENTITY

Ryuichi Shirasaki and Samuel L. Pfaff
Gene Expression Laboratory, The Salk Institute for Biological Studies, La Jolla, California 92037; email: shirasaki@salk.edu; pfaff@salk.edu

Key Words LIM code, motor neuron subtypes, axon pathfinding, topography, spinal cord

■ **Abstract** The topographic assembly of neural circuits is dependent upon the generation of specific neuronal subtypes, each subtype displaying unique properties that direct the formation of selective connections with appropriate target cells. Studies of motor neuron development in the spinal cord have begun to elucidate the molecular mechanisms involved in controlling motor projections. In this review, we first describe the actions of transcription factors within motor neuron progenitors, which initiate a cascade of transcriptional interactions that lead to motor neuron specification. We next highlight the contribution of the LIM homeodomain (LIM-HD) transcription factors in establishing motor neuron subtype identity. Importantly, it has recently been shown that the combinatorial expression of LIM-HD transcription factors, the LIM code, confers motor neuron subtypes with the ability to select specific axon pathways to reach their distinct muscle targets. Finally, the downstream targets of the LIM code are discussed, especially in the context of subtype-specific motor axon pathfinding.

INTRODUCTION

One of the fundamental issues in the field of developmental neuroscience is to understand the mechanisms that control the identity of distinct classes of neurons located at defined positions within the nervous system. This has been the goal of considerable research effort in vertebrates and invertebrates alike, for the ability of an animal to accomplish its behavioral repertoires depends critically on the generation of appropriate neuronal types and the establishment of specific neuronal connections. Importantly, in adult organisms, the mature identity of a neuron, which underlies the functional unit of a neural circuit, is represented by its characteristic features that include the soma position, the axonal projection pattern, the elaboration of dendritic morphology, the expression of specific ion channels and neurotransmitter receptors, and the production of appropriate neurotransmitter. Many of these characteristic traits are coordinately regulated and acquired during development (for reviews, see Edlund & Jessell 1999, Jessell 2000). Recent genetic studies of vertebrates and invertebrates have identified a number of genes that regulate the expression of these unique phenotypes, and further suggested that the ultimate outcome of neuronal identity is under the control of complex regulatory networks of transcription factors acting in a sequential fashion (for recent reviews, see Jurata et al. 2000, Lee & Pfaff 2001). It should be noted, however, that many of these transcription factors act in the progenitor cells for neurons prior to cell cycle exit (for reviews, see Bang & Goulding 1996, Doe & Skeath 1996). Therefore, for the most part, mutation of progenitor cell transcription factors alters the expression of the downstream transcription factors driving neural development, leading to a change in expression of the terminal differentiation genes in postmitotic neurons. This results in the alteration of the mature phenotype of neurons. Therefore, whereas progenitor cell transcription factors are important for establishing neuronal cell identity, it is unlikely that they directly control the expression of unique characteristics of postmitotic neurons.

Axon pathfinding is one aspect of the neuronal differentiation program that occurs in postmitotic neurons. We now understand that the correct wiring of the nervous system depends upon the ability of individual axons to recognize specific guidance cues during their growth toward their final cellular targets (for reviews, see Tessier-Lavigne & Goodman 1996, Mueller 1999). In addition, the topographic organization and elaboration of neural networks in higher organisms is dependent upon the generation of related, but distinct, neuronal subclasses (or subtypes) from broad classes of neurons during development. In general, the axonal trajectory of a single class of neurons is composed of several discrete pathways, some of which are shared by multiple subtypes and others are unique pathways used by individual subtypes. The classical experimental studies using avian embryos have provided a wealth of information on these axonal pathfinding steps (for a review, see Landmesser 1992). The emerging view is that each neuronal subtype possesses an intrinsic capacity to detect its own unique path soon after it becomes postmitotic (for a review, see Eisen 1994). This suggests that, before they reach their specific pathfinding decision points, they have already been genetically programmed to be able to sense their unique pathway-specific guidance cues. Accumulating evidence now suggests that the intrinsic genetic programs are in most cases encoded not by a single but rather by unique combinatorial arrays of certain classes of transcription factors (for a review, see Pfaff & Kintner 1998).

This review focuses primarily on the role of postmitotically expressed transcription factors that ultimately control axon pathfinding during the development of vertebrates and invertebrates. In particular, we describe the role of transcription factors in the development of a single class of neurons in the spinal cord, motor neurons, from the generation of motor neuron progenitors to the acquisition of specific subtype properties. In recent years, numerous reviews have covered portions of this work, most of which focused on inductive signals and the control of neuronal patterning along the dorsoventral axis of the spinal cord as well as the transcriptional mechanisms for determination of neuronal progenitor identity. Thus, rather than providing a comprehensive review on the transcriptional mechanisms for motor neuron fate specification, we mainly discuss the role of one class of postmitotically expressed transcription factors, the LIM homeodomain

(LIM-HD) transcription factors, in fate determination of motor neuron subtypes and in subtype-specific motor axon pathfinding, especially in light of intrinsic genetic regulators of axon guidance.

SPECIFICATION OF PROGENITOR CELL IDENTITY BY HOMEODOMAIN TRANSCRIPTION FACTORS

We first describe the early patterning events that confer the identity of neuronal progenitors in the spinal cord. During the early development of the vertebrate central nervous system (CNS), dividing progenitor cells in the ventricular zone acquire regionally restricted characteristics that later impose distinct identities on their neuronal progeny. The properties of progenitor cells are defined by secreted inductive signals, which diffuse from their cellular source to form a concentration gradient. For example, in the dorsal spinal cord, bone morphogenetic proteins, which are secreted from the surface ectoderm and the roof plate, control the specification of dorsal cell types such as neural crest cells and dorsal sensory interneurons (for a review, see Lee & Jessell 1999). In contrast, in the ventral spinal cord, Sonic Hedgehog (Shh), a glycoprotein secreted from the notochord and the floor plate, specifies the pattern of generation of motor neurons and certain classes of ventral interneurons in a concentration-dependent manner (for a recent review, see Briscoe & Ericson 2001). Indeed, in mice deficient in Shh function, motor neurons and ventral interneurons fail to develop (Chiang et al. 1996).

Progenitor cells respond to the graded inductive signals by translating the specific concentration of the signals into the patterned expression of transcription factors (for a recent review, see Gurdon & Bourillot 2001). Recent studies have identified a set of homeodomain transcription factors expressed by ventral neuronal progenitors that function as intermediaries in interpreting graded Shh signaling. The combinatorial expression of these homeodomain transcription factors subdivides the ventral spinal cord into five progenitor domains, each of which gives rise to a distinct class of postmitotic neurons (Figure 1) (Briscoe et al. 2000). These homeodomain transcription factors can be categorized into class I and class II factors based on their mode of regulation by Shh. The class I proteins (Pax7, Dbx1, Dbx2, Irx3, and Pax6) are repressed by Shh signaling, whereas the class II proteins (Nkx6.2, Nkx6.1, Olig2, Nkx2.2, and Nkx2.9) depend on Shh signaling for their expression (Briscoe et al. 2000, Mizuguchi et al. 2001, Novitch et al. 2001, Vallstedt et al. 2001). It should be noted that Olig2 is a basic helix-loop-helix (bHLH) transcription factor (see below), unlike the other factors which contain homeodomains.

Evidence has been provided that cross-repressive interactions between neighboring class I and class II factors subsequently act to refine and sharpen the boundaries of these progenitor domains (Briscoe et al. 2000), thus ensuring that progenitor cells within individual domains express distinct combinations of these homeodomain transcription factors. Importantly, it has also been shown that most of these homeodomain transcription factors, which share a motif related to the eh1 region of the Engrailed repressor domain (Smith & Jaynes 1996), act directly as

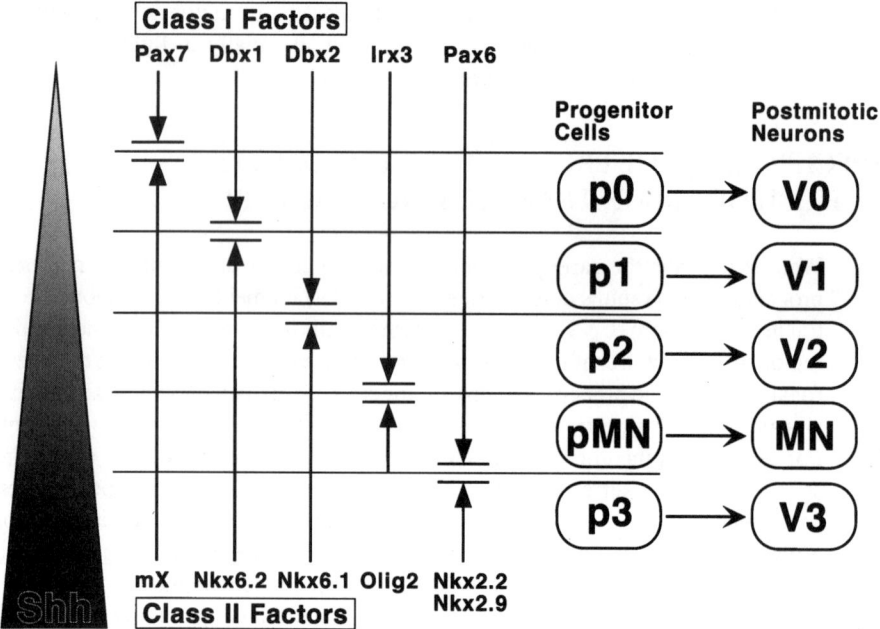

Figure 1 A combinatorial code of homeodomain transcription factors and basic helix-loop-helix transcription factors specifies the identity of neuronal progenitor cells in the ventral neural tube. Graded Sonic Hedgehog (Shh) signaling establishes distinct ventral progenitor domains by regulating the expression of transcription factors, which are subdivided into class I and class II factors based on whether their expression is repressed or induced by Shh signaling. These transcription factors act as intermediaries in Shh-dependent neural patterning. Crossrepressive interactions between class I and class II factors, which abut a common progenitor domain, contribute to establishing individual progenitor domains (p3-p0) with sharp boundaries. Each progenitor domain gives rise to a specific class of postmitotic neurons. The identity of potential class II factors (mX) counteracting the activity of the Pax7 class I factor remains unknown. In this model, for example, the boundaries of the motor neuron (MN) progenitor domain are delineated ventrally by Pax6/Nkx2.2 and dorsally by Irx3/Olig2. Nkx6.2 expression overlaps Nkx6.1 ventrally in the chick but is restricted to the p1 domain in the mouse.

transcriptional repressors through the recruitment of Groucho-TLE corepressors (Muhr et al. 2001). Consistent with the idea that these transcription factors participate in a code for establishing neuronal identity, genetic studies have shown that the pattern of generation of certain ventral neuronal classes is impaired in rodents carrying mutations in these genes such as *Pax6*, *Dbx1*, *Nkx2.2*, *Nkx6.1*, and *Nkx6.2* (Burrill et al. 1997, Ericson et al. 1997, Osumi et al. 1997, Briscoe et al. 1999, Sander et al. 2000, Pierani et al. 2001, Vallstedt et al. 2001).

Together, these findings suggest that mutual repression of class I and class II transcriptional repressors at progenitor domain boundaries, as well as the combinatorial expression of these factors within each progenitor domain, plays an instructive role in the specification of neuronal progenitor fate in the developing spinal cord (Figure 1). Furthermore, the regulation of neuronal fate by negatively acting transcription factors in progenitor cells implicates a derepression strategy for neuronal fate determination. This model therefore predicts that cell identity within each domain is achieved through active suppression of all the other possible neuronal fates directed by other transcription factors (see below).

NEURONAL SPECIFICATION AND CELL CYCLE EXIT BY bHLH TRANSCRIPTION FACTORS

Recent studies have now provided evidence that bHLH transcription factors in vertebrates can also impose some aspects of neuronal identity, in addition to their more general roles in neurogenesis (for reviews, see Kageyama & Nakanishi 1997, Lee 1997, Guillemot 1999; see also Gowan et al. 2001). For example, in the peripheral nervous system, *Mash1*, a mammalian homolog of *Drosophila achaete-scute*, is expressed in progenitor cells of autonomic neurons, where it is required for the acquisition of a noradrenergic neurotransmitter phenotype (Hirsch et al. 1998). In addition, bHLH transcription factors have the ability to direct the exit of neural progenitors from the cell cycle (Morrow et al. 1999, Farah et al. 2000, Mizuguchi et al. 2001, Novitch et al. 2001). In contrast, none of the class I/II homeodomain transcription factors described above have so far been shown to regulate cell cycle exit in the ventricular zone of the neural tube.

In the developing spinal cord, bHLH transcription factors are expressed in a number of discrete domains along the dorsoventral axis of the neural tube. Some of these bHLH transcription factors, such as the *Drosophila biparous/tap*-related genes *Neurogenins* (*Ngns*) and the *atonal*-related gene *Math1*, show expression patterns restricted to specific progenitor domains of the spinal cord (Akazawa et al. 1995, Ben-Arie et al. 1996, Gradwohl et al. 1996, Sommer et al. 1996, Ma et al. 1997, Helms & Johnson 1998, Lee et al. 1998, Gowan et al. 2001). Indeed, like the action of homeodomain transcription factors in the ventral spinal cord described above, crossinhibitory regulation between Neurogenin1, Math1, and Mash1 has been shown to determine discrete progenitor domains and neuronal fate in the dorsal spinal cord (Gowan et al. 2001).

Recently, another family of bHLH transcription factors has been identified and two of these members, Olig1 and Olig2, show a restricted pattern of expression within the ventral spinal cord (Lu et al. 2000, Takebayashi et al. 2000, Zhou et al. 2000). Furthermore, functional studies using ectopic expression of these genes have suggested an instructive role in oligodendrocyte differentiation (Lu et al. 2000; Zhou et al. 2000, 2001). Importantly, the onset of expression of these genes precedes the appearance of oligodendrocyte progenitors and, in particular, they are precisely expressed in the progenitor domain for motor neurons in the ventral spinal cord (Mizuguchi et al. 2001, Novitch et al. 2001), suggesting an additional role for these bHLH transcription factors in motor neuron fate specification. Indeed, misexpression of Olig2 in the chick neural tube generated ectopic motor neurons. The ectopic cells formed at the expense of more dorsal neuronal types are correlated with Olig2's ability to repress the expression of the class I homeodomain factor Irx3 (Mizuguchi et al. 2001, Novitch et al. 2001).

The involvement of Neurogenin2 (Ngn2), another member of the bHLH transcription factor family, in neuronal fate specification has been suggested from the regulatory interactions between *Ngn2* and *Pax6*, revealed by the analysis of mice carrying mutations in these genes (Scardigli et al. 2001). Interestingly, consistent with the temporal and spatial expression pattern of Olig2 and Ngn2 in the ventral spinal cord, ectopic expression of Olig2 in the neural tube induced the upregulation of Ngn2 (Novitch et al. 2001) and, in some regions of the neural tube, triggered expression of homeodomain transcription factors such as MNR2 which is involved in motor neuron specification (Mizuguchi et al. 2001, Novitch et al. 2001). This therefore suggests that Olig2 regulates factors for establishing cell identity and neurogenesis in a coordinated manner. In support of this notion, mice deficient in *Ngn2* function have reduced motor neuron numbers and altered molecular marker profiles (Scardigli et al. 2001). Because Ngns have been shown to regulate neurogenesis and promote cell cycle exit, the role of Olig2 appears not just to activate specific motor neuron determinants but also to restrict the number of cell cycles available to motor neuron progenitors and activate pan-neuronal traits via the activation of Ngn2.

MOLECULAR PATHWAY FOR SPECIFICATION OF MOTOR NEURON IDENTITY BY COMBINATORIAL ACTIONS OF HOMEODOMAIN AND bHLH TRANSCRIPTION FACTORS

The combinatorial expression of homeodomain and bHLH transcription factors thus delineates distinct progenitor domains in the ventricular zone of the spinal cord, each of which is responsible for the fate determination of their neuronal progeny. Here, we describe in detail a model of the action of these transcription factors on motor neuron fate specification.

The motor neuron progenitor (pMN) domain is demarcated by the V2 interneuron progenitor (p2) domain dorsally and by the V3 interneuron progenitor (p3) domain ventrally (Figure 1). The combinatorial actions of the class I/II factors, Pax6/Nkx2.2 and Irx3/Olig2, define the boundaries of the pMN domain. Within this pMN domain itself, where progenitors express Nkx6.1 and Nkx6.2, the activities of these two Nkx6 repressor proteins inhibit the expression of other cell determinants while allowing Olig2 to become expressed for the coordinate regulation of factors involved in motor neuron differentiation such as Ngn2, MNR2, and HB9 (Briscoe et al. 2000, Sander et al. 2000, Vallstedt et al. 2001). These findings raise the issue of whether Olig2 acts in parallel with or independently of Nkx6 and Ngn2 in this molecular pathway. Gain-of-function and loss-of-function studies have demonstrated that Olig2 acts downstream of Nkx6 proteins (Novitch et al. 2001).

Taken together, these findings establish a molecular hierarchy of transcription factors in motor neuron fate specification, i.e., within the pMN domain, Nkx6 class proteins initiate the generation of motor neuron progenitors by upregulating the expression of Olig2, which then promotes the expression of dedicated determinants such as MNR2 that impose specific motor neuron identity on cells, in parallel with the promotion of Ngn2 expression to drive motor neuron progenitors to leave the cell cycle and acquire panneuronal traits characteristic of postmitotic neurons.

TRANSCRIPTIONAL CONTROL OF MOTOR NEURON FATE BY HOMEODOMAIN FACTORS, MNR2 AND HB9

The onset of expression of MNR2 and subsequent expression of HB9 correlates with the progression of progenitor cells toward a motor neuron fate. In the chick embryo, MNR2 is first expressed by motor neuron progenitors during their final division cycle at the time when motor neurons attain independence from Shh signaling (Ericson et al. 1996, Tanabe et al. 1998), and the expression persists transiently in postmitotic motor neurons. Functional significance for the MNR2 homeodomain factor is provided by the finding that ectopic expression of MNR2 is sufficient to activate a program of somatic (skeletal muscle–innervating) motor neuron differentiation, characterized by the expression of motor neuron–specific homeodomain factors [e.g., Islet-1 (Isl1), Islet-2 (Isl2), and Lhx3 (also termed Lim3 or P-Lim)] and the enzyme choline acetyltransferase (ChAT) for neurotransmitter synthesis, and by the extension of axons that exit from the ventral spinal cord (Tanabe et al. 1998). In addition, the observation that MNR2 can positively regulate its own expression (Tanabe et al. 1998) points to a molecular basis for the transition from an Shh-dependent to an Shh-independent state, thus contributing to an irreversible commitment for the specification of somatic motor neuron differentiation (Tanabe et al. 1998).

HB9 (Harrison et al. 1994), a homeobox gene whose expression in the CNS is also restricted to motor neurons (Pfaff et al. 1996, Saha et al. 1997, Arber et al. 1999, Thaler et al. 1999), possesses a homeodomain almost identical to that of *MNR2* and shares with MNR2 the ability to induce motor neurons when ectopically expressed in the chick neural tube (Tanabe et al. 1998). In contrast to MNR2, chick HB9 appears only in postmitotic motor neurons (Tanabe et al. 1998). Interestingly, mice seem to lack *MNR2*, but the expression pattern of HB9 in mice shows a pattern that approximates a composite of MNR2 and HB9 in chick (Arber et al. 1999, Thaler et al. 1999). In mice lacking *HB9* function, motor neurons are generated on schedule and in normal number. However, soon after motor neurons have left the cell cycle, they transiently express the Chx10 homeodomain transcription factor ordinarily characteristic of V2 interneurons.

Although this aberrant regulation of the gene profile in motor neurons does not cause a complete fate conversion of motor neurons to V2 interneurons, this abnormal transcriptional code is accompanied by topological disorganization of motor columns and marked errors in the pattern of axonal projections of somatic motor neurons in the periphery (Arber et al. 1999, Thaler et al. 1999). Thus, these studies show that, although HB9 is not required for motor neuron generation, unlike other factors such as the postmitotically expressed Isl1 (Pfaff et al. 1996), it is indispensable for facilitating a normal program of motor neuron differentiation and for suppressing expression of inappropriate interneuron genes. Indeed, the repressive role of HB9 deduced from the above loss-of-function studies is complementary to the MNR2 gain-of-function study in which ectopic expression of MNR2 suppresses the generation of V2 interneurons in the chick (Tanabe et al. 1998).

It is important to discuss why motor neurons, in the absence of *HB9* function, preferentially upregulate V2 interneuron characteristics rather than other interneuron genes. Cell lineage–tracing studies in the chick and the zebrafish have suggested that a single clonal progenitor can generate both motor neurons and ventral interneurons (Leber et al. 1990, Kimmel et al. 1994). More recently, it has been shown in mice that neuronal progenitors of motor neurons and V2 interneurons express a related profile of homeodomain transcription factors, including Pax6 and Lhx3/Lhx4 (Ericson et al. 1997, Sharma et al. 1998). These findings therefore suggest that motor neurons and V2 interneurons have genetically related progenitor cells. In addition, it has been shown in the chick that ectopic expression of Lhx3 is sufficient to induce the V2 interneuron marker Chx10 in the absence of MNR2 (Tanabe et al. 1998). Thus, the emerging view from the analysis of *HB9* knockout mice is that motor neurons inherently possess the potential to initiate genetic programs representative of V2 interneurons. In this context, active suppression of V2 interneuron genetic programs within developing motor neurons is essential to consolidate and establish the identity of postmitotic motor neurons, because Lhx3 is shared by both V2 interneurons and motor neurons.

Thus, like other homeodomain transcription factors, MNR2 and HB9 contribute to the determination of neuronal fate by restricting the intrinsic cellular potential to express conflicting genetic programs within a developing neuron (e.g., Tanabe et al. 1998, Arber et al. 1999, Siegler & Jia 1999, Thaler et al. 1999, Moran-Rivard et al. 2001, Pierani et al. 2001).

TRANSCRIPTION FACTORS EXPRESSED IN POSTMITOTIC NEURONS: LIM-HD TRANSCRIPTION FACTORS

The initiation of axon outgrowth and subsequent pathfinding steps by a neuron is one characteristic feature of the neuronal differentiation program that occurs postmitotically. However, as described above, many of the transcription factors that have critical roles in neuronal fate specification act prior to or around the final cell division of progenitor cells, making it unlikely that they directly regulate the onset of the specific events occurring in postmitotic neurons. In this context, transcription factors expressed in postmitotic neurons are considered to have key functions to activate axon guidance programs for specific pathway choices and target recognition.

Among transcription factors categorized in this way, LIM-HD transcription factors are a family of genes in which all members identified to date in higher vertebrates are expressed in distinct subsets of postmitotic neurons (for a recent review, see Hobert & Westphal 2000). Apart from a DNA-binding homeodomain, one of the unique properties of LIM-HD transcription factors is their structure with two copies of a specialized zing-finger motif called the LIM domain, after its original discovery in two *Caenorhabditis elegans* gene products, lin-11 and mec-3, and the insulin-enhancer-binding protein Isl1 (Way & Chalfie 1988, Freyd et al. 1990, Karlsson et al. 1990). The LIM domain is thought to mediate protein-protein interactions, and, in particular, all LIM-HD transcription factor interactions reported to date are dependent upon the LIM domains (for reviews, see Dawid et al. 1998, Jurata & Gill 1998), except for the interaction between the MEC-3 and the POU homeodomain (POU-HD) transcription factor UNC-86 (Xue et al. 1993). It has also been shown that NLI (also termed Ldb or CLIM), a widely expressed nuclear protein with intrinsic dimerization capacity, binds with high affinity to the LIM domains of all LIM-HD transcription factors (Agulnick et al. 1996, Jurata et al. 1996, Jurata & Gill 1997, Bach et al. 1997) and interestingly, it is mainly expressed in postmitotic cells in the developing CNS (Jurata et al. 1996). Importantly, evidence exists that the nuclear interactor NLI can assemble different LIM-HD transcription factors into a single complex (Figure 2) (Jurata et al. 1998).

Thus, the unique combinatorial arrays of LIM-HD transcription factors in postmitotic neurons have the potential to assemble homomeric and heteromeric complexes in a neuron-specific fashion. It seems likely that the formation of such higher-order complexes allows LIM-HD transcription factors to engage in activation of terminal differentiation genes responsible for the expression of unique properties of postmitotic neuronal subclasses (J. P. Thaler, S.-K. Lee, L. W. Jurata, G. N. Gill, S. L. Pfaff, submitted).

FUNCTIONS OF EXEMPLARY MEMBERS OF LIM-HD TRANSCRIPTION FACTORS

Before proceeding to the role of LIM-HD transcription factors in the determination of motor neuron fate, we describe briefly the functions of exemplary members of this class of transcription factors, *lin-11*, *mec-3*, *apterous*, and *Isl1*.

lin-11

The complete genome sequence of *C. elegans* reveals seven LIM-HD genes, which are expressed in largely nonoverlapping sets of cell types (for a review, see Hobert & Westphal 2000). *lin-11* was first identified on the basis of its role in regulating asymmetric cell divisions in vulval cell lineages (Ferguson & Horvitz 1985,

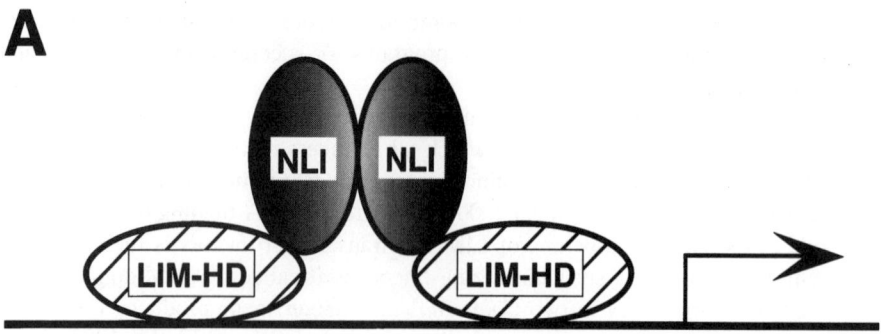

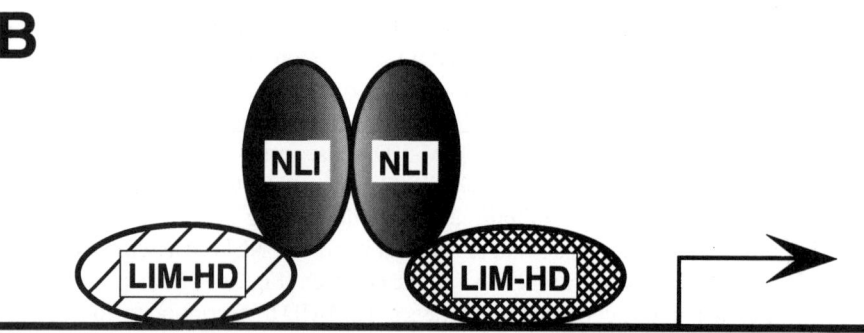

Figure 2 The formation of NLI-mediated LIM-HD transcription factor complexes. NLI specifically interacts with the LIM domain of LIM-HD transcription factors. The intrinsic dimerization capacity of NLI allows any given LIM-HD transcription factors to form homomeric (*A*) and heteromeric (*B*) complexes with other LIM-HD transcription factors. Thus, combinations of LIM-HD transcription factors are thought to assemble into unique complexes for the regulation of effector genes in postmitotic neurons responsible for the characteristic properties of individual neuronal subclasses.

Ferguson et al. 1987, Freyd et al. 1990). Subsequent work has shown that postmitotic AIZ interneurons, which are a major component of the thermoregulatory neural network, express *lin-11*. The gene activity is not required for the generation of these neurons per se, but has a role in specifying the function of these cells (Hobert et al. 1998). *lin-11* is also expressed in ventral cord motor neurons and the gene activity is not necessary for the axonal outgrowth of these motor neurons but is required for them to fasciculate correctly (Hobert et al. 1998). Recently, it has been shown that *lin-11* is also necessary for the functional specification of the AWA olfactory neurons by regulating expression of the AWA-specific *odr-7* nuclear hormone receptor gene (Sarafi-Reinach et al. 2001), which promotes the expression of AWA-specific characteristics and suppresses the fate of sibling AWC olfactory neurons (Sengupta et al. 1994, Sagasti et al. 1999). It should be noted that the vertebrate *lin-11* homolog *Lim1* (also termed *Lhx1*) is also expressed in a subset of motor neurons in the spinal cord (Tsuchida et al. 1994, Kania et al. 2000).

mec-3

The *mec-3* gene is required for the terminal differentiation of the six mechanosensory neurons (the touch receptor neurons) in *C. elegans* (Way & Chalfie 1988). In *mec-3* mutants, the expression of *mec-4*, which encodes a putative ion channel subunit necessary for touch sensitivity (Driscoll & Chalfie 1991), is absent and the expression of mec-7, which encodes a β-tubulin present in the processes of touch cells (Savage et al. 1989), is greatly reduced (Hamelin et al. 1992, Mitani et al. 1993). The POU-HD transcription factor *unc-86* is necessary for the generation of touch cells and continues to be expressed by postmitotic touch cells (Finney et al. 1988, Finney & Ruvkun 1990). Subsequent molecular analyses have suggested that MEC-3 and UNC-86 bind cooperatively to sites in the *mec-3* promoter region and can synergistically activate transcription from the *mec-3* promoter (Xue et al. 1993, Lichtsteiner & Tjian 1995). Recently, it has been shown that MEC-3 binds poorly to *mec-4* and *mec-7* promoters in the absence of UNC-86, and the hetero-oligomer complexes of UNC-86 and MEC-3 directly activate two downstream genes, *mec-4* and *mec-7*, that are necessary for the specific function of touch cells (Duggan et al. 1998). Combinatorial transcription control by LIM-HD and POU-HD transcription factors can also be seen in development of the mammalian pituitary. For example, the LIM-HD transcription factor Lhx3 and POU-HD transcription factor Pit-1 synergistically activate several downstream target genes, such as the prolactin and thyroid-stimulating hormone β-subunit genes (Bach et al. 1995).

apterous

The *apterous* gene is expressed in small subsets of postmitotic interneurons and muscle precursors during embryonic development of *Drosophila* (Bourgouin et al. 1992, Lundgren et al. 1995). It has been shown that *apterous* controls the ability of these neurons to make their proper pathway choices and selectively fasciculate with one another (Lundgren et al. 1995). *apterous* is also expressed in dorsal cells of the wing imaginal disc and is required for cells of the dorsal compartment to assume a dorsal identity (Cohen et al. 1992, Diaz-Benjumea & Cohen 1993, Blair et al. 1994). Interestingly, ectopic expression of *apterous* induces ectopic ventral expression of PS1 integrin, which is normally expressed in a dorsal-specific pattern. In contrast, loss of apterous causes the ectopic dorsal expression of PS2 integrin, a ventral-specific characteristic (Blair et al. 1994). These results suggest that cell adhesion molecules are included among the downstream target genes of LIM-HD transcription factors.

Isl1

Isl1 was originally identified as a protein that binds to enhancer elements in the rat insulin gene (Karlsson et al. 1990). In the adult rat, Isl1 is expressed in all pancreatic islet cell types and in a variety of other polypeptide hormone-producing cells of the endocrine system (Thor et al. 1991). In addition, Isl1 is also expressed in a subset of neurons, including motor neurons in the spinal cord and the brain stem (Thor et al. 1991). The expression pattern of Isl1 in the vertebrate CNS, together with the findings from genetic studies in *C. elegans* and *Drosophila* described above, raised the intriguing possibility that LIM-HD transcription factors may contribute to the specification of neuronal cell fate in the vertebrate nervous system. This idea gained credence when it was found in the chick spinal cord that Isl1 is expressed in all motor neurons immediately after they leave the cell cycle and before they initiate the expression of other differentiated properties (Ericson et al. 1992). Similar observations have been made in studies of spinal motor neurons in the mouse (Pfaff et al. 1996), of zebrafish primary motor neurons (Korzh et al. 1993, Inoue et al. 1994), and of *Drosophila* motor neurons and interneurons (Thor & Thomas 1997). In *Isl1* knockout mice and in chick embryo spinal cord explants cultured in the presence of *Isl1* antisense oligonucleotides, the generation of motor neurons does not occur, leading to the appearance of apoptotic cells at the stage when Isl1 is first expressed (Pfaff et al. 1996). In contrast, *Drosophila islet* is not required for the generation of motor neurons and interneurons. Instead, in *islet* mutants, these neurons show axon pathfinding errors and fail to exhibit their proper neurotransmitter phenotype (Thor & Thomas 1997). Recently, the role of zebrafish *Isl2*, an ortholog of *Drosophila islet*, in neuronal differentiation has been examined by disrupting the heteromeric complex of Isl2 and NLI dimers and through the use of antisense morpholino oligonucleotides against *Isl2* mRNA (Segawa et al. 2001). This functional repression of Isl2 causes abnormal soma settling within the spinal cord, elimination of ventrally projecting axons, and alteration in neurotransmitter phenotype by subsets of primary motor neurons. Interestingly, in these embryos, the peripheral projections of the Isl2-positive primary sensory neurons are never formed.

MOTOR NEURON SUBTYPE IDENTITY IN VERTEBRATES: CELLULAR ORGANIZATION OF MOTOR NEURON SUBTYPES

Motor neurons located at different positions in the spinal cord project their axons in a highly stereotyped manner to innervate distinct targets in the periphery (Figure 3, see color insert). This high degree of spatial order establishes a topographic neural map. The topographic and functional organization of the spinal cord motor projections is a consequence of the generation of subtypes of motor neurons during development (for a review, see Pfaff & Kintner 1998). Motor neuron subtypes become evident when their axons select specific pathways to their muscle targets, and their cell bodies settle into longitudinally aligned columns within the spinal

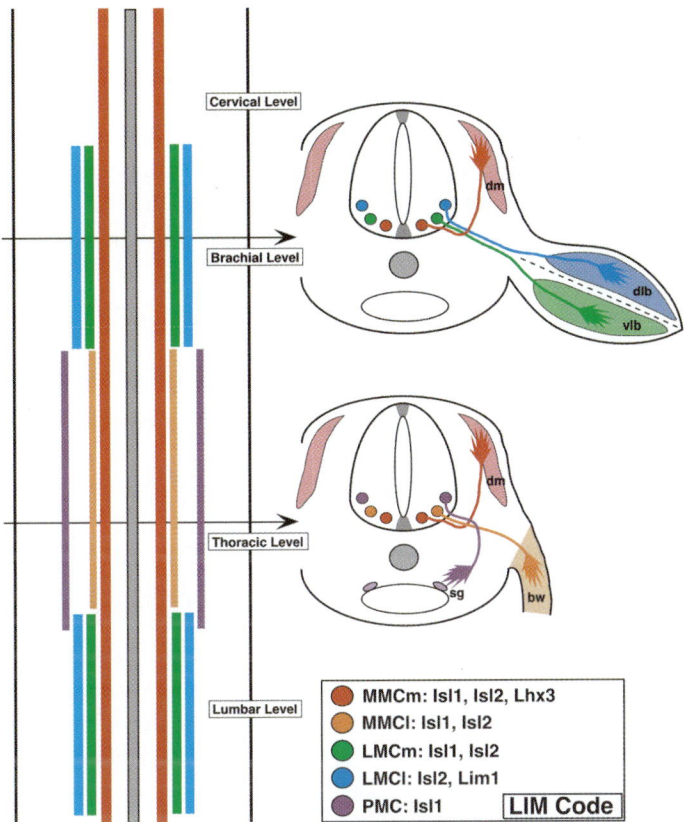

Figure 3 Columnar organization of motor neuron subtypes in the chick spinal cord and the target specificity of motor neuron subtypes are defined by distinct combinations of LIM-HD transcription factors. In the spinal cord, motor neurons are organized into longitudinal columns, as shown in the open-book whole-mount view on the *left*. In this preparation, the center is the floor plate (*grey*) at the ventral midline of the neural tube. The axonal pathways of motor neuron subtypes are represented in transverse sections at brachial and thoracic levels on the *right*, respectively. The expression of LIM-HD transcription factors in individual motor neuron subtypes, i.e., the LIM code for motor neuron subtypes, is shown by color coding. bw, body wall musculature; dlb, dorsal limb bud musculature; dm, dermomyotome; sg, neurons of the sympathetic ganglia; vlb, ventral limb bud musculature. LMCl (*blue*), lateral half of lateral motor column; LMCm (*green*), median half of lateral motor column; MMCl (*orange*), lateral half of medial motor column; MMCm (*red*), median half of medial motor column; PMC (*purple*), preganglionic motor column (also called the column of Terni in chick). *Lhx4* (also termed *Gsh-4*) is a gene highly related to *Lhx3* in mouse (Li et al. 1994). Because chick and mouse LIM-HD genes typically exhibit similar expression, *Lhx4* is likely expressed in chick MMCm neurons as well.

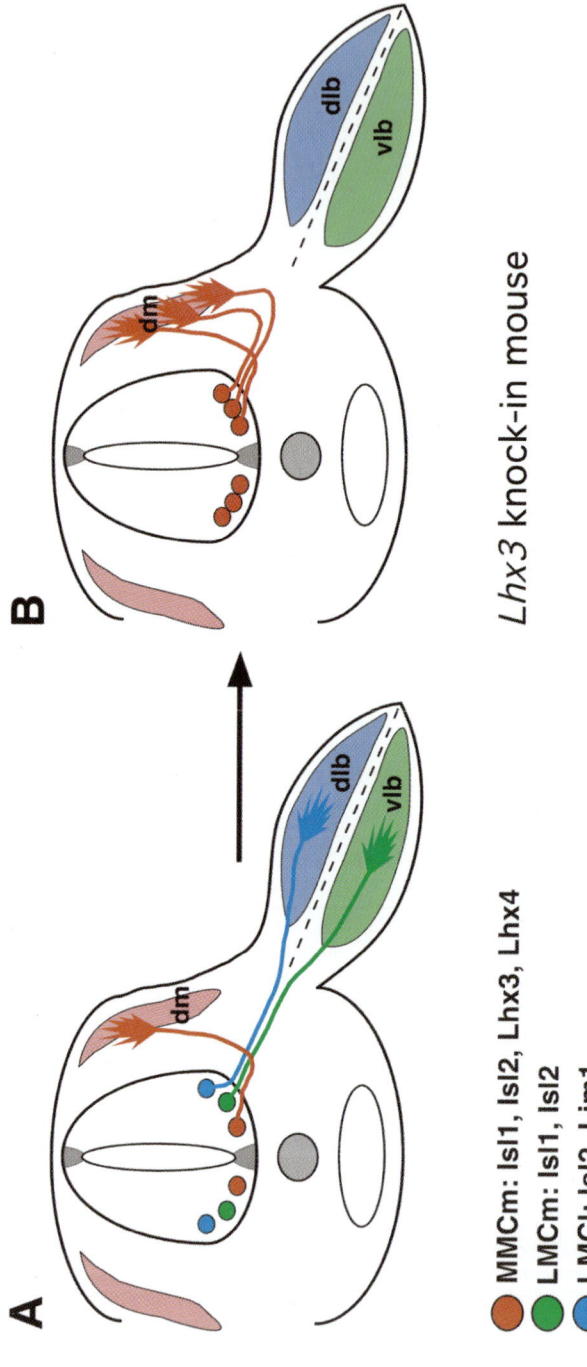

Figure 5 A LIM code for MMCm neurons confers axons of this subtype with the ability to select a specific pathway toward their muscle targets, the dermomyotome. (*A*) A LIM code for MMCm neurons is defined by the unique expression of Lhx3 and redundant factor Lhx4 in combination with Isl1/2. (*B*) To test whether the subtype-specific LIM code controls motor axon pathway selection in the periphery according to the LIM code hypothesis, all motor neurons in mice were genetically forced to stably express Lhx3 in order to establish an MMCm-like code. This genetic alteration is sufficient to convert the cell body settling pattern, gene-expression profile and axonal projection patterns of motor neurons to that of the MMCm subtype. However, owing to a competitive interaction in embryos with large numbers of MMCm cells, some axons are diverted from the dermomyotome (not shown).

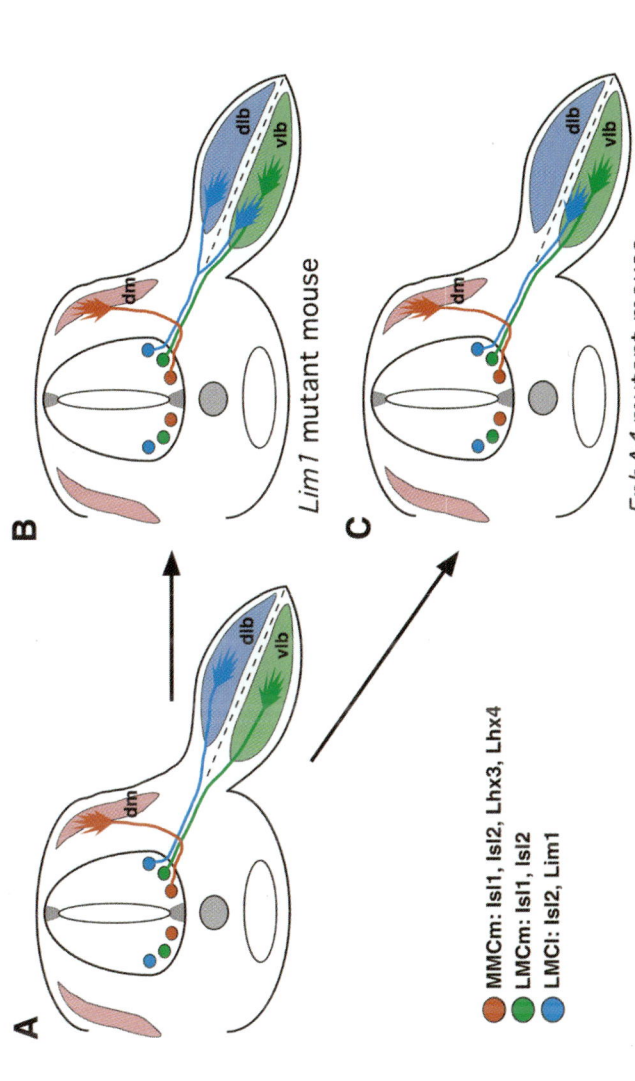

Figure 6 LMCl axon pathfinding at the base of the limb. (*A*) Lim1 marks the LMCl subtype, and EphA4 is preferentially expressed in LMCl axons when they extend into the dorsal limb. At the bifurcation point, LMCl axons (*blue*) select a dorsal pathway, while LMCm axons (*green*) select a ventral trajectory to enter the limb bud. (*B*) In *Lim1* mutant mice, LMCl axons project into the dorsal and ventral pathways within the limb at equal incidence. In this mutant, the projections of LMCm axons are not affected. Thus, the LIM code for LMCl neurons composed of Lim1 and Isl2 contributes to the fidelity of a binary choice in axonal trajectory at the base of the limb. (*C*) In *EphA4* mutant mice, LMCl axons fail to invade the dorsal compartment of the limb and instead select a ventral pathway to enter the ventral limb mesenchyme.

cord (for a review, see Landmesser 1980). In higher vertebrates, the arrangement of this organization is hierarchical and composed of several layers. First, motor neurons that project to a single muscle are clustered together in a longitudinal column called a "pool." Second, pools are grouped into larger columns on the basis of their specific target location in the periphery. Individual columns occupy characteristic positions along the rostrocaudal axis of the spinal cord. Third, the limb- and axial muscle–innervating motor columns are further subdivided into the medial and the lateral compartments corresponding to a dorsoventral subdivision in the position of their respective targets. In addition, motor neurons are also categorized into two classes based on whether their axons emerge dorsally or ventrally from the neural tube, as well as on the cell type that they innervate (i.e., somatic or visceral).

In the chick spinal cord (Figure 3, see color insert), for example, somatic motor neurons that innervate trunk muscles form a medial motor column (MMC), which is continuous along the length of the spinal cord. In contrast, somatic motor neurons that innervate limb muscles are present only at brachial and lumbar levels, where they constitute a discontinuous lateral motor column (LMC). At thoracic levels, visceral motor neurons that innervate sympathetic neurons form a preganglionic motor column of Terni (purple, Figure 3, see color insert). Furthermore, the MMC and LMC can be subdivided into two groups, respectively. A medial group of MMC, MMCm, extends along the entire rostrocaudal length of the spinal cord and projects axons to axial muscles (red, Figure 3, see color insert). A lateral group, MMCl, found only at thoracic levels, projects axons to muscles of the ventral body wall (orange, Figure 3, see color insert). Within the LMC, a medial group, LMCm, projects to ventral limb muscles (green, Figure 3, see color insert), whereas a lateral group, LMCl, projects to dorsal limb muscles (blue, Figure 3, see color insert).

DISTINCT INTRINSIC PROPERTIES OF MOTOR NEURON SUBTYPES

The specificity of motor projections to individual muscle targets at different peripheral locations depends critically on the pathfinding decisions made by axonal growth cones of each subtype of motor neuron during embryonic development (for reviews, see Landmesser 1992, Eisen 1994). A remarkable feature of axon pathfinding by motor neuron subtypes is that after leaving the spinal cord, they initially follow a common pathway to grow ventrally, at which point they diverge to select a subtype-specific pathway leading toward their appropriate targets. For example, the axons of MMCm neurons break away from the common path, the ventral root, and form a nerve branch (termed the dorsal ramus) to innervate the dermomyotome, the final target of MMCm axons (Tosney & Landmesser 1985a,b). In contrast, the axons of LMCm and LMCl neurons appear to ignore the choice point of MMCm axons and instead continue to grow ventrolaterally to reach the base of the limb. At this point, LMCm axons project exclusively into the ventral limb mesenchyme, whereas LMCl axons project only to the dorsal limb mesenchyme (Landmesser 1978; Tosney & Landmesser 1985a,b). Importantly, motor neurons develop their target specificity prior to their target innervation, as suggested by the analyses of motor axon projection patterns after a variety of surgical manipulations of chick and zebrafish embryos (for reviews, see Landmesser 1992, Eisen 1994; see also Matise & Lance-Jones 1996). Furthermore, in embryonic zebrafish, it has been shown by transplanting identified primary motor neurons to new spinal cord locations that the identity of motor neuron subtypes is determined just prior to the onset of axon outgrowth (Eisen 1991).

COMBINATORIAL EXPRESSION OF LIM-HD TRANSCRIPTION FACTORS MARKS MOTOR NEURON SUBTYPES: A LIM CODE

Tsuchida et al. (1994) identified a family of chick LIM-HD transcription factors and found that four members, *Isl1*, *Isl2*, *Lim1*, and *Lhx3*, are expressed in motor neurons in the embryonic chick spinal cord. Interestingly, the expression of a single LIM-HD transcription factor was not sufficient to distinguish individual motor columns, nor did it obviously delineate individual motor pools. However, the combinatorial expression of these genes (LIM code) defines subtypes of motor neurons that occupy different columns in the spinal cord, select specific axonal pathways in the periphery, and innervate distinct targets (Figure 3, see color insert). Furthermore, each subtype of motor neuron expresses its own LIM code prior to the segregation of motor neurons into columns and before distinct axonal pathways are established in the periphery. Thus, a LIM code expressed in motor neurons accurately marks a motor neuron subtype long before they reach their own muscle targets. Subsequent studies have shown that the patterned expression of LIM-HD transcription factors correlates with the axonal trajectory of primary motor neurons in embryonic zebrafish as well (Appel et al. 1995, Tokumoto et al. 1995). In addition, the transplantation of zebrafish primary motor neurons to new spinal cord positions initiated a new LIM code in these motor neurons, accompanied by axonal projection patterns appropriate for the new position (Appel et al. 1995), suggesting a tight correlation between the LIM-HD factors and the commitment to a particular motor neuron fate represented by subtype-specific axonal projection patterns.

Together, these findings raise the possibility that the LIM code confers subtypes of motor neurons with the ability to select distinct axonal pathways and to recognize specific muscle targets in the periphery. In this context, one function of the LIM code may be to regulate the expression of downstream genes that determine motor axons' responsiveness to guidance cues specific for a subtype of motor neurons.

DYNAMIC EXPRESSION OF LIM-HD TRANSCRIPTION FACTORS IN MOTOR NEURONS

The expression patterns of LIM-HD transcription factors change rapidly in motor neurons during the initial period in which they are generated (Tsuchida et al. 1994, Appel et al. 1995, Sharma et al. 1998), suggesting that precise temporal regulation of LIM-HD transcription factor expression potentially expands the LIM code. Using CRE-mediated lineage tracing in mice, Sharma et al. (1998) found that, for a brief period around the time when motor neurons are born, *Lhx3* and the redundant gene *Lhx4* are expressed in all motor neuron subtypes that extend axons ventrally from the neural tube (Figure 4). In contrast, motor neurons whose axons emerge dorsally from the neural tube never express these genes. Surprisingly, when ventrally exiting motor neurons become topographically organized into columns and extend axons in the periphery, they rapidly downregulate the expression of Lhx3 and Lhx4, except for MMCm neurons that maintain expression of these factors (Figure 4). Because all motor neurons begin to express Isl1 soon after they exit the cell cycle irrespective of the axonal trajectory (Ericson et al. 1992), the dynamic expression of Lhx3 and Lhx4 in combination with Isl1 is found to define a transient LIM code that is predicted to assign the axonal exit point selected by motor neurons. To test whether these LIM-HD transcription factors regulate axonal exit point identity, Sharma et al. (1998) analyzed mice deficient in both *Lhx3* and *Lhx4*. They found that, although motor neuron differentiation proceeds, ventrally exiting motor neurons switch their subtype identity to become motor neurons that extend axons dorsally from the neural tube. Conversely, they showed using chick embryo electroporation that ectopic expression of Lhx3 in dorsally exiting motor neurons is sufficient to reorient their axonal projections ventrally from the neural tube.

Thus, at the birth of motor neurons, Isl1 and Lhx3/Lhx4 act together in a binary fashion during a brief period to direct motor axons ventrally from the neural tube, providing the first functional evidence that the LIM code plays a role in axon pathfinding in vertebrates. Notably, this "early" LIM code is very dynamic and cells rapidly switch to new LIM codes as motor column identity begins to be established.

FUNCTIONAL TESTS OF THE LIM CODE FOR MOTOR AXON PATHWAY SELECTION IN THE PERIPHERY

The LIM code model, as first conceived (Figure 3, see color insert) (Tsuchida et al. 1994), predicted that axonal projections in the periphery of the embryo were regulated by the combinatorial activities of the LIM-HD transcription factors expressed by a given motor neuron subtype. For example, if *Lhx3/Lhx4* were knocked out, cells fated to become MMCm neurons were predicted to express an Isl1/Isl2 LIM code characteristic of MMCl and LMCm neurons, and were therefore predicted to project axons to the targets of these motor neuron subtypes. However, because Lhx3 and Lhx4 have an extremely dynamic expression pattern and thus seem to act at several distinct phases of motor neuron development (Figure 4), previous studies using *Lhx3*- and *Lhx4*-deficient mice were not able to reveal the role of these factors in motor axon pathfinding in the periphery owing to the cell fate conversion to a dorsally exiting motor neuron identity (Sharma et al. 1998). To define the later functions of these factors in motor axon guidance, Sharma et al. (2000) therefore

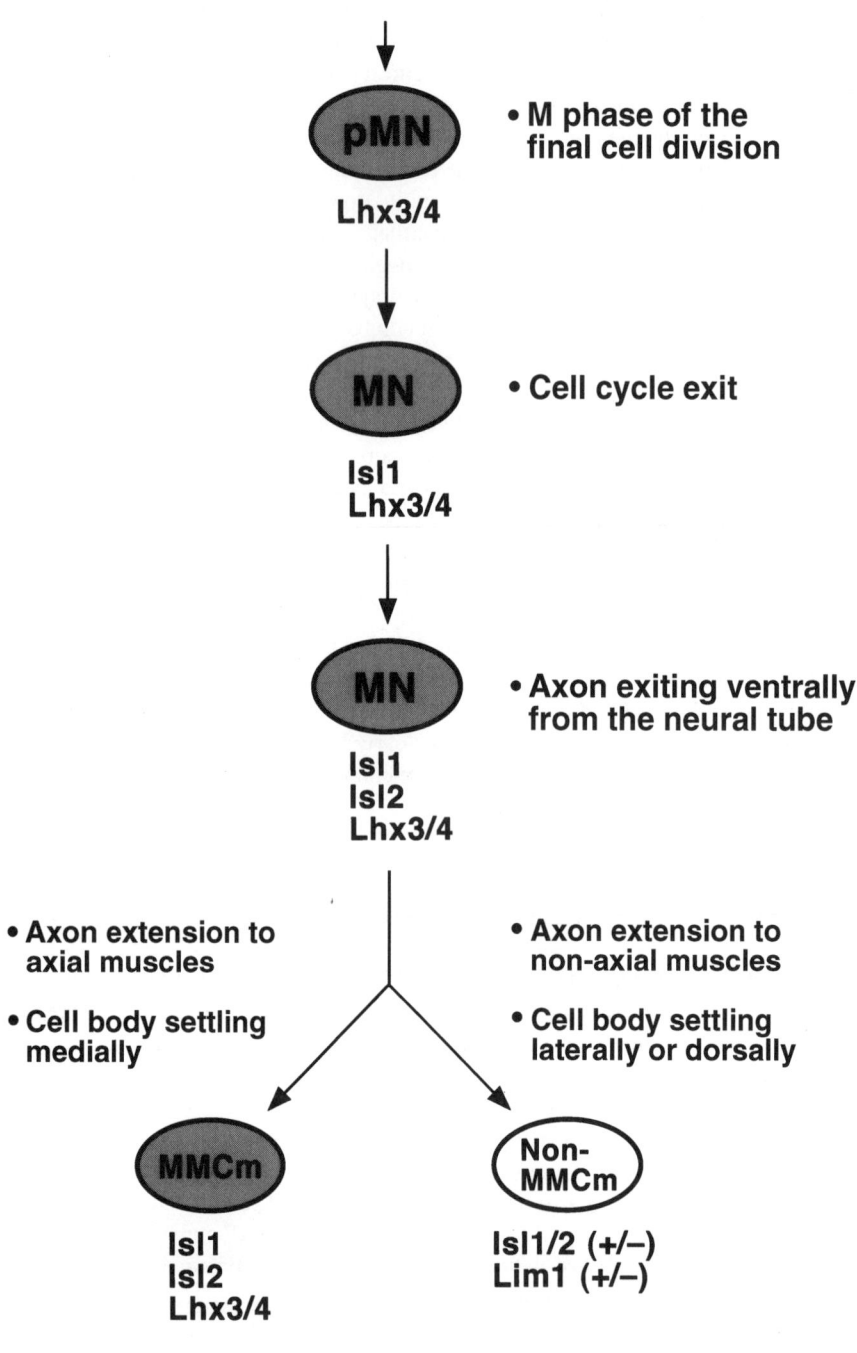

employed a mice knock-in strategy to ectopically and stably express Lhx3 in all somatic motor neurons. This genetic alteration was sufficient to convert motor neuron subtypes to the MMCm identity, as defined by the cell body settling pattern within the spinal cord and gene expression profile. Importantly, this resulted in a dramatic increase, but not total conversion, in the number of axonal projections toward axial muscles, the normal targets of MMCm neurons (Figure 5, see color insert). Thus, during the later period in which motor axons extend in the periphery, a LIM code specific for MMCm neurons (i.e., Isl1/Isl2 and Lhx3/Lhx4) now controls motor axon pathway selection and target recognition, presumably by regulating MMCm axons' responsiveness to guidance cues that promote and direct their axonal growth toward axial muscles.

Another line of evidence for a role of the LIM code in the control of motor axon pathfinding in vertebrates was provided by focusing on the selective expression of Lim1 in LMCl neurons (Kania et al. 2000). A characteristic feature of LMCl neurons is that, at the base of the limb, the LMCl axons project into dorsal limb mesenchyme. This is in marked contrast to the axons of LMCm neurons which select a ventral trajectory within the limb (Figure 6A, see color insert). By introducing an axonal tracer tau-LacZ into the *Lim1* locus, Kania et al. analyzed the axonal trajectory of LMCl neurons within the limb. They found that, in mice deficient in *Lim1* function, the specification of LMCl neurons, such as the cell migration pattern in the spinal cord and appearance of characteristic molecular attributes, proceeds normally, but the LMCl axons project into the dorsal and ventral halves of the limb at equal incidence, thus randomizing the selection of a dorsal or ventral pathway by these axons (Figure 6B, see color insert). In addition, in *Lim1* mutants, LMCl axons are unable to extend into the distal part of the limb. These results suggest that, although *Lim1* is not essential for dorsally directed growth of LMCl axons, a LIM code for LMCl neurons (i.e., Isl2 and Lim1) may control the ability of LMCl axons to respond to specific guidance cues which ensure that a correct pathfinding decision is made at the bifurcation point.

Figure 4 Dynamic expression of LIM-HD transcription factors during motor neuron development in mice. The onset of expression of Lhx3 and Lhx4 occurs in the final cell division of ventrally exiting motor neuron progenitors. Lhx3 and Lhx4 are not detected in dorsally exiting motor neuron progenitors. When motor neurons exit the cell cycle, they begin to express Isl1. Although MMCm neurons maintain the expression of Lhx3 and Lhx4 even when their axons exit ventrally from the neural tube to project dorsally toward their axial muscles, other motor neuron subtypes rapidly downregulate the expression of these factors when they begin to form specific motor columns within the spinal cord and extend axons ventrally from the neural tube to grow toward their nonaxial muscle targets. LIM-HD gene expression is dynamic and produces several transient combinations of factors that contribute to motor neuron development in a sequential fashion.

The action of the LIM code for axon pathfinding decisions seems to be conserved through evolution. In *Drosophila* embryos, motor neurons that project axons through the b branch of the intersegmental nerve (ISNb) express both *islet* and *lim3*, a homolog of vertebrate *Lhx3*, whereas motor neurons that project axons through the d branch (ISNd) express only *islet* (Thor et al. 1999). As in vertebrates, subtypes of motor neurons in *Drosophila* therefore express their unique combinations of LIM-HD transcription factors. Thor et al. (1999) found that, in *lim3* mutants, the axons of ISNb motor neurons behave like those of ISNd motor neurons. Conversely, ectopic expression of *lim3* in ISNd motor neurons now causes the axons of these neurons to behave as those of ISNb motor neurons in a predictable way.

Together, the findings from these genetic studies therefore establish the role of the LIM code in motor axon pathway selection, which represents a critical step for generating the topographic organization of motor projections. More generally, the LIM code expressed in postmitotic neurons is likely to participate in determining the fates and axonal projections of subtypes of neuronal classes throughout the CNS (for a recent review, see Hobert & Westphal 2000; see also Varela-Echavarría et al. 1996, Marín et al. 2000, Nakagawa & O'Leary 2001).

DOWNSTREAM TARGETS OF THE LIM CODE FOR SUBTYPE-SPECIFIC MOTOR AXON GUIDANCE

The downstream targets of the LIM code that control motor axon pathway selection in vertebrates as well as the subtype-specific axon guidance cues remain poorly characterized. Here, we discuss candidate guidance cues and their motor neuron subtype-specific receptors that may be regulated by the LIM code.

Exit Point from the Neural Tube: Ventral or Dorsal

Analysis of the fate of motor neurons in mice deficient in both *Lhx3* and *Lhx4* revealed that the combinatorial expression of Isl1 and Lhx3/Lhx4 specifies motor neuron axonal exit point identity (Sharma et al. 1998). The finding that the absence of Lhx3 and Lhx4 causes motor axons to exit more dorsally from the neural tube suggests that the status of these factors may control the responsiveness of motor axons to either attractants or repellents secreted from the region outside the motor column. It has been shown, for example, that the floor plate secretes diffusible chemorepellents, such as netrin-1 and semaphorins, for axons that show dorsally directed growth away from the ventral midline (Colamarino & Tessier-Lavigne 1995, Guthrie & Pini 1995, Tamada et al. 1995, Shirasaki et al. 1996, Varela-Echavarría et al. 1997). To date, involvement of netrin-1 and semaphorins in the formation of the initial ventral trajectory seems unlikely, for in mice deficient in the netrin receptor *DCC* and semaphorin receptors *neuropilins*, motor axon exit from the ventral neural tube is not perturbed (Fazeli et al. 1997, Kitsukawa et al. 1997, Chen et al. 2000, Giger et al. 2000). However, among the repellents expressed in floor plate cells, Slit2 is able to repel spinal motor axons at a distance in vitro (Brose et al. 1999). Because spinal motor neurons express the Slit receptor *Robo* around the time when these axons exit the neural tube (Kidd et al. 1998, Brose et al. 1999, Yuan et al. 1999), it is possible that Slit/Robo signaling in some way contributes to the guidance of motor axons from the neural tube. Alternatively, Lhx3 and Lhx4 in motor neurons might regulate the expression of receptors for chemorepellents secreted from the dorsal neural tube. If this is the case, the source of chemorepellents might exclude the roof plate because this tissue did not repel spinal motor axons in vitro (Augsburger et al. 1999). The release of repellent activity from the dorsal neural tube for subsets of cranial motor axons has recently been reported (see Caton et al. 2000).

Growth Away from the Neural Tube

It has been observed that motor neuron subtypes other than MMCm neurons rapidly downregulate the expression of Lhx3 and Lhx4 after their axons exit the neural tube (Figure 4) (Sharma et al. 1998). Interestingly, in the developing neural tube, commissural axons have been shown to rapidly change their responsiveness to guidance cues upon encountering the floor plate, an intermediate target of these axons (Shirasaki et al. 1998, Shirasaki & Murakami 2001, Zou et al. 2000). These studies suggest that growing axons still possess an ability to change their initial navigation program during their growth toward their final targets. Thus, the rapid change in the status of Lhx3/Lhx4 within motor neurons might be relevant to the change in their initial axon-navigation program, which may enable axons of motor neuron subtypes other than MMCm to extend ventrally away from the neural tube. On the other hand, the maintenance of Lhx3/Lhx4 in MMCm motor neurons as they extend axons in the periphery raises the intriguing possibility that a related guidance cue is used both for directing motor axons ventrally from the neural tube and for their selection of the dorsal ramus pathway in the periphery. Finally, it should be noted that the involvement of HGF/c-met signaling in directed growth of LMC axons toward the base of the limb is still controversial, for the HGF receptor c-met has not been observed to be expressed in LMC neurons when they first grow toward the limb (Ebens et al. 1996, Novak et al. 2000), and the axonal projections toward the limb are unaffected in *HGF* mutant mice (Ebens et al. 1996), suggesting another signal may be responsible.

Directed Growth Toward Axial Muscles

It has been shown that ablation of the dermomyotome, the precursor of axial muscles, in the chick embryo prevents the dorsal deviation of MMCm axons (Tosney 1987, 1988), suggesting the possibility that the dermomyotome secretes a chemoattractant for the axons of MMCm neurons. Similarly, evidence has been provided that Xenopus somitic myoblasts can attract neurites from Xenopus neural tube explants in culture (McCaig 1986). Furthermore, in the zebrafish mutant *spadetail*, which is deficient in trunk myotome development (Kimmel et al. 1989), the projection patterns of motor axons toward the myotome develop abnormally (Eisen & Pike 1991). Although the hypothesis of the existence of a dermomyotome-derived chemoattractant for MMCm axons has fascinated developmental neurobiologists for many years, direct evidence has yet to be provided.

Tessier-Lavigne and colleagues have attempted to detect such a chemoattractive activity from the dermomyotome using a collagen-gel coculture assay, but were unable to detect its activity (Ebens et al. 1996). In addition, they have also tested whether candidate secreted molecules expressed in the dermomyotome could promote the outgrowth of motor axons from ventral spinal cord explants. However, the factors that they have tested such as netrin-1, basic FGF, and TGFβ-1 failed to promote motor axon outgrowth (Ebens et al. 1996). Consistent with this, most of the known secreted molecules expressed in the dermomyotome do not show outgrowth-promoting activity for MMCm axons (R. Shirasaki & S. L. Pfaff, unpublished data). Possible explanations for these failures are that the cultured target tissue fails to differentiate properly or motor axons lose responsiveness to this putative chemoattractant when placed in vitro (Ebens et al. 1996).

Most importantly, however, the observation that, in mice in which Lhx3 is ectopically expressed in all motor neurons, many motor axons that have first bypassed the normal choice point of MMCm axons now are dramatically redirected to grow back toward axial muscles (Figure 5, see color insert) (Sharma et al. 2000) provides further in vivo evidence to support the notion that the axial muscle, the target of MMCm axons, secretes a long-range chemoattractant specific for MMCm axons. To date, the nature of the chemoattractive activity specific for MMCm axons as well as the identity of the chemoattractant receptor remain unknown.

Choice of Axonal Entry into the Limb: Ventral or Dorsal

The cellular environment of the dorsal and ventral halves of the limb mesenchyme is molecularly distinguished by the restricted dorsal expression of LIM-HD transcription factor Lmx1b (Riddle et al. 1995, Vogel et al. 1995). Kania et al. (2000) found that, in *Lmx1b* mutants, the axons of both LMCl and LMCm neurons randomly project into the dorsal and ventral limb, suggesting that LMCl and LMCm axons can no longer sense the distinction of guidance cues that direct these axons toward their appropriate territory. Together with the complementary results obtained in *Lim1* mutants described above (Figure 6B, see color insert) (Kania et al. 2000), these findings suggest that one of the downstream genes regulated by *Lim1* in LMCl neurons may encode a receptor that recognizes the guidance cues preferentially expressed in the dorsal or the ventral half of the limb mesenchyme. In addition, at the base of the limb, the axons of LMCl and LMCm neurons segregate immediately to enter either the dorsal or ventral limb mesenchyme (Tosney & Landmesser 1985a), suggesting that the guidance cues that control this pathfinding decision may operate at short range.

Recent studies have shown that the Eph family of tyrosine kinase receptors and their membrane-bound ligands, the ephrins, mediate axon guidance by a contact-dependent mechanism (for a recent review, see Wilkinson 2001). Indeed, some members of the Eph/ephrin signaling system are differentially and transiently expressed in subsets of motor neurons (Kilpatrick et al. 1996, Ohta et al. 1996, Araujo et al. 1998, Iwamasa et al. 1999). Among these, it has been shown that in chick and mouse embryos, the EphA4 protein is predominantly expressed in the axons of LMCl neurons during their pathfinding decision at the base of the limb (Ohta et al. 1996, Helmbacher et al. 2000). Moreover, the ventral limb mesenchyme appears to express higher levels of two EphA4 ligands, ephrin-A2 and ephrin-A5 proteins, as LMC axons reach this choice point (Ohta et al. 1997, Eberhart et al. 2000). Strikingly, in mice deficient in *EphA4* function, the LMCl axons fail to enter the dorsal limb and instead select a ventral trajectory (Figure 6C, see color insert) (Helmbacher et al. 2000). Because ephrin-A2 and ephrin-A5 have an inhibitory effect on axonal outgrowth of motor neurons expressing EphA4 (Ohta et al. 1997), the misprojection by LMCl axons in *EphA4* mutant mice could be due to the failure of these axons to respond to the repellent ligands in the ventral limb.

Although the phenotype of LMCl axon trajectories in *EphA4* mutants does not perfectly correspond to that in *Lim1* mutants (compare Figure 6B with 6C, see color insert), the finding that both of these molecules are critically involved in the binary choice in axonal trajectory at their specific decision point provides a potential link between the downstream genes of a LIM code involving Lim1 and the Eph/ephrin signaling through EphA4 within the LMCl subtype.

CONTRIBUTION OF THE HOX CODE AND THE ETS CODE

Although the LIM code defines the columnar identity of motor neurons in the spinal cord, the currently known LIM-HD transcription factors are insufficient to account for the entire range of motor neuron subtypes. For example, the LIM code alone does not distinguish the motor pool identity of limb-innervating LMC neurons. Lin et al. (1998) found that specific motor pools are defined by the combinatorial expression of LIM-HD transcription factors and two closely-related ETS transcription factors. In addition, the two ETS transcription factors, ER81 and PEA3, are also expressed in subsets of muscle sensory afferent neurons. Strikingly, they found a high degree of concordance in the profile of ETS protein expression by sensory and motor neurons that project to the same muscle target (Lin et al. 1998), thus revealing a molecular matching between functionally related sensory and motor neurons. It is important to note that the onset of expression of ETS transcription factors occurs at a relatively late stage when LMC axons reach the base of the limb (Lin et al. 1998). It has recently been shown that individual motor pools exhibit their characteristic features before LMC axons enter the limb (Milner et al. 1998, Milner & Landmesser 1999). Thus, it is likely that specification of motor pool identity occurs prior to the initiation of ETS gene expression. Indeed, in *Er81* mutant mice, the pool identity as well as the columnar identity of LMC neurons do not appear to be perturbed, although Ia sensory afferents fail to establish their characteristic termination zone within the ventral spinal cord (Arber et al. 2000).

Several lines of evidence suggest that the *Hox* genes, which have been implicated in specifying positional identities along the anteroposterior axis of the caudal neural tube (for a review, see Lumsden & Krumlauf 1996), may contribute to the establishment of the pool identity of motor neurons. Members of *Hoxc* and *Hoxd* clusters have been shown to be expressed by subsets of LMC neurons (Ensini et al. 1998, Tiret et al. 1998, Lance-Jones et al. 2001). In addition, the specification of the motor pool seems to occur prior to the onset of expression of the LIM code (Matise & Lance-Jones 1996), and the reprogramming of motor neuron subtype properties is associated with the reconfiguration of *Hox* gene profiles (Lance-Jones et al. 2001). In support of a role for *Hox* genes in motor neuron pool development, targeted disruptions of *Hoxd9* and *Hoxd10* cause alterations in peripheral nerve projections from specific motor pools (Carpenter et al. 1997, de la Cruz et al. 1999). Furthermore, in *Hoxc8* deficient mice, the topographic organization of motor pools that innervate forelimb distal muscles becomes disorganized (Tiret et al. 1998). Thus, these findings support the idea that the action of the Hox code within motor neurons is a part of the mechanism that imposes on motor neurons the patterning information required for the fine tuning of subtype identity necessary to establish connections between motor pools and muscle targets.

CONCLUSIONS

The ultimate outcome of neuronal fate specification is the accomplishment of synapse formation with appropriate target cells and the acquisition of appropriate physiological traits. Among a series of stages of the neuronal differentiation program, axon pathfinding that occurs in postmitotic neurons plays a fundamental role in the formation of functional neural circuits. Classical embryological studies focusing on topographic projections of spinal motor axons have suggested the existence of intrinsic differences between motor neuron subtypes before these axons select their specific pathways to reach their proper target muscles. One of the key discoveries during the past decade is that unique combinatorial arrays of LIM-HD transcription factors, the LIM code, define the target specificity of individual motor neuron subtypes. Recent molecular and genetic studies have provided compelling evidence to support the notion that the LIM code confers motor neuron subtypes with the ability to select specific axon pathways to reach their targets.

However, a number of important questions remain to be addressed. For example, it is still unclear to what extent the initial intrinsic genetic programs can control the later aspects of motor neuron development, such as positionally selective synaptogenesis by motor pools onto skeletal muscles (Laskowski & Sanes 1987, Feng et al. 2000) and specific expression of ion channels on motor nerve endings. Notably, it has recently been shown that individual motor pools exhibit their characteristic patterned bursting activity even while motor axons are growing toward their muscle targets (Milner & Landmesser 1999). In addition, motor neurons within a single pool can be further subdivided into two classes, based on whether they innervate fast or slow muscle fiber types. Recent studies have shown that, long before motor axons reach their muscle targets, fast- and slow-muscle innervating motor neurons are molecularly distinct in their axonal surface properties required for their selective fasciculation and target recognition (Rafuse et al. 1996, Milner et al. 1998). These findings suggest that, although the LIM code explains neither pool-specific nor subpool-specific properties, certain genetic programs involved in regulation of their characteristic membrane surface properties on their growing axons must be active prior to target innervation.

Finally, an important challenge is now to identify the downstream genes that are directly regulated by axon pathfinding-related transcriptional codes. Implicit in the suggestion from the LIM code hypothesis is that the code determines the differential responsiveness of motor axons to guidance cues specific for a subtype of motor neurons. In this context, it is reasonable to speculate that the receptor or receptor-associated cofactor for sensing such guidance cues is a prime candidate to be among the downstream targets. Because the transcriptional codes seem to be conserved through evolution, the identification of such key players will surely lead to the opening of a new and exciting chapter in this field of biology.

ACKNOWLEDGMENTS

Our research is supported by the National Institutes of Health, the Christopher Reeves Paralysis Foundation, the Human Frontier Science Program, and the Muscular Dystrophy Association. RS is supported by the Japan Society for the Promotion of Science Postdoctoral Fellowships for Research Abroad and SLP by the Pew Charitable Trusts and the Mathers Foundation.

The *Annual Review of Neuroscience* is online at http://neuro.annualreviews.org

LITERATURE CITED

Agulnick AD, Taira M, Breen JJ, Tanaka T, Dawid IB, Westphal H. 1996. Interactions of the LIM-domain binding factor Ldb1 with LIM homeodomain proteins. *Nature* 384:270–72

Akazawa C, Ishibashi M, Shimizu C, Nakanishi S, Kageyama R. 1995. A mammalian helix-loop-helix factor structurally related to the product of Drosophila proneural gene atonal is a positive transcriptional regulator expressed in the developing nervous system. *J. Biol. Chem.* 270:8730–38

Appel B, Korzh V, Glasgow E, Thor S, Edlund T, et al. 1995. Motoneuron fate specification revealed by patterned LIM homeobox gene expression in embryonic zebrafish. *Development* 121:4117–25

Araujo M, Piedra ME, Herrera MT, Ros MA, Nieto MA. 1998. The expression and regulation of chick *EphA7* suggests roles in limb patterning and innervation. *Development* 125:4195–204

Arber S, Han B, Mendelsohn M, Smith M, Jessell TM, Sockanathan S. 1999. Requirement for the homeobox gene *Hb9* in the consolidation of motor neuron identity. *Neuron* 23:659–74

Arber S, Ladle DR, Lin JH, Frank E, Jessell TM.

2000. ETS gene *Er81* controls the formation of functional connections between group Ia sensory afferents and motor neurons. *Cell* 101:485–98

Augsburger A, Schuchardt A, Hoskins S, Dodd J, Butler S. 1999. BMPs as mediators of roof plate repulsion of commissural neurons. *Neuron* 24:127–41

Bach I, Carriere C, Ostendorff HP, Andersen B, Rosenfeld MG. 1997. A family of LIM domain-associated cofactors confer transcriptional synergism between LIM and Otx homeodomain proteins. *Genes Dev.* 11:1370–80

Bach I, Rhodes SJ, Pearse RV 2nd, Heinzel T, Gloss B, et al. 1995. P-Lim, a LIM homeodomain factor, is expressed during pituitary organ and cell commitment and synergizes with Pit-1. *Proc. Natl. Acad. Sci. USA* 92:2720–24

Bang AG, Goulding MD. 1996. Regulation of vertebrate neural cell fate by transcription factors. *Curr. Opin. Neurobiol.* 6:25–32

Ben-Arie N, McCall AE, Berkman S, Eichele G, Bellen HJ, Zoghbi HY. 1996. Evolutionary conservation of sequence and expression of the bHLH protein Atonal suggests a conserved role in neurogenesis. *Hum. Mol. Genet.* 5:1207–16

Blair SS, Brower DL, Thomas JB, Zavortink M. 1994. The role of *apterous* in the control of dorsoventral compartmentalization and PS integrin gene expression in the developing wing of *Drosophila*. *Development* 120:1805–15

Bourgouin C, Lundgren SE, Thomas JB. 1992. *apterous* is a Drosophila LIM domain gene required for the development of a subset of embryonic muscles. *Neuron* 9:549–61

Briscoe J, Ericson J. 2001. Specification of neuronal fates in the ventral neural tube. *Curr. Opin. Neurobiol.* 11:43–49

Briscoe J, Pierani A, Jessell TM, Ericson J. 2000. A homeodomain protein code specifies progenitor cell identity and neuronal fate in the ventral neural tube. *Cell* 101:435–45

Briscoe J, Sussel L, Serup P, Hartigan-O'Connor D, Jessell TM, et al. 1999. Homeobox gene *Nkx2.2* and specification of neuronal identity by graded sonic hedgehog signalling. *Nature* 398:622–27

Brose K, Bland KS, Wang KH, Arnott D, Henzel W, et al. 1999. Slit proteins bind Robo receptors and have an evolutionarily conserved role in repulsive axon guidance. *Cell* 96:795–806

Burrill JD, Moran L, Goulding MD, Saueressig H. 1997. PAX2 is expressed in multiple spinal cord interneurons, including a population of EN1+ interneurons that require PAX6 for their development. *Development* 124:4493–503

Carpenter EM, Goddard JM, Davis AP, Nguyen TP, Capecchi MR. 1997. Targeted disruption of *Hoxd-10* affects mouse hindlimb development. *Development* 124:4505–14

Caton A, Hacker A, Naeem A, Livet J, Maina F, et al. 2000. The branchial arches and HGF are growth-promoting and chemoattractant for cranial motor axons. *Development* 127:1751–66

Chen H, Bagri A, Zupicich JA, Zou Y, Stoeckli E, et al. 2000. Neuropilin-2 regulates the development of selective cranial and sensory nerves and hippocampal mossy fiber projections. *Neuron* 25:43–56

Chiang C, Litingtung Y, Lee E, Young KE, Corden JL, et al. 1996. Cyclopia and defective axial patterning in mice lacking Sonic hedgehog gene function. *Nature* 383:407–13

Cohen B, McGuffin ME, Pfeifle C, Segal D, Cohen SM. 1992. *apterous*, a gene required for imaginal disc development in Drosophila encodes a member of the LIM family of developmental regulatory proteins. *Genes Dev.* 6:715–29

Colamarino SA, Tessier-Lavigne M. 1995. The axonal chemoattractant netrin-1 is also a chemorepellent for trochlear motor axons. *Cell* 81:621–29

Dawid IB, Breen JJ, Toyama R. 1998. LIM domains: multiple roles as adapters and functional modifiers in protein interactions. *Trends Genet.* 14:156–62

de la Cruz CC, Der-Avakian A, Spyropoulos DD, Tieu DD, Carpenter EM. 1999. Targeted disruption of *Hoxd9* and *Hoxd10* alters locomotor behavior, vertebral identity, and peripheral nervous system development. *Dev. Biol.* 216:595–610

Diaz-Benjumea FJ, Cohen SM. 1993. Interaction between dorsal and ventral cells in the imaginal disc directs wing development in Drosophila. *Cell* 75:741–52

Doe CQ, Skeath JB. 1996. Neurogenesis in the insect central nervous system. *Curr. Opin. Neurobiol.* 6:18–24

Driscoll M, Chalfie M. 1991. The *mec-4* gene is a member of a family of Caenorhabditis elegans genes that can mutate to induce neuronal degeneration. *Nature* 349:588–93

Duggan A, Ma C, Chalfie M. 1998. Regulation of touch receptor differentiation by the Caenorhabditis elegans *mec-3* and *unc-86* genes. *Development* 125:4107–19

Ebens A, Brose K, Leonardo ED, Hanson MG Jr, Bladt F, et al. 1996. Hepatocyte growth factor/scatter factor is an axonal chemoattractant and a neurotrophic factor for spinal motor neurons. *Neuron* 17:1157–72

Eberhart J, Swartz M, Koblar SA, Pasquale EB, Tanaka H, Krull CE. 2000. Expression of EphA4, ephrin-A2 and ephrin-A5 during axon outgrowth to the hindlimb indicates potential roles in pathfinding. *Dev. Neurosci.* 22:237–50

Edlund T, Jessell TM. 1999. Progression from extrinsic to intrinsic signaling in cell fate specification: a view from the nervous system. *Cell* 96:211–24

Eisen JS. 1991. Determination of primary motoneuron identity in developing zebrafish embryos. *Science* 252:569–72

Eisen JS. 1994. Development of motoneuronal phenotype. *Annu. Rev. Neurosci.* 17:1–30

Eisen JS, Pike SH. 1991. The *spt-1* mutation alters segmental arrangement and axonal development of identified neurons in the spinal cord of the embryonic zebrafish. *Neuron* 6:767–76

Ensini M, Tsuchida TN, Belting HG, Jessell TM. 1998. The control of rostrocaudal pattern in the developing spinal cord: specification of motor neuron subtype identity is initiated by signals from paraxial mesoderm. *Development* 125:969–82

Ericson J, Morton S, Kawakami A, Roelink H, Jessell TM. 1996. Two critical periods of Sonic Hedgehog signaling required for the specification of motor neuron identity. *Cell* 87:661–73

Ericson J, Rashbass P, Schedl A, Brenner-Morton S, Kawakami A, et al. 1997. Pax6 controls progenitor cell identity and neuronal fate in response to graded Shh signaling. *Cell* 90:169–80

Ericson J, Thor S, Edlund T, Jessell TM, Yamada T. 1992. Early stages of motor neuron differentiation revealed by expression of homeobox gene *Islet-1*. *Science* 256:1555–60

Farah MH, Olson JM, Sucic HB, Hume RI, Tapscott SJ, Turner DL. 2000. Generation of neurons by transient expression of neural bHLH proteins in mammalian cells. *Development* 127:693–702

Fazeli A, Dickinson SL, Hermiston ML, Tighe RV, Steen RG, et al. 1997. Phenotype of mice lacking functional deleted in colorectal cancer (*Dcc*) gene. *Nature* 386:796–804

Feng G, Laskowski MB, Feldheim DA, Wang H, Lewis R, et al. 2000. Roles of ephrins in positionally selective synaptogenesis between motor neurons and muscle fibers. *Neuron* 25:295–306

Ferguson EL, Horvitz HR. 1985. Identification and characterization of 22 genes that affect the vulval cell lineages of the nematode Caenorhabditis elegans. *Genetics* 110:17–72

Ferguson EL, Sternberg PW, Horvitz HR. 1987. A genetic pathway for the specification of the vulval cell lineages of Caenorhabditis elegans. *Nature* 326:259–67

Finney M, Ruvkun G. 1990. The *unc-86* gene product couples cell lineage and cell identity in C. elegans. *Cell* 63:895–905

Finney M, Ruvkun G, Horvitz HR. 1988. The C. elegans cell lineage and differentiation gene *unc-86* encodes a protein with a homeodomain and extended similarity to transcription factors. *Cell* 55:757–69

Freyd G, Kim SK, Horvitz HR. 1990. Novel

cysteine-rich motif and homeodomain in the product of the Caenorhabditis elegans cell lineage gene lin-11. *Nature* 344:876–79

Giger RJ, Cloutier JF, Sahay A, Prinjha RK, Levengood DV, et al. 2000. Neuropilin-2 is required in vivo for selective axon guidance responses to secreted semaphorins. *Neuron* 25:29–41

Gowan K, Helms AW, Hunsaker TL, Collisson T, Ebert PJ, et al. 2001. Crossinhibitory activities of Ngn1 and Math1 allow specification of distinct dorsal interneurons. *Neuron* 31:219–32

Gradwohl G, Fode C, Guillemot F. 1996. Restricted expression of a novel murine atonal-related bHLH protein in undifferentiated neural precursors. *Dev. Biol.* 180:227–41

Guillemot F. 1999. Vertebrate bHLH genes and the determination of neuronal fates. *Exp. Cell Res.* 253:357–64

Gurdon JB, Bourillot P-Y. 2001. Morphogen gradient interpretation. *Nature* 413:797–803

Guthrie S, Pini A. 1995. Chemorepulsion of developing motor axons by the floor plate. *Neuron* 14:1117–30

Hamelin M, Scott IM, Way JC, Culotti JG. 1992. The *mec-7* beta-tubulin gene of Caenorhabditis elegans is expressed primarily in the touch receptor neurons. *EMBO J.* 11:2885–93

Harrison KA, Druey KM, Deguchi Y, Tuscano JM, Kehrl JH. 1994. A novel human homeobox gene distantly related to proboscipedia is expressed in lymphoid and pancreatic tissues. *J. Biol. Chem.* 269:19968–75

Helmbacher F, Schneider-Maunoury S, Topilko P, Tiret L, Charnay P. 2000. Targeting of the EphA4 tyrosine kinase receptor affects dorsal/ventral pathfinding of limb motor axons. *Development* 127:3313–24

Helms AW, Johnson JE. 1998. Progenitors of dorsal commissural interneurons are defined by MATH1 expression. *Development* 125:919–28

Hirsch MR, Tiveron MC, Guillemot F, Brunet JF, Goridis C. 1998. Control of noradrenergic differentiation and Phox2a expression by MASH1 in the central and peripheral nervous system. *Development* 125:599–608

Hobert O, D'Alberti T, Liu Y, Ruvkun G. 1998. Control of neural development and function in a thermoregulatory network by the LIM homeobox gene *lin-11*. *J. Neurosci.* 18:2084–96

Hobert O, Westphal H. 2000. Functions of LIM-homeobox genes. *Trends Genet.* 16:75–83

Inoue A, Takahashi M, Hatta K, Hotta Y, Okamoto H. 1994. Developmental regulation of islet-1 mRNA expression during neuronal differentiation in embryonic zebrafish. *Dev. Dyn.* 199:1–11

Iwamasa H, Ohta K, Yamada T, Ushijima K, Terasaki H, Tanaka H. 1999. Expression of Eph receptor tyrosine kinases and their ligands in chick embryonic motor neurons and hindlimb muscles. *Dev. Growth Differ.* 41:685–98

Jessell TM. 2000. Neuronal specification in the spinal cord: inductive signals and transcriptional codes. *Nat. Rev. Genet.* 1:20–29

Jurata LW, Gill GN. 1997. Functional analysis of the nuclear LIM domain interactor NLI. *Mol. Cell. Biol.* 17:5688–98

Jurata LW, Gill GN. 1998. Structure and function of LIM domains. *Curr. Top. Microbiol. Immunol.* 228:75–113

Jurata LW, Kenny DA, Gill GN. 1996. Nuclear LIM interactor, a rhombotin and LIM homeodomain interacting protein, is expressed early in neuronal development. *Proc. Natl. Acad. Sci. USA* 93:11693–98

Jurata LW, Pfaff SL, Gill GN. 1998. The nuclear LIM domain interactor NLI mediates homo- and heterodimerization of LIM domain transcription factors. *J. Biol. Chem.* 273:3152–57

Jurata LW, Thomas JB, Pfaff SL. 2000. Transcriptional mechanisms in the development of motor control. *Curr. Opin. Neurobiol.* 10:72–79

Kageyama R, Nakanishi S. 1997. Helix-loop-helix factors in growth and differentiation of the vertebrate nervous system. *Curr. Opin. Genet. Dev.* 7:659–65

Kania A, Johnson RL, Jessell TM. 2000. Coordinate roles for LIM homeobox genes in directing the dorsoventral trajectory of motor axons in the vertebrate limb. *Cell* 102:161–73

Karlsson O, Thor S, Norberg T, Ohlsson H, Edlund T. 1990. Insulin gene enhancer binding protein Isl-1 is a member of a novel class of proteins containing both a homeo- and a Cys-His domain. *Nature* 344:879–82

Kidd T, Brose K, Mitchell KJ, Fetter RD, Tessier-Lavigne M, et al. 1998. Roundabout controls axon crossing of the CNS midline and defines a novel subfamily of evolutionarily conserved guidance receptors. *Cell* 92:205–15

Kilpatrick TJ, Brown A, Lai C, Gassmann M, Goulding M, Lemke G. 1996. Expression of the *Tyro4/Mek4/Cek4* gene specifically marks a subset of embryonic motor neurons and their muscle targets. *Mol. Cell. Neurosci.* 7:62–74

Kimmel CB, Kane DA, Walker C, Warga RM, Rothman MB. 1989. A mutation that changes cell movement and cell fate in the zebrafish embryo. *Nature* 337:358–62

Kimmel CB, Warga RM, Kane DA. 1994. Cell cycles and clonal strings during formation of the zebrafish central nervous system. *Development* 120:265–76

Kitsukawa T, Shimizu M, Sanbo M, Hirata T, Taniguchi M, et al. 1997. Neuropilin-semaphorin III/D-mediated chemorepulsive signals play a crucial role in peripheral nerve projection in mice. *Neuron* 19:995–1005

Korzh V, Edlund T, Thor S. 1993. Zebrafish primary neurons initiate expression of the LIM homeodomain protein Isl-1 at the end of gastrulation. *Development* 118:417–25

Lance-Jones C, Omelchenko N, Bailis A, Lynch S, Sharma K. 2001. *Hoxd10* induction and regionalization in the developing lumbosacral spinal cord. *Development* 128:2255–68

Landmesser L. 1978. The development of motor projection patterns in the chick hind limb. *J. Physiol.* 284:391–414

Landmesser LT. 1980. The generation of neuromuscular specificity. *Annu. Rev. Neurosci.* 3:279–302

Landmesser LT. 1992. Growth cone guidance in the avian limb: a search for cellular and molecular mechanisms. In *The Nerve Growth Cone*, ed. PC Letourneau, SB Kater, ER Macagno. pp. 373–85. New York: Raven

Laskowski MB, Sanes JR. 1987. Topographic mapping of motor pools onto skeletal muscles. *J. Neurosci.* 7:252–60

Leber SM, Breedlove SM, Sanes JR. 1990. Lineage, arrangement, and death of clonally related motoneurons in chick spinal cord. *J. Neurosci.* 10:2451–62

Lee JE. 1997. Basic helix-loop-helix genes in neural development. *Curr. Opin. Neurobiol.* 7:13–20

Lee KJ, Mendelsohn M, Jessell TM. 1998. Neuronal patterning by BMPs: a requirement for GDF7 in the generation of a discrete class of commissural interneurons in the mouse spinal cord. *Genes Dev.* 12:3394–407

Lee KJ, Jessell TM. 1999. The specification of dorsal cell fates in the vertebrate central nervous system. *Annu. Rev. Neurosci.* 22:261–94

Lee S-K, Pfaff SL. 2001. Transcriptional networks regulating neuronal identity in the developing spinal cord. *Nat. Neurosci.* 4(Suppl.):1183–91

Lichtsteiner S, Tjian R. 1995. Synergistic activation of transcription by UNC-86 and MEC-3 in *Caenorhabditis elegans* embryo extracts. *EMBO J.* 14:3937–45

Li H, Witte DP, Branford WW, Aronow BJ, Weinstein M, et al. 1994. *Gsh-4* encodes a LIM-type homeodomain, is expressed in the developing central nervous system and is required for early postnatal survival. *EMBO J.* 13:2876–85

Lin JH, Saito T, Anderson DJ, Lance-Jones C, Jessell TM, Arber S. 1998. Functionally related motor neuron pool and muscle sensory afferent subtypes defined by coordinate *ETS* gene expression. *Cell* 95:393–407

Lu QR, Yuk D-I, Alberta JA, Zhu Z, Pawlitzky I, et al. 2000. Sonic hedgehog-regulated oligodendrocyte lineage genes encoding bHLH proteins in the mammalian central nervous system. *Neuron* 25:317–29

Lumsden A, Krumlauf R. 1996. Patterning the vertebrate neuraxis. *Science* 274:1109–15

Lundgren SE, Callahan CA, Thor S, Thomas JB. 1995. Control of neuronal pathway selection by the Drosophila LIM homeodomain gene apterous. *Development* 121:1769–73

Ma Q, Sommer L, Cserjesi P, Anderson DJ. 1997. Mash1 and neurogenin1 expression patterns define complementary domains of neuroepithelium in the developing CNS and are correlated with regions expressing notch ligands. *J. Neurosci.* 17:3644–52

Marin O, Anderson SA, Rubenstein JL. 2000. Origin and molecular specification of striatal interneurons. *J. Neurosci.* 20:6063–76

Matise MP, Lance-Jones C. 1996. A critical period for the specification of motor pools in the chick lumbosacral spinal cord. *Development* 122:659–69

McCaig CD. 1986. Myoblasts and myoblast-conditioned medium attract the earliest spinal neurites from frog embryos. *J. Physiol.* 375:39–54

Milner LD, Landmesser LT. 1999. Cholinergic and GABAergic inputs drive patterned spontaneous motoneuron activity before target contact. *J. Neurosci.* 19:3007–22

Milner LD, Rafuse VF, Landmesser LT. 1998. Selective fasciculation and divergent pathfinding decisions of embryonic chick motor axons projecting to fast and slow muscle regions. *J. Neurosci.* 18:3297–313

Mitani S, Du H, Hall DH, Driscoll M, Chalfie M. 1993. Combinatorial control of touch receptor neuron expression in *Caenorhabditis elegans*. *Development* 119:773–83

Mizuguchi R, Sugimori M, Takebayashi H, Kosako H, Nagao M, et al. 2001. Combinatorial roles of Olig2 and Neurogenin2 in the coordinated induction of pan-neuronal and subtype-specific properties of motoneurons. *Neuron* 31:757–71

Moran-Rivard L, Kagawa T, Saueressig H, Gross MK, Burrill J, Goulding M. 2001. *Evx1* is a postmitotic determinant of V0 interneuron identity in the spinal cord. *Neuron* 29:385–99

Morrow EM, Furukawa T, Lee JE, Cepko CL. 1999. NeuroD regulates multiple functions in the developing neural retina in rodent. *Development* 126:23–36

Mueller BK. 1999. Growth cone guidance: first steps towards a deeper understanding. *Annu. Rev. Neurosci.* 22:351–88

Muhr J, Andersson E, Persson M, Jessell TM, Ericson J. 2001. Groucho-mediated transcriptional repression establishes progenitor cell pattern and neuronal fate in the ventral neural tube. *Cell* 104:861–73

Nakagawa Y, O'Leary DDM. 2001. Combinatorial expression patterns of LIM-homeodomain and other regulatory genes parcellate developing thalamus. *J. Neurosci.* 21:2711–25

Novak KD, Prevette D, Wang S, Gould TW, Oppenheim RW. 2000. Hepatocyte growth factor/scatter factor is a neurotrophic survival factor for lumbar but not for other somatic motoneurons in the chick embryo. *J. Neurosci.* 20:326–37

Novitch BG, Chen AI, Jessell TM. 2001. Coordinate regulation of motor neuron subtype identity and pan-neuronal properties by the bHLH repressor Olig2. *Neuron* 31:773–89

Ohta K, Iwamasa H, Drescher U, Terasaki H, Tanaka H. 1997. The inhibitory effect on neurite outgrowth of motoneurons exerted by the ligands ELF-1 and RAGS. *Mech. Dev.* 64:127–35

Ohta K, Nakamura M, Hirokawa K, Tanaka S, Iwama A, et al. 1996. The receptor tyrosine kinase, Cek8, is transiently expressed on subtypes of motoneurons in the spinal cord during development. *Mech. Dev.* 54:59–69

Osumi N, Hirota A, Ohuchi H, Nakafuku M, Iimura T, et al. 1997. Pax-6 is involved in the specification of hindbrain motor neuron subtype. *Development* 125:2961–72

Pfaff S, Kintner C. 1998. Neuronal diversification: development of motor neuron subtypes. *Curr. Opin. Neurobiol.* 8:27–36

Pfaff SL, Mendelsohn M, Stewart CL, Edlund T, Jessell TM. 1996. Requirement for LIM homeobox gene *Isl1* in motor neuron generation reveals a motor neuron-dependent step in interneuron differentiation. *Cell* 84:309–20

Pierani A, Moran-Rivard L, Sunshine MJ, Littman DR, Goulding M, Jessell TM. 2001. Control of interneuron fate in the developing spinal cord by the progenitor homeodomain protein Dbx1. *Neuron* 29:367–84

Rafuse VF, Milner LD, Landmesser LT. 1996. Selective innervation of fast and slow muscle regions during early chick neuromuscular development. *J. Neurosci.* 16:6864–77

Riddle RD, Ensini M, Nelson C, Tsuchida T, Jessell TM, Tabin C. 1995. Induction of the LIM homeobox gene *Lmx1* by WNT7a establishes dorsoventral pattern in the vertebrate limb. *Cell* 83:631–40

Sagasti A, Hobert O, Troemel ER, Ruvkun G, Bargmann CI. 1999. Alternative olfactory neuron fates are specified by the LIM homeobox gene *lim-4*. *Genes Dev.* 13:1794–806

Saha MS, Miles RR, Grainger RM. 1997. Dorsal-ventral patterning during neural induction in *Xenopus*: assessment of spinal cord regionalization with xHB9, a marker for the motor neuron region. *Dev. Biol.* 187:209–23

Sander M, Paydar S, Ericson J, Briscoe J, Berber E, et al. 2000. Ventral neural patterning by *Nkx* homeobox genes: *Nkx6.1* controls somatic motor neuron and ventral interneuron fates. *Genes Dev.* 14:2134–39

Sarafi-Reinach TR, Melkman T, Hobert O, Sengupta P. 2001. The *lin-11* LIM homeobox gene specifies olfactory and chemosensory neuron fates in *C. elegans*. *Development* 128:3269–81

Savage C, Hamelin M, Culotti JG, Coulson A, Albertson DG, Chalfie M. 1989. *mec-7* is a beta-tubulin gene required for the production of 15-protofilament microtubules in *Caenorhabditis elegans*. *Genes Dev.* 3:870–81

Scardigli R, Schuurmans C, Gradwohl G, Guillemot F. 2001. Crossregulation between *Neurogenin2* and pathways specifying neuronal identity in the spinal cord. *Neuron* 31:203–17

Segawa H, Miyashita T, Hirate Y, Higashijima S, Chino N, et al. 2001. Functional repression of Islet-2 by disruption of complex with Ldb impairs peripheral axonal outgrowth in embryonic zebrafish. *Neuron* 30:423–36

Sengupta P, Colbert HA, Bargmann CI. 1994. The *C. elegans* gene *odr-7* encodes an olfactory-specific member of the nuclear receptor superfamily. *Cell* 79:971–80

Sharma K, Leonard AE, Lettieri K, Pfaff SL. 2000. Genetic and epigenetic mechanisms contribute to motor neuron pathfinding. *Nature* 406:515–19

Sharma K, Sheng HZ, Lettieri K, Li H, Karavanov A, et al. 1998. LIM homeodomain factors Lhx3 and Lhx4 assign subtype identities for motor neurons. *Cell* 95:817–28

Shirasaki R, Katsumata R, Murakami F. 1998. Change in chemoattractant responsiveness of developing axons at an intermediate target. *Science* 279:105–7

Shirasaki R, Mirzayan C, Tessier-Lavigne M, Murakami F. 1996. Guidance of circumferentially growing axons by netrin-dependent and -independent floor plate chemotropism in the vertebrate brain. *Neuron* 17:1079–88

Shirasaki R, Murakami F. 2001. Crossing the floor plate triggers sharp turning of commissural axons. *Dev. Biol.* 236:99–108

Siegler MV, Jia XX. 1999. Engrailed negatively regulates the expression of cell adhesion molecules connectin and neuroglian in embryonic *Drosophila* nervous system. *Neuron* 22:265–76

Smith ST, Jaynes JB. 1996. A conserved region of engrailed, shared among all en-, gsc-, Nk1-, k Nk2- and msh-class homeoproteins, mediates active transcriptional repression in vivo. *Development* 122:3141–50

Sommer L, Ma Q, Anderson DJ. 1996. Neurogenins, a novel family of atonal-related bHLH transcription factors, are putative mammalian neuronal determination genes that reveal progenitor cell heterogeneity in the developing CNS and PNS. *Mol. Cell. Neurosci.* 8:221–41

Takebayashi H, Yoshida S, Sugimori M, Kosako H, Kominami R, et al. 2000. Dynamic expression of basic helix-loop-helix

Olig family members: implication of Olig2 in neuron and oligodendrocyte differentiation and identification of a new member, Olig3. *Mech. Dev.* 99:143–48

Tamada A, Shirasaki R, Murakami F. 1995. Floor plate chemoattracts crossed axons and chemorepels uncrossed axons in the vertebrate brain. *Neuron* 14:1083–93

Tanabe Y, William C, Jessell TM. 1998. Specification of motor neuron identity by the MNR2 homeodomain protein. *Cell* 95:67–80

Tessier-Lavigne M, Goodman CS. 1996. The molecular biology of axon guidance. *Science* 274:1123–33

Thaler J, Harrison K, Sharma K, Lettieri K, Kehrl J, Pfaff SL. 1999. Active suppression of interneuron programs within developing motor neurons revealed by analysis of homeodomain factor HB9. *Neuron* 23:675–87

Thor S, Andersson SG, Tomlinson A, Thomas JB. 1999. A LIM-homeodomain combinatorial code for motor-neuron pathway selection. *Nature* 397:76–80

Thor S, Ericson J, Brannstrom T, Edlund T. 1991. The homeodomain LIM protein Isl-1 is expressed in subsets of neurons and endocrine cells in the adult rat. *Neuron* 7:881–89

Thor S, Thomas JB. 1997. The Drosophila *islet* gene governs axon pathfinding and neurotransmitter identity. *Neuron* 18:397–409

Tiret L, Le Mouellic H, Maury M, Brulet P. 1998. Increased apoptosis of motoneurons and altered somatotopic maps in the brachial spinal cord of *Hoxc-8*-deficient mice. *Development* 125:279–91

Tokumoto M, Gong Z, Tsubokawa T, Hew CL, Uyemura K, et al. 1995. Molecular heterogeneity among primary motoneurons and within myotomes revealed by the differential mRNA expression of novel islet-1 homologs in embryonic zebrafish. *Dev. Biol.* 171:578–89

Tosney KW. 1987. Proximal tissues and patterned neurite outgrowth at the lumbosacral level of the chick embryo: deletion of the dermamyotome. *Dev. Biol.* 122:540–58

Tosney KW. 1988. Proximal tissues and patterned neurite outgrowth at the lumbosacral level of the chick embryo: partial and complete deletion of the somite. *Dev. Biol.* 127:266–86

Tosney KW, Landmesser LT. 1985a. Development of the major pathways for neurite outgrowth in the chick hindlimb. *Dev. Biol.* 109:193–214

Tosney KW, Landmesser LT. 1985b. Growth cone morphology and trajectory in the lumbosacral region of the chick embryo. *J. Neurosci.* 5:2345–58

Tsuchida T, Ensini M, Morton SB, Baldassare M, Edlund T, et al. 1994. Topographic organization of embryonic motor neurons defined by expression of LIM homeobox genes. *Cell* 79:957–70

Vallstedt A, Muhr J, Pattyn A, Pierani A, Mendelsohn M, et al. 2001. Different levels of repressor activity assign redundant and specific roles to *Nkx6* genes in motor neuron and interneuron specification. *Neuron* 31:743–55

Varela-Echavarria A, Pfaff SL, Guthrie S. 1996. Differential expression of LIM homeobox genes among motor neuron subpopulations in the developing chick brain stem. *Mol. Cell. Neurosci.* 8:242–57

Varela-Echavarria A, Tucker A, Puschel AW, Guthrie S. 1997. Motor axon subpopulations respond differentially to the chemorepellents netrin-1 and semaphorin D. *Neuron* 18:193–207

Vogel A, Rodriguez C, Warnken W, Izpisua Belmonte JC. 1995. Dorsal cell fate specified by chick Lmx1 during vertebrate limb development. *Nature* 378:716–20

Way JC, Chalfie M. 1988. *mec-3*, a homeobox-containing gene that specifies differentiation of the touch receptor neurons in C. elegans. *Cell* 54:5–16

Wilkinson DG. 2001. Multiple roles of EPH receptors and ephrins in neural development. *Nat. Rev. Neurosci.* 2:155–64

Xue D, Tu Y, Chalfie M. 1993. Cooperative interactions between the Caenorhabditis elegans homeoproteins UNC-86 and MEC-3. *Science* 261:1324–28

Yuan W, Zhou L, Chen JH, Wu JY, Rao Y, Ornitz DM. 1999. The mouse SLIT family: secreted ligands for ROBO expressed in patterns that suggest a role in morphogenesis and axon guidance. *Dev. Biol.* 212:290–306

Zhou Q, Choi G, Anderson DJ. 2001. The bHLH transcription factor Olig2 promotes oligodendrocyte differentiation in collaboration with Nkx2.2. *Neuron* 31:791–807

Zhou Q, Wang S, Anderson DJ. 2000. Identification of a novel family of oligodendrocyte lineage-specific basic helix-loop-helix transcription factors. *Neuron* 25:331–43

Zou Y, Stoeckli E, Chen H, Tessier-Lavigne M. 2000. Squeezing axons out of the grey matter: a role for slit and semaphorin proteins from midline and ventral spinal cord. *Cell* 102:363–75

THE ROLE OF HYPOCRETINS (OREXINS) IN SLEEP REGULATION AND NARCOLEPSY

Shahrad Taheri, Jamie M. Zeitzer, and Emmanuel Mignot

Stanford University Center for Narcolepsy, 701 Welch Road B, Basement, Palo Alto, California 94304-5742; email: staheri@stanford.edu; jzeitzer@stanford.edu; mignot@stanford.edu

Key Words lateral hypothalamus, arousal, appetite, neuroendocrine, energy expenditure

■ **Abstract** The hypocretins (orexins) are two novel neuropeptides (Hcrt-1 and Hcrt-2), derived from the same precursor gene, that are synthesized by neurons located exclusively in the lateral, posterior, and perifornical hypothalamus. Hypocretin-containing neurons have widespread projections throughout the CNS with particularly dense excitatory projections to monoaminergic centers such as the noradrenergic locus coeruleus, histaminergic tuberomammillary nucleus, serotoninergic raphe nucleus, and dopaminergic ventral tegmental area. The hypocretins were originally believed to be primarily important in the regulation of appetite; however, a major function emerging from research on these neuropeptides is the regulation of sleep and wakefulness. Deficiency in hypocretin neurotransmission results in the sleep disorder narcolepsy in mice, dogs, and humans. The hypocretins are also uniquely positioned to link sleep, appetite, and neuroendocrine control. The aim of this review is to describe and discuss the current knowledge regarding the hypocretin neurotransmitter system in narcolepsy and normal sleep.

INTRODUCTION

Sleep is a highly complex and regulated state that remains as one of the great mysteries in neuroscience. The importance of sleep to survival is emphasized by studies demonstrating that total sleep deprivation in animals results in multiple pathophysiological alterations that can threaten life. Furthermore, the presence of sleep/rest in multiple species indicates an evolutionary advantage. However, despite these studies and suggestions that sleep is restorative to brain metabolism and/or is required for memory consolidation, the precise physiological reason why most human beings spend about a third of their life sleeping remains unknown.

Electrophysiological studies have demonstrated that sleep is not a quiescent state. Normal sleep has a highly predictable architecture that can be divided into two states that alternate with a cyclicity of about ninety minutes in humans: rapid eye movement (REM) sleep and nonrapid eye movement (NREM) sleep. NREM sleep is associated with a synchronized rhythm on the electroencephalogram (EEG) and can be further subdivided into four stages: stages I and II (light NREM sleep) and stages III and IV (deep slow-wave sleep). During NREM sleep, muscle tone is maintained allowing changes in posture. REM sleep is associated with a low amplitude, high frequency desynchronized EEG pattern that resembles the awake state; therefore, REM sleep is also called paradoxical sleep. In REM sleep, there is phasic activity of eye muscles, increased activity of the sympathetic nervous system, and loss of muscle tone. Individuals are more likely to report a dream if they are awoken during a period of REM sleep. Other influences determine the timing, duration, and intensity of sleep (circadian and homeostatic mechanisms). Sleep is also associated with a variety of sleep-state–specific and circadian-controlled physiological parameters such as changes in hormone release, cardiovascular control, regulation of breathing, convulsive thresholds, and gastrointestinal physiology.

The EEG is generated through reciprocal neuronal connections between the thalamus and cortex (thalamocortical loops). A widely accepted neuroanatomical model for the regulation of sleep and EEG involves reciprocal inhibitory interactions between cholinergic (laterodorsal tegmental [LDT]) area and pedunculopontine nuclei [PPT]) and monoaminergic neurons (adrenergic locus coeruleus [LC], serotoninergic raphe nucleus [RN], and histaminergic tuberomammillary nucleus [TMN]). Projections from these neurons mediate EEG changes by regulating the activity of thalamocortical loops. During wakefulness, monoaminergic tone is high (resulting in EEG desynchronization), though across the sleep cycle, this tone decreases (resulting in EEG synchronization). The decrease in monoaminergic tone disinhibits cholinergic neurons in the latter part of the sleep cycle thus generating REM sleep and EEG desynchronization.

Narcolepsy is a unique model for dysfunction in mechanisms that regulate wakefulness and the transition between NREM, wake, and REM sleep. The narcolepsy syndrome is characterized by the narcolepsy tetrad: excessive daytime sleepiness (EDS), cataplexy (sudden loss of muscle tone in response to strong emotion such as laughter or anger), hypnagogic hallucinations (dream-like experiences occurring at sleep onset), and sleep paralysis (the inability to move while falling asleep or upon awakening). Narcolepsy therefore appears to consist of two major problems: first, an inability to maintain wakefulness, and second, intrusion of REM sleep into wakefulness or at sleep onset resulting in hallucinations, sleep paralysis, and possibly cataplexy. The total daily amount of sleep and REM sleep, however, is little different from those of healthy subjects. The current hypothesis regarding the etiology of narcolepsy is that it is an autoimmune disorder because of its strong association with the human leukocyte antigen (HLA) system (HLA-DR2 and HLA-DQB1*0602).

Until recently, little was known about the basic pathophysiology of narcolepsy, and most studies concentrated on abnormalities in classical neurotransmitter circuits. The only animal model available was narcolepsy in dogs. Genetic studies of canine models of narcolepsy and targeted gene deletion and neuronal degeneration studies in mice supported by human studies have now identified the lateral hypothalamic peptidergic hypocretin (orexin) neurons as key targets in the pathogenesis of narcolepsy. Familial cases of canine narcolepsy are associated with mutations in one of the receptors for hypocretins (hcrtr2), though sporadic cases are deficient in the hypocretin precursor, prepro-hypocretin. Mice with targeted deletion of the prepro-hypocretin gene and mice with targeted degeneration of hypocretin neurons display phenotypes similar to humans with narcolepsy. The majority of HLA-DQB1*0602 positive patients with narcolepsy-cataplexy have undetectable hypocretins in their cerebrospinal fluid (CSF), and the few postmortem brains from narcolepsy subjects examined to date reveal absence or profound decrease of hypocretins. This discovery has revolutionized our understanding of narcolepsy and the regulation of sleep and wakefulness. With the importance of the hypocretins in sleep and arousal becoming increasingly apparent, it is likely that these peptides are altered by or contribute to other neurological and psychiatric disorders. The aim of this review is to describe and discuss the current knowledge regarding the hypocretin neurotransmitter system in narcolepsy and normal sleep.

THE ROLE OF THE HYPOTHALAMUS IN THE REGULATION OF SLEEP

Key mechanisms involved in the maintenance of homeostasis are located within the hypothalamus. Neural circuits within the hypothalamus regulate temperature, heart rate, blood pressure, plasma osmolality, food and water intake, pituitary hormone secretion, and the sleep-wake cycle. A variety of experimental approaches have identified multiple areas within the hypothalamus as major centers in the regulation of circadian rhythms, sleep, and wakefulness. One such area, the suprachiasmatic nucleus (SCN), contains neurons with a genetically identified biological clock that oscillates with a periodicity of approximately 24 hours. These neurons primarily send projections to other hypothalamic nuclei and convey circadian rhythmicity to a variety of physiological functions and behaviors including sleep.

Experiments by Nauta suggested that the hypothalamic preoptic area (POA) is a sleep center because lesioning of this area results in insomnia (Nauta 1946). Sleep-active neurons in the POA and basal forebrain have been identified; these begin their firing when drowsiness begins and are maximally active during NREM sleep. Sleep-active neurons in the ventrolateral preoptic area (VLPO) (Sherin et al. 1996), containing the inhibitory neurotransmitters GABA and galanin, project to wake-active histaminergic TMN and brainstem monoaminergic neurons (Szymusiak et al. 2001). These neurons, therefore, may promote sleep through inhibition of wake-active neurons. Unlike damage to the POA, damage to the posterior hypothalamus and peduncular area immediately anterior to the occulomotor nerve, as observed during the viral encephalitis epidemic in the early twentieth century (see Mignot 2001a), results in somnolence, which suggests that this area serves as a wake center. The recent discovery of hypocretin neurons in the lateral and posterior hypothalamus has confirmed this brain region as an important center for the regulation of sleep and wakefulness.

THE DISCOVERY OF HYPOCRETINS

The hypocretins were discovered in search of the most abundant and exclusively expressed mRNAs in the rat hypothalamus using directional tag polymerase chain reaction subtraction. This search resulted in the identification of a novel mRNA that appeared to be exclusively expressed in the lateral, posterior, and perifornical hypothalamus (de Lecea et al. 1998). Based on the structure of this mRNA, the structure of its prepropeptide was determined. The prepropeptide consisted of a signal peptide and cleavage sites that suggested the existence of at least two peptide products named hypocretins (1 and 2) from the combination of hypothalamic and incretin (based on weak structural homology with the peptide secretin, which belongs to the incretin superfamily of peptides). The structure of endogenous hypocretin 2 was accurately determined, but the full structure of endogenous hypocretin 1 remained unknown. In the same year, another group, using the reverse pharmacology technique, independently discovered the same peptides (Sakurai et al. 1998). The reverse pharmacology approach aims to identify endogenous ligands for orphan G protein–coupled receptors (GPCRs). This strategy involves challenging cultured cells expressing the orphan receptor on their surface with chromatographically purified extracts from tissues and monitoring receptor activation through measurement of changes in secondary messengers, ionic fluxes, or pH. Transfectant cell lines, expressing the orphan GPCR HFGAN72 (now orexin-1/hypocretin-1 receptor, OX_1R/hcrtr1), were challenged with purified rat brain extracts. The fractions that elicited an increase in cytoplasmic Ca^{2+} levels in these cells were further purified to isolate the peptide orexin A (corresponding to hypocretin-1, Hcrt-1). A smaller activity peak was further purified to obtain orexin B (corresponding to hypocretin-2, Hcrt-2) and resulted in the discovery of another orphan GPCR, the orexin-2/hypocretin-2 receptor (OX_2R/hcrtr2). The name orexin (from *orexis*, Greek for appetite) was selected based on the anatomical location of these neurons synthesizing the peptides in the lateral hypothalamus (feeding center), and the observations that administration of these peptides into the cerebral ventricle of rats (intracerebroventricular, ICV) potently increased food intake while the mRNA for peptide precursor (prepro-orexin) was upregulated by prolonged fasting (Sakurai et al. 1998).

The orexins and the hypocretins are the same peptides, but, confusingly, both names continue to be used. Some have argued that the name hypocretins is inaccurate because these peptides bear closer structural homology to the bombesin family of peptides (rather than the incretin family) (Willie et al. 2001), whereas others have argued that the name orexins is too restrictive because these peptides have multiple functions. Gene databases use the name hypocretin based on the first report published, though most publications to date, focusing on feeding aspects, have used the name orexins. Perhaps the less functionally specific name hypocretin is now preferable because these peptides have many other functions beside possible involvement in the regulation of appetite. For clarity, the name hypocretin is used in this review.

HYPOCRETIN GENE AND PEPTIDE STRUCTURE

The human prepro-hypocretin gene, spanning 1432 base pairs, consists of 2 exons and 1 intron and is located on chromosome 17q21 (Sakurai et al. 1999). Preprohypocretin mRNA encodes a 131–amino acid human (130–amino acid rat) precursor prepro-hypocretin (Figure 1). The structures of Hcrt-1 and -2 are well conserved between species suggesting important physiological roles. Hcrt-1 is a 33–amino acid carboxy-amidated peptide with an N-terminal pyroglutamyl residue and two intra-chain disulphide bonds (Figure 1). The originally reported structure of Hcrt-1 was based on the structure of prepro-hypocretin mRNA; it is therefore longer than the endogenous peptide, lacks the intrachain disulfide bonds and carboxy-terminal amidation, and, not surprisingly, is inactive at hypocretin receptors. Human Hcrt-1 is identical to the mouse, rat, bovine, and porcine Hcrt-1. Hcrt-2 is a C-terminally amidated, linear peptide of 28 amino acids (Figure 1). Human Hcrt-2 has two amino acid substitutions compared with rodent Hcrt-2 and one substitution compared to porcine and canine Hcrt-2. Hcrt-1 has been reported to be more lipophilic than Hcrt-2 and to weakly diffuse passively across the blood-brain barrier (Kastin & Akerstrom 1999); however, this observation has not been confirmed.

THE ANATOMY OF HYPOCRETIN NEUROTRANSMISSION

Hypocretin immunoreactive cell bodies have been observed mainly in the lateral, posterior, and perifornical hypothalamus. A small proportion of hypocretin neurons are immunopositive for galanin (Hakansson et al. 1999), though essentially all hypocretin neurons contain dynorphin (Chou et al. 2001). Hypocretin immunoreactivity and immunoreactive fibers are widely distributed throughout the central nervous system (Figure 2) (Peyron et al. 1988, Taheri et al. 1999) but are particularly

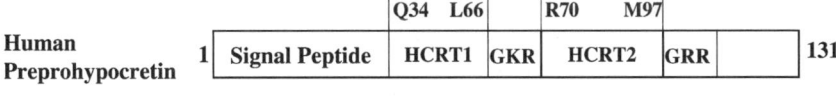

Figure 1 The structures of human prepro-hypocretin (orexin), hypocretin 1 (orexin A) and hypocretin 2 (orexin B). <E = pyroglutamyl residue. Note the C-terminal amidation of both peptides, essential for biological activity.

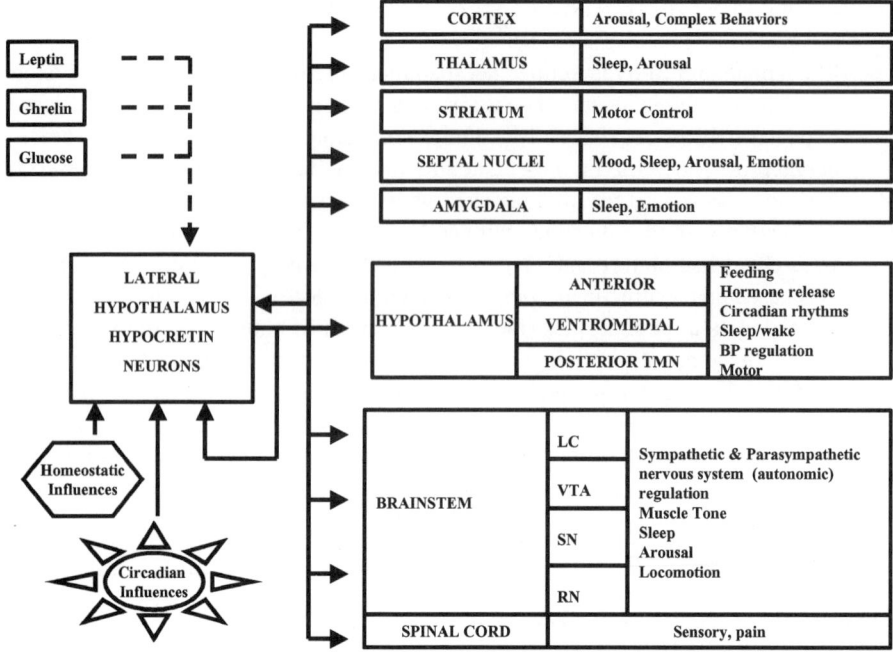

Figure 2 Relative position of hypocretins (orexins) in neural physiology. Hypocretin neurons in the lateral hypothalamus receive neural input from both circadian (SCN) and homeostatic mechanisms, as well as being connected with cortical, thalamic, striatal, septal, amygdalar, hypothalamic, brainstem, and spinal nuclei. Hypocretin neurons may also receive, either directly or indirectly, information from circulating leptin, ghrelin, and glucose, which may influence their capacity as modulators of metabolism and energy expenditure. Other proposed mechanisms of hypocretinergic influence include modulation of sleep/wake activity (thalamus, septum, amygdala, hypothalamus, brainstem), arousal/affect (cortex, septum, amygdala, thalamus, hypothalamus, brainstem), motor activity (striatum, hypothalamus, brainstem), feeding and neuroendocrine (hypothalamus), and autonomic physiology (hypothalamus, brainstem, spinal cord). LC = locus coeruleus, VTA = ventral tegmental area, SN = substantia nigra, RN = raphe nucleus.

concentrated in the hypothalamus, amygdala, nucleus accumbens, septum, and monoaminergic centers such as the noradrenergic LC, histaminergic TMN, serotoninergic RN, and dopaminergic ventral tegmental area (VTA). Based on these projection sites, hypocretins are likely to have multiple functions (Figure 2), but they appear to have significant inputs to areas involved in the regulation of the sleep/wake cycle.

Two previously orphan G protein–coupled receptors for the hypocretins have been reported: hcrtr1 and hcrtr2. Hcrt-1 has greater affinity than Hcrt-2 for the human hcrtr1 receptor, but both peptides have similar affinities for the human hcrtr2 receptor; this has made it difficult to identify the individual functions of these peptides. Hcrtr1 mRNA has been detected in the ventromedial hypothalamus, tenia tecta, hippocampus, dorsal raphe, and LC. Hcrtr2 mRNA is expressed in the paraventricular hypothalamic nucleus (PVN), subthalamic and thalamic nuclei, septal nuclei, cerebral cortex, nucleus accumbens, anterior pretectal nucleus, and several regions in the medulla oblongata (Trivedi et al. 1998, Marcus et al. 2001). It is important to note that noradrenergic LC neurons are densely packed with hcrtr1 but do not contain hcrtr2, as opposed to the histaminergic TMN, which contains hcrtr2 but not hcrtr1 receptors. The serotoninergic raphe and dopaminergic VTA and substantia nigra (SN) contain both receptor subtypes. Hypocretin receptor mRNA expression has also been reported in the pituitary gland (Blanco et al. 2001), adrenal gland (Lopez et al. 1999, Malendowicz et al. 1999, Randeva et al. 2001), enteric nervous system, and pancreas (Kirchgessner & Liu 1999).

Elevated Fos immunoreactivity (a measure of neuronal activation) in response to hypocretins (Date et al. 1999, Mullett et al. 2000) has been detected in hypothalamic areas involved in appetite regulation, neuroendocrine regulation, and sleep and circadian rhythms (lateral hypothalamus, posterior and dorsomedial hypothalamus, anterior hypothalamus, perifornical area, arcuate and paraventricular nuclei), as well as in the lateral septal area, central nucleus of the amygdala, shell of the nucleus accumbens, bed nucleus of the stria terminalis, and nucleus of the solitary tract (involved in visceral and autonomic regulation).

In all electrophysiological experiments reported to date, hypocretins are excitatory. Excitatory effects have been reported in the LC, RN, VTA, SN, and TMN (Hagan et al. 1999, Sutcliffe & de Lecea 2000, Uramura et al. 2001, Brown et al. 2001, Eriksson et al. 2001). In vitro, both hcrtr1 and hcrtr2 can couple to Gq and mobilize intracellular calcium. Hcrtr2 may couple with either Gq or Go/Gi in some cases; using isolated pro-opiomelanocortin (POMC) neurons and fura-2 fluorescence imaging, Yada and colleagues have observed a pertussis-sensitive decrease in intracellular calcium after application of hypocretins (T. Yada, personal communication). Others have proposed the existence of hcrtr2-inhibitory autoreceptors on hypocretin neurons. The finding that the TMN neurons are strongly excited by hypocretins (Eriksson et al. 2001) indicates that hcrtr2 has excitatory effects at least on histaminergic neurotransmission.

CLINICAL ASPECTS OF NARCOLEPSY

Epidemiology

Narcolepsy is a much more common disorder than generally recognized. This is because, up to now, its diagnosis has very much depended on clinical history and examination with few objective measures. The prevalence of narcolepsy has been reported to be 20–60 per 100,000 population in Western countries (Overeem et al. 2001). This prevalence is comparable to multiple sclerosis and Parkinson's disease. Narcolepsy usually begins in the second to third decade of life, with both sexes being equally affected. Narcolepsy in children comprises only 10% of cases. Life expectancy and sexual reproductive capacity are unaltered in narcolepsy. However, taking into account that the onset of narcolepsy is usually in adolescence, when social interaction and academic achievement become increasingly important, and then continues throughout life and is compounded by treatments with multiple side effects, morbidity is great. In fact, with the exception of driving, narcolepsy has been reported to have a greater adverse impact on quality of life than epilepsy (Broughton & Broughton 1994).

Only 10% to 15% of patients experience the full tetrad of abnormalities. The majority (approximately 70%) have cataplexy, the most specific symptom of narcolepsy. About a third of patients suffer from hypnagogic hallucinations while a quarter of patients have sleep paralysis. The presenting symptom in the majority (90%) is excessive sleepiness alone or in combination with hypnagogic hallucinations and/or sleep paralysis. Cataplexy usually appears later, being the presenting symptom in less than 10% of patients. Symptoms usually have a gradual onset. Once symptoms occur, their severity usually remains stable after several decades. Cataplexy and sleep paralysis may diminish or even disappear in some patients; this may reflect neural plasticity or may be due to reduced attention paid to symptoms with aging. It is exceptional for new symptoms to appear ten years or more after disease onset.

Symptoms of Narcolepsy

Gelineau first used the term narcolepsy by combining the Greek for "somnolence" and "to seize" (see Mignot 2001a). Yoss & Daly introduced the narcolepsy tetrad as excessive daytime sleepiness (EDS), cataplexy, hypnagogic hallucinations, and sleep paralysis (see Mignot 2001a). Because disturbed nighttime sleep invariably accompanies narcolepsy, it was added to these symptoms by other clinicians (narcolepsy pentad).

EDS manifests in two ways: a continuous feeling of sleepiness and irresistible sleep attacks. Most patients experience both symptoms, but one may have a greater impact in an individual patient. Sleep attacks/naps, lasting usually less than 20 minutes, start gradually with drowsiness, which may be associated with diplopia, blurred vision, and ptosis. The occurrence of these naps is dependent on circumstances; passive activities such as watching television or being a passenger in a car are usual triggers. Patients can usually be easily awakened from naps. Although naps are refreshing, irresistible sleepiness recurs within several hours. Despite excessive sleepiness during the day, the total daily amount of sleep or REM sleep in narcolepsy is not different from normal individuals.

Cataplexy (Greek: to strike down) usually consists of sudden, bilateral loss of muscle tone without loss of consciousness. It is triggered by any situation that requires sudden action or strong emotion (laughing and joking are commonly reported triggers). Attacks may be partial or complete (causing total collapse onto the floor) and may involve some or, less frequently, all muscles. The knees (giving way) and head and neck muscles (sagging of the jaw, inclination of the head, slurred speech) are commonly involved in partial attacks. Injury is uncommon with cataplexy, which usually lasts for seconds (rarely minutes). Partial attacks may be very subtle with the patient, relatives, and physicians interpreting them as normal. If the patient is neurologically examined during an attack, atonia and areflexia may be observed. Although cataplexy is the most specific symptom in narcolepsy, it has also been observed or suggested (sometimes difficult to distinguish from atonic seizures or other forms of hypotonia) in several other disorders including Prader Willi syndrome, Niemann Pick disease Type C, Norrie disease, and with diencephalic tumors; the majority of these disorders are associated with hypothalamic dysfunction.

Hypnagogic hallucinations are dream-like episodes occurring at the time of going to sleep. These hallucinations are rarely pleasant and when combined with sleep paralysis can be highly distressing. Occasionally, vivid hypnagogic hallucinations are mistaken for psychosis and schizophrenia. Both hypnagogic hallucinations and sleep paralysis are not specific to narcolepsy and are commonly reported in the general population. Other symptoms of narcolepsy include disturbed nighttime sleep secondary to excessive sleep during the day, periodic leg movements (nocturnal myoclonus) during sleep, automatic behavior, sleep talking, and more rarely, somnambulism. Patients with narcolepsy tend to have a significantly higher body mass index (BMI) compared to age and sex matched controls (Schuld et al. 2000). This may be related to lower metabolic rate (see below). Patients with narcolepsy also have a greater propensity toward type 2 diabetes mellitus (Honda et al. 1986).

Genetic Aspects and HLA Association of Human Narcolepsy

The majority of human narcolepsy cases are due to a multigenic, environmentally determined disorder (Mignot 1998). However, first-degree relatives have a 20- to 40-fold increased risk of developing narcolepsy-cataplexy. Approximately 25% to 35% of reported monozygotic twins are concordant for narcolepsy. Although a tight association between narcolepsy and HLA DR2 has been reported, recent studies have shown that HLA-DQB1*0602 is the unifying susceptibility HLA allele across ethnic groups; other HLA alleles further modulate susceptibility (Mignot et al. 2001). It is interesting that a hierarchy of HLA alleles may be predisposing and protective, as reported for other HLA-associated diseases. The most important effect is carried by DQB1*0602 as opposed to alleles such as DQB1*0601 and DQB1*0501, which may be protective. Of note, DQB1*0601 is very similar to DQB1*0602 yet is a protective allele, which suggests that very minute changes in the peptide binding pockets of these molecules may dramatically alter their effect in narcolepsy (Lin et al. 2001). HLA-DQB1*0602 positivity is especially high (90% to 100%) in patients with definite cataplexy but decreases dramatically with atypical or no cataplexy (40%). However, approximately 12% of Japanese, 25% of Caucasian, and 38% of African American controls are DQB1*0602 positive (Mignot et al. 2001). Up to a third of familial cases are DQB1*0602 negative, which suggests the contribution of other genes.

The established association between HLA-DQB1*0602 and narcolepsy suggests an autoimmune etiology. However, to date, no concrete evidence has been reported in support of this hypothesis. This could be because, up to now, the neuronal target for such an autoimmune attack was unknown; autoimmune attack could involve a small discrete set of neurons, such as hypocretin cells in the lateral and perifornical hypothalamus. Further, because patients are diagnosed much later than the actual disease onset, any evidence of an inflammatory insult may have disappeared.

Diagnosis of Narcolepsy

In most international classifications, narcolepsy is defined by the presence of sleepiness and cataplexy or by the polysomnographic documentation of REM sleep abnormalities (ASDA 1997). The most commonly accepted diagnostic test is the multiple sleep latency test (MSLT). In the MSLT, nocturnal sleep polysomnography is first performed, followed the next day by 4 to 5 daytime naps during which sleep latency is measured. Untreated narcoleptic patients typically display short mean sleep latency (MSL) (≤ 8 minutes) and 2 or more sleep onset REM periods (SOREMPs) during naps. Using the polysomnographic definition, narcolepsy contrasts with idiopathic hypersomnia, a more common condition (affecting approximately 4% of the general population) that is characterized by extended sleep time, no REM-related symptoms, and a short MSL without SOREMPs (Bassetti & Aldrich 1997). The MSLT is time-consuming, expensive, and not readily available to all physicians. Also, false-negatives and false-positives are common in clinical practice. HLA typing is only supportive of a diagnosis. With the key pathophysiology of narcolepsy being abnormal hypocretin neurotransmission, measurement of CSF hypocretin levels is likely to be an important diagnostic test for narcolepsy, especially in difficult cases with multiple sleep abnormalities or equivocal MSLT. The majority of HLA-DQB1*0602 positive patients with narcolepsy-cataplexy have low hypocretin levels in their CSF.

Treatment of Human Narcolepsy

Current treatments are directed toward the alleviation of symptoms rather than the disease pathology itself (Nishino & Mignot 1998, Overeem et al. 2001). Behavioral modifications are helpful, but the majority of patients (over 95%) require pharmacological treatment. EDS, the most disabling symptom, is treated with amphetamine-like stimulants or modafinil. These compounds act by stimulating dopamine release (e.g. dextroamphetamine) and/or inhibiting dopamine reuptake (modafinil) (Nishino et al. 1998, Wisor et al. 2001). Amphetamine-like stimulants are generally more effective than modafinil but produce more side effects, are more difficult to prescribe, and have the dangers of addiction and abuse. Stimulant therapy is relatively ineffective for cataplexy and other symptoms.

Cataplexy, the second most disabling symptom, is usually treated using tricyclic antidepressants or higher doses of serotonin reuptake inhibitors (e.g. fluoxetine).

Adrenergic reuptake inhibition is most effective in reducing cataplexy in animal models (Mignot et al. 1993a), but few antidepressants targeting the adrenergic transporter site are available, and at present, clinical experience with these compounds is limited. Sleep paralysis and hypnagogic hallucinations, two symptoms of dissociated REM sleep, respond partially to antidepressants. Disturbed nocturnal sleep can be treated using benzodiazepine hypnotics, but this may aggravate daytime sleepiness. A novel hypnotic, gammahydroxybutyrate (GHB), a former anesthetic agent, is now undergoing clinical trial for the treatment of narcolepsy (Nishino & Mignot 1997, Bernasconi et al. 1999). GHB consolidates nocturnal sleep by increasing deep slow-wave sleep and to a lesser extent REM sleep while decreasing daytime sleepiness and cataplexy. This compound, however, has a very short half-life requiring it to be administered twice in the night, and it is also a drug of abuse. Because GHB promotes slow-wave sleep, resulting in growth hormone release, it has been abused by athletes. The compound has also been abused as a rape drug or in association with other drugs to induce euphoria at rave parties. The exact mode of action of the GHB is still unknown but may involve specific GHB receptors and/or stimulation of GABA-B receptors (Bernasconi et al. 1999). GHB administration dramatically decreases the firing rate of dopaminergic cells and dopamine release in awake animals. Dopamine synthesis is uncoupled and brain dopamine content increases as a result. It is hypothesized that the increased dopamine store could alleviate daytime sleepiness in the following day.

PHARMACOLOGICAL STUDIES IN CANINE NARCOLEPSY

Canine narcolepsy was first reported in a dachshund and in a poodle (see Mignot 2001a). In these breeds, as in most human cases, narcolepsy is a polygenic and/or an environmentally influenced disorder. However, litters with several affected animals were discovered in Doberman pinschers and Labrador retrievers; these were used to establish a narcoleptic dog-breeding colony at Stanford in 1976. Genetic transmission in these two breeds was determined to be autosomal recessive with full penetrance through a single gene called *canarc-1*.

The validity of the canine model has been established through neurophysiological and neuropharmacological similarities with the human disorder. As in human narcolepsy, affected dogs display emotionally triggered cataplexy, fragmented sleep, and a short sleep latency. Familial cases are due to hcrtr2 mutations as opposed to sporadic cases, which are hypocretin-deficient but have no hypocretin gene mutations (Lin et al. 1999, Ripley et al. 2001a). Cataplexy can be induced easily in these animals by presenting them with food or by play. Cataplexy is generally more severe and onset of symptoms more variable in prepro-hypocretin–deficient dogs. Considering the polyphasic nature of sleep in canines, sleepiness is more difficult to document in narcoleptic animals. Twenty-four–hour sleep recordings have indicated that narcoleptic dogs, whether prepro-hypocretin ligand– or receptor-deficient, have disrupted sleep patterns and reduced REM sleep latency (Nishino & Mignot 1997).

The effects of over 200 pharmacological agents have been studied since the establishment of a canine narcolepsy breeding colony at Stanford (Baker & Dement 1985, Nishino & Mignot 1997). Unlike humans, where cataplexy may be difficult to diagnose and evaluate, canine cataplexy can be reliably induced using the food-elicited cataplexy test (FECT). In general, the pharmacological control of cataplexy has been found to be similar to that of REM sleep and consistent with the aminergic-cholinergic reciprocal interaction model (McCarley & Massaquoi 1992). In this model, high monoaminergic and cholinergic activity leads to wakefulness and EEG activation. In NREM sleep, monoaminergic and cholinergic activity decreases in association with EEG synchronization. During REM sleep, cholinergic systems are active, though monoaminergic neuronal activity is almost absent. This pattern of activity is most striking for adrenergic, histaminergic, and serotoninergic tone. Dopaminergic neurons in the VTA and SN, however, do not significantly change their firing rates during the sleep cycle. Drugs that either increase cholinergic tone or decrease monoaminergic activity increase cataplexy while anticholinergic and monoaminergic-enhancing drugs reduce it (Baker & Dement 1985, Mignot et al. 1993b, Nishino & Mignot 1997). Adrenergic uptake inhibition is more effective than serotonin or dopamine reuptake inhibition in reducing cataplexy (Mignot et al. 1993a). As serotonin reuptake inhibitors also reduce REM sleep, this may suggest a preferential adrenergic control of REM atonia. It is surprising that at small doses, D2/D3 dopaminergic drugs have important effects on cataplexy, an effect that contrasts with the lack of effect of dopaminergic reuptake blockers on this symptom (Nishino & Mignot 1997). This effect is likely to be mediated by D2/D3 autoreceptors, with stimulation and blockade exacerbating and reducing the symptoms, respectively (Nishino & Mignot 1997).

Fewer pharmacological studies have been performed to evaluate drug effects on sleepiness. A number of studies have shown that amphetamine-like stimulants and modafinil produce wakefulness by presynaptic enhancement of dopaminergic neurotransmission, either via dopamine reuptake inhibition and/or by enhancing dopamine release (Nishino et al. 1998, Wisor et al. 2001).

HYPOCRETIN DEFICIENCY IN ANIMALS

Although sleep disturbances have been noted with lesions of the lateral hypothalamus, these lesions do not produce a narcolepsy phenotype, which suggests that destruction of discrete neurons is necessary. The importance of the hypocretin neurotransmitter system in narcolepsy was recognized through the use of state-of-the art molecular biology techniques by two simultaneously published studies. First, after a decade's search for gene abnormalities in canine narcolepsy, an autosomal recessive condition inherited through the *canarc-1* gene, it was discovered that this model was associated with mutations in the hcrtr2 receptor gene such that this receptor was not functional in narcoleptic dogs (Lin et al. 1999). Three Hcrtr2 mutations have been identified in labradors, dobermans, and dachshunds. Two of these are exon-skipping mutations while the third involves a single-base pair change. Second, an abnormal phenotype was noted in mice with targeted deletion of the prepro-hypocretin gene (Chemelli et al. 1999). Because it was originally believed that the hypocretins have an important role in the regulation of appetite, and rodents tend to consume the majority of their daily food in the dark (active phase), infrared cameras were installed to further characterize the feeding phenotype of prepro-hypocretin gene knockout mice. Unexpectedly, these mice suffered from periods of freezing, totter, and atonia. These periods of behavioral arrest were similar to cataplexy observed in human and canine narcolepsy. EEG studies confirmed significant disruptions of sleep-wakefulness cycling during the dark phase. In the same report, modafinil was shown to activate hypocretin neurons in wild-type mice, and important projections of hypocretin neurons to cholinergic and monoaminergic centers involved in the regulation of the sleep/wake cycle were demonstrated. Prepro-hypocretin gene knockout mice have abnormalities present from early development, but human narcolepsy is rarely an inherited condition. Sakurai and colleagues, therefore, ingeniously inserted a truncated Machado-Joseph disease gene product (ataxin-3) with an expanded polyglutamine stretch into the prepro-hypocretin gene (Hara et al. 2001). This resulted in postnatal degeneration of hypocretin neurons that was complete by 15 weeks of age. These mice also displayed behavioral abnormalities, similar to human narcolepsy, and abnormal transitions to REM sleep. Others have used the ribosomal toxin saporin to selectively destroy hypocretin neurons in rats: Although lateral hypothalamic neurons apart from hypocretin neurons were also eliminated, increased slow-wave sleep, rapid-eye movement (REM) sleep, and sleep-onset REM sleep periods were observed (Gerashchenko et al. 2001a). Hypocretin deficiency also exists in sporadic cases of canine narcolepsy where cataplexy is more severe than in dogs with hcrtr-2 gene mutations. These sporadic cases have low or undetectable Hcrt-1 in their CSF (Ripley et al. 2001a).

The role of individual hypocretin receptors has been difficult to investigate owing to the lack of specific receptor antagonists. ICV administration of Hcrt-1 on sleep is ineffective in hcrtr2 gene–mutated dogs (Fujuki et al. 2001), which suggests a primary effect of hcrtr2 in the promotion of wakefulness. Further information has been obtained through targeted deletion of the genes for hypocretin receptors hcrtr1 and hcrtr2. Hcrtr2 gene knockout mice have features of narcolepsy, whereas there is milder sleep disturbance in hcrtr1 gene knockouts (Chemelli et al. 2000). The narcolepsy phenotype appears to be most strongly displayed in prepro-hypocretin (ligand) gene knockouts and double receptor gene knockouts; both hypocretin receptors are therefore important in sleep-wake mechanisms. Hcrtr1 may, therefore, play a modulatory role in the control of sleep and wakefulness. This is to be expected because the locus coeruleus, a major target for hypocretin neurotransmission and an important brainstem nucleus for the induction of arousal and regulation of muscle tone, contains only hcrtr1.

HYPOCRETIN DEFICIENCY IN HUMAN NARCOLEPSY

All the above animal studies demonstrate the unprecedented progress made since the discovery of hypocretins to a possible link to a human disease condition. Also, the study of hypocretins has shown the value of novel molecular techniques to answer important biological questions. Though animal studies emphasized the importance of hypocretin neurotransmission in narcolepsy, it remained to be shown that the hypocretins also had a pivotal role in human narcolepsy. When Hcrt-1 immunoreactivity was measured in the CSF of patients with narcolepsy, it was observed that 7 out of 9 patients with narcolepsy had undetectable Hcrt-1 compared to controls (Nishino et al. 2000). This study has been extended to more patients and comparisons have been made to other neurological disorders (Ripley et al. 2001b, Bassetti et al. 2001). The majority of HLA positive patients with narcolepsy-cataplexy have low or undetectable Hcrt-1 levels in their CSF. The original study suggested that high hypocretin levels (2 out of 9 patients) also occur in narcolepsy, but this is unlikely to be specific because high levels are also observed in controls and patients with other neurological disorders. More than 10% of narcolepsy-cataplexy patients have normal CSF hypocretin-1 levels, which suggests that multiple pathophysiologies may result in a narcolepsy phenotype. These cases may still have abnormalities in hypocretin neurotransmission (e.g., hypocretin receptor or terminals at target sites may be affected). Only a few patients with other neurological disorders have undetectable CSF Hcrt-1 levels (Ripley et al. 2001b). It is interesting that most of these patients have disorders associated with abnormal hypothalamic function.

In a recent study, narcolepsy patients were screened for mutations in preprohypocretin and the hypocretin receptor genes. Unlike familial canine narcolepsy, no hypocretin receptor mutations were observed. One severely affected patient with an unusually early onset narcolepsy (cataplexy at the age of 6 months) had a prepro-hypocretin gene mutation resulting in abnormal trafficking of the mutant peptide precursor. In the same study, prepro-hypocretin mRNA was undetectable in the post-mortem hypothalami of narcoleptic subjects and immunoreactive peptide levels were undetectable in brain regions (Peyron et al. 2000). These findings were confirmed independently by another post-mortem study (Thannickal et al. 2000). Together, these studies suggest that even in the absence of a specific mutation, narcolepsy is still associated with a deficiency in the hypocretin system. It is at present unclear how the selective loss of hypocretin neurons occurs in narcolepsy. In the above studies, neighboring neurons containing melanin-concentrating hormone (MCH) were observed to be intact. There was little evidence of an inflammatory insult to hypocretin neurons, although the study by Thannickal et al. suggested residual gliosis and a few remaining hypocretin positive neurons. It is possible that the lack of evidence for inflammation is because of few and/or older patients. The autoimmune insult, suspected from the HLA association, and any concomitant degeneration may have occurred years before.

The available evidence strongly suggests that hypocretin neurotransmission is excitatory to monoaminergic systems and possibly cholinergic systems. In narcolepsy, symptoms may be explained by baseline monoaminergic hypoactivity, hyperreactivity of cholinergic systems, and cholinergic receptor hypersensitivity (Baker & Dement 1985, Nishino & Mignot 1997). Canine narcolepsy is associated with decreased histamine levels in the brain (Nishino et al. 2001), a system that appears to have exclusive hcrtr2 localization. Other studies have reported elevated dopamine and 3,4-dihydroxy-phenylacetic acid (DOPAC, a dopamine metabolite) content in the amygdala (Miller et al. 1990), which suggests a secondary dysregulated dopaminergic tone. Dopamine dysregulation could either primarily contribute to excessive daytime sleepiness or could represent a compensatory mechanism. Abnormal dopaminergic neurotransmission could also explain periodic leg movements in narcolepsy (Mignot 2001b)—a symptom currently treated using dopamine D2/D3 receptor agonist.

HYPOCRETINS AND OTHER SLEEP DISORDERS

If hypocretins are important in the regulation of the sleep/wake cycle, then it would be expected that hypocretin neurotransmission would be abnormal in other sleep disorders, especially the narcolepsy spectrum: idiopathic hypersomnia, secondary hypersomnia, periodic hypersomnia, and/or atypical narcolepsy. However, only a few patients with these disorders have abnormally low CSF Hcrt-1 levels (Bassetti et al. 2001). An explanation may be that the etiology of these cases does not involve abnormal hypocretin neurotransmission and that narcolepsy with typical cataplexy is a distinct disease entity. It also indicates that separating narcolepsy without cataplexy and idiopathic hypersomnia into two distinct disease entities may not be justified; this view has been taken by others on clinical grounds (Honda & Matsuki 1990, Aldrich 1998). Another view is that hypersomnia and narcolepsy-cataplexy are part of the same disease continuum: Some narcolepsy cases may have hypocretin deficiency in projection areas important for sleep regulation but have normal CSF hypocretin-1 levels, especially in younger HLA-DQB1*0602 negative subjects. Partial hypocretin deficiency would be associated with less severe symptoms and less frequent cataplexy.

HYPOCRETINS AND NEUROLOGICAL DISORDERS

Based on studies in narcolepsy, it is clear that absence of CSF Hcrt-1 accurately reflects lack of hypocretins in the brain and will have a major role to play in the clinical diagnosis of narcolepsy. There have been case histories of hypocretin deficiency in patients with sleepiness and hypothalamic lesions. Only a few patients with other neurological disorders have undetectable CSF Hcrt-1 levels. Three patients with severe quadriplegic Guillain-Barré syndrome (an autoimmune condition) have been observed to have undetectable CSF Hcrt-1 levels (Ripley et al. 2001b). It remains to be determined whether this is due to alterations in CSF flow, blood brain barrier permeability, and/or directly associated disease pathology. Lower than normal but detectable hypocretin-1 levels have been observed in subjects with Guillain-Barré syndrome, head trauma, and encephalitis. These disorders may alter hypothalamic function (Ripley et al. 2001b) and can result in temporary or occasionally long-lasting sleepiness (Guilleminault et al. 1983, Guilleminault & Mondini 1986, Mignot 2001a) and rarely narcolepsy-cataplexy (Adie 1926, Van Economo 1930, Guilleminault et al. 1983, Lankford et al. 1994, Mignot 2001a).

HYPOCRETIN MEASUREMENT IN HUMAN CSF AND PLASMA

The absence of Hcrt-1 in the CSF was the first report to link deficient hypocretin neurotransmission to human narcolepsy. To date, it has not been possible to measure Hcrt-2 in the CSF; this could be due to the sensitivity of the available assays, but it is unlikely because most assays have been able to detect higher hypocretin-2 levels in brain tissues. A more probable explanation is that Hcrt-2 is rapidly degraded (neuropeptide neurotransmitters rarely have uptake mechanisms) or is unstable in CSF. The source of CSF Hcrt-1 requires further study. Hypocretin immunoreactive fibers have been detected in the vicinity of the cerebral ventricles, and one report suggests that hypocretins may be synthesized locally (Kummer et al. 2001). Human CSF Hcrt-1 only shows a modest diurnal variation and, surprisingly, is highest during the night.

There have been several reports of measurement of Hcrt-1 in the plasma. Most of these studies, giving disparate results, have used commercially available radioimmunoassay (RIA) kits and have not been accompanied by chromatographic analysis of the detected radioimmunoreactivity. It is possible that interfering substances in the plasma or those produced through plasma extraction procedures have resulted in incorrect detection in the various RIAs. Using the same commercial kits and independently developed RIAs, we were not able to consistently detect hypocretins in the plasma. Hcrt-1 may circulate at very low concentrations requiring extraction from large volumes of plasma; however, its source is unclear (apart from the hypothalamus, prepro-hypocretin mRNA has only been consistently reported in the testis). More rigorous examination using different extraction procedures and RIAs is required before the presence of hypocretins in the plasma can be confidently accepted.

HYPOCRETINS IN NORMAL SLEEP AND AROUSAL

It appeared from early research on hypocretins that they may be involved in the regulation of sleep/wake physiology. One early observation, now replicated in a number of studies, was that ICV injection of either Hcrt-1 or -2 increases locomotor activity (Ida et al. 1999, Hagan et al. 1999, Piper et al. 2000). Only one group has also reported that Hcrt-1 increases wakefulness with a concomitant decrease in REM sleep and, sometimes at high doses, a decrease in slow-wave sleep (Hagan et al. 1999, Piper et al. 2000). This effect was relatively minor compared to our unpublished observations. The question that has not been addressed is the relation between the effects on locomotion and wakefulness. The increase in locomotion may also lead to a secondary increase in body temperature (Yoshimichi et al. 2001), but such a change in body temperature has not been observed consistently (Hagan et al. 1999, Jones et al. 2001). When examining the stimulated locomotor behavior more closely, increases in stress-related behaviors or stereotypy are most common following ICV Hcrt-1 or -2; this implies the possible involvement of dopaminergic neurotransmission (Duxon et al. 2001, Nakamura et al. 2000, Ida et al. 1999, Jones et al. 2001, Sunter et al. 2001). Indeed, dopamine antagonists (also serotonin antagonists) block hypocretin-induced motor behaviors (Nakamura et al. 2000, Duxon et al. 2001).

Two processes are believed to regulate sleep and wakefulness in humans: a circadian and a homeostatic process (Dijk & Czeisler 1994). The former is a daily process that occurs with a 24-hour periodicity, irrespective of state. The latter is an appetitive process, such that the longer the time spent awake, the greater the drive for sleep, and the longer the time spent asleep, the less the drive for sleep. Using a forced desynchrony protocol in which the circadian and homeostatic components of sleep/wake regulation can be separated, Dantz and colleagues have demonstrated that in narcolepsy, both the circadian pacemaker and the homeostat are normal but the circadian alertness signal is inadequate (Dantz et al. 1994). As narcolepsy is characterized by a deficient hypocretin system, it is possible that hypocretins are involved in the circadian alertness signal. The circadian alertness signal peaks during the late evening hours (late in the active portion of the daily cycle) in humans (Dijk & Czeisler 1994). This is the same circadian time at which Hcrt-1 peaks in the CSF (Fujiki et al. 2001), pons (Taheri et al. 2000), lateral hypothalamus (Yoshida et al. 2001), and medial thalamic area (Yoshida et al. 2001) of rats (late in the active portion of the daily cycle). Only in the preoptic area (Taheri et al. 2000) has hypocretin been found to peak at a different time (early during the inactive portion of the daily cycle). Fos immunoreactivity in hypocretin neurons in the rat also has a diurnal rhythm, with the peak occurring during the active portion of the daily cycle (Estabrooke et al. 2001). There are direct projections from the SCN to hypocretin neurons (Abrahamson et al. 2001), which suggests that hypocretin activity may be modulated by the biological clock. Hypocretin levels in the CSF, or measured using in vivo microdialysis, increase during the second half of the active period in rats under entrained conditions (Yoshida et al. 2001). Additional studies in constant darkness and in SCN-lesioned animals are required to establish if this fluctuation is driven by the SCN.

The role of hypocretins in the regulation of the homeostatic aspects of sleep remains to be explored. Fos experiments have indicated that hypocretin neurons may be activated by enforced wakefulness (Estabrooke et al. 2001). Short-term sleep deprivation does not alter prepro-hypocretin mRNA levels, but transcription may not reflect alterations in extracellular peptide release. A recent study using in vivo microdialysis has suggested that sleep deprivation increases hypocretin-1 release (Yoshida et al. 2001). It is interesting to note that sleep deprivation has antidepressant effects (Gillin 1983), which could reflect activated serotoninergic and adrenergic transmission through increased hypocretin tone. Prolonged sleep deprivation is associated with hyperphagia, hyperthermia, and a paradoxical increase in energy expenditure (Rechtschaffen & Bergmann 1995)—symptoms that may be secondary to enhanced hypocretin neurotransmission (Mignot 2001b).

One of the brain nuclei thought to be intimately involved in the regulation of alertness is the monoaminergic locus coeruleus (Aston-Jones et al. 1999). LC neurons in the rat have a higher discharge rate during the dark portion of their daily cycle, irrespective of locomotor activity (Aston-Jones et al. 2001). It is likely that this cyclicity is controlled by the hypothalamic SCN via the dorsomedial hypothalamus (DMH) and, possibly, hypocretin neurons (Aston-Jones et al. 2001, Abrahamson et al. 2001). The LC receives a dense hypocretinergic innervation (Date et al. 1999, Peyron et al. 1998, McGranaghan & Piggins 2001, Hagan et al. 1999, Horvath et al. 1999a, Moore et al. 2001). Most, if not all, LC neurons express hcrtr1, while hcrtr2 has not been found in the LC (Trivedi et al. 1998, Greco & Shiromani 2001, Bourgin et al. 2000). Both hypocretins can excite LC neurons in vitro (Hagan et al. 1999, Ivanov & Aston-Jones 2000, Horvath et al. 1999a). Microinjection of Hcrt-1 or -2 directly onto the LC in vivo also causes an increase in LC firing activity (Kiyashchenko et al. 2001, Bourgin et al. 2000). In the same way, ICV administration of either hypocretin can increase Fos protein expression in the LC (Date et al. 1999). At a systems level, one study has shown that direct microinjection of Hcrt-1 onto the LC leads to an acute increase in wakefulness with a corresponding decrease in active sleep in rats (Bourgin et al. 2000).

The LC is not the only neural structure involved in sleep/wake regulation that receives hypocretinergic innervation. Histaminergic (TMN), serotoninergic (dorsal and medial raphe nuclei—DR, MR), dopaminergic (VTA, SN), and cholinergic (basal forebrain, LDT, PPT) cell groups in the brain all receive hypocretinergic innervation to varying degrees (Date et al. 1999, Peyron et al. 1998, McGranaghan & Piggins 2001, Nakamura et al. 2000, Moore et al. 2001, Horvath et al. 1999a) and have hypocretin receptors (Marcus et al. 2001, Trivedi et al. 1998). Compared to the LC, the precise role of hypocretins in modulating the activities of these other nuclei has been less well studied. Microinjection of Hcrt-1 onto neurons in the preoptic area, a sleep-active group of neurons that project to the histaminergic TMN (Sherin et al. 1996), increases wakefulness and decreases both REM and slow-wave sleep (Methippara et al. 2000). Application of Hcrt-1 directly onto TMN neurons, which express only hcrtr2 (Marcus et al. 2001), results in an increase in wakefulness with a concomitant decrease in slow-wave and REM sleep in rats (Huang et al. 2001). The wake-promoting effect of Hcrt-1 infusion in the CSF is absent in histamine-1 receptor gene knockout mice, which suggests important downstream mediation by the histaminergic system (Huang et al. 2001). A significant decrease in histamine content in the cortex and thalamus, two structures important for histamine-mediated cortical arousal, has been found in hcrtr-2–mutated narcoleptic

Dobermans (Nishino et al. 2001). However, selective lesioning of TMN neurons that express hcrtr2 caused no change in sleep architecture (Gerashchenko et al. 2001a). Furthermore, mice that lack production of the histamine-1 receptor have normal sleep (Huang et al. 2001). Though histamine may have a role in regulating the sleep/wake cycle and hypocretin may stimulate histaminergic neurons, the interaction does not appear to be necessary for the occurrence of normal sleep.

Serotoninergic DR and MR neurons are well innervated by hypocretin neurons, and they express both hypocretin receptors. Hypocretins appear to have a direct, post-synaptic excitatory effect on raphe neurons (Brown et al. 2001). Furthermore, ICV administration of hypocretin peptides increases Fos immunoreactivity in the DR (Date et al. 1999). Hypocretins may also be able to act on downstream release of serotonin because ICV hypocretin injection can increase serotonin turnover in the hypothalamus and striatum (Jones et al. 2001). As with the serotoninergic raphe, the dopaminergic VTA and SN also receive hypocretin innervation (Horvath et al. 1999a) and express both hypocretin receptors (Marcus et al. 2001). Both hypocretin peptides evoke an increase in the activity of VTA neurons (Nakamura et al. 2000). Hypocretin may be involved in the regulation of dopaminergic control of locomotor activity because ICV Hcrt-1 results in stereotypy that is blocked by D2 receptor antagonists (Nakamura et al. 2000).

Another intriguing area of the brain involved in sleep/wake regulation and which receives substantial hypocretinergic innervation is the cholinergic septal area and LDT/PPT. Application of hypocretin-1 onto neurons of either the LDT in cats (Xi et al. 2001) or the horizontal limb of the diagonal band (septal) in rats (Thakkar et al. 2001) increases wakefulness and decreases in REM sleep, with a decrease in slow-wave sleep also occurring after septal application. Selective ablation of medial septal area cholinergic and noncholinergic neurons that express hypocretin receptors, however, did not significantly change sleep architecture or delta power during sleep or recovery sleep (delta power is frequently used as a measure of the strength of the sleep homeostat) (Gerashchenko et al. 2001b). There was, however, a complete abolition of the hippocampal theta rhythm during both REM sleep and exploratory wake behavior. As the theta rhythm is considered a function of memory track consolidation (McNaughton 2000), the hypocretins may modulate memory consolidation.

Initial evidence indicates that hypocretin is an excitatory neuropeptide able to activate multiple neuromodulatory systems involved in the regulation of sleep/wake behavior. The peak of the hypocretinergic stimulation appears to occur late in the activity cycle, a time at which the circadian alertness drive is at its peak. Recent models of sleep regulation (Kilduff & Peyron 2000, Hungs & Mignot 2001) have placed hypocretinergic innervation in a position such that it activates neuromodulatory cell groups during wakefulness and activates them less so during slow-wave sleep. The models differ in the role of hypocretin during REM sleep, with one model characterizing hypocretin neurons as nearly quiescent during REM sleep (Hungs & Mignot 2001) and the other characterizing them as active during REM sleep (Kilduff & Peyron 2000). A recent study reported that during carbachol-induced REM sleep, Fos immunoreactivity in the hypocretin neurons of the cat hypothalamus is very low, as compared with active wakefulness; this indicates that these cells may have reduced activity during REM sleep (Torterolo et al. 2001). Another provocative finding from this study was that Fos immunoreactivity in hypocretin neurons was nearly absent during quiet wakefulness. This would imply that hypocretin neurons may be involved more in the stimulation of activity than in the stimulation of wakefulness per se. If this were the case, the increase in hypocretin activity would lead to an increase in behavioral activity, possibly through dopaminergic, histaminergic, or other mechanisms, whose feedback would help to maintain wakefulness. This possibility, however, remains to be examined.

HYPOCRETINS AND DEPRESSION

Narcolepsy can often be misdiagnosed as depression for two main reasons: First, these two conditions share some of the symptoms, and second, up to a third of patients with narcolepsy suffer from depression. Depression is associated with shortened REM sleep latency and suppressed slow-wave sleep. It is interesting that sleep deprivation has antidepressant actions. Studies of the neurochemical basis of depression have concentrated on the role of classical neurotransmitters, especially the noradrenergic and serotoninergic pathways. However, the importance of neuropeptides in depression is increasingly recognized. It can be speculated that hypocretin neurons, with their dense innervation of brain monoaminergic systems, may have a role in the pathophysiology of human depression. Sleep deprivation also increases hypocretin levels in animals (Yoshida et al. 2001). The Wistar-Kyoto (WKY) rat model is hypoactive with reduced exploratory activity, hyponeophagic, and resistant to antidepressants. WKY rats also have increased REM sleep present from early in development (Dugovic & Turek 2001). This animal model of depression has global brain hypocretin deficiency that is especially marked in brain regions involved in the regulation of sleep and emotion (Taheri et al. 2001). These observations remain to be confirmed in other animal models of depression and in human cases. CSF hypocretin levels have been measured in a few human cases with depression and are not different from normal controls. However, it is possible that the hypocretins may be involved in particular types of depression such as seasonal depression. It would also be interesting to see whether CSF hypocretin levels are elevated in mania.

THE ROLE OF HYPOCRETINS IN FOOD INTAKE AND ENERGY EXPENDITURE

The location of hypocretin neurons in the lateral hypothalamus and early experiments with the hypocretins suggested that they have an important role in appetite regulation and energy expenditure. Lesioning of the lateral hypothalamus results in hypophagia (and weight loss) and hypodipsia. With larger lesions, death by starvation occurs in animals, but recovery is possible after smaller lesions. Though lesioning experiments may damage fiber bundles running through the lateral hypothalamus, it is likely that lateral hypothalamic neurons are involved in the regulation of food intake. This hypothalamic area contains not only hypocretin neurons but also distinct neurons expressing the orexigenic peptide melanin-concentrating hormone (MCH). When hypocretins were administered into the cerebral ventricle of rats (ICV), they dose-dependently stimulated food intake, and prepro-hypocretin mRNA was upregulated with fasting (Sakurai et al. 1998). There is also evidence for the inhibition of food intake by an hcrtr-1 antagonist, 1-(2-methylbenzoxazol-6-yl)-3-[1,5] naphthyridin-4-yl urea hydrochloride (SB-334867-A) (Haynes et al. 2000). The orexigenic effect of hypocretins may be through neuropeptide Y (NPY) (Jain et al. 2000, Dube et al. 2000, Yamanaka et al. 2000). Hypocretin neurons have reciprocal connections with NPY neurons in the hypothalamic arcuate nucleus (Horvath et al. 1999b).

Prepro-hypocretin gene knockout mice (Chemelli et al. 1999) consume 15% less food than wild-type mice but are not lean. It is interesting that knockout mice do not have abnormalities in basal metabolic rate, but they have decreased oxygen consumption during the dark phase. When degeneration of hypocretin neurons was induced by the insertion of polyglutamine repeats (ataxin-3 mutants) into the prepro-hypocretin gene in mice, this resulted in hypophagia; however, paradoxically, it also resulted in weight gain (Hara et al. 2001). This may be due to reduced energy expenditure, but further reports are awaited. It has also been proposed that because hypocretin neurons also produce the peptide dynorphin (an orexigenic peptide), the lack of dynorphin may be responsible for the hypophagia observed in this model (Scammell 2001).

Despite the above evidence, the importance of hypocretins in food intake is not entirely clear. Edwards et al. (1999) observed that ICV administration of Hcrt-1 stimulated food intake in rats but not as potently as NPY. Hcrt-2 is only occasionally (if at all) orexigenic. Though Hcrt-1 may increase food intake in the first 4 h after ICV administration, it actually decreases food intake in the subsequent 20 h (Taheri et al. 2000). Prepro-hypocretin mRNA is elevated with prolonged fasting, but immunoreactive peptide levels are unaltered (Taheri et al. 1999). Chronic ICV administration of Hcrt-1 does not result in obesity. Because the orexigenic activity of Hcrt-1 is mainly observed in the light (inactive) phase in rodents, it is possible that its orexigenic effects are more related to an increase in activity and arousal. It is also possible that the hypocretins are important in food intake only in particular circumstances, such as in response to hypoglycemia (Griffond et al. 1999, Cai et al. 1999). It has been proposed that hypocretin neurons are the previously reported, yet unidentified, glucose-sensitive neurons in the lateral hypothalamus. However, it should be borne in mind that most studies of the interaction between hypocretins and hypoglycemia have used very high doses of insulin to induce hypoglycemia, enough to activate the stress axis (see below).

Using indirect calorimetry, Hcrt-1, but not Hcrt-2, increases metabolic rate (Lubkin & Stricker-Krongrad 1998). Mice with targeted disruption of the prepro-hypocretin are hypophagic but are prone to diet-induced obesity, which could be due to their lower metabolic rate (Willie et al. 2001). However, these mice also have episodes of behavioral arrest (such as freezing, totter, and atonia), which may interfere with feeding and exercise. Although hypocretin neurons have receptors for the adipocyte-derived cytokine hormone leptin and are immunoreactive for STAT3, a transcription factor activated by leptin (Hakansson et al. 1999), the interaction of hypocretins and leptin is unlike that observed for other orexigenic peptides (Taheri & Bloom 2001). NPY mRNA is upregulated in leptin-deficient ob/ob and leptin receptor–deficient db/db mice, while prepro-orexin/hypocretin mRNA is downregulated but increases with starvation in these animals (Yamamoto et al. 2000). The possible interaction between the hypocretins and the gastric-derived hormone ghrelin (Yamanaka et al. 2001) is interesting and requires further investigation.

THE NEUROENDOCRINOLOGY OF HYPOCRETINS

Hypocretin fibers innervate hypothalamic regions involved in the regulation of pituitary hormone release and have a stimulatory effect on neuroendocrine and neuropeptide hypothalamic neurons in vitro (van Den Pol et al. 1998, Taheri & Bloom 2001). ICV administration of both hypocretin peptides dramatically decreases plasma prolactin and growth hormone (GH) while increasing plasma ACTH and corticosterone. These peptides may also alter pituitary thyroid stimulating hormone (TSH) secretion (Mitsuma et al. 1999, Jones et al. 2001), but this remains to be further investigated. Although Hcrt-2 has not been studied as rigorously as Hcrt-1, most studies to date report similar actions for the two peptides on pituitary hormone secretion. The inhibition of plasma prolactin by ICV Hcrt-1 appears to be partially independent of dopamine (Russell et al. 2000). Both hypocretins dose-dependently increase plasma ACTH and corticosterone. There is increased Fos mRNA in the PVN of the hypothalamus where corticotrophin-releasing factor (CRF) neurons are located (Hagan et al. 1999, Kuru et al. 2000, Ida et al. 2000). An in situ hybridization study has shown that in response to ICV Hcrt-1, both CRF and AVP mRNA levels are significantly increased in the parvocellular cells of the hypothalamic PVN (Al-Barazanji et al. 2001). Others have suggested that the effect of ICV hypocretins on the hypothalamo-pituitary-adrenal axis is through activation of NPY (Jaszberenyi et al. 2001). It is therefore likely that the hypocretins regulate the hypothalamo-pituitary-adrenal (HPA) axis, which is consistent with effects on increasing arousal. Preliminary results show that adrenalectomy or treatment with dexamethasone does not alter brain hypocretin content (S. Taheri, unpublished data). Hypocretins may not be part of the day-to-day regulation of the HPA axis but may be important in particular stress situations (e.g., hypoglycaemia or high arousal states). It remains to be determined whether the activation of the HPA axis is directly through activation of the PVN or indirectly via monoaminergic centers. The mechanism through which the hypocretins inhibit GH secretion is as yet undefined but could be a nonspecific response to the activation of the stress axis.

Hypocretin neurons may regulate LH secretion via gonadotropin-releasing hormone (GnRH) neurons in the hypothalamus. Pu et al. (1998) studied the effects of hypocretins on LH secretion in ovariectomized (ovx) and ovarian steroid-treated ovx rats. ICV administration of Hcrt-1 or Hcrt-2 rapidly stimulated LH secretion in ovx animals pretreated with estradiol and progesterone. However, these peptides inhibited LH release in unprimed ovx rats. The actions of the hypocretins on the gonadal axis are therefore dependent on the hormonal milieu (Taheri & Bloom 2001).

No anatomical definition for the neuroendocrine effects of the hypocretins has yet been carried out except for hypocretin and GnRH neurons in the ovine hypothalamus (Iqbal et al. 2001). Most studies to date have concentrated on ICV administration of pharmacological doses of hypocretins, which results in hypocretin action at several nuclei and does not determine the specificity of hypocretin action. Further, the physiological circumstances in which hypocretins regulate the neuroendocrine system and the existence of any negative or positive feedback loops remain to be studied. Also, these hormonal effects are not consistent with available human data that indicate decreased prolactin and growth hormone release during sleep but show relatively unaffected cortisol levels in human narcolepsy (Higushi et al. 1979). Reproductive capacity is also unaltered in human narcolepsy, and no reproductive problems have been reported in prepro-hypocretin knockout mice. It should be borne in mind that only a few small studies have investigated any possible association of narcolepsy with altered hormone secretion. These studies were also carried out at a time when the diagnosis of narcolepsy was perhaps less precise. It now remains for similar investigations to be carried out in patients with documented hypocretin deficiency. Also, because cataplexy occurs at times when immediate action is required or there is a high level of arousal, it would be interesting to know whether the stress response is altered with hypocretin deficiency. Reports are awaited regarding endocrine abnormalities (if any) in prepro-hypocretin and hypocretin receptor–gene knockout mice, and in ataxin-3 mice.

OTHER BIOLOGICAL ACTIONS OF HYPOCRETINS

The hypocretins have been implicated in the regulation of vagally mediated gastric acid secretion, sympathetic nervous system activation, cardiovascular function, and drinking behavior. ICV Hcrt-1 increases water intake and prepro-hypocretin mRNA is upregulated by water deprivation (Kunii et al. 1999). However, osmotic sensitivity of hypocretin neurons has not been investigated. Both hypocretin peptides, when injected ICV, increase blood pressure and heart rate in rats (Shirasaka et al. 1999, Samson et al. 1999). This is consistent with the arousal effect of these peptides. In the spinal cord, the hypocretins may have a modulatory role in the sensory and autonomic nervous systems; in particular, they may be involved in pain mechanisms (van Den Pol 1999, Bingham et al. 2001). There are several reports on the effect of hypocretins on peripheral tissues such as the gastrointestinal tract, the pancreas, and adrenal gland (Kirchgessner & Liu 1999, Malendowicz et al. 1999,

Nowak et al. 2000, Kirchgessner 2002). However, although hypocretin receptors have been detected in several tissues (Johren et al. 2001, Kirchgessner 2002), the presence of hypocretin peptides has only occasionally been reported outside the CNS.

CONCLUDING REMARKS

The development of the hypocretin/narcolepsy field has been extraordinary in many respects. Only three years have elapsed between the discovery of these peptides and our understanding of the cause of a human sleep disorder that had puzzled clinicians for over a century. Hundreds of genes and peptides are currently known, yet their functional characterization is lagging behind or, if available, is often not shedding light on any specific physiology. The hypocretins are clearly outstanding molecules with strong physiological effects whose importance is only beginning to be appreciated. The combination of persistence in defining the phenotype of the prepro-hypocretin gene knockout mouse and discovering the cause of a canine model of narcolepsy culminated in the discovery of the basis of human narcolepsy, bringing together scientists with different interests and technical expertise.

Many important clinical questions remain unanswered. The true cause of narcolepsy—whether it is an autoimmune disorder—remains to be established. Hypocretin-based therapies also need to be explored. Some investigators have suggested limited penetration of hypocretins in the CNS after peripheral administration, so high doses of peptides may have rescuing properties in ligand-deficient animals. Hypocretin receptor agonists could be an excellent treatment, but agonists for peptide receptors have proved notoriously difficult to develop. Gene-based therapies and cell transplantation may be viable treatment options, depending on whether distant hypocretin projections are needed for functional rescue. The role of hypocretins in other disorders also warrants further investigations. For example, the findings of lower CSF levels in head trauma and encephalitis suggest additional therapeutic possibilities. It is also possible that hypocretin receptor antagonists will have excellent hypnotic properties or that hypocretins, with their intimate connection to monoaminergic systems, will be involved in depression.

At the more basic level, the field now needs to shift emphasis from appetite regulation to the regulation of sleep and energy metabolism. Studies are now indicating changes of hypocretin transmission with circadian timing and sleep deprivation. Other investigators have suggested increased transmission during active wakefulness, suggesting that hypocretin may also be involved in promoting brain activity even within the wakefulness state. Hypocretins promote energy consumption, produce hyperthermia, and stimulate sympathetic tone after central administration. These observations, together with previous work, indicate that the hypocretins are controlled at least partially by peripheral nutritional signals and suggest an integration of sleep and energy metabolism in the hypothalamus. In contrast to nutritional signals, however, very little is known about what is signaling sleep-deprivation and circadian-related effects on hypocretin neurotransmission.

Studies also need to be conducted that aim to understand how hypocretins have such a wide range of pharmacological effects. Most studies have used ICV administration of high doses that may not reflect physiological actions. The majority of the work has focused on few output structures—most notably the LC and other monoaminergic nuclei. It is unlikely that all the effects of hypocretins on sleep and wakefulness are mediated by a single downstream neurotransmitter system such as dopamine, norepinephine, or histamine. Clearly, the effects of other downstream systems in the hypothalamus and elsewhere need to be defined.

At the practical level, a revolution is now occurring in the diagnosis of narcolepsy. Some patients can now be diagnosed early with high accuracy and low cost, which opens the possibility of intervention to stop progression of the disease in these cases. The finding of such a simple process (the lack of a few thousand cells among billions) causing such a profound disease condition uncovers the possibility that similar discoveries will be made in the future in the vast field of neuropsychiatric disorders, many of which have a complex genetic basis like that of human narcolepsy.

The *Annual Review of Neuroscience* is online at http://neuro.annualreviews.org

LITERATURE CITED

Abrahamson EE, Leak RK, Moore RY. 2001. The suprachiasmatic nucleus projects to posterior hypothalamic arousal systems. *Neuro. Rep.* 12(2):435–40

Adie WJ. 1926. Idiopathic narcolepsy: a disease sui generis; with remarks on the mechanisms of sleep. *Brain* 49:257–306

Al-Barazanji KA, Wilson S, Baker J, Jessop DS, Harbuz MS. 2001. Central orexin-A activates hypothalamic-pituitary-adrenal axis and stimulates hypothalamic corticotropin releasing factor and arginine vasopressin neurones in conscious rats. *J. Neuroendocrinol.* 13(5):421–24

Aldrich MS. 1998. Diagnostic aspects of narcolepsy. *Neurology* 50(2 Suppl 1):S2–7

American Sleep Disorders Association (ASDA). 1997. *The International Classification of Sleep Disorders, Revised: Diagnostic and Coding Manual.* Rochester: Am. Sleep Disord. Assoc.

Aston-Jones G, Chen S, Zhu Y, Oshinsky ML. 2001. A neural circuit for circadian regulation of arousal. *Nat. Neurosci.* 4:732–38

Aston-Jones G, Rajkowski J, Cohen J. 1999. Role of locus coeruleus in attention and behavioral flexibility. *Biol. Psych.* 46:1309–20

Baker TL, Dement WC. 1985. Canine narcolepsy-cataplexy syndrome: evidence for an inherited monoaminergic-cholinergic imbalance. In *Brain Mechanisms of Sleep*, ed. DJ McGinty, R Drucker-Colin, A Morisson, PL Parmengiani, pp. 199–233. New York: Raven

Bassetti C, Aldrich MS. 1997. Idiopathic hypersomnia. A series of 42 patients. *Brain* 120 (Pt. 8):1423–35

Bassetti CL, Gugger M, Mathis J, Sturznegger C, Radanov B, et al. 2001. Cerebrospinal fluid levels of hypocretin (orexin) in hypersomnolent patients without cataplexy. *Actas Fisiolog.* 7:205

Bernasconi R, Mathivet P, Bischoff S, Marescaux C. 1999. Gamma-hydroxybutyric acid: an endogenous neuromodulator with abuse potential? *Trends Pharmacol. Sci.* 20(4): 135–41

Bingham S, Davey PT, Babbs AJ, Irving EA, Sammons MJ, et al. 2001. Orexin-A, an hypothalamic peptide with analgesic properties. *Pain* 92(1–2):81–90

Blanco M, Lopez M, Garcia-Caballero T, Gallego R, Vazquez-Boquete A, et al. 2001. Cellular localization of orexin receptors in human pituitary. *J. Clin. Endocrinol. Metab.* 86(7):1616–19

Bourgin P, Huitrón-Reséndiz S, Spier AD, Fabre V, Morte B, et al. 2000. Hypocretin-1 modulates rapid eye movement sleep through activation of locus coeruleus neurons. *J. Neurosci.* 20:7760–65

Broughton WA, Broughton RJ. 1994. Psychosocial impact of narcolepsy. *Sleep* 17(Suppl): S45–49

Brown RE, Sergeeva O, Eriksson KS, Haas HL. 2001. Orexin A excites serotonergic neurons in the dorsal raphe nucleus of the rat. *Neuropharmacology* 40:457–59

Cai XJ, Widdowson PS, Harrold J, Wilson S, Buckingham RE, et al. 1999. Hypothalamic orexin expression: modulation by blood glucose and feeding. *Diabetes* 48(11):2132–37

Chemelli RM, Sinton CM, Yanagisawa M. 2000. Polysomnographic characterization of Orexin-2 receptor knockout mice. *Sleep* 23(Suppl. 2):A296–29

Chemelli RM, Willie JT, Sinton CM, Elmquist JK, Scammell T, et al. 1999. Narcolepsy in orexin knockout mice: molecular genetics of sleep regulation. *Cell* 98:437–51

Chou TC, Lee CE, Lu J, Elmquist JK, Hara J, et al. 2001. Orexin (hypocretin) neurons contain dynorphin. *J. Neurosci.* 21(19):RC168

Dantz B, Edgar DM, Dement WC. 1994. Circadian rhythms in narcolepsy: studies on a 90 minute day. *Electroencephalogr. Clin. Neurophysiol.* 90:24–35

Date Y, Ueta Y, Yamashita H, Yamaguchi H, Matsukura S, et al. 1999. Orexins, orexigenic hypothalamic peptides, interact with autonomic, neuroendocrine and neuroregulatory systems. *Proc. Natl. Acad. Sci. USA* 96:748–53

de Lecea L, Kilduff TS, Peyron C, Gao X, Foye PE, et al. 1998. The hypocretins: hypothalamus-specific peptides with neuroexcitatory activity. *Proc. Natl. Acad. Sci. USA* 95:322–27

Dijk DJ, Czeisler CA. 1994. Paradoxical timing of the circadian rhythm of sleep propensity serves to consolidate sleep and wakefulness in humans. *Neurosci. Lett.* 166:63–68

Dube MG, Horvath TL, Kalra PS, Kalra SP. 2000. Evidence of NPY Y5 receptor involvement in food intake elicited by orexin A in sated rats. *Peptides* 21(10):1557–60

Dugovic C, Turek FW. 2001. Similar genetic mechanisms may underlie sleep-wake states in neonatal and adult rats. *Neuro. Rep.* 12 (14):3085–89

Duxon MS, Stretton J, Starr K, Jones DNC, Holland V, et al. 2001. Evidence that orexin-A-evoked grooming in the rat is mediated by orexin-1 (OX_1) receptors, with downstream $5-HT_{2C}$ receptor involvement. *Psychopharmacology* 153:203–9

Edwards CM, Abusnana S, Sunter D, Murphy KG, Ghatei MA, Bloom SR. 1999. The effect of the orexins on food intake: comparison with neuropeptide Y, melanin-concentrating hormone and galanin. *J. Endocrinol.* 160: R7–12

Eriksson KS, Sergeeva O, Brown RE, Haas HL. 2001. Orexin/hypocretin excites the histaminergic neurons of the tuberomammillary nucleus. *J. Neurosci.* 21(23):9273–79

Estabrooke IV, McCarthy MT, Ko E, Chou TC, Chemelli RM, et al. 2001. Fos expression in orexin neurons varies with behavioral state. *J. Neurosci.* 21(5):1656–62

Fujiki N, Yoshida Y, Ripley B, Honda K, Mignot E, Nishino S. 2001. Changes in CSF hypocretin-1 (orexin A) levels in rats across 24 hours and in response to food deprivation. *Neuro. Rep.* 12(5):993–97

Gerashchenko D, Kohls MD, Greco MA, Waleh NS, Salin-Pascual R, et al. 2001a. Hypocretin-2-saporin lesions of the lateral hypothalamus produce narcoleptic-like sleep behavior in the rat. *J. Neurosci.* 21:7273–83

Gerashchenko D, Salin-Pascual R, Shiromani PJ. 2001b. Effects of hypocretin-saporin injections into the medial septum on sleep and hippocampal theta. *Brain Res.* 913:106–15

Gillin JC. 1983. The sleep therapies of depression. *Prog. Neuropsychopharmacol. Biol. Psychiatry* 7(2–3):351–64

Greco MA, Shiromani PJ. 2001. Hypocretin receptor protein and mRNA expression in the dorsolateral pons of rats. *Mol. Brain Res.* 88:176–82

Guilleminault C, Faull KF, Miles L, van den Hoed J. 1983. Posttraumatic excessive daytime sleepiness: a review of 20 patients. *Neurology* 33(12):1584–89

Guilleminault C, Mondini S. 1986. Mononucleosis and chronic daytime sleepiness. A long-term follow-up study. *Arch. Intern. Med.* 146(7):1333–35

Hagan JJ, Leslie RA, Patel S, Evans ML, Wattam TA, et al. 1999. Orexin A activates locus coeruleus cell firing and increases arousal in the rat. *Proc. Natl. Acad. Sci. USA* 96:10911–16

Hakansson M, de Lecea L, Sutcliffe JG, Yanagisawa M, Meister B. 1999. Leptin receptor- and STAT3-immunoreactivities in hypocretin/orexin neurons of the lateral hypothalamus. *J. Neuroendocrinol.* 11:653–63

Hara J, Beuckmann CT, Nambu T, Willie JT, Chemelli RM, et al. 2001. Genetic ablation of orexin neurons in mice results in narcolepsy, hypophagia, and obesity. *Neuron* 30(2):345–54

Haynes AC, Jackson B, Chapman H, Tadayyon M, Johns A, et al. 2000. A selective orexin-1 receptor antagonist reduces food consumption in male and female rats. *Regul. Pept.* 96(1–2):45–51

Higushi T, Takahashi Y, Takahashi K, Nimi Y, Miyasita A. 1979. Twenty-four hour secretatory patterns of growth hormone, prolactin, and cortisol in narcolepsy. *J. Clin. Endocrinol. Metab.* 49(2):197–204

Honda Y, Doi Y, Ninomiya R, Ninomiya C. 1986. Increased frequency of non-insulin-dependent diabetes mellitus among narcoleptic patients. *Sleep* 9:254–59

Honda Y, Matuski K. 1990. Genetic aspects of narcolepsy. In *Handbook of Sleep Disorders*, ed. MJ Thorpy, pp. 217–34. New York: Dekker

Horvath TL, Diano S, van Den Pol AN. 1999b. Synaptic interaction between hypocretin (orexin) and neuropeptide Y cells in the rodent and primate hypothalamus: a novel circuit implicated in metabolic and endocrine regulations. *J. Neurosci.* 19:1072–87

Horvath TL, Peyron C, Diano S, Ivanov A, Aston-Jones G, et al. 1999a. Hypocretin (orexin) activation and synaptic innervation of the locus coeruleus noradrenergic system. *J. Comp. Neurol.* 415:145–59

Huang Z-L, Qu W-M, Li W-D, Mochizuki T, Eguchi N, et al. 2001. Arousal effect of orexin A depends on activation of the histaminergic system. *Proc. Natl. Acad. Sci. USA* 98:9965–70

Hungs M, Mignot E. 2001. Hypocretin/orexin, sleep and narcolepsy. *Bioessays* 23(5):397–408

Ida T, Nakahara K, Katayama T, Murakami N, Nakazato M. 1999. Effect of lateral cerebroventricular injection of the appetite-stimulating neuropeptide, orexin and neuropeptide Y, on the various behavioral activities of rats. *Brain Res.* 821(2):526–29

Ida T, Nakahara K, Murakami T, Hanada R, Nakazato M, et al. 2000. Possible involvement of orexin in the stress reaction in rats. *Biochem. Biophys. Res. Commun.* 270:318–23

Iqbal J, Pompolo S, Sakurai T, Clarke IJ. 2001. Evidence that orexin-containing neurones provide direct input to gonadotropin-releasing hormone neurons in the ovine hypothalamus. *J. Neuroendocrinol.* 13(12):1033–41

Ivanov A, Aston-Jones G. 2000. Hypocretin/orexin depolarizes and decreases potassium conductance in locus coeruleus neurons. *Neuro. Rep.* 11:1755–58

Jain MR, Horvath TL, Kalra PS, Kalra SP. 2000. Evidence that NPY Y1 receptors are involved in stimulation of feeding by orexins (hypocretins) in sated rats. *Regul. Pept.* 87(1–3):19–24

Jaszberenyi M, Bujdoso E, Telegdy G. 2001. The role of neuropeptide Y in orexin-induced hypothalamic-pituitary-adrenal activation. *J. Neuroendocrinol.* 13(5):438–41

Johren O, Neidert SJ, Kummer M, Dendorfer A, Dominiak P. 2001. Prepro-orexin and orexin receptor mRNAs are differentially expressed in peripheral tissues of male and female rats. *Endocrinology* 142(8):3324–31

Jones DNC, Gartlon J, Parker F, Taylor SG, Routledge C, et al. 2001. Effects of centrally administered orexin-B and orexin-A: a role for orexin-1 receptors in orexin-B-induced hyperactivity. *Psychopharmacology* 153:210–18

Kastin AJ, Akerstrom V. 1999. Orexin A but not orexin B rapidly enters brain from blood by simple diffusion. *J. Pharmacol. Exp. Ther.* 289:219–23

Kilduff TS, Peyron C. 2000. The hypocretin/orexin ligand-receptor system: implications for sleep and sleep disorders. *Trends Neurosci.* 23:359–65

Kirchgessner AL. 2002. Orexins in the brain-gut axis. *Endocr. Rev.* 23(1):1–15

Kirchgessner AL, Liu M. 1999. Orexin synthesis and response in the gut. *Neuron* 24:941–51

Kiyashchenko LI, Mileykovskiy BY, Lai Y-Y, Siegel JM. 2001. Increased and decreased muscle tone with orexin (hypocretin) microinjections in the locus coeruleus and pontine inhibitory area. *J. Neurophysiol.* 85:2008–16

Kummer M, Neidert SJ, Johren O, Dominiak P. 2001. Orexin (hypocretin) gene expression in rat ependymal cells. *Neuro. Rep.* 12(10):2117–20

Kunii K, Yamanaka A, Nambu T, Matsuzaki I, Goto K, Sakurai T. 1999. Orexins/hypocretins regulate drinking behaviour. *Brain Res.* 842:256–61

Kuru M, Ueta Y, Serino R, Nakazato M, Yamamoto Y, et al. 2000. Centrally administered orexin/hypocretin activates HPA axis in rats. *Neuro. Rep.* 11:1977–80

Lankford DA, Wellman JJ, O'Hara C. 1994. Posttraumatic narcolepsy in mild to moderate closed head injury. *Sleep* 17(8 Suppl.):S25–28

Lin L, Faraco J, Li R, Kadotani H, Rogers W, et al. 1999. The sleep disorder canine narcolepsy is caused by a mutation in the hypocretin (orexin) receptor 2 gene. *Cell* 98:365–76

Lin L, Hungs M, Mignot E. 2001. Narcolepsy and the HLA region. *J. Neuroimmunol.* 117(1–2):9–20

Lopez M, Senaris R, Gallego R, Garcia-Caballero T, Lago F, et al. 1999. Orexin receptors are expressed in the adrenal medulla of the rat. *Endocrinology* (12):5991–94

Lubkin M, Stricker-Krongrad A. 1998. Independent feeding and metabolic actions of orexins in mice. *Biochem. Biophys. Res. Commun.* 253:241–45

Marcus JN, Aschkenasi CJ, Lee CE, Chemelli RM, Saper CB, et al. 2001. Differential expression of orexin receptors 1 and 2 in the rat brain. *J. Comp. Neurol.* 435:6–25

McCarley RW, Massaquoi SG. 1992. Neurobiological structure of the revised limit cycle reciprocal interaction model of REM cycle control. *J. Sleep Res.* 1(2):132–37

McGranaghan PA, Piggins HD. 2001. Orexin A-like immunoreactivity in the hypothalamus and thalamus of the Syrian hamster (*Mesocricetus auratus*) and Siberian hamster (*Phodopus sungorus*), with special reference to circadian structures. *Brain Res.* 904:234–44

McNaughton BL. 2000. Memory trace reactivation in hippocampal and neocortical neuronal ensembles. *Curr. Opin. Neurobiol.* 10:180–86

Malendowicz LK, Tortorella C, Nussdorfer GG. 1999. Orexins stimulate corticosterone secretion of rat adrenocortical cells, through the activation of the adenylate cyclase-dependent signaling cascade. *J. Steroid Biochem. Mol. Biol.* 70:185–88

Methippara MM, Alam MN, Szymusiak R, McGinty D. 2000. Effects of lateral preoptic area application of orexin-A on sleep-wakefulness. *Neuro. Rep.* 11:3423–26

Mignot E. 1998. Genetic and familial aspects of narcolepsy. *Neurology* 50(Suppl. 1): S16–22

Mignot E. 2001a. A hundred years of narcolepsy research. *Arch. Ital. Biol.* 139(3):207–20

Mignot E. 2001b. A commentary on the neurobiology of the hypocretin/orexin system. *Neuropsychopharmacology* 25(5 Suppl. 1): S5–13

Mignot E, Lin L, Rogers W, Honda Y, Qiu X, et al. 2001. Complex HLA-DR and -DQ interactions confer risk of narcolepsy-cataplexy in three ethnic groups. *Am. J. Hum. Genet.* 68(3):686–99

Mignot E, Nishino S, Sharp LH, Arrigoni J, Siegel JM, et al. 1993b. Heterozygosity at the canarc-1 locus can confer susceptibility for narcolepsy: induction of cataplexy in heterozygous asymptomatic dogs after administration of a combination of drugs acting on monoaminergic and cholinergic systems. *J. Neurosci.* 13(3):1057–64

Mignot E, Renaud A, Nishino S, Arrigoni J, Guilleminault C, Dement WC. 1993a. Canine cataplexy is preferentially controlled by adrenergic mechanisms: evidence using monoamine selective uptake inhibitors and release enhancers. *Psychopharmacology* (Berl.) 113(1):76–82

Miller JD, Faull KF, Bowersox SS, Dement WC. 1990. CNS monoamines and their metabolites in canine narcolepsy: a replication study. *Brain Res.* 509(1):169–71

Mitsuma T, Hirooka Y, Mori Y, Kayama M, Adachi K, et al. 1999. Effects of orexin A on thyrotropin-releasing hormone and thyrotropin secretion in rats. *Horm. Metab. Res.* 31(11):606–9

Moore RY, Abrahamson EA, van den Pol A. 2001. The hypocretin neuron system: an arousal system in the human brain. *Arch. Ital. Biol.* 139:195–205

Mullett MA, Billington CJ, Levine AS, Kotz CM. 2000. Hypocretin I in the lateral hypothalamus activates key feeding-regulatory brain sites. *Neuro. Rep.* 11:103–8

Nakamura T, Uramura K, Nambu T, Yada T, Goto K, et al. 2000. Orexin-induced hyperlocomotion and stereotypy are mediated by the dopaminergic system. *Brain Res.* 873:181–87

Nauta JH. 1946. Hypothalamic regulation of sleep in rats. An experimental study. *J. Neurophysiol.* 9:285–316

Nishino S, Fujiki N, Ripley B, Sakurai E, Kato M, et al. 2001. Decreased brain histamine content in hypocretin/orexin receptor-2 mutated narcoleptic dogs. *Neurosci. Lett.* 313(3):125–28

Nishino S, Mao J, Sampathkumaran R, Shelton J, Mignot E. 1998. Increased dopaminergic transmission mediates the wake-promoting effects of CNS stimulants. *Sleep Res. Online* 1(1):49–61

Nishino S, Mignot E. 1997. Pharmacological aspects of human and canine narcolepsy. *Prog. Neurobiol.* 52:27–78

Nishino S, Ripley B, Overeem S, Lammers GJ, Mignot E. 2000. Hypocretin (orexin) deficiency in human narcolepsy. *Lancet* 355:39–40

Nowak KW, Mackowiak P, Switonska MM, Fabis M, Malendowicz LK. 2000. Acute orexin effects on insulin secretion in the rat: in vivo and in vitro studies. *Life Sci.* 66(5):449–54

Overeem S, Mignot E, Gert van Dijk J, Lammers GJ. 2001. Narcolepsy: clinical features, new pathophysiologic insights, and future perspectives. *J. Clin. Neurophysiol.* 18(2):78–105

Peyron C, Faraco J, Rogers W, Ripley B, Overeem S, et al. 2000. A mutation in a case of early onset narcolepsy and a generalized absence of hypocretin peptides in human narcoleptic brains. *Nat. Med.* 6(9):991–97

Peyron C, Tighe DK, van Den Pol AN, de Lecea L, Heller HC, et al. 1998. Neurons containing hypocretin (orexin) project to multiple neuronal systems. *J. Neurosci.* 18:9996–10015

Piper DC, Upton N, Smith MI, Hunter AJ. 2000. The novel brain neuropeptide, orexin-A, modulates the sleep-wake cycle of rats. *Eur. J. Neurosci.* 12:726–30

Pu S, Jain MR, Kalra PS, Kalra SP. 1998. Orexins, a novel family of hypothalamic neuropeptides, modulate pituitary luteinizing hormone secretion in an ovarian steroid-dependent manner. *Regul. Pept.* 78:133–36

Randeva HS, Karteris E, Grammatopoulos D, Hillhouse EW. 2001. Expression of orexin-A and functional orexin type 2 receptors in the human adult adrenals: implications for adrenal function and energy homeostasis. *J. Clin. Endocrinol. Metab.* 86(10):4808–13

Rechtschaffen A, Bergmann BM. 1995. Sleep

deprivation in the rat by the disk-over-water method. *Behav. Brain Res.* 69(1–2):55–63

Ripley B, Fujiki N, Okura M, Mignot E, Nishino S. 2001a. Hypocretin levels in sporadic and familial canine narcolepsy. *Neurobiol. Dis.* 8(3):525–34

Ripley B, Overeem S, Fujiki N, Nevsimalova S, Uchino M, et al. 2001b. CSF hypocretin/orexins levels in narcolepsy and other neurological conditions. *Neurology* 7(12):2253–58

Russell SH, Kim MS, Small CJ, Abbott CR, Morgan DG, et al. 2000. Central administration of orexin A suppresses basal and domperidone stimulated plasma prolactin. *J. Neuroendocrinol.* 12(12):1213–18

Sakurai T, Amemiya A, Ishii M, Matsuzaki I, Chemelli RM, et al. 1998. Orexins and orexin receptors: a family of hypothalamic neuropeptides and G protein-coupled receptors that regulate feeding behavior. *Cell* 92:573–85

Sakurai T, Moriguchi T, Furuya K, Kajiwara N, Nakamura T, et al. 1999. Structure and function of human prepro-orexin gene. *J. Biol. Chem.* 274:17771–76

Samson WK, Gosnell B, Chang JK, Resch ZT, Murphy TC. 1999. Cardiovascular regulatory actions of the hypocretins in brain. *Brain Res.* 831:248–53

Scammell TE. 2001. Wakefulness: an eye-opening perspective on orexin neurons. *Curr. Biol.* 11(19):R769–71

Schuld A, Hebebrand J, Geller F, Pollmacher T. 2000. Increased body-mass index in patients with narcolepsy. *Lancet* 355:1274–75

Sherin JE, Shiromani PJ, McCarley RW, Saper CB. 1996. Activation of ventrolateral preoptic neurons during sleep. *Science* 271:216–19

Shirasaka T, Nakazato M, Matsukura S, Takasaki M, Kannan H. 1999. Sympathetic and cardiovascular actions of orexins in conscious rats. *Am. J. Physiol. Regul. Integr. Comp. Physiol.* 277:R1780–85

Sunter D, Morgan I, Edwards CMB, Dakin CL, Murphy KG, et al. 2001. Orexins: effects on behavior and localisation of orexin receptor 2 messenger ribonucleic acid in the rat brainstem. *Brain Res.* 907:27–34

Sutcliffe JG, de Lecea L. 2000. The hypocretins: excitatory neuromodulatory peptides for multiple homeostatic systems, including sleep and feeding. *J. Neurosci. Res.* 62(2):161–68

Szymusiak R, Steininger T, Alam N, McGinty D. 2001. Preoptic area sleep-regulating mechanisms. *Arch. Ital. Biol.* 139(1–2):77–92

Taheri S, Bloom S. 2001. Orexins/hypocretins: waking up the scientific world. *Clin. Endocrinol. (Oxf)* 54(4):421–29

Taheri S, Gardiner J, Hafizi S, Murphy K, Dakin C, et al. 2001. Orexin A immunoreactivity and preproorexin mRNA in the brain of Zucker and WKY rats. *Neuro. Rep.* 12(3):459–64

Taheri S, Mahmoodi M, Opacka-Juffry J, Ghatei MA, Bloom SR. 1999. Distribution and quantification of immunoreactive orexin A in rat tissues. *FEBS Lett.* 457:157–61

Taheri S, Sunter D, Dakin C, Moyes S, Seal L, et al. 2000. Diurnal variation in orexin A immunoreactivity and prepro-orexin mRNA in the rat central nervous system. *Neurosci. Lett.* 279:109–12

Thakkar MM, Ramesh V, Strecker RE, McCarley RW. 2001. Microdialysis perfusion of orexin-A in the basal forebrain increases wakefulness in freely behaving rats. *Arch. Ital. Biol.* 139:313–28

Thannickal TC, Moore RY, Nienhuis R, Ramanathan L, Gulyani S, et al. 2000. Reduced number of hypocretin neurons in human narcolepsy. *Neuron* 27:469–74

Torterolo P, Yamuy J, Sampogna S, Morales FR, Chase MH. 2001. Hypothalamic neurons that contain hypocretin (orexin) express *c-fos* during active wakefulness and carbachol-induced active sleep. *Sleep Res. Online* 4:25–32

Trivedi P, Yu H, MacNeil DJ, Van der Ploeg LH, Guan XM. 1998. Distribution of orexin receptor mRNA in the rat brain. *FEBS Lett.* 438:71–75 van Den Pol AN. 1999. Hypothalamic hypocretin (orexin): robust innervation of the spinal cord. *J. Neurosci.* 19:3171–82

van Den Pol AN, Gao XB, Obrietan K, Kilduff TS, Belousov AB. 1998. Presynaptic and postsynaptic actions and modulation of neuroendocrine neurons by a new hypothalamic peptide, hypocretin/orexin. *J. Neurosci.* 18:7962–71

van Economo C. 1930. Sleep as a problem of localization. *J. Nerv. Ment. Dis.* 71:249–59

Willie JT, Chemelli RM, Sinton CM, Yanagisawa M. 2001. To eat or to sleep? Orexin in the regulation of feeding and wakefulness. *Annu. Rev. Neurosci.* 24:429–58

Wisor JP, Nishino S, Sora I, Uhl GH, Mignot E, Edgar DM. 2001. Dopaminergic role in stimulant-induced wakefulness. *J. Neurosci.* 21(5):1787–94

Xi M-C, Morales FR, Chase MH. 2001. Effects on sleep and wakefulness of the injection of hypocretin-1 (orexin-A) into the laterodorsal tegmental nucleus of the cat. *Brain Res.* 901:259–64

Yamamoto Y, Ueta Y, Serino R, Nomura M, Shibuya I, Yamashita H. 2000. Effects of food restriction on the hypothalamic prepro-orexin gene expression in genetically obese mice. *Brain Res. Bull.* 51:515–21

Yamanaka A, Kunii K, Nambu T, Tsujino N, Sakai A, et al. 2000. *Brain Res.* 859(2):404–9

Yamanaka A, Tsujino M, Masu M, Yamamoto M, Katsumoto T, et al. 2001. The electrophysiological analysis of orexins neurons. *Soc. Neurosci.* (Abstr.) Vol. 27 Program No. 635.11

Yoshida Y, Fujiki N, Nakajima T, Ripley B, Matsumura H, et al. 2001. Fluctuation of extracellular hypocretin-1 (orexin A) levels in the rat in relation to the light-dark cycle and sleep-wake activities. *Eur. J. Neurosci.* 14:1075–81

Yoshimichi G, Yoshimatsu H, Masaki T, Sakata T. 2001. Orexin-A regulates body temperature in coordination with arousal status. *Exp. Biol. Med.* 226:468–76

# A DECADE OF MOLECULAR STUDIES OF FRAGILE X SYNDROME

William T. O'Donnell and Stephen T. Warren
Howard Hughes Medical Institute and Departments of Human Genetics, Pediatrics, and Biochemistry, Emory University School of Medicine, Atlanta, Georgia 30322; email: wodonne@learnlink.emory.edu; swarren@emory.edu

Key Words mental retardation, trinucleotide repeat, RNA binding, dendritic spine, synaptic plasticity

■ **Abstract** Fragile X syndrome is one of the most common forms of inherited mental retardation. In most cases the disease is caused by the methylation-induced transcriptional silencing of the *fragile X mental retardation 1* (*FMR1*) gene that occurs as a result of the expansion of a CGG repeat in the gene's 5′UTR and leads to the loss of protein product fragile X mental retardation protein (FMRP). FMRP is an RNA binding protein that associates with translating polyribosomes as part of a large messenger ribonucleoprotein (mRNP) and modulates the translation of its RNA ligands. Pathological studies from the brains of patients and from *Fmr1* knockout mice show abnormal dendritic spines implicating FMRP in synapse formation and function. Evidence from both in vitro and in vivo neuronal studies indicates that FMRP is located at the synapse and the loss of FMRP alters synaptic plasticity. As synaptic plasticity has been implicated in learning and memory, analysis of synapse abnormalities in patients and *Fmr1* knockout mice should prove useful in studying the pathogenesis of fragile X syndrome and understanding learning and cognition in general.

If an appreciable portion of the total variance (in IQ) is due to sex linked genes, it is of more importance that a boy should have a clever mother than a clever father.

Hogben 1932 (quoted in Lehrke 1974)

INTRODUCTION

Evidence that mental retardation has a sex-linked component came as early as the turn of the twentieth century. In the population at large and in prison populations, Johnson (1897), Penrose (1938), and others found an excess of males with mental retardation compared to females (reviewed in Lehrke 1974). At the time, this preponderance of males was thought to be due to ascertainment bias because males were more likely to be institutionalized and the expectations placed upon them by society were higher. There was evidence of an excess of males with high IQ, of an excess of retarded male sibships, and that the IQ of the mother has more effect

0147-006X/02/0721-0315$14.00

on mental retardation in sons than does the IQ of the father. These data led Lehrke (1974) to argue that genes affecting intelligence are located on the X chromosome and to estimate that one fourth of mental retardation can be traced to X-linked factors. Indeed, as early as 1943, a pedigree reported by Martin & Bell showed mental retardation segregating as an X-linked recessive gene, and many others showed the same (Losowsky 1961, Dunn et al. 1963, Lubs 1969).

Lubs (1969) noticed a peculiar chromosomal variant that segregated with mental retardation over three generations. He described a constriction near the end of the long arm of the X chromosome apparent in metaphase spreads from four mentally retarded males and one normal female. This variant would later be localized to Xq27.3 (Harrison et al. 1983) and become known as a fragile X chromosome. Some early studies confirmed the link between the fragile X and mental retardation (Giraud et al. 1976, Harvey et al. 1977), but many others were negative. Sutherland (1977, 1979) showed that culture media deficient in folic acid and thymidine are required for fragile site expression. He called for repetition of previous cytological studies of families with X-linked mental retardation. Males from the original Martin & Bell pedigree were found to express the fragile site (Richards et al. 1981), and mental retardation associated with a fragile X chromosome was then termed the Martin-Bell syndrome, now known as fragile X syndrome.

With the advent of a cytogenetic marker for some cases of X-linked mental retardation, and as a number of families with the fragile X chromosome were identified, the syndromic nature of the disorder became more apparent. Studies in the late 1970s and 1980s led to the current clinical picture of fragile X syndrome. The primary attributes of an affected male are moderate to severe mental retardation (Bennetto & Pennington 1996), macroorchidism (Turner et al. 1980), and a connective tissue dysplasia leading to a characteristic yet mild physical appearance of a long, narrow face and large ears (Opitz et al. 1984, Hagerman et al. 1984). Other clinical signs, presumably due to the connective tissue disorder, include velvet-like skin, finger-joint hyperextensibility, recurrent otitis media, aortic root dilatation, and mitral valve prolapse (Loehr 1986, Hagerman 1996). Patients often display autistic features ranging from shyness, poor eye contact, and social anxiety in less affected individuals to hyperactivity, hand flapping, hand biting, and perserverative speech in the severely affected (Merenstein et al. 1996), as well as seizures and EEG findings consistent with epilepsy (Musumeci et al. 1999, Sabaratnam et al. 2001). Some obligate carrier women are also affected (Nielsen et al. 1981, Fryns 1986, Hagerman et al. 1992). In general, affected females have a less severe phenotype than do males, and the severity of dysfunction is correlated to the degree of X-inactivation on the abnormal chromosome (Abrams et al. 1994, Sobesky et al. 1996).

Along with the existence of affected females heterozygous for the fragile X, a number of pedigrees also show unaffected males who transmit the marker to their daughters (Martin & Bell 1943, Losowsky 1961, Nielsen et al. 1981). These two results point to a mode of inheritance more complicated than a simple X-linked recessive model. Because of these irregularities in the set of reported pedigrees,

Sherman et al. (1984, 1985) performed a large-scale segregation analysis on 206 fragile X syndrome pedigrees. They too saw a significant number of asymptomatic males and affected females and put forward a model of X-linked dominant inheritance with reduced penetrance (79% for males and 35% for females). Sherman et al. also noted the lack of sporadic cases and thus argued that each affected male received the gene from his mother. Finally, they noticed that an asymptomatic carrier male is more likely to have grandsons with the disorder than to have brothers with the disorder. Therefore, the penetrance for the disease increases in succeeding generations of a pedigree—an observation referred to as the Sherman paradox. The mechanism responsible for the Sherman paradox became clear in 1991 with the cloning of the gene defective in fragile X syndrome.

FMR1

Before the gene responsible for fragile X syndrome could be cloned, a great deal of both genetic and physical mapping was done. Although the fragile site cosegregated with the syndrome phenotype, it was unknown whether or not the syndrome was caused by the fragile site itself or a closely linked causal mutation. Pedigree analysis localized both the causal locus and the fragile site to a 22 cM region on the X chromosome between the factor IX gene and marker St14 (Oberle et al. 1986), and further studies revealed a number of linked markers that reduced the interval to 1–2 Mb and strengthened the localization of the disease locus to the fragile site (Suthers et al. 1989, 1990; Hirst et al. 1991; Rousseau et al. 1991). Warren et al. (1987, 1990) devised a method using somatic cell hybrid lines to mark the fragile site itself by selecting for X chromosome breakage and translocation generated when the fragility was induced.

Methylation-sensitive restriction fragment digests separated by pulsed-field gel electrophoresis and probed with markers flanking the fragile site identified a region methylated in fragile X patients but not in unaffected carriers or normal males (Vincent et al. 1991, Bell et al. 1991). Furthermore, fragments abnormally methylated in affected individuals were found to be unstable and to increase in size when transmitted through a pedigree (Oberle et al. 1991, Yu et al. 1991). Heitz et al. (1991) identified a yeast artificial chromosome (YAC) containing two markers known to flank the fragile site hybrid breakpoints and showed that this YAC contains a CpG island aberrantly methylated in fragile X patients. Finally, Verkerk et al. (1991) independently used the same YAC to probe a human brain cDNA library and clone the gene responsible for the disease: *fragile X mental retardation 1* (*FMR1*). Expression of *FMR1* is absent in the majority of fragile X patients (Pieretti et al. 1991), and the number of patients with *FMR1* deletions proves the syndrome is caused by the loss of *FMR1* function (Gedeon et al. 1992, Wohrle et al. 1992).

The gene itself spans 38 kb and encodes a 4.4 kb transcript consisting of 17 exons (Figure 1*A*, see color insert) (Eichler et al. 1993, Ashley et al. 1993a). *FMR1* has two autosomal paralogs (*FXR1* and *FXR2*; Siomi et al. 1995, Zhang et al. 1995) and highly conserved orthologs in mammals, chickens, and fruit flies (Ashley et al. 1993a, Price et al. 1996, Wan et al. 2000). Full-length fragile X mental retardation protein (FMRP), the protein encoded by *FMR1*, has a molecular weight of 69 kDa, but extensive alternate splicing of exons produces a number of protein isoforms (Eichler et al. 1993, Ashley et al. 1993a; the structure and function of FMRP are discussed below). Consistent with the primary features of the phenotype, *FMR1* mRNA and protein are highly expressed in testis and in fetal and adult brain, with the majority of signal localized to neurons (Abitbol et al. 1993, Devys et al. 1993). *FMR1* harbors a novel, dynamic mutation that accounts for both the fragile site and the genetic peculiarities observed to be associated with the region.

Both the hybrid breakpoints and the unstable DNA fragments map to a $(CGG)_n$ repeat (Kremer et al. 1991, Verkerk et al. 1991), later shown to be in the 5' untranslated region of *FMR1* (Ashley et al. 1993a). The repeat is polymorphic in the general population with a range of 6–60 and a mode of 30 (Fu et al. 1991, Snow et al. 1993). Examination of fragile X pedigrees reveals two other classes of alleles: nonpenetrant premutations with 60–200 repeats and completely penetrant full mutations with >200 repeats, which often are in the thousands (Figure 1*B*, see color insert) (Fu et al. 1991, Snow et al. 1993). Premutation alleles are unstable and tend to expand when transmitted. A premutation can undergo a small expansion to another, usually slightly larger allele in the premutation range, or it can undergo massive expansion to a full mutation. This massive expansion to a full mutation occurs only when the premutation is transmitted from a female with male spermatogenesis apparently unable to maintain the long full-mutation repeat (Malter et al. 1997). Furthermore, the risk of expansion to the full mutation is determined by the size of the premutation: The larger the premutation is, the more likely it will expand to the full mutation. Because premutations increase through a pedigree, the risk of expansion to full mutation also increases—an observation that resolves the Sherman paradox (Fu et al. 1991, Heitz et al. 1992).

Massive CGG expansion is the causative mutation in >95% of patients with fragile X syndrome (Warren & Sherman 2001). As a result of expansion, the repeat and the upstream CpG island are methylated, and *FMR1* expression is silenced (Pieretti et al. 1991, Sutcliff et al. 1992, Hornstra et al. 1993). Treatment of full-mutation cell lines with methylation inhibitors reactivates a low level of *FMR1* expression, which indicates that methylation causes the gene silencing (Chiurazzi et al. 1998). Methyl-C binding protein MeCP2 binds to methylated DNA and recruits histone deacetylases. These deacetylases induce chromatin condensation and prevent transcription machinery from binding to a gene's promoter (Razin 1998). Indeed, the association of acetylated histones with *FMR1* is reduced in full-mutation cell lines and is restored with methylation inhibitor treatment (Coffee et al. 1999). Furthermore, treatment of full-mutation lines with histone deacetylase inhibitors along with methylation inhibitors reactivates the normally silent *FMR1* more than does treatment with methylation inhibitors alone (Chiurazzi et al. 1999). However, treatment with histone deacetylase inhibitors alone induces little or no increase in gene expression, which indicates that at least to some extent methylation alone is sufficient for gene silencing. In vivo footprinting analysis of the *FMR1* promoter region shows a number of putative transcription factor binding sites that are protected in normal cells but are absent in methylated patient cell lines (Schwemmle et al. 1997, Drouin et al. 1997). Kumari & Usdin (2001) have shown that transcription factors USF1, USF2, and alpha-Pal/Nrf-1 bind to the *FMR1* promoter and are required for reporter gene expression driven by the *FMR1* promoter. Methylating the reporter constructs suppresses gene expression, but not fully. Thus, it appears that both histone-dependent and histone-independent effects are at work in methylation-dependent gene silencing at the *FMR1* promoter. Though it may appear that reactivation of *FMR1* in patients is an ideal approach to therapy, current data suggest translation of FMRP is suppressed in large full-mutation transcripts (Feng et al. 1995). Thus, reactivating transcription may not appreciably increase cellular FMRP levels.

Repeat Expansion

The elucidation of the exact timing and mechanism for both modest and massive expansions of premutation alleles has been a major goal in fragile X research over recent years. Initially, the findings of somatic mosaicism in repeat size in full-mutation males and of the absence of full-mutation alleles in the sperm of those males led to a model of postzygotic expansion occurring after germline differentiation (Rousseau et al. 1991, Reyniers et al. 1993). However, Moutou et al. (1997) argues that in this postzygotic model, the degree of mosaicism should be inversely proportional to the length of maternal premutation; this prediction, in fact, is not the case. Thus, the mosaicism seen in full-mutation patients is likely due to variable contraction of somatic full-mutation alleles and not variable expansion of maternal premutation alleles. More support of a prezygotic model came from studies of fetal gonadal tissue. Malter et al. (1997) found only full-mutation alleles in the ovaries of a 16-week-old fetus, which suggests that the timing of expansion must be before germline differentiation. Furthermore, testis tissue from a 13-week-old full-mutation fetus showed only full mutations in the germ cells while testis tissue from an older fetus showed evidence of both full and premutations in the germ cells (Malter et al. 1997). These observations, taken together with the lack of full-mutation alleles in adult sperm, led Malter and his colleagues to argue that spermatogenesis is unable to maintain full mutations. Similar conclusions were reached for myotonic dystrophy, a disease in which males do not transmit long CTG repeats (Jansen et al. 1994). While these studies point to a prezygotic expansion, expansion occurring very early in embryogenesis (before day 3–5) cannot be ruled out (Malter et al. 1997, Moutou et al. 1997). To definitively time the expansion, repeat lengths in the oocytes of premutation females must be measured to determine if expansion occurs during oogenesis—a difficult prospect in humans. It is clear that animal models for expansion will be essential to determine both the timing and the mechanism of expansion.

For the most part, attempts to recapitulate expansion in the mouse have been ineffective (Bontekoe et al. 1997; Lavedan et al. 1997, 1998). While some modest expansions have recently been reported in a mouse containing a human $(CGG)_{98}$ repeat recombined into the endogenous mouse *Fmr1* locus (Bontekoe et al. 2001), to date there are no reports of massive expansion occurring in the mouse. It will be interesting to see if the repeat in the mice of Bontekoe and colleagues continues to expand with time and eventually reaches a threshold for massive expansion. Though the lack of a good animal model for expansion is disappointing, the apparent stability of the CGG repeat in mouse should prove useful in identifying both *cis* and *trans* acting factors in the mechanism of expansion. The loss of repeat stability in knockin mice generated with increasing amounts of human *FMR1* sequence can be used as an assay for factors that influence the repeat in *cis*, and mutational analysis may identify genes expressed in the mouse and not humans that when mutated alter the stability of premutation-sized alleles.

Studies in humans tend to point to the structure of the repeat itself as the most important variable in expansion, although certain haplotypes of polymorphic markers in and around the *FMR1* locus are overabundant in fragile X chromosomes, which indicates the possibility of other *cis* acting sequences (Eichler et al. 1996, Gunter et al. 1998). Single AGG triplets that are variable in both number and location interrupt the CGG repeat of *FMR1* in normal chromosomes (Verkerk et al. 1991, Kunst & Warren 1994). The ancestral sequence is 5′— $(CGG)_9AGG(CGG)_9AGG(CGG)_9$—3′ (Eichler et al. 1995), and variation of repeat length occurs almost exclusively at the 3′ end of the repeat (Kunst & Warren 1994, Eichler et al. 1994). Furthermore, the length of pure CGG tracts predicts repeat stability with a threshold for expansion of 34–38 uninterrupted repeats. Unstable premutations arise either from the gradual expansion of the 3′ end or by the loss of one or more AGG interruptions (Eichler et al. 1996).

The polarity of expansion of the CGG repeat in humans mimics the orientation-dependent instability of CGG repeats cloned into *E. coli* and *S. cerevisiae* (Hirst & White 1998, White et al. 1999). This finding suggests that a difference in leading-versus lagging-strand synthesis may be involved in repeat expansion. Because trinucleotide repeats have been shown to form secondary structures whose stability increases with length and decreases with AGG interruption (Gacy et al. 1995, Gacy & McMurray 1998, Pearson et al. 1998), many models for repeat expansion evoke hairpin formation and subsequent slippage during replication. One model for repeat expansion that is rapidly gaining acceptance involves the abnormal processing of repeat containing Okazaki fragments during lagging strand synthesis (Gordenin et al. 1997). During normal replication, the synthesis of the newly formed, upstream Okazaki fragment displaces the 5′ end of the downstream fragment to facilitate RNA primer removal by FEN1 endonuclease (Budd et al. 1995, Budd & Campbell 1997). If the downstream Okazaki fragment was initiated in the repeat, the displaced strand could fold back upon itself and form a stable secondary structure. The ability of FEN1 to cleave these secondary structures is reduced and is inversely proportional to the length of the foldback (Henricksen et al. 2000).

Supporting this model, yeast mutant for RAD27, the ortholog of FEN1, show a tenfold increase in repeat expansion (White et al. 1999).

Prevalence and the Premutation Revisited

With the discovery of the causal, dynamic mutation in *FMR1*, a more sensitive and specific DNA-based test supplanted cytology in the diagnosis of fragile X syndrome (Rousseau et al. 1991). Numerous studies have found fragile X syndrome in every ethnic group, with current estimates putting the prevalence at 1 out of 4500 males and 1 out of 9000 females (reviewed in Warren & Sherman 2001). With the advent of a DNA-based diagnostic test, the study of premutation carriers has increased. The prevalence of the premutation is estimated to be 1 out of 1000 males and 1 out of 400 females, consistent with the dynamic nature of the mutation (Warren & Sherman 2001). Some premutation individuals are not truly unaffected, as once thought, but exhibit subtle fragile X–like features (Hull & Hagerman 1993, Loesch et al. 1994, Riddle et al. 1998) as well as premature ovarian failure in females and Parkinsonism in elderly males (Allingham-Hawkins et al. 1999, Uzielli et al. 1999, Hagerman et al. 2001). These latter two features are remarkable in that they are unique to the premutation, as full-mutation individuals are not affected. Premutation carriers express *FMR1* mRNA at higher levels and FMRP at lower levels than do normal controls (Tassone et al. 2000, Kenneson et al. 2001). This observation led to the suggestion that higher level of *FMR1* mRNA may be responsible for the unique features in premutation individuals (Tassone et al. 2000).

FROM GENOTYPE TO PHENOTYPE: THE CAUSE OF COGNITIVE IMPAIRMENT IN FRAGILE X SYNDROME

As the primary characteristic of fragile X syndrome is mental retardation, an obvious point of focus for much of the work over the past years is how the loss of FMRP leads to abnormal cognitive function. Studies addressing this issue include: (*a*) anatomical ones using cadaveric tissue as well as CT and MRI images from affected patients to more fully characterize the fragile X phenotype, (*b*) the creation and characterization of a mouse model for the syndrome that is amenable to more rigorous laboratory manipulation than are human models, (*c*) biochemical studies to determine the function of FMRP, and (*d*) studies to determine how loss of FMRP function in neurons leads to cognitive impairment.

Anatomical Studies

Brains taken at autopsy from a total of six fragile X males show no gross abnormalities (Rudelli et al. 1985, Hinton et al. 1991, Wisniewski et al. 1991, Irwin et al. 2001). Brain weight and structure appear normal with the exception of mild cortical atrophy in one patient and ventricular enlargement in four. CT scanning confirmed this ventricular enlargement in seven of eight live patients (Wisniewski et al. 1991). In a number of studies comparing MRI images of fragile X patients with those of normal controls, Reiss and his colleagues have not only confirmed the increase of lateral ventricular volume (Reiss et al. 1995, Eliez et al. 2001), but also have revealed a number of other, somewhat subtle structural differences in patients. In both male and female patients, the posterior vermis of the cerebellum is decreased (Reiss et al. 1991a,b, Mostofsky et al. 1998) and the caudate nucleus is increased (Reiss et al. 1995, Eliez et al. 2001). Furthermore, the posterior vermis, caudate nucleus, and lateral ventricle volumes appear to correlate with cognitive function, and the authors suggest that the vermis and caudate sizes correlate with *FMR1* expression, which indicates that the observed anatomical differences in patients are due to the lack of FMRP (Reiss et al. 1995, Mostofsky et al. 1998). Reiss also found the hippocampus to be enlarged in fragile X children and young adults (Reiss et al. 1994, Kates et al. 1997), but Jakala et al. (1997) did not reproduce this finding in adults. The difference in size of all of these structures is small, however, and there is substantial overlap in each case between controls and patients.

Though it is tempting to speculate on how these changes relate to the neurological deficits seen in fragile X syndrome, it is important to remember that while these descriptive studies point to structures that may be involved in the syndrome, they cannot address whether any anatomical change is causative for—or is merely caused by—the neurological deficits seen in patients. The anatomical studies also do not begin to address a molecular mechanism for the cognitive dysfunction seen in patients—to do so requires the study of the neurons themselves. The limitations to performing such research in the brains of humans require that an animal model be developed to further the neurobiological research into FMRP.

The FMRP Knockout Mouse

Though the need for a mouse model for fragile X syndrome is unquestionable, the validity and applicability of any animal model must be examined. *FMR1* is highly conserved between human and mouse with a nucleotide and amino acid identity of 95% and 97% respectively, including the CGG repeat in the 5'UTR (Ashley et al. 1993a). Furthermore, in situ hybridization and *Fmr1*: β-galactosidase transcriptional fusion experiments have shown the expression pattern of murine *Fmr1* to be similar to the human version in both tissue specificity and time of expression (Hinds et al. 1993, Hergersberg 1995). Finally, though the mouse repeat does not appear to undergo expansion and thus cannot mimic the timing of methylation and inactivation of *FMR1* seen in humans, the existence of deletion and point mutations patients with fragile X syndrome shows that the lack of FMRP throughout development is sufficient to cause the phenotype. Thus, because the knockout animal never has functional FMRP, it should provide an accurate molecular model for the human condition.

With this reasoning in mind, a consortium of the Oostra and Willems labs created an *Fmr1* knockout mouse by the insertion of a neomycin cassette into exon 5 of the murine gene (Dutch-Belgian Fragile X Consortium 1994). This mouse has no functional FMRP in any of the tissues assayed by Western blotting. As with the human syndrome, the phenotype of the mutant mouse is mild: No major neurological deficits are found, no obvious abnormalities of the major organ systems are revealed by pathological examination, and reproductive fitness is not reduced. Also consistent with the human disease, adult mutant mice show significant macroorchidism, undergo audiogenic seizures, and exhibit subtle behavioral abnormalities (Dutch-Belgian Fragile X Consortium 1994, Musumeci et al. 2000, Chen & Toth 2001). While it is impossible to give mice the equivalent of an IQ test, a number of paradigms have been designed that test different aspects of cognition and behavior in mice (Crawley & Paylor 1997). With these paradigms, *Fmr1* knockout mice exhibit increased exploratory and motor activity, deficits in spatial learning ability, and decreased anxiety-related responses (Dutch-Belgian Consortium 1994, D'Hooge et al. 1997, Van Dam et al. 2000, Peier et al. 2000), although there has been some difficulty replicating these results owing to possible strain-dependent effects (Paradee et al. 1999, Fisch et al. 1999). The creation of an *Fmr1* knockout mouse that has phenotypic characteristics consistent with the human disease has proven valuable for the functional and neurophysiologic studies performed in order to zero in on the cause of fragile X syndrome.

Biochemical Studies

Because the lack of functional FMRP is both necessary and sufficient for the development of fragile X syndrome, elucidation of the function of FMRP in normal cells is imperative for understanding the pathogenesis of the disease. FMRP has functional domains in common with proteins known to form large ribonucleoprotein (RNP) complexes in vivo. By sequence analysis, Siomi et al. (1993b) and Ashley et al. (1993b) independently identified three RNA binding domains in FMRP: two KH domains (KH1, KH2) that show homology to hnRNP K (Siomi et al. 1993a), and an RGG box similar to hnRNP U (Kiledjian & Dreyfuss 1992). Both in vitro translated and purified FMRP bind RNA homopolymers and 4% of fetal brain messages in vitro, which confirms that FMRP is indeed an RNA binding protein (Siomi et al. 1993b, Ashley et al. 1993b, Brown et al. 1998). Recent studies have identified both the in vivo ligands of FMRP and the RNA motif to which FMRP binds. In vitro, the RGG box of FMRP binds strongly to an intramolecular stem-loop structure that is termed a G-quartet (Figure 2) (Schaeffer et al. 2001, Darnell et al. 2001). Furthermore, this G-quartet is found in approximately 70% of transcripts that immunoprecipitate with FMRP, which indicates this structure is indeed the in vivo target for FMRP binding (Brown et al. 2001).

When FMRP is sedimented through sucrose gradients, it associates with translating polyribosomes as part of a large mRNP complex (Eberhart et al. 1996, Khandjian et al. 1996, Corbin et al. 1997, Feng et al. 1997a). This mRNP is >660 kDa in size and contains a number of proteins, including both FMRP autosomal homologs (Feng et al. 1997a, Ceman et al. 1999, 2000). The functional importance of this polyribosome association of FMRP is demonstrated by a severely

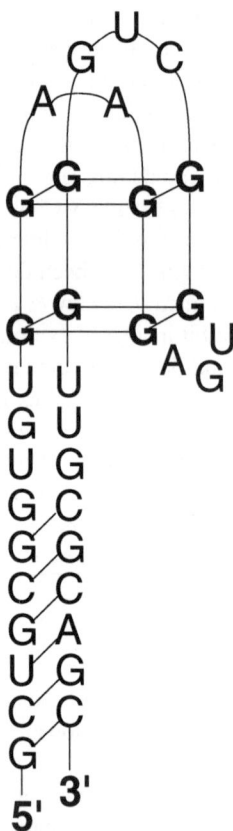

Figure 2 G-quartet structure mediates FMRP binding. Consensus stem-loop and G-quartet structure found in RNA species purified in vitro by multiple rounds of selection for FMRP binding. The G-quartet binds to the RGG box of FMRP (Darnell et al. 2001). This structure is also present in 70% of RNAs that coimmunoprecipitate with FMRP from mouse brain lysates, which indicates that the G-quartet may indeed be the in vivo target of FMRP binding (Brown et al. 2001).

affected fragile X syndrome patient who has a missense mutation in the KH2 domain of FMRP (De Boulle et al. 1993). Crystal structure analysis of highly conserved KH domains in other proteins suggests that this isoleucine to asparagine substitution (I304N) compromises the RNA binding ability of the KH2 domain (Musco et al. 1996, Lewis et al. 2000). However, absolute RNA binding of the I304N mutant FMRP is not affected (Feng et al. 1997a, Brown et al. 1998). Instead, the mutant protein no longer associates with translating polyribosomes, but is part of a smaller, abnormal mRNP (Feng et al. 1997a). Thus, the loss of proper mRNP formation and polyribosome association causes this severe fragile X phenotype in this patient.

In the brain, FMRP is located primarily in the cytoplasm of neurons (Devys et al. 1993), but a small amount of the protein has been found in the nucleus by both light and electron microscopy (Verheij et al. 1993, Eberhart et al. 1996, Feng et al. 1997b). The study of various FMRP truncation proteins has led to the identification of both a nuclear localization signal and a nuclear export signal in FMRP, leading to the hypothesis that FMRP shuttles into and out of the nucleus (Eberhart et al. 1996). The paralogs of FMRP (FXR1P and FXR2P) also shuttle into and out of the nucleus, however it appears that unlike FMRP, which shuttles to the nucleoplasm, FXR2P and certain isoforms of FXR1P localize to the nucleolus (Tamanini et al. 1999, 2000). Thus, while the FXR proteins are associated in the cytoplasm with FMRP in an mRNP, they may be playing different functional roles in the nucleus. One model of FMRP function taking into account the data discussed above is that FMRP is transported into the nucleus via its nuclear localization signal. Once in the nucleus it associates with a number of other proteins and a distinct population of mRNAs to form an mRNP complex. This complex is then exported to the cytoplasm via the NES, where it associates with ribosomes and influences translation of FMRP ligand mRNAs (Jin & Warren 2000).

A recent study lends credence to the influence of FMRP on translation. Brown et al. (2001) determined the identities of in vivo mRNA ligands that were selectively immunoprecipitated from mouse brain with the FMRP mRNP. A substantial proportion of these transcripts show a change in translation status by a shift on polyribosome profiles in full-mutation lymphoblastoid cells compared to normal cells. Since in vitro studies have shown that the addition of purified FMRP to translation mixes suppresses translation (Laggerbauer et al. 2001, Li et al. 2001), one might expect that cells deficient in FMRP would exhibit increased polysome loading of certain messages. However, Brown et al. observe that both increased and decreased polysome loading of specific messages occurs in fragile X syndrome cells. Clearly the cellular effects are more complex than simple in vitro translation assays; however, one may speculate on the mechanism(s). It is unclear whether all FMRP-containing mRNPs are identical, except for the bound mRNA, or if there are differences in associated proteins among them. If there are distinct FMRP complexes, each may respond differently to the absence of FMRP. Alternatively, it is possible that mRNAs interact with the FMRP complex differently depending upon specific sequence signals on the messages. For example, the RGG domain of FMRP recognizes a G-quartet structure (Darnell et al. 2001); the other RNA-binding KH domains of FMRP may recognize a distinct structure. The presence of one or both of these structures may influence the translational fate of specific messages in the absence of FMRP. Finally, the loss of FMRP could simply shift the delicate balance of mRNA translation, so that each message may be regarded as having a variable capacity to be translated and all messages therefore compete with one another for limited ribosome pools. If FMRP indeed suppresses translation, the absence of FMRP could lead to polyribosome loading of normally FMRP-associated messages, competing off ribosomes those messages with less translational attractiveness.

Neuronal Studies

A neuronal phenotype has been evident in fragile X syndrome since the first autopsy studies by Rudelli et al. (1985). While neuron counts are normal, qualitative analysis of Golgi-impregnated dendrites in parieto-occipital cortex sections show abnormal dendritic spines—the postsynaptic protrusions from dendrites at which the vast majority of excitatory synapses occur (Harris & Kater 1994). Spines are sensitive to their environment and change density and morphology to a number of stimuli (Yuste & Bonhoeffer 2001), and spine abnormalities have long been associated with mental retardation of unknown etiology (Purpura 1974), as well as with Down's and Rett syndromes (Kaufmann & Moser 2000). Whether these abnormalities reflect a foundation in the pathogenesis of the mental retardation or are merely a by-product of that mental retardation is not known.

A number of spines in one patient at autopsy were long and thin with prominent heads resembling immature spine-like structures called filopodia (Rudelli et al. 1985). This observation was confirmed in two other patients (Hinton et al. 1991) and was recently quantified in three additional patients (Irwin et al. 2001). In studies of sections from the temporal and visual cortex, Irwin and his colleagues found that three adult fragile X males had significantly more long, immature spines and fewer short, mature spines than did controls. Spine density in these patients was also increased compared to control males of the same age, which led the authors to postulate a defect in spine maturation and elimination in fragile X patients. The number of patients studied is still very small and finding adequate controls is nearly impossible, so more studies must be done to address the possibility of other variables causing the observed spine phenotype, as spines are so mutable. For example, the patient in the original study by Rudelli and colleagues was blind for the last 15 years of his life, presented with seizures, and was on antipsychotic medication at the time of his death—all variables that conceivably could alter spine morphology in the cortex independently of FMRP absence. Detailed histories are not provided for the other patients examined, so it remains to be determined whether lack of FMRP is the sole cause of the spine phenotype in those patients.

Spine study results in the *Fmr1* knockout mouse, while somewhat conflicting, do support a role for spine abnormalities in fragile X syndrome. In a Golgi impregnation study of layer V pyramidal neurons in sections from the visual cortex of wild-type and *Fmr1* knockout mice, Comery et al. (1997) found that spine length and density was increased in knockout dendrites compared to those in wild type. This finding is consistent with the findings in humans and further supports the hypothesis of a maturation and pruning defect. Unfortunately, the studies of Comery et al. were performed on *Fmr1* knockout mice that were homozygous for a gene causing retinal degeneration, and thus it is possible a portion of the knockout mice studied were blind. Because slices were taken from the visual cortex, the blindness of the mice may have substantially affected spine phenotype in an FMRP-independent manner. Subsequent studies, comparing congenic littermates to determine the degree of spine phenotype in adult animals related to FMRP, are underway and appear to confirm at least the immaturity if not the increased density of knockout (KO) spines (Irwin et al. 1999, 2000a).

Other studies point to a transient spine phenotype in developing knockout mice. Using two-photon microscopy to visualize enhanced green fluorescence protein EGFP in individual neurons in the barrel region of somatosensory cortex, Nimchinsky et al. (2001) showed that knockouts have significantly longer spines, more long spines, and fewer short spines than do wild-type animals. However, this effect was seen only on postnatal days 7 and 14 and was absent by postnatal day 28. Thus, it appears that in barrel cortex, knockout spines show a transient delay in maturation that resolves with time. Consistent with a transient difference, cultured hippocampal neurons from knockout animals show a similar delay in spine maturation at 14 and 21 days that eventually subsides by 35 days (E. R. Torre, personal communication), and electrophysiology experiments in cultured hippocampal neurons indicate that there is also a delay in establishing synaptic connections that resolves with time (Braun & Segal 2000).

The spine phenotype in fragile X patients and *Fmr1* knockout mice suggests FMRP may be involved in synaptogenesis early in development. One current model of spine formation in synaptogenesis requires initial axon-dendrite contact to occur at filopodia. A subset of these filopodia then becomes mature spines (Dailey & Smith 1996, Ziv & Smith 1996, Fiala et al. 1998). If the long, thin spine-like structures seen in fragile X syndrome are really immature filopodia as their morphology suggests, their prolonged presence could reflect a defect in normal maturation of synapse formation in the absence of FMRP. Even a transient delay may affect the neural circuitry enough to produce the cognitive abnormalities associated with fragile X syndrome. Because the spine phenotype is seen concurrently with the critical period of experience-dependent synapse development in the barrel region (Lendvai et al. 2000), one intriguing possibility is that the maturational delay seen in knockout spines in the barrel cortex is present at all new synapses. Parallel studies of spine morphology in different brain regions during their critical periods of development as well as denervation/reinnervation experiments would determine the scope of FMRP involvement in spine maturation.

Synaptic plasticity—a long-term change in synaptic strength after stimulation—is thought to be the mechanism of information storage in learning and memory (Steward & Schuman 2001). While the discovery of a spine morphological phenotype indicates a possible defect in synaptic plasticity in fragile X syndrome, altered spine morphology alone does not elucidate a cause of mental retardation. As spines are thought to play a pivotal role in synaptic plasticity, the role of FMRP at the synapse has been a central question. Currently, evidence of FMRP involvement in synaptic plasticity is circumstantial. In vivo, FMRP levels increase in animals raised in a complex environment (Irwin et al. 2000b) and in response to sensory stimulation (Todd & Mack 2000); both paradigms are thought to induce synaptic plasticity (Greenough et al. 1985, Lendvai et al. 2000). *FMR1* mRNA is present in synaptoneurosome preparations thought to be highly enriched for synapses (Weiler et al. 1997). Furthermore, stimulation with mGluR-specific agonists causes FMRP levels to increase significantly in these preparations. It is unclear whether this increase is specific for FMRP or if all mRNAs present at the synapse are translated after stimulation. While these data point to the involvement of FMRP in synaptic plasticity, one cannot yet infer a causative role.

The most direct evidence of FMRP's role in synaptic plasticity comes from recent studies comparing wild-type and knockout mice using two synaptic models for information storage: long-term depression (LTD) and long-term potentiation (LTP). LTD is a decrease in the strength of the same synapses after prolonged, low-frequency stimulation (Bear & Abraham 1996). One form of LTD is mGluR-dependent and requires protein synthesis (Huber et al. 2000). This form of LTD is enhanced in *Fmr1* KO mice (M.F. Bear, personal communication), constituting direct evidence that the absence of FMRP alters synaptic plasticity. LTP is a long-term increase in synaptic strength in response to high-frequency stimulation, and it is thought to have an early phase requiring neither transcription nor protein synthesis, an intermediate—mGluR-dependent—phase requiring only protein synthesis, and a late phase requiring transcription and protein synthesis (Bliss & Collingridge 1993, Raymond et al. 2000). *Fmr1* knockout mice show no difference in either the early- or late-phase LTP (Godfraind et al. 1996, Paradee et al. 1999), but it remains to be seen if the intermediate, mGluR-dependent phase is affected. This evidence of altered synaptic plasticity along with spine maturation abnormalities in knockout mice, and both the RNA binding and protein translational capacities of FMRP lead to intriguing possibilities for the role of FMRP in synaptic function.

Protein synthesis has long been considered a necessary component of synaptic plasticity (Steward & Schuman 2001), and for years there has been mounting evidence that specificity of synaptic plasticity is gained in part through local protein synthesis at individual synapses in dendrites. All the necessary machinery for protein synthesis is found at the site of synaptic contact: free polyribosomes and rough endoplasmic recticula are found in dendrites at the base of spines as are a number mRNAs (Steward & Levy 1982, Steward & Reeves 1988, Miyashiro et al. 1994, Kuhl & Skehel 1998). Furthermore, local protein synthesis occurs in live, transected dendrites (Torre & Steward 1992, Aakalu et al. 2001), in vitro in response to mGluR activation (Weiler & Greenough 1993), and in both LTP and LTD (Kang & Schuman 1996, Huber et al. 2000). FMRP has been detected in dendrites and dendritic spines by immunogold labeling and electron microscopy, and it is thought to be associated with translating polyribosomes in dendrites (Feng et al. 1997b). Taken together the above observations are compelling evidence that FMRP plays a role in protein synthesis–dependent synaptic plasticity.

FMRP's role in synaptic plasticity could take a number of turns after the FMRP mRNP leaves the nucleus. FMRP could be involved in the dendritic localization of its ligands and could influence the translation of those mRNAs either at the site of synapse or in the cell body (for a model of FMRP function in the neuron see Figure 3, in color insert). A recent study indicates that the dendritic mRNAs MAP2, CAMII kinase, dendrin, and ARC are not abnormally localized in *Fmr1* KO mice (Steward et al. 1998); however, these mRNAs are not ligands for FMRP RNA binding (Brown et al. 2001). Thus, though FMRP is not involved in localization of all dendritic mRNAs, it may still localize its specific ligands to the dendrites. In vivo ligands that are translationally altered by FMRP include a number of transcripts involved in synaptic function in both the pre- and postsynaptic cells (Brown et al. 2001). For example, translation of both UNC-13 and SAPAP4 is downregulated in patient cell lines, and both genes appear to play a role in synaptic function. However, UNC-13 is involved in presynaptic vesicle fusion, whereas SAPAP4 is associated with PSD-95 at the postsynaptic density. Further study of the effects of altered translation of these transcripts and others should prove useful in elucidating the molecular mechanisms involved in the pathogenesis of fragile X syndrome.

The subtle effect of FMRP on translation and the transient nature of the spine phenotype are consistent with the phenotype seen in patients. It is clear that there is no global and fundamental defect in synapses of fragile X patients, because for the most part nervous system function in patients is normal. Furthermore, both patients and KO mice can learn to perform simple tasks with practice, so it appears that although higher-order thinking is impaired, basic cognitive function still remains. That *Fmr1* KO mice show a change in one type of synaptic plasticity (LTD) and not in another (LTP) indicates that the loss of FMRP may not even affect an individual synapse under all conditions. Identifying at which synapses and under what conditions the loss of FMRP affects synaptic function should provide insight not only into the pathogenesis of fragile X, but also into learning and cognition in general.

TREATMENT

Current treatment modalities for fragile X syndrome are palliative and involve individually tailored behavioral and cognitive therapy designed to help each patient reach his maximum potential in conjunction with symptom-specific therapy for medical problems (for reviews see Hagerman & Cronister 1996). Unfortunately, no molecular-based approach exists, and the difficulties of finding one are numerous. Because phenotypic consequences of the absence of FMRP are apparent so early after birth (Nimchinsky et al. 2001), and these defects in synaptic plasticity could produce long-term neural wiring effects, any molecular treatment may need to be applied early. It also appears that any therapeutic approach may need to target downstream effects of FMRP function. Though full-mutation *FMR1* expression can be reactivated with the use of methylation and histone deacetylase inhibitors, reactivation is slight, cellular toxicity is substantial, and transcripts with large repeats are not translated well (Feng et al. 1995, Chiurazzi et al. 1999, Coffee et al. 1999, Kenneson et al. 2001). Thus, reactivating transcription of *FMR1* may not be enough to restore FMRP levels. FMRP replacement therapy may also have its downside, as overexpression of human FMRP in the mouse appears to cause a phenotype of its own (Peier et al. 2000). The identification of the in vivo ligands of FMRP by Brown et al. (2001) opens the door for novel therapeutic approaches targeted downstream of FMRP that may avoid these difficulties. Translational changes, and any subsequent downstream effects, may prove to be more assayable markers for the fragile X phenotype than transient spine abnormalities. If a high-throughput assay of the phenotype can be established, one can screen combinatorial libraries for small molecules that rescue this phenotype and thus identify potential therapeutic agents for the treatment of fragile X syndrome (Lam 1997).

The *Annual Review of Neuroscience* is online at http://neuro.annualreviews.org

LITERATURE CITED

Aakalu G, Smith WB, Nguyen N, Jiang C, Schuman EM. 2001. Dynamic visualization of local protein synthesis in hippocampal neurons. *Neuron* 30:489–502

Abitbol M, Menini C, Delezoide AL, Rhyner T, Vekemans M, Mallet J. 1993. Nucleus basalis magnocellularis and hippocampus are the major sites of FMR-1 expression in the human fetal brain. *Nat. Genet.* 4:147–53

Abrams MT, Reiss AL, Freund LS, Baumgardner TL, Chase GA, Denckla MB. 1994. Molecular-neurobehavioral associations in females with the fragile X full mutation. *Am. J. Med. Genet.* 51:317–27

Allingham-Hawkins DJ, Babul-Hirji R, Chitayat D, Holden JJ, Yang KT, et al. 1999. Fragile X premutations is a significant risk factor for premature ovarian failure: the international collaborative POF in fragile X study—preliminary data. *Am. J. Med. Genet.* 83:322–25

Ashley CT, Sutcliff JS, Kunst CB, Leiner HA, Eichler EE, et al. 1993a. Human and murine FMR-1: alternate splicing and translational initiation downstream of the CGG-repeat. *Nat. Genet.* 4:244–51

Ashley CT, Wilkinson KD, Reines D, Warren ST. 1993b. FMR1 protein: conserved RNP family domains and selective RNA binding. *Science* 262:563–66

Bear MF, Abraham WC. 1996. Long-term depression in hippocampus. *Annu. Rev. Neurosci.* 19:437–62

Bell MV, Hirst MC, Nakahori Y, MacKinnon RN, Roche A, et al. 1991. Physical mapping across the fragile X: hypermethylation and clinical expression of the fragile X syndrome. *Cell.* 64:861–66

Bennetto L, Pennington BF. 1996. The neuropsychology of fragile X syndrome. See Hagerman & Cronister 1996, pp. 210–48

Bliss TV, Collingridge GL. 1993. A synaptic model of memory: long-term potentiation in the hippocampus. *Nature* 361:31–39

Bontekoe CJ, Bakker CE, Nieuwenhuizen IM, van der Linde H, Lans H, et al. 2001. Instability of a $(CGG)_{98}$ repeat in the Fmr1 promoter. *Hum. Mol. Genet.* 10:1693–99

Bontekoe CJ, de Graaff E, Nieuwenhuizen IM, Willemsen R, Oostra BA. 1997. FMR1 premutation allele $(CGG)_{81}$ is stable in mice. *Eur. J. Hum. Genet.* 5:293–98

Braun K, Segal M. 2000. FMRP involvement in formation of synapses among cultured hippocampal neurons. *Cereb. Cortex* 10:1045–52

Brown V, Jin P, Ceman S, Darnell JC, O'Donnell WT, et al. 2001. Microarray identification of FMRP-associated brain mRNAs and altered mRNA translational profiles in fragile X syndrome. *Cell* 107:477–87

Brown V, Small K, Lakkis L, Feng Y, Gunter C, et al. 1998. Purified recombinant Fmrp exhibits selective RNA binding as an intrinsic property of the fragile X mental retardation protein. *J. Biol. Chem.* 273:15521–27

Budd ME, Campbell JL. 1997. A yeast replicative helicase, Dna2 helicase, interacts with yeast FEN-1 nuclease in carrying out its essential function. *Mol. Cell. Biol.* 17:2136–42

Budd ME, Choe WC, Campbell JL. 1995. DNA2 encodes a DNA helicase essential for replication of eukaryotic chromosomes. *J. Biol. Chem.* 270:26766–69

Ceman S, Brown V, Warren ST. 1999. Isolation of an FMRP-associated messenger ribonucleoprotein particle and identification of nucleolin and the fragile X related proteins as components of the complex. *Mol. Cell. Biol.* 19:7925–32

Ceman S, Nelson R, Warren ST. 2000. Identification of mouse YB1/p50 as a component of the FMRP-associated mRNP particle. *Biochem. Biophys. Res. Commun.* 29:904–8

Chen L, Toth M. 2001. Fragile X mice develop sensory hyperreactivity to auditory stimuli. *Neuroscience* 4:1043–50

Chiurazzi P, Pomponi MG, Pietrobono R, Bakker CE, Neri G, Oostra BA. 1999. Synergistic effect of histone hyperacetylation and DNA demethylation in the reactivation of the FMR1 gene. *Hum. Mol. Genet.* 8:2317–23

Chiurazzi P, Pomponi MG, Willemsen R, Oostra BA, Neri G. 1998. In vitro reactivation of the FMR1 gene involved in fragile X syndrome. *Hum. Mol. Genet.* 7:109–13

Coffee B, Zhang F, Warren ST, Reines D. 1999. Acetylated histones are associated with FMR1 in normal but not fragile X-syndrome cells. *Nat. Genet.* 22:98–101

Comery TA, Haris JB, Willems PJ, Oostra BA, Irwin SA, et al. 1997. Abnormal dendritic spines in fragile X knockout mice: maturation and pruning deficits. *Proc. Natl. Acad. Sci. USA* 94:5401–4

Corbin F, Bouillon M, Fortin A, Morin S, Rousseau F, Khandjian EW. 1997. The fragile X mental retardation protein is associated with poly(A)$^+$ mRNA in actively translating polyribosomes. *Hum. Mol. Genet.* 6:1465–72

Crawley JN, Paylor R. 1997. A proposed test battery and constellations of specific behavioral paradigms to investigate the behavioral phenotypes of transgenic and knockout mice. *Horm. Behav.* 31:197–211

Dailey ME, Smith SJ. 1996. The dynamics of dendritic structure in developing hippocampal slices. *J. Neurosci.* 16:2983–94

Darnell JC, Jensen KB, Brown V, Jin P, Warren ST, Darnell RB. 2001. Fragile X mental retardation protein mRNA targets harboring intramolecular G-quartets encode proteins related to synaptic function. *Cell* 107:489–99

De Boulle K, Verkerk AJ, Reyniers E, Vits L, Hendrickx J, et al. 1993. A point mutation in the FMR-1 gene associated with fragile X mental retardation. *Nat. Genet.* 3:31–35

Devys D, Lutz Y, Rouyer N, Bellocq JP, Mandel JL. 1993. The FMR-1 protein is cytoplasmic, most abundant in neurons and appears normal in carriers of a fragile X premutation. *Nat. Genet.* 4:335–40

D'Hooge R, Nagels G, Franck F, Bakker CE, Reyniers E, et al. 1997. Mildly impaired water maze performance in male Fmr1 knockout mice. *Neuroscience* 76:367–76

Drouin R, Angers M, Dallaire N, Rose TM, Khandjian EW, Rousseau F. 1997. Structural and functional characterization of the human FMR1 promoter reveals similarities with the hnRNP-A2 promoter region. *Hum. Mol. Genet.* 6:2051–60

Dunn HG, Renpenning H, Gerrard JW, Miller JR, Tabata T, Federoff S. 1963. Mental retardation as a sex-linked defect. *Am. J. Ment. Defic.* 67:827–46

Dutch-Belgian Fragile X Consortium. 1994. Fmr1 knockout mice: a model to study fragile X mental retardation. *Cell* 78:23–33

Eberhart DE, Malter HE, Feng Y, Warren ST. 1996. The fragile X mental retardation protein is a ribonucleoprotein containing both nuclear localization and nuclear export signals. *Hum. Mol. Genet.* 5:1083–91

Eichler EE, Holden JJ, Popovich BW, Reiss AL, Snow K, et al. 1994. Length of uninterrupted CGG repeats determines instability in the FMR1 gene. *Nat. Genet.* 8:88–94

Eichler EE, Kunst CB, Lugenbeel KA, Ryder OA, Davison D, et al. 1995. Evolution of the cryptic FMR1 CGG repeat. *Nat. Genet.* 11:301–8

Eichler EE, Macpherson JN, Murray A, Jacobs PA, Chakravarti A, Nelson DL. 1996. Haplotype and interspersion analysis of the FMR1 CGG repeat identifies two different mutational pathways for the orgin of the fragile X syndrome. *Hum. Mol. Genet.* 5:319–30

Eichler EE, Richards S, Gibbs RA, Nelson DL. 1993. Fine structure of the human FMR1 gene. *Hum. Mol. Genet.* 2:1147–53

Eliez S, Blasey CM, Freund LS, Hastie T, Reiss AL. 2001. Brain anatomy, gender and IQ in children and adolescents with fragile X syndrome. *Brain* 124:1610–18

Feng Y, Absher D, Eberhart DE, Brown V, Malter HE, Warren ST. 1997a. FMRP associates with polyribosomes as an mRNP and the I304N mutation of severe fragile X syndrome abolishes this association. *Mol. Cell* 1:109–118

Feng Y, Gutekunst CA, Eberhart DE, Yi H, Warren ST, Hersch SM. 1997b. Fragile X mental retardation protein: nucleocytoplasmic shuttling and association with somatodendritic ribosomes. *J. Neurosci.* 17:1539–47

Feng Y, Zhang F, Lokey LK, Chastain JL, Lakkis L, et al. 1995. Translational suppression by trinucleotide repeat expansion at FMR1. *Science* 268:731–34

Fiala JC, Feinberg M, Popov V, Harris KM. 1998. Synaptogenesis via dendritic filopodia in developing hippocampal area CA1. *J. Neurosci.* 18:8900–11

Fisch GS, Hao HK, Bakker C, Oostra BA. 1999. Learning and memory in the FMR1 knockout mouse. *Am. J. Med. Genet.* 84:277–82

Fryns JP. 1986. The female and the fragile X: a study of 144 obligate female carriers. *Am. J. Med. Genet.* 23:157–69

Fu YH, Kuhl DP, Pizzuti A, Pieretti M, Sutcliffe JB, et al. 1991. Variation of the CGG repeat at the fragile site results in genetic instability: resolution of the Sherman paradox. *Cell* 67:1047–58

Gacy AM, Goellner G, Juranic N, Macura S, McMurray CT. 1995. Trinucleotide repeats that expand in human disease form hairpin structures in vitro. *Cell* 81:533–40

Gacy AM, McMurray CT. 1998. Influence of hairpins on template reannealing at trinucleotide repeat duplexes: a model for slipped DNA. *Biochemistry* 37:9426–34

Gedeon AK, Baker E, Robinson H, Partington MW, Gross B, et al. 1992. Fragile X syndrome without CCG amplification has an FMR1 deletion. *Nat. Genet.* 1:341–44

Giraud F, Ayme S, Mattei JF, Mattei MG. 1976. Constitutional chromosomal breakage. *Hum. Genet.* 34:125–36

Godfraind JM, Reyniers E, De Boulle K, D'Hooge R, De Deyn PP, et al. 1996. Long-term potentiation in the hippocampus of fragile X knockout mice. *Am. J. Med. Genet.* 64:246–51

Gordenin DA, Kunkel TA, Resnick MA. 1997. Repeat expansion—all in a flap? *Nat. Genet.* 16:116–18

Greenough WT, Hwang HM, Gorman C. 1985. Evidence for active synapse formation or altered postsynaptic metabolism in visual cortex of rats reared in complex environments. *Proc. Natl. Acad. Sci. USA* 82:4549–52

Gunter C, Paradee W, Crawford DC, Meadows KA, Newman J, et al. 1998. Re-examination of factors associated with expansion of CGG repeats using a single nucleotide polymorphism in FMR1. *Hum. Mol. Genet.* 7:1935–46

Hagerman RJ, Cronister A, eds. 1996. *Fragile X Syndrome Diagnosis, Treatment, and Research.* Baltimore, MD: Johns Hopkins Univ. Press. 481pp.

Hagerman RJ, Jackson C, Amiri K, Silverman AC, O'Connor R, Sobesky W. 1992. Girls with fragile X syndrome: physical and neurocognitive status and outcome. *Pediatrics* 89:395–400

Hagerman RJ, Leehey M, Heinrichs W, Tassone F, Wilson R, et al. 2001. Intention tremor, Parkinsonism, and generalized brain atrophy in male carriers of fragile X. *Neurology.* 57:127–30

Hagerman RJ, Van Housen K, Smith AC, McGavran L. 1984. Consideration of connective tissue dysfunction in the fragile X syndrome. *Am. J. Med. Genet.* 17:111–21

Hagerman RJ. 1996. Physical and behavioral phenotype. See Hagerman & Cronister 1996, pp. 3–87

Harris KM, Kater SB. 1994. Dendritic spines: cellular specializations imparting both stability and flexibility to synaptic function. *Annu. Rev. Neurosci.* 17:341–71

Harrison CJ, Jack EM, Allen TD, Harris R. 1983. The fragile X: a scanning electron microscopic study. *J. Med. Genet.* 20:280–85

Harvey J, Judge C, Wiener S. 1977. Familial

X-linked mental retardation with an X chromosome abnormality. *J. Med. Genet.* 14:46–50

Heitz D, Devys D, Imbert G, Kretz C, Mandel JL. 1992. Inheritance of the fragile X syndrome: size of the fragile X premutation is a major determinant of the transition to full mutation. *J. Med. Genet.* 29:794–801

Heitz D, Rousseau F, Devys D, Saccone S, Abderrahim H, et al. 1991. Isolation of sequences that span the fragile X and identification of a fragile-X–related CpG island. *Science* 251:1236–39

Henricksen LA, Tom S, Liu Y, Bambara RA. 2000. Inhibition of flap endonuclease 1 by flap secondary structure and relevance to repeat expansion. *J. Biol. Chem.* 22:16420–27

Hergersberg M, Matsuo K, Gassmann M, Schaffner W, Luscher B, et al. 1995. Tissue-specific expression of a FMR1/β-galactosidase fusion gene in transgenic mice. *Hum. Mol. Genet.* 4:359–66

Hinds HL, Ashley CT, Sutcliff JS, Nelson DL, Warren ST, et al. 1993. Tissue specific expression of FMR-1 provides evidence for a functional role in fragile X syndrome. *Nat. Genet.* 3:36–43

Hinton VJ, Brown WT, Wisniewski K, Rudelli RD. 1991. Analysis of neocortex in three males with the fragile X syndrome. *Am. J. Med. Genet.* 41:289–94

Hirst MC, Roche A, Flint TJ, MacKinnon RN, Bassett JH, et al. 1991. Linear order of new and established DNA markers around the fragile site at Xq27.3. *Genomics* 10:243–49

Hirst MC, White PJ. 1998. Cloned human FMR1 trinucleotide repeats exhibit a length- and orientation-dependent instability suggestive of in vivo lagging strand secondary structure. *Nucleic Acids Res.* 26:235–38

Hornstra IK, Nelson DL, Warren ST, Yang TP. 1993. High resolution methylation analysis of the *FMR1* gene trinucleotide repeat region in fragile X syndrome. *Hum. Mol. Genet.* 2:1659–65

Huber KM, Kayser MS, Bear MF. 2000. Role for rapid dendritic protein synthesis in hippocampal mGluR-dependent long-term depression. *Science* 288:1254–56

Hull C, Hagerman RJ. 1993. A study of the physical, behavioral, and medical phenotype, including anthropometric measures, of females with fragile X syndrome. *Am. J. Dis. Child.* 147:1236–41

Irwin SA, Galvez R, Greenough WT. 2000a. Dendritic spine structural anomalies in fragile X mental retardation syndrome. *Cereb. Cortex* 10:1038–44

Irwin SA, Idupulapati M, Mehta AB, Crisostomo RA, Rogers EJ, et al. 1999. Abnormal dendritic and dendritic spine characteristics in fragile-X patients and the mouse model of fragile-X syndrome. *Soc. Neurosci. Abstr.* 25:2548

Irwin SA, Patel B, Idupulapati M, Harris JB, Crisostomo RA, et al. 2001. Abnormal dendritic spine characteristics in the temporal and visual cortices of patients with fragile-X syndrome. *Am. J. Med. Genet.* 98:161–67

Irwin SA, Swain RA, Christmon CA, Chakravarti A, Weiler IJ, Greenough WT. 2000b. Evidence for altered fragile-X mental retardation protein expression in response to behavioral stimulation. *Neurobiol. Learn. Mem.* 74:87–93

Jakala P, Hanninen T, Ryynanen M, Laakso M, Partanen K, et al. 1997. FragileX: neuropsychological test performance, CGG triplet repeat lengths and hippocampal volumes. *J. Clin. Invest.* 100:331–38

Jansen G, Willems P, Coerwinkel M, Nillesen W, Smeets H, et al. 1994. Gonosomal mosaicism in myotonic dystrophy patients: involvement of mitotic events in $(CTG)_n$ repeat variation and selection against extreme expansion in sperm. *Am. J. Hum. Genet.* 54:575–85

Jin P, Warren ST. 2000. Understanding the molecular basis of fragile X syndrome. *Hum. Mol. Genet.* 9:901–8

Johnson GE. 1897. Contribution to the psychology and pedagogy of feeble-minded children. *J. Psycho-Asthen.* 2:26–32

Kang H, Schuman EM. 1996. A requirement

for local protein synthesis in neurotropin-induced hippocampal synaptic plasticity. *Science* 273:1402–6

Kates WR, Abrams MT, Kaufmann WE, Breiter SN, Reiss AL. 1997. Reliability and validity of MRI measurement of the amygdala and hippocampus in children with fragile X syndrome. *Psychiatry Res.* 75:31–48

Kaufmann WE, Moser HW. 2000. Dendritic anomalies in disorders associated with mental retardation. *Cereb. Cortex* 10:981–91

Kenneson A, Zhang F, Hagedorn CH, Warren ST. 2001. Reduced FMRP and increased FMR1 transcription is proportionally associated with CGG repeat number in intermediate-length and premutation carriers. *Hum. Mol. Genet.* 10:1449–54

Khandjian E, Corbin F, Woerly S, Rousseau F. 1996. The fragile X mental retardation protein is associated with ribosomes. *Nat. Genet.* 12:91–93

Kiledjian M, Dreyfuss G. 1992. Primary structure and binding activity of the hnRNP U protein: binding RNA through RGG box. *EMBO J.* 11:2655–64

Kremer EJ, Pritchard M, Lynch M, Yu S, Holman K, et al. 1991. Mapping of DNA instability at the fragile X to a trinucleotide repeat sequence p(CGG)$_n$. *Science* 252:1711–14

Kuhl D, Skehel P. 1998. Dendritic localization of mRNAs. *Curr. Opin. Neurobiol.* 8:600–6

Kumari D, Usdin K. 2001. Interaction of the transcription factors USF1, USF2, and α-Pal/Nrf-1 with the FMR1 promoter. *J. Biol. Chem.* 276:4357–64

Kunst CB, Warren ST. 1994. Cryptic and polar variation of the fragile X repeat could result in predisposing normal alleles. *Cell* 77:853–61

Laggerbauer B, Ostareck D, Keidel EM, Ostareck-Lederer A, Fischer U. 2001. Evidence that fragile X mental retardation protein is a negative regulator of translation. *Hum. Mol. Genet.* 10:329–38

Lam KS. 1997. Application of combinatorial library methods in cancer research and drug discovery. *Anticancer Drug Des.* 12:145–67

Lavedan C, Grabczyk E, Usdin K, Nussbaum RL. 1998. Long uninterrupted CGG repeats within the first exon of the human FMR1 gene are not intrinsically unstable in transgenic mice. *Genomics* 50:229–40

Lavedan CN, Garrett L, Nussbaum RL. 1997. Trinucleotide repeats (CGG)22TGG(CGG) 43TGG(CGG)21 from the fragile X gene remain stable in transgenic mice. *Hum. Genet.* 100:407–14

Lehrke RG. 1974. X-linked mental retardation and verbal disability. *Birth Defects* 10:1–100

Lendvai B, Stern EA, Chen B, Svoboda K. 2000. Experience-dependent plasticity of dendritic spines in the developing rat barrel cortex in vivo. *Nature* 404:876–81

Lewis HA, Musunuru K, Jensen KB, Edo C, Chen H, et al. 2000. Sequence-specific RNA binding by a Nova KH domain: implications for paraneoplastic disease and the fragile X syndrome. *Cell* 100:323–32

Li Z, Zhang Y, Ku L, Wilkinson KD, Warren ST, Feng Y. 2001. The fragile X mental retardation protein inhibits translation via interacting with mRNA. *Nucleic Acids Res.* 29: 2276–83

Loehr JP, Synhorst DP, Wolfe RR, Hagerman RJ. 1986. Aortic root dilation and mitral valve prolapse in the fragile X syndrome. *Am. J. Med. Genet.* 23:189–94

Loesch DZ, Hay DA, Mulley J. 1994. Transmitting males and carrier females in fragile X—revisited. *Am. J. Med. Genet.* 51:392–99

Losowsky MS. 1961. Hereditary mental defect showing the pattern of sex influence. *J. Ment. Defic. Res.* 5:60

Lubs HA. 1969. A marker X chromosome. *Am. J. Hum. Genet.* 21:231–44

Malter HE, Iber JC, Willemsen R, de Graaff E, Tarleton JC, et al. 1997. Characterization of the full fragile X syndrome mutation in fetal gametes. *Nat. Genet.* 15:165–69

Martin J, Bell J. 1943. A pedigree of mental defect showing sex-linkage. *Arch. Neurol. Psychiat.* 6:154–57

Merenstein SA, Sobesky WE, Taylor AK, Riddle JE, Tran HX, Hagerman RJ. 1996. Molecular-clinical correlations in males with

A

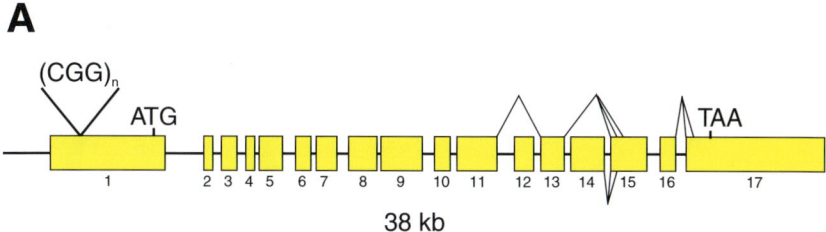

B

Normal

(CGG)$_{6-60}$ ATG

phenotype: normal
transmission: stable

methylation: no
transcription: yes

Premutation

(CGG)$_{60-200}$ ATG

phenotype: largely normal
transmission: unstable, prone to expansion

methylation: no
transcription: yes

Full mutation

(CGG)$_{>200}$ ATG

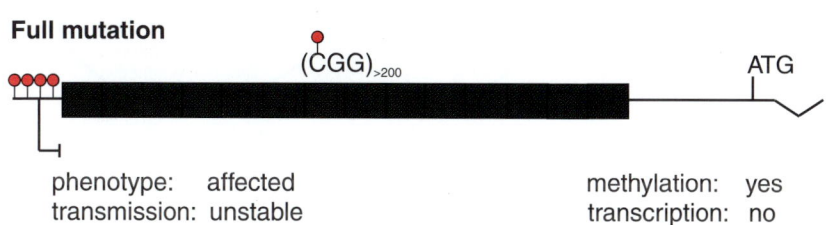

phenotype: affected
transmission: unstable

methylation: yes
transcription: no

Figure 1 Structure of *FMR1* and repeat expansion. (*A*) The 38 kb *FMR1* gene has 17 exons that undergo substantial alternate splicing. Note the CGG repeat in the 5′UTR of the gene. (*B*) Schematic representation of the repeat expansion seen in premutation and full mutation patients. Expansion over 200 repeats leads to promoter methylation and transcriptional silencing. *Black bar* represents the CGG repeat; *red lollipops* represent cytosine methylation of the repeat and promoter.

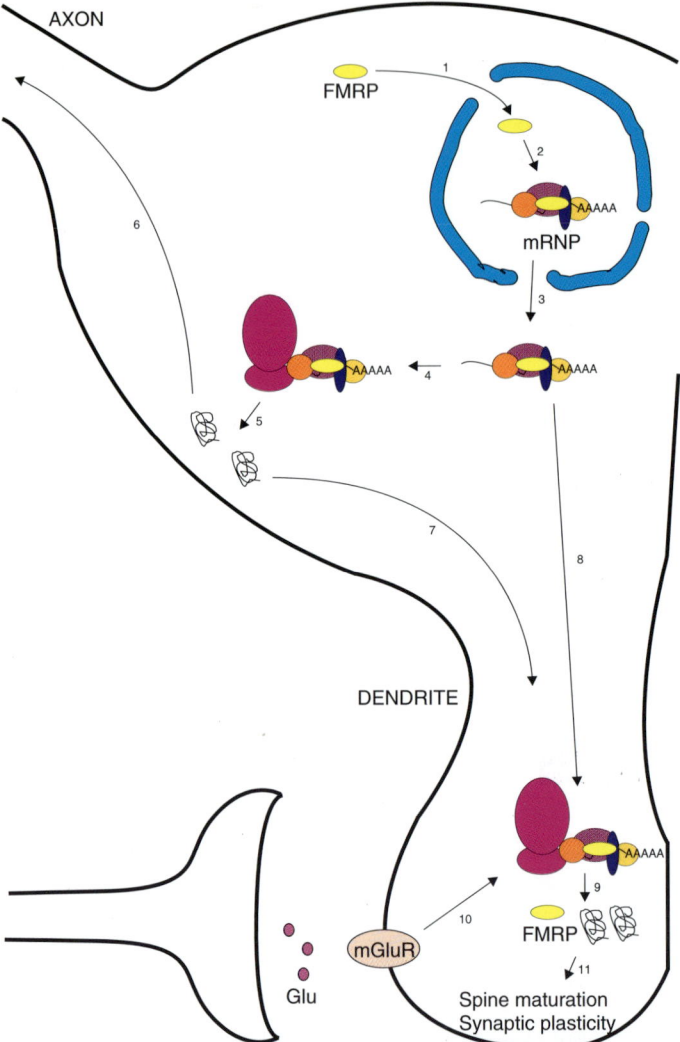

Figure 3 Model of FMRP function in the neuron. FMRP is transported into the nucleus of neurons via its nuclear localization signal (1). Once in the nucleus FMRP associates with 4% of cellular mRNAs as well as a number of proteins to form a large RNP (2) that is subsequently transported out of the nucleus via the nuclear export signal of FMRP (3). Once in the cytoplasm the FMRP mRNP can either associate with ribosomes in the cell body (4) producing proteins (5) that can then be transported into the axon (6) or dendrites (7), or the FMRP mRNP can itself be transported into the dendrites (8) and associate with ribosomes to produce proteins, including its own message and others in response in part to mGluR activation (10). Both the local translation of protein in the dendrite and transport of proteins from the cell body mediate synaptic plasticity and spine maturation (11).

an expanded FMR1 mutation. *Am. J. Med. Genet.* 64:388–94

Miyashiro K, Dichter M, Eberwine J. 1994. On the nature and differential distribution of mRNAs in hippocampal neurites: implications for neuronal functioning. *Proc. Natl. Acad. Sci. USA* 91:10800–4

Mostofsky SH, Mazzocco MM, Aakalu G, Warsofsky IS, Denckla MB, Reiss AL. 1998. Decreased cerebellar posterior vermis size in fragile X syndrome. *Neurology* 50:121–30

Moutou C, Vincent MC, Biancalana V, Mandel JL. 1997. Transition from premutation to full mutation in fragile X syndrome is likely to be prezygotic. *Hum. Mol. Genet.* 6:971–79

Musco G, Stier G, Joseph C, Antonietta M, Morelli C, et al. 1996. Three-dimensional structure and stability of the KH domain: molecular insights into the fragile X syndrome. *Cell* 85:237–45

Musumeci SA, Bosco P, Calabrese G, Bakker C, De Sarro GB, et al. 2000. Audiogenic seizures susceptibility in transgenic mice with fragile X syndrome. *Epilepsia* 41:19–23

Musumeci SA, Hagerman RJ, Ferri R, Bosco P, Dalla Bernardina B, et al. 1999. Epilepsy and EEG findings in males with fragile X syndrome. *Epilepsia* 40:1092–99

Nielsen KB, Tommerup N, Poulsen H, Mikkelsen M. 1981. X-linked mental retardation with fragile X. *Hum. Genet.* 59:23–25

Nimchinsky EA, Oberlander AM, Svoboda K. 2001. Abnormal development of dendritic spines in FMR1 knock-out mice. *J. Neurosci.* 21:5139–46

Oberle I, Heilig R, Moisan JP, Kloepfer C, Mattei GM, et al. 1986. Genetic analysis of the fragile-X mental retardation syndrome with two flanking polymorphic DNA markers. *Proc. Natl. Acad. Sci. USA* 83:1016–20

Oberle I, Rousseau F, Heitz D, Kretz C, Devys D, et al. 1991. Instability of a 550–base pair DNA segment and abnormal methylation in fragile X syndrome. *Science* 252:1097–102

Opitz JM, Westphal JM, Daniel A. 1984. Discovery of a connective tissue dysplasia in the Martin-Bell syndrome. *Am. J. Med. Genet.* 17:101–9

Paradee W, Melikian HE, Rasmussen DL, Kenneson A, Conn PJ, Warren ST. 1999. Fragile X mouse: strain effects of knockout phenotype and evidence suggesting deficient amygdala function. *Neuroscience* 94:185–92

Pearson CE, Eichler EE, Lorenzetti D, Kramer SF, Zoghbi HY, et al. 1998. Interruptions in the triplet repeats of SCA1 and FRAXA reduce the propensity and complexity of slipped strand DNA (S-DNA) formation. *Biochemistry* 37:2701–8

Peier AM, McIlwain KL, Kenneson A, Warren ST, Paylor R, Nelson DL. 2000. (Over) correction of FMR1 deficiency with YAC transgenics: behavioral and physical features. *Hum. Mol. Genet.* 9:1145–59

Penrose L S. 1938. A clinical and genetic study of 1,280 cases of mental cases of mental defect. *Spec. Rep. Ser.* 299. London: Med. Res. Counc.

Pieretti M, Zhang FP, Fu YH, Warren ST, Oostra BA, et al. 1991. Absence of expression of the FMR-1 gene in fragile X syndrome. *Cell* 66:817–22

Price DK, Zhang F, Ashley CT, Warren ST. 1996. The chicken FMR1 gene is highly conserved with a CCT 5' untranslated repeat and encodes an RNA-binding protein. *Genomics* 31:3–12

Purpura DP. 1974. Dendritic spine "dysgenesis" and mental retardation. *Science* 186:1126–28

Raymond CR, Thompson VL, Tate WP, Abraham WC. 2000. Metabotropic glutamate receptors trigger homosynaptic protein synthesis to prolong long-term potentiation. *J. Neurosci.* 20:969–76

Razin A. 1998. CpG methylation, chromatin structure and gene silencing—a three-way connection. *EMBO J.* 17:4905–8

Reiss AL, Abrams MT, Greenlaw R, Freund L, Denckla MB. 1995. Neurodevelopmental effects of the FMR-1 full mutation in humans. *Nat. Med.* 1:159–67

Reiss AL, Aylward E, Freund LS, Joshi PK, Bryan RN. 1991a. Neuroanatomy of fragile X syndrome: the posterior fossa. *Ann. Neurol.* 29:26–32

Reiss AL, Freund L, Tseng JE, Joshi PK. 1991b.

Neuroanatomy in fragile X females: the posterior fossa. *Am. J. Hum. Genet.* 49:279–88

Reiss AL, Lee J, Freund L. 1994. Neuroanatomy of fragile X syndrome: the temporal lobe. *Neurology* 44:1317–24

Reyniers E, Vits L, De Boulle K, Van Roy B, Van Velzen D, et al. 1993. The full mutation in the FMR-1 gene of male fragile X patients is absent in their sperm. *Nat. Genet.* 4:143–46

Richards BW, Sylvester PE, Brooker C. 1981. Fragile X-linked mental retardation: the Martin-Bell syndrome. *J. Ment. Defic. Res.* 25:253–56

Riddle JE, Cheema A, Sobesky WE, Gardner SC, Taylor AK, et al. 1998. Phenotypic involvement in females with FMR1 gene mutation. *Am. J. Ment. Retard.* 102:590–601

Rousseau F, Heitz D, Biancalana V, Blumenfeld S, Kretz C, et al. 1991. Direct diagnosis by DNA analysis of the fragile X syndrome of mental retardation. *N. Engl. J. Med.* 325:1673–81

Rudelli RD, Brown WT, Wisniewski K, Jenkins EC, Laure-Kamionowska M. 1985. Adult fragile X syndrome: clinico-neuropathologic findings. *Acta Neuropathol.* 67:289–95

Sabaratnam M, Vroegop PG, Gangadharan SK. 2001. Epilepsy and EEG findings in 18 males with fragile X syndrome. *Seizure* 10:60–63

Schaeffer C, Bardoni B, Mandel JL, Ehresmann B, Ehresmann C, Moine H. 2001. The fragile X mental retardation protein binds specifically to its mRNA via a purine quartet motif. *EMBO J.* 20:4803–13

Schwemmle S, de Graaff E, Deissler H, Glaser D, Wohrle D, et al. 1997. Characterization of FMR1 promoter elements by in vivo–footprinting analysis. *Am. J. Hum. Genet.* 60:1354–63

Sherman SL, Jacobs PA, Morton NE, Froster-Iskenius U, Howard-Peebles PN, et al. 1985. Further segregation analysis of the fragile X syndrome with special reference to transmitting males. *Hum. Genet.* 69:289–99

Sherman SL, Morton NE, Jacobs PA, Turner G. 1984. The marker (X) syndrome: a cytogenetic and genetic analysis. *Ann. Hum. Genet.* 48:21–37

Siomi H, Matunis MJ, Michael WM, Dreyfuss G. 1993a. The pre-mRNA binding K protein contains a novel evolutionarily conserved motif. *Nucleic Acids Res.* 21:1193–98

Siomi H, Siomi MC, Nussbaum RL, Dreyfuss G. 1993b. The protein product of the fragile X gene, FMR1, has characteristics of an RNA-binding protein. *Cell* 74:291–98

Siomi MC, Siomi H, Sauer WH, Srinivasan S, Nussbaum RL, Dreyfuss G. 1995. FXR1, an autosomal homolog of the fragile X mental retardation gene. *EMBO J.* 14:2401–8

Snow K, Doud LK, Hagerman R, Pergolizzi RG, Erster SH, Thibodeau SN. 1993. Analysis of a CGG sequence at the FMR-I locus in fragile X families and in the general population. *Am. J. Hum. Genet.* 53:1217–28

Sobesky WE, Taylor AK, Pennington BF, Bennetto L, Porter D, et al. 1996. Molecular/clinical correlations in females with fragile X. *Am. J. Med. Genet.* 64:340–45

Steward O, Bakker CE, Willems PJ, Oostra BA. 1998. No evidence for disruption of normal patterns of mRNA localization in dendrites or dendritic transport of recently synthesized mRNA in FMR1 knockout mice, a model for human fragile-X mental retardation syndrome. *Neuroreport* 9:477–81

Steward O, Levy WB. 1982. Preferential localization of polyribosomes under the base of dendritic spines in granule cells of the dentate gyrus. *J. Neurosci.* 2:284–91

Steward O, Reeves TM. 1988. Protein-synthetic machinery beneath postsynaptic sites on CNS neurons: association between polyribosomes and other organelles at the synaptic site. *J. Neurosci.* 8:176–84

Steward O, Schuman EM. 2001. Protein synthesis at synaptic sites on dendrites. *Annu. Rev. Neurosci.* 24:299–325

Sutcliffe JS, Nelson DL, Zhang F, Pieretti M, Caskey CT, et al. 1992. DNA methylation represses FMR-1 transcription in fragile X syndrome. *Hum. Mol. Genet.* 1:397–400

Sutherland GR. 1977. Fragile sites on human chromosomes: demonstration of their dependence on the type of tissue culture medium. *Science* 197:265–66

Sutherland GR. 1979. Heritable fragile sites on human chromosomes. *Hum. Genet.* 53:23–27

Suthers GK, Callen DF, Hyland VJ, Kozman HM, Baker E, et al. 1989. A new DNA marker tightly linked to the fragile X locus (FRAXA). *Science* 246:1298–1300

Suthers GK, Hyland VJ, Callen DF, Oberle I, Rocchi M, et al. 1990. Physical mapping of new DNA probes near the fragile X mutation (FRAXA) by using a panel of cell lines. *Am. J. Hum. Genet.* 47:187–95

Tamanini F, Bontekoe C, Bakker CE, van Unen L, Anar B, et al. 1999. Different targets for the fragile X-related proteins revealed by their distinct nuclear localizations. *Hum. Mol. Genet.* 8:863–69

Tamanini F, Kirkpatrick LL, Schonkeren J, van Unen L, Bontekoe C, et al. 2000. The fragile X-related proteins FXR1p and FXR2P contain a functional nucleolar-targeting signal equivalent to the HIV-1 regulatory proteins. *Hum. Mol. Genet.* 9:1487–93

Tassone F, Hagerman RJ, Taylor AK, Gane LW, Godfrey TE, Hagerman PJ. 2000. Elevated levels of FMR1 mRNA in carrier males: a new mechanism of involvement in the fragile-X syndrome. *Am. J. Hum. Genet.* 66:6–15

Todd PK, Mack KJ. 2000. Sensory stimulation increases cortical expression of the fragile X mental retardation protein in vivo. *Mol. Brain Res.* 80:17–25

Torre ER, Steward O. 1992. Demonstration of local protein synthesis within dendrites using a new cell culture system that permits the isolation of living axons and dendrites from their cell bodies. *J. Neurosci.* 12:762–72

Turner G, Daniel A, Frost M. 1980. X-linked mental retardation, macro-orchidism, and the Xq27 fragile site. *J. Pediatr.* 96:837–41

Uzielli ML, Guarducci S, Lapi E, Cecconi A, Ricci U, et al. 1999. Premature ovarian failure (POF) and fragile X premutation females: from POF to fragile X carrier identification, from fragile X carrier diagnosis to POF association data. *Am. J. Med. Genet.* 84:300–3

Van Dam D, D'Hooge R, Hauben E, Reyniers E, Gantois I, et al. 2000. Spatial learning, contextual fear conditioning and conditioned emotional response in Fmr1 knockout mice. *Behav. Brain Res.* 117:127–36

Verheij C, Bakker CE, de Graaff E, Keulemans J, Willemsen R, et al. 1993. Characterization and localization of the FMR-1 gene product associated with fragile X syndrome. *Nature* 363:722–24

Verkerk AJ, Pieretti M, Sutcliff JS, Fu YH, Kuhl DP, Pizzuti A, et al. 1991. Identification of a gene (FMR1) containing a CGG repeat coincident with a breakpoint cluster region exhibiting length variation in fragile X syndrome. *Cell* 65:905–14

Vincent A, Heitz D, Petit C, Kretz C, Oberle I, Mandel JL. 1991. Abnormal pattern detected in fragile-X patients by pulsed-field gel electrophoresis. *Nature* 349:624–26

Wan L, Dockendorff TC, Jongens TA, Dreyfuss G. 2000. Characterization of dFMR1, a *Drosophila melanogaster* homolog of the fragile X mental retardation protein. *Mol. Cell. Biol.* 20:8536–47

Warren ST, Knight SJ, Peters JF, Stayton CL, Consalez GG, Zhang F. 1990. Isolation of the human chromosomal band Xq28 within somatic cell hybrids by fragile site breakage. *Proc. Natl. Acad. Sci. USA* 87:3856–60

Warren ST, Sherman SL. 2001. The fragile X syndrome. In *The Metabolic and Molecular Basis of Inherited Disease*, ed. CR Scriver, AL Beaudet, WS Sly, D Valle, 8:1257–89. New York: McGraw Hill

Warren ST, Zhang F, Licameli GR, Peters JF. 1987. The fragile X site in somatic cell hybrids: an approach for molecular cloning of fragile sites. *Science* 237:420–23

Weiler IJ, Greenough WT. 1993. Metabotropic glutamate receptors trigger postsynaptic protein synthesis. *Proc. Natl. Acad. Sci. USA* 90:7168–71

Weiler IJ, Irwin SA, Klintsova AY, Spencer CM, Brazelton AD, et al. 1997. Fragile X mental retardation protein is translated near synapses in response to neurotransmitter activation. *Proc. Natl. Acad. Sci. USA* 94:5395–400

White PJ, Borts RH, Hirst MC. 1999. Stability of the human fragile X $(CGG)_n$ triplet repeat array in *Saccharomyces cerevisiae* deficient in aspects of DNA metabolism. *Mol. Cell Biol.* 19:5675–84

Wisniewski KE, Segan SM, Miezejeski CM, Sersen EA, Rudelli RD. 1991. The Fra(X) syndrome: neurological, electrophysiological, and neuropathological abnormalities. *Am. J. Med. Genet.* 38:476–80

Wohrle D, Kotzot D, Hirst MC, Manca A, Korn B, et al. 1992. A microdeletion of less that 250kb, including the proximal part of the FMR-I gene and the fragile site, in a male with the clinical phenotype of fragile-X syndrome. *Am. J. Hum. Genet.* 51:299–306

Yu S, Pritchard M, Kremer E, Lynch M, Nancarrow J, et al. 1991. Fragile X genotype characterized by an unstable region of DNA. *Science* 252:1179–81

Yuste R, Bonhoeffer T. 2001. Morphological changes in dendritic spines associated with long-term synaptic plasticity. *Annu. Rev. Neurosci.* 24:1071–89

Zhang Y, O'Connor JP, Siomi M, Srinivasan S, Dutra A, et al. 1995. The fragile X mental retardation syndrome protein interacts with novel homologs FXR1 and FXR2. *EMBO J.* 14:5358–66

Ziv NE, Smith SJ. 1996. Evidence for a role of dendritic filopodia in synaptogenesis and spine formation. *Neuron* 17:91–102

CONTEXTUAL INFLUENCES ON VISUAL PROCESSING

Thomas D. Albright and Gene R. Stoner

Howard Hughes Medical Institute, Systems Neurobiology Laboratories, The Salk Institute for Biological Studies, La Jolla, California 92037; email: tom@salk.edu; gene@salk.edu

Key Words occlusion, depth-ordering, figure-ground interpretation, filling-in, visual cortex

■ **Abstract** The visual image formed on the retina represents an amalgam of visual scene properties, including the reflectances of surfaces, their relative positions, and the type of illumination. The challenge facing the visual system is to extract the "meaning" of the image by decomposing it into its environmental causes. For each local region of the image, that extraction of meaning is only possible if information from other regions is taken into account. Of particular importance is a set of image cues revealing surface occlusion and/or lighting conditions. These information-rich cues direct the perceptual interpretation of other more ambiguous image regions. This context-dependent transformation from image to perception has profound—but frequently under-appreciated—implications for neurophysiological studies of visual processing: To demonstrate that neuronal responses are correlated with perception of visual scene properties, rather than visual image features, neuronal sensitivity must be assessed in varied contexts that differentially influence perceptual interpretation. We review a number of recent studies that have used this context-based approach to explore the neuronal bases of visual scene perception.

INTRODUCTION

The *Oxford English Dictionary* defines "context" as "the parts which immediately precede or follow any particular passage or 'text' and determine its meaning." More generally, context is the "whole situation, background, or environment relevant to a particular event, etc.," which reveals its meaning. These definitions, of course, beg the question of "meaning," to which there are no simple or all-encompassing answers. For our purposes, the definition of meaning is implicit in the central problems of vision, which are identifying the environmental causes of the patterns of light falling on the retinae, and the behavioral significance of those causes for the observer. Accordingly, the meaning of a contour of light on the retina includes, among other things, the particular object in the scene that reflected that pattern and the information that the object conveys about the world. Context, in turn, includes sensory cues that enable the image feature to be assigned to an object,

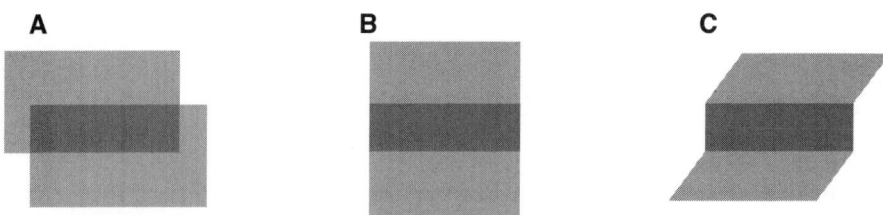

Figure 1 Illustration of the influence of local sensory context on visual perception. Each image contains a horizontal dark-gray rectangle. Although the rectangles are physically identical, the surrounding features differ in the three images. As a result, the rectangle is perceptually attributed to different environmental causes in the three instances, i.e., it conveys different meanings. (*A*) The rectangle appears to result from overlap of two surfaces, one of which is transparent (e.g., a piece of tinted glass). (*B*) The rectangle appears to result from a variation in surface reflectance (e.g., a stripe painted across a flat canvas). (*C*) The rectangle appears to result from variation in the angle of the surface with respect to the illumination source. These markedly different perceptual interpretations argue for the existence of different neuronal representations.

the observer's history with the object, his motivational state, and the reward value of the object to the observer. As a practical matter, extraction of meaning can be identified with perception, and context is the sensory/behavioral/cognitive milieu that influences the way each sensory feature is perceived.

Contextual influences on perception are necessarily manifested as interactions, rather than passive associations, between sensory, behavioral, or cognitive elements. For example, the percepts elicited by the images in Figures 1*A–C* are not merely of three different sets of features. On the contrary, the meaning of the dark gray bar is changed dramatically through its interactions with surrounding features. The perceptual whole is more than the sum of the sensory parts.

With these definitions as a foundation, we begin our review with a brief historical account of views on the role of context in vision. This history originates with philosophical doctrine and empirical psychology and is offered with the belief that past controversies, misunderstandings, and insights on this topic can guide and inform modern-day visual neuroscience. Indeed, the growing bond between psychology and neuroscience has led to remarkable new discoveries regarding the role of context in the neuronal bases of perception, as well as paradigmatic changes in the way neuronal representations are viewed and approached experimentally. These changes, and the data that support them, are the primary focus of our review.

HISTORICAL CONTEXT

The Reductionist Psychology of Perception

Discussions leading to our present-day understanding of the role of context in perception can be traced to the concept of "associationism" in the seventeenth-century philosophy of John Locke (1690). The crux of Locke's view was that perception results from a passive association of "ideas," of which immediate sensation is the primary source. Associationism remained a dominant theme in philosophy well into the nineteenth century. Its early proponents adopted the metaphor of "mechanical compounding" to characterize their conclusion that sensations are linked noninteractively to render the objects of perceptual experience.

In the mid–nineteenth century, associationism was embraced by many of the great figures in the emerging field of experimental psychology. Wilhelm Wundt (1902), in particular, pressed further the deconstruction of psychological reality by advancing "elementism." Whereas the associationists believed percepts to be the product of linked sensations, elementism held the inverse to be true: Any percept may be reduced to independent internal states (sensations) elicited by individual elements of the proximal stimulus (such as luminance or chrominance). Wundt's arguments were highly influential and fitted squarely with the emerging physiology of his time, specifically Johannes Müller's (1838) law of specific nerve energies and the growing belief that different sensory pathways convey specific and distinct—perhaps elemental—types of information.

Despite the prevalence of the Wundtian view, there were dissenting voices. William James (1890) was an associationist to be sure, but he promoted the position that perception is an emergent product of interactions between associated sensations. The Austrian physicist Ernst Mach held similar beliefs. In *Contributions to the Analysis of Sensations* (1886), Mach discussed context explicitly and provided striking demonstrations of its role in visual perception, which anticipated the Gestalt movement in experimental psychology.

Gestalt Psychology

Flawed by its failure to account for emergent properties of perception, elementism fell from favor and Gestalt psychology rose in its place. Max Wertheimer [1961 (1912)] founded the Gestalt movement and rejected elementism following his studies of "phi motion," now commonly known as "apparent motion." This phenomenon is the illusory percept of smooth or continuous motion that results when a stimulus (a spot of light, for example) is displaced by discrete intervals over space and time. (This is, of course, the basis for motion picture photography.) Elementism argues that this motion percept should be reducible to the elemental sensations, which consist of static points of light occurring at different points in space and time. This reduction is impossible, Wertheimer observed; perceived motion under these conditions is thus an emergent property that comes about through the relations between discrete sensations—again, the whole is more than the sum of the parts.

The Gestaltists offered a set of laws to characterize the sensory interactions underlying "perceptual organization." Despite the empirical validity of these laws and the many compelling demonstrations of contextual influences on visual processing, the Gestalt school did not exert a pervasive influence on the developing field of neurobiology in the latter half of the twentieth century. There are several reasons for this, including the heavy emphasis placed by the Gestaltists on phenomenology and their attendant failure to build firm neurobiological and mechanistic foundations for their beliefs. As a consequence, the Gestalt doctrine was rapidly eclipsed by a new reductionism, the neurophysiology of vision.

Neurophysiology of Vision: A Neurobiological Elementism?

With the development of techniques for recording neuronal action potentials and the subsequent application of these techniques to the study of the mammalian visual system, the field of visual science underwent a paradigm shift. Fundamental to the conduct and interpretation of neurophysiological experiments has been the "classical" concept of the receptive field (RF) as a discrete, spatially restricted, representational unit (Hartline & Graham 1932). Implicit in this concept is an assumption of functional independence: Once the information represented by individual RFs is known, it should be possible to deduce how neurons collectively represent the objects of perceptual experience. This assumption of independence, in turn, has fostered a localizationist view, in which different sensory elements are thought to be represented in different brain regions that are mechanistically and functionally independent of one another (e.g., Livingstone & Hubel 1987).

Conjunctively—and in spite of the many merits of the neurophysiological approach—these premises are a throwback to the nineteenth-century views of the associationists and the elementists. An association of sensory attributes represented by functionally independent neurons is reminiscent of "mechanical compounding," and the notion that different brain regions represent different attributes smells of Wundtian elementism. Neurophysiological approaches shackled by these premises cannot reveal in toto the neural substrates of perceptual experience—for the same reasons that psychological elementism cannot explain the contextual phenomena highlighted by the Gestaltists.

As a simple illustration, consider an orientation-selective neuron in primary visual cortex (Hubel & Wiesel 1968). The RF of the neuron illustrated in Figure 2 was characterized without contextual manipulations, and the data clearly reveal how the neuron represents the proximal (retinal) stimulus elements of orientation and direction of motion. From such data, however, it is frankly impossible to know whether this neuron would respond differentially to locally identical image features viewed in different contexts, such as the dark gray bars in Figures 1A–C. In other words, the full meaning of the pattern of responses in Figure 2 is unclear, as that pattern is not sufficient to reveal what the neuron conveys about the visual scene.

The Emergence of a Neurobiology of Perception

The ready ability to record neuronal activity in alert nonhuman primates has allowed one of the most notable achievements of modern neuroscience: the establishment of direct links between the behavior of single neurons and that of the whole organism (e.g., Mountcastle et al. 1972, Newsome et al. 1989; see also Barlow 1972).

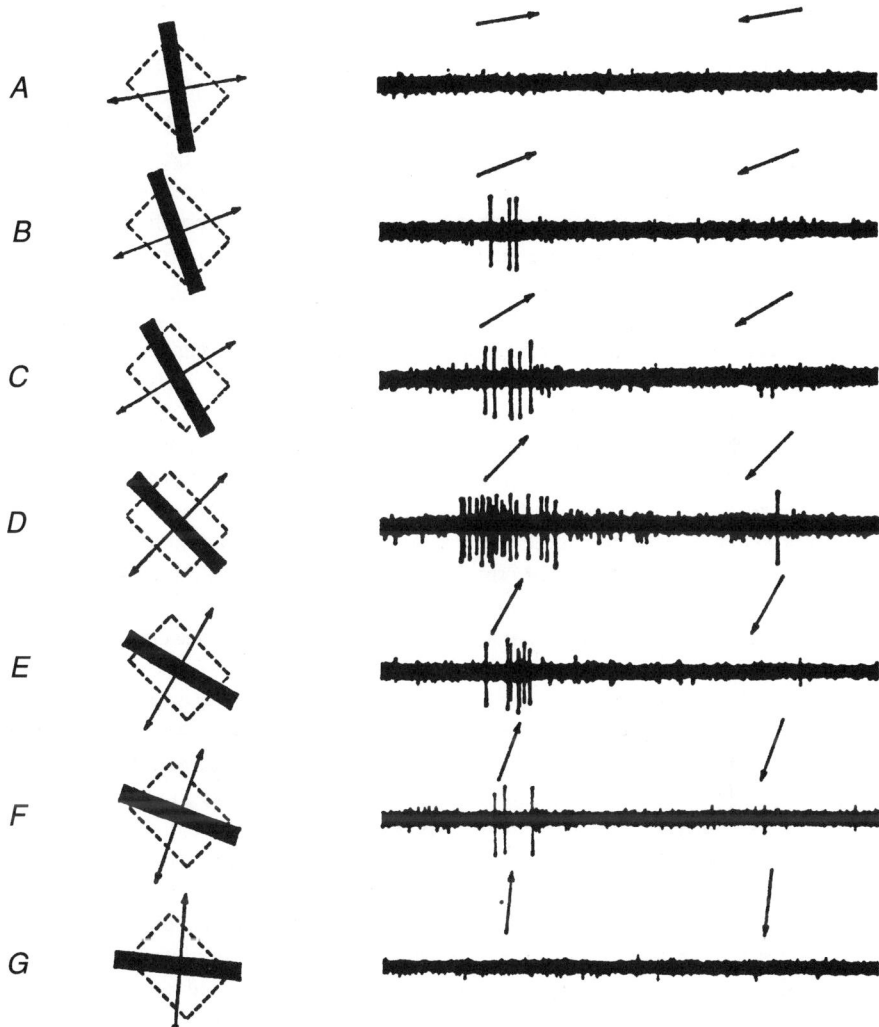

Figure 2 Neuronal directional selectivity, as first observed by Hubel & Wiesel (1968) in primary visual cortex (area V1) of rhesus monkey. The neuronal receptive field is indicated by broken rectangles in the left column. The visual stimulus was moved in each of 14 directions (rows A–G, opposing directions indicated by arrows) through receptive field. Recorded traces of cellular activity are shown at the right, in which the horizontal axis represents time (2 s/trace) and each vertical line represents an action potential. This neuron responds most strongly to motion up and to the right (row D). (From Hubel & Wiesel 1968. Permission from The Physiological Society.)

As impressive as such links are, their relevance to the extraction of "meaning" from visual images depends upon the meaningfulness of the stimuli being perceived. Unless neurophysiologists use stimuli embodying the "semantics" of natural images, they will advance—with or without the inclusion of behavioral links—only a few fledgling steps toward an understanding of the neuronal bases of perception.

To illustrate this assertion, imagine attempting to understand the function of language areas of the human brain armed only with nonlanguage stimuli such as frequency-modulated audible noise. Imagine further that, using these stimuli, you make the remarkable discovery that the responses of neurons within Wernicke's area (for example) are correlated with behavioral judgments (of, for instance, whether the frequency modulation was high-to-low or low-to-high). Although this finding would offer an interesting (and perhaps satisfyingly parametric) link between single neurons and perceptual decisions, it seems clear that stimuli constructed of words and sentences would yield results more likely to illuminate language processing. Just as we can only progress so far using nonsense sounds to explore language function, so are we constrained by visual stimuli lacking the rich cue interdependencies that permit the perceptual interpretation of natural scenes.

In recognition of this limitation, some have advocated an approach in which neuronal activity is recorded while animals are viewing natural scenes (Gallant et al. 1998, Stanley et al. 1999, Vinje & Gallant 2000). Clearly the sensory features of such scenes are replete with contextual cues, but disentangling the roles of specific cues from among the many that exert parallel influences on neurons and perception is a formidable challenge.

Much promise lies in a simpler and more controlled method inspired by the well-documented effects of local context on visual perception—such as the influence of one image region on the surface color appearance of an adjacent region (e.g., Shevell 1978, von Helmholtz 2000 [1860/1924], Wachtler et al. 2001) or the influence of depth cues on surface brightness (e.g., Figure 1) (e.g., Gilchrist 1977, Adelson 1993, Purves et al. 1999) and perceived direction of motion (Shimojo et al. 1989, Duncan et al. 2000). By systematically varying one contextual cue at a time, neuronal-perceptual links can be established that reveal how the visual system uses these cues to recover the properties of more complex natural scenes.

Using this controlled contextual approach, several recent studies have obtained evidence for parallel influences on neuronal and perceptual sensitivity. Before considering these studies in detail, we note a development that has profoundly impacted the evolution of a context-based approach: the revamping of the concept of the RF.

Influences from Beyond the Classical Receptive Field

Interest in neuronal effects of context became focused with the discovery of modulatory influences from beyond the classical RF. There were several early indications of such effects (e.g., Barlow 1953, McIlwain 1964, Hubel & Wiesel 1965, Nelson & Frost 1978), suggesting that the classical RF concept was flawed and perhaps limited efforts to understand neural substrates of perception. By the mid-1980s, however, the accumulation of evidence for response modulation via the "nonclassical RF," or RF "surround," reached a critical level, and the topic found a broad audience (Allman et al. 1985).

Perhaps the most well-documented nonclassical RF effect is the modulation, by motion in the surround, of neuronal responses in the middle temporal visual area (area MT) of primate cerebral cortex (Figure 3, see color insert). MT neurons are well known to exhibit a high degree of selectivity for the direction of motion of a stimulus within the classical RF (e.g., Albright 1984). The presence of motion outside the classical RF is, by definition, incapable of eliciting a response. Nonclassical RF motion leads, however, to marked modulation of the response to a stimulus within the classical RF (Figure 4) (Allman et al. 1985, Tanaka et al. 1986, Xiao et al. 1997). Similar surround effects have been reported for other forms of feature contrast, including line orientation (e.g., Gilbert & Wiesel 1990; Knierim & van Essen 1992; Kapadia et al. 2000; Li et al. 2000, 2001) and binocular disparity (Bradley & Andersen 1998).

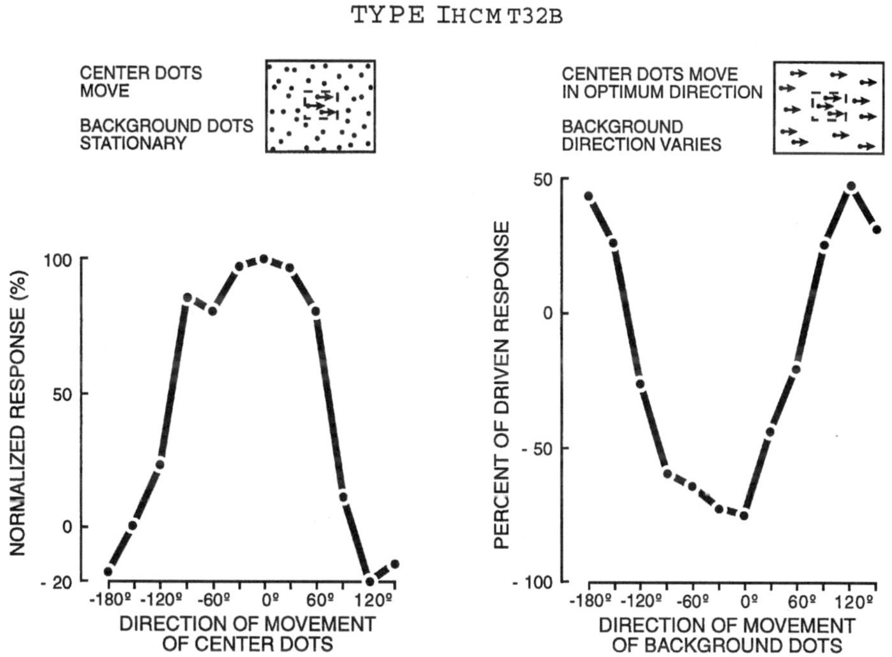

TYPE IHCM T32B

Figure 4 Responses to stimulus motion in the classical receptive field (RF) of a middle temporal (MT) neuron are modulated by motion in the RF "surround." (*Left*) Direction tuning for a moving random dot pattern confined to the classical RF. (*Right*) Modulation of the response to classical RF motion in the preferred direction as a function of the direction of surround motion. When center and surround motion were in the same direction (0°), the response to classical RF motion was suppressed. When center and surround moved in opposite directions, the response to classical RF motion was facilitated. (From Allman et al. 1985. Permission from the Annual Review of Neuroscience, Volume 8, © 1985 by Annual Reviews.)

Although these findings suggested a neuronal substrate through which perceptual context effects might be exerted, there exists an important difference between the definition of context as it has been applied to neuronal RF surround effects and the perceptual definition we have adopted. The neuronal definition is based strictly upon an interesting spatial interaction in the RF, i.e., contextual information from the surround modulates the neuronal response to the classical RF stimulus. By contrast, the perceptual definition of context is based on the ability of contextual information to disambiguate the environmental origins of a sensory stimulus. As we show below, there are phenomena that seem to satisfy both definitions, suggesting that surround modulation may indeed underlie specific types of perceptual disambiguation. However, surround modulation is surely not the only means by which perceptual context effects are implemented neuronally. For example, because surround modulation takes on a very specific spatial form (by definition), it is difficult to imagine how it might account for perceptual effects that occur when contextual information is spatially comingled with stimulus features that are disambiguated by that context [e.g., the ability of a feature's color to disambiguate, both perceptually and neuronally, the motion of that feature (Croner & Albright 1999a; for review see Croner & Albright 1999b).

ON THE VARIETIES OF CONTEXTUAL INFLUENCES

Broadly defined—as sources of information used to identify the meaning of a sensory stimulus—the scope of contextual influences on visual processing is vast and may include stimuli present at other points in space or time, as well as effects of attention, memory, and self-movement. For practical reasons, and in an effort to present our arguments in a coherent framework, we have chosen to concentrate on spatial contextual interactions in which information in one region of the visual image influences the interpretation of another region. Except where noted otherwise, the perceptual and neuronal phenomena we have reviewed are based on observations made in primates (humans and monkeys).

There is also a common functional theme fundamental to the context-based processes we consider: recovery of the multiple visual scene attributes associated with each single point in the visual image. The optical projection of a three-dimensional world onto a two-dimensional retina reduces the information at each spatial location to a single image value, which characterizes the light emanating from that location. Thus, although each value is a direct product of specific surface and lighting conditions in the visual scene, it is only possible to identify those conditions by taking into account the context in which the image value occurs. That context, of course, consists of other values in the image. But how is this context used to infer visual scene content? What are its critical attributes, and how do those attributes reveal surface and lighting relationships?

One way to approach these questions is to rephrase the problem as that of recovering "missing" information. Indeed, there exists a variety of circumstances, mostly intrinsic to the visual scene, but some intrinsic to the visual system itself, that cause attributes of the visual scene to be literally, fully or partially, missing from view. As an extreme example—but one of great functional significance—consider what happens when one opaque surface partially blocks the observer's view of another surface. In such cases a single image value arising from the occluder reflects only the properties of that surface. Contextual cues from surrounding spatial regions, however, lead to perceptual decomposition or "scission" of the image value into visible and implied surfaces and to perceptual interpolation ("filling-in" or "completion") of the "missing" attributes of the occluded surface.

A similar recovery of missing information occurs when the foreground surface is not opaque but transparent. Under these conditions each image value from the foreground surface is a physical composite of the separate reflectance and transmission properties of the overlapping surfaces, i.e., the actual properties of those surfaces are missing in the mix. Once again, surrounding contextual cues promote a scission of each image value into perceptually distinct properties of the transparent foreground and background surfaces (see Figure 1A).

A precisely analogous recovery process is engaged more generally—and ubiquitously—owing to the fact that each image value is a product of surface reflectance and illumination. Here also contextual cues promote scission: Just as context enables filling-in of surface properties missing behind an opaque or transparent occluder, context enables the viewer to see "behind" the illuminant to recover the surface reflectance properties of objects (see Figure 1C).

There are several computational steps that are broadly associated with these (and other related) context-mediated recovery processes. These include the interpretation of surface "depth-ordering," "boundary assignment," and filling-in. How these processes operate, what specific contextual cues are utilized, and what neuronal substrates and mechanisms are employed are the primary topics addressed in the research summarized below.

Although the visual scene information supplied by every visual image value is incomplete, this incompleteness, and hence the need for contextual resolution, is nowhere greater than when information about a part of the visual scene is absent altogether. We turn first to a discussion of how these informational gaps are filled-in using contextual cues.

FILLING-IN I: USING CONTEXT TO PREDICT MISSING INFORMATION

Filling-In the Blind Spot

Large enough to conceal 76 full moons (Campbell & Andrews 1992), the "blind spot" went unnoticed until the seventeenth century (Mariotte 1668). The blind spot is a roughly elliptical ($\sim$5° wide by $\sim$8° high) hole in the photoreceptor mosaic through which information exits (via the optic nerve) and nourishment enters (via blood vessels). The trunks of blood vessels arborize atop each retina, casting shadows and creating "angioscotomas" (Le Gros 1967). The blind spot and angioscotoma form a large, continuous, irregularly shaped interruption in the central visual field of each eye (sparing the fovea), which, in collaboration with other minor scotomas and those caused by neuronal pathologies, conspire to prevent us from seeing all that lies within our visual field. That the blind spot and scotomas go unnoticed is testament to the importance of filling-in of these visual holes (Ramachandran & Gregory 1991).

A question of great importance to our understanding of neuronal representation is whether filling-in is an active process in which spatial context explicitly provides the missing information, or simply ignorance of the fact that anything is absent (e.g., von Helmholtz 2000 [1860/1924], Dennet 1991, Churchland & Ramachandran 1993, Ramachandran 1994, De Weerd et al. 1995). About the blind spot, Helmholtz concluded (contradicting many of his contemporaries), "Nothing bright or coloured or dark is to be seen in the gap in the visual field. What we see there is literally nothing—a nothing that is not even a hole..." (von Helmholtz 2000 [1860/1924], 3:208). Equally emphatically, he asserts "... and *especially* [our italics]... no sensations whatever are transferred from the surrounding neighborhood to the gap."

Helmholtz's denial of a role for context in filling-in has not prevailed. On the contrary, in recent years the field has witnessed several intriguing perceptual demonstrations of phenomenal filling-in of the blind spot with information from adjacent regions of the visual field (e.g., Ramachandran & Gregory 1991, Ramachandran 1992), as well as quantitative psychophysical support for an active process (Paradiso & Hahn 1996). For example, He & Davis (2001) showed that filled-in content from the blind spot locus of one eye could suppress real image content from the corresponding visual field locus of the other eye. These investigators reasoned that such binocular rivalry favors an active filling-in mechanism because ignored (i.e., nonactively processed) information should not be able to compete with actively processed information. The most direct evidence for active context-based filling-in comes, however, from neurophysiological studies of the blind spot representation.

The blind spots of the two retinae do not correspond to the same parts of visual space, and hence the blind spot "representation" in primary visual cortex (V1) (see Figure 3 in color insert) is a large [about 5×10 mm in humans (Tong & Engel 2001)] monocular island within a larger sea of neurons fed by both eyes. Fiorani et al. (1992) found that a bar swept across the blind spot of anesthetized Cebus monkeys (monocular viewing) elicited a response from neurons in these blind spot zones. Neurons with this "completion property" required stimulation of opposite sides of the blind spot to be activated. Similarly, Komatsu et al. (2000) found that uniform rectangles, with edges well outside the blind spot, elicited responses in the blind spot representation in alert rhesus monkeys. Together these studies indicate that perceptual filling-in of the blind spot results from an active neuronal process fed by informational content present in surrounding regions of the visual field. Furthermore, filling-in occurs for both oriented features (such as surface boundaries) and homogenous image regions (such as surface interiors). From a functional perspective, this type of process is scarcely surprising, as local spatial context offers the best available clues to the missing content from the visual scene, and thus provides a meaningful resolution to sensory gaps.

Filling-In of Acquired or Induced Visual Gaps

Monocular retinal lesions, like the blind spot, rob information from a localized region of one eye. Whereas the visual system has had plenty of time to evolve adaptations to the visual gap created by the optic disk, the same is not true of pathological scotomas. Evidence nonetheless suggests that these visual holes are filled-in in humans (Gerrits & Timmerman 1969) and in monkeys with experimentally induced retinal lesions (Murakami et al. 1997).

Gerrits et al. (1966) asked whether filling-in extended to "artificial scotomas," created by temporarily removing localized information from the retina. These artificial scotomas were formed from a homogenous black patch that was positioned over a restricted region of the retina and surrounded by light. When eye position was fixed relative to the patch, the patch faded over the course of a few seconds and was replaced by a percept of the surrounding light. In subsequent experiments in which the patches were surrounded by dynamic colored texture, Ramachandran & Gregory (1991) found that the color and dynamic properties of the contextual texture filled-in, but color did so faster, implicating the involvement of different mechanisms and perhaps different cortical areas (De Weerd et al. 1998).

To explore the neuronal bases of these filling-in effects, De Weerd et al. (1995) positioned similarly constructed artificial scotomas over the classical receptive fields of neurons in cortical areas V1, V2, and V3 (see Figure 3) of alert rhesus monkeys. Following stimulus onset, neuronal activity in areas V2 and V3 (but not V1) gradually increased as if responding to the missing texture. The time course of this "climbing activity" correlated well with perceptual filling-in, suggesting that these neuronal events in V2 and V3 may underlie the perceptual effect.

In a study bearing on mechanism, Pettet & Gilbert (1992) found that, when presented with artificial scotomas, V1 RFs increased in size up to fivefold. Neurons, deprived of information within their classical RF, extend their sensitivity to surrounding regions of visual space. This enlarged field could serve to fill-in the scotoma with information from the surrounding spatial field, just as sensitivity to regions surrounding the blind spot permits neurons in the blind spot representation to fill-in that visual hole. Similar changes in RF size have been seen minutes after focal retinal lesions (Chino et al. 1992, Gilbert & Wiesel 1992). Although longer-term plasticity had been documented in adult animals after restricted deafferentation (e.g., Kaas 1991, Garraghty & Kaas 1992), the results of Pettet & Gilbert's study suggest that perceptual filling-in reflects a form of neuronal plasticity in which visual input alters RF structure and cortical topography from moment to moment. This remarkable suggestion challenges the long-standing view that a neuron's visual RF is a stable entity within the healthy adult.

It should be pointed out that the time scale for perceptual filling-in of artificial scotomas (De Weerd et al. 1998) is not as fast as the seemingly instantaneous perceptual filling-in of the blind spot, nor is it as fast as measured for other types of filling-in (Paradiso & Hahn 1996, Paradiso & Nakayama 1991). This suggests that different neuronal mechanisms are involved in these different types of filling-in (De Weerd et al. 1998). Furthermore, the neuronal effects observed by Pettet & Gilbert (1992) occurred over a period of several minutes, not seconds, and hence the exact relationship between their observed RF changes and perceptual filling-in remains to be determined. Taken together, the different time courses observed for different types of neuronal and perceptual filling-in suggest the involvement of multiple mechanisms.

FILLING-IN II: SURFACE OCCLUSION

Intrinsic, pathological, and artificial scotomas of the sort discussed above are all characterized by loss of visual information from a fixed retinal locus—i.e., the gap moves with the eye or is present only when the eye is in a fixed position. There are, however, other circumstances of transient visual information "loss" that are caused by surface occlusion in the visual scene. These circumstances are, of course, endemic to normal visual experience, and the resolution of information loss is made possible by context in a manner that is functionally—and perhaps mechanistically—similar to the resolution of loss from scotomas.

Although we provide evidence for common principles, occlusion-based informational gaps differ importantly from those caused by defects intrinsic to the visual system. Occluders elicit a sense of depth-ordered surfaces in the visual scene. Gaps caused by the optic disk and retinal vasculature do not. Accordingly, an observer cannot usually perceive his/her own optic disk or retinal vasculature but generally will see occluding surfaces.

It is useful to distinguish two context-based processes in the resolution of information gaps caused by surface occlusion. The first involves establishing which image regions constitute "figure" and "ground" (occluder and occluded) using a set of characteristic image cues. The second process, which is intimately tied to the first, involves filling-in or perceptual completion of occluded content from the visual scene and depends upon local contextual cues that offer evidence of that missing content.

Surface Depth-Ordering, Figure-Ground Interpretation, and Border Ownership

Figure-ground interpretation is coupled to the establishment of the three-dimensional spatial arrangement of surfaces; image regions in the foreground depth plane are generally perceived as figure. There are, of course, well-known sources of depth information (as a scalar quantity) present in the visual image, such as binocular disparity and motion parallax, which serve this function. In addition, there are image consequences that are specific to surface occlusion, which supply reliable contextual cues for surface depth-ordering (an ordinal quantity). One of the most common of such cues is due to the horizontal displacement of the two eyes, which enables one eye to see part(s) of the visual scene that a foreground surface occludes from the other eye. Human observers are particularly sensitive to this monocular occlusion cue, which is known as Da Vinci stereopsis (Anderson 1994, Nakayama & Shimojo 1990).

Occlusion produces several other types of depth-ordering cues that emerge from the distinctive characteristics of overlapping surfaces. One important cue consists of spatially defined "X-" and "T-junctions," which are stereotyped contour arrangements in visual images that generally occur where edges of two surfaces overlap (Kanizsa 1979, Beck et al. 1984). A second cue consists of temporally defined "accretion-deletion," which results from the progressive uncovering and/or covering of a background surface by a foreground surface (Kaplan 1969, Stoner & Albright 1995). A third cue consists of the coincident alignment of unconnected line endings or edges, which is indicative of occlusion by a common surface (Kanizsa 1979). The presence of any of these contextual cues generally biases perceived depth-ordering and figure-ground interpretation in human observers.

The Gestalt psychologists emphasized a number of additional image cues that lead to figure-ground interpretation but do not offer explicit evidence of occlusion. For example, regions within a closed boundary usually appear as figure (Koffka 1935). Furthermore, if the boundary resembles the profile of a known object, the image region corresponding to the implied object will tend to be seen as figure (Rubin 1921, Peterson 2002).

In some of the earliest neurophysiological studies of this topic, Lamme (1995) and Zipser et al. (1996) searched for correlates of figure-ground interpretation in area V1. Stimuli consisting of texture-defined rectangles (perceptual figures defined by boundary closure) were positioned such that individual texture elements drawn from figure or ground were within the classical RF. V1 responses to figure elements were reported to be significantly larger than responses to ground elements. This result implies that figure versus ground is encoded by differential response rate, and that larger spatial context elicits this differential. Using stimuli similar to those of Lamme (1995), however, a more recent study (Rossi et al. 2001) failed to find any response differential in V1 for figure versus ground elements in the classical RF (see also Cumming & Parker 1998). On the other hand, using stimuli consisting of two overlapping rectangles, which possess robust occlusion-based depth-ordering cues (T-junctions), Chang et al. (1999, 2001) detected small populations of both V1 and V2 neurons that responded differentially to the interiors of figure versus ground rectangles. Although differential responses to figure versus ground interiors could constitute the neuronal substrate for the heightened salience of figure versus ground (e.g., Kanizsa 1979), further evidence is needed to resolve the discrepancies between these different studies.

The attribution of figures to one side or the other of an image boundary is commonly referred to as assigning "border ownership" (Koffka 1935, Rubin 1921,

Zhou et al. 2000). Border ownership identifies which side of the boundary is seen as the occluder and which side as the occluded background. Several recent studies have found evidence of neuronal correlates of this perceptual process at early stages within the visual pathway. Zhou et al. (2000) stimulated V1, V2, and V4 (see Figure 3, see color insert) neurons using stimuli that consisted of single rectangles (figure defined by closure) and overlapping pairs of rectangles (figure defined by occlusion-based depth-ordering cues, i.e., T-junctions). These investigators placed the border between figure and ground within the classical RF and varied border ownership, such that the figure lay on one side of the RF or the other. Slightly more than half of the recorded neurons in areas V2 and V4 exhibited significantly different responses for contrast edges as a function of border ownership (Figure 5). A smaller fraction of V1 neurons showed similar effects. The spatial integration of these contextual cues was found to extend to at least 20°. Moreover, they emerged almost immediately following stimulus onset, suggesting a rapid conveyance of contextual information from the nonclassical RF. Zhou et al. also found that different cues for border ownership (closure, T-junctions, relative size) yielded similar effects, which indicates that these neurons were encoding border ownership per se, rather than a particular cue for border ownership.

Chang et al. (2001) used a different strategy to differentiate selectivity for border ownership from sensitivity to incidental aspects of stimulus geometry. Specifically, stimuli with occlusion cues (T-junctions) yielding ambiguous figure-ground interpretation were compared with those with unambiguous interpretation. The unambiguous stimuli were overlapping rectangles similar to those of Zhou et al. This enlarged stimulus set allowed identification of neurons responding merely to the presence of T-junctions, rather than to depth-ordering per se. With the incorporation of these experimental controls, Chang et al. (2001) confirmed that responses correlated with border ownership existed within V2, although these effects were modest in magnitude and prevalence and were rarely seen in V1.

Two additional studies sought neuronal correlates of border ownership using stimuli with occlusion cues local to the stimulating border and thus present in the classical RF. Baumann et al. (1997) stimulated neurons in areas V2 and V3/V3A with spatially defined occlusion cues and observed responses that were consistent with border ownership. Stoner et al. (1998) stimulated neurons in area MT using temporally defined occlusion cues (i.e., accretion-deletion) and found evidence for figure-ground selectivity. Much more work is needed to establish the precise role(s) of these different cortical areas in figure-ground interpretation. The existing neurophysiological evidence suggests, nonetheless, that neuronal representations of figure-ground relationships are pervasive and are manifested early in the visual pathway.

Completion of Occluded Information

As we have seen, figure-ground interpretation follows the occlusion of one surface by another. Figure-ground interpretation itself leads to another important process: the filling-in of occluded features. There are two types of filling-in associated

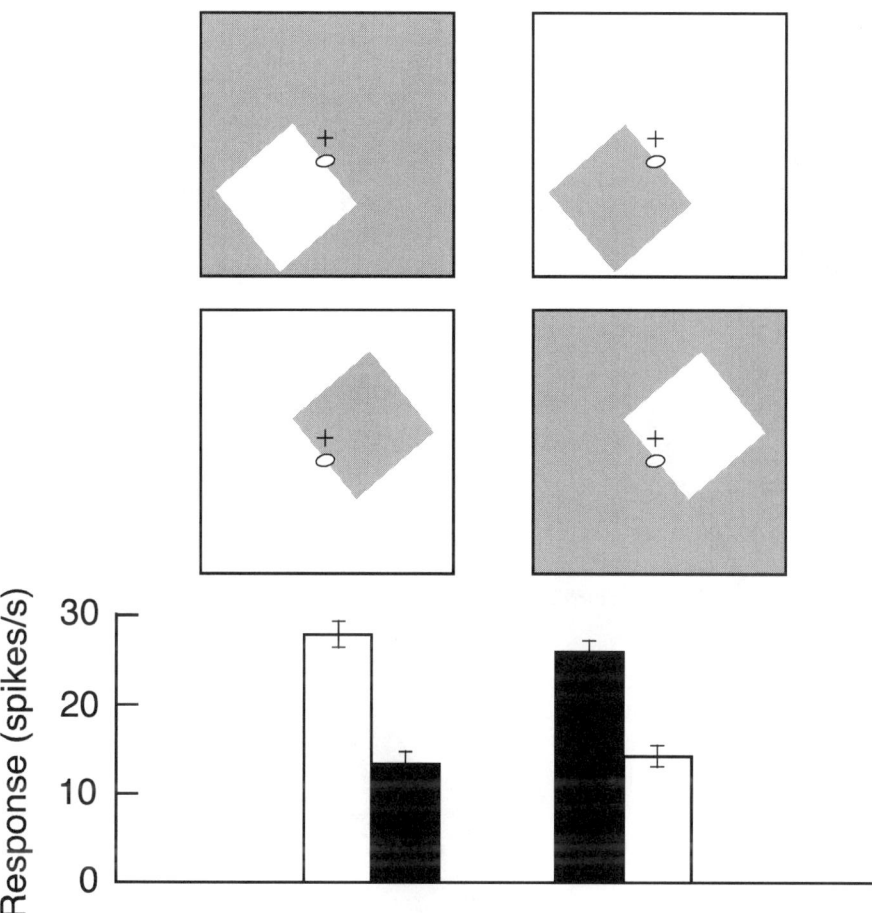

Figure 5 Border ownership responses from a V2 neuron. Responses to white squares on gray background are indicated by open bars. Responses to gray squares on white background are indicated by filled bars. Stimuli vary in luminance contrast polarity (i.e., white vs. gray or gray vs. white) across columns and implied border ownership across rows. This cell preferred the border ownership implied by the top row of stimuli in which figures were on the bottom left side of the receptive field (RF). Ellipses indicate classical RF (note that the preferred orientation is that of the short axis of the ellipse). This cell differentiated displays that were identical in an 8 × 16° region around the RF. (From Zhou et al. 2000. Copyright 2000 by the Society for Neuroscience.)

with occlusion, termed "modal" and "amodal" completion (Michotte et al. 1964). Modal completion refers to illusory image features that appear to have resulted from direct stimulation of the visual modality (Kanizsa 1979). The classic Kanizsa triangle (Figure 6) elicits illusory contours, which comprise a case of modal completion driven by contextual cues for surface occlusion. As illustrated in the

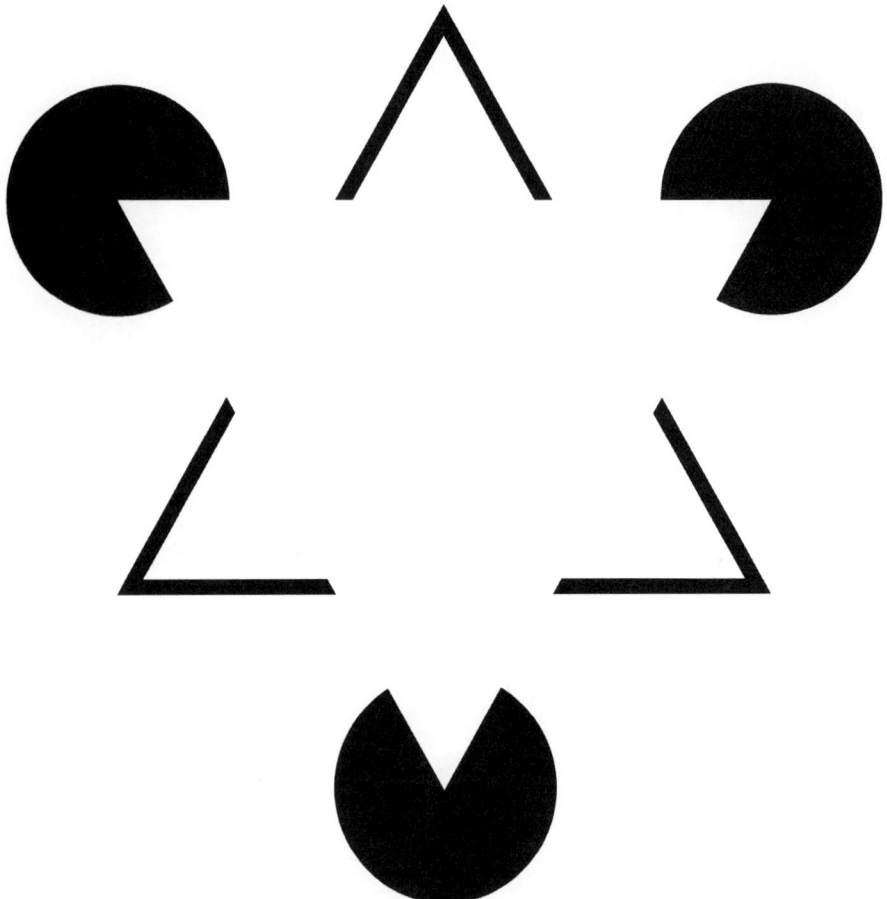

Figure 6 The Kanizsa triangle illustrates two forms of filling-in—modal and amodal completion—elicited by contextual cues for surface occlusion. Modal completion: Coincident alignment of contrast edges and line endings allows for perceptual interpolation of a foreground occluding triangle. Although the inferred foreground occluder and background are physically indistinct, contextual cues enable this "missing" boundary information to be filled-in. The resulting illusory contours have a sensory-like quality or "modal presence." Amodal completion: The existence of a triangular foreground occluder is implied, in part, by the V-shaped segments missing from each of three black features. The occluder implies, in turn, that the black features are actually partially occluded full circles. Similarly, the three black-outlined Vs along the edges of the foreground occluding surface appear to compose a second, partially occluded, triangle. Although observers do not perceive the filled-in occluded features of these objects with the verisimilitude of their unoccluded portions, they are nonetheless fully and directly aware of the presence of occluded features. Moreover, the filled-in occluded features possess very specific forms, which are largely dictated by contextual cues, and observers can readily direct actions toward filled-in features, suggesting that they are explicitly represented in the brain. (From Kanizsa 1979.)

Kanizsa triangle, modal completion usually requires that foreground and occluded background have the same apparent intensity—a relatively rare occurrence in natural scenes. [The illusory "phi motion" discovered by Wertheimer [1961 (1912)] and perceptual filling-in of the blind spot/scotomas (all reviewed above) are also striking examples of modal completion, though unrelated to surface occlusion.]

In seminal experiments von der Heydt et al. (1984) explored neuronal correlates of modal completion. Stimuli consisted of illusory contours, which were induced by contextual cues for the presence of a foreground occluding surface (analogous to the case illustrated in Figure 6). The illusory contours were moved through the classical RFs of V1 and V2 neurons by moving the inducing context through the RF surround. Remarkably, when contextual cues were configured to elicit a percept of illusory contours, they also elicited neuronal responses in V2 (but rarely V1) that were comparable to responses elicited by real contours (Figure 7)—despite the fact that the illusory contours presented no physical contrast within the classical RF! Redies et al. (1986) obtained similar results from cat cortical areas 17 and 18.

Although these data support a model in which modal completion evolves in V2 from real contour signals present in V1, the V1-V2 distinction made by von der Heydt and colleagues has been challenged. Grosof et al. (1993) and Lee & Nguyen (2001) found that a significant fraction of V1 neurons responded to illusory contours (see also Sheth et al. 1996, Ramsden et al. 2001). In addition, the neuronal blind spot filling-in effects reported in V1 by Komatsu et al. (2000) (see above) suggest the existence of a mechanism for contour completion in V1. Lee & Nguyen reported, however, that V1 responses to illusory contours lag behind responses to real contours by about 55 ms, whereas this difference is only ~30 ms in area V2. They suggest that this lag could reflect feedback latencies from area V2 or from horizontal connections intrinsic to V1. As we shall see, the debate about the contribution of V1 to modal contour completion is further complicated by recent evidence that V1 neurons represent amodal contours.

Amodal completion, which is the second type of filling-in, refers to perceptual interpolation of features seen to lie behind a foreground occluding surface [termed amodal, according to Kanizsa (1979), because the completed information "is not verified by any sensory modality"]. Matching the ubiquity of occlusion in the visual environment, amodal completion is much more common, and hence arguably of greater functional significance than the modal variety.

The specific features that are amodally completed are determined by contextual cues surrounding the occluding surface. Given that we do not actually "see" amodal features (unlike modally completed features), it might be thought that they are not explicitly represented. Counter to this supposition, however, are the observations that amodal features have a specific form, surface character, and extent, and are not easily modifiable by cognitive intent (e.g., try imagining the occluded triangle contours in Figure 6 as anything other than straight lines). Moreover, an observer can easily direct a motor action to amodal features (as one might reach behind a book to grab the edge of a desk it partially occludes). Quantitative psychophysical evidence supports these anecdotal phenomena (e.g., Nakayama & Shimojo 1990)—all of which argues that amodal features are explicitly represented by the brain.

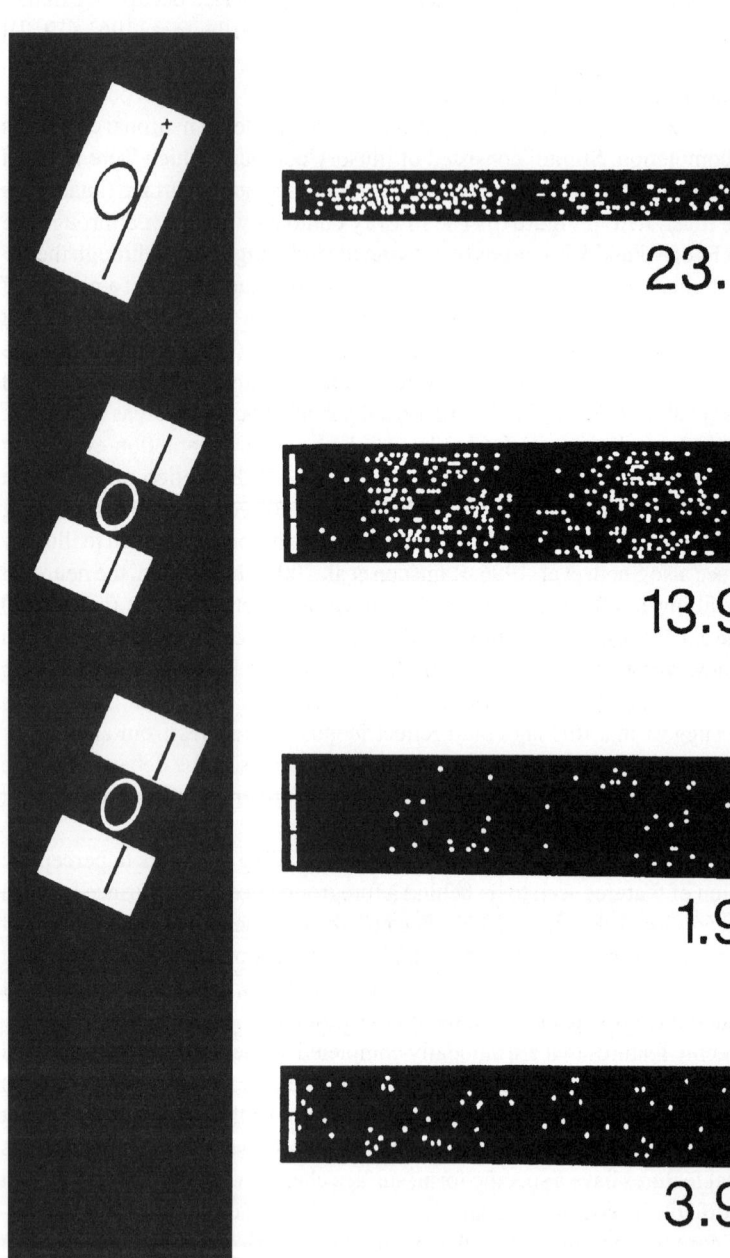

A 23.1

B 13.9

C 1.9

D 3.9

Even if explicitly represented, the phenomenally "nonvisual" character of amodal features would seem to place them in the domain of higher-order visual areas. This appears not to be the case. Sugita (1999) reported neuronal correlates of amodal completion within area V1. When presented with two collinear line segments placed on opposite sides of the classical RF, over 10% of V1 neurons responded if the gap between the segments was placed stereoscopically near (hence consistent with occlusion) (Figure 8). By contrast, these same neurons failed to respond if the gap was positioned stereoscopically behind the line segments, such that amodal completion was prohibited. Sugita found that response latencies to amodal contours were similar to those elicited by real contours presented within the classical RF. On this basis, Sugita concluded that these responses are likely to be mediated by horizontal connections within V1 or perhaps by short-latency feedback connections from V2.

The sensitivity of V1 to amodal features implies neuronal "X-ray vision"; with the help of contextual cues, area V1 constructs a representation of what lies hidden behind occluders. The apparent qualitative similarity between neuronal representations of amodal, modal, and real contours seems to conflict, however, with their differing perceptual qualities. On the other hand, as we have noted, amodally completed features are tangible enough to expect some processing overlap with real features. The degrees of processing differences and overlap are sure to emerge as research begins to focus on the cellular mechanisms by which context gives rise to these forms of filling-in.

INTERRUPTED CONTOURS: FACILITATION AND GROUPING

As we have seen, contours can be interrupted in the visual image by a variety of means, including surface occlusion and shadows. Depending upon the visual configuration, contours may be perceptually completed across these gaps (modally or amodally). Two related behavioral paradigms have also been used to explore

---

Figure 7 Responses of a V2 neuron to modal (illusory) contours. Stimuli are shown at left. Ellipses indicate the classical receptive field (RF). Crosses indicate the position of the monkey's fixation point. Each row of dots in the right column corresponds to a forward and backward sweep across the neuronal RF. Each dot represents an action potential. (*A*) Responses to a real contour. (*B*) Responses to a modally completed (illusory) contour were qualitatively similar, even though no physical contrast was present in its classical RF. (*C*) Placing thin intersection lines between black contours and intervening black patch prevented modal completion. Consistent with perception, no response was observed. (*D*) Spontaneous activity. Numbers below each panel of spike trains indicate mean spike counts per stimulus cycle. (From Peterhans & von der Heydt 1989. Copyright 1989 by the Society for Neuroscience.)

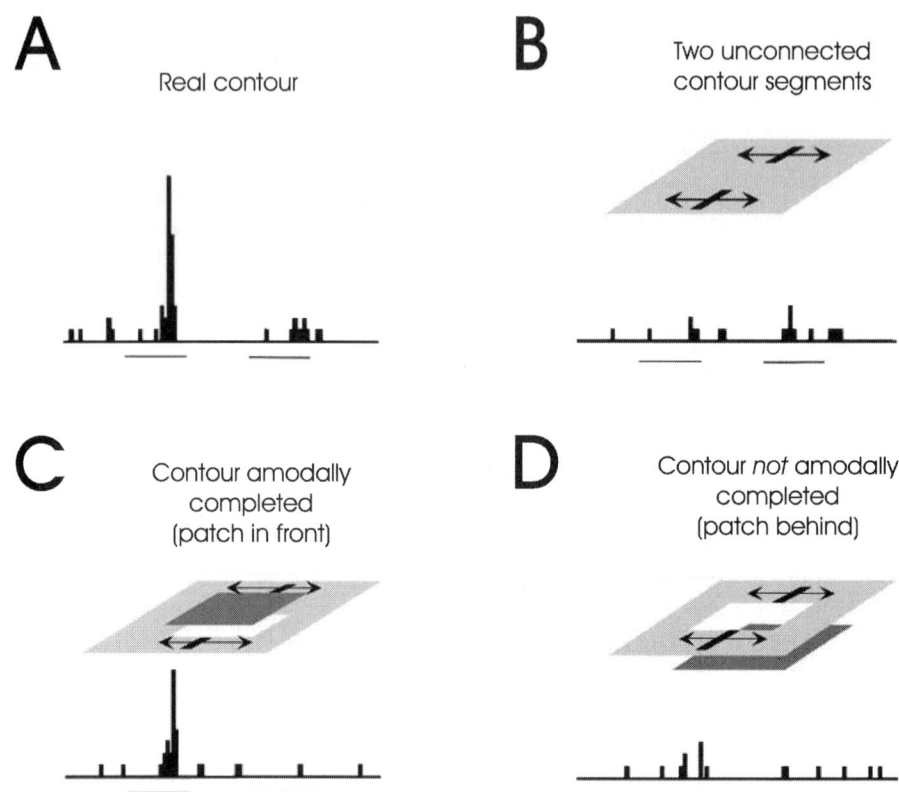

Figure 8 Responses of a V1 neuron to amodal contours. (*A*) Responses to a real contour. (*B*) Responses to two unconnected line segments (no stimulus in classical receptive field). (*C*) Responses to two collinear segments with a patch placed stereoscopically in front. (*D*) Responses to two collinear segments with a patch placed stereoscopically behind. Each stimulus type was visible during periods indicated by bars under each histogram and was swept in each of two directions (left bar, preferred direction; right bar, not preferred direction). This neuron exhibited directionally selective responses to both real (*A*) and amodally completed (*C*) contours. Little response was observed for unconnected segments in which amodal completion was not engaged by the presence of a foreground occluding surface (*B* and *D*). (From Sugita 1999.)

contextual interactions between contour fragments. The first has been used to show that collinear contour fragments tend to be perceptually "grouped" (Field et al. 1993). The second paradigm of this genre has been used to demonstrate that the processing of oriented contour fragments can be facilitated by flanking collinear contours (Polat & Sagi 1993, 1994; Kapadia et al. 1995).

Neurophysiological experiments that aim to identify neuronal correlates of grouping and contour facilitation require assessment of the degree to which the response to an oriented RF stimulus is modulated by contours in the RF surround. Many studies have documented surround effects in area V1 using oriented stimuli

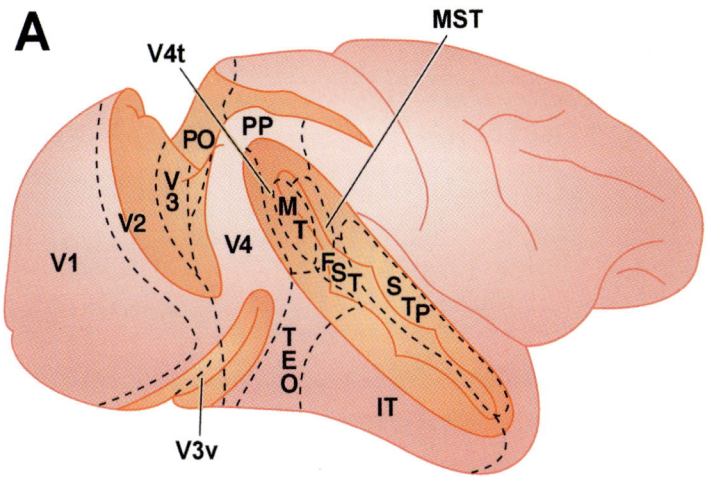

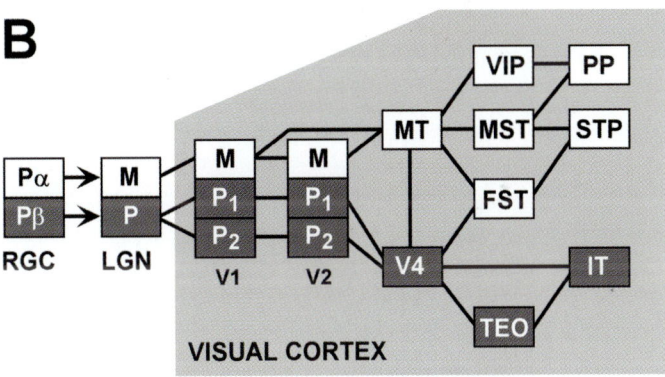

Figure 3 Organization of visual cortex in the rhesus monkey (*Macaca mulatta*), which has been the primary subject in studies of the neuronal substrates of contextual influences on visual perception. (*A*) Lateral view of the monkey brain, which illustrates the locations of some of the known visual areas. Primary visual cortex (area V1) is located on the occipital pole (*left*); "extrastriate" cortical visual areas extend anteriorally (*rightward*) and are labeled by their commonly used abbreviations. Indicated borders of visual areas (*dashed lines*) are approximate. Some sulci have been partially opened (*yellow regions*) to show visual areas that lie buried within these sulci. PO, parieto-occipital sulcus; ST, superior temporal sulcus. (*B*) Visual areas are hierarchically organized. The image illustrates some of the anatomical connections known to exist from the retina through visual cortex. Except where indicated by arrows, anatomical connections are known to be reciprocal. RGC, retinal ganglion cell layer; LGN, lateral geniculate nucleus of the thalamus; M, magnocellular subdivisions; P_1 and P_2, parvocellular subdivisions; MT, middle temporal; MST, medial superior temporal; FST, fundus superior temporal; PP, posterior parietal cortex; VIP, ventral intraparietal; STP, superior temporal polysensory.

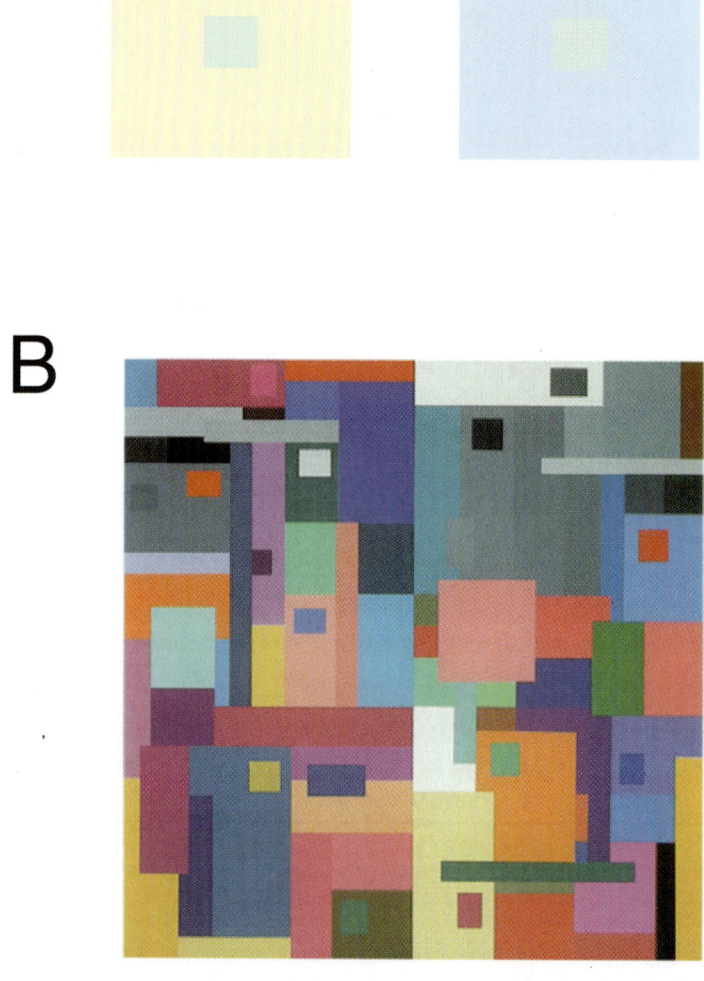

Figure 13 Contextual influences on color appearance. (*A*) The colors of the two central squares are identical, but the one on the right appears greener than the one on the left. This difference in appearance of the central squares is caused by the physical difference between the larger background contexts. This illusion, which has been known for centuries, is caused by the contrast difference between the two image regions and is a product of a mechanism designed to recover the reflectance of surfaces. Neurons in area V1 exhibit context-induced shifts in their sensitivity to receptive field stimuli, which mirror changes in color appearance (Rossi et al. 1996, Wachtler et al. 1999, MacEvoy & Paradiso 2001). (*B*) "Mondrian" stimulus used to examine influence of context on perceptual and neuronal sensitivity to color (Land & McCann 1971, Zeki 1983a).

(e.g., Hubel & Wiesel 1965; Bishop et al. 1973; Maffei & Fiorentini 1976; Nelson & Frost 1985; Gulyas et al. 1987; Gilbert & Wiesel 1990; Knierim & van Essen 1992; Li & Li 1994; Kapadia et al. 1995, 2000; Sillito et al. 1995; Polat et al. 1998; Li et al. 2000, 2001). A subset of these have manipulated contour collinearity and have yielded evidence of response facilitation when discontinuous contours are placed collinearly along the long axis of a RF (Maffei & Fiorentini 1976, Nelson & Frost 1985, Kapadia et al. 1995). Most notably, Kapadia et al. (1995, 1999) found that the responses of V1 and V2 neurons to a suprathreshold line within the classical RF can be greatly increased by adding a flanking collinear line in the RF surround.

Contour response facilitation could represent a neuronal correlate of perceptual continuity in the face of contour fragmentation caused by occlusion. In an elegant examination of this hypothesis, Bakin et al. (2000) obtained evidence suggesting that the neuronal collinear facilitation effect is associated with amodal completion (see Figure 9). To control the degree of amodal completion, these investigators placed additional lines orthogonally between the RF stimulus and flanking stimuli, and manipulated the relative depth of these lines using binocular disparity cues. The effects of these manipulations on neuronal response facilitation appeared strikingly consistent with amodal completion: Facilitation was blocked by placing orthogonal bars in a depth plane behind the collinear fragments, in a manner suggesting that the fragments lie on separate foreground surfaces (no amodal completion). Facilitation recovered when the orthogonal bars were placed in the near depth plane, in a manner that allows the fragments to lie on the same background surface (amodal completion).

This differential effect of orthogonal bar depth was seen by Bakin et al. (2000) in half of the V2 neurons so tested but was rarely seen in area V1. Although interesting, these results are somewhat at odds with those of Sugita (1999). Unlike Bakin et al., Sugita found neuronal correlates of amodal completion in V1. This discrepancy may simply be due to different sampling of the visual field (Sugita's sample was limited to the central 2°, whereas that of Bakin et al. extended 3–6°). Somewhat more puzzling, perhaps, is the fact that Bakin et al. observed neuronal contour facilitation under one condition in which neuronal (and indeed perceptual) amodal completion does not appear to occur (Sugita 1999): two interrupted fragments with no intervening surface (compare the stimulus configuration of Figure 8B with that of 9C). This observation casts doubt on the conclusion that contour facilitation depends upon amodal completion. This intriguing incongruity might best be resolved by examining neuronal correlates of amodal completion and flank facilitation in tandem while simultaneously monitoring perceptual interpretation.

These neuronal findings of Gilbert and colleagues (Bakin et al. 2000; Kapadia et al. 1995, 1999) also appear to have a connection to the behavioral phenomena of contrast facilitation (Dresp 1993; Polat & Sagi 1993, 1994; Zenger & Sagi 1996) and contour grouping (Field et al. 1993). Plausible candidates for mediation of these varied contour phenomena are long-range horizontal connections within area V1 (Rockland & Lund 1982, Gilbert & Wiesel 1983, Martin & Whitteridge 1984). These connections are anisotropic and preferentially connect neurons that have collinear RFs (Bosking et al. 1997, Schmidt et al. 1997).

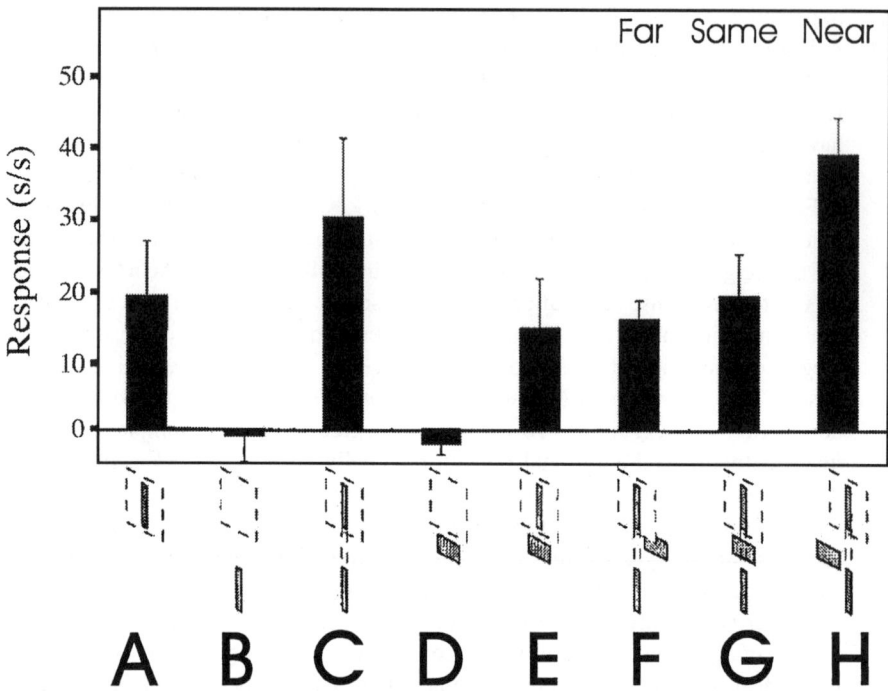

Figure 9 Collinear contextual response facilitation in area V2. Collinear contours in nonclassical receptive field (RF) facilitate responses to RF contour, but not if contextual conditions are inconsistent with amodal completion (but see text). Flanking contour facilitates responses (compare A with C). An orthogonal bar placed in either the same plane (G) as the flank and target stimulus or in the far depth plane (F; 0.16° uncrossed disparity) blocked the flank-induced facilitation of the response to the target stimulus. Conversely, placing the orthogonal bar in the near depth plane (H; 0.16° crossed disparity) did not block this facilitation. (From Bakin et al. 2000. Copyright 2000 by the Society for Neuroscience.)

We have discussed evidence of neuronal filling-in/completion of oriented features and surface properties. We next discuss how completion of moving surfaces might manifest itself as a contextual effect arising from the nonclassical RF of motion-detecting neurons.

OCCLUSION AND VISUAL MOTION PROCESSING

Although occlusion presents a ubiquitous challenge in the recovery of visual scene properties, it poses a special challenge to the recovery of visual motion. Where occluded, the edge of a moving surface appears terminated in the visual image, yielding a feature that traces an unambiguous trajectory across the image. Because such features and their trajectories are defined, however, not by the moving object, but by the object's coincidental intersection with a foreground occluder (Figure 10), they provide spurious information about the motion of the object in the scene. For that reason, the visual motion system must distinguish the false motion information provided by features "extrinsic" to a moving object from the valid information provided by features "intrinsic" to that object (Nakayama et al. 1989). Shimojo et al. (1989) demonstrated that human subjects utilize depth-ordering cues to distinguish intrinsic and extrinsic features, which in turn enables recovery of the motions of partially occluded objects.

Duncan et al. (2000) devised a variant of the classic barber-pole illusion (Wallach 1935) to explore the neuronal substrates of contextual depth-motion interactions. Their "barber-diamond" stimuli consisted of a moving grating framed by a static diamond-shaped aperture (Figure 11). Two of four textured panels that defined the aperture were placed in front of the grating via stereoscopic depth cues, and the other two were placed behind. These depth manipulations simulated partial occlusion of the grating. The features formed by termination ("terminators") of the grating stripes at the far panels were seen as intrinsic, and those formed at the near panels were seen as extrinsic to the grating. When the grating was moved horizontally, human observers reported that the grating moved diagonally upwards or downwards, in the direction of the intrinsic terminators. The depth cues at the margins of the moving stimulus thus provided a context for perceptual disambiguation of motion within the interior. [Similar perceptual motion effects can be obtained using simulated shadows, rather than binocular disparity, to elicit depth-ordering (G.R. Stoner, unpublished observations).]

Duncan et al. (2000) compared these perceptual effects to the responses of directionally selective neurons in cortical area MT. Barber-diamond stimuli were positioned such that the aperture edges (and hence depth-ordering cues) were outside of the classical RF. Many MT neurons exhibited directionally selective responses consistent with perceived direction of motion, i.e., depth-ordering cues that were restricted to the RF surround predictably altered responses to moving features within the classical RF (Figure 12).

These findings challenge previous characterizations of nonclassical receptive field (RF) surround effects within area MT, which have stressed relatively simple feature contrast effects. As noted above, the responses of many MT neurons to motion in the classical RF are modulated by motion in the surround (Figure 4) (Allman et al. 1985). A similar modulation has been reported for binocular disparity (Bradley & Andersen 1998). These two types of "intra-modal" (motion-motion and depth-depth) interactions may be helpful in signaling cue-specific image discontinuities (Bradley & Andersen 1998) or extracting depth variation based on either motion parallax or binocular disparity (Buracas & Albright 1996, Liu & Kersten 1998, Xiao et al. 1997). Bradley & Andersen (1998) found few interactions between depth and motion, however, which reinforced a view of surrounds as exerting intra-modal and mostly antagonistic modulation of responses to stimulation of the classical RF.

In the Bradley & Andersen study, depth and motion were independent stimulus attributes, neither bearing on the interpretation of the other. By contrast, in the barber-diamond stimuli employed by Duncan et al. (2000) (as in natural scenes)

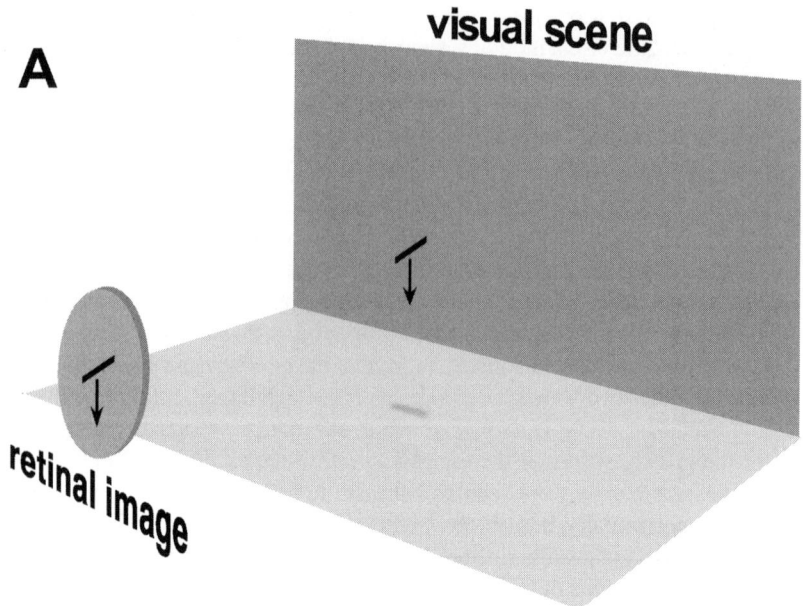

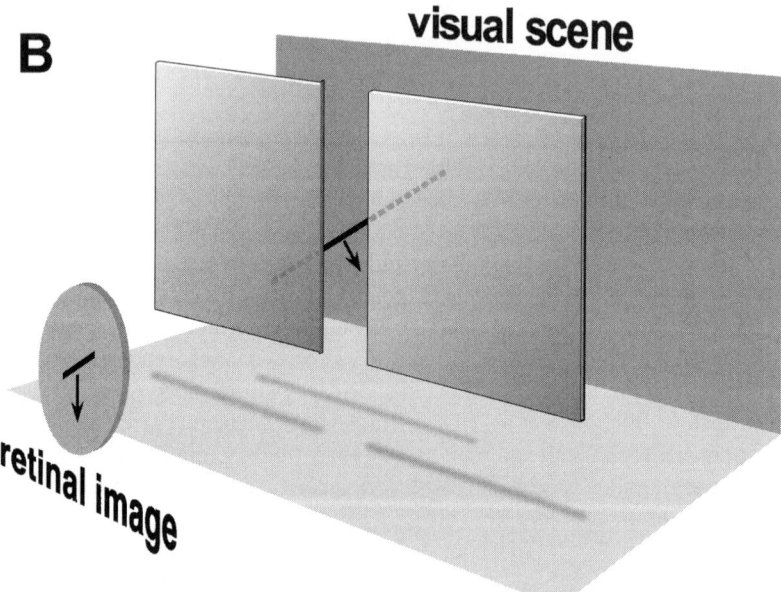

depth cues provide a context that triggers recovery of occluded scene elements, which in turn modulates the representation of motion. Thus, as was the case for many of the contextual effects reviewed herein, it was the incorporation of occlusion cues that uncovered the sophisticated RF surround interactions observed by Duncan et al.

Although the precise mechanisms responsible for the depth-motion interactions of Duncan et al. are unknown, there are some plausible alternatives. One possibility is that the occlusion cues elicit a neuronal representation of the amodally completed grating behind the near panels, as suggested by the findings of Sugita (1999) (reviewed above). These completed contours might then lead to the recruitment of additional motion detectors, which may in turn shift the balance in favor of motion in the direction of intrinsic terminators. A second possible mechanism involves selective pooling of motion signals determined to be intrinsic to the same surface (Stoner & Albright 1993). For example, pooling of the signals arising from the horizontally moving gratings with the diagonally oriented far edge (intrinsic to the grating) would yield the observed diagonal motion of the barber-diamond. A third mechanism involves selective suppression of motion signals arising from extrinsic features (Liden & Pack 1999), which would tilt the balance in favor of intrinsic terminator motions. These three potential mechanisms are, of course, not mutually exclusive.

FILLING-IN REVISITED: NORMAL SURFACE PERCEPTION

> At any one time, only a small proportion of cells (in visual cortex) are likely to be influenced (activated or suppressed), since contours of inappropriate orientation and diffuse light (within the receptive field) have little or no effect on a cell (Quotation from Hubel & Wiesel 1968).

Figure 10 Illustration of the contextual dependence of motion perception. Two simple visual scenes are shown, along with image motions they give rise to. (*a*) A single moving object oriented obliquely and moving downward. Image motion reflects object motion. (*b*) A single obliquely oriented bar moving to lower right (orthogonal to orientation). The scene also contains occluders, which block the observer's view of the ends of the bar. Resulting visual image motion in (*b*) is identical to that caused by the scene in (*a*), although the visual scene motions are clearly different. Observers perceive the different object motions correctly, and they do so because of different spatial contexts, which elicit different interpretations of the terminations of the moving line in the visual image. Terminators in (*a*) are interpreted as ends of an elongated object, i.e., they are intrinsic features of that object and their downward motions genuinely reflect object motion. Terminators in (*b*), by contrast, are extrinsic to the moving line (an accident of occlusion), and their downward motions are deemed spurious. Thus, terminator interpretation determines how the image motions are perceived.

Hubel & Wiesel's observation implies that visual neurons only respond to image contrast and that every homogenous region in the visual image is, in effect, a blind spot. Surfaces with uniform reflectance are common and, hence, so are uniform image regions. This poses a profound computational problem and a mystery: How is it that we are not blind to much of the world?

Part of the answer might be thought to lie in the functions of "luxotonic" neurons, which do respond to diffuse sustained light. However, the importance of these neurons is called into question by their relative rarity in the species in which they have been detected (DeYoe & Bartlett 1980, Kayama et al. 1979) and their absence in other species (Kahrilas et al. 1980). Moreover, as others have noted (Walls 1954, Gerrits & Vendrik 1970), the computational problem posed

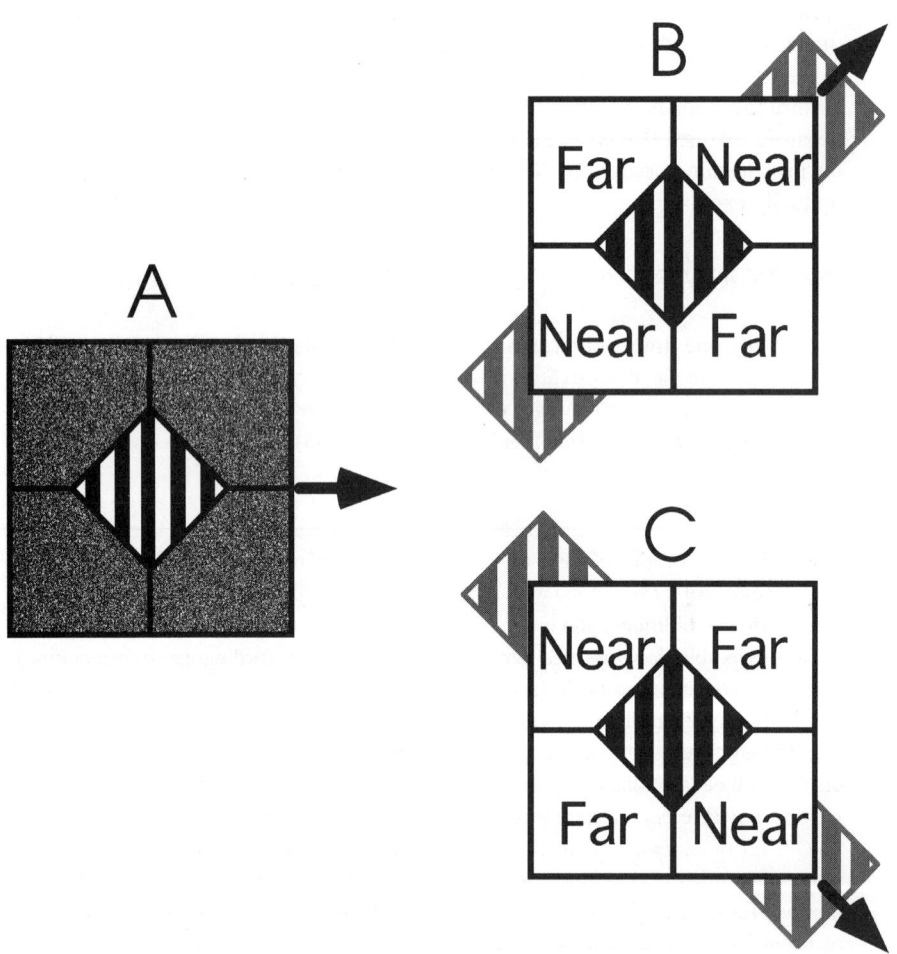

by uniform regions in the visual image has an obvious parallel with that posed by the blind spot and scotomas, and the same mechanistic considerations and potential neuronal solutions apply. In particular, one can ask whether filling-in of homogenous image regions is an active or passive process.

As we have seen, perceptual completion of image gaps caused by the blind spot and scotomas, as well as by surface occlusion, involves an active neuronal process in which the missing features are explicitly reconstructed. It is possible, nonetheless, that our ability to perceive areas of constant coloration is accomplished implicitly (i.e., passively). For example, imagine a gray surface atop a white background. Neurons encoding one spatial location detect one surface border (light-dark); neurons encoding another location detect the complementary discontinuity (dark-light). According to an implicit encoding scheme, border-detecting neurons signal that the surface is darker than the background, while uniformity of

←───

Figure 11 Schematic depiction of "barber-diamond" stimuli used to study the influence of context on perceived motion and its neuronal correlates (a demonstration can be seen at http://www.cnl.salk.edu/~gene/). (*a*) Stimuli consisted of a moving square-wave grating framed by a static, diamond-shaped aperture. The grating itself was placed in the plane of ocular fixation while four textured panels that defined the aperture were independently positioned in depth using binocular disparity cues. The grating was moved either leftward or rightward on each trial. Each of the four stimulus conditions used was a unique conjunction of direction of grating motion (right vs. left) and depth-ordering configuration. The two conditions illustrated (*b* and *c*) were created using rightward-moving gratings; two additional conditions (not shown) were created using leftward-moving gratings. "Near" and "Far" identify the depth-ordering of the textured panels relative to the plane of the grating. (*b*) Upper-right and lower-left panels were placed in the near depth plane, while the upper-left and lower-right panels were in the far depth plane. Line terminators formed at the boundary of near surfaces and grating are classified as extrinsic features resulting from occlusion, and the grating is amodally completed behind the near surface. (Note that gray stripes are not part of the stimulus and are used solely to illustrate amodal completion.) Conversely, line terminators formed at the boundary of the far surfaces and the grating are classified as intrinsic, i.e., they appear to result from the physical termination of the surface upon which the stripes are "painted." As a result of this depth-ordering manipulation and ensuing feature interpretation, observers typically perceive the moving grating in (*b*) as belonging to a surface that slides behind the near panels and across the far panels, i.e., to the upper-right. This direction is identified with motions of intrinsic terminators. The condition shown in (*c*) contains the same rightward grating motion but employs the depth-ordering configuration that is complementary to (*b*). In this case observers typically perceive the grating belonging to a surface that is drifting to the lower right. (After Duncan et al. 2000. Copyright 2000 by the Society for Neuroscience.)

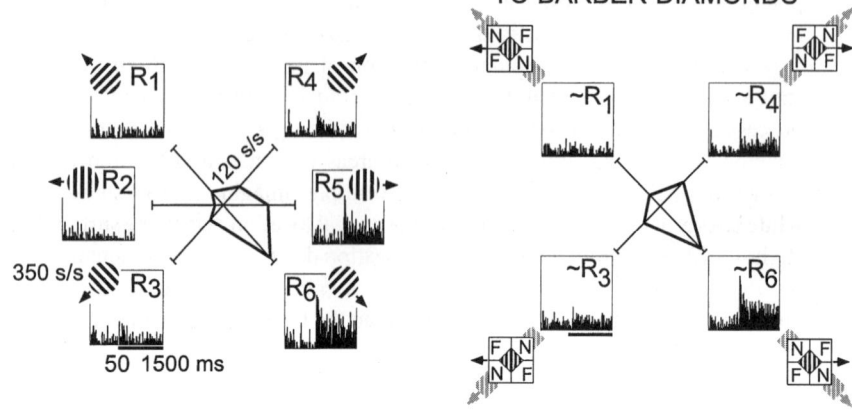
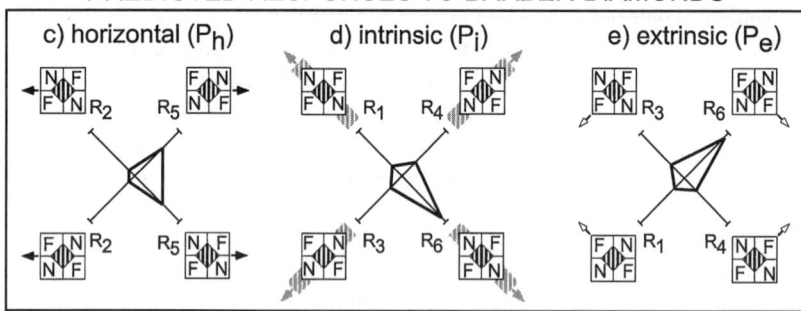

Figure 12 Influence of depth-ordering on direction selectivity of an MT neuron. (*a*) Neuronal responses to circular gratings moved in each of six directions. These responses (R1–R6) were used to form three different predictions for barber-diamond stimuli (*c, d* and *e*). (*b*) Actual responses to barber-diamond stimuli. Icons illustrate the stimulus configuration for each of four experimental conditions. (*c*) If depth-ordering in the receptive field surround has no effect, neuronal responses will be influenced solely by the horizontal motion (black arrows) of the grating and, hence, be of the same relative magnitude as responses to circular gratings moved leftward (R2) and rightward (R5), regardless of depth-ordering configuration. (*d*) If, as expected, neuronal responses are associated with the direction of intrinsic terminator motion (gray arrows), they should be of the same relative magnitude as responses to circular grating moving in the corresponding oblique directions (R1, R3, R4, and R6). (*e*) If neuronal responses are associated with the direction of extrinsic terminator motion (white arrows), they should be of the same relative magnitude as the intrinsic motion prediction but flipped about the horizontal axis. Observed responses (*b*) of this neuron to barber-diamond stimuli were more closely correlated with the intrinsic motion prediction than with either the horizontal or extrinsic motion prediction. (From Duncan et al. 2000. Copyright 2000 by the Society for Neuroscience.)

the interior is signaled by the lack of activity of neurons with RFs positioned over intermediate spatial locations. An analogous proposal can be made for our sense of a temporally continuous world (for related discussion see O'Regan 1992). Say your eyes fixate a blue surface, then a red one; neurons sensitive to that chromatic change will respond transiently, and this activation renders a sense of redness even after the response has ended. Taken together, these considerations suggest a brain very stingy with its action potentials, none being wasted on signaling that things have not changed over space or time.

Although it is appealingly economical, arguments and evidence against the sufficiency of this implicit encoding scheme are not hard to find. One empirical argument comes from studies of brightness induction. The interior of a gray patch will appear to undergo a brightness change if the surrounding area is temporally modulated from light to dark. If the frequency of that modulation is varied sinusoidally, it will be discovered that this induction effect disappears above about 2.5 Hz (De Valois et al. 1986, Rossi & Paradiso 1996). This temporal cutoff is much lower than that at which the border between the gray patch and surround appear unchanging. The perception of edges and interior brightness can thus be decoupled and we can conclude that the two cannot be mediated by the same pattern of neuronal activity.

A more general problem with the implicit coding scheme is that contrast measures at the nearest borders are simply not sufficient to support color and brightness perception of interior zones: As discussed below, the visual system integrates diverse types of information from large regions of visual space to infer the reflectance of a given region of visual space (e.g., Land & McCann 1971, Adelson 1993, Brown & MacLeod 1997, Barnes et al. 1999, Purves et al. 1999, Shevell & Wei 1998, Wachtler et al. 2001) (Figure 13B, see color insert). Very complicated nontopographic schemes involving the neurons responding to these diverse information sources could, of course, be advanced, but they hardly seem parsimonious. Fortunately, whether there exist topographic neuronal responses related to brightness perception is not a question whose answer is hostage to theory.

Paradiso and colleagues (Rossi et al. 1996, Rossi & Paradiso 1999, MacEvoy & Paradiso 2001) have examined neuronal correlates of brightness induction. In one study they recorded from retina, thalamic lateral geniculate nucleus, and area V1 of anesthetized cats. The experiments were designed to determine whether neurons respond to brightness changes caused by either "direct" or contextual ("induced") means. Direct brightness changes were produced by varying the intensity of a central patch of light superimposed on the classical RF. Induced brightness changes were caused by holding the classical RF stimulus constant and varying light intensity in the RF surround. A substantial fraction of V1 neurons—unlike those at earlier processing stages—exhibited responses that covaried with the brightness of the classical RF stimulus, regardless of whether those brightness changes were caused directly or induced.

The findings of Paradiso and colleagues and those from recent related studies (Hung et al. 2001) suggest a topographically explicit, rather than implicit, representation of brightness. This conclusion does not refute the observations of Hubel &

Wiesel (1968)—cortical neurons do, in fact, respond to luminance contrast within the classical RF but weakly or not at all to "diffuse light." Some cortical neurons are, however, sensitive to luminance contrast cues outside the classical field, and that sensitivity permits them to, in effect, encode the ambiguous properties of homogenous regions within that field. Evidence indicates that the observed topographic neuronal representation of uniform image regions develops progressively from border to interior. In human observers this progressive completion can be observed directly and its time course measured (Paradiso & Hahn 1996, Paradiso & Nakayama 1991). Neuronal evidence from "artificial scotoma" experiments (reviewed above) (De Weerd et al. 1995) also supports progressive completion, although it is important to note that these artificial scotomas were fixed (at least transiently) relative to eye position, unlike uniform surfaces under normal viewing conditions.

RECOVERING SURFACE REFLECTANCE: USING CONTEXT TO "SEE BENEATH" THE ILLUMINANT

Having reviewed evidence for an active neuronal filling-in of uniform image regions, we now turn to function: What is completed, and why? We begin by emphasizing a resemblance to the "recovery" of visual information lost owing to occlusion. As we have seen, amodal completion answers the question "What lies behind?" by predicting the content of occluded surfaces. Similarly, completion of uniform image regions identifies what is in front. Although we have portrayed this latter form of completion as propagation of information from edges to interior, the computation required is generally far more complex. As amodal completion must "see beneath" occluding surfaces, so must the mechanisms of foreground surface completion see beneath the light illuminating the scene in order to recover surface reflectance.

To appreciate the magnitude and significance of this problem, recall that the light reaching the retina is a generally a product of two factors: surface reflectance and illumination. Reflectance relates the quantity of light reflected to its wavelength. Reflectance is constant over time (though it may vary over space) for most surfaces and is thus a valuable source of information for object recognition. Illumination is characterized by its intensity as a function of wavelength. Unlike reflectance, however, the illuminant can and often does vary over space and time. Because surface reflectance and illumination properties are confounded in the light reaching the retina, it is impossible to deduce the reflectance of a surface given only the light arising from that surface. Despite this limitation, the primate visual system is quite good at recovering surface reflectance in natural scenes even when faced with significant changes in illumination. (For example, the cover of this book will appear red under many different lighting conditions.) This invariance of color appearance is known as "color constancy." Brightness constancy is an achromatic analogue of color constancy, which is likely to result from similar neuronal substrates and mechanisms.

Vision scientists have puzzled over color constancy for centuries. Thomas Young (1807) observed, "When a room is illuminated by either the yellow light of a candle, or the red light of a fire, a sheet of writing paper still appears to retain its whiteness." Helmholtz (2000) also recognized the phenomenon and its functional significance: "Colors are mainly important for us as properties of objects and as means of identifying objects.... we constantly aim to reach a judgment on the object colors and to eliminate differences of illumination." Helmholtz largely attributed the observer's ability to discount the illuminant, and thereby recover surface reflectance, to prior experience with objects in different lighting conditions. Many of Helmholtz's contemporaries and followers offered explanations rooted solely in spatial context (e.g., Hering 1964 [1920], Mach 1886).

Contextual Influences on Color Appearance

The influence of context on color appearance has long been a popular topic in the field of visual psychophysics, and there now exists an enormous body of data characterizing human sensitivity. One of the most robust effects of context in this domain—one that has been known for centuries—is the dependency of perceived color on the color of adjacent regions (Figure 13A, see color insert). From a functional perspective, simple context-induced shifts in color appearance can be viewed as a by-product of a system designed to recover surface reflectance in the face of changing illumination: While an illumination change will alter the spectrum of light arising from a given surface, it does so for every surface so illuminated. Therefore, as noted by the psychologist Hans Wallach (1948), the ratios of spectra from adjacent surfaces generally remain unchanged. Invariant color appearance should thus result from stimuli that possess spectral ratios that are invariant between adjacent image regions. Conversely, in the absence of ratio invariance, perceived color should change, as it indeed does.

Neuronal Correlates of Color Appearance

Until recently, very little was known of the neuronal bases of color appearance. Although there have been several detailed neuropsychological studies of acquired deficiencies of human color vision (achromatopsia) (for review see Zeki 1990), it has been difficult to discover from these studies brain regions specifically involved in representing perceived color, as opposed to low-level color detection and analysis. The past four decades have also seen numerous electrophysiological studies of neurons in the retina, thalamus, and visual cortex that respond to chrominance and/or luminance contrast within their classical RFs (for review see Lennie & D'Zmura 1988). However, most of these studies did not include contextual manipulations that alter color appearance, such as those shown in Figure 13A in color insert. Without such manipulations it is impossible to determine whether a neuronal response covaries with the spectral power of the light in the classical RF or, alternatively, with perceived color/brightness. We now turn to a few recent studies that have included appropriate contextual manipulations and have obtained intriguing results.

ACHROMATIC COLOR APPEARANCE (BRIGHTNESS)   The complexity of color appearance phenomena is reduced when considering achromatic stimuli, but similar contextual effects and mechanistic interpretations apply. We have already discussed the brightness induction effects reported for neurons in cat area V1 by Paradiso and colleagues (Rossi et al. 1996, Rossi & Paradiso 1999). Not only might these effects underlie perceptual filling-in of the interiors of uniform image regions, but they satisfy one of the criteria for a mechanism that recovers surface reflectance. Specifically, the neuronal response changes when the ratio of intensities changes between adjacent surfaces. If these neurons do, in fact, represent surface reflectance, we should also expect them to exhibit an invariant response under conditions that mimic variations in diffuse illumination, i.e., where contrast remains unchanged. Recent evidence is consistent with this requirement (MacEvoy & Paradiso 2001).

CHROMATIC COLOR APPEARANCE   Zeki (1983a,b) was the first to attempt to distinguish between neuronal representations of chromatic spectral composition and perceived color. Zeki adopted a stimulus configuration (Figure 13*B*, see color insert) that had been a favorite tool of the color theorist Edwin Land for manipulating color appearance (Land & McCann 1971). The stimulus is known as a "Mondrian" for its resemblance to the style of the painter. These stimuli possess rich chromatic structure and they elicit robust color constancy effects.

In Zeki's experiments, one of the colored panels in the Mondrian was placed over the classical RF of a cortical neuron, and the remaining panels were placed in the RF surround. When the color of the illuminant changed, the responses of neurons in area V1 changed in accordance with the wavelength of light falling in the classical RF. By contrast, the responses of a subset of neurons in area V4 (see Figure 3 in color insert) were unaltered by illumination changes and thus paralleled the perceived constancy of the RF stimulus. Zeki suggested, on the basis of these results, that V4 may be the site at which chromatic context is integrated with the RF stimulus in order to achieve a representation of perceived color.

Using simpler stimulus manipulations of the type shown in Figure 13*A* (see color insert), Schein & Desimone (1990) also examined the influence of chromatic context on the color selectivity of V4 neurons. Figure 14 illustrates one of their findings. The response of this neuron to the "preferred" chromatic stimulus placed in the classical RF was dependent upon the color of the surround. To a first approximation, these results are consistent with a mechanism in which the output is a function of chromatic contrast between the classical RF stimulus and the surround. But what is the function of such cells? Schein & Desimone speculated that they may signal the average intensity of the illuminant in specific wave-bands. The response of the cell illustrated, for example, should decline as the redness of the illuminant increases. Cells that serve as "wave-band intensity gauges" could, by explicitly recovering the properties of the illuminant, permit the illuminant to be discounted and surface reflectance computed (Land 1986).

Wachtler et al. (1999) used similarly simple stimuli to re-examine the role of area V1 in color appearance. In contrast to Schein & Desimone (1990) and Rossi

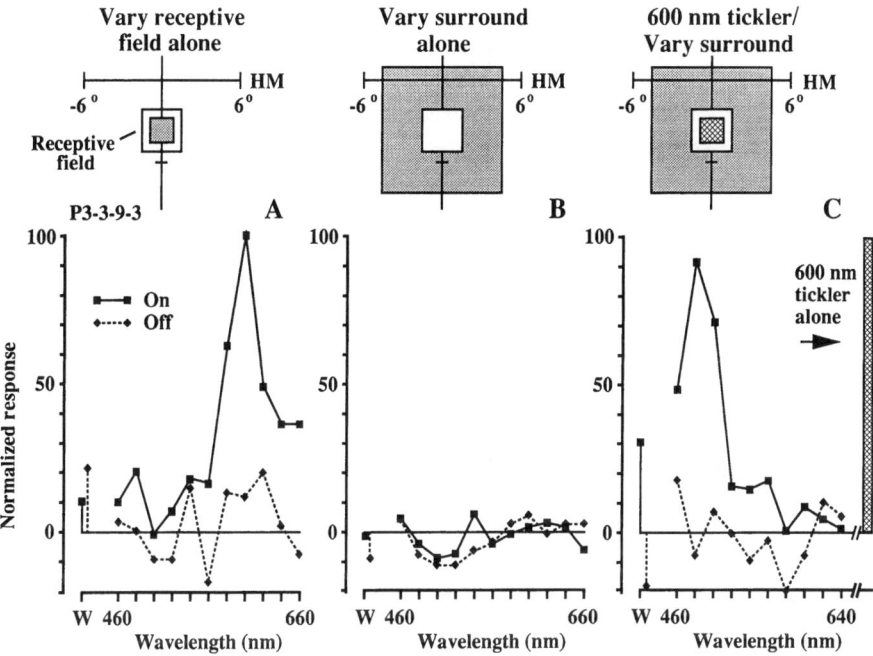

Figure 14 Chromatic selectivity of a V4 neuron is influenced by chromatic context. (*Left*) Wavelength tuning reveals that the neuron responds optimally to a stimulus of 600 nm in the classical receptive field (RF). (*Center*) The cell failed to respond to a chromatic background that extended around but did not intrude upon the classical RF. (*Right*) The chromatic background did, however, modulate the response to the preferred stimulus in the classical RF. To a first approximation, these results are consistent with a mechanism in which the output is a function of chromatic contrast between adjacent stimuli. Moreover, the modulation of response to the classical RF stimulus parallels the context-induced shift in color appearance that occurs for such stimuli. (From Schein & Desimone 1990. Copyright 1990 by the Society for Neuroscience.)

et al. (1996), Wachtler et al. evaluated the effects of color changes in the RF surround on the full chromatic tuning to the classical RF stimulus. For many V1 neurons the effects were striking. For example, introduction of a reddish surround to a red-preferring neuron had the effect of shifting the chromatic tuning curve away from red and toward the green side of color space. Significantly, the same contextual manipulation causes the central square to appear more greenish to an observer (Wachtler et al. 2001). Interestingly, this type of neuronal response shift is similar to what one would expect from double-opponent color-selective neurons, which have been thought to play a role in color constancy (Conway 2001). The important difference, however, is that the contextual manipulations of Wachtler et al. were highly effective even though they were restricted to the RF surround.

These results suggest that at least some V1 neurons use context to achieve a representation of perceived color. Although this conclusion mirrors that of Rossi et al. (1996) regarding the role of V1 in brightness perception, it is at variance with Zeki's (1983a,b) claim that such contextual effects are not seen until V4. This difference is difficult to reconcile, although it is noteworthy that Wachtler et al. studied V1 responses in alert monkeys, whereas Zeki's subjects were anesthetized.

Context and Color: Beyond Spatial Effects

We have thus far considered the simple contextual conditions in which the color appearance of a surface is influenced by light arising from adjacent (or nearby) coplanar surfaces. In most real scenes, however, the light arising from surfaces is determined not only by their reflectance and the properties of the illuminant, but also generally by the angle of the surface relative to the observer and the illumination source. One important consequence is that color ratios between surfaces, which provide evidence of reflectance and illumination, can serve as meaningful cues only if interpreted in the additional context of three-dimensional surface configuration. Striking demonstrations and psychophysical confirmation of this important caveat have been provided by Adelson (1993) and Purves et al. (1999) (see also Mach 1886), but nothing is currently known of the neuronal bases of these effects. Although this is an important area for future research, the number and complexity of contextual cue interactions required present a formidable challenge for controlled neurophysiological experiments.

CONCLUDING REMARKS

Computational considerations and a substantial body of psychophysical and physiological evidence indicate that the visual system employs a well-characterized set of contextual cues to identify the visual scene origins—i.e., the meaning—of visual image features. A common theme in this review is the use of contextual information to recover visual scene properties lost owing to the superimposition of causes in the visual image. We have, for example, discussed the contextual cues used by the visual system to decompose images into multiple overlapping surfaces and/or lighting conditions. Moreover, we have shown that individual neurons at early stages of cortical processing (e.g., V1 and V2) are involved in the detection of these contextual cues and the representation of recovered information.

This recent progress and the attention directed to the role of context represents a critical transition in efforts to understand the neuronal bases of perception. Historically, the reductionism of single-cell physiology has fostered a view of visual neurons as functionally independent windows on the world. In practice, this view has led to an experimental approach that neglects context. The result is that most of the reported responses of visual neurons (e.g., Figure 2) can only be described as neural correlates of the visual image; the full contributions of these neurons to visual scene perception remain unknown.

The transition toward a context-based approach has been inspired, in part, by the discovery of modulatory influences from outside the classical RF. Of even greater importance, perhaps, has been the increased willingness of visual physiologists to turn to experimental psychology for guidance, inspiration, and tools. Embedded in this approach is a renewed focus on function. Accordingly, many of the investigations of neuronal representation reviewed herein are motivated not simply by the question, "How does the visual system work?" but also by the question, "What is the visual system 'designed' to do?"

The studies we have reviewed are only the tip of an iceberg. The contextual influences we have omitted include, among others, attention and memory, which have been topics of many neuroscientific studies in recent years. There is also a vast realm of contextual effects that are closely related to those we have reviewed, whose manifestations are termed "perceptual constancies." These constancies—color (reviewed above), size, viewing angle, etc.—reflect the recovery of invariant attributes of the visual scene in the face of changing image properties. Further identification of neural correlates of perceptual constancies is an important goal for the future.

Finally, in considering future directions, we emphasize that discovery of neuronal correlates is only one step toward understanding the neuronal bases of perception. Evidence of such correlates is now seemingly in abundance but, by contrast, we know very little about how these interesting neuronal properties are implemented. For example, horizontal intrinsic cortical connections are implicated in many phenomena of spatial context, such as filling-in and color appearance effects, but few firm conclusions can be reached. The complexity of cortical wiring and synaptic connectivity/plasticity is indeed daunting. New technologies that enable monitoring of global patterns of neuronal activity and manipulation of that activity within complex circuits nonetheless offer a promising armament for future efforts to reveal the mechanisms underlying contextual influences on visual processing.

ACKNOWLEDGMENTS

We thank Charles Gilbert, Greg Horwitz, Mike Paradiso, and John Reynolds for stimulating discussions and comments on the manuscript. The authors have been supported by grants from the NEI. TDA is an Investigator of the Howard Hughes Medical Institute.

The *Annual Review of Neuroscience* is online at http://neuro.annualreviews.org

LITERATURE CITED

Adelson EH. 1993. Perceptual organization and the judgment of brightness. *Science* 262: 2042–44

Albright TD. 1984. Direction and orientation selectivity of neurons in visual area MT of the macaque. *J. Neurophysiol.* 52:1106–30

Allman J, Miezin F, McGuinness E. 1985. Stimulus specific responses from beyond the classical receptive field: neurophysiological mechanisms for local-global comparisons in visual neurons. *Annu. Rev. Neurosci.* 8:407–30

Anderson BL. 1994. The role of partial occlusion in stereopsis. *Nature* 367:365–68

Bakin JS, Nakayama K, Gilbert CD. 2000. Visual responses in monkey areas V1 and V2 to three-dimensional surface configurations. *J. Neurosci.* 20:8188–98

Barlow HB. 1953. Summation and inhibition in the frog's retina. *J. Physiol.* 119:69–88

Barlow HB. 1972. Single units and sensation: a neuron doctrine for perceptual psychology? *Perception* 1:371–94

Barnes CS, Wei J, Shevell SK. 1999. Chromatic induction with remote chromatic contrast varied in magnitude, spatial frequency, and chromaticity. *Vis. Res.* 39:3561–74

Baumann R, ver der Zwan R, Peterhans E. 1997. Figure-ground segregation at contours: a neural mechanism in the visual cortex of the alert monkey. *Eur. J. Neurosci.* 9:1290–303

Beck J, Prazdny K, Ivry R. 1984. The perception of transparency with achromatic colors. *Percept. Psychophys.* 35:407–22

Bishop PO, Coombs JS, Henry GH. 1973. Receptive fields of simple cells in the cat striate cortex. *J. Physiol.* 231:31–60

Bosking WH, Zhang Y, Schofield B, Fitzpatrick D. 1997. Orientation selectivity and the arrangement of horizontal connections in tree shrew striate cortex. *J. Neurosci.* 17:2112–27

Bradley DC, Andersen RA. 1998. Center-surround antagonism based on disparity in primate area MT. *J. Neurosci.* 18:7552–65

Brown RO, MacLeod DI. 1997. Color appearance depends on the variance of surround colors. *Curr. Biol.* 7:844–49

Buracas GT, Albright TD. 1996. Contribution of area MT to perception of three-dimensional shape: a computational study. *Vis. Res.* 36:869–87

Campbell FW, Andrews PR. 1992. Motion reveals spatial visual defects. *Ophthalmic Physiol. Optics* 12:131–32

Chang G, Stoner GR, Albright TD. 1999. Neuronal correlates for depth-ordering in area V2. *Invest. Opthalmol. Vis. Sci.* 40(4): S640

Chang GC, Stoner GR, Albright TD. 2001. Figure-ground selectivity in macaque V2. *Soc. Neurosci. Abstr.* 27:432

Chino YM, Kaas JH, Smith EL, Langston AL, Cheng H. 1992. Rapid reorganization of cortical maps in adult cats following restricted deafferentation in retina. *Vis. Res.* 32:789–96

Churchland PS, Ramachandran VI. 1993. Filling-in: why Dennett is wrong. In *Dennett and His Critics: Demystifying Mind*, ed. B Dahlbom, pp. vii, 247. Oxford/Cambridge, MA: Blackwell

Conway BR. 2001. Spatial structure of cone inputs to color cells in alert macaque primary visual cortex (V-1). *J. Neurosci.* 21:2768–83

Croner LJ, Albright TD. 1999a. Segmentation by color influences responses of motion-sensitive neurons in the cortical middle temporal visual area. *J. Neurosci.* 19:3935–51

Croner LJ, Albright TD. 1999b. Seeing the big picture: integration of image cues in the primate visual system. *Neuron* 24:777–89

Cumming BG, Parker AJ. 1998. Eye movements explain contextual modulation by binocular disparity in primate V1. *Soc. Neurosci. Abstr.* 24:1980

De Valois RL, Webster MA, De Valois KK, Lingelbach B. 1986. Temporal properties of brightness and color induction. *Vis. Res.* 26:887–97

De Weerd P, Desimone R, Ungerleider LG. 1998. Perceptual filling-in: a parametric study. *Vis. Res.* 38:2721–34

De Weerd P, Gattass R, Desimone R, Ungerleider LG. 1995. Responses of cells in monkey visual cortex during perceptual filling-in of an artificial scotoma. *Nature* 377:731–34

Dennett DC. 1991. *Consciousness Explained.* New York: Little Brown

DeYoe EA, Bartlett JR. 1980. Rarity of luxotonic responses in cortical visual areas of the cat. *Exp. Brain. Res.* 39:125–32

Dresp B. 1993. Bright lines and edges facilitate the detection of small light targets. *Spatial Vis.* 7:213–25

Duncan RO, Albright TD, Stoner GR. 2000.

Occlusion and the interpretation of visual motion: perceptual and neuronal effects of context. *J. Neurosci.* 20:5885–97

Field DJ, Hayes A, Hess RF. 1993. Contour integration by the human visual system: evidence for a local "association field." *Vis. Res.* 33:173–93

Fiorani JM, Rosa MG, Gattass R, Rocha-Miranda CE. 2001. Dynamic surrounds of receptive fields in primate striate cortex: a physiological basis for perceptual completion? *Proc. Natl. Acad. Sci. USA* 89:8547–51

Gallant JL, Connor CE, Van Essen DC. 1998. Neural activity in areas V1, V2 and V4 during free viewing of natural scenes compared to controlled viewing. *Neuroreport* 9:1673–78

Garraghty PE, Kaas JH. 1992. Dynamic features of sensory and motor maps. *Curr. Opin. Neurobiol.* 2:522–27

Gerrits HJ, Timmerman GJ. 1969. The filling-in process in patients with retinal scotomata. *Vis. Res.* 9:439–42

Gerrits HJM, de Haan B, Vendrick AJH. 1966. Experiments with retinal stabilized images. Relations between the observations and neural data. *Vision Res.* 6:427–440

Gerrits HJ, Vendrik AJ. 1970. Simultaneous contrast, filling-in process and information processing in man's visual system. *Exp. Brain. Res.* 11:411–30

Gilbert CD, Wiesel TN. 1983. Clustered intrinsic connections in cat visual cortex. *J. Neurosci.* 3:1116–33

Gilbert CD, Wiesel TN. 1990. The influence of contextual stimuli on the orientation selectivity of cells in primary visual cortex of the cat. *Vis. Res.* 30:1689–701

Gilbert CD, Wiesel TN. 1992. Receptive field dynamics in adult primary visual cortex. *Nature* 356:150–52

Gilchrist AL. 1977. Perceived lightness depends on perceived spatial arrangement. *Science* 195:185–87

Grosof DH, Shapley RM, Hawken MJ. 1993. Macaque V1 neurons can signal 'illusory' contours. *Nature* 365:550–52

Gulyas B, Orban GA, Duysens J, Maes H. 1987. The suppressive influence of moving textured backgrounds on responses of cat striate neurons to moving bars. *J. Neurophysiol.* 57:1767–91

Hartline HK, Graham CH. 1932. Nerve impulses from single receptors in the eye. *J. Cell Comp. Physiol.* 1:277–95

He S, Davis WL. 2001. Filling-in at the natural blind spot contributes to binocular rivalry. *Vis. Res.* 41:835–40

Hering E. 1964 (1920). *Outlines of a Theory of the Light Sense.* Cambridge, MA: Harvard Univ. Press

Hubel DH, Wiesel TN. 1965. Receptive fields and functional architecture in two nonstriate visual areas (18 and 19) of the cat. *J. Neurophysiol.* 28:228–89

Hubel DH, Wiesel TN. 1968. Receptive fields and functional architecture of monkey striate cortex. *J. Physiol.* 195:215–43

Hung CP, Ramsden BM, Chen LM, Roe AW. 2001. Building surfaces from borders in areas 17 and 18 of the cat. *Vis. Res.* 41:1389–407

James W. 1890. *Principles of Psychology.* New York: Holt

Kaas JH. 1991. Plasticity of sensory and motor maps in adult mammals. *Annu. Rev. Neurosci.* 14:137–67

Kahrilas PJ, Doty RW, Bartlett JR. 1980. Failure to find luxotonic responses for single units in visual cortex of the rabbit. *Exp. Brain Res.* 39:11–16

Kanizsa G. 1979. *Organization in Vision: Essays on Gestalt Perception.* New York: Praeger

Kapadia MK, Ito M, Gilbert CD, Westheimer G. 1995. Improvement in visual sensitivity by changes in local context: parallel studies in human observers and in V1 of alert monkeys. *Neuron* 15:843–56

Kapadia MK, Westheimer G, Gilbert CD. 1999. Dynamics of spatial summation in primary visual cortex of alert monkeys. *Proc. Natl. Acad. Sci. USA* 96:12073–78

Kapadia MK, Westheimer G, Gilbert CD. 2000. Spatial distribution of contextual interactions in primary visual cortex and in visual perception. *J. Neurophysiol.* 84:2048–62

Kaplan GA. 1969. Kinetic disruption of optical

texture: the perception of depth at an edge. *Percept. Psychophys.* 6:193–98

Kayama Y, Riso RR, Bartlett JR, Doty RW. 1979. Luxotonic responses of units in macaque striate cortex. *J. Neurophysiol.* 42:1495–517

Knierim JJ, van Essen DC. 1992. Neuronal responses to static texture patterns in area V1 of the alert macaque monkey. *J. Neurophysiol.* 67:961–80

Koffka K. 1935. *Principles of Gestalt Psychology.* New York: Harcourt Brace

Komatsu H, Kinoshita M, Murakami I. 2000. Neural responses in the retinotopic representation of the blind spot in the macaque V1 to stimuli for perceptual filling-in. *J. Neurosci.* 20:9310–19

Lamme VAF. 1995. The neurophysiology of figure-ground segregation in primary visual cortex. *J. Neurosci.* 15:1605–15

Land EH. 1986. Recent advances in retinex theory. *Vis. Res.* 26:7–21

Land EH, McCann JJ. 1971. Lightness and retinex theory. *J. Opt. Soc. Am.* 61:1–11

Le Gros. 1967. Entopic phemonena. In *Le Grand, Yves, 1908—Form and Space Vision.* Transl. M Millodot, GG Heath, pp. 146–58. Bloomington: Indiana Univ. Press (From French)

Lee TS, Nguyen M. 2001. Dynamics of subjective contour formation in the early visual cortex. *Proc. Natl. Acad. Sci. USA* 98:1907–11

Lennie P, D'Zmura M. 1988. Mechanisms of color vision. *Crit. Rev. Neurobiol.* 3:333–400

Li CY, Li W. 1994. Extensive integration field beyond the classical receptive field of cat's striate cortical neurons—classification and tuning properties. *Vis. Res.* 34:2337–55

Li W, Thier P, Wehrhahn C. 2000. Contextual influence on orientation discrimination of humans and responses of neurons in V1 of alert monkeys. *J. Neurophysiol.* 83:941–54

Li W, Thier P, Wehrhahn C. 2001. Neuronal responses from beyond the classic receptive field in V1 of alert monkeys. *Exp. Brain Res.* 139:359–71

Liden L, Pack C. 1999. The role of terminators and occlusion cues in motion integration and segmentation: a neural network model. *Vis. Res.* 39:3301–20

Liu Z, Kersten D. 1998. 2D observers for human 3D object recognition? *Vis. Res.* 38:2507–19

Livingstone MS, Hubel DH. 1987. Psychophysical evidence for separate channels for the perception of form, color, movement, and depth. *J. Neurosci.* 7:3416–68

Locke J. 1690. *An Essay Concerning Human Understanding.* London: Basset

MacEvoy SP, Paradiso MA. 2001. Lightness constancy in primary visual cortex. *Proc. Natl. Acad. Sci. USA* 98:8827–31

Mach E. 1886. *Contributions to the Analysis of Sensations.* Trans. CM Williams, 1910. Chicago: Open Court

Maffei L, Fiorentini A. 1976. The unresponsive regions of visual cortical receptive fields. *Vis. Res.* 16:1131–39

Mariotte E. 1668. *Nouvelle Découverte Touchant la Veve.* Paris: Frederic Leonard. 27 pp.

Martin KA, Whitteridge D. 1984. The relationship of receptive field properties to the dendritic shape of neurons in the cat striate cortex. *J. Physiol.* 356:291–302

McIlwain JT. 1964. Receptive fields of optic tract axons and lateral geniculate cells: peripheral extent and barbiturate sensitivity. *J. Neurophysiol.* 27:1154–73

Michotte AE, Thinès GO, Crabbé G. 1964. *Les Compléments Amodaux des Structures Perceptives.* Louvain: Inst. Psychol. Univ. Louvain. 56 pp.

Mountcastle VB, LaMotte RH, Carli G. 1972. Detection thresholds for stimuli in humans and monkeys: comparison with threshold events in mechanoreceptive afferent nerve fibers innervating the monkey hand. *J. Neurophysiol.* 35:122–36

Müller J. 1838. *Elements of Physiology.* London: Printed for Taylor & Walton. 2 vols.

Murakami I, Komatsu H, Kinoshita M. 1997. Perceptual filling-in at the scotoma following a monocular retinal lesion in the monkey. *Vis. Neurosci.* 14:89–101

Nakayama K, Shimojo S. 1990. Toward a neural

understanding of visual surface representation. *Cold Spring Harbor Symp. Quant. Biol.* LV: 911–24

Nakayama K, Shimojo S, Silverman GH. 1989. Stereoscopic depth: its relation to image segmentation, grouping, and the recognition of occluded objects. *Perception* 18:55–68

Nelson JI, Frost BJ. 1978. Orientation-selective inhibition from beyond the classic visual receptive field. *Brain Res.* 139:359–65

Nelson JI, Frost BJ. 1985. Intracortical facilitation among co-oriented, co-axially aligned simple cells in cat striate cortex. *Exp. Brain Res.* 61:54–61

Newsome WT, Britten KH, Movshon JA. 1989. Neuronal correlates of a perceptual decision. *Nature* 341:52–54

O'Regan JK. 1992. Solving the "real" mysteries of visual perception: the world as an outside memory. *Can. J. Psychol.* 46:461–88

Paradiso MA, Hahn S. 1996. Filling-in percepts produced by luminance modulation. *Vision Res.* 36:2657–63

Paradiso MA, Nakayama K. 1991. Brightness perception and filling-in. *Vis. Res.* 31:1221–36

Peterhans E, von der Heydt R. 1989. Mechanisms of contour perception in monkey visual cortex. II. Contours bridging gaps. *J. Neurosci.* 9:1749–63

Peterson MA. 2002. On figures, grounds, and varieties of surface completion. In *The 31st Carnegie Symposium: Perceptual Organization in Vision: Behavioral and Neural Perspectives*, ed. M Behrmann, R Kimchi, C Olson. Mahwah, NJ: Earlbaum. In press

Pettet MW, Gilbert CD. 1992. Dynamic changes in receptive-field size in cat primary visual cortex. *Proc. Natl. Acad. Sci. USA* 89:8366–70

Polat U, Mizobe K, Pettet MW, Kasamatsu T, Norcia AM. 1998. Collinear stimuli regulate visual responses depending on cell's contrast threshold. *Nature* 391:580–84

Polat U, Sagi D. 1993. Lateral interactions between spatial channels: suppression and facilitation revealed by lateral masking experiments. *Vis. Res.* 33:993–99

Polat U, Sagi D. 1994. The architecture of perceptual spatial interactions. *Vis. Res.* 34:73–78

Purves D, Shimpi A, Lotto RB. 1999. An empirical explanation of the cornsweet effect. *J. Neurosci.* 19:8542–51

Ramachandran VS. 1992. Blind spots. *Sci. Am.* 266:86–91

Ramachandran VS. 1994. Phantom limbs, neglect syndromes, repressed memories, and Freudian psychology. *Int. Rev. Neurobiol.* 37:291–333

Ramachandran VS, Gregory RL. 1991. Perceptual filling in of artificially induced scotomas in human vision. *Lett. Nat.* 350:699–702

Ramsden BM, Hung CP, Roe AW. 2001. Real and illusory contour processing in area V1 of the primate: a cortical balancing act. *Cereb. Cortex* 11:648–65

Redies C, Crook JM, Creutzfeldt OD. 1986. Neuronal responses to borders with and without luminance gradients in cat visual cortex and dorsal lateral geniculate nucleus. *Exp. Brain Res.* 61:469–81

Rockland KS, Lund JS. 1982. Widespread periodic intrinsic connections in the tree shrew visual cortex. *Science* 215:1532–34

Rossi AF, Desimone R, Ungerleider LG. 2001. Contextual modulation in primary visual cortex of macaques. *J. Neurosci.* 21:1698–709

Rossi AF, Paradiso MA. 1996. Temporal limits of brightness induction and mechanisms of brightness perception. *Vis. Res.* 36:1391–98

Rossi AF, Paradiso MA. 1999. Neural correlates of perceived brightness in the retina, lateral geniculate nucleus, and striate cortex. *J. Neurosci.* 19:6145–56

Rossi AF, Rittenhouse CD, Paradiso MA. 1996. The representation of brightness in primary visual cortex. *Science* 273:1104–7

Rubin E. 1921. *Visuell Wahrgenommene Figuren; Studien in Psychologischer Analyse.* Kobenhavn: Gyldendalske Boghandel

Schein SJ, Desimone R. 1990. Spectral properties of V4 neurons in the macaque. *J. Neurosci.* 10:3369–89

Schmidt KE, Kim DS, Singer W, Bonhoeffer T, Lowel S. 1997. Functional specificity of long-range intrinsic and interhemispheric connections in the visual cortex of strabismic cats. *J. Neurosci.* 17:5480–92

Sheth BR, Sharma J, Rao SC, Sur M. 1996. Orientation maps of subjective contours in visual cortex. *Science* 274:2110–15

Shevell SK. 1978. The dual role of chromatic backgrounds in color perception. *Vis. Res.* 18:1649–61

Shevell SK, Wei J. 1998. Chromatic induction: border contrast or adaptation to surrounding light? *Vis. Res.* 38:1561–66

Shimojo S, Silverman GH, Nakayama K. 1989. Occlusion and the solution to the aperture problem for motion. *Vis. Res.* 29:619–26

Sillito AM, Grieve KL, Jones HE, Cudeiro J, Davis J. 1995. Visual cortical mechanisms detecting focal orientation discontinuities. *Nature* 378:492–96

Stanley GB, Li FF, Dan Y. 1999. Reconstruction of natural scenes from ensemble responses in the lateral geniculate nucleus. *J. Neurosci.* 19:8036–42

Stoner GR, Albright TD. 1993. Image segmentation cues in motion processing: implications for modularity in vision. *J. Cogn. Neurosci.* 5:129–49

Stoner GR, Albright TD. 1995. Depth from occlusion and non-Fourier motion. *Soc. Neurosci. Abstr.* 21:511

Stoner GR, Duncan RO, Albright TD. 1998. Area MT responses to contrast-modulated stimuli may encode depth-from-dynamic occlusion. *Invest. Opthalmol. Vis. Sci.* 39:S904

Sugita Y. 1999. Grouping of image fragments in primary visual cortex. *Nature* 401:269–72

Tanaka K, Hikosaka K, Saito H, Yukie M, Fukada Y, Iwai E. 1986. Analysis of local and wide-field movements in the superior temporal visual areas of the macaque monkey. *J. Neurosci.* 6:134–44

Tong F, Engel SA. 2001. Interocular rivalry revealed in the human cortical blind-spot representation. *Nature* 411:195–99

Vinje WE, Gallant JL. 2000. Sparse coding and decorrelation in primary visual cortex during natural vision. *Science* 287:1273–76

von der Heydt R, Peterhans E, Baumgartner G. 1984. Illusory contours and cortical neuron responses. *Science* 224:1260–62

von Helmholtz H. 2000 (1860/1924). *Treatise on Psychological Optics*, trans. JPC Southall. Rochester, NY: Opt. Soc. Am.

Wachtler T, Albright TD, Sejnowski TJ. 2001. Nonlocal interactions in color perception: nonlinear processing of chromatic signals from remote inducers. *Vis. Res.* 41:1535–46

Wachtler T, Sejnowski TJ, Albright TD. 1999. Responses of cells in macaque V1 to chromatic stimuli are compatible with human color constancy. *Soc. Neurosci. Abs.* 25(S7.8):4

Wallach H. 1935. Uber visuell wahrgenommene Bewegungsrichtung. *Psychol. Forsch.* 20:325–80

Wallach H. 1948. Brightness constancy and the nature of achromatic colors. *J. Exp. Psychol.* 38:310–24

Walls G. 1954. The filling-in process. *Am. J. Optom.* 31:329–40

Wertheimer M. 1961 (1912). *Classics in Psychology*, ed. T Shipley, pp. 1032–89. New York: Phil. Library

Wundt WM. 1902. *Principles of Physiological Psychology*. Trans. EB Titchener. New York: Macmillan (from German)

Xiao DK, Marcar VL, Raiguel SE, Orban GA. 1997. Selectivity of macaque MT/V5 neurons for surface orientation in depth specified by motion. *Eur. J. Neurosci.* 9:956–64

Young T. 1807. *A Course of Lectures on Natural Philosophy and the Mechanical Arts*, 2 vols. London: Johnson

Zeki S. 1983a. Colour coding in the cerebral cortex: the reaction of cells in monkey visual cortex to wavelengths and colours. *Neuroscience* 9:741–65

Zeki S. 1983b. Colour coding in the cerebral cortex: the responses of wavelength-selective and colour-coded cells in monkey visual cortex to changes in wavelength composition. *Neuroscience* 9:767–81

Zeki S. 1990. Colour vision and functional specialisation in the visual cortex. *Discuss. Neurosci.* 6:11–64

Zenger B, Sagi D. 1996. Isolating excitatory and inhibitory nonlinear spatial interactions involved in contrast detection. *Vis. Res.* 36:2497–513

Zhou H, Friedman HS, von der Heydt R. 2000. Coding of border ownership in monkey visual cortex. *J. Neurosci.* 20:6594–611

Zipser K, Lamme VA, Schiller PH. 1996. Contextual modulation in primary visual cortex. *J. Neurosci.* 16:7376–89

LARGE-SCALE SOURCES OF NEURAL STEM CELLS

David I. Gottlieb
Department of Anatomy and Neurobiology, Washington University School of Medicine, St. Louis, Missouri; email: gottlied@pcg.wustl.edu

Key Words neural cell lines, neural development, in vitro neural differentiation

■ **Abstract** Large-scale sources of neural stem cells are crucial for both basic research and novel approaches toward treating neurological disorders. Three sources that produce neural cells closely resembling their normal counterparts are now available: oncogene immortalized stem cells, neurospheres, and embryonic stem cell (ES)-derived neural cells. Cells including multiple subtypes of CNS and PNS neurons, as well as oligodendrocytes, Schwann cells, and astrocytes, are modeled by these large-scale sources. Although most cell lines were originally from rodents, their human counterparts are being discovered and characterized.

INTRODUCTION

There is currently a high level of interest in neural stem cells. A number of distinct aspects of these cells is responsible. One is the intrinsic importance of stem cells in the process of normal development. Another is the recent appreciation that stem cells occur in the adult as well as the developing embryo. Stem cells as the basis of cell lines provide valuable research tools for general molecular and cellular neuroscience. Finally, stem cells are seen as having significant potential in therapy for a wide range of clinical conditions that involve loss of neural cells. The general biology of neural stem cells has been covered in several recent reviews (Gage 2000, Panicker & Rao 2001, Temple 2001). This review focuses on neural cell lines derived from stem cells. Progress in this area has been rapid, and a clearer view of neural cell lines and their relationship to normal stem cells is beginning to emerge. Also, it is cell lines, rather than stem cells themselves, that are likely to be used in most transplantation research.

Neural cell lines may be defined as clonally derived cells capable of indefinite replication in tissue culture and differentiation into neural cells. This definition is broad enough to include embryonic stem cells and teratocarcinomas, which are covered in the review. Some lines are capable of differentiating into multiple neural cell types, whereas others have a more restricted fate. There have been two distinct motives for developing neural cell lines. The first is the continuing need for tractable cell systems for analyzing developmental and physiological mechanisms in neural cells. The intact nervous system consists of a large variety of cell types that are spatially intermingled; most are postmitotic and cannot be expanded in tissue culture. Many experimental approaches require large numbers of homogeneous cells. It is often important that these cells be adapted to the simplified environment of tissue culture. Cell lines meet these requirements and are widely used to study mechanisms at the molecular level. The current use of the PC12 neural cell line, derived by Greene & Tischler in 1976 to analyze mechanisms of transmitter release, nerve growth factor (NGF)-related signal transduction pathways, as well as a variety of other mechanisms in neuronal cell biology, underscores the importance of cell lines in neuroscience. The second motive for studying neural cell lines is their potential for neural transplantation. The basic premise that some types of neurological disease may be treatable by cell transplantation is attractive, but for many years the hope of finding a practical source of transplantable cells seemed remote. Recent discoveries in the area of neural cell lines are changing this picture rapidly.

In this article I highlight the principles involved in the production and use of neural cell lines. Next, I describe progress using each of the major categories of cell lines. These sections are intended not to be comprehensive but rather to illustrate the strengths of the various strategies employed. Finally, the current status of basic conceptual issues relating to cell lines is discussed, and issues for future research are outlined.

Fundamental Features of Neural and Pluripotent Cell Lines

Neural and pluripotent cell lines may be understood with respect to a set of defining properties that includes (*a*) phenotypic resemblance to a particular type of stem cell; (*b*) a mechanism for overcoming natural limits to replication; (*c*) cellular memory; and (*d*) differentiation in response to extrinsic cues. This set of properties provides a framework for thinking about cell lines, though the level of understanding varies widely among the cell lines reviewed.

It is likely that each neural or pluripotent cell line is derived from a particular type of stem cell. In some cases, such as lines derived from spontaneous tumors, the originating stem cell can be inferred only from the phenotype of the line. In other cases (ES cells, lines derived by transformation with oncogenes, and certain types of neurospheres), the original stem cell is known more precisely. In spite of significant gaps in our knowledge, the view that the phenotype of individual cell lines is determined in part by the stem cell from which it arose seems valid and is a useful organizing principle.

In the natural setting of the nervous system, stem cell division is tightly regulated. Even in the early embryonic nervous system, purely exponential cell division occurs only during a brief time window. Therefore, during most of development, mechanisms that constrain cell division are in constant operation. All cell lines overcome these constraints and exhibit continuous cell division. The precise mechanisms that limit cell division are likely to vary among different stem cell types. Perhaps some day cell lines may be classified according to the mechanism that permits continual cell division. However, at present we are unable to do this for even a single cell type.

Neural and pluripotent (as well as most other) cell lines exhibit the fascinating property of cellular memory. Cell lines are routinely propagated in tissue culture for many generations. Compared with the normal embryo, the tissue culture environment is vastly simplified. Nevertheless, each line retains its essential phenotype as if it were locked in and independent of the specialized environment in the embryo from which it arose. One hypothesis is that cellular memory reflects constant signaling through pathways that start with extracellular effectors to maintain expression of cell-type determining genes (Blau & Baltimore 1991). An alternative is that cellular memory arises from a self-propagating epigenetic state. Repression of the X chromosome in female somatic tissues (the Lyon effect) is one example of this kind of epigenetic state. It is presently unclear whether cellular memory is due to one or the other of these mechanisms or to the combined action of both. This question is made all the more fascinating by the discovery that cell lines can, under special circumstances, transdifferentiate into other lineages. In summary, the very existence of cell lines is based on mechanisms that may be inferred but are not understood in any detail.

Finally, neural cell lines are capable of switching from cell division to differentiation. In a few cases, a defined inducing agent is sufficient. For example, NGF causes neuronal differentiation of PC12 cells, and retinoic acid induces neural differentiation of a variety of pluripotent cells. At the other end of the spectrum, differentiation of some cell lines is induced by transplanting into the nervous system to provide an appropriate microenvironment. It is likely that the next few years will see the discovery of other defined inducing signals.

Organization

In this review, cell lines are grouped into the following categories: cell lines derived from tumors; lines derived by oncogene transformation; neurospheres and related cell lines; and pluripotent cell lines (ES cells and teratocarcinomas). This scheme simplifies description and comparison of lines and is somewhat related to the mechanism by which cells overcome barriers to cell division. However, it is likely that as we understand more about stem cells a more natural classification will be possible. Because of the breadth of coverage, a comprehensive review of each type is not feasible. In the case of older cell lines, only selected new references are cited to indicate current activity in the field. For the newer systems, papers are chosen to illustrate basic principles of selection and use of the lines.

TUMOR-DERIVED CELL LINES

The use of cell lines as models of neural cells became widespread in the mid-1970s. Among those used were neuroblastomas, glioblastomas (reviewed in Brismar 1995), the rat pheochromocytoma PC12 (Greene & Tischler 1976), and induced tumors of the CNS (Schubert et al. 1974). Of these, PC12 cells and the C6 glioblastoma are currently in wide use and best illustrate the advantages of this type of cell line.

PC12 Cells

The PC12 line originated as a spontaneous tumor of the rat adrenal medulla, from which a clonal cell line adapted to growth in culture was established (Greene & Tischler 1976). Because the tumor was spontaneous, there is no direct knowledge of how PC12 cells overcome the limit to continued cell division. PC12 cells synthesize acetylcholine and norepinephrine; both transmitters are packaged into synaptic vesicles and can be released in a voltage- and calcium-dependent manner. Epidermal growth factor (EGF) is a mitogen for PC12 cells, whereas NGF inhibits cell division and induces terminal postmitotic differentiation. The hallmark of PC12 differentiation is a prodigious outgrowth of neurites.

PC12 cells illustrate the advantages of a cell line that divides as well as differentiates as a homogeneous population of cells. Cultures containing large numbers of cells are readily available. Thus experiments requiring enough cells for biochemical analysis are feasible. Also feasible are experiments that require large numbers of precise replicate cultures. Examples of such applications include determining concentration curves of growth factor action or the time course of steps in regulatory cascades. In addition, differentiation is induced by nerve growth factor (NGF), which allows the kinetics of this process to be followed. PC12 cells are also genetically manipulatable as one can create stable clones of transgenic cells. These properties have made PC12 cells a popular model for analyzing the mechanisms of synaptic vesicle exocytosis (reviewed in Burgoyne & Morgan 1998, Chen et al. 2001, and Sugita et al. 1999). Studies of the signal transduction pathways linking NGF-receptor activation to differentiation are also being intensively studied in PC12 cells (Boss et al. 2001, Persengiev et al. 2001, Wu et al. 2001). The above detailed examples provide only a small sample of the extensive current use of the PC12 line.

C6 Glioma Cells

The C6 glioma cell line was one of the first neural cell lines to be studied and is still a significant tool (reviewed in Brismar 1995). Like PC12 cells, C6 cells are homogeneous both during division and differentiation. C6 cells have many of the characterisitcs of astrocytes, though they differ from primary astrocytes in important respects. An advantage of C6 cells is that it is possible to select sublines with interesting biological properties. An instructive example is provided by C6-R, a new subline of C6 that resembles radial glia (Hormigo et al. 2001). C6-R cells express RC1, an antigen characteristic of radial glia. They also promote neuronal migration in vitro, as do normal radial glia in the developing brain. During development, radial glia regress and are thus not present in the adult brain. When transplanted to the adult brain or spinal cord, C6-R cells migrate in white matter tracks. Embryonic rat neurons coinjected into the adult brain with C6-R cells migrate much farther than neurons injected without C6-R cells. Thus transplanted C6-R cells are able to restore one feature of an embryonic microenvironment in the adult brain. C6-R cells are thus a valuable model for investigating cellular mechanisms that promote migration of transplanted neurons. Understanding such mechanisms is an important goal in transplantation research.

CELL LINES DERIVED BY ONCOGENE EXPRESSION

For a stem cell to become the clonal progenitor of a cell line, the intrinsic barriers to continued cell division must be bypassed. In tumor-derived cell lines, the mechanism(s) responsible is not identified. In the 1980s a number of laboratories developed strategies that exploit oncogenes to promote continuous cell division (see Whittemore & Snyder 1996, for review of early work). The essential idea was to transfect embryonic brain cells (which are enriched in stem cells) grown in tissue culture with an oncogene with the aim of creating immortalized cell lines. This approach has several fundamental advantages. The region and developmental stage of the target population of stem cells can be chosen at will. One can obtain a large number of continuously replicating clones and select those with favorable properties. Finally, reversible oncogenes can be utilized that allow the replicative signal to be turned off at will. There are still two limitations to the technique. First, while the identity of the target cell is narrowed, it is not fully defined. Embryonic neural tissues still contain multiple types of stem cells including multipotent progenitors, restricted neuronal progenitors, and restricted glial progenitors (Panicker & Rao 2001, Temple 2001). Each of these might be transformed by expression of an oncogene. Second, insertion of the transforming oncogene into the genome is random. Randomly inserted transgenes are expressed in patterns strongly influenced by the site of insertion. Thus two cell lines with insertion of the identical oncogene at different chromosomal sites could have different phenotypes.

HiB5 and C17 Cells

Two classic examples illustrate the features of this approach. In one, a stem-cell line termed HiB5 was derived from the rat embryonic hippocampal formation (Renfranz et al. 1991). Cultured hippocampal cells were transformed with the large T antigen of SV40; the particular allele was temperature sensitive being functional at 33°C (permissive temperature) but inactive at 39°C, the body temperature of the rat. At the permissive temperature in culture, HiB5 cells replicate and express phenotypes of neural stem cells. At 39°C in culture, they showed some signs of differentiation. Differentiation was not complete as cells continued to express nestin, a marker for stem cells. The cell's full capacity for differentiation was shown by transplantation into the neonatal rat brain. Cells transplanted into the hippocampal formation integrated into the dentate gyrus and differentiated into cells with the morphology of dentate granule neurons. When transplanted to the cerebellum, they differentiated into cerebellar granule cells and Bergmann glial cells. One explanation for the ability of HiB5 cells to adopt widely different fates in response to the transplant site is that the founder cells immortalized by the virus had yet to be specified as to a final neural fate. Alternatively, the process of transformation may have actually reversed the specification of the target cells. In either case these experiments provide very strong evidence that the final pathway of neural differentiation is determined by local cues provided by the host tissue. They also illustrate the potential power of neural stem cells in therapeutic applications where the brain is damaged by disease or trauma.

Analogous results were obtained using another oncogene and population of stem cells (Snyder et al. 1992). Neonatal mouse cerebellar cells were transformed with the v-myc oncogene and clones with continuously dividing stem cells selected; a clone designated C17 has been studied in detail. In vitro, these cells were able to partially differentiate into cells expressing markers of oligodendrocytes and neurons. When transplanted into the cerebellum of neonatal mice, they differentiated into cells with the morphology of cerebellar granule cells. Electron microscopy (EM) analysis revealed synapses formed upon the transplanted cells. More recent data show that v-myc-immortalized C17 cells can also differentiate into cortical pyramid-like neurons (Snyder et al. 1997). The adult mouse cortex was lesioned by a method that causes selective degeneration of layer II/III pyramidal neurons (Macklis 1993). Progenitors injected into the midst of the lesion differentiated into cells with the morphology of pyramidal neurons. EM analysis indicates that some of the transplant-derived neurons received synapses from host neurons. Progenitors injected into control cortex differentiated into glia but not neurons. These experiments demonstrate that highly restricted lesions in the cortex create a microenvironment that promotes a very specific type of neuronal differentiation and that virally immortalized progenitor cells have the ability to respond appropriately.

Parkinson's disease is caused by loss of midbrain dopaminergic neurons innervating the striatum. Transplant therapy for Parkinson's disease is being intensively investigated. A suitable large-scale source of dopaminergic neurons for transplantation therapy in Parkinson's disease is an urgent priority. C17 cells demonstrate one promising approach to this goal (Wagner et al. 1999). Using normal differentiation protocols, the cells do not differentiate into dopaminergic neurons. However, additional signals can induce a high percentage of C17 cells to differentiate into dopaminergic neurons. A subline of C17 cells expressing the dopaminergic-associated nuclear receptor Nurr-1 was selected. A panel of purified growth factors failed to induce these cells to become dopaminergic. However, coculture with ventral mesencephalon cells did induce a small percentage to differentiate into dopamine-producing cells. Further analysis revealed that coculture with mesencephalon-derived astrocytes induced 80% of the cells to become dopaminergic. The differentiated cells synthesize, package, and release dopamine. Though the factor(s) secreted by ventral mesencephalon astrocytes has not been identified, these experiments demonstrate that C17 cells are capable of choosing a dopaminergic fate when given proper extracellular signals. The approach of inducing stem cells to differentiate into other desired neuronal subtypes with extracellular signals will be an important new chapter in this field.

RN33B Cells

It is predictable that oncogene-immortalized cell lines derived from separate brain regions will differ significantly in many properties. Potentially they might vary in their behavior after transplantation. Results with the RN33B cell line show this principle in action. RN33B is a cell line obtained by immortalizing cells of the raphe nucleus with the temperature-sensitive SV40 T oncogene (Onifer et al. 1993, Whittemore & White 1993). Upon transplant into the normal adult cortex, RN33B cells integrate and form neurons with the morphology of pyramidal neurons (Shihabuddin et al. 1995). They also form appropriate-looking neurons in the adult hippocampal formation. The results in the cortex are in striking contrast to C17 cells, which do not form neurons in the uninjured adult cortex. Also, in contrast to C17 cells, transplanted RN33B cells did not differentiate into glial cells. The mechanisms that instruct RN33B cells to undergo neuron birth and differentiation in the adult brain are not known. One possibility is that RN33B cells secrete some of the molecules that cells in the injured cortex use to establish a permissive niche for pyramidal cell differentiation. Alternatively, RN33B cells may secrete factors that maintain them in a juvenile state after transplant to an adult cortex site. Whatever the operative mechanism, these results demonstrate the important point that the adult cortex and hippocampus can be permissive environments for neural differentiation and growth.

ST14A Cells

The results cited above show that stem cell lines are plastic with respect to final fate choice and can use extracellular cues at a transplant site to guide terminal differentiation. A study of striatal-derived precursors shows that they can also retain region-specific characteristics during repeated passage in tissue culture (Ehrlich et al. 2001). Cells from the embryonic rat striatum were immortalized with the temperature-sensitive SV40 T oncogene. At the nonpermissive temperature, cells differentiated into neuron-like cells. The striatal marker DARPP-32 is expressed and dopamine-receptor–activated phosphorylation of CREB occurs. The cells contain gamma amino butyric acid (GABA) as do most striatal neurons. The phenotype of the neurons most resembles medium-sized spiny neurons of the striatum. Thus, immortalized cell lines have the potential to express phenotypes associated with a particular region of the central nervous system (CNS). Discovering the basis of this identity is an exciting prospect for future work.

PNS-Derived Lines

Immortalized lines can also be derived from the peripheral nervous system (PNS) (Rao & Anderson 1997). Prior studies had established conditions in which cultures of the E8.5 mouse spinal cord were enriched in neural crest progenitors. Such cultures were transfected with a retrovirus carrying the v-myc oncogene. The clonal cell line MONC-1 was selected from a large number of colonies of dividing cells. Normal neural crest progenitors can be induced to differentiate into neurons, Schwann cells, or melanocytes by culture in specialized media. MONC-1 cells also differentiate appropriately in each type of specialized medium. These results show that viral immortalized cells may be directed along natural pathways in ways highly similar to their normal counterparts. Oncogene immortalization of PNS progenitors has been extended to the human (Raymon et al. 1999). Dorsal root ganglia (DRG) were cultured from first-trimester human embryos and transfected with a retroviral vector carrying v-myc under control of a tetracycline-repressible promoter. Cells expressing v-myc divided indefinitely. Upon repression of v-myc, many cells exhibited a neuronal morphology and stained with neuronal markers. It is significant that some neurons exhibited capsaicin sensitivity, a feature seen in a subset of normal DRG neurons.

The Schwann cells of the peripheral nervous system arise from mitotically active Schwann cell precursor cells (SCPCs). Oncogene immortalization has been used to generate a cell line with many characteristics of SCPCs (Lobsiger et al. 2001). Cultures of embryonic rat SCPCs were transfected with a retrovirus carrying a transgene for a fusion protein consisting of the extracellular ligand-binding domain of the epidermal growth factor (EGF) receptor fused to the intracellular part of the c-neu tyrosine kinase. This fusion protein provides a mitogenic signal in the presence of EGF. Transfected cells were cultured in the presence of EGF, and candidate SCPC-immortalized clones were selected. Clone SpL201 expresses a number of marker genes of the Schwann cell lineage. The line has the functional capacity to myelinate axons in vitro and also in demyelinated CNS axon tracts. Given the complexity of myelin formation, this is a stringent test that these cells have a physiologically relevant phenotype.

Oncogene immortalization has also been used to create cell lines with key characteristics of olfactory receptor neurons (ORNs). ORNs are a propitious cell type because their progenitors divide throughout the life of the animal and have a significant ability to divide in culture (Mumm et al. 1996, reviewed in Calof 1998). Early postnatal nasal epithelial cells were transfected with a retrovirus carrying a temperature-sensitive SV40 T oncogene (Murrell & Hunter 1999). A cell line (termed odora) was cloned. After differentiation, cells express several marker proteins associated with olfactory sensory neurons. Upon transfection with a cDNA, the cells express the exogenous olfactory receptor protein and give a physiological response to odorant. In a related approach, an immortalized ORN line was obtained by culturing olfactory cells from the so-called immortamouse (Barber et al. 2000). This mouse carries a transgene in which the temperature-sensitive SV40 T oncogene is driven by an interferon (IFN)-responsive promoter. Thus at physiological temperatures in the absence of IFN, functional T antigen is not expressed. Early postnatal nasal epithelium was explanted and cultured at permissive temperatures in the presence of IFN. Many clonal lines grew, one of which was studied in detail. This line expresses a number of olfactory receptor neuron (OFN)-specific genes. Differentiated cells are responsive to odorant. The spectrum of effective odorants suggests that a wide range of odorant receptors is expressed in the population. These lines will be important for studying structure-function relationships in the olfactory transduction pathway, as well as for deciphering mechanisms that control olfactory receptor gene expression at the molecular level.

Astrocyte Cell Lines

Oncogene immortalization can also produce astrocyte cell lines (Whittemore et al. 1994). Adult human spinal-cord cells were transfected with temperature-sensitive SV40 T oncogene. Two lines were obtained that had an astrocyte phenotype when the oncogene was inactive. Specifically the cells had an astrocytic morphology, glutamine synthetase activity, Na^+-dependent glutamate uptake, and the ability to support embryonic neurite outgrowth.

Summary

Recent work has widened the scope and power of oncogene immortalization of neural stem cells. Cells produced by this approach show phenotypes closely resembling those of normal neurons and glia. Exciting results are being obtained with both CNS-derived and PNS-derived lines. Strategies for precise control of oncogene expression, including inducible promoters (Raymon et al. 1999) or oncogenes activated by an extracellular ligand such as EGF (Lobsiger et al. 2001), will increase the utility of this approach for basic studies and transplantation. It is notable that as yet only a handful of oncogenes have been used as immortalizing agents. Currently the control of the cell cycle in neural stem cells is only partially understood. As new regulatory components are discovered, the exciting possibility of additional useful immortalized lines will be realized.

NEUROSPHERES AND RELATED CELL LINES

Rodent Neurospheres

An alternative method of obtaining neural cell lines is based on the existence of neural stem cells that divide continuously when cultured in a serum-free medium. In the paper that defined this approach, the striatum of adult mice was dissociated enzymatically and cultured as a cell monolayer in defined serum-free medium (Reynolds & Weiss 1992). In cultures with EGF, a small percentage of cells detached and formed multicellular spheres that quickly increased in size owing to cell division; these were called neurospheres. When neurospheres were dissociated, some of the resulting single cells would divide to form new spheres; serial propagation by repetition of this cycle was possible. Upon plating on an adhesive substrate, neurospheres differentiate into mixed cultures containing cells with the morphology and antigen-expression profile of neurons and astrocytes. Several aspects of the results were unexpected and novel. First, they demonstrated that a cell type with the capacity to divide exists in a region of the adult CNS not considered to be a proliferative center. They also showed that long-term proliferation could be established by the simple expedient of culturing cells in a serum-free medium with an established mitogen. The original results with adult cells were extended to the embryonic striatum (Reynolds & Weiss 1996). A careful analysis of isolated individual cells formally showed that neurospheres contain an EGF-responsive stem cell capable of either dividing or giving rise to neurons, astrocytes, and oligodendrocytes. Other studies suggest that striatal stem cells responsive to bFGF are present before the EGF-responsive cells and that bFGF-responsive cells give rise to the EGF-responsive cells (Ciccolini & Svendsen 1998).

Other regions of the adult CNS not considered to be proliferative contain progenitors similar to those of the striatum. For example, neurospheres have been derived from the adult eye (Tropepe et al. 2000). Cells cultured from the ciliary margin in defined medium formed spheres that could be propagated. It is interesting that no exogenous growth factors were required to obtain continuously dividing cells. Upon plating, the spheres give rise to neurons with some of the characteristics of rod photoreceptors and bipolar neurons. It remains to be demonstrated that functional photoreceptors differentiate from this adult-derived stem cell. New neuronal production does not occur in the adult spinal cord. Nevertheless, continuously dividing stem cells can be cultured from this region using a method similar to that for neurospheres (Shihabuddin et al. 1997). Based on these results in the retina and spinal cord, it is likely that most regions of the CNS contain stem cells that can give rise to cell lines.

Human Neurospheres

Given the potential that continuously dividing stem cells have for therapeutic transplantation, there has been great interest in obtaining such lines from the human nervous system. The methodology for generating rodent neurospheres has been extended to the human (Carpenter et al. 1999). First trimester fetal forebrain was dissociated by mechanical shearing to yield living dispersed cells. These were cultured in N2-defined medium with added basic fibroblast growth factor (bFGF), EGF, and leukemia inhibitory factor (LIF). Neurospheres formed readily. When left in suspension, cells in neurospheres proliferated vigorously and could be serially passaged; a 1000-fold increase in cell number was obtained. Upon plating on an adhesive substrate, neurospheres gave rise to oligodendrocytes, astrocytes, and neurons. Among the neurons, some expressed GABA and tyrosine hydroxylase (TH). In a similar study, continuous passage of stem cells from human fetal spinal cord resulted in loss of ability to differentiate into neurons; differentiation into astrocytes was unaffected by multiple divisions of the line (Quinn et al. 1999). In contrast, a line isolated from the embryonic human diencephalon retained the capacity to differentiate into neurons after repeated passages (Vescovi et al. 1999). The reason for these divergent results has not been established. There may be intrinsic differences between stem cells of different regions of the human CNS.

Alternatively, small methodological differences may influence whether cells remain multipotent.

Neurospheres from adult human brains have been cultured (Kukekov et al. 1999). The subependymal zone or hippocampal tissue removed from adult brains during surgery for epilepsy was dissociated enzymatically and cultured in a serum-containing medium. Spheres formed that gave rise to neurons and glia when plated on an adhesive substrate. The extent of proliferation in these experiments is not documented. Other studies on human neurospheres (Uchida et al. 2000, Caldwell et al. 2001) are cited below.

Monolayer Cultures of Progenitors

An important variant method for culturing progenitors that uses monolayer culture instead of spheres has been developed (Ray et al. 1993). In this approach, embryonic rat hippocampus is dissociated mechanically and plated on an adhesive substrate in a defined medium. In the absence of bFGF, many cells died or were unhealthy. Addition of bFGF had a survival effect on neurons and also caused stem cells to proliferate. Stem cells generated in culture could differentiate into neurons. In contrast to neurospheres, no glia were present. This system was used to identify and isolate a novel proliferation-stimulating protein (Taupin et al. 2000). The factor is a glycosylated form of cystatin C called CCg. It has replication-stimulating activity both in vivo and in vitro.

A similar approach produces continuously replicating cell lines from adult hippocampal progenitors (Palmer et al. 1997). Genetic marking of single cells shows that true multipotential stem cell lines are produced. One of the lines derived by this method has been analyzed by single-cell physiological recording. Neuron-like cells fired action potentials and formed both GABAergic and glutaminergic synapses with each other in monolayer culture (Toda et al. 2000). This documents that a key aspect of terminal differentiation is possible by adult stem-cell–derived neurons.

Deriving Neurospheres from Identified Progenitors

In the literature cited above, spheres arise from complex mixtures of dissociated brain cells. Such experiments give no direct information on the cell type within the originating tissue that gives rise to neurospheres. Without such information two divergent ideas regarding the origin of neurospheres are tenable. One is that a large subset of brain cells is capable of giving rise to neurospheres but does so with a very low probability. Alternatively, a minor subpopulation could give rise to neurospheres with a high probability. New evidence strongly supports the latter. It has been proposed that astrocytes of the subventricular zones are the true stem cells of the adult nervous system. Clonally derived cultures of isolated subventricular zone (SVZ) astrocytes give rise to neurospheres (Doetsch et al. 1999). In another study, subventricular and ependymal cells of the adult mouse brain were dissociated and analyzed by fluorescence-activated cell sorter (FACS) for size and expression of two surface markers: PNA (peanut agglutinin) and HSA (heat-stable antigen) (Rietze et al. 2001). A small (~.27%) fraction of the unsorted cell population derived from brain were PNA-low + HSA-low and greater than 12 μm in diameter. Almost all these cells could give rise to neurospheres. About 60% of the sphere-forming ability of the unsorted starting population was present in this fraction. This purification is an important achievement because it will allow a detailed examination of the events leading to sphere formation.

Parallel efforts to directly isolate stem cells from human fetal brain have been successful (Uchida et al. 2000). A panel of antibodies directed to cell-surface antigens was surveyed on enzymatically dissociated human fetal brain cells to identify antibodies that delineate neurosphere-forming stem cells. About 4% of cells were positive for antibody CD133. CD133+ cells were sorted and tested in a clonogenic assay. CD133+ cells are highly enriched (about 20-fold relative to the starting population of brain cells) in cells capable of forming neurospheres. These neurospheres could be expanded and were capable of differentiating into neurons and glia. The developmental potential of expanded cells was tested by transplantation into neonatal mouse brain. Transplanted cells disseminated widely and engrafted in multiple regions of the brain where they differentiate into neurons and glia. On an interesting note, cells that had engrafted into the SVZ continue long-term proliferation; tumor formation was not observed. These studies establish a system for systematically investigating the factors involved in establishing long-term proliferating stem cells from adult brain.

Other studies showing that neurospheres (NS) may arise directly from ES cells (Tropepe et al. 2001) are discussed below. An approach utilizing transfected cells has also been used to identify and purify progenitor cells directly (Roy et al. 2000a,b). Adult human hippocampi were obtained from patients undergoing surgery for intractable epilepsy and were dissociated by standard techniques. Cells were then transfected with a plasmid with the nestin promoter–driving green fluorescent protein (GFP). The nestin promoter is selectively expressed in neural progenitors. Upon culture, a minor population of cells express GFP. Nestin/GFP-positive cells were capable of division and subsequent differentiation. Their ability to undergo long-term proliferation was not investigated.

Directed Differentiation of Neurospheres

In most studies to date, differentiation of neurosphere cells into neurons and glia occurs spontaneously when they are plated onto an adhesive substrate. It would be desirable to control the differentiation step especially in ways that might induce desired subpopulations of neural cells. One potential method is by addition of relevant signaling molecules; work along this line has begun. Differentiation of human neurospheres into neurons can be significantly increased by addition of NT3, NT4, and platelet-derived growth factor (PDGF) (Caldwell et al. 2001). It is likely that these are acting as survival factors rather than differentiation factors. Neurospheres derived from embryonic or adult rodent brain show limited differentiation into TH neurons under normal growth conditions. However, upon culture with glial-conditioned medium and bFGF, there is a large increase in the number of TH-positive cells (Daadi & Weiss 1999). It is likely that these early results will be quickly extended by the discovery of many extracellular factors that induce or modulate neuronal differentiation.

Neurospheres as a Model for Stem-Cell Gene Screening

Large-scale sources of stem cells are ideally suited for biochemical applications that would be difficult with the small numbers of stem cells that can be obtained directly from brain. An example is a recent study using neurospheres to screen for genes expressed in neural stem cells and then downregulated in their differentiated derivatives (Geschwind et al. 2001). cDNA prepared from undifferentiated spheres (NS) were subtracted with cDNA from differentiated cultures (DC). Candidate regulated genes were further tested on cDNA arrays from NS and DC cultures to identify differentially regulated genes. Differential regulation was confirmed by northern blots. A number of known and unknown genes selectively expressed in NS cells were identified. Finally, their differential expression was confirmed by in situ hybridization in the developing and adult brain. As expected many of the genes were selectively expressed in the ventricular layer of the embryonic brain thus confirming their stem-cell specificity. It is interesting that some genes were also enriched in hematopoietic stem cells. Further characterization of genes using this approach will contribute toward a more exact definition of neural stem cells in terms of expressed genes. It will also shed light on the important basic issue of whether stem cells of different lineages such as neural and hematopoietic are strongly related.

Transdifferentiation of Neurosphere Cells

Until very recently it was assumed that stem cells committed to the neural lineage could not differentiate into cells of other lineages of the body. This basic assumption is being called into question by new evidence that indicates that neurospheres have a wider differentiation potential than previously assumed (Clarke et al. 2000). Neurospheres were obtained from the adult brain of the ROSA26 mouse strain, which carries a lacz transgene expressed in all tissues. The early embryo has microenvironments capable of instructing cells to choose ectodermal, mesodermal, or endodermal fates. The response of neurosphere cells to these cues was tested by injecting them into the early chick embryo. Neurosphere cells integrated into most organs of the developing chick including the nervous system, stomach, liver, and kidney. In each case, transplanted cells acquired the cell morphology appropriate to the host tissue. The analogous experiment was done in the mouse by transplanting neurosphere cells into the blastocyst. Again neurosphere cells integrated into host tissues to produce a chimera. Nonneural tissues such as liver and heart had significant populations of transplanted cells. These studies concluded that neurospheres contain cells that respond to the embryonic environment by differentiating into nonneural tissues. The relative abundance of cells with this property within the neurosphere is not clear; potentially it might be a property of a minority. It also hasn't been established whether neurosphere-derived nonneural cells possess all the functional capacities of their new lineage. Nevertheless, the results show that some neurosphere cells are capable of differentiating into multiple nonneural lineages. It has also been shown that cells in neurospheres can transdifferentiate into blood cells (Bjornson et al. 1999). ROSA26-derived neurospheres were injected into sublethally irradiated hosts. Grafted cells differentiated into multiple types of hematopoietic cells and gave a partial functional reconstitution of host marrow.

A Neural Precursor with Unlimited Proliferative Capacity

A recent paper indicates that normal oligodendrocyte precursor cells (OPCs) have unlimited replicative capacity (Tang et al. 2001). OPCs were cultured from the postnatal rat optic nerve in a serum-free medium containing PDGF. It is surprising that the cells proliferated continuously for over 60 days without exhibiting any signs of senescence. Even though they avoid senescence in routine cultures, cells retain the capacity to undergo senescence when it is deliberately induced. Addition of serum caused them to stop dividing and express senescence-associated β-galactosidase. Thus some normal neural progenitors can undergo long-term proliferation without senescence. Furthermore, serum normally used in tissue culture medium is implicated as a cause of replicative senescence. The capacity of these cells to undergo normal differentiation was not tested, so it is not clear if long-term proliferation was accompanied by loss of the original OPC identity.

EMBRYONIC STEM CELLS AND TERATOCARCINOMAS

Introduction

Embryonic stem (ES) cells are totipotent cells derived from the early embryo. ES cells were first isolated from the mouse (Evans & Kaufman 1981, Martin 1981). More recently ES cells have been isolated from the monkey (Thomson et al. 1995) and human (Thomson et al. 1998). Mouse ES cells most closely resemble cells of the inner cell mass, a transient structure present in the 3–4-day embryo. Mouse ES cells have three cardinal properties: they are totipotent, genetically normal, and divide without limit. Developmental totipotency is demonstrated by transplanting ES cells to the inner cell mass of a host embryo. If the embryo is then implanted into a foster mother, it develops into an outwardly normal mouse called a chimera, in which all tissues have cells derived from both the ES cells and the host. Thus the transplanted ES cells respond to appropriate cues and differentiate into all cell types of the normal body. Chimeric male mice are fertile and have sperm derived from ES cells. These are competent to provide the paternal genome to a new generation, thus proving the genetic integrity of the ES cell genome. ES cells are capable of continuous replication in tissue culture. Even after 140 cycles of division, they remain genetically normal (Suda et al. 1987). Why ES cells have this natural immortality is not understood in detail. The cytokine LIF plays a key role (reviewed in Smith 2001). Using a signaling pathway through STAT3, LIF prevents ES cells from differentiating. ES cells also maintain a high level of telomerase. ES cells are used as the starting point for creating knockout mice. Genes introduced into ES cells are able to undergo homologous recombination with the corresponding normal gene and can thus introduce mutations into that gene. Homologous recombination in ES cells will be extremely useful as a tool in the in vitro systems described below.

Neural Differentiation of ES Cells

The experiments with chimeric mice show that ES cells have the intrinsic capacity to form all cell types of the nervous system when they are provided with the appropriate developmental cues. This raises the exciting possibility of reconstructing the neural development pathway starting with ES cells in vitro (see earlier reviews: Gottlieb & Huettner 1999, O'Shea 1999).

If ES cells are cultured on a nonadhesive substrate in the absence of LIF, they form small aggregates of cells termed embryoid bodies (EBs). After 4 days of culture, the original stem cells differentiate into a complex mixture of ectodermal and mesodermal progenitors surrounded by a single layer of extraembryonic membrane-like cells. If cultured without added factors, this mixture of cells gives rise to a complex mixture of terminally differentiated cells including heart, muscle, blood, and neurons among others. Addition of retinoic acid (RA) at 4 days has a dramatic effect. After 4 days of additional culture in the presence of RA, the great majority of cells differentiate into neural lineage cells (Bain et al. 1995, 1998). Cells derived by this and related methods will be termed ESNLCs (ES cell-derived neural lineage cells).

When RA-induced EBs are dissociated and plated on an adhesive substrate, cells differentiate further into neurons, astrocytes, and oligodendrocytes (Bain et al. 1995, Finley et al. 1996, Liu et al. 2000). The neurons are postmitotic and have axons and dendrites closely resembling those in primary neuronal cultures. The resting potential, sodium action potentials, potassium and calcium currents in ES cell-derived neurons also closely resemble those in dispersed primary neuronal cultures. Dendrites stain for MAP2, whereas axons stain for synaptophysin. Most significantly, the axons and dendrites form functional synapses with each other. Single-cell recordings reveal that the synapses are indistinguishable from those in dispersed primary cultures of embryonic CNS (Finley et al. 1996). Approximately 70% of the neurons are glutaminergic, 25% GABAergic, and 5% glycinergic. Other protocols utilizing RA but differing in detail also produce neurons and glia (Fraichard et al. 1995, Strubing et al. 1995).

Neural cells can also be derived from ES cells by a selective method (Okabe et al. 1996). ES cells are first cultured as EBs for 4 days. EBs are then plated on an adhesive substrate in serum-free medium. A large proportion of cells die, but surviving cells are enriched in neural progenitors. Further culture for 7 days in the presence of basic fibroblast growth factor (bFGF) causes division and expansion of this population. Withdrawal of bFGF triggers differentiation into a mixture of neurons and glial cells.

ESNLCs have been transplanted into a rat model of spinal-cord injury (McDonald et al. 1999). Rats received a weight drop injury to the thoracic spinal cord. Nine days after injury, when a cavity had developed, ESNLCs were transplanted into the cavity. ESNLCs integrated into the injured cord and differentiated into neurons, astrocytes, and oligodendrocytes. A modest but significant behavioral recovery relative to controls resulted from the ESNLC transplant. Rats receiving transplants were able to support weight on their hindlimbs better than control rats. However, coordinate movement of hindlimbs and forelimbs was not restored by the transplants. The basic culture techniques for producing ESNLCs have been extended to generate cultures enriched in oligodendrocytes (Brustle et al. 1999, Liu et al. 2000). These oligodendrocytes express characteristic marker genes and myelinate axons in vitro. When transplanted into either a genetic model or an injury model of demyelinating disease, the ES cell–derived oligodendrocytes integrate into the tissue and form myelin around available axons. These results suggest that ES cell-derived oligodendrocytes will be useful for therapy in demyelinating conditions. Transplantation of ES cells into the striatum is capable of generating a population of serotonergic and catecholaminergic neurons (Deacon et al. 1998).

Developmental Pathway from ES Cells to Neural Cells

An interesting question is whether the pathway from ES cells to ESNLCs in vitro has the same intermediate steps as normal neural development. This question was addressed by following the expression of early neural markers in ES cells undergoing in vitro differentiation (Bain et al. 1996). Within 2 days of RA addition, the early embryonic markers MASH-1 and WNT-1 are expressed. It is interesting to note, mesodermal markers such as brachyury, zeta globin, and cardiac actin are repressed by RA, which suggests that RA causes a switch in fate from mesodermal to neural. This action of RA has a parallel in the normal development of the mouse (Shum et al. 1999). Mice pregnant with E 9.5-day embryos were injected with RA. Treated embryos had ectopic neural tubes in the posterior half of the body, which had apparently grown at the expense of mesenchyme. Another parallel between the pathway leading to ESNLCs and normal neural development is that BMP4 (bone morphogenetic factor 4) acts as a negative signal in both (Finley et al. 1999).

One model of normal early CNS development suggests that multipotent neuroepithelial cell precursors (NEPs) first differentiate into neural cell-restricted and glial cell-restricted precursors (NRPs and GRPs, respectively). Cells with the functional properties of NRPs and GRPs may be isolated from the mouse neural tube by immunopanning and use of selective media. Analogous cells can also be isolated from ES cells differentiating in vitro (Mujtaba et al. 1999). These results give substantial evidence that ES cells proceed through the same intermediates as they differentiate into ESNLCs. Another line of evidence—that ES cells differentiate through intermediates that resemble normal stem cells—depends on gene targeting.

Sox2 is one of the earliest genes expressed throughout the normal neural plate. Gene targeting was used to introduce the bifunctional selection marker/reporter gene β-*geo* into the *Sox2* gene (Li et al. 1998). When ES cells with this construct were differentiated, they became G418-resistant, demonstrating that the *Sox2* gene was transcribed as it would be in normal early neural cells.

Phenotypic characterization of ESNLCs in vitro still leaves open the question of whether these cells can recapitulate the entire developmental program of neural stem cells. The behavior of ESNLCs transplanted into the embryonic brain suggests they can and strengthens the idea that ES cells differentiate in vitro through a normal pathway (Brustle et al. 1997). ESNLCs were derived in vitro by the selection method described above and injected into the ventricles of E16-E18 rat embryos in utero. Embryos were allowed to survive to 2 weeks after birth. Many of the injected cells entered the telencephalon, diencephalon, and mesencephalon. Transplanted cells differentiated into neurons, astrocytes, and oligodendocytes. Neurons appeared remarkably like adjacent host neurons. For instance, in the cortex there were transplant-derived pyramidal neurons with characteristic dendritic trees and axonal projections. In conclusion, it appears that stem cells derived from ES cells and normal stem cells respond to local developmental cues in a strikingly similar manner.

Being totipotent, ES cells have the potential to become any cell type in the nervous system. Conversely, any cell type of the nervous system is derived from an ES cell–like inner cell mass (ICM) cell. An exciting prospect is to develop systems in which ES cells are converted efficiently into particular types of neural cells in vitro. Recent work has begun to realize this potential. Liu et al. (2000) used several cycles of dissociation and reassociation of EBs to create cultures highly enriched in oligodendrocytes. A successful attempt to induce differentiation of dopaminergic neurons was carried out by Lee et al. (2000). In explanted embryonic midbrain cultures, FGF8 and sonic hedgehog promote dopaminergic neuron differentiation. Cultures of neurally differentiated ES cells were exposed to FGF8 and sonic hedgehog; 15% of the neurons were dopaminergic and capable of synthesizing and releasing dopamine. Although ESNLCs express several transcription factors associated with motorneurons, (Renoncourt et al. 1998) fully differentiated motorneurons have not been found. No doubt the approach of using instructive signals will make it possible to go from ES cells to enriched cultures of many neuronal types. The ability to do gene targeting is an important feature of this system that will greatly facilitate analysis of developmental pathways. For example a double knockout of the GD3 synthase gene blocks ganglioside production but, contrary to expectation, does not disrupt normal differentiation (Kawai et al. 1998).

Until recently all protocols for neural induction of ES cells began by forming EBs. Two reports show that neurogenesis may be uncoupled from EB formation. Tropepe et al. (2001) cultured undifferentiated ES cells on an adhesive substratum at low density in serum-free medium. A small percentage spontaneously transformed into neurospheres. Addition of BMP4 inhibits the formation of neurospheres. This suggests that neurosphere induction has parallels to early neuroectoderm specification in *Xenopus*, where BMP4 also inhibits ectoderm differentiation. In the presence of bFGF, ES cell–derived neurosphere cells proliferate. Upon plating on an adhesive substrate, they formed mixed cultures of neurons and glia. Neurons in this system have not yet been characterized physiologically. In another method that bypasses EB formation, Kawasaki et al. (2000) plated ES cells on a feeder layer of mesenchymal stromal cells. ES cells proliferated and then spontaneously differentiated into neural cells. Many of the neurons were dopaminergic.

Human ES Cells

The application of ES cells to human neuroscience and medicine appeared remote until 1995 when rhesus monkey ES cells were isolated (Thomson et al. 1995). This was followed by isolation of human ES cells (Thomson et al. 1998). A closely related human cell type, the embryonic germ cell (EG cell), was also discovered (Shamblott et al. 1998). Human ES cells are capable of extended replication in culture. They retain a normal karyotype after 6 months of continuous culture. This is a reasonable standard of genetic normality, although clearly not as rigorous as that possible with mouse ES cells. A crucial question is whether human ES cells are totipotent. When injected into immunodeficient mice they form teratomas with complex histology. Many tissue types are represented in these teratomas including rosettes of cells that resemble the neural tube (Thomson et al. 1998). Because ethical considerations preclude testing the developmental potential of human ES cells by chimera formation, teratoma formation and in vitro differentiation are the most stringent ways to assess developmental totipotency. Human ES cells differentiate in vitro into complex mixtures of cells (Schuldiner et al. 2000). Gene markers for ectoderm, mesoderm, and endoderm are expressed in the cultures as determined by analysis of RNA from whole cultures. The expression profile in culture changes with addition of numerous growth factors. RA upregulates the expression of a heavy chain of neurofilament protein (NF-H), which suggests a regulation with similarities to mouse ES cells. A more-detailed study of neuronal differentiation of human ES cells has been carried out (Schuldiner et al. 2001). ES cells cultured as EBs in the presence of RA and subsequently plated on an adhesive substrate differentiated into cells with extensive neuronal networks. Related studies show that human ES cells are readily transfectable with foreign DNA (Eiges et al. 2001). Additional features of the human ES cell–derived neural cells are detailed in a second study (Carpenter et al. 2001). Both male and female ES cell lines were maintained as dividing cells for 6 months, while remaining karyotypically stable, before analyzing their ability to differentiate. Efficient differentiation into PS-NCAM and A2B5-positive progenitor-like cells was obtained by induction with RA. Upon plating, many of these cells exhibited the morphology and antigen expression profile of mature neurons. Some of these were capable of generating sodium-dependent action potentials. Finally, an efficient protocol for obtaining enriched cultures of human neurons based on differential adhesion has been developed (Zhang et al. 2001). In conclusion, many of the approaches for deriving neural cells from mouse ES cells are likely to be applicable to human ES cells.

P19 and Ntera2 Cells

P19 and NT-2 cells are tumor-derived pluripotent cell lines that share some biological properties with ES cells. The P19 line was derived by transplanting an E7 mouse embryo beneath the testis capsule, which resulted in tumor formation (McBurney & Rogers 1982). A clonal line was then established from the tumor. Under standard tissue culture conditions, P19 cells have the phenotype of embryonal carcinoma cells; these strongly resemble cells of the E4–5-day embryo. Culture of P19 cells as aggregates in the presence of retinoic acid induces neural differentiation (Jones-Villeneuve et al. 1983). Induced cultures consist mostly of neurons and glia. The neurons are postmitotic and have axons and dendrites. Crucially, they are able to form functional synapses among themselves (Finley et al. 1996). Both glutaminergic and GABAergic neurons are present. Thus the P19 cells are able to differentiate from a pluripotent, early embryonic cell type to cells with the defining phenotype of CNS neurons. Undifferentiated P19 cells are readily transfectable with DNA; therefore, the P19 system allows a combination of genetic, biochemical, and cell biological investigation. These properties have been exploited to screen for novel genes regulated during early neural development (Bouillet et al. 1997; Oulad-Abdelghani et al. 1997, 1996). The system is also useful for analyzing gene regulatory elements (Tamura et al. 2001).

NTERA-2 cells are human cells that strongly resemble P19 cells. They were derived from a human germ-cell tumor (Andrews et al. 1984, Pleasure et al. 1992). Undifferentiated cells express many markers characteristic of early human embryonic cells. NTERA-2 cells are readily transfected with DNA, and stable transgenic lines expressing transgenes can be obtained. Thus genetic approaches to developmental and physiological questions are feasible. In the presence of RA, NTERA-2 cells differentiate into a population enriched in neurons. These cells have many of the characteristics of postmitotic CNS neurons. These include firing tetrodotoxin-sensitive spikes (Rendt et al. 1989) and glutamate receptors (Younkin et al. 1993). They also form functional synapses (Hartley et al. 1999). The developmental sequence from pluripotent, undifferentiated stem cells to differentiated neurons recapitulates three phases of gene expression seen in the normal nervous system (Przyborski et al. 2000). In phase 1, there is an increase in nestin mRNA. In phase 2, nestin mRNA levels fall and neuroD1 levels rise; there is also a rise of A2B5 surface antigen expression. In phase 3, levels of mRNA for synaptophysin and neuron-specific enolase, markers of terminally differentiated neurons, rise.

SUMMARY IDEAS

There has been great progress toward obtaining large-scale sources of neural cells over the past five years, and three distinct approaches have taken shape. A number of basic issues cut across the entire field; consideration of these issues will help us define the challenges for the immediate future.

How Real Are the Cells?

The first neural cell lines were derived from tumors. Differentiated cells from these lines were manifestly abnormal and only approximated the phenotype of normal neural cells. Now the gap between normal neural cells and line-derived cells has narrowed dramatically. For example, myelin formation is a highly complex process that may be regarded as a stringent end-point for judging differentiation. ES cell–derived neural cells are capable of forming myelin in vitro and after transplantation (Brustle et al.1999, Liu et al. 2000). A virally immortalized cell line can also form myelin (Lobsiger et al. 2001). Synapse formation is another extremely complex process that requires large numbers of genes to be appropriately expressed. A neurosphere-like cell line forms synapses (Toda et al. 2000). ES cell–derived neurons do so as well (Finley et al. 1996). Complex neuronal morphology is another valid criterion of differentiation. Oncogene immortalized cells transplanted into the developing or adult brain readily differentiate into cells with appropriate neuronal morphology (Renfranz et al. 1991, Snyder et al. 1992, Shihabuddin et al. 1995, Snyder et al. 1997). ES cell-derived neural cells do the same (Brustle et al. 1997). In conclusion, properties we associate with the most specific aspects of neural differentiation are expressed in cell-line systems.

As compelling as these examples are, there are important remaining questions about the phenotypes of cell line–derived neural cells. One regards the extent and rate of differentiation. Thus far most of the literature has focused on the fact that interesting processes such as synaptogenesis or myelination actually occur. It is not known how efficient these processes are compared with those in normal neural cells. In future studies it will be essential to determine the percentage of cells that carry out a particular process and the rate and extent to which it happens. With such data in hand, comparisons with normal cells will be far more meaningful. Another challenge is to apply the new generation of gene expression profiling technologies to this problem. These measure the expression levels of large numbers of mRNA or proteins simultaneously. It will be informative to compare and contrast the expression profiles of normal and cell-line–derived neural cells. Together, studies of specialized processes and expression profiling will give detailed answers to the question of how close to normal cell-line–derived neural cells truly are.

Mechanisms Leading to Lines

The lines reviewed were derived at a time when we had only partial knowledge of the most basic relevant control processes. All cell lines overcome the natural barriers to continued cell division. The particular technical steps used were arrived at by trial and error. How oncogenes such as v-myc and SV40 T override restraining mechanisms to immortalize neural progenitors is not understood in detail. It is equally unclear why culture of progenitors in serum-free–defined medium in the presence of EGF or bFGF allows neurospheres to form. In a similar manner, the practical steps needed to convert ICM cells of the early embryo to ES cells are known, but the mechanisms underlying the transformation are only now beginning to be unraveled (Smith 2001). We are now poised to learn much more about the regulatory systems that determine the choice between continued division and differentiation of neural progenitors. This knowledge can then be used to explain in detail how lines arose. We will also gain new insight into all of the consequences of overcoming barriers to division. This new understanding will provide a needed bridge between normal stem cells and lines derived from them.

The phenotypic stability of neural stem cells and lines is a very important current topic. Under some conditions, lines can be remarkably stable even after multiple rounds of cell division. Other environments allow cells to transdifferentiate into nonneural lineages. The rules and mechanisms responsible for stability on the one hand and transdifferentiation on the other are presently unknown. Here too, new understanding of mechanisms will help unify our perception of normal stem cells and cell lines.

Neural Cell Lines and Transplantation Research

Large-scale sources of neural stem cells have opened up new possibilities for transplantation research. There are a large variety of medical conditions where the common denominator is the loss of neural cells. These range from loss due to trauma to degenerative diseases. The idea that these conditions can be treated by cell transplantation so as to replace lost neural cells is intrinsically appealing. The cell lines discussed in this review are having a powerful impact on neural transplantation research in animal models. They provide a practical and affordable source of the large numbers of cells needed to perform transplantation experiments. Their theoretical potential in this arena is large. It is very likely that the tissue culture methods of today will be quickly extended so that major subsets of neural cells can be identified and physically purified. This will open the door to a systematic exploration of which cells are optimal in a particular disease model. It will be possible to test each lineage to determine which is best. Likewise, the early, middle, and late stages of a particular lineage will be available and can be tested by transplantation.

A key advantage of cell lines is that they are amenable to genetic engineering. In particular, ES cells are suitable for gene targeting and thus allow extremely sophisticated manipulation of genes. Genetically engineered cell lines will bring additional power to transplantation research. Cells may be engineered so as to overexpress growth factors to determine which may be helpful for transplants. It is possible to engineer cells so that they are resistant to apoptosis and thus determine if this improves survival. Gene-knockout experiments can be performed to learn if removing particular signaling pathways enhances the effectiveness of transplantation. We may discover in the long run that normal neural cells are far less effective than genetically engineered derivatives. No one can foresee exactly which manipulations will lead to improved results. What is certain is that there are now powerful tools to actively investigate the issue.

Until recently transplantation research was confined to animal models. Each of the cell-line types we have been discussing was first obtained from rodents. Now there are human counterparts of virtually all of them. There is much to be done with animal models. As these studies reveal therapeutic possibilities, the new human cell lines will be available to move this research into human trials.

ACKNOWLEDGMENTS

I thank Ben Barres, Jim Skeath, and Scott Whittemore for comments on the manuscript. I also thank Larry Adams and Hai-Qing Xian, members of my laboratory, for valuable discussions, and Cheryl Rivers for help preparing the manuscript. Work in the author's laboratory was supported by P01 NS39577 from the NIH and a grant from the Pharmacia/Washington University Biomedical Research Program.

The *Annual Review of Neuroscience* is online at http://neuro.annualreviews.org

LITERATURE CITED

Bain G, Kitchens D, Yao M, Huettner JE, Gottlieb DI. 1995. Embryonic stem cells express neuronal properties in vitro. *Dev. Biol.* 168:342–57

Bain G, Ray WJ, Yao M, Gottlieb DI. 1994. From embryonal carcinoma cells to neurons: the P19 pathway. *BioEssays* 16:343–48

Bain G, Ray WJ, Yao M, Gottlieb DI. 1996. Retinoic acid promotes neural and represses mesodermal gene expression in mouse embryonic stem cells in culture. *Biochem. Biophys. Res. Commun.* 223:691–94

Bain G, Yao M, Huettner J, Finley M, Gottlieb D. 1998. Neuronlike cells derived in culture from P19 embryonal carcinoma and embryonic stem cells. In *Culturing Nerve Cells*, ed. G Banker, K Goslin, pp. 189–212. Cambridge, MA: MIT

Barber RD, Jaworsky DE, Yau KW, Ronnett GV. 2000. Isolation and in vitro differentiation of conditionally immortalized murine olfactory receptor neurons. *J. Neurosci.* 20:3695–704

Bjornson CR, Rietze RL, Reynolds BA, Magli MC, Vescovi AL. 1999. Turning brain into blood: a hematopoietic fate adopted by adult neural stem cells in vivo. *Science* 283:534–37

Blau HM, Baltimore D. 1991. Differentiation requires continuous regulation. *J. Cell Biol.* 112:781–83

Boss V, Roback J, Young A, Roback L, Weisenhorn D. 2001. Nerve growth factor, but not epidermal growth factor, increases Fra-2 expression and alters Fra-2/JunD binding to AP-1 and CREB binding elements in pheochromocytoma (PC12) cells. *J. Neurosci.* 21:18–26

Bouillet P, Sapin V, Chazaud C, Messaddeq N, Decimo D, et al. 1997. Developmental expression pattern of Stra6, a retinoic acid-responsive gene encoding a new type of membrane protein. *Mech. Dev.* 63:173–86

Brismar T. 1995. Physiology of transformed glial cells. *Glia* 15:231–43

Brustle O, Jones KN, Learish RD, Karram K, Choudhary K, et al. 1999. Embryonic stem cell-derived glial precursors: a source of myelinating transplants. *Science* 285:754–56

Brustle O, Spiro AC, Karram K, Choudhary K, Okabe S, McKay RD. 1997. In vitro-generated neural precursors participate in mammalian brain development. *Proc. Natl. Acad. Sci. USA* 94:14809–14

Burgoyne RD, Morgan A. 1998. Analysis of regulated exocytosis in adrenal chromaffin cells: insights into NSF/SNAP/SNARE function. *BioEssays* 20:328–35

Caldwell MA, He X, Wilkie N, Pollack S, Marshall G, et al. 2001. Growth factors regulate the survival and fate of cells derived from human neurospheres. *Nat. Biotechnol.* 19:475–79

Calof AL, Mumm JS, Rim PC, Shou J. 1998. The neuronal stem cell of the olfactory epithelium. *J. Neurobiol.* 36:190–205

Carpenter MK, Cui X, Hu ZY, Jackson J, Sherman S, et al. 1999. In vitro expansion of a multipotent population of human neural progenitor cells. *Exp. Neurol.* 158:265–78

Carpenter MK, Inokuma MS, Denham J, Mujtaba T, Chiu CP, Rao MS. 2001. Enrichment of neurons and neural precursors from human embryonic stem cells. *Exp. Neurol.* 172:383–97

Chen YA, Scales SJ, Scheller RH. 2001. Sequential SNARE assembly underlies priming and triggering of exocytosis. *Neuron* 30:161–70

Ciccolini F, Svendsen CN. 1998. Fibroblast growth factor 2 (FGF-2) promotes acquisition of epidermal growth factor (EGF) responsiveness in mouse striatal precursor cells: identification of neural precursors responding to both EGF and FGF-2. *J. Neurosci.* 18:7869–80

Clarke DL, Johansson CB, Wilbertz J, Veress B, Nilsson E, et al. 2000. Generalized potential of adult neural stem cells. *Science* 288:1660–63

Daadi MM, Weiss S. 1999. Generation of tyrosine hydroxylase-producing neurons from precursors of the embryonic and adult forebrain. *J. Neurosci.* 19:4484–97

Deacon T, Dinsmore J, Costantini L, Ratliff J, Isacson O. 1998. Blastula-stage stem cells can differentiate into dopaminergic and serotonergic neurons after transplantation. *Exp. Neurol.* 149:28–41

Doetsch F, Caille I, Lim DA, Garcia-Verdugo JM, Alvarez-Buylla A. 1999. Subventricular zone astrocytes are neural stem cells in the adult mammalian brain. *Cell* 97:703–16

Ehrlich ME, Conti L, Toselli M, Taglietti L, Fiorillo E, et al. 2001. ST14A cells have properties of a medium-size spiny neuron. *Exp. Neurol.* 167:215–26

Eiges R, Schuldiner M, Drukker M, Yanuka O, Itskovitz-Eldor J, Benvenisty N. 2001. Establishment of human embryonic stem cell-transfected clones carrying a marker for undifferentiated cells. *Curr. Biol.* 11:514–18

Evans M, Kaufman M. 1981. Establishment in culture of pluripotent cells from mouse embryos. *Nature* 292:154–56

Finley MF, Devata S, Huettner JE. 1999. BMP-4 inhibits neural differentiation of murine embryonic stem cells. *J. Neurobiol.* 40:271–87

Finley MF, Kulkarni N, Huettner JE. 1996. Synapse formation and establishment of neuronal polarity by P19 embryonic carcinoma cells and embryonic stem cells. *J. Neurosci.* 16:1056–65

Fraichard A, Chassande O, Bilbaut G, Dehay C, Savatier P, Samarut J. 1995. In vitro differentiation of embryonic stem cells into glial cells and functional neurons. *J. Cell Sci.* 108:3181–88

Gage FH. 2000. Mammalian neural stem cells. *Science* 287:1433–38

Geschwind DH, Ou J, Easterday MC, Dougherty JD, Jackson RL, et al. 2001. A genetic analysis of neural progenitor differentiation. *Neuron* 29:325–39

Gottlieb D, Huettner JE. 1999. An in vitro pathway from embryonic stem cells to neurons and glia. *Cells Tissues Organs* 165:165–72

Greene LA, Tischler AS. 1976. Establishment of a noradrenergic clonal line of rat adrenal pheochromocytoma cells which respond to nerve growth factor. *Proc. Natl. Acad. Sci. USA* 73:2424–28

Hartley RS, Margulis M, Fishman PS, Lee VM, Tang CM. 1999. Functional synapses are formed between human NTera2 (NT2N, hNT) neurons grown on astrocytes. *J. Comp. Neurol.* 407:1–10

Hormigo A, McCarthy M, Nothias JM, Hasegawa K, Huang W, et al. 2001. Radial glial cell line C6-R integrates preferentially in adult white matter and facilitates migration of coimplanted neurons in vivo. *Exp. Neurol.* 168:310–22

Jones-Villeneuve EM, Rudnicki MA, Harris JF, McBurney MW. 1983. Retinoic acid-induced neural differentiation of embryonal carcinoma cells. *Mol. Cell Biol.* 3:2271–79

Kawai H, Sango K, Mullin KA, Proia RL. 1998. Embryonic stem cells with a disrupted GD3 synthase gene undergo neuronal differentiation in the absence of b-series gangliosides. *J. Biol. Chem.* 273:19634–38

Kawasaki H, Mizuseki K, Nishikawa S, Kaneko S, Kuwana Y, et al. 2000. Induction of midbrain dopaminergic neurons from ES cells by stromal cell-derived inducing activity. *Neuron* 28:31–40

Kukekov VG, Laywell ED, Suslov O, Davies K, Scheffler B, et al. 1999. Multipotent stem/progenitor cells with similar properties arise from two neurogenic regions of adult human brain. *Exp. Neurol.* 156:333–44

Lee SH, Lumelsky N, Studer L, Auerbach JM, McKay RD. 2000. Efficient generation of midbrain and hindbrain neurons from mouse embryonic stem cells. *Nat. Biotechnol.* 18:675–79

Li M, Pevny L, Lovell-Badge R, Smith A. 1998. Generation of purified neural precursors from embryonic stem cells by lineage selection. *Curr. Biol.* 8:971–74

Liu S, Qu Y, Stewart T, Howard M, Chakrabortty S, et al. 2000. Embryonic stem cells differentiate into oligodendrocytes and myelinate in culture and after spinal cord transplantation. *Proc. Natl. Acad. Sci. USA* 97:6126–31

Lobsiger CS, Smith PM, Buchstaller J, Schweitzer B, Franklin RJ, et al. 2001. SpL201: a conditionally immortalized Schwann cell precursor line that generates myelin. *Glia* 36:31–47

Macklis J. 1993. Transplanted neocortical neurons migrate selectively into regions of neuronal degeneration produced by chromophore-targeted laser photolysis. *J. Neurosci.* 13:3848–63

Martin GR. 1981. Isolation of a pluripotent cell line from early mouse embryos cultured in medium conditioned by teratocarcinoma stem cells. *Proc. Natl. Acad. Sci. USA* 78:7634–38

McBurney MW, Rogers BJ. 1982. Isolation of male embryonal carcinoma cells and their chromosome replication patterns. *Dev. Biol.* 89:503–8

McDonald J, Liu X, Qu Y, Liu S, Mickey S, et al. 1999. Transplanted embryonic stem cells survive, differentiate, and promote recovery in injured rat spinal cord. *Nat. Med.* 5:1410–12

Mujtaba T, Piper DR, Kalyani A, Groves AK, Lucero MT, Rao MS. 1999. Lineage-restricted neural precursors can be isolated from both the mouse neural tube and cultured ES cells. *Dev. Biol.* 214:113–27

Mumm JS, Shou J, Calof AL. 1996. Colony-forming progenitors from mouse olfactory epithelium: evidence for feedback regulation of neuron production. *Proc. Natl. Acad. Sci. USA* 93:11167–72

Murrell JR, Hunter DD. 1999. An olfactory sensory neuron line, odora, properly targets olfactory proteins and responds to odorants. *J. Neurosci.* 19:8260–70

Okabe S, Forsberg-Nilsson K, Spiro AC, Segal M, McKay RD. 1996. Development of neuronal precursor cells and functional postmitotic neurons from embryonic stem cells in vitro. *Mech. Dev.* 59:89–102

Onifer SM, Whittemore SR, Holets VR. 1993. Variable morphological differentiation of a raphe-derived neuronal cell line following transplantation into the adult rat CNS. *Exp. Neurol.* 122:130–42

O'Shea KS. 1999. Embryonic stem cell models of development. *Anat. Rec.* 257:32–41

Oulad-Abdelghani M, Bouillet P, Chazaud C, Dolle P, Chambon P. 1996. AP-2.2: a novel AP-2-related transcription factor induced by retinoic acid during differentiation of P19 embryonal carcinoma cells. *Exp. Cell. Res.* 225:338–47

Oulad-Abdelghani M, Chazaud C, Bouillet P, Sapin V, Chambon P, Dolle P. 1997. Meis2, a novel mouse Pbx-related homeobox gene induced by retinoic acid during differentiation of P19 embryonal carcinoma cells. *Dev. Dyn.* 210:173–83

Palmer TD, Takahashi J, Gage FH. 1997. The adult rat hippocampus contains primordial neural stem cells. *Mol. Cell. Neurosci.* 8: 389–404

Panicker M, Rao M. 2001. Stem cells and neurogenesis. In *Stem Cell Biology*, ed. D Marshak, R Gardner, D Gottlieb, pp. 399–438. Cold Spring Harbor, NY: Cold Spring Harbor

Persengiev SP, Li J, Poulin ML, Kilpatrick DL. 2001. E2F2 converts reversibly differentiated PC12 cells to an irreversible, neurotrophin-dependent state. *Oncogene* 20:5124–31

Pleasure SJ, Page C, Lee VM. 1992. Pure, postmitotic, polarized human neurons derived from NTera 2 cells provide a system for expressing exogenous proteins in terminally differentiated neurons. *J. Neurosci.* 12:1802–15

Quinn SM, Walters WM, Vescovi AL, Whittemore SR. 1999. Lineage restriction of neuroepithelial precursor cells from fetal human spinal cord. *J. Neurosci. Res.* 57:590–602

Rao MS, Anderson DJ. 1997. Immortalization and controlled in vitro differentiation of murine multipotent neural crest stem cells. *J. Neurobiol.* 32:722–46

Ray J, Peterson D, Schinstine M, Gage F. 1993. Proliferation, differentiation, and long-term culture of primary hippocampal neurons. *Proc. Natl. Acad. Sci. USA* 90:3602–6

Raymon HK, Thode S, Zhou J, Friedman GC, Pardinas JR, et al. 1999. Immortalized human dorsal root ganglion cells differentiate into neurons with nociceptive properties. *J. Neurosci.* 19:5420–28

Rendt J, Erulkar S, Andrews PW. 1989. Presumptive neurons derived by differentiation of a human embryonal carcinoma cell line exhibit tetrodotoxin-sensitive sodium currents and the capacity for regenerative responses. *Exp. Cell Res.* 180:580–84

Renfranz PJ, Cunningham MG, McKay RD. 1991. Region-specific differentiation of the hippocampal stem cell line HiB5 upon implantation into the developing mammalian brain. *Cell* 66:713–29

Renoncourt Y, Carroll P, Filippi P, Arce V, Alonso S. 1998. Neurons derived in vitro from ES cells express homeoproteins characteristic of motoneurons and interneurons. *Mech. Develop.* 79:185–97

Reynolds BA, Weiss S. 1992. Generation of neurons and astrocytes from isolated cells of the adult mammalian central nervous system. *Science* 255:1707–10

Reynolds BA, Weiss S. 1996. Clonal and population analyses demonstrate that an EGF-responsive mammalian embryonic CNS precursor is a stem cell. *Dev. Biol.* 175:1–13

Rietze RL, Valcanis H, Brooker GF, Thomas T, Voss AK, Bartlett PF. 2001. Purification of a pluripotent neural stem cell from the adult mouse brain. *Nature* 412:736–39

Roy NS, Benraiss A, Wang S, Fraser RA, Goodman R, et al. 2000a. Promoter-targeted selection and isolation of neural progenitor cells from the adult human ventricular zone. *J. Neurosci. Res.* 59:321–31

Roy NS, Wang S, Jiang L, Kang J, Benraiss A, et al. 2000b. In vitro neurogenesis by progenitor cells isolated from the adult human hippocampus. *Nat. Med.* 6:271–77

Schubert D, Heinemann S, Carlisle W, Tarikas H, Kimes B, et al. 1974. Clonal cell lines from the rat central nervous sytem. *Nature* 249:224–27

Schuldiner M, Eiges R, Eden A, Yanuka O, Itskovitz-Eldor J, et al. 2001. Induced neuronal differentiation of human embryonic stem cells. *Brain Res.* 913:201–5

Schuldiner M, Yanuka O, Itskovitz-Eldor J, Melton DA, Benvenisty N. 2000. From the cover: effects of eight growth factors on the differentiation of cells derived from human embryonic stem cells. *Proc. Natl. Acad. Sci. USA* 97:11307–12

Shamblott M, Axelman J, Wang S, Bugg E, Littlefield J, et al. 1998. Derivation of pluripotent stem cells from cultured human primordial germ cells. *Proc. Natl. Acad. Sci. USA* 95:13726–31

Shihabuddin LS, Hertz JA, Holets VR, Whittemore SR. 1995. The adult CNS retains the potential to direct region-specific differentiation of a transplanted neuronal precursor cell line. *J. Neurosci.* 15:6666–78

Shihabuddin L, Ray J, Gage F. 1997. FGF-2 is sufficient to isolate progenitors found in adult mammalian spinal cord. *Exp. Neurol.* 148:577–86

Shum AS, Poon LL, Tang WW, Koide T, Chan BW, et al. 1999. Retinoic acid induces downregulation of Wnt-3a, apoptosis and diversion of tail bud cells to a neural fate in the mouse embryo. *Mech. Dev.* 84:17–30

Smith A. 2001. Embryonic stem cells. In *Stem Cell Biology*, ed. D Marshak, R Gardner, D Gottlieb, pp. 205–30. Cold Spring Harbor, New York: Cold Spring Harbor

Snyder EY, Deitcher DL, Walsh C, Arnold-Aldea S, Hartwieg EA, Cepko CL. 1992. Multipotent neural cell lines can engraft and participate in development of mouse cerebellum. *Cell* 68:33–51

Snyder EY, Yoon C, Flax JD, Macklis JD. 1997. Multipotent neural precursors can differentiate toward replacement of neurons undergoing targeted apoptotic degeneration in adult mouse neocortex. *Proc. Natl. Acad. Sci. USA* 94:11663–68

Strubing C, Ahnert-Hilger G, Shan J, Wiedenmann B, Hescheler J, Wobus AM. 1995. Differentiation of pluripotent embryonic stem cells into the neuronal lineage in vitro gives rise to mature inhibitory and excitatory neurons. *Mech. Dev.* 53:275–87

Suda Y, Suzuki M, Ikawa Y, Aizawa S. 1987. Mouse embryonic stem cells exhibit indefinite proliferative potential. *J. Cell Physiol.* 133:197–201

Sugita S, Janz R, Sudhof TC. 1999. Synaptogyrins regulate Ca^{2+}-dependent exocytosis in PC12 cells. *J. Biol. Chem.* 274:18893–901

Tamura T, Hashimoto M, Aruga J, Konishi Y, Nakagawa M, et al. 2001. Promoter structure and gene expression of the mouse inositol 1,4,5-trisphosphate receptor type 3 gene. *Gene* 275:169–76

Tang DG, Tokumoto YM, Apperly JA, Lloyd AC, Raff MC. 2001. Lack of replicative senescence in cultured rat oligodendrocyte precursor cells. *Science* 291:868–71

Taupin P, Ray J, Fischer WH, Suhr ST, Hakansson K, et al. 2000. FGF-2-responsive neural stem cell proliferation requires CCg, a novel autocrine/paracrine cofactor. *Neuron* 28:385–97

Temple S. 2001. The development of neural stem cells. *Nature* 414:112–17

Thomson JA, Kalishman J, Golos TG, Durning M, Harris CP, et al. 1995. Isolation of a primate embryonic stem cell line. *Proc. Natl. Acad. Sci. USA* 92:7844–48

Thomson JA, Itskovitz-Eldor J, Shapiro SS, Waknitz MA, Swiergiel JJ, et al. 1998. Embryonic stem cell lines derived from human blastocysts. *Science* 282:1145–47

Thomson JA, Marshall VS, Trojanowski JQ. 1998. Neural differentiation of rhesus embryonic stem cells. *Apmis* 106:149–57

Toda H, Takahashi J, Mizoguchi A, Koyano K, Hashimoto N. 2000. Neurons generated from adult rat hippocampal stem cells form functional glutamatergic and GABAergic synapses in vitro. *Exp. Neurol.* 165:66–76

Tropepe V, Coles BL, Chiasson BJ, Horsford DJ, Elia AJ, et al. 2000. Retinal stem cells in the adult mammalian eye. *Science* 287:2032–36

Uchida N, Buck DW, He D, Reitsma MJ, Masek M, et al. 2000. Direct isolation of human central nervous system stem cells. *Proc. Natl. Acad. Sci. USA* 97:14720–25

Vescovi AL, Parati EA, Gritti A, Poulin P, Ferrario M, et al. 1999. Isolation and cloning of multipotential stem cells from the embryonic human CNS and establishment of transplantable human neural stem cell lines by epigenetic stimulation. *Exp. Neurol.* 156:71–83

Wagner J, Akerud P, Castro DS, Holm PC, Canals JM, et al. 1999. Induction of a midbrain dopaminergic phenotype in Nurr1-overexpressing neural stem cells by type 1 astrocytes. *Nat. Biotechnol.* 17:653–59

Whittemore SR, Snyder EY. 1996. Physiological relevance and functional potential of central nervous system-derived cell lines. *Mol. Neurobiol.* 12:13–38

Whittemore SR, White LA. 1993. Target regulation of neuronal differentiation in a temperature-sensitive cell line derived from medullary raphe. *Brain Res.* 615:27–40

Whittemore SR, Neary JT, Kleitman N, Sanon HR, Benigno A, et al. 1994. Isolation and characterization of conditionally immortalized astrocyte cell lines derived from adult human spinal cord. *Glia* 10:211–26

Wu C, Lai CF, Mobley WC. 2001. Nerve growth factor activates persistent Rap1 signaling in endosomes. *J. Neurosci.* 21:5406–16

Younkin DP, Tang CM, Hardy M, Reddy UR, Shi QY, et al. 1993. Inducible expression of neuronal glutamate receptor channels in the NT2 human cell line. *Proc. Natl. Acad. Sci. USA* 90:2174–78

Zhang SC, Wernig M, Duncan ID, Brustle O, Thomson JA. 2001. In vitro differentiation of transplantable neural precursors from human embryonic stem cells. *Nat. Biotechnol.* 19:1129–33

SCHIZOPHRENIA AS A DISORDER OF NEURODEVELOPMENT

David A. Lewis[1] and Pat Levitt[2,3]

[1]Departments of Psychiatry and Neuroscience and [2]Department of Neurobiology, University of Pittsburgh School of Medicine, University of Pittsburgh, Pittsburgh, Pennsylvania 15213; email: lewisda@msx.upmc.edu
[3]J. F. Kennedy Center for Human Development, Vanderbilt University, Nashville, Tennessee 37203; email: pat.levitt@vanderbilt.edu

Key Words neuropsychiatric disorder, genetics, environment, psychosis, polygenic

■ **Abstract** A combination of genetic susceptibility and environmental perturbations appear to be necessary for the expression of schizophrenia. In addition, the pathogenesis of the disease is hypothesized to be neurodevelopmental in nature based on reports of an excess of adverse events during the pre- and perinatal periods, the presence of cognitive and behavioral signs during childhood and adolescence, and the lack of evidence of a neurodegenerative process in most individuals with schizophrenia. Recent studies of neurodevelopmental mechanisms strongly suggest that no single gene or factor is responsible for driving a highly complex biological process. Together, these findings suggest that combinatorial genetic and environmental factors, which disturb a normal developmental course early in life, result in molecular and histogenic responses that cumulatively lead to different developmental trajectories and the clinical phenotype recognized as schizophrenia.

INTRODUCTION

Schizophrenia is a severe brain disorder that usually produces a lifetime of disability and emotional distress for affected individuals (Lewis & Lieberman 2000). The diagnostic clinical features of the disorder typically appear in the late second to third decade of life, with the average age of onset generally about five years earlier in males than in females. Although schizophrenia afflicts approximately 1% of the population throughout the world, the specific factors that give rise to the illness remain elusive. A number of studies have focused on identifying genetic and environmental components that separately, or in combination, may be causative of the disease. This literature is particularly complex, replete with both enticing original observations and subsequent studies that fail to replicate findings of genetic linkage or environmental risk factors. These apparent discrepancies may, in fact, reflect the complex nature of schizophrenia, in which apparently subtle brain abnormalities produce an illness that has its onset in early adulthood and then proceeds with apparently minimal degenerative changes over the lifetime of the individual.

Like many other human diseases, the clinical syndrome recognized as schizophrenia may represent the termination of multiple different pathogenetic paths. Attempts to produce a unifying concept of the etiology of schizophrenia have posited biological mechanisms that have their origins in developmental processes that transpire prior to the onset of clinical symptoms. Although agreement has not yet been achieved regarding the specific causal factors and the timeframe during which they may act, the neurodevelopmental nature of schizophrenia appears to be a particularly attractive concept. In this chapter we provide a brief overview of current literature that relates to possible developmental origins of the illness and then provide a framework for a developmental perspective that takes into account the contributions of both genetic susceptibility and environmental factors to produce the complex phenotype of schizophrenia.

GENETIC RISK AND ENVIRONMENTAL CONTRIBUTIONS

The risk of developing schizophrenia is directly associated with the degree of biological relatedness to an affected individual (Gottesman 1991); that is, first-degree relatives of an affected individual have a higher risk of manifesting the illness than do second-degree relatives, and a monozygotic twin of an individual with schizophrenia is at greater risk than a dizygotic twin. In addition, when the biological children of individuals with schizophrenia are adopted, their risk of developing schizophrenia remains elevated, as expected for first-degree relatives, and is much higher than the general population rates of schizophrenia exhibited by their adoptive families (Kety et al. 1971, Ingraham & Kety 2000). Furthermore, the offspring of identical twins discordant for schizophrenia have elevated rates of the disorder, independent of whether the parent was affected or unaffected (Gottesman & Bertelsen 1989).

The genetic liability to schizophrenia appears to be transmitted in a polygenic, non-Mendelian fashion (Risch & Baron 1984, Risch 2000). A number of loci demonstrate associations with schizophrenia, including 22q11–13, 6p, 13q, and 1q21–22 (Pulver 2000, Brzustowicz et al. 2000). Interestingly, several of these loci contain genes that have well-delineated neurobiological functions. However, these observations have not yet been converted into reliable single gene findings through the use of positional cloning approaches. In addition, although linkage studies, in general, have proven difficult to replicate in subsequent cohorts, this phenomenon may highlight the genetic complexity of schizophrenia, with different subtypes or etiologies of schizophrenia produced by different fundamental molecular defects.

Although the etiology of schizophrenia clearly involves genetic factors, with the heritability of the disorder estimated to range from 70–85%, about 60% of all persons with schizophrenia have neither a first- nor second-degree relative with the disorder (Gottesman & Erlenmeyer-Kimling 2001). In addition, given that the degree of concordance for schizophrenia among monozygotic twins only approaches 50% (Gottesman 1991), genetic liability alone is not sufficient for the clinical manifestation of the illness. These observations suggest both that the genetic predisposition to schizophrenia is complex and that sporadic forms of the illness may exist.

Given that inheritance alone cannot account for schizophrenia, considerations of the etiology of the illness have also included the role of environmental factors. The importance of these factors is demonstrated by the fact that in twin studies the nonshared environment accounts for almost all of the liability for schizophrenia attributable to environmental effects (Tsuang et al. 2001). Indeed, it has been suggested recently that just as the entire genome may need to be searched for genes that convey susceptibility to schizophrenia, so the entire "envirome" must be examined for environmental risk factors (Tsuang et al. 2001). Consequently, most models of the etiology of schizophrenia propose additive and/or interactive effects between multiple susceptibility genes and environmental factors. Of the environmental events that have been associated with an increased risk of schizophrenia, many occur during the prenatal or perinatal periods of life, long before the typical onset, in late adolescence or early adulthood, of the psychotic symptoms that are required for the diagnosis of schizophrenia. This delay between environmental events of possible etiological significance and the appearance of clinical illness has played a major role in the idea that schizophrenia may be a disorder of neural development.

The idea of a developmental etiology has also been fueled by negative data. Specifically, the majority of postmortem studies have failed to find evidence of gliosis, whether examined by Nissl staining or immunoreactivity for glial fibrillary acidic factor, in the brains of subjects with schizophrenia (Roberts & Harrison 2000). Furthermore, glial membrane turnover does not appear to be increased, as assessed by magnetic resonance spectroscopy, at illness onset or in subjects with chronic schizophrenia (Bertolino et al. 1998). This apparent absence of gliosis has been interpreted to exclude typical neurodegenerative processes as operative in schizophrenia. Given the epidemiological data for perinatal brain damage as a precursor of the illness (discussed below), the absence of gliosis may also be surprising from a neurodevelopmental perspective because the brain may be able to mount a gliotic response as early as the twentieth week of gestation. However, neuronal loss or other types of disturbances can certainly occur in either the developing or adult brain without a sustained glial reaction (Milligan et al. 1991), so the absence of gliosis may only be informative about the nature of brain abnormalities in schizophrenia, rather than their timing.

CONCEPTUALIZATION OF SCHIZOPHRENIA AS A NEURODEVELOPMENTAL DISORDER

In the broadest sense the view of schizophrenia as a neurodevelopmental disorder posits that pathogenetic biological events or characteristics are present much earlier in life than the onset of the features of the illness (e.g., psychosis) required for diagnosis (Chua & Murray 1996). This general idea has a long history, dating back to the initial descriptions of the disorder a century ago. For example, Hecker proposed that schizophrenia occurs at puberty in individuals who are "slightly retarded" in physical and mental development (see Takei & Murray 1998); Kraeplin observed that premorbid signs could be detected in early childhood (see Marenco & Weinberger 2000); and Bleuler noted that many individuals who developed schizophrenia could be found earlier in life to exhibit "a tendency to seclusion, withdrawal, together with moderate or severe degrees of irritability" (as quoted in Malmberg et al. 1998).

Specific forms of a neurodevelopmental conceptualization of schizophrenia began to appear in the 1980s. In a highly influential paper, Weinberger (1987) proposed that schizophrenia is a "neurodevelopmental disorder in which a fixed brain lesion from early in life interacts with certain normal maturational events that occur much later." This hypothesis was based on the idea that a "brain lesion can remain clinically silent until normal developmental processes bring the structures affected by the lesion 'on line'" (Marenco & Weinberger 2000). Other investigators, while endorsing the basic tenets of this view, suggested that it was applicable to only a subset of individuals with schizophrenia. For example, Murray and colleagues (1992) proposed that the age of appearance of the first behavioral abnormalities be used to delineate the following three types of schizophrenia: "congenital," adult onset, and late onset. The congenital or neurodevelopmental form of schizophrenia applies to those individuals who have a brain abnormality from pre- or perinatal life that is manifest as subtle behavioral disturbances prior to late adolescence, when the full syndrome of the illness appears. Compared with other types of schizophrenia, congenital schizophrenia was thought to be more common in males, more severe, and more likely to be associated with soft neurological signs and cognitive impairments (Pilowsky et al. 1993).

In contrast, Feinberg (1982), impressed by the substantial changes in brain function and structure, such as sleep architecture and cortical synaptic density that occur during normal adolescence, suggested that schizophrenia was the result of a disturbance of late developmental events. Specifically, he proposed that altered cortical synaptic pruning during adolescence was the central pathogenetic process.

Thus, at the extremes, neurodevelopmental views of schizophrenia have posited that the illness results from either (a) an early (pre- or perinatal), static brain lesion with a long latency until the appearance of clinical signs and symptoms, or (b) a late (adolescence) brain disturbance of limited duration and short latency. However, there is no clear evidence against a pathophysiological process, with cumulative potency, operating across infancy and adolescence and possibly through the initial phases of the illness (Marenco & Weinberger 2000). For example, computer simulations have been used to propose a model of schizophrenia in which reduced synaptic connectivity results from disturbances in neural development active during both perinatal and adolescent periods (McGlashan & Hoffman 2000).

In the following section we review a variety of observations, including environmental events and behavioral abnormalities, that have been reported to occur prior to the onset of the clinical features of schizophrenia. We then consider the possible pathogenetic significance of these data, discuss how an understanding of neurodevelopmental mechanisms can inform these considerations, and conclude with a proposal for and illustration of an integrated model of the etiopathogenesis of schizophrenia.

FINDINGS ACROSS DEVELOPMENT IN INDIVIDUALS WITH SCHIZOPHRENIA

Prenatal Period

MATERNAL NUTRITION   Individuals conceived at the height of the Dutch Hunger Winter of 1944–1945 showed a twofold increase in the risk for schizophrenia in both male and female offspring (Susser et al. 1996). In contrast, lesser famine exposure in early pregnancy or severe famine in later pregnancy were not associated with increased risk. Furthermore, the increased risk associated with severe malnutrition during early pregnancy appeared to be somewhat specific to schizophrenia relative to other psychiatric disorders. However, it is unclear whether individuals who conceived during the height of the famine might have had other factors that predisposed their offspring to schizophrenia. In addition, the unique circumstances that made this study possible have rendered independent replications difficult.

MATERNAL INFECTION   Beginning in 1988 (Mednick et al. 1988) a series of studies reported a greater than expected incidence of schizophrenia in the offspring of women who were in their second trimester of gestation at the peak of the 1957 influenza epidemic. However, some subsequent studies failed to detect an association between the timing of pregnancy and influenza epidemics (Morgan et al. 1997, Selten et al. 1998, Jablensky 2000). Furthermore, a number of methodological concerns were raised about the reports with positive findings (Crow 1994, McGrath & Castle 1995). Most importantly, two investigations that examined the offspring of women known to have had influenza during the 1957 epidemic failed to find an increased risk of schizophrenia (Cannon et al. 1996, Grech et al. 1997), although the sample sizes were relatively small. More recent findings, such as an increased risk of schizophrenia in offspring following maternal rubella infections (Brown et al. 2000) and elevated serum levels of IgG and IgM class immunoglobulins found prior to delivery in mothers of individuals who subsequently developed schizophrenia (Buka et al. 2001), provide interesting leads, but conclusive evidence of an in utero infectious etiology of schizophrenia remains elusive.

SEASON OF BIRTH   A large number of studies, in both the northern and southern hemispheres, have replicated the observation that compared with the general population, schizophrenia is associated with a 5–8% excess of births in the winter and spring months (Torrey et al. 1997). However, a similar excess of winter-spring births is also seen in individuals with certain other psychiatric conditions such as bipolar disorder and major depression. In addition, whether these observations reflect the action of an intrauterine factor that varies seasonally, such as respiratory viral infections (or the typical treatments for these infections), the procreational habits of individuals who carry the risk genes for schizophrenia, or some other factor has not yet been determined.

URBAN BIRTH   A significant positive relationship between the size of urban places of birth and the incidence of schizophrenia and other psychoses has been reported (Marcelis et al. 1998, Mortensen et al. 1999). Interestingly, although a history of schizophrenia in a first-degree relative is associated with a much higher relative risk for schizophrenia, on a population basis, place and season of birth may account for many more causes of illness (Mortensen et al. 1999). In addition, a recent study suggests that, independent of place of birth, urban residence during upbringing is associated with an increased risk of schizophrenia (Pedersen & Mortensen 2001). However, it is not clear whether this association with schizophrenia reflects greater exposure to other factors, such as infection, toxins, or malnutrition, that may be more common in urban populations.

SMALL HEAD SIZE   Several studies have reported reduced head circumference at birth, relative to controls, in individuals who later developed schizophrenia (see McNeil et al. 2000a, Kunugi et al. 2001 for reviews). In contrast, 78% of 50 magnetic resonance imaging studies failed to find statistically significant reductions in total brain volume in adults with schizophrenia (Shenton et al. 2001). However, given the substantial variation in head/brain size among the general population and the host of potential covariates influencing these measures, the possibilities that reductions in brain size are subtle (i.e., leading to small effect sizes) in schizophrenia, or present in only a subset of affected individuals, cannot yet be excluded.

MINOR PHYSICAL ANOMALIES   So-called minor physical anomalies (MPAs), slight deviations in external physical characteristics (e.g., low set ears, furrowed tongue, high arched palate, curved fingers, adherent earlobes) are considered to be the consequence of disturbed prenatal development of the ectoderm. Because the central nervous system also develops from the ectodermal germ layer, MPAs are thought to be associated with abnormal development of the brain. MPAs have been found with increased prevalence in individuals with schizophrenia and other psychiatric disorders (Lane et al. 1996), and also in individuals with other developmental disabilities such as Down syndrome and pervasive mental retardation. Unfortunately, many of the studies that focused on schizophrenia were limited by a scorer bias because it is difficult to assess MPAs in adults in a fashion blind to the presence of a psychiatric diagnosis. However, in individuals at increased genetic risk for schizophrenia by virtue of having an affected parent, but assessed for MPAs before the clinical onset of symptoms, high MPA scores were associated with a significantly greater likelihood of developing schizophrenia-spectrum disorders (Schiffman et al. 2001). In addition, MPAs do not appear to be associated with genetic liability for schizophrenia in either high-risk offspring or unaffected siblings of schizophrenic subjects, suggesting that the appearance of MPAs may be independent of genetic risk, and thus perhaps a marker of a "second" prenatal hit that increases vulnerability to the illness (Green et al. 1994). However, the extent to which MPAs actually inform the neurodevelopmental basis for schizophrenia is not clear because not only are MPAs associated with other disorders, but the frequency of MPAs in normal subjects has raised the question of whether they can truly be considered abnormalities (Marenco & Weinberger 2000, McNeil et al. 2000a).

Perinatal Period

OBSTETRICAL COMPLICATIONS   Schizophrenia has frequently been associated with an increase in obstetrical complications, with such problems reported in the medical histories of 20% or more of subjects with schizophrenia (Cannon 1997, McNeil et al. 2000a). In a meta-analysis (Geddes & Lawrie 1995) the pooled odds ratio of the effect of exposure to obstetrical complications on the subsequent development of schizophrenia was 2.0 (95% confidence interval 1.6–2.4), indicating that individuals with obstetrical complications are twice as likely to develop schizophrenia. However, interpreting the significance of this finding is hampered by the fact that the definition of obstetrical complications differs across studies, variably including events during pregnancy, at the time of delivery, and during the first postnatal month.

In general, labor-delivery complications (LDCs) appear as risk factors in a larger proportion of subjects with schizophrenia than do pregnancy complications (including viral exposure) or signs of fetal maldevelopment (Cannon 1997). After taking into account family history and gender, Verdoux et al. (1997) found in a meta-analysis that LDCs were particularly associated with an increased risk of early-onset schizophrenia. For example, onset before age 22 was associated with 2.7-fold increase in abnormal presentation at birth and 10-fold greater likelihood of a complicated Caesarean section. This association does not appear to be a consequence of selective maternal recall because it has also been found in studies that examined original birth records (see Takei & Murray 1998 for details). However, in population-based studies 97% of those with LDCs do not develop schizophrenia (Done et al. 1991, Buka et al. 1993), indicating both that LDCs have a very low predictive value for the appearance of schizophrenia and that their presence is not sufficient for the illness.

The impact of obstetric complications has also been assessed in twin studies in which differences between monozygotic twins discordant for schizophrenia must be considered to reflect the disease process and/or environmental factors, rather than genetic predisposition. Interestingly, LDCs, but not events during pregnancy, characterized monozygotic twins of which one or both was affected with schizophrenia but not twins of which neither was affected (McNeil et al. 2000b). Furthermore, for discordant twin pairs, when the twin affected with schizophrenia was born second, there were very high rates of prolonged labor and lower rates of complications earlier during pregnancy, whereas the opposite was true of twins in which the individual with schizophrenia was born first. Although these findings are interesting, they are limited by the small sample size (n = 22 twin pairs) and the fact that the history of obstetrical complications was based on retrospective maternal reports.

Through what mechanisms might LDCs act to increase the risk of schizophrenia? In a large Finnish cohort Jones and colleagues (1998) found that low birth weight alone and combined with premature birth were more common among schizophrenic subjects, findings suggestive of abnormalities in prenatal growth. This type of observation raises the possibility that LDCs reflect abnormal fetal development rather than representing an independent risk factor for schizophrenia; however, other lines of investigation have failed to provide support for this interpretation (McNeil et al. 2000a). Some studies suggest that LDCs are not a consequence of genetic predisposition. For example, neither the unaffected siblings nor the offspring of individuals with schizophrenia appear to be more likely to have LDCs than the general population (see Cannon 1997 for review). However, the association of LDCs with schizophrenia may be greater among those at increased genetic risk for the specific illness, suggesting that predisposing genes may make the developing brain more susceptible to the effects of LDCs, such as hypoxemia (Cannon 1997). Consistent with a model of gene-environment interactions, the offspring born with LDCs of individuals with schizophrenia may be more likely to develop schizophrenia than the offspring born without LDCs, whereas the same degree of LDCs does not increase risk of schizophrenia in the offspring of control subjects (Cannon 1997).

Childhood and Early Adolescence

MOTOR ABNORMALITIES   In a clever use of home movies Walker and colleagues (1994) found that clinicians were able to identify children who subsequently developed schizophrenia on the basis of abnormal movements, especially choreoathetoid movements and posturing of the upper limbs, and poorer motor skills. The differences in these children, compared with their unaffected siblings, other subjects who later developed a mood disorder and their unaffected siblings, and children from families without mental illness, were greatest when evaluated before 2 years of age. During this developmental time frame there is a rapid increase in motor skills such as manual manipulation and locomotion. In addition, in the subjects with schizophrenia, these early childhood motor defects were related to greater ventricular size in adulthood (Walker et al. 1996). Similarly, the high risk offspring of subjects with schizophrenia were reported to exhibit delayed development of certain motor skills, such as posture control, standing, and walking (Fish et al. 1992, Marcus et al. 1993). Finally, in a large cohort of subjects born during a single week in 1946, the individuals who subsequently developed schizophrenia achieved milestones of motor development, especially walking, later than expected (Jones et al. 1994).

Thus, at least some individuals who later develop schizophrenia exhibit motor abnormalities during childhood, especially during the first 2 years of life. The strength of these observations rests, in part, on the fact that different study designs, which do not share the same limitations, point to the same conclusion. Although these findings are nonspecific and may appear in association with other psychiatric conditions, they do suggest that disturbances in brain function are present at an early age in individuals with schizophrenia. However, whether these observations represent early signs of the illness, indicative of its neurodevelopmental pathogenesis, or risk factors that may help translate a genetic/environmental predisposition into the illness, remains to be determined.

SOCIAL ABNORMALITIES   In the same cohort of subjects born in 1946, the individuals who subsequently developed schizophrenia exhibited the following characteristics: at ages 4 and 6 they were observed to be more likely to play alone; at age 13 they rated themselves as less socially confident; and at age 15 they were rated by teachers as being more anxious in social situations (Jones et al. 1994). Similarly, in a 1958 British birth cohort children who later developed schizophrenia were more likely to be rated by teachers as exhibiting socially maladaptive behavior as early as age 7 (Done et al. 1994). Boys were more overactive at ages 7 and 11, whereas girls were more withdrawn at age 11. In contrast, children who eventually developed affective psychoses did not differ from controls at either age 7 or 11. In addition, the Danish High Risk Project (Olin & Mednick 1996) found that teachers rated girls who later developed schizophrenia as more withdrawn and nervous and boys as more inappropriate and disruptive, than other children.

In a study of over 100,000 conscripts to the Israeli Army, poor social relations were evident at ages 16–17 in individuals who subsequently developed schizophrenia (Davidson et al. 1999). Similarly, in a study of 50,000 18-year-old Swedish Army conscripts, those who later developed schizophrenia had fewer close friends, preferred to socialize in small groups, were more sensitive, and were less likely to have a girlfriend (Malmberg et al. 1998). The combination of these four factors was associated with a strikingly increased risk of schizophrenia, but 79% of the sample reported experiencing at least one of these four factors.

These abnormalities in social factors may not be simply a prodromal feature of schizophrenia (i.e., a herald of incipient psychosis) because they were present in some individuals more than 10 years prior to illness onset. However, it is unclear whether these social abnormalities reflect the actual disease process or whether they represent psychological risk factors, in the sense that those who are less socially engaged have fewer opportunities for reality testing and correction of paranoid thinking.

IMPAIRMENTS IN IQ AND SCHOOL PERFORMANCE   Both IQ scores, assessed as early as age 8, and educational achievements have been reported to be lower in children who subsequently manifest the clinical features of schizophrenia than in comparison subjects, such as individuals who later develop mood disorders (reviewed in Chua & Murray 1996). In the New York High Risk Project, the offspring of individuals with schizophrenia had lower IQ and reduced general cognitive functioning than children with normal parents or with parents who had an affective disorder (Ott et al. 1998). As a group, preschizophrenic individuals achieve lower educational qualifications than individuals who later develop mood disorders (Isohanni et al. 1998). In addition, the highest premorbid occupational level of subjects with schizophrenia tends to be lower than their fathers; in contrast, the occupational performance of individuals with a mood psychosis does not differ from their fathers (Jones et al. 1993). In Israeli army conscripts IQ was lower at age 16–17 in the individuals who subsequently developed schizophrenia than in control subjects who attended the same high schools (Davidson et al. 1999). Similarly, the risk of schizophrenia in 18-year-old Swedish army conscripts increased linearly with decrement in IQ, although lower IQ was also present in those who developed a nonschizophrenic psychosis (David et al. 1997). Finally, some studies suggest that a decline in childhood IQ may be predictive of psychosis in adulthood (Kremen et al. 1998). Specifically, a decline in IQ between ages 4 and 7 was associated with a 10-fold increase in risk of psychosis at age 23, findings that may be consistent with a progressive disease process.

These findings indicate that intellectual performance is impaired many years before the onset of psychosis, suggesting that a decrement in cognitive function does not simply represent the prodromal phase of schizophrenia. Lower intellectual function may reflect a causal factor for the psychotic manifestations of schizophrenia, predisposing an individual to the development of false beliefs and perceptions. For example, the reduction in cognitive function in Alzheimer's disease is associated with an increased risk of psychosis. However, IQ does not appear to be outside the normal range in preschizophrenic individuals; the differences from comparison subjects are small; and, in isolation, the positive predictive value of low IQ is very modest, predicting only 3% of cases of schizophrenia (David et al. 1997).

IMPLICATIONS FOR NEURODEVELOPMENTAL MODELS OF SCHIZOPHRENIA

The studies reviewed above converge on several conclusions. First, individuals who develop schizophrenia appear to be more likely than comparison subjects to have experienced one or a combination of a variety of potentially adverse events during pre- or perinatal life. Although a substantial literature, including replicated findings, supports this conclusion, these events generally have low predictive value; that is, most individuals who experience such an event do not manifest the clinical features of schizophrenia later in life, and the absence of such an event does not preclude the later development of schizophrenia. Of these early events, the increased risk associated with LDCs may be the most robust finding. Second, disturbances (usually subtle) in a variety of types of behaviors (e.g., motor, cognitive, or social) may be evident in individuals with schizophrenia years or decades before the diagnostic features of the illness appear. Whether these observations represent (*a*) epiphenomena associated with other risk factors for schizophrenia, (*b*) independent risk factors that contribute to the causality of the illness, or (*c*) different aspects of a spectrum of age-specific manifestations of schizophrenia remain critical questions. In any case the nature and number of these observations strongly suggest that disturbances in brain function are present very early in life in at least some individuals who subsequently develop schizophrenia.

As noted earlier, one view of schizophrenia as a neurodevelopmental disorder posits that (*a*) the primary pathogenetic event is a disturbance in brain development during pre- or perinatal life, (*b*) the action of the causative agent is relatively short in duration, such that it and the resulting brain lesion are static, and (*c*) the behavioral consequences remain in large part latent for a substantial time after the action of the primary cause has ceased. The data summarized above on pre- and perinatal abnormalities and insults, and recent experimental data in animal models, are consistent with an early onset of the pathophysiological process, although clearly the process need not be limited to a specific developmental time period.

This particular view of schizophrenia as an early-onset neurodevelopmental disorder, however, is difficult to reconcile with several other observations. For example, Woods (1999) has noted that one of the structural brain features of schizophrenia, an excess of extraventricular cerebral spinal fluid (CSF), reflecting diminished brain volume, may not be explicable by an early, static lesion. Because brain growth drives intracranial cavity growth, the brain grows outward from the ventricles and intracranial cavity size is not reducible after skull sutures fuse, any loss of brain tissue after brain growth reaches its maximum will produce increased CSF equivalently in both ventricles and extracerebral spaces. However, diffuse loss of brain tissue in the pre- or perinatal period should produce a smaller cranial cavity and a persistent increase in ventricular size, but not an increase in extracerebral CSF. Thus, the observations of smaller head size at birth in schizophrenia and larger extracerebral CSF volumes would seem to require both an early lesion and later volume loss. However, it remains unclear whether the increased extracerebral CSF, which may be evident at the onset of clinical illness, represents a fixed loss of brain tissue, the effect of physiological factors (e.g., hydration, nutritional status, medications) that can reversibly affect brain tissue volume (Marenco & Weinberger 2000), or genetically controlled changes in the expression of molecules that regulate cellular components that produce changes in brain volume (see example in the following section).

A second potential problem with the early, static lesion model has been the absence of convincing data from postmortem studies of a brain abnormality in schizophrenia that could only be explained by an early lesion. In this regard reports of cytoarchitectural disturbances in the entorhinal cortex of schizophrenic subjects (Jakob & Beckmann 1986, Arnold et al. 1991) attracted a great deal of attention because the reported findings were strongly suggestive of an abnormality in neuronal migration (Weinberger 1999). However, subsequent studies have both failed to confirm these reports and provided likely methodological explanations for the initial findings (Heinsen et al. 1996, Akil & Lewis 1997, Krimer et al. 1997). The report of an altered distribution of interstitial neurons in cortical white matter also was conceptually very attractive because it strongly suggested an early developmental lesion (Akbarian et al. 1993). However, the majority of subjects with schizophrenia appear to lack such abnormalities (Akbarian et al. 1996, Anderson et al. 1996). Abnormalities in the expression of *reelin*, a gene that encodes an extracellular matrix protein known to play a critical role in neuronal migration in the cerebral cortex, also have been reported in postmortem studies of schizophrenia (Fatemi et al. 2000, Guidotti et al. 2000), but disturbances in brain cytoarchitecture that even marginally mimic those seen in *reelin*-null mice have not been observed. Although reelin probably participates in multiple developmental processes, particularly in light of the recent discovery that the protein exhibits metalloprotease activity (Quattrocchi et al. 2001), the extent to which the findings in schizophrenia represent an early developmental, potentially causal, disturbance versus a later event that reflects a consequence of the illness, has not been determined. Thus, the field still awaits the identification of a robust brain abnormality in schizophrenia that unequivocally implicates a disturbance (static lesion) in early development.

Models of schizophrenia incorporating late developmental changes have focused on maturational events that occur during adolescence. This period of development has been viewed as critical for the appearance of the clinical syndrome of schizophrenia for a number of reasons, such as the considerable "environmental and psychological adventure" and social stress that occurs during these years (Weinberger 1987) and the maturation of certain complex brain regions and functions that are critical to meeting the challenges of adult life. For example, the number of excitatory synapses in the cerebral cortex declines markedly in the peri-adolescent period (Huttenlocher 1979, Rakic et al. 1986), and reciprocal changes in gray and white matter volumes occur (i.e., gray matter volume normally decreases from childhood, beginning at around 5 years of age, whereas white matter volume increases throughout childhood and adolescence (Bartzokis et al. 2001).

The timing of these biological events raises the possibility that alterations in these processes may contribute to the pathogenesis of schizophrenia. The presence in schizophrenia of progressive increases in ventricular volume and decreases in whole or regional brain volumes might argue for a late or at least an ongoing pathophysiological process in schizophrenia. Recent studies have demonstrated accelerated reductions in cortical gray matter in childhood-onset schizophrenia (Thompson et al. 2001). In addition, longitudinal MRI studies of brain volume measures in adults with schizophrenia have revealed evidence of progressive changes, especially in the frontal lobes (see Shenton et al. 2001 for review), although additional studies are needed to confirm these findings. It should also be noted that an absence of structural changes after illness onset does not exclude the possibility of progressive alterations prior to the clinical appearance of the illness. Thus, observable progression, suggestive of abnormalities in late developmental processes, may be limited to the initial phase of illness and/or to those with an early-onset form of the illness.

NEURODEVELOPMENTAL PROCESSES AND THE PATHOGENESIS OF SCHIZOPHRENIA

Clearly, a current challenge is to more precisely define the developmental mechanisms that are critical to the pathogenesis of schizophrenia. This task has not been straightforward, in part because of the problems involved in understanding the causal nature of developmental perturbations that typically fail to produce overt pathophysiological consequences until brain maturation approaches completion. In one sense, because development of the central nervous system is characterized by the general spatial and temporal segregation (with some overlap) of specific histogenic events that are readily traced through specific pre- and postnatal epochs, studies have tended to focus on individual aspects of development that may account for the disease. However, this approach has problems that, in some sense, parallel the attempts to identify a specific chromosomal locus associated with schizophrenia. For example, the attempt to link schizophrenia to a disturbance in the prenatal migration of a specific population of neurons has not been integrated with the abnormal functioning of specific circuits that produce psychosis or cognitive deficits. Although, as noted above, the evidence for altered neuronal migration in schizophrenia remains controversial, such a finding highlights a particular problem that investigators face in reconciling one perspective of the illness with other replicated observations of altered cortical and subcortical structure and organization.

As a discipline, developmental neurobiology has made great strides in achieving a detailed understanding of the sequence of events and the molecular machinery that lead to the assembly of brain systems (see Tessier-Lavigne & Goodman 1996, Jessell 2000 for reviews). It is clear that rather than an independent collection of histogenic events, the assembly of brain circuitry occurs through a continuum of processes, which change over time and which are highly dependent upon previous events to produce ultimately a normal functional state (Figure 1a). Neuronal and glial proliferation, cell migration, morphological and biochemical differentiation, and circuit formation rely on complex intracellular and cell-environmental interactions that control particular developmental processes. Effector signaling systems such as transcription factors, growth factors, and guidance molecules that mediate attractive and repulsive growth have been delineated. Such detailed data afford the opportunity to identify the different ways in which disturbances at one point in development may establish an altered trajectory for subsequent events, eventually resulting in a dysfunctional state.

Developmental neurogenetic studies commonly have revealed single gene mutations that can affect key processes, resulting in major, permanent changes in brain structure. For example, mutations of *LIS1* result in altered cell migration (Feng & Walsh 2001, Leventer et al. 2001), and mutations in *cpp32* produce abnormal increases in early cell survival (Kuida et al. 1996). In both conditions major central nervous system structural defects are evident from the outset, and as such, early and persistent evidence for dysfunction, such as mental retardation, reflects a profound deviation from a normal developmental trajectory (Figure 1b).

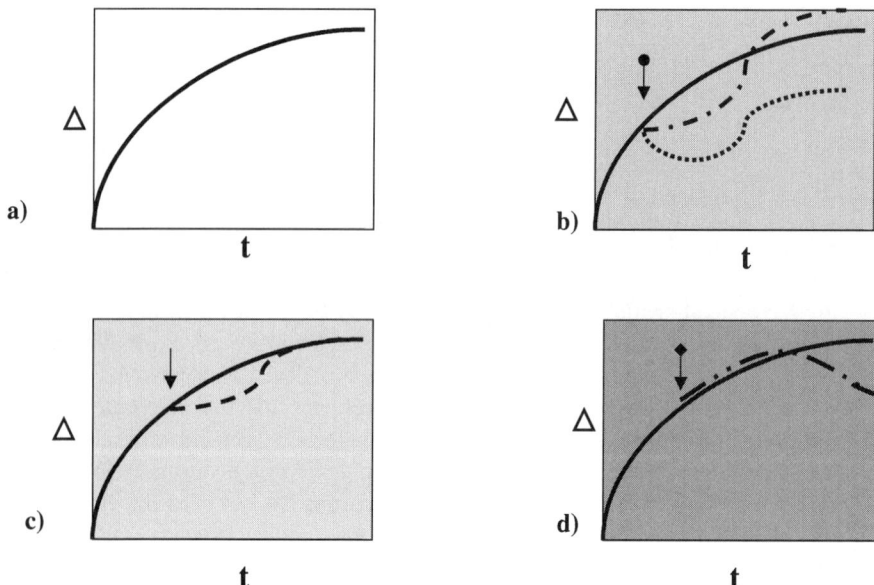

Figure 1 The histogenic events that underlie development proceed along a defined trajectory that changes over time, eventually reaching a mature state of brain function (arrows). (*a*) Responses to adverse events vary, depending upon the timing and type of perturbation and the genetic susceptibility of the individual (*shaded background*). (*b*) In certain conditions development proceeds abnormally, with the individual attaining a dysfunctional state. (*c*) Development may deviate for a brief period and then recover to a normal state owing to adaptive mechanisms. (*d*) Adaptive changes may be initially subtle, with development appearing to continue on a normal trajectory for a period of time, after which an altered functional state is expressed.

Devastating defects, however, are not the sole outcome of altered genetic and epigenetic regulation of neural development. Among the most essential issues that have emerged from basic developmental studies is the idea that specific aspects of brain formation are regulated by quantitatively precise molecular cues, with the same signaling system used to control a variety of developmental processes. Consequently, quantitative modification to a signaling system could produce different effects on distinct developmental processes. A number of recent examples in experimental models demonstrate the susceptibility of developing systems to quantitative (rather than qualitative) perturbations. Graded reduction of *LIS1* expression results in either delayed or severely disrupted neuronal migration (Hirotsune et al. 1998). Cellular responsiveness to epidermal growth factor changes considerably over developmental time, depending upon the levels of receptor expression. Low levels of epidermal growth factor receptor can affect gene expression of early developing neurons (Ferri & Levitt 1995), intermediate and high levels can modulate cell migration (Caric et al. 2001), and high levels of receptor and ligand influence differentiation into glial cells (Lillien 1995, Burrows et al. 1997). The impact of quantitative alterations to developmental cues can lead to permanent changes in developmental trajectory (Figure 1b). For example, what appears to be a simple disruption of a gradient of fibroblast growth factor 8 (fgf8) in the developing dorsal telencephalon dramatically alters cortical map formation (Fukuchi-Shimogori & Grove 2001) and the expression patterns of molecular cues that are used during the assembly of normal circuitry. Alternatively, quantitative changes can lead to developmental defects that induce adaptive processes, resulting in partial or complete recovery (Figure 1c). For example, deletion of the cyclin kinase inhibitors Ink4d and Kip1 produces a dramatic, aberrant proliferation of neurons throughout the brain that is balanced by an increase in apoptosis (Zindy et al. 1999), resulting in no obvious changes in brain cytoarchitecture. In addition, partial reduction of *LIS1* expression causes delayed neuronal migration in the cortex, but this process subsequently recovers to yield normal laminar patterns (Hirotsune et al. 1998).

Perhaps the most difficult developmental defects to reconcile are those that fail to produce immediate changes but result in dysfunction at a later stage of life (Figure 1d). This particular trajectory may represent the essence of understanding how polygenic and environmental factors combine to produce the developmental changes that are eventually expressed as the characteristic pathophysiology of schizophrenia. Both genetic and lesion studies in animals demonstrate that such delayed manifestations of changes earlier in development can occur. For example, the transforming growth factor-α (TGF-α) hypomorphic mutation, *waved-1 (wa-1)* causes a gradual reduction in growth factor levels postnatally (Weickert & Blum 1995) that culminates in adolescent-onset changes in brain structure, including an enlargement of lateral ventricles, and defects in behavioral performance (Burrows et al. 2000). However, no changes in brain structure or behavior are evident prior to the adolescent period. A pleiotrophic molecule such as TGF-α could impact a variety of late-onset developmental events, such as synapse maturation, astroglial differentiation, and brain vasculature remodeling. In a model using a static lesion, structural damage of the ventral hippocampus in rats during the first postnatal week causes significant, early structural changes, but alterations in certain measures of brain chemistry and performance on specific behavioral tasks appear only after puberty (Wood et al. 1997, Lipska & Weinberger 2000). The underlying mechanisms of such delayed changes (Figure 1d) remain a mystery, but these findings suggest that late developmental events, such as synapse formation, myelination, or biochemical maturation, may be the targets of developmental disturbances that occur much earlier in life.

COMBINATORIAL COMPONENTS THAT MAY CONTRIBUTE TO SCHIZOPHRENIA

A unifying hypothesis of the pathogenesis of schizophrenia may include components of developmental disturbances that encompass different adaptive or maladaptive responses (see Figure 1). Genetic susceptibility would, of course, be present prenatally, but the impact of such combinatorial mutations and/or allelic variations (particularly if they lie in regulatory regions of genes), in concert with the occurrence of an environmental trigger, may lead the developmental trajectory towards a critical threshold that when crossed produces the clinical syndrome at a later stage in life. This scenario could result in progressive alterations in brain circuitry without a persistent gliotic response. In addition, the rate at which these progressive alterations occur may be tied to the normal developmental time course of the affected neural systems, with the consequences of the disturbed trajectory most evident at the age at which the system normally changes most rapidly. Such a scenario may explain why differences in motor function between preschizophrenic and unaffected individuals are most prominent prior to age two (Walker 1994), when a number of motor skills are rapidly acquired, whereas abnormalities that reflect dysfunction of higher-order cortical systems (e.g., delusions) frequently first appear during adolescence and early adulthood, when synaptic pruning and myelination in those brain areas are transpiring at maximal rates.

We suggest that a comprehensive view of schizophrenia as a neurodevelopmental disorder may need to include the following components:

1. Genetic predisposition. Affected individuals may need to carry both gene mutations that are responsible for the disease phenotype and some combination of allelic variations in other genes whose protein products normally interact to mediate a particular brain function (Mirnics & Lewis 2001). The presence of either the mutated genes or predisposing alleles alone would not be sufficient to manifest the illness but would contribute to its heritability. This notion has a parallel in animal models, in which the genetic background (allelic variation) of different mouse strains has a powerful impact on baseline phenotype (Crawley et al. 2000) or on an abnormal phenotype caused by the complete elimination of a particular gene by homologous recombination (Crawley 1999). Examples of behavioral and early brain patterning phenotypes that are "lost" when the same mutation is placed on a different background strain also have been reported (Qiu et al. 1996, Crawley 1999, Rakic 2000, Weher et al. 2001, Bolivar et al. 2001).

2. Gene-environmental interactions. This combination of genetic factors subsequently results in the altered expression of genes (those that mediate the inherited susceptibility and perhaps other genes that are downstream of their control) whose protein products play an essential role in brain function. These altered patterns of gene expression may be influenced, or even triggered, by environmental phenomena during a particular period of ontogeny or perinatally. Thus, the resultant altered course of development would be due to a combination of genetic changes, which when combined with particular stressors during development, drive the system towards a critical threshold for an altered phenotype. This notion suggests that the individuality exhibited in the biological response (such as disease expression) to a particular environmental stimulus would be dependent upon the genetic susceptibility inherent in the mutations and allelic variations that are associated with the

disease. For example, this interaction has been reported in coronary artery disease in which the combination of Apo E genotype and area of residence influence the expression of the pathophysiology (Stengard et al. 1999). Although this has not been investigated directly in the context of behavioral and neurochemical defects, identical environmental perturbations can have widely divergent effects on developing systems, depending upon genetic background (Reik et al. 1993, Schumacher et al. 2000).

3. Cascade effects of altered development. These altered patterns of gene expression have a cumulative effect that leads to additional disturbances in subsequent developmental processes. These disturbances depend both upon the timing of normal maturational events and circuitry-specific susceptibility. Unfortunately, animal models have not yet examined in detail the manner in which subtle changes in developmental course might culminate in more dramatic alterations later in life. However, the *wa-1* example cited above may be reflective of such a mechanism, in that gradual changes in growth factor expression lead to subtle changes in glial cell numbers (Weickert & Blum 1995), which ultimately may impact on synaptic and vascular maturation during adolescence.

4. Steady-state outcome. Once developmental processes approach completion, the impact of these alterations in gene expression and their sequelae becomes relatively stabilized, being only modestly affected by adult levels of plasticity but more subject than the normal brain to further adverse functional consequences from aging-related changes. Although we are not aware of specific examples from the experimental literature that have directly examined this issue, we predict that an animal model with a combination of genetic alterations and early environmental events that affect adult behavior would also exhibit distinctive neurobiological and behavioral changes during senescence.

TOWARDS AN INTEGRATED MODEL

Our laboratories recently performed gene-expression studies of the cerebral cortex in postmortem brain specimens from adults with schizophrenia. The findings of these investigations provide a novel framework upon which this four-component view of the developmental basis of the illness may be applied (Mirnics et al. 2001a). We found a consistent decrease across subjects with schizophrenia in expression of the gene encoding the regulator of G-protein signaling 4 (*RGS4*), one of a family of proteins that controls the duration and timing of synaptic responsiveness for Gi/Go and Gq-coupled neurotransmitter systems (Mirnics et al. 2001b). In addition, *RGS4* mRNA levels appear to decrease across highly divergent cortical regions in subjects with schizophrenia in a diagnosis-specific manner; the *RGS4* protein modulates the function of neurotransmitter systems implicated in the pathophysiology and pharmacological treatment of schizophrenia; and the gene for *RGS4* is found at a chromosomal locus (1q21–22), which is significantly associated with familial schizophrenia (Brzustowicz et al. 2000). Thus, *RGS4* represents a putative susceptibility gene for schizophrenia, a view supported by preliminary linkage and association studies (Chowdari et al. 2002). Consequently, allelic variations in *RGS4* may be one of several inherited susceptibility factors for schizophrenia. The resulting functional outcome would be alterations in the timing of information transfer (Doupnik et al. 1997) at a subset of synapses, given that only subpopulations of neurons express *RGS4*.

One prediction arising from these findings is that such an alteration in synaptic function would subsequently unleash a cascade of changes in the expression of genes whose protein products contribute to synaptic function. Such a response would be especially likely if the altered synaptic function was present during developmental periods in which neural connections are particularly sensitive to the experience-driven activity required for both functional and structural synaptic maturation. Consistent with this prediction, we found that as a group, the genes encoding for proteins involved in the mechanics of presynaptic function showed decreased expression in the prefrontal cortex of subjects with schizophrenia (Mirnics et al. 2000). Changes in expression of the components of glutamate and GABA neurotransmission also have been well-documented (see Lewis et al. 1999, Meador-Woodruff & Healy 2000 for reviews), providing additional evidence of altered synaptic function in specific circuitry. In addition, this evidence of impaired synaptic function was associated with diminished expression of genes that regulate specific aspects of metabolic activity (Middleton et al. 2002). These findings are convergent with clinical and imaging studies of altered synaptic function and metabolism in the prefrontal cortex in subjects with schizophrenia. In a neurodevelopmental context this constellation of altered gene expression would be expected to be present from early in life and might contribute to the disturbances in brain function that account for the types of motor, cognitive, and behavioral findings observed during childhood of individuals who later develop schizophrenia.

The available evidence supports the idea that certain circuits (e.g., the connections between the mediodorsal thalamus and the dorsal prefrontal cortex) are preferentially affected in schizophrenia (Lewis 2000). The impaired synaptic function in these and other (e.g., corticocortical) circuits would render these connections more vulnerable to additional alterations as a result of later developmental events. For example, the adolescence-related pruning of cortical excitatory synapses (Huttenlocher 1979, Rakic et al. 1986) would not only reduce the total complement of cortical synapses, but may do so to a degree that is more severe than in the normal state because of the diminished function of the affected circuits (Mirnics et al. 2001a). In this regard the evidence of cognitive abnormalities in childhood may be considered to reflect both the disease process and a risk factor for later developmental disturbances required for the clinical manifestations of the illness. Once pruning is complete, normally during the second decade of life (although the temporal trajectory could itself be altered in schizophrenia), the manifestations of the illness would be expected to be relatively stable, but the effects of normal aging might be more severe, perhaps accounting for the non-Alzheimer's dementia seen in elderly individuals with schizophrenia (Arnold et al. 1998).

CONCLUSIONS

In recent years, we have witnessed an improving ability to extrapolate backwards from adult and postmortem studies to developmental time points when the combinatorial effects of genetic mutations and environmental insults may establish the trajectory that eventually leads to the clinical syndrome of schizophrenia. As a point of convergence, studies of neurodevelopmental mechanisms have departed from attempting to identify the key, singular factor driving highly complex processes, and recent investigations of schizophrenia have clearly moved in the same direction. Instead, the polygenic nature of the disease, in combination with perhaps multiple, subtle environmental determinants, have parallels in developmental concepts. That is, current data indicate that multiple molecular cues control individual histogenic processes, which when altered result in adaptive or maladaptive changes that lead to different developmental trajectories and unique end states. Technical advances will continue to facilitate more integrated studies of the combinatorial nature of the biological processes that lead to schizophrenia.

ACKNOWLEDGMENTS

The interactions between our laboratories that are reflected in this chapter were supported by NIMH Conte Center Grant MH45156. We are indebted to our colleagues, Drs. Karoly Mirnics, Frank Middleton, Joseph Pierri, and Gregg Stanwood, for critical discussions and for leading key research efforts in our studies of schizophrenia.

The *Annual Review of Neuroscience* is online at http://neuro.annualreviews.org

LITERATURE CITED

Akbarian S, Bunney WE Jr, Potkin SG, Wigal SB, Hagman JO, et al. 1993. Altered distribution of nicotinamide-adenine dinucleotide phosphate-diaphorase cells in frontal lobe of schizophrenics implies disturbances of cortical development. *Arch. Gen. Psychiatry* 50:169–77

Akbarian S, Kim JJ, Potkin SG, Hetrick WP, Bunney WE Jr, Jones EG. 1996. Maldistribution of interstitial neurons in prefrontal white matter of the brains of schizophrenic patients. *Arch. Gen. Psychiatry* 53:425–36

Akil M, Lewis DA. 1997. The cytoarchitecture of the entorhinal cortex in schizophrenia. *Am. J. Psychiatry* 154:1010–12

Anderson S, Volk DW, Lewis DA. 1996. Increased density of microtubule-associated protein 2-immunoreactive neurons in the prefrontal white matter of schizophrenic subjects. *Schizophr. Res.* 19:111–19

Arnold SE, Hyman BT, Van Hoesen GW, Damasio AR. 1991. Some cytoarchitectural abnormalities of the entorhinal cortex in schizophrenia. *Arch. Gen. Psychiatry* 48:625–32

Arnold SE, Trojanowski JQ, Gur RE, Blackwell P, Han L-Y, Choi C. 1998. Absence of neurodegeneration and neural injury in the cerebral cortex in a sample of elderly patients with schizophrenia. *Arch. Gen. Psychiatry* 55:225–32

Bartzokis G, Beckson M, Lu PH, Neuchterlein KH, Edwards N, et al. 2001. Age-related changes in frontal and temporal lobe volumes in men: a magnetic resonance imaging study. *Arch. Gen. Psychiatry* 58:461–65

Bertolino A, Callicott JH, Elman I, Mattay VS,

Tedeschi G, et al. 1998. Regionally specific neuronal pathology in untreated patients with schizophrenia: a proton magnetic resonance spectroscopic imaging study. *Biol. Psychiatry* 43:641–48

Bolivar VJ, Cook MN, Flaherty L. 2001. Mapping of quantitative trait loci with knockout/congenic strains. *Genome Res.* 11:1549–52

Brown AS, Cohen P, Greenwald S, Susser E. 2000. Nonaffective psychosis after prenatal exposure to rubella. *Am. J. Psychiatry* 157:438–43

Brzustowicz LM, Hodgkinson KA, Chow EWC, Honer WG, Bassett AS. 2000. Location of a major susceptibility locus for familial schizophrenia on chromosome 1q21–q22. *Science* 288:678–82

Buka SL, Tsuang MT, Lipsitt LP. 1993. Pregnancy/delivery complications and psychiatric diagnosis. A prospective study. *Arch. Gen. Psychiatry* 50:151–56

Buka SL, Tsuang MT, Torrey EF, Klebanoff MA, Bernstein D, Yolken RH. 2001. Maternal infections and subsequent psychosis among offspring. *Arch. Gen. Psychiatry* 58:1032–37

Burrows RC, Levitt P, Shors TJ. 2000. Postnatal decrease in TGFα is associated with enlarged ventricles, deficient amygdaloid vasculature and performance deficits. *Neuroscience* 96:825–36

Burrows RC, Wancio D, Levitt P, Lillien L. 1997. Response diversity and the timing of progenitor cell maturation are regulated by developmental changes in EGFR expression in the cortex. *Neuron* 19:251–67

Cannon M, Cotter D, Coffey VP, Sham PC, Takei N, et al. 1996. Prenatal exposure to the 1957 influenza epidemic and adult schizophrenia: a follow-up study. *Br. J. Psychiatry* 168:368–71

Cannon TD. 1997. On the nature and mechanisms of obstetric influences in schizophrenia: a review and synthesis of epidemiologic studies. *Int. Rev. Psychiatry* 9:387–97

Caric D, Raphael H, Viti J, Feathers A, Wancio D, Lillien L. 2001. EGFRs mediate chemotactic migration in the developing telencephalon. *Development* 128:4203–16

Chowdari KW, Mirnics K, Semwal P, Wood J, Lawrence E, et al. 2002. Association and linkage analyses of *RGS4* polymorphisms in schizophrenia. *Hum. Mol. Genet.* In press

Chua SE, Murray RM. 1996. The neurodevelopment theory of schizophrenia: evidence concerning structure and neuropsychology. *Ann. Med.* 28:547–55

Crawley JN. 1999. Behavioral phenotyping of transgenic and knockout mice: experimental design and evaluation of general health, sensory functions, motor abilities, and specific behavioral tests. *Brain Res.* 835:18–26

Crawley JN, Belknap JK, Collins A, Crabbe JC, Frankel W, et al. 2000. Behavioral phenotypes of inbred mouse strains: implications and recommendations for molecular studies. *Psychopharmacology (Berlin)* 132:107–24

Crow TJ. 1994. Prenatal exposure to influenza as a cause of schizophrenia: There are inconsistencies and contradictions in the evidence. *Br. J. Psychiatry* 164:586–92

David AS, Malmberg A, Brandt L, Allebeck P, Lewis G. 1997. IQ and risk for schizophrenia: a population-based cohort study. *Psychol. Med.* 27:1311–23

Davidson M, Reichenberg A, Rabinowitz J, Weiser M, Kaplan Z, Mark M. 1999. Behavioral and intellectual markers for schizophrenia in apparently healthy male adolescents. *Am. J. Psychiatry* 156:1328–35

Done DJ, Crow TJ, Johnstone EC, Sacker A. 1994. Childhood antecedents of schizophrenia and affective illness: social adjustment at ages 7 and 11. *Br. Med. J.* 309:699–703

Done DJ, Johnstone EC, Frith CD, Golding J, Shepherd PM, Crow TJ. 1991. Complications of pregnancy and delivery in relation to psychosis in adult life: data from the British perinatal mortality survey sample. *Br. Med. J.* 302:1576–80

Doupnik CA, Davidson N, Lester HA, Kofuji P. 1997. RGS proteins reconstitute the rapid gating kinetics of G$\beta\gamma$-activated inwardly rectifying K+ channels. *Proc. Natl. Acad. Sci. USA* 94:10461–66

Fatemi SH, Earle JA, McMenomy T. 2000. Reduction in reelin immunoreactivity in hippocampus of subjects with schizophrenia, bipolar disorder and major depression. *Mol. Psychiatry* 5:654–63

Feinberg I. 1982. Schizophrenia: caused by a fault in programmed synaptic elimination during adolescence? *J. Psychiatry Res.* 17:319–34

Feng Y, Walsh CA. 2001. Protein-protein interactions, cytoskeletal regulation and neuronal migration. *Nat. Rev. Neurosci.* 2:408–16

Ferri RT, Levitt P. 1995. Regulation of regional differences in the differentiation of cerebral cortical neurons by EGF family-matrix interactions. *Development* 121:1151–60

Fish B, Marcus J, Hans SL, Auerbach JG, Perdue S. 1992. Infants at risk for schizophrenia: sequelae of a genetic neurointegrative defect. *Arch. Gen. Psychiatry* 49:221–35

Fukuchi-Shimogori T, Grove EA. 2001. Neocortex patterning by the secreted signaling molecule FGF8. *Science* 294:1071–74

Geddes JR, Lawrie SM. 1995. Obstetric complications and schizophrenia: a meta-analysis. *Br. J. Psychiatry* 167:786–93

Gottesman II. 1991. *Schizophrenia Genesis: The Origins of Madness*. New York: Freeman

Gottesman II, Bertelsen A. 1989. Confirming unexpressed genotypes for schizophrenia. Risks in the offspring of Fischers's Danish identical and fraternal discordant twins. *Arch. Gen. Psychiatry* 46:867–72

Gottesman II, Erlenmeyer-Kimling L. 2001. Family and twin strategies as a head start in defining prodromes and endophenotypes for hypothetical early-interventions in schizophrenia. *Schizophr. Res.* 51:93–102

Grech A, Takei N, Murray RM. 1997. Maternal exposure to influenza and paranoid schizophrenia. *Schizophr. Res.* 26:121–25

Green MF, Satz P, Christenson C. 1994. Minor physical anomalies in schizophrenia patients, bipolar patients, and their siblings. *Schizophr. Bull.* 20:433–40

Guidotti A, Auta J, Davis JM, Gerevini VD, Dwivedi Y, et al. 2000. Decrease in reelin and glutamic acid decarboxylase$_{67}$ (GAD_{67}) expression in schizophrenia and bipolar disorder. *Arch. Gen. Psychiatry* 57:1061–69

Heinsen H, Gössmann E, Rüb U, Eisnemenger W, Bauer M, et al. 1996. Variability in the human entorhinal region may confound neuropsychiatric diagnoses. *Acta Anat.* 157:226–37

Hirotsune S, Fleck MW, Gambello MJ, Bix GJ, Chen A, et al. 1998. Graded reduction of Pafah1b1 (Lis1) activity results in neuronal migration defects and early embryonic lethality. *Nat. Genet.* 19:333–39

Huttenlocher PR. 1979. Synaptic density in human frontal cortex: developmental changes and effects of aging. *Brain Res.* 163:195–205

Ingraham LJ, Kety SS. 2000. Adoption studies of schizophrenia. *Am. J. Med. Genet.* 97:18–22

Isohanni I, Järvelin M-R, Nieminen P, Jones P, Rantakallio P, et al. 1998. School performance as a predictor of psychiatric hospitalization in adult life. A 28-year follow-up in the Northern Finland 1966 birth cohort. *Psychol. Med.* 28:967–74

Jablensky A. 2000. Epidemiology of schizophrenia: the global burden of disease and disability. *Eur. Arch. Psychiatry Clin. Neurosci.* 250:274–85

Jakob H, Beckmann H. 1986. Prenatal developmental disturbances in the limbic allocortex in schizophrenics. *J. Neural Transm.* 65:303–26

Jessell TM. 2000. Neuronal specifications in the spinal cord: inductive signals and transcriptional codes. *Nat. Rev. Genet.* 1:20–29

Jones P, Rodgers B, Murray R, Marmot M. 1994. Child development risk factors for adult schizophrenia in the British 1946 birth cohort. *Lancet* 344:1398–402

Jones PB, Bebbington P, Foerster A, Lewis SW, Murray RM, et al. 1993. Premorbid social underachievement in schizophrenia. Results from the Camberwell Collaborative Psychosis Study. *Br. J. Psychiatry* 162:65–71

Jones PB, Rantakallio P, Hartikainen A-L, Isohanni M, Sipila P. 1998. Schizophrenia as a long-term outcome of pregnancy, delivery, and perinatal complications: a 28-year follow-up of the 1966 North Finland general population birth cohort. *Am. J. Psychiatry* 155:355–64

Kety SS, Rosenthal D, Wender PH, Schulsinger F. 1971. Mental illness in the biological and adoptive families of adopted schizophrenics. *Am. J. Psychiatry* 128:82–86

Kremen WS, Buka SL, Seidman LJ, Goldstein JM, Koren D, Tsuang MT. 1998. IQ decline during childhood and adult psychotic symptoms in a community sample: a 19-year longitudinal study. *Am. J. Psychiatry* 155:672–77

Krimer LS, Herman MM, Saunders RC, Boyd JC, Hyde TM, et al. 1997. A qualitative and quantitative analysis of the entorhinal cortex in schizophrenia. *Cereb. Cortex* 7:732–39

Kuida K, Zheng TS, Na S, Kuan C, Yang D, et al. 1996. Decreased apoptosis in the brain and premature lethality in CPP32-deficient mice. *Nature* 384:368–72

Kunugi H, Nanko S, Murray RM. 2001. Obstetric complications and schizophrenia: prenatal underdevelopment and subsequent neurodevelopmental impairment. *Br. J. Psychiatry* 178(S40):S25–29

Lane A, Larkin C, Waddington JL, O'Callaghan E. 1996. Dysmorphic features and schizophrenia. In *The Neurodevelopmental Basis of Schizophrenia*, ed. JL Waddington, PF Buckley, pp. 79–94. Georgetown, TX: Landes

Leventer RJ, Cardoso C, Ledbetter DH, Dobyns WB. 2001. LIS1: from cortical malformation to essential protein of cellular dynamics. *Trends Neurosci.* 24:439–92

Lewis DA. 2000. Is there a neuropathology of schizophrenia? *Neuroscientist* 6:208–18

Lewis DA, Lieberman JA. 2000. Catching up on schizophrenia: natural history and neurobiology. *Neuron* 28:325–34

Lewis DA, Pierri JN, Volk DW, Melchitzky DS, Woo T-U. 1999. Altered GABA neurotransmission and prefrontal cortical dysfunction in schizophrenia. *Biol. Psychiatry* 46:616–26

Lillien L. 1995. Changes in retinal cell fate induced by overexpression of EGF receptor. *Nature* 377:158–62

Lipska BK, Weinberger DR. 2000. To model a psychiatric disorder in animals: schizophrenia as a reality test. *Neuropsychopharmacology* 23:223–39

Malmberg A, Lewis G, David A, Allebeck P. 1998. Premorbid adjustment and personality in people with schizophrenia. *Br. J. Psychiatry* 172:308–13

Marcelis M, Navarro-Mateau F, Murray R, Selten J-P, van Os J. 1998. Urbanization and psychosis: a study of 1942–1978 birth cohorts in The Netherlands. *Psychol. Med.* 28:871–79

Marcus J, Hans SL, Auerbach JG, Auerbach AG. 1993. Children at risk for schizophrenia: the Jerusalem infant development study. II. Neurobehavioral deficits at school age. *Arch. Gen. Psychiatry* 50:797–809

Marenco S, Weinberger DR. 2000. The neurodevelopmental hypothesis of schizophrenia: following a trail of evidence from cradle to grave. *Dev. Psychopathol.* 12:501–27

McGlashan TH, Hoffman RE. 2000. Schizophrenia as a disorder of developmentally reduced synaptic connectivity. *Arch. Gen. Psychiatry* 57:637–48

McGrath J, Castle DJ. 1995. Does influenza cause schizophrenia? A five year review. *Aust. NZ J. Psychiatry* 29:23–31

McNeil TF, Cantor-Graae E, Ismail B. 2000a. Obstetric complications and congenital malformation in schizophrenia. *Brain Res. Rev.* 31:166–78

McNeil TF, Cantor-Graae E, Weinberger DR. 2000b. Relationship of obstetric complications and differences in size of brain structures in monozygotic twin pairs discordant for schizophrenia. *Am. J. Psychiatry* 157:203–12

Meador-Woodruff JH, Healy DJ. 2000. Glutamate receptor expression in schizophrenic brain. *Brain Res. Rev.* 31:288–94

Mednick SA, Machon RA, Huttunen MO, Bonett D. 1988. Adult schizophrenia following prenatal exposure to an influenza epidemic. *Arch. Gen. Psychiatry* 45:189–92

Middleton FA, Mirnics K, Pierri JN, Lewis DA,

Levitt P. 2002. Gene expression profiling reveals alterations of specific metabolic pathways in schizophrenia. *J. Neurosci.* 22:2718–29

Milligan CE, Levitt P, Cunningham TJ. 1991. Brain macrophages and microglia respond differently to lesions of the developing and adult visual system. *J. Comp. Neurol.* 314:136–46

Mirnics K, Lewis DA. 2001. Genes and subtypes of schizophrenia. *Trends Mol. Med.* 7:281–83

Mirnics K, Middleton FA, Lewis DA, Levitt P. 2001a. Analysis of complex brain disorders with gene expression microarrays: schizophrenia as a disease of the synapse. *Trends Neurosci.* 24:479–86

Mirnics K, Middleton FA, Marquez A, Lewis DA, Levitt P. 2000. Molecular characterization of schizophrenia viewed by microarray analysis of gene expression in prefrontal cortex. *Neuron* 28:53–67

Mirnics K, Middleton FA, Stanwood GD, Lewis DA, Levitt P. 2001b. Disease-specific changes in regulator of G-protein signaling 4 (RGS4) expression in schizophrenia. *Mol. Psychiatry* 6:293–301

Morgan V, Castle D, Page A, Fazia S, Gurrin L, et al. 1997. Influenza epidemics and incidence of schizophrenia, affective disorders and mental retardation in Western Australia: no evidence of a major effect. *Schizophr. Res.* 26:25–39

Mortensen PB, Pedersen CB, Westegaard T, Wohlfahrt J, Ewald H, et al. 1999. Effects of family history and place and season of birth on the risk of schizophrenia. *N. Engl. J. Med.* 340:603–8

Murray RM, O'Callaghan E, Castle DJ, Lewis SW. 1992. A neurodevelopmental approach to the classification of schizophrenia. *Schizophr. Bull.* 18:319–32

Olin SS, Mednick SA. 1996. Risk factors of psychosis: identifying vulnerable populations premorbidly. *Schizophr. Bull.* 22:223–40

Ott SL, Spinelli S, Rock D, Roberts S, Amminger GP, Erlenmeyer-Kimling L. 1998. The New York high-risk project: social and general intelligence in children at risk for schizophrenia. *Schizophr. Res.* 31:1–11

Pedersen CB, Mortensen PB. 2001. Evidence of a dose-response relationship between urbanicity during upbringing and schizophrenia risk. *Arch. Gen. Psychiatry* 58:1039–46

Pilowsky LS, Kerwin RW, Murray RM. 1993. Schizophrenia: a neurodevelopmental perspective. *Neuropsychopharmacology* 9:83–91

Pulver AE. 2000. Search for schizophrenia susceptibility genes. *Biol. Psychiatry* 47:221–30

Qiu M, Anderson S, Chen S, Meneses JJ, Hevner R, et al. 1996. Mutation of Emx-1 homeobox gene disrupts the corpus callosum. *Dev. Biol.* 178:174–78

Quattrocchi CC, Wannasnes F, Persico AM, Ciafre SA, D'Arcangelo G, et al. 2001. Reelin is a serine protease of the extracellular matrix. *J. Biol. Chem.* 277:303–9

Rakic P. 2000. From spontaneous to induced neurological mutations: a personal witness of the ascent of the mouse model. *Results Probl. Cell Differ.* 30:1–19

Rakic P, Bourgeois J-P, Eckenhoff MF, Zecevic N, Goldman-Rakic PS. 1986. Concurrent overproduction of synapses in diverse regions of the primate cerebral cortex. *Science* 232:232–35

Reik W, Romer I, Barton SC, Surani MA, Howlett SK, et al. 1993. Adult phenotype in the mouse can be affected by epigenetic events in the early embryo. *Development* 119:933–42

Risch N, Baron M. 1984. Segregation analysis of schizophrenia and related disorders. *Am. J. Hum. Genet.* 36:1039–59

Risch NJ. 2000. Searching for genetics determinants in the new millennium. *Nature* 405:847–56

Roberts GW, Harrison PJ. 2000. Gliosis and its implications for the disease process. In *The Neuropathology of Schizophrenia: Progress and Interpretation*, ed. PJ Harrison, GW Roberts, pp.137–50. New York: Oxford Univ. Press

Schiffman J, Ekstrom M, LaBrie J, Schulsinger F, Sorensen H, Mednick SA. 2002. Minor physical anomalies and schizophrenia-spectrum disorders: a prospective investigation. *Am. J. Psychiatry.* 159:238–43

Schumacher A, Koetsier PA, Hertz J, Doerfler W. 2000. Epigenetic and genotype-specific effects on the stability of de novo imposed methylation patterns in transgenic mice. *J. Biol. Chem.* 275:37915–21

Selten JP, Slaets J, Kahn R. 1998. Prenatal exposure to influenza and schizophrenia in Surinamese and Dutch Antillean immigrants to The Netherlands. *Schizophr. Res.* 30:101–3

Shenton ME, Dickey CC, Frumin M, McCarley RW. 2001. A review of MRI findings in schizophrenia. *Schizophr. Res.* 49:1–52

Stengard JH, Kardi SL, Tervahauta M, Ehnholm C, Nissinen A, et al. 1999. Utility of the predictors of coronary heart disease mortality in a longitudinal study of elderly Finnish men aged 65 to 84 years is dependent on the context defined by Apo E genotype and area of residence. *Clin. Genet.* 56:367–77

Susser E, Neugebauer R, Hoek HW, Brown AS, Lin S, et al. 1996. Schizophrenia after prenatal famine: further evidence. *Arch. Gen. Psychiatry* 53:25–31

Takei N, Murray RM. 1998. The current status of the neurodevelopmental hypothesis of schizophrenia. *Int. Med. J.* 5:13–20

Tessier-Lavigne M, Goodman CS. 1996. The molecular biology of axon guidance. *Science* 274:1123–33

Thompson PM, Vidal C, Giedd JN, Gochman P, Blumenthal J, et al. 2001. Mapping of adolescent brain change reveals dynamic wave of accelerated gray matter loss in very early-onset schizophrenia. *Proc. Natl. Acad. Sci. USA* 98:11650–55

Torrey EF, Miller J, Rawlings R, Yolken RH. 1997. Seasonality of births in schizophrenia and bipolar disorder: a review of the literature. *Schizophr. Res.* 28:1–38

Tsuang MT, Stone WS, Faraone SV. 2001. Genes, environment and schizophrenia. *Br. J. Psychiatry* 40S:18–24

Verdoux H, Geddes JR, Takei N, Lawrie SM, Bovet P, et al. 1997. Obstetric complications and age at onset in schizophrenia: an international collaborative meta-analysis of individual patient data. *Am. J. Psychiatry* 154:1220–27

Walker EF. 1994. Developmentally moderated expressions of the neuropathology underlying schizophrenia. *Schizophr. Bull.* 20:453–80

Walker EF, Lewine RRJ, Neumann C. 1996. Childhood behavioral characteristics and adult brain morphology in schizophrenia. *Schizophr. Res.* 22:93–101

Walker EF, Savoie T, Davis D. 1994. Neuromotor precursors of schizophrenia. *Schizophr. Bull.* 20:441–51

Weher JM, Radcliffe RA, Bowers BJ. 2001. Quantitative genetics and mouse behavior. *Annu. Rev. Neurosci.* 24:845–67

Weickert CS, Blum M. 1995. Striatal TGF-alpha: postnatal developmental expression and evidence for a role in the proliferation of subependymal cells. *Dev. Brain Res.* 86:203–16

Weinberger DR. 1987. Implications of normal brain development for the pathogenesis of schizophrenia. *Arch. Gen. Psychiatry* 44:660–69

Weinberger DR. 1999. Cell biology of the hippocampal formation in schizophrenia. *Biol. Psychiatry* 45:395–402

Wood GK, Lipska BK, Weinberger DR. 1997. Behavioral changes in rats with early ventral hippocampal damage vary with age at damage. *Dev. Brain Res.* 101:17–25

Woods BT. 1999. Is schizophrenia a progressive neurodevelopmental disorder? Toward a unitary pathogenetic mechanism. *Am. J. Psychiatry* 155:1661–70

Zindy F, Cunningham JJ, Sherr CJ, Jogal S, Smeyne RJ, et al. 1999. Postnatal neuronal proliferation in mice lacking Ink4d and Kip1 inhibitors of cyclin-dependent kinases. *Proc. Natl. Acad. Sci. USA* 96:13462–67

THE CENTRAL AUTONOMIC NERVOUS SYSTEM: Conscious Visceral Perception and Autonomic Pattern Generation

Clifford B. Saper

Department of Neurology and Program in Neuroscience, Harvard Medical School, Beth Israel Deaconess Medical Center, 330 Brookline Avenue, Boston, Massachusetts 02215; email: csaper@caregroup.harvard.edu

Key Words sympathetic, parasympathetic, hypothalamus, thalamus, cortex

■ **Abstract** The overall organization of the peripheral autonomic nervous system has been known for many decades, but the mechanisms by which it is controlled by the central nervous system are just now coming to light. In particular, two major issues have seen considerable progress in the past decade. First, the pathways that provide visceral sensation to conscious perception at a cortical level have been elucidated in both animals and humans. The nociceptive system runs in parallel to the pathways carrying visceral sensation from the cranial nerves and may be considered in itself a component of visceral sensation. Second, structures in the central nervous system that generate patterns of autonomic response have been identified. These pattern generators are located at multiple levels of the central nervous system, and they can be combined in temporal and spatial patterns to subserve a wide range of behavioral needs.

INTRODUCTION

Although the basic plan of the peripheral components of the autonomic nervous system has been known since the late nineteenth century, and its neurotransmitters were characterized more than 50 years ago, the organization of central control of the autonomic nervous system has only become clear in the past two decades, and several aspects remain controversial. Rather than attempt to review the entire subject, much of which has not progressed substantially since other comprehensive reviews were published (Saper 1995, Loewy & Spyer 1990), this review focuses on two key elements of the central autonomic system that remain controversial: how visceral sensory information reaches conscious appreciation and how patterns of autonomic response are generated by the brain. We attempt to put these areas into historical perspective and to provide a theoretical framework for considering current and future advances.

HOW DOES VISCERAL SENSORY INFORMATION REACH CONSCIOUS APPRECIATION?

Historical Perspective

The mechanisms for emotional expression were explored in the nineteenth century by Charles Darwin in his classic work, *The Expression of the Emotions in Man and Animals* (Darwin 1873). Although perhaps not as influential as his better-known work on natural selection, Darwin used the same approach of observation of animals in natural situations to demonstrate that different patterns of facial expression were common in animals of different species, including humans, who were facing similar behavioral situations, including anger, fear, happiness, and jealousy (Figure 1). These observations had profound effects on subsequent thought concerning the mechanisms of emotion and their obvious autonomic concomitants. First, Darwin legitimized the use of animal models for human emotional and autonomic expression. Subsequently, the use of animal experimentation became an accepted approach to understanding the physiological basis for such responses. A second consequence of Darwin's arguments was his recognition that autonomic responses were an intrinsic part of the emotional process. Although Darwin took pains to emphasize that he did not have the tools to address emotional autonomic responses, he explicitly recognized Claude Bernard's work on the subject, and thus prepared the field for later workers.

As tools for measuring autonomic responses improved, physiologists and psychologists in the 1880s began to approach the problems of emotional expression by evaluating autonomic alterations that accompany them. The Danish physiologist Lange (1885) proposed that autonomic responses occurred as reflex reactions during behaviors, and that the basis for emotional sensation was the perceptions that arose from such cardiovascular adjustments (e.g., flushing during embarrassment, pelvic blood flow during sexual arousal, elevated blood pressure and heart rate during anger, etc.).

There was considerable debate at the time whether cardiovascular sensation could account for the entire range of sensory experience associated with emotion. William James, professor of psychology at Harvard University, proposed a variation on Lange's model, usually called the James-Lange theory (James 1884). In this view, emotional experience was the sum product of the entire range of autonomic responses that accompany behavior (not just the cardiovascular response). Hence, the James-Lange theory allowed for additional sensory experiences reaching conscious perception, such as queasiness from abdominal sensations, or breathlessness during a panic attack.

This theory was criticized by Cannon (1929), who recognized his debt to Darwin in identifying patterns of motor response during emotions but sought to carry that approach forward by observing patterns of autonomic response in different emotional states (see "How are Different Patterns of Autonomic Output Generated?" below). Cannon argued that visceral sensation itself could not account

Fig. 14. Head of snarling Dog. From life, by Mr. Wood.

Fig 17 The same, when pleased by being caressed.

Figure 1 Comparisons of animal and human faces exhibiting snarling (*upper row*) or pleased (*lower row*) facial expressions. The similarity of facial musculature engaged in these expressions across different species provided evidence for Darwin's hypothesis that neural control of emotional response involved common pattern generators in different mammals. These observations also provided support for the use of animal models to study emotional expression. Reproduced from Darwin (1873) with permission.

for the full range of emotional experience. He pointed to experiments of nature or medical conditions in which the spinal cord or the vagus nerves had been transected in humans, who retained most if not all of their emotional sensation. Hence, Cannon argued for a central system for emotional experience and behavior that was separable from the brain system for appreciation of visceral sensation.

The first hints of a forebrain sensory representation devoted to visceral sensation came from experiments in the mid-twentieth century in which the newly developed tool of evoked potentials was used. Stimulation of the proximal end of the cut cervical vagus nerve allowed neurophysiologists to map the fields in the forebrain in monkeys, cats, and rats that received vagal afferent input (Dell & Olson 1951, Bailey & Bremer 1938, Ogawa et al. 1990, Yamamoto et al. 1980). These experiments demonstrated a region of vagal receptive cortex, corresponding in each species to the insular cortex.

The importance of the vagal sensory cortex to conscious appreciation of visceral sensation became clear in studies performed by the neurosurgeons Wilder Penfield and Theodore Rasmussen in the 1950s (Penfield & Faulk 1955, Penfield & Rasmussen 1950), in which they used electrical stimulation to map the cerebral cortex in patients undergoing neurosurgical procedures. The surgery was done under local anesthesia so patients could report their responses to the surgeon when he stimulated different cortical areas. This approach allowed the surgeons to produce functional maps of the cerebral cortex in individual patients, and these were used as a guide during subsequent surgery to resect tumors, arteriovenous malformations, or sites of epileptic electrical activity.

As Penfield moved his stimulating electrode ventrally along the primary sensory cortex, he identified a region extending just beyond the tongue somatosensory area into the opercular cortex overlying the insula, where electrical stimulation produced taste sensation. When he moved his electrode further into the insula, his subjects reported that stimulation produced oropharyngeal, esophageal, or even gastrointestinal sensation. In his classic maps, Penfield positioned the gastrointestinal tract as part of his sensory homunculus, running ventrally from the tongue sensory area into the operculum and insula.

It is important to note that although Penfield's subjects volunteered a variety of descriptions about their visceral sensory experiences, none of them felt complete emotional responses from stimulation in the insular region. This experience was in contradistinction to experiments in which depth electrodes were used to stimulate the medial temporal lobe, also in patients with underlying epilepsy (Gloor et al. 1982). These patients instead reported emotional reactions, such as feelings of fear, or complex experiential phenomena with attendant emotions (such as seeing an old boyfriend, hearing a song on a guitar, or balancing on the edge of a fountain) with stimulation of the amygdala or hippocampal formation. Thus, although the visceral sensory cortex does relay information to medial temporal lobe structures where it may be integrated with emotional experience, it is unlikely that the insular cortex itself functions as an emotional integrator, or that visceral sensation is equated by the brain with the emotional experience that often accompanies it.

The mechanisms by which such conscious perceptions arise from activity of the internal organs remain poorly understood. We review the underlying anatomy and physiology of the visceral sensory pathways in a systematic way and attempt to arrive at a modern synthesis for understanding the conscious appreciation of visceral sensation. Note that we do not consider mechanisms of visceral sensory input in reflex pathways or even in highly coordinated and complex responses (e.g., digestion, cardiovascular response to muscle activity) that do not reach the level of conscious perception. For this information the reader is referred to the classic reviews by Sato & Schmidt (Sato & Schmidt 1973, Sato et al. 1997) on somato-autonomic reflexes and the thorough reviews in the volume by Loewy & Spyer (1990).

Two main sensory systems provide information to the brain about the state of the internal organs. The afferents provided by the cranial nerves, sometimes called the parasympathetic afferent system, carry mainly mechanoreceptor and chemosensory information. By contrast, afferents that arrive via spinal nerves, often called the sympathetic afferent system, convey mainly sensations related to temperature and impending or ongoing tissue injury, of either mechanical, chemical, or thermal origin (Figure 2, see after section on the gustatory system). We first consider the cranial nerve afferent system and then review the spinal afferent system, mainly in the context of its relationship to the cranial nerve system.

The Cranial Nerve (Parasympathetic) Visceral Sensory System

Visceral sensory information enters the brain via four cranial nerves: the trigeminal nerve, which conveys internal face and head visceral sensation; the facial nerve, which provides visceral sensory information (taste) from the tongue; the glossopharyngeal nerve, which conveys visceral sensory input (including taste) from the hard palate and upper part of the oropharynx as well as the carotid body; and the vagus nerve, which provides visceral sensation (including taste) from the lower part of the oropharynx, as well as supplying the larynx, trachea, esophagus, and thoracic and abdominal organs, with the exception of the pelvic viscera (Kerr 1961, Altschuler et al. 1989, Contreras et al. 1982). The pelvic viscera are innervated by nerves from the second through fourth sacral spinal segments. Although their patterns of spinal cord termination have been studied in detail (Morgan et al. 1981), what little is known about their central pathways or projections is similar to other spinal visceral afferent pathways, which are covered below.

VISCERAL AFFERENTS FROM ALL FOUR CRANIAL NERVES TERMINATE TOPOGRAPHICALLY IN THE NUCLEUS OF THE SOLITARY TRACT The abbreviation NTS, which is commonly used, stands for the Latin term, *nucleus tracti solitarii*. Other corruptions of the Latin term are frequently seen, such as *nucleus tractus solitarius* or *nucleus tractus solitarii*. The visceral sensory axons form a bundle, the solitary tract, which is analogous to the spinal trigeminal tract or Lissauer's tract and conveys sensory fibers along the length of the NTS.

Studies of the distribution of these sensory nerves along the length of the NTS show that they terminate in a strongly topographic pattern (Altschuler et al. 1989). The rostral tip of the NTS, which is in close proximity to the tongue region of the spinal trigeminal nucleus, is innervated by axons concerned with taste (a chemosensory modality) from the anterior two thirds of the tongue. Gustatory fibers from the posterior third of the tongue and the surrounding posterior oropharynx end more caudally in the rostral NTS (Contreras et al. 1982). At the level where the NTS begins to shift dorsomedially, to abut the floor of the fourth ventricle, fibers from the esophagus terminate in a tight cluster, the central NTS subnucleus (Altschuler

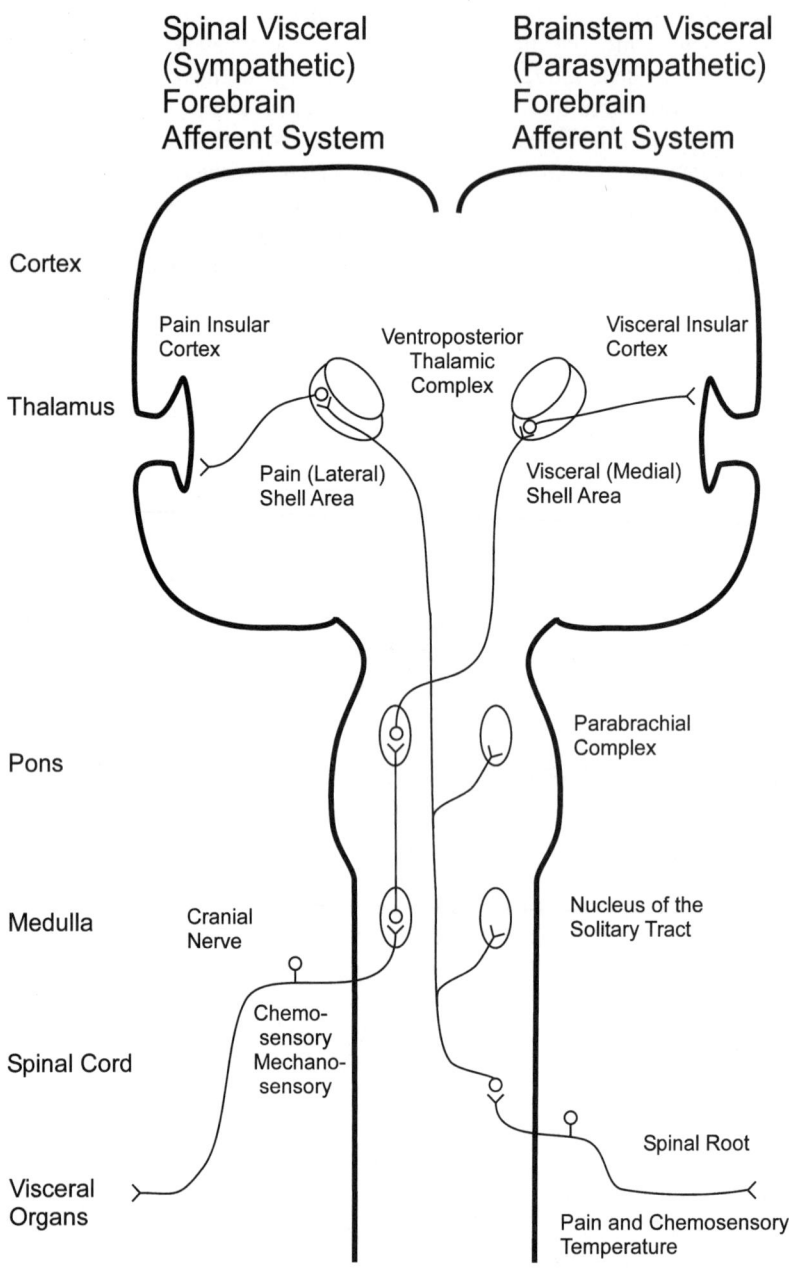

et al. 1989). Axons from parts of the vagus nerve that innervate progressively more caudal parts of the gastrointestinal tract end more caudally in the medial part of the NTS.

Axons from cardiovascular structures, including the carotid body and aortic arch baroreceptors, terminate within the dorsomedial part of the NTS, from just rostral to the level of the obex caudally through the commissural subnucleus of the NTS (Panneton & Loewy 1980, Ciriello 1983). Respiratory chemoreceptors are distributed in a similar pattern, but laryngeal, bronchial, tracheal, and pulmonary mechanoreceptor afferents synapse in the lateral part of the NTS, including the ventrolateral, interstitial, intermediate, and commissural nuclei (Kalia & Richter 1988).

THE PARABRACHIAL NUCLEUS IS THE MAJOR RELAY FOR ASCENDING VISCERAL INPUTS FROM THE NTS TO THE FOREBRAIN Because all of the cranial visceral afferents are relayed via a single cell group, the NTS, it is possible to study their further processing in the brain by examining the outputs from the NTS (Ricardo & Koh 1978, Beckstead 1980). In general, the most massive site for termination of fibers from the NTS is the parabrachial nucleus. The parabrachial nucleus consists of at least 13 separate subnuclei, which in turn provide extensive projections to a wide range of sites in the brainstem, hypothalamus, basal forebrain, thalamus, and cerebral cortex (Fulwiler & Saper 1984, Chamberlin & Saper 1994, Bester et al. 1997, Moga et al. 1990, Herbert et al. 1990), some of which are augmented by smaller direct projections from the NTS. Because this is a complex topic and conscious appreciation of visceral sensation depends upon the visceral sensory cortex (see below), we focus on the ways in which information from the NTS and parabrachial nucleus may reach the visceral sensory thalamus and hence the cerebral cortex.

Figure 2 A schematic drawing illustrating the spinal (sympathetic) and brainstem (parasympathetic) visceral sensory pathways to the thalamus and cerebral cortex. Note that the spinal afferent pathway crosses in the spinal cord; it projects bilaterally in the brainstem to the nucleus of the solitary tract and the parabrachial nucleus, but only ipsilateral projections are shown for clarity of illustration. Spinal visceral afferents relay in the lateral part of the "shell" of the ventroposterior thalamic complex, along with other spinothalamic afferents, especially nociception. The pathway taken by visceral afferents conveyed in the facial, glossopharyngeal, and vagus nerves relays in the nucleus of the solitary tract and the parabrachial nucleus, then crosses to innervate the medial part of the shell of the contralateral ventroposterior thalamic complex. The brainstem visceral sensory pathway terminates in the anterior part of the insular cortex, in comparison with the spinal pathway, which terminates more caudally in the insular area.

THE GUSTATORY SYSTEM PROVIDES A MODEL FOR VISCERAL SENSORY ACCESS TO CONSCIOUS APPRECIATION   To understand the relays responsible for conscious perception of visceral stimuli, it is instructive to consider the pathways that convey taste, a special visceral sensory modality that is appreciated in detail at a conscious level. As Herrick (1905) first pointed out in fishes, taste afferents from the NTS relay through the parabrachial region. In mammals inputs from the rostral part of the NTS are widely distributed in the parabrachial nucleus, including the medial subnucleus, external medial subnucleus, and ventral lateral subnucleus (Norgren 1978, Ricardo & Koh 1978). Large lesions of the parabrachial nucleus eliminate salt appetite, prevent conditioned taste avoidance in rats (Flynn et al. 1991), and cause deficits of taste discrimination in humans (Sato & Nitta 2000, Comvarros et al. 2000, Kokima & Hirano 1999, Onoda & Ikeda 1999, Nakjima et al. 1983). However, the location within the parabrachial nucleus of neurons that give rise to conscious taste perception has been controversial.

One approach to identifying the parabrachial region responsible for taste perception has been to determine the inputs to the thalamocortical taste system. The taste cortex has been mapped in rodents and primates by single-unit recording of neurons that respond to the application of gustatory stimuli applied to the tongue. The taste cortex in primates includes the inner, upper lip of the opercular cortex (closest to the insula) and the adjacent anterior insular region (Pritchard et al. 1986, Scott et al. 1986, Yaxley et al. 1990, Ito & Ogawa 1991, Baylis et al. 1995, Scott & Plata-Salaman 1999, Ito et al. 2001). In rodents a homologous taste area is found just dorsal to the rhinal sulcus, just rostral to where it is crossed by the middle cerebral artery (Kosar et al. 1986a, Cechetto & Saper 1987). The dysgranular field of the insular cortex in rats corresponds to this functionally defined region. The thalamic relay nucleus for the taste cortex is the most medial part of the ventroposterior parvicellular nucleus [VPpc, sometimes called the ventroposterior medial parvicellular nucleus (VPMpc) and sometimes ventromedial basal nucleus (VMb)] (Kosar et al. 1986b, Flynn et al. 1991, Cechetto & Saper 1987). Studies of lesions in humans have demonstrated that the representation of taste at a cortical or thalamic level is contralateral to the tongue (Onoda & Ikeda 1999a). Hence, by studying the contralateral brainstem inputs to the VPpc it should be possible to determine the brainstem relay for conscious appreciation of taste.

The inputs to the medial part of the VPpc have been examined using both anterograde and retrograde tracing studies in rodents (Cechetto & Saper 1987, Yasui et al. 1989, Bester et al. 1999). Retrograde transport studies are complicated by the fact that the small size of the VPpc makes it almost impossible to confine tracer injections to this nucleus, and hence the adjacent intralaminar and midline nuclei are virtually always involved by the injections (Cechetto & Saper 1987, Yasui et al. 1989). Retrogradely labeled neurons were found on the ipsilateral side of the brain in a wide range of parabrachial subnuclei, including the medial, external medial, and ventral lateral groups. However, on the contralateral side of the brain there were few labeled neurons in the medial and ventral lateral parabrachial subnuclei, whereas the external medial subnucleus contained about three to four times as many retrogradely labeled neurons as on the ipsilateral side. Hence, the projection from the external medial subnucleus is the only parabrachial-thalamic projection that is predominantly contralateral.

Anterograde tracing studies demonstrated a pattern of labeling consistent with these findings (Bester et al. 1999; D. F. Cechetto & C. B. Saper, unpublished). Efferents from the medial and ventral lateral subnuclei innervated predominantly the midline and intralaminar nuclei on the ipsilateral side, with a smaller contralateral projection, and only a light projection to the VPpc. By contrast, after tracer injections into the external medial nucleus, the bulk of the projection was to the contralateral VPpc, with only a light ipsilateral projection and relatively few axons in the intralaminar and midline nuclei.

This point is made most clearly by examining the pattern of immunostaining for calcitonin gene-related peptide (CGRP) in the thalamus (Yasui et al. 1989). Among the thalamically projecting neurons in the parabrachial complex, only the external medial subnucleus contains CGRP. The CGRP terminal field from this projection clearly outlines the VPpc in rats. De Lacalle & Saper (2000) recently described a similar CGRP projection in human brains.

Norgren and colleagues (Beckstead et al. 1980, Norgren 1984) have put forward an alternative view of the ascending taste pathway in primates. They noted that following injections of tritiated amino acids into the most rostral pole of the NTS in monkeys, they could trace a direct projection to the ipsilateral VPMpc and that only their more caudal NTS injections labeled substantial projections to the parabrachial nucleus.

On the other hand, in humans lesions of the dorsolateral pons, including the parabrachial nucleus, cause an ipsilateral taste deficit, whereas thalamic lesions cause a contralateral taste deficit (Sato & Nitta 2000, Combarros et al. 2000, Kojima & Hirano 1999, Onoda & Ikeda 1999, Nakajima et al. 1983). Thus, the ipsilateral projection from the NTS to the thalamus described by Beckstead and colleagues (1980) in monkeys is unlikely to account for conscious taste perception, and the parabrachial relay to the contralateral thalamus seems to be a critical component of taste perception in humans.

OTHER VISCERAL SENSORY MODALITIES ARE ORGANIZED IN A TOPOGRAPHIC PATTERN THAT PARALLELS THE TASTE PATHWAYS The thalamocortical localization of other visceral sensory modalities has been studied most carefully in rats, in which Cechetto & Saper (1987) systematically mapped cardiovascular baroreceptor, carotid chemoreceptor, pulmonary inflation, gastric stretch, and taste responses in the insular cortex. These experiments identified a topographic pattern of visceral sensory responses in the dysgranular and granular insular region of rats. As in the NTS, taste was located most rostrally; taste neurons were found mainly in the dysgranular insular field. Neurons responding to gastrointestinal sensation were found just caudal to the taste area, and most were in the more dorsal granular insular field. There was some overlap of the locations of neurons responding to taste and to gastrointestinal inputs, and occasional cells responded to both modalities.

Neurons responding to cardiovascular and respiratory afferents were located further caudally in the insular region. Again, there was spatial overlap of the neurons responding to these modalities, and occasional neurons responded to both types of inputs. Retrograde tracer, injected from the recording pipettes at the precise sites of the cortical visceral sensory responses, demonstrated a topographic input from the VPpc into the insular cortex, with the most medial tip of the VPpc innervating the dysgranular taste cortex, the midpart of the VPpc sending inputs to the rostral granular gastrointestinal insular sensory field, and the most lateral part of the VPpc supplying the more caudal granular cardiovascular part of the insular visceral sensory cortex. Preliminary recordings of responses of single neurons in the thalamus to these different stimuli confirmed the medial to lateral topography of the VPpc (D. F. Cechetto & C. B. Saper, unpublished observations). Electrical stimulation of the homologous region in the human thalamus produces sensations both of taste and of gastric fullness (Lenz et al. 1997), suggesting that the special (taste) and general visceral relay nuclei are similarly organized in humans.

Single unit recordings from monkeys support a similar topographic organization of visceral sensory cortex in primates. The taste area in primates actually occupies two separate representations: a primary taste area in the inner lip of the rostral opercular cortex extending into the insular region, which constitutes a dysgranular insular field; and a taste association area in the orbitofrontal cortex, which also receives olfactory inputs (Pritchard et al. 1986, Scott et al. 1986, Yaxley et al. 1990, Ito & Ogawa 1991, Baylis et al. 1995, Scott & Plata-Salaman 1999, Ito et al. 2001, Ongur & Price 2000). One study of single neurons in the insular area of awake monkeys reported responses to stimulating the cervical vagus nerve, suggesting a general visceral input to the insular region caudal to the taste area (Radna & MacLean 1981). Recent functional magnetic resonance imaging studies in humans demonstrated increased blood flow in the insular cortex, just caudal to the gustatory cortex, with visceral stimuli such as air hunger, maximal inspiration, valsalva maneuver, or physical manipulations such as hand grip or cold application to the forehead or hands, which elevate blood pressure and heart rate (King et al. 1999, Harper et al. 2000, Banzett et al. 2000). Taken as a whole, these studies suggest that the insular cortex of primates, including humans, contains a visceral sensory representation. The more rostral representation of taste, compared with general visceral modalities, also suggests that the topographic ordering of visceral sensations are preserved throughout the system, from the visceral sensory "homunculus" in the NTS all the way through to the cerebral cortex.

OTHER POSSIBLE ROUTES BY WHICH VISCERAL SENSORY AFFERENTS MAY REACH THE CEREBRAL CORTEX   In addition to the NTS-parabrachial-thalamic-cortical relay, the NTS and parabrachial nucleus project to other targets that might provide access to cortical appreciation of visceral sensation. One possible route is a direct projection from the parabrachial nucleus to the cerebral cortex (Saper 1982). This pathway originates from neurons in the medial part of the medial parabrachial subnucleus, in a region that receives gustatory visceral afferents. Some of these direct parabrachio-cortical axons innervate the insular visceral sensory field (the dysgranular and granular cortex), but most project to the agranular insular cortex and the lateral prefrontal cortex, regions in which the neurons do not respond to discrete visceral stimuli (Saper 1982, Allen et al. 1991). Hence, this direct parabrachio-cortical pathway probably does not play a major role in conscious visceral perception, but it may instead serve a secondary role, e.g., as an arousing influence, perhaps to direct behavior towards a food source.

The NTS also projects to a number of other sites with forebrain projections, which therefore could potentially contribute to visceral perception, including the ventrolateral medulla, the hypothalamus, and the amygdala/bed nucleus of the stria terminalis (Ricardo & Koh 1978). A small percentage of neurons in the ventrolateral medulla project to sites such as the locus coeruleus, lateral hypothalamus, and midline thalamic nuclei, each of which has direct but diffuse cortical projections (Otake et al. 1994, Woulfe et al. 1990, Aston-Jones et al. 1986, Saper 1985). Neurons in the lateral hypothalamus that project diffusely to the cerebral cortex also could potentially relay inputs directly from the NTS. However, because these projections are diffuse, they are more likely to be important in arousal than in conveying specific sensory information. Other basal forebrain targets of the NTS, such as other hypothalamic targets or the bed nucleus of the stria terminalis and the central nucleus of the amygdala, do not have direct cortical projections and hence probably do not contribute in a major way to conscious appreciation of visceral sensation.

SUMMARY Current evidence suggests that, in mammals from rats to humans, the pathway by which cranial nerve visceral afferents reach conscious appreciation involves a relay from the NTS to the ipsilateral external medial parabrachial nucleus. This cell group, many of whose neurons contain the peptide CGRP, then relays visceral afferents to the contralateral VPpc, which in turn innervates a dysgranular and granular anterior insular sensory field that acts as a visceral sensory cortex. This pathway appears to maintain its organotopic ordering, in roughly the same pattern as in the NTS, at a cortical level. There is less evidence for the maintenance of this topopraphic ordering at thalamic and parabrachial levels, but it is likely that the viscerotopic map is maintained throughout the system. Visceral sensations may then be relayed to other structures, including the medial temporal lobe, where they may become part of a more complex emotional experience. However, discrete visceral sensations appear to depend upon the integrity of the visceral sensory cortex and thalamus.

The Spinal (Sympathetic) Visceral Afferent System

Sensory afferents from the internal organs also enter the central nervous system via the spinal nerves, and these inputs have been called sympathetic visceral sensory afferents. This term is confusing, however, as visceral sensory afferents (e.g., concerned with muscle chemosensation during exercise) may arise at all spinal levels, not just the thoracic levels at which sympathetic preganglionic neurons are found (Kalia et al. 1981). In addition, dorsal horn sensory neurons that respond to visceral organs may be located at spinal levels quite distant from the level at which the visceral sensory afferents enter the spinal cord (e.g., dorsal horn neurons that respond to cardiac pain are found from at least the C2 to the T5 spinal levels) (see Blair et al. 1982, Chandler et al. 2000). Conversely, pelvic sensory afferents from spinal segments S2–4 are important for the regulation of sacral parasympathetic outflow, but there is little evidence that the information they convey is handled differently from inputs entering at other spinal levels. Thus, we consider spinal visceral afferent systems as a whole.

Our concepts concerning which spinal visceral afferents reach conscious perception are shaped by reports from patients with spinal cord transections, in whom the contribution from spinal as opposed to cranial nerve visceral inputs can be most readily observed. In general, visceral afferents that enter via spinal nerves convey information concerned with temperature as well as nociceptive visceral inputs related to mechanical, chemical, or thermal stimulation (Adelson et al. 1997, Ammons 1992), and the spinal afferents seem to be the principal source for these modalities reaching conscious perception. For example, patients with spinal transection may have a vague sense of vagally mediated fullness after eating a hot meal, but they do not report abdominal warmth or discomfort of acid reflux or colicky pain owing to a distended or obstructed viscus; the absence of these sensations in patients with spinal transection presents a continuous danger to survival (Strauther et al. 1999, Juler & Eltorai 1985). It is possible to show that these sensations do arrive at the spinal cord but are not relayed to the brain, because spinal reflexes due to visceral afferents remain active in patients with spinal cord transections and in fact may be exaggerated. For example, spinal patients often become aware of an overdistended bladder when they experience episodes of diffuse sympathetic activation, including sweating and hypertension (Silver 2000, Giannantoni et al. 1998).

PRIMARY PATHWAYS TAKEN BY ASCENDING SPINAL VISCERAL AFFERENTS   The primary pathways taken by ascending spinal visceral afferents are controversial. Although most spinal visceral afferents are thought to converge with musculoskeletal and cutaneous afferents and ascend via the spinothalamic and spinoreticular tracts (Foreman 1999, Weiss & Chowdhury 1998), there is some evidence for ascending fibers taking a dorsal column trajectory. A few visceral fibers enter the dorsal columns directly (Knuepfer & Schramm 1985), but most terminate either in lamina I or in the deep layers of the dorsal horn (IV and V) or the intermediate gray matter (layers VII and X), where a few may reach as far as the preganglionic neurons in the intermediolateral column (Kuo et al. 1983; Roppolo et al. 1985; Sugiura et al. 1989, 1993).

Some cells in layer X, around the central canal, send ascending afferents concerned with visceral pain (e.g., colorectal distention) through the dorsal columns (Willis et al. 1999). These neurons are particularly numerous at sacral levels, and their axons ascend along the medial septum of the dorsal columns. In addition, smaller numbers of cells in layer X at thoracic levels project to the dorsal column nuclei, by sending their axons along the intermediate septum. These afferents synapse in the dorsal column nuclei, and the postsynaptic neurons project to the contralateral ventroposterior thalamic complex. A majority, although not all, ascending afferents concerned with pelvic visceral pain appear to take this pathway; the importance of this route for other visceral afferents is not as clear (Willis et al. 1999).

Neurons from laminae I, IV, and V that respond to visceral stimuli generally also receive nociceptive inputs from cutaneous sensory fields (Foreman 1999). Hence, from the level of the of the first synapse in the spinal cord, the ascending visceral sensory pathways from these laminae converge with and are therefore essentially identical to the classic spinoreticular and spinothalamic tracts.

CONVERGENCE WITH CRANIAL NERVE VISCERAL SENSORY PATHWAYS   A key feature of these ascending pathways is that they provide collaterals that converge with the cranial nerve visceral sensory pathways at virtually every level (Mehler et al. 1960, Saper 2000). For example, neurons in laminae I, V, VII, and X innervate the nucleus of the solitary tract (Menetrey & Basbaum 1987; Gamboa-Esteves et al. 2001a,b), and there is also a projection from visceroceptive neurons in laminae I, V, and VII of the spinal dorsal horn to the ventrolateral medulla (Ammons 1988). Some of these afferents may be responsible for autonomic reflex responses to visceral stimuli, including visceral pain.

As many as 80% of lamina I spinothalamic axons in rats send collaterals to the parabrachial nucleus (Hylden et al. 1989), and recordings in monkeys from spinal dorsal horn neurons that receive visceral sensory input show that many can be antidromically activated from parabrachial nucleus (Foreman 1999). The spinal inputs to the parabrachial nucleus are rather complex, with afferents to different subnuclei arising from distinct laminae and levels of the spinal cord (Cechetto et al. 1985, Feil & Herbert 1995, Bernard et al. 1995, Panneton & Burton 1985). Many of these afferents end in subnuclei that primarily innervate the medulla, hypothalamus, or amygdala and hence for reasons discussed above are not likely to contribute to conscious appreciation of visceral sensation (Chamberlin & Saper 1994, Moga et al. 1990). However, other afferents end in the internal lateral parabrachial subnucleus, which provides a diffuse input to the intralaminar thalamic nuclei and may be involved in arousal responses to visceral stimuli (Cechetto et al. 1985, Feil & Herbert 1995, Fulwiler & Saper 1984, Bourgeais et al. 2001). Modest numbers of spinal afferents also terminate in the external medial nucleus (Feil & Herbert 1995, Bernard et al. 1995) and hence may contribute to conscious appreciation of visceral sensation via the visceral sensory thalamic relay nucleus and cortex.

Smaller numbers of spinal nociceptive neurons in laminae I, IV, V, VII, and X directly innervate the hypothalamus, and a few even reach the amygdala and medial prefrontal cortex (Burstein & Potrebic 1993, Cliffer et al. 1991, Burstein et al. 1996). However, because none of these latter targets specifically innervates the visceral sensory thalamus or cortex, they are not likely to bring visceral sensation to the level of conscious perception.

Within the thalamus the spinothalamic tract provides a medial projection to the intralaminar nuclei, as well as a lateral projection to the ventroposterior thalamic complex. The terminal field for the lateral projection in rats, cats, and monkeys has been characterized as a "shell" area along the ventral and lateral margins of the complex, just lateral to the VPpc (Horie & Yokota 1990, Koyama et al. 1998, Kobayashi 1998; C.B. Saper, unpublished observations). In monkeys this region has been termed the posterior part of the ventromedial nucleus (Craig & Dostrovsky 2001, Blomqvist et al. 2000) and constitutes a calbindin-immunoreactive terminal field. The cortical projection from this region innervates the posterior insular cortex in monkeys, and in humans pain activates a homologous region just caudal to the primary visceral sensory cortex (Casey et al. 1996, 2001; Craig et al. 2000).

Neurons in the shell region in cats respond not only to cutaneous pain but also to visceral sensory stimuli (Horie & Yokota 1990). In humans neuronal responses to cardiovascular stimuli can be recorded in a similar region, and many of these neurons also have cutaneous sensory fields in the parts of the neck and arm associated with referred cardiac pain (Oppenheimer 1998). Thus, the spinal visceral sensory system is maintained as a component of the spinothalamic system all the way from the first synapse in the spinal cord to the thalamus.

SUMMARY   In summary, a remarkable feature of the spinal visceral sensory system is its close relationship with the cranial nerve visceral sensory system at every level. At brainstem levels collaterals from the spinal system converge extensively with the cranial nerve sensory system in the nucleus of the solitary tract, ventrolateral medulla, and parabrachial nucleus. At the level of the forebrain the spinal visceral sensory system constitutes a posterolateral continuation of the cranial nerve visceral sensory thalamus and cortex (Saper 2000).

This relationship of pain with visceral sensation may seem surprising. However, pain is at its root a visceral sensory modality: the sensation that arises from mechanical or thermal stress that threatens tissue integrity. The tissue usually tested when examining a patient for pain perception is the cutaneous surface, which has resulted in the conceptualization of a spinothalamic system of cutaneous sensation representing pain and temperature, compared to the dorsal column system representing position sense and fine cutaneous discrimination. Another way to view the dichotomy is that discriminative sensations carried by the dorsal columns arise from a variety of tissues, both superficial and deep (including both skin deformation and joint receptors, for example), that are related to fine discrimination at body surfaces, muscles, and joints. These sensations are externally directed, i.e., they are concerned with exploring the external world and discerning the relationship of the body with external space. Nociceptive sensations, by contrast, are related to mechanical and thermal stresses of both deep and superficial tissues. Because they monitor tissue integrity, these sensations are internally directed, i.e., they are concerned ultimately with the state of the body itself, as opposed to its relationship with the external world. When viewed from this perspective, it is reasonable to describe pain per se as a visceral modality.

HOW ARE DIFFERENT PATTERNS OF AUTONOMIC OUTPUT GENERATED?

Historical Perspective

In addition to noting the marked similarities in facial expression in animals and humans experiencing related emotions, Darwin was aware that patterns of autonomic response also occur during different emotional states. For example, he noted that during rage "the action of the heart is much accelerated, or it may be much disturbed. The face reddens, or it becomes purple ... or it may turn deadly pale" (Darwin 1873). He also commented upon changes in piloerection and sweating in different emotional states. However, despite the progress that he cited in autonomic physiology owing to the efforts of contemporary physiologists such as Claude Bernard, Darwin did not have adequate methods at the time to pursue this level of analysis.

Walter B. Cannon, the eminent physiology professor at Harvard Medical School in the first half of the twentieth century, was heavily influenced by Darwin's observations and spent much of his career examining the physiological responses associated with emotions (Cannon 1929). Cannon focused on the secretion of adrenal catecholamines, as assessed by bioassay, and also examined changes in blood sugar (which are associated with adrenal secretion) as measures of sympathetic response during emotional states.

Cannon chose as his model for evaluating sympathoadrenal responses a condition known as "sham rage." Dusser de Barenne (1920) had discovered in the early 1920s that following decortication, a cat could respond to such innocuous stimuli as stroking the fur with a generalized "rage" response, characterized by somatomotor activity including arching of the back, extension of the claws, hissing, and spitting. These animals would also display autonomic responses such as retraction of the nictitating membrane, elevations of blood pressure and heart rate, and increased adrenal secretion similar to those seen in an intact cat that is threatened. However, in the absence of a functioning cerebral cortex, Cannon characterized this state as sham rage (Cannon & Britton 1927, Cannon 1929).

Cannon hypothesized that the diencephalon was the origin of the coordinated emotional response. His student Philip Bard (1928) demonstrated by means of serial transections of the remaining neuraxis in decorticate cats that severing the connections from the diencephalon to the midbrain eliminated the coordinated sham rage response (Figure 3). These experiments provided the basis for placing a pattern generator for both the somatomotor and the autonomic responses associated with this "fight-or-flight" response in the diencephalon.

Perhaps as a consequence of the simple models and measures available to him, Cannon viewed sympathetic response as a generalized reaction to environmental stimuli, which he characterized as "stressful." Interestingly, this unitary conception of stress influenced Selye (1975), who later equated stress with conditions that elevate adrenal corticosteroid levels. Cannon's view of sympathoadrenal response

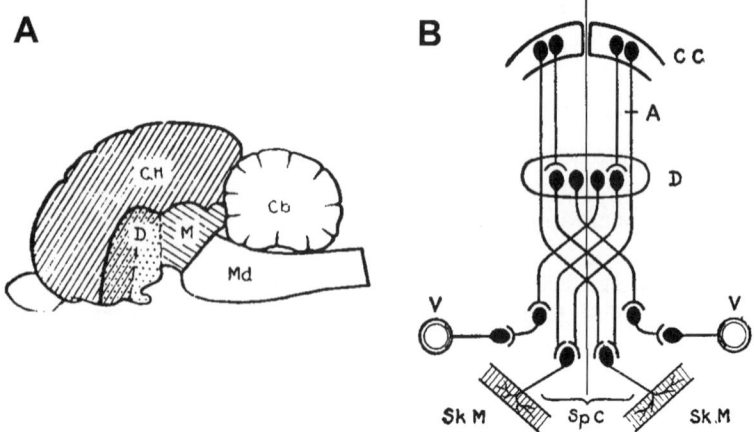

Figure 3 Two drawings from the work of Walter B. Cannon illustrating his ideas on the organization of sympathetic pattern generators. (*A*) The experiments of Cannon & Britton (1927) showed that after removal of the cortical mantle (CM) from a cat, even innocuous stimuli such as stroking the fur would result in fits of "sham rage," including all of the autonomic and somatomotor components of a rage attack. Subsequently, Cannon's student Bard placed transections through the diencephalon (D) and midbrain (M), demonstrating that a fully developed rage attack required the intact diencephalon. Md, medulla; Cb, cerebellum. (*B*) Cannon explained that although the cerebral cortex (CC) projects directly to the spinal cord (SpC) and controls skeletal muscle (SkM) responses, it also projects to the diencephalon, where it inhibits neurons that can produce patterned motor and autonomic responses. V, blood vessel. In the absence of cortical input, these diencephalic pattern generators become hypersensitive to other inputs, so that cutaneous stimuli that would ordinarily produce little if any response can result in a rage attack. From Cannon (1929) with permission.

as a monolithic reaction to a wide range of environmental stressors may have been a natural conclusion, given the limited range of autonomic functions he could measure in unanesthetized animals. However, subsequent work has demonstrated that sympathetic responses are highly patterned and differentiated.

Patterns of Autonomic Response

DEFENSE REACTION   For example, the pattern of sympathoadrenal response studied by Cannon has been described as a "defense reaction." It can be elicited by electrical stimulation in the lateral hypothalamus or periaqueductal gray matter, and hence has been studied in great detail in anesthetized animals (Hilton & Spyer 1980, Yardley & Hilton 1987, Arthur et al. 1991, Schadt & Hasser 2001). Although there is an increase in blood pressure and heart rate during the defense reaction, the use of miniature doppler flow probes has demonstrated sympathetically mediated vasoconstriction and reduction in blood flow to cutaneous beds or to renal or splanchnic vessels but vasodilation and increased flow in vessels carrying blood to the hindlimbs. The hindlimb vasodilation may be due to activation of sympathetic cholinergic vasodilation in cats (i.e., still a product of diffuse sympathetic activation) (Morrison 2001), but in rats, which lack cholinergic vasodilation, the increased blood flow to hindlimb skeletal muscles is due to withdrawal of sympathetic vasoconstriction (Yardley & Hilton 1987). This pattern of response is clearly adaptive (i.e., it is useful to have maximal tissue perfusion available to the hindlimbs to provide energy to fight or flee and to reduce blood flow to other vascular beds in case of injury), but even the fight-or-flight response is not a uniform pattern of sympathetic activation as Cannon imagined.

Other studies of sympathetic nerve responses under a wide range of physiological conditions have demonstrated highly differentiated patterns of activity, consistent with the need for modulation of different organ functions. This topic has recently been reviewed in detail by Morrison (2001) and is therefore discussed only briefly here.

THERMOREGULATION Thermoregulation requires differential control of sudomotor versus cutaneous vasoconstrictor and pilomotor sympathetic outflow, i.e., during heating sweating is required, whereas during cooling it is necessary to reduce cutaneous blood flow and increase heat retention of the fur by piloerection. At the same time, it is necessary to dilate deeper vascular beds to increase heat retention. This patterning has been identified in humans, in which studies of single sympathetic fibers in peripheral nerves can distinguish sudomotor, cutaneous vasoconstrictor, and muscle vasoconstrictor fibers (Janig & McLachlan 1992). In rats recordings have been made from tail artery and brown fat sympathetic afferents (Johnson & Gilbey 1994, Rathner & McAllen 1999, Morrison et al. 1999). Because the tail skin in rats is not covered by fur, it radiates heat efficiently and thus is used as a thermoregulatory organ. Brown adipose tissue contains massive numbers of mitochondria and high levels of uncoupling protein, which allow the mitochondria to burn calories to produce heat rather than synthesize ATP. The level of this thermogenic activity is under sympathetic control via the $\beta 3$ adrenergic receptor (Hamann et al. 1998). Exposure to cooling increases sympathetic discharge to both brown adipose tissue and the tail artery (Morrison 2001). During fever responses owing to immune system activation, body temperature is increased by sympathetic responses similar to those seen during cooling (see Zhang et al. 2000 for review).

METABOLIC CHALLENGE Metabolic challenges also produced highly patterned responses. Hypoglycemia, which can be reproduced by challenge with insulin or 2-deoxyglucose, causes increased sympathetic outflow to the liver and adrenal glands (resulting in increased gluconeogenesis), muscle vascular beds (thus reducing glucose utilization), and sweat glands (thus lowering body temperature and hence metabolic rate) (Niijima 1975, Sacca et al. 1977, Brodows et al. 1975, Medvedev et al. 1988). At the same time there is a decrease in sympathetic discharge to brown adipose tissue and cutaneous blood vessels (thus promoting cooling) and little effect on cardiac or renal sympathetic outflow.

DEHYDRATION   Dehydration is associated with an increase in adrenal, renal, and cardiac sympathetic activation, all of which are necessary to support blood pressure (Gharbi et al. 1999). However, there is a reduction in sweating, thus conserving fluid volume (Grucza et al. 1987, Baker 1989).

SENSORY STIMULI   Sensory stimuli produce a wide range of patterned autonomic responses, owing to local spinal reflexes as well as spino-bulbo-spinal pathways (Sato & Schmidt 1973, Sato et al. 1997). As a result, pain can produce different patterns of response depending upon the location of the stimulus. Painful cutaneous stimuli to the body, for example, cause increases in sympathetic outflow to the heart and adrenal glands, as well as causing vasoconstriction of cutaneous, renal, and mesenteric beds (Shimoda et al. 1998, Yamaguchi et al. 2001). On the other hand, trigeminal stimuli or deep stimuli, such as muscle pain, may provoke a fall in blood pressure and decreased sympathetic outflow to the heart, visceral, and muscle vascular beds (Keay et al. 2000). Vestibular stimulation also produces a differentiated pattern of sympathetic response, which includes increased blood flow to the hindlimbs but decreased perfusion of the face and forelimbs (Kerman et al. 2000).

Evidence for Organotopic Organization of Neuronal Pools that Generate Patterned Sympathetic Responses

One approach to identifying the origins of the different patterns of autonomic activation in varying physiological states has been to look for topographic organization of the premotor cell groups. Finding an organotopic map, along the lines of the somatic sensory or motor systems, would not necessarily identify the sources of pattern generation, but it would facilitate the search for inputs that result in patterned responses (e.g., finding the motor homunculus in the primary motor cortex did not solve the problem of motor pattern generators, but it did help elucidate the organization of motor control).

RETROGRADE TRACER STUDIES   Early experiments were unable to identify a somatotopic patern of organization in the central sympathetic regulatory system. Because sympathetic preganglionic neurons for many target organs tend to be located at specific spinal levels, e.g., pupillary neurons at T1–2, cardiac neurons at T2–5, adrenal neurons at T7–11, tail artery vasoconstrictor neurons at T13–L1, etc., (see Tucker & Saper 1985 for review), it was tempting to speculate that inputs to the sympathetic preganglionic neurons might be topographically ordered. However, regardless of which spinal level was injected, retrograde transport of tracers from different spinal levels of the sympathetic preganglionic column identified the same set of central structures, including the paraventricular nucleus, the retrochiasmatic area, the lateral hypothalamus, the parabrachial nucleus, the A5 area, the rostral ventrolateral medulla, the medullary raphe, and the nucleus of the solitary tract (NTS) (Tucker & Saper 1985). There was no obvious topography of retrograde labeling within any brain structure that contained labeled presympathetic cells. In double-labeling studies, using two distinguishable retrograde tracers to identify inputs to two spinal levels simultaneously, mixed collections of cells were seen with every pair of injections, with only occasional doubly labeled cells. These observations suggested that there is a fair degree of specificity of projection (i.e., that most presympathetic neurons do not project diffusely to every level of the spinal cord). At the same time, the studies failed to find any organizational principles that could explain how the projections were ordered or any mechanism for generating patterns of response.

VIRAL TRANSNEURONAL TRACER STUDIES The introduction of viral transneuronal tracers has allowed the extension of this strategy to looking for topographic organization of the inputs to sympathetic preganglionic neurons regulating different organ systems. A key issue in considering these findings is the extent to which the virus crosses between neurons only at synapses (transsynaptically) vs. crossing between nearby neurons that may not establish synaptic contacts. When pseudorabies virus of the Bartha strain is applied at peripheral tissues, the retrograde labeling in the sympathetic ganglia corresponds closely to the pattern of conventional retrograde tracers (i.e., the virus does not leak out of ganglion cells and infect neighboring cell bodies that innervate other tissues) (Strack & Loewy 1990). Furthermore, the pattern of transneuronal labeling in the spinal sympathetic preganglionic column is highly characteristic for each peripheral tissue but often different from other tissues served by the same ganglion. For example, the sympathetic preganglionic neurons labeled from the eye are in the T1–T3 segments, but those labeled from the pinna are in the T2–T5 segments (Strack & Loewy 1990). Because the two tissues are targets of independent but intermixed populations of sympathetic neurons in the superior cervical ganglion, the transfer of virus appears to be highly selective (i.e., to the presynaptic preganglionic population for each cell type). Thus, the transneuronal labeling at the level of the sympathetic preganglionic neurons is likely to be transsynaptic.

However, the specificity of the transynaptic transfer after successive waves of transport, replication, and release within the CNS is less well established. Card and colleagues (1993) carefully documented the passage of pseudorabies virus at an ultrastructural level through infected neurons in the dorsal motor vagal nucleus. They found that viral particles were packaged into bilaminar membranous particles in the Golgi apparatus and that the viral particles were preferentially released near the sites of afferent synaptic content. Nonsynaptic transfer was limited by astrocytic processes. Even after a severely infected cell died, local astrocytes and macrophages tended to limit the spread of infective particles. However, such studies are not capable of proving that viral transfer within the CNS occurs only at synapses.

This issue becomes important because injections of pseudorabies virus into a wide variety of tissues produce retrograde labeling in a very similar set of CNS structures. Careful studies of the transneuronal labeling after injections of pseudorabies virus have included, for example, skeletal muscle (Rotto-Percelay et al. 1992), kidney (Schramm et al. 1993, Huang & Weiss 1999), stellate ganglion (Jansen et al. 1995), pancreas (Jansen et al. 1997), tail artery (Smith et al. 1998), and spleen (Cano et al. 2001). These experiments virtually all demonstrated retrograde labeling of neurons in four structures: the C1 adrenergic neurons in the ventrolateral medulla, the medullary raphe nuclei and adjacent ventromedial medulla, the A5 noradrenergic neurons in the ventrolateral pons, and the paraventricular nucleus and adjacent lateral hypothalamus. Furthermore, the pattern was similar regardless of whether the labeling was restricted to parasympathetic or sympathetic afferents to the target tissue. Interestingly, large numbers of cells in those structures were often labeled from quite restricted injections into target tissues, suggesting a substantial amount of divergence of output from individual cells in these structures to multiple tissue types. Another interpretation, however, might be that virus is transferred nonsynaptically in the advanced stages of the infection, when several CNS synapses have been crossed.

Evidence for somatotopic ordering of transneuronal labeling within specific hypothalamic nuclei argues against this interpretation. For example, Strack and colleagues (Strack et al. 1989) demonstrated that injections into different sympathetic ganglia produced slightly different patterns of retrograde labeling in the paraventricular nucleus of the hypothalamus and that only injections of the stellate ganglion demonstrated retrograde labeling in the lateral hypothalamic area. Sved and coworkers (Sved et al. 2001) also noted some organotopic organization in the paraventricular nucleus, with cells retrogradely labeled from brown fat clustering mainly in the anterior and dorsal parvicellular subnuclei, whereas neurons in the lateral parvicellular subnucleus were most prominently retrogradely labeled from spleen and pineal gland injections. However, within the dorsal and ventral parvicellular subnuclei, no topography was apparent.

An alternative approach to determining whether the projections to the different tissues actually emerge from the same central neurons was provided by the use of two different strains of pseudorabies virus to label simultaneously the inputs to the superior cervical ganglion and adrenal medulla (Jansen et al. 1995). Antibodies were used to distinguish the two viral strains and demonstrated that a large percentage of infected CNS cells, particularly in the brainstem catecholaminergic fields and in the paraventricular nucleus of the hypothalamus, contained both viruses. This overlap could still be interpreted as nonspecific transfer of virus (i.e., the proof of specificity would be in the maintenance of distinct channels of communication rather than the presence of double labeling). However, Loewy and coworkers interpreted the convergence of these different pathways at the same central cells as evidence for "command neurons" in the central autonomic control system, with wide-ranging outputs that could produce patterns of autonomic response in different tissues. They therefore interpreted their findings as providing a possible basis for Cannon's predictions of a diencephalic pattern generator for sympathetic response in fight-or-flight situations (i.e., single hypothalamic neurons could contact a wide range of sympathetic preganglionic neurons concerned with producing an integrated response involving multiple tissues). However, these studies do not provide insight into how the very different patterns of sympathetic response required by varying physiological conditions may be generated, nor do they pinpoint which neurons are critically involved in generating the fight-or-flight response, which is just one of a large array of hypothalamic response patterns.

Evidence for Functional Organization of Neuronal Pools that Generate Patterned Autonomic Responses

A different perspective on the organization of sympathetic pattern generation has come from recent studies of the functional neuroanatomy of presympathetic neurons at several levels of the central autonomic control system outlined by the viral tracer studies. Evidence now is available for the generation of specific patterned autonomic responses by identified populations of neurons in the ventrolateral medulla, the rostral medullary raphe, the periaqueductal gray matter, and the hypothalamus.

VENTROLATERAL MEDULLA   Neurons in the ventrolateral medullary reticular formation are thought to play a key role in producing the pattern of autonomic and endocrine response necessary to maintain adequate arterial perfusion of the body's tissues (Morrison 2001). Information from baroreceptors that monitor vascular wall stretch in the aortic arch and the carotid sinus converges in the nucleus of the solitary tract and is conveyed from there to the ventrolateral medulla. A fall in arterial pressure causes reduced excitatory drive from the nucleus of the solitary tract to cardiovagal motor neurons in the compact formation of the nucleus ambiguus, which slows vagal firing, thus increasing heart rate (Beiger & Hopkins 1987, Ross et al. 1985). There is also reduced excitation of inhibitory interneurons in the caudal part of the ventrolateral medulla that project to a population of neurons in the rostral ventrolateral medulla that regulate vasoconstrictor sympathetic tone. The disinhibition of these rostral ventrolateral medullary neurons increases their excitation of vasoconstrictor sympathetic preganglionic neurons, resulting in elevated arterial blood pressure (Elliott et al. 1985). Other neurons in the caudal ventrolateral medulla project to the forebrain, where they regulate release of vasopressin. A fall in baroreceptor input results in the release of vasopressin, thus increasing fluid retention and causing vasoconstriction (Blessing & Willoughby 1985). The net effect of these three limbs of the baroreceptor response is to increase cardiac output, blood pressure, and heart rate in the short term and to increase blood volume in the longer term. This pattern of response appears to be intrinsic to the connections of the baroresponsive neurons of the ventrolateral medulla.

There is some evidence for topographic ordering of the sympathetic vasoconstrictor responses elicited from the rostral ventrolateral medulla in cats (see

Morrison 2001). Interestingly, the topographic arrangement is more along the lines of functional groupings than body map. Hence, neurons controlling vasocontrictor tone to muscles were more caudolateral, those regulating cutaneous vasoconstriction were more medial, and those involved with increasing sympathetic tone to renal, cardiac, and splanchnic beds were rostromedial. Other classes of sympathetic response such as pupillomotor responses, cutaneous vasoconstriction, and brown fat activation apparently were not represented, lending further support to the functional specificity of this region.

ROSTRAL MEDULLARY RAPHE   Recent evidence has identified a different patterned sympathetic response generated by the rostral part of the raphe pallidus formation in the rostral ventromedial medulla. Neurons in this parapyramidal region (between the pyramidal tracts at the level of the caudal part of the facial nucleus) show Fos expression and increased firing rates in response to external cooling (Morrison et al. 1999, Rathner et al. 2001). Injection of GABA agonists into the parapyramidal region inhibits cooling-induced sympathetic outflow to brown adipose tissue (Morrison et al. 1999), and injection of excitatory amino acids at the same site causes increased sympathetic outflow to the tail artery in rats. Because sympathetic stimulation of brown adipose tissue results in thermogenesis, and vasoconstriction of the tail artery limits passive heat loss, the rostral medullary raphe region has been considered a pattern generator for thermogenic responses (Morrison 2001). These responses are important not only for allowing small animals to survive in a harsh climate but also are involved in energy balance and metabolism, as thermogenesis is metabolically expensive (Lowell & Flier 1997).

PERIAQUEDUCTAL GRAY MATTER   Stimulation in discrete regions of the periaqueductal gray matter may produce quite distinct and highly stereotyped patterns of both somatic and autonomic response in cats (Zhang et al. 1990, Bandler & Shipley 1994). For example, excitatory amino acid stimulation in the ventrolateral periaqueductal gray matter caused motor quiescence and a fall in blood pressure, whereas stimulation in the lateral periaqueductal region produced a flight reaction with an accompanying increase in blood pressure. The fall in blood pressure with stimulation of the rostral part of the ventrolateral column was associated with a fall in heart rate as well as a reduction in vasoconstrictor tone to the hindlimbs, but not the kidneys; stimulation of the caudal part of the ventrolateral column resulted in bradycardia plus renal, but not hindlimb, vasodilation (Carrive & Bandler 1991). Interestingly, there was a topographically ordered projection from these sites to the ventrolateral medulla, with the rostral part of the ventrolateral periaqueductal gray matter projecting to the caudal part of the rostral ventrolateral medulla, which controls blood flow to muscles, and the caudal part of the ventrolateral periaqueductal gray matter sending efferents to the part of the rostral ventrolateral medulla, which regulates blood flow in renal and other visceral vascular beds (Carrive & Bandler 1991, Morrison 2001). This functional organization suggests that higher levels of the neuraxis may produce progressively more complex and highly integrated response patterns by means of their activation of multiple, more elemental response generators.

HYPOTHALAMUS The hypothalamus contains several distinct sets of neuronal populations that innervate the parasympathetic and sympathetic preganglionic populations: the paraventricular nucleus, the lateral hypothalamic area, and the arcuate nucleus and adjacent retrochiasmatic area (Saper et al. 1976, Swanson & Sawchenko 1983, Cechetto & Saper 1988). Within the paraventricular nucleus, neurons that project to the spinal cord are found in the dorsal, ventral, and lateroposterior parvicellular subnuclei. The dominant neurotransmitters appear to be oxytocin and vasopressin (Cechetto & Saper 1988, Hallbeck & Blomqvist 1999, Hallbeck et al. 2001), although many neurons also contain either dynorphin or enkephalin, and many of these same cells may employ excitatory neurotransmitters as well. In the lateral hypothalamic area, neurons that project to the spinal cord are found at levels roughly coextensive with the ventromedial nucleus. They are most dense in the perifornical region but spill over medially into the lateral edge of the dorsomedial nucleus and dorsally into the zona incerta and reach laterally to the edge of the cerebral peduncle. The lateral hypothalamic presympathetic neurons include cells containing orexin/hypocretin and melanin-concentrating hormone (van den Pol 1999, Bittencourt & Elias 1993) but also encompass many other neurons whose neurotransmitter specificity is not known. The arcuate neurons are also located at the level of the ventromedial nucleus, and they spill out laterally into the region ventral to the ventromedial nucleus (the retrochiasmatic area). They include many neurons that express the pro-opiomelanocortin gene, and hence make α-melanocyte-stimulating hormone (α-MSH) (Cechetto & Saper 1988), as well as cocaine and amphetamine-regulated transcript (CART) (Elias et al. 1998).

Evidence has accumulated within the past few years that specific subsets of neurons within the hypothalamo-spinal system may be engaged by distinct physiological processes. For example, expression of Fos protein is often used to determine whether neurons are engaged by a particular physiological stimulus. Administration of lipopolysaccharide, a bacterial cell wall component, produces a vigorous CNS response that augments the immune response to this stimulus (see Elmquist et al. 1996). The CNS components include fever, secretion of corticosteroids, and sickness behavior (Figure 4). Fever, in turn, is mainly due to engaging a series of sympathetic responses, including increased activity of brown adipose tissue (thus generating heat), shunting of blood flow to deep from cutaneous vascular beds (particularly the tail, which radiates heat in rats), and increased cardiac output and heart rate, as well as increased adrenalin secretion (see Zhang et al. 2000 for review). After lipopolysaccharide treatment, Fos immunoreactive neurons are found in a range of central autonomic structures including the ventrolateral medulla, the nucleus of the solitary tract, the ventromedial preoptic nucleus, and the paraventricular nucleus and lateral hypothalamic area (Elmquist et al. 1996). By combining retrograde tracing from the sympathetic preganglionic column with Fos

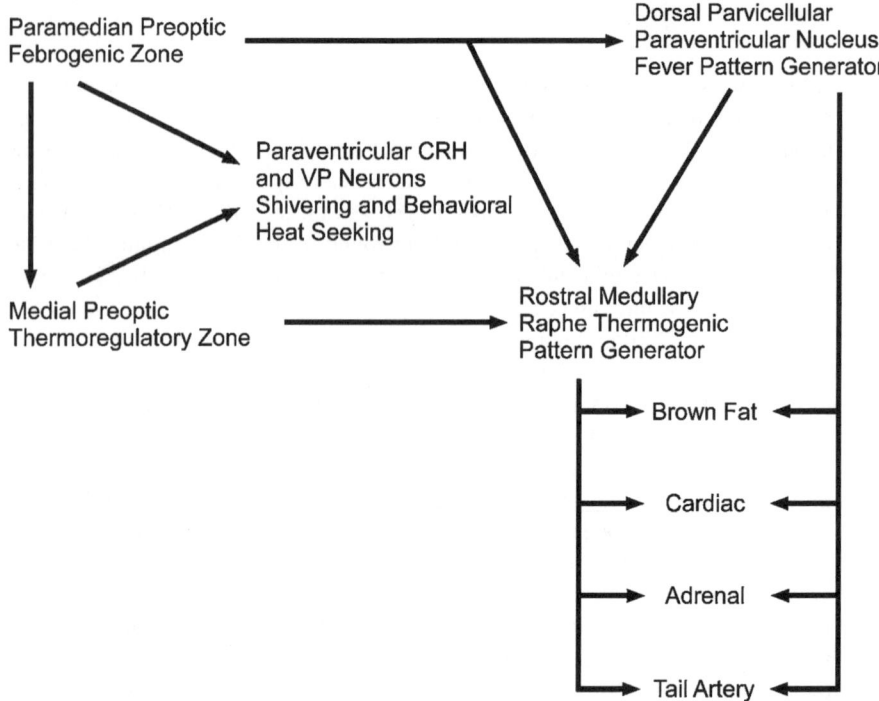

Figure 4 A schematic outline of the possible interactions among pattern generators for thermogenesis, thermoregulation, and fever in rats. Under normal circumstances, the medial preoptic area contains thermosensitive neurons that regulate body temperature around a set point. The medial preoptic area can elicit appropriate endocrine and behavioral patterns of response necessary to increase or decrease body temperature. It also controls a rostral medullary thermogenic pattern generator. This region, in the rostral medullary raphe, controls the sympathetic input to both brown adipose tissue (which results in thermogenesis in small animals) and the tail artery (which regulates heat radiation in rats). Sympathetic cardiac outflow and adrenal outflow may also be elevated during thermogenesis. The paramedian preoptic region, surrounding the anterior tip of the third ventricle, contains neurons that respond to prostaglandins elaborated during an immune stimuli, by producing a fever. Prostaglandin E2 activates the fever response, which includes the thermogenic pathways that are engaged during thermoregulation in the cold. In addition, the paraventricular nucleus is also engaged, including neurons in its medial part that secrete corticotropin-releasing hormone (CRH) and thyrotropin-releasing hormone and neurons in the dorsal parvicellular subnucleus, which contains vasopressin (VP) and contributes to thermogenesis by means of their pattern of projections to sympathetic preganglionic neurons, as well as to the rostral medullary raphe.

immunostaining, it has been possible to identify the hypothalamo-spinal neurons that demonstrate Fos activation after lipopolysaccharide administration.

Surprisingly, although Fos-immunoreactive neurons are found throughout the range of the hypothalamo-spinal cell groups, the only population that shows doubly labeled neurons is the dorsal parvicellular part of the paraventricular nucleus (Zhang et al. 2000). Fos-immunoreactive neurons in this one cluster project to every level of the intermediolateral cell column, and few other hypothalamo-spinal neurons are doubly labeled. Interestingly, retrograde transneuronal tracing from the rat tail artery using pseudorabies virus results in intensive labeling in the dorsal parvicellular part of the paraventricular nucleus (Smith et al. 1998). Lesions of the paraventricular nucleus attenuate fever responses to lipopolysaccharide but do not affect circadian cycles of body temperature or the thermoregulatory response to cooling (Horn et al. 1994, Zhang et al. 2000). Thus, the dorsal parvicellular part of the paraventricular nucleus may be a pattern generator for fever responses during immune stimulation, and it may work at least in part by its projections to the spinal cord. However, it may also activate the rostral medullary raphe region that generates sympathetic patterns resulting in thermogenesis. The rostral medullary raphe, on the other hand, may play a broader role in thermogenesis related to cooling or energy metabolism, responses regulated by the medial preoptic area (Chen et al. 1998, Zhang et al. 1997) and in which the dorsal parvicellular paraventricular nucleus may have little role.

Similarly, after intravenous administration of leptin, a hormone made by white fat cells during times of high levels of substrate availability, there is a restricted pattern of Fos expression in the hypothalamus, including neurons in the paraventricular nucleus, lateral hypothalamus, and arcuate/retrochiasmatic groups (Elmquist et al. 1997, Elias et al. 2000). After retrograde labeling from the sympathetic preganglionic column (to identify leptin-activated sources of sympathetic activation), doubly labeled neurons were found in the arcuate/retrochiasmatic area but not in the other sources of hypothalamo-spinal projections, regardless of which spinal levels were injected (Elias et al. 1998). A high percentage of the doubly labeled neurons also contained the neuropeptides a-MSH (a derivative of pro-opiomelanocortin) and CART.

These observations suggest that specific patterns of sympathetic response, associated with discrete stimuli, may differentially activate small populations of neurons within the hypothalamo-autonomic projection system. These neurons, which may be marked by chemical (neurotransmitter phenotype) as well as functional and anatomical specificity, appear as a group to project to the entire length of the sympathetic preganglionic cell column. It is not yet clear whether individual hypothalamic neurons may contact sympathetic preganglionic cells at multiple spinal levels. Double-label retrograde tracer studies suggest that there is relatively little collateralization of axons from individual hypothalamic neurons to different spinal levels (Tucker & Saper 1985), with only 1–2% of cells being doubly labeled after injections of different colored fluorescent tracers at two distinct spinal levels. However, such studies may understate the degree of collateralization, if each axon contacts a small subset of (e.g., two or three) spinal levels rather than branching diffusely over the entire column. (For example, a single combination of injections at the T2 and T10 levels of the intermediolateral column would identify only about 8% (one twelfth) of the neurons projecting to either the T2 or T10 level that also innervate a second level of the sympathetic preganglionic column). Superficially, it might seem that the command neurons, described by Loewy and colleagues (Jansen et al. 1995), would demand a much higher degree of collateralization. However, the preganglionic neurons projecting to any ganglion or peripheral tissue are generally scattered across at least four to five spinal levels, so that a high degree of double-labeling is possible in their double-virus experiments, even if each neuron innervates only two or three spinal levels.

In this view, it would be possible for a population of neurons, each of which innervates only discrete portions of the preganglionic column, as a whole to undertake the entire range of sympathetic preganglionic projections necessary to produce a patterned response. The pattern of sympathetic (as well as parasympathetic and endocrine) response that is generated would be intrinsic to the specificity of the connections of this set of functionally defined neurons.

Integration Across Multiple Levels of Autonomic Pattern Generators

The evidence reviewed here indicates that neurons at multiple levels of the CNS are capable of generating specific patterns of sympathetic response but that in each case the anatomical organization is along the lines of functionally related cell groups rather than somatotopic maps. The sympathetic response patterns organized at different brainstem levels may in turn be integrated at those sites with simple parasympathetic, endocrine, and behavioral components. For example, neurons in the ventrolateral medulla defend against low blood pressure with a sympathetic response (increased vasoconstrictor tone) that is tightly linked to parasympathetic activity (withdrawal of vagal cardiac inhibition resulting in tachycardia) and endocrine response (secretion of vasopressin). Neurons in the periaqueductal gray matter produce combinations of autonomic and behavioral patterns such as quiet coping (which is linked to withdrawal of vasoconstrictor tone, resulting in hypotension) or flight responses (which are accompanied by hindlimb vasodilation but splanchnic vasoconstriction, to provide maximal blood flow to the hindlimbs).

These specific functional patterns may, in turn, be incorporated into larger-scale response patterns. As in the examples of the fight-or-flight response or the thermoregulatory systems (Figures 4 and 5), the different pattern generators at different levels of the neuraxis are organized in a hierarchical manner that allows individual response patterns to become parts of larger responses. For example, fever responses engage sympathetic pattern generators at both the hypothalamic (paraventricular nucleus) and medullary (rostral medullary raphe) levels. Both of these sites engage the sympathetic preganglionic neurons directly, and the integrity of both sites is necessary to produce elevation of body temperature.

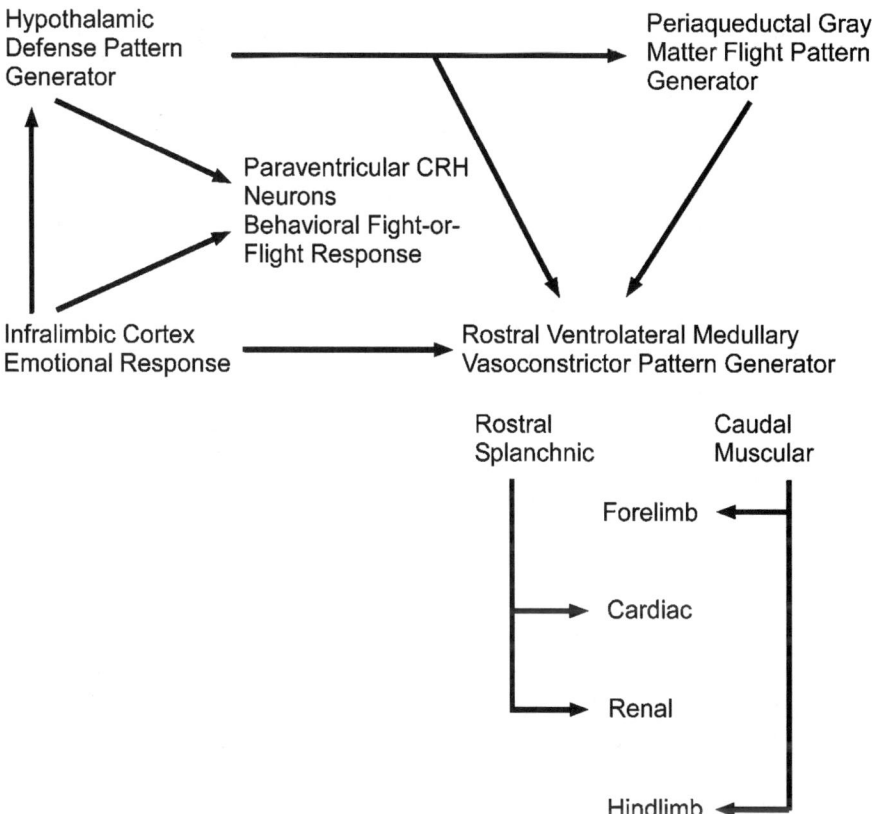

Figure 5 Schematic diagram of possible interactions among pattern generators for vascular responses to emotional stress and fight-or-flight reactions. The infralimbic cortex serves as an emotional motor cortex. It receives inputs from the prelimbic and cingulate areas and insures that the emotional reaction is linked to autonomic and endocrine responses by means of its projections both to the hypothalamus and to the ventrolateral medulla. The hypothalamic pattern generators for fight or flight can activate a coordinated pattern of autonomic, endocrine, and behavioral response associated with either fighting or fleeing. The flight response is due in part to hypothalamic activation of a region in the lateral part of the periaqueductal gray matter, which contains a pattern generator for flight behavior including associated autonomic responses. These responses include vasodilation of arterial supply to the hindlimb muscles but vasoconstriction of blood supply to the splanchnic vascular beds, along with increased cardiac output and adrenal secretion. This pattern of sympathetic response, in turn, is due to activation of distinct rostral and caudal components of the ventrolateral medullary pattern generator that differentially regulates tissue perfusion in splanchnic and muscular beds. CRH, corticotropin-releasing hormone.

At the same time, an individual response pattern may be used for multiple different larger-scale responses. The rostral medullary raphe, for example, is a critical component of thermogenesis in defending body temperature against hypothermia, by increasing heat generation by brown adipose tissue and by shunting blood flow from cutaneous (tail artery) to deep vascular beds. At the same time, these responses are incorporated into fever responses, and brown adipose activity is used to balance energy metabolism (Lowell & Flier 1997, Hamann et al. 1998).

Highly integrated patterns of response, as first noted by Cannon (1929) and Bard (1928), are generally organized at a hypothalamic level. These integrated responses, which maintain energy metabolism, body temperature, and fluid and electrolyte balance, and allow for reproduction and defense against attack, involve autonomic, endocrine, and behavioral components that are played out on a temporal and spatial sequence as a combination of more limited patterned responses, organized at other levels of the basal forebrain, brainstem, and spinal cord. Thus, while the hypothalamus may determine the overall "composition" of the response (and how it will fit with other ongoing needs), subsidiary pattern generators may each produce a series of "chords," or more elementary response patterns, which are in turn composed of "keys" consisting of individual autonomic (and endocrine and motor) actions. When engaged in different combinations by different compositions, these pattern generators can produce the entire range of highly differentiated and complex patterned responses necessary to maintain homeostasis, defend against threat, and ultimately pass on genetic information to the next generation.

SUMMARY   The highly interconnected nature of the central autonomic control system has for many years served as an impediment to assigning responsibility for specific autonomic patterns to identified neuronal populations. Although network properties of a system are a convenient explanation for complex responses, they tell us little about how they actually work, and the concept tends to stifle exploration for more parsimonious explanations. Recent data indicate that it is possible to correlate certain discrete patterns of autonomic response with identified neuronal sets, whose integrity may be necessary for that function. The challenge of identifying these pattern generators will lead in turn to determining their connections, neurotransmitters, and physiological activities. Such a prospect offers hope of controlling, or even reversing, maladaptive patterns of autonomic response such as occur during psychogenically elicited cardiac arrhythmias or gastrointestinal disease.

CONCLUSION

Although the two sections of this review may seem disparate in their intent, they both follow from the key issue of the specificity of organization of visceral control, a debate that dates back to the nineteenth century. The autonomic nervous system has often been viewed as lacking the finely regulated control that is inherent in somatic sensory and motor systems. While this view is undoubtedly correct in a temporal sense (i.e., somatomotor responses are generally much faster and contain rapidly changing temporal patterns that autonomic responses lack), it has obscured the importance of fine discrimination within the visceral system, both at the level of conscious appreciation of visceral sensation, and in the recognition of fine gradations and differentiation of patterns of autonomic motor response.

Recent work makes it clear that the visceral sensory and motor systems contain elements of organization that have long been accepted in the somatomotor systems but have been neglected in central autonomic control. The ascending visceral sensory system, in addition to providing for short- and long-loop reflexes, also contributes a pathway that parallels the classical spinothalamic system, including a visceral sensory thalamic relay nucleus and visceral sensory cortex. This close relationship underscores the similarity in the way the brain handles visceral vs. cutaneous pain, and it suggests that pain itself may be thought of as a visceral modality, as it is concerned with maintenance of tissue integrity, rather than exploration of the external world (except in the sense that it threatens tissue integrity).

The emergence of evidence for pattern generators in the visceral motor system also parallels the somatomotor systems, where such pattern generators have long been accepted and studied. Recent evidence has pinpointed a number of sites that appear to serve this role for discrete autonomic patterns associated with specific physiological responses. The challenge over the next few years will be to test this hypothesis critically and to begin working out the relationships of these pattern generation modules with each other, and with forebrain endocrine and behavioral systems.

The "missing link" in autonomic research has long been the gap between understanding visceral sensory systems and their role in emotion on the one hand, and the ways in which emotional and cognitive responses impact autonomic function on the other. The advances described here narrow this gap and provide some anchor points for future investigators wishing to establish the links.

ACKNOWLEDGMENTS

The author thanks Dr. Rami Burstein for constructive discussions and criticism. The work was supported in part by US PHS grant NS33987.

The *Annual Review of Neuroscience* is online at http://neuro.annualreviews.org

LITERATURE CITED

Adelson DW, Wei JY, Kruger L. 1997. Warm-sensitive afferent splanchnic C-fiber units in vitro. *J. Neurophysiol.* 77:2989–3002

Allen GV, Saper CB, Hurley KM, Cechetto DF. 1991. Organization of visceral and limbic connections in the insular cortex of the rat. *J. Comp. Neurol.* 311:1–16

Altschuler SM, Bao X, Bieger D, Hopkins DA, Miselis RR. 1989. Viscertopic representation of the upper alimentary tract in the rat: sensory ganglia and nuclei of the solitary and spinal trigeminal tracts. *J. Comp. Neurol.* 283:248–68

Ammons WS. 1988. Renal and somatic input to spinal neurons antidromically activated from the ventrolateral medulla. *J. Neurophysiol.* 60:1967–81

Ammons WS. 1992. Bowditch lecture. Renal afferent inputs to ascending spinal pathways. *Am. J. Physiol.* 262:R165–76

Arthur JM, Bonham AC, Gutterman DD, Gebhart GF, Marcus ML, Brody MJ. 1991. Coronary vasoconstriction during stimulation in hypothalamic defense region. *Am. J. Physiol.* 260:R335–45

Aston-Jones G, Ennis M, Pieribone VA, Nickell WT, Shipley MT. 1986. The brain nucleus locus coeruleus: restricted afferent control of a broad efferent network. *Science* 234:734–37

Bailey P, Bremer F. 1938. A sensory cortical representation of the vagus nerve with a note on the effects of low blood pressure on the cortical electrogram. *J. Neurophysiol.* 1:405–12

Baker MA. 1989. Effects of dehydration and rehydration on thermoregulatory sweating in goats. *J. Physiol.* 417:421–35

Bandler R, Shipley MT. 1994. Columnar organization in the midbrain periaqueductal gray: modules for emotional expression? *Trends Neurosci.* 17:379–89

Banzett RB, Mulnier HE, Murphy K, Rosen SD, Wise RJ, Adams L. 2000. Breathlessness in humans activates insular cortex. *Neuroreport* 11:2117–20

Bard P. 1928. A diencephalic mechanism for the expression of rage with special reference to the sympathetic nervous system. *Am. J. Physiol.* 84:490–515

Baylis LL, Rolls ET, Baylis GC. 1995. Afferent connections of the caudolateral orbitofrontal cortex taste area of the primate. *Neuroscience* 64:801–12

Beckstead RM, Morse JR, Norgren R. 1980. The nucleus of the solitary tract in the monkey: projections to the thalamus and brain stem nuclei. *J. Comp. Neurol.* 190:259–82

Bernard JF, Dallel R, Raboisson P, Villanueva L, Le Bars D. 1995. Organization of the efferent projections from the spinal cervical enlargement to the parabrachial area and periaqueductal gray: a PHA-L study in the rat. *J. Comp. Neurol.* 353:480–505

Bester H, Besson JM, Bernard JF. 1997. Organization of efferent projections from the parabrachial area to the hypothalamus: a phaseolus vulgaris-leucoagglutinin study in the rat. *J. Comp. Neurol.* 383:245–81

Bester H, Bourgeais L, Villanueva L, Besson JM, Bernard JF. 1999. Differential projections to the intralaminar and gustatory thalamus from the parabrachial area: a PHA-L study in the rat. *J. Comp. Neurol.* 405:421–49

Bieger D, Hopkins DA. 1987. Viscerotopic representation of the upper alimentary tract in the medulla oblongata in the rat: the nucleus ambiguus. *J. Comp. Neurol.* 262:546–62

Bittencourt JC, Elias CF. 1993. Diencephalic origins of melanin-concentrating hormone immunoreactive projections to medial septum/diagonal band complex and spinal cord using two retrograde fluorescent tracers. *Ann. NY Acad. Sci.* 680:462–65

Blair RW, Weber RN, Foreman RD. 1982. Responses of thoracic spinothalamic neurons to intracardiac injection of bradykinin in the monkey. *Circ. Res.* 51:83–94

Blessing WW, Willoughby JO. 1985. Inhibiting the rabbit caudal ventrolateral medulla prevents baroreceptor-initiated secretion of vasopressin. *J. Physiol.* 367:253–65

Blomqvist A, Zhang ET, Craig AD. 2000. Cytoarchitectonic and immunohistochemical characterization of a specific pain and temperature relay, the posterior portion of the ventral medial nucleus, in the human thalamus. *Brain* 123:601–19

Bourgeais L, Monconduit L, Villanueva L, Bernard JF. 2001. Parabrachial internal lateral neurons convey nociceptive messages from the deep laminas of the dorsal horn to the intralaminar thalamus. *J. Neurosci.* 21:2159–65

Brodows RG, Pi S, Campbell RG. 1975. Sympathetic control of hepatic glycogenolysis during glucopenia in man. *Metabolism* 24:617–24

Burstein R, Falkowsky O, Borsook D, Strassman A. 1996. Distinct lateral and medial projections of the spinohypothalamic tract of the rat. *J. Comp. Neurol.* 373:549–74

Burstein R, Potrebic S. 1993. Retrograde labeling of neurons in the spinal cord that project directly to the amygdala or the orbital cortex in the rat. *J. Comp. Neurol.* 335:469–85

Cannon WB. 1929. *Bodily Changes in Pain, Hunger, Fear and Rage.* New York: Littleton. 404 pp.

Cannon WB, Britton SW. 1927. Pseudoaffective medulliadrenal secretion. *Am. J. Physiol.* 79:433–65

Cano G, Sved AF, Rinaman L, Rabin BS, Card JP. 2001. Characterization of the central nervous system innervation of the rat spleen using viral transneuronal tracing. *J. Comp. Neurol.* 439:1–18

Card JP, Rinaman L, Lynn RB, Lee BH, Meade RP, et al. 1993. Pseudorabies virus infection of the rat central nervous system: ultrastructural characterization of viral replication, transport, and pathogenesis. *J. Neurosci.* 13:2515–39

Carrive P, Bandler R. 1991. Viscerotopic organization of neurons subserving hypotensive reactions within the midbrain periaqueductal grey: a correlative functional and anatomical study. *Brain Res.* 541:206–15

Casey KL, Minoshima S, Morrow TJ, Koeppe RA. 1996. Comparison of human cerebral activation pattern during cutaneous warmth, heat pain, and deep cold pain. *J. Neurophysiol.* 76:571–81

Casey KL, Morrow TJ, Lorenz J, Minoshima S. 2001. Temporal and spatial dynamics of human forebrain activity during heat pain: analysis by positron emission tomography. *J. Neurophysiol.* 85:951–59

Cechetto DF, Saper CB. 1987. Evidence for a viscerotopic sensory representation in the cortex and thalamus in the rat. *J. Comp. Neurol.* 262:27–45

Cechetto DF, Saper CB. 1988. Neurochemical organization of the hypothalamic projection to the spinal cord in the rat. *J. Comp. Neurol.* 272:579–604

Cechetto DF, Saper CB. 1990. Role of the cerbral cortex in autonomic function. In *Central Regulation of Autonomic Functions*, ed. AD Loewy, KD Spyer, pp. 208–23. New York: Oxford Univ. Press

Cechetto DF, Standaert DG, Saper CB. 1985. Spinal and trigeminal dorsal horn projections to the parabrachial nucleus in the rat. *J. Comp. Neurol.* 240:153–60

Chamberlin NL, Saper CB. 1994. Topographic organization of respiratory responses to glutamate microstimulation of the parabrachial nucleus in the rat. *J. Neurosci.* 14:6500–10

Chandler MJ, Zhang J, Qin C, Yuan Y, Foreman RD. 2000. Intrapericardiac injections of algogenic chemicals excite primate C1-C2 spinothalamic tract neurons. *Am. J. Physiol Regul. Integr. Comp. Physiol.* 279:R560–68

Chen XM, Hosono T, Yoda T, Fukuda Y, Kanosue K. 1998. Efferent projection from the preoptic area for the control of non-shivering thermogenesis in rats. *J. Physiol.* 512:883–92

Ciriello J. 1983. Brainstem projections of aortic baroreceptor afferent fibers in the rat. *Neurosci. Lett.* 36:37–42

Cliffer KD, Burstein R, Giesler GJ Jr. 1991. Distributions of spinothalamic, spinohypothalamic, and spinotelencephalic fibers revealed by anterograde transport of PHA-L in rats. *J. Neurosci.* 11:852–68

Combarros O, Sanchez-Juan P, Berciano J, De Pablos C. 2000. Hemiageusia from an ipsilateral multiple sclerosis plaque at the midpontine tegmentum. *J. Neurol. Neurosurg. Psychiatry* 68:796

Contreras RJ, Beckstead RM, Norgren R. 1982. The central projections of the trigeminal, facial, glossopharyngeal and vagus nerves: an autoradiographic study in the rat. *Brain Res. Bull.* 6:303–22

Craig AD, Chen K, Bandy D, Reiman EM. 2000. Thermosensory activation of insular cortex. *Nat. Neurosci.* 3:184–90

Craig AD, Dostrovsky JO. 2001. Differential projections of thermoreceptive and nociceptive lamina I trigeminothalamic and spinothalamic neurons in the cat. *J. Neurophysiol.* 86:856–70

Darwin C. 1873. *The Expression of the Emotions in Man and Animals.* New York: Appleton. 374 pp.

de Lacalle S, Saper CB. 2000. Calcitonin gene-related peptide-like immunoreactivity marks putative visceral sensory pathways in human brain. *Neuroscience* 100:115–30

Dell P, Olson R. 1951. Projections thalamiques corticales et cérébelleuses des afférences viscérales vagales. *C. R. Soc. Seances Soc. Biol.* 145:1084–88

Dusser de Barenne JG. 1920. Recherches expérimentales sur les fonctions du système nerveux central, faites en particulier sur deux chats donc le néopallium a eté enlevé. *Arch. Neurol. Physiol.* 4:31–123

Elias CF, Kelly JF, Lee CE, Ahima RS, Drucker DJ, et al. 2000. Chemical characterization of leptin-activated neurons in the rat brain. *J. Comp. Neurol.* 423:261–81

Elias CF, Lee C, Kelly J, Aschkenasi C, Ahima RS, et al. 1998. Leptin activates hypothalamic CART neurons projecting to the spinal cord. *Neuron* 21:1375–85

Elliott JM, Kapoor V, Cain M, West MJ, Chalmers JP. 1985. The mechanism of hypertension and bradycardia following lesions of the caudal ventrolateral medulla in the rabbit: the role of sympathetic nerves, circulating adrenaline, vasopressin and renin. *Clin. Exp. Hypertens. A* 7:1059–82

Elmquist JK, Ahima RS, Maratos-Flier E, Flier JS, Saper CB. 1997. Leptin activates neurons in ventrobasal hypothalamus and brainstem. *Endocrinology* 138:839–42

Elmquist JK, Scammell TE, Jacobson CD, Saper CB. 1996. Distribution of Fos-like immunoreactivity in the rat brain following intravenous lipopolysaccharide administration. *J. Comp. Neurol.* 371:85–103

Feil K, Herbert H. 1995. Topographic organization of spinal and trigeminal somatosensory pathways to the rat parabrachial and Kolliker-Fuse nuclei. *J. Comp. Neurol.* 353:506–28

Flynn FW, Grill HJ, Schulkin J, Norgren R. 1991. Central gustatory lesions: II. Effects on sodium appetite, taste aversion learning, and feeding behaviors. *Behav. Neurosci.* 105:944–54

Foreman RD. 1999. Mechanisms of cardiac pain. *Annu. Rev. Physiol.* 61:143–67

Fulwiler CE, Saper CB. 1984. Subnuclear organization of the efferent connections of the parabrachial nucleus in the rat. *Brain Res. Rev.* 7:229–59

Gamboa-Esteves FO, Kaye JC, McWilliam PN, Lima D, Batten TF. 2001b. Immunohistochemical profiles of spinal lamina I neurones retrogradely labelled from the nucleus tractus solitarii in rat suggest excitatory projections. *Neuroscience* 104:523–38

Gamboa-Esteves FO, Tavares I, Almeida A, Batten TF, McWilliam PN, Lima D. 2001a. Projection sites of superficial and deep spinal dorsal horn cells in the nucleus tractus solitarii of the rat. *Brain Res.* 921:195–205

Gharbi N, Somody L, El Fazaa S, Kamoun A, Gauquelin-Koch G, Gharib C. 1999. Tissue norepinephrine turnover and cardiovascular responses during intermittent dehydration in the rat. *Life Sci.* 64:2401–10

Giannantoni A, Di Stasi SM, Scivoletto G, Mollo A, Silecchia A, et al. 1998. Autonomic dysreflexia during urodynamics. *Spinal Cord* 36:756–60

Gloor P, Olivier A, Quesney LF, Andermann F, Horowitz S. 1982. The role of the limbic system in experiential phenomena of temporal lobe epilepsy. *Ann. Neurol.* 12:129–44

Grucza R, Lecroart JL, Carette G, Hauser JJ, Houdas Y. 1987. Effect of voluntary dehydration on thermoregulatory responses to heat in men and women. *Eur. J. Appl. Physiol. Occup. Physiol.* 56:317–22

Hallbeck M, Blomqvist A. 1999. Spinal cord-projecting vasopressinergic neurons in the rat paraventricular hypothalamus. *J. Comp. Neurol.* 411:201–11

Hallbeck M, Larhammar D, Blomqvist A. 2001. Neuropeptide expression in rat paraventricular hypothalamic neurons that project to the spinal cord. *J. Comp. Neurol.* 433:222–38

Hamann A, Flier JS, Lowell BB. 1998. Obesity after genetic ablation of brown adipose tissue. *Z. Ernährungswiss.* 37(Suppl. 1):1–7

Harper RM, Bandler R, Spriggs D, Alger JR.

2000. Lateralized and widespread brain activation during transient blood pressure elevation revealed by magnetic resonance imaging. *J. Comp. Neurol.* 417:195–204

Herbert H, Moga MM, Saper CB. 1990. Connections of the parabrachial nucleus with the nucleus of the solitary tract and the medullary reticular formation in the rat. *J. Comp. Neurol.* 293:540–80

Herrick CJ. 1905. The central gustatory paths in the brains of bony fishes. *J. Comp. Neurol.* 15:375–456

Hilton SM, Spyer KM. 1980. Central nervous regulation of vascular resistance. *Annu. Rev. Physiol.* 42:399–441

Horie H, Yokota T. 1990. Responses of nociceptive VPL neurons to intracardiac injection of bradykinin in the cat. *Brain Res.* 516:161–64

Horn T, Wilkinson MF, Landgraf R, Pittman QJ. 1994. Reduced febrile responses to pyrogens after lesions of the hypothalamic paraventricular nucleus. *Am. J. Physiol.* 267:R323–28

Huang J, Weiss ML. 1999. Characterization of the central cell groups regulating the kidney in the rat. *Brain Res.* 845:77–91

Hylden JL, Anton F, Nahin RL. 1989. Spinal lamina I projection neurons in the rat: collateral innervation of parabrachial area and thalamus. *Neuroscience* 28:27–37

Ito S, Ogawa H. 1991. Cytochrome oxidase staining facilitates unequivocal visualization of the primary gustatory area in the fronto-operculo-insular cortex of macaque monkeys. *Neurosci. Lett.* 130:61–64

Ito SI, Ohgushi M, Ifuku H, Ogawa H. 2001. Neuronal activity in the monkey fronto-opercular and adjacent insular/prefrontal cortices during a taste discrimination GO/NOGO task: response to cues. *Neurosci. Res.* 41:257–66

James W. 1884. What is emotion? *Mind* 19:188–205

Janig W, McLachlan EM. 1992. Specialized functional pathways are the building blocks of the autonomic nervous system. *J. Auton. Nerv. Syst.* 41:3–13

Jansen AS, Hoffman JL, Loewy AD. 1997. CNS sites involved in sympathetic and parasympathetic control of the pancreas: a viral tracing study. *Brain Res.* 766:29–38

Jansen AS, Nguyen XV, Karpitskiy V, Mettenleiter TC, Loewy AD. 1995. Central command neurons of the sympathetic nervous system: basis of the fight-or-flight response. *Science* 270:644–46

Johnson CD, Gilbey MP. 1994. Sympathetic activity recorded from the rat caudal ventral artery in vivo. *J. Physiol.* 476:437–42

Juler GL, Eltorai IM. 1985. The acute abdomen in spinal cord injury patients. *Paraplegia* 23:118–23

Kalia M, Mei SS, Kao FF. 1981. Central projections from ergoreceptors (C fibers) in muscle involved in cardiopulmonary responses to static exercise. *Circ. Res.* 48:I48–62

Kalia M, Richter D. 1988. Rapidly adapting pulmonary receptor afferents: I. arborization in the nucleus of the tractus solitarius. *J. Comp. Neurol.* 274:560–73

Keay KA, Li QF, Bandler R. 2000. Muscle pain activates a direct projection from ventrolateral periaqueductal gray to rostral ventrolateral medulla in rats. *Neurosci. Lett.* 290:157–60

Kerman IA, Yates BJ, McAllen RM. 2000. Anatomic patterning in the expression of vestibulosympathetic reflexes. *Am. J. Physiol. Regul. Integr. Comp. Physiol.* 279:R109–17

Kerr FWL. 1961. Facial, vagal, and glossopharyngeal nerves in the cat. Afferent connections. *Arch. Neurol.* 6:264–81

King AB, Menon RS, Hachinski V, Cechetto DF. 1999. Human forebrain activation by visceral stimuli. *J. Comp. Neurol.* 413:572–82

Knuepfer MM, Schramm LP. 1985. Properties of renobulbar afferent fibers in rats. *Am. J. Physiol.* 248:R113–19

Kobayashi Y. 1998. Distribution and morphology of spinothalamic tract neurons in the rat. *Anat. Embryol.* 197:51–67

Kojima Y, Hirano T. 1999. A case of gustatory disturbance caused by ipsilateral pontine hemorrhage. *Rinsho Shinkeigaku* 39:979–81

Kosar E, Grill HJ, Norgren R. 1986a. Gustatory cortex in the rat. I. Physiological properties and cytoarchitecture. *Brain Res.* 379:329–41

Kosar E, Grill HJ, Norgren R. 1986b. Gustatory cortex in the rat. II. Thalamocortical projections. *Brain Res.* 379:342–52

Kuo DC, Nadelhaft I, Hisamitsu T, De Groat WC. 1983. Segmental distribution and central projections of renal afferent fibers in the cat studied by transganglionic transport of horseradish peroxidase. *J. Comp. Neurol.* 216:162–74

Koyama N, Nishikawa Y, Yokota T. 1998. Distribution of nociceptive neurons in the ventrobasal complex of macaque thalamus. *Neurosci. Res.* 31:39–51

Lange KG. 1885. *Om Sindsbevagelser, et Psyko-Fysiologisk Studie.* Kjobenhavn: Lund. 91 pp.

Lenz FA, Gracely RH, Zirh TA, Leopold DA, Rowland LH, Dougherty PM. 1997. Human thalamic nucleus mediating taste and multiple other sensations related to ingestive behavior. *J. Neurophysiol.* 77:3406–9

Loewy AD, Spyer KM. 1990. *Central Regulation of Autonomic Functions.* New York: Oxford Univ. Press. 390 pp.

Lowell BB, Flier JS. 1997. Brown adipose tissue, beta 3-adrenergic receptors, and obesity. *Annu. Rev. Med.* 48:307–16

Medvedev OS, Delle M, Thoren P. 1988. 2-Deoxy-D-glucose-induced central glycopenia differentially influences renal and adrenal nerve activity in awake SHR rats. *Clin. Exp. Hypertens. A* 10(Suppl. 1):375–81

Mehler WR, Feferman ME, Nauta WJH. 1960. Ascending axon degeneration following anterolateral cordotomy. An experimental study in the monkey. *Brain* 83:718–50

Menetrey D, Basbaum AI. 1987. Spinal and trigeminal projections to the nucleus of the solitary tract: a possible substrate for somatovisceral and viscerovisceral reflex activation. *J. Comp. Neurol.* 255:439–50

Moga MM, Herbert H, Hurley KM, Yasui Y, Gray TS, Saper CB. 1990. Organization of cortical, basal forebrain, and hypothalamic afferents to the parabrachial nucleus in the rat. *J. Comp. Neurol.* 295:624–61

Morgan C, Nadelhaft I, De Groat WC. 1981. The distribution of visceral primary afferents from the pelvic nerve to Lissauer's tract and the spinal gray matter and its relationship to the sacral parasympathetic nucleus. *J. Comp. Neurol.* 201:415–40

Morrison SF. 2001. Differential control of sympathetic outflow. *Am. J. Physiol Regul. Integr. Comp. Physiol.* 281:R683–98

Morrison SF, Sved AF, Passerin AM. 1999. GABA-mediated inhibition of raphe pallidus neurons regulates sympathetic outflow to brown adipose tissue. *Am. J. Physiol. Regul. Integr. Comp. Physiol.* 276:R290–97

Nakajima Y, Utsumi H, Takahashi H. 1983. Ipsilateral disturbance of taste due to pontine haemorrhage. *J. Neurol.* 229:133–36

Niijima A. 1975. The effect of 2-deoxy-D-glucose and D-glucose on the efferent discharge rate of sympathetic nerves. *J. Physiol.* 251:231–43

Norgren R. 1978. Projections from the nucleus of the solitary tract in the rat. *Neuroscience* 3:207–18

Norgren R. 1984. Central neural mechanisms of taste. In *Handbook of Physiology, Section I: the Nervous System, Vol. III, Sensory Processes*, ed. ID Smith, pp. 1087–128. Bethesda: Am. Physiol. Soc.

Ogawa H, Ito S, Murayama N, Hasegawa K. 1990. Taste area in the granular and dysgranular insular cortices in the rat identified by stimulation of the entire oral cavity. *Neurosci. Res.* 9:196–201

Ongur D, Price JL. 2000. The organization of networks within the orbital and medial prefrontal cortex of rats, monkeys and humans. *Cereb. Cortex* 10:206–19

Onoda K, Ikeda M. 1999. Gustatory disturbance due to cerebrovascular disorder. *Laryngoscope* 109:123–28

Otake K, Reis DJ, Ruggiero DA. 1994. Afferents to the midline thalamus issue collaterals to the nucleus tractus solitarii: an anatomical basis for thalamic and visceral reflex integration. *J. Neurosci.* 14:5694–707

Panneton WM, Burton H. 1985. Projections from the paratrigeminal nucleus and the medullary and spinal dorsal horns to the peribrachial area in the cat. *Neuroscience* 15:779–97

Panneton WM, Loewy AD. 1980. Projections of the carotid sinus nerve to the nucleus of the solitary tract in the cat. *Brain Res.* 191:239–44

Penfield W, Faulk ME Jr. 1955. The insula: further observations on its function. *Brain* 78:445–70

Penfield W, Rasmussen T. 1950. *The Cerebral Cortex of Man*, pp. 78–79. New York: MacMillan.

Pritchard TC, Hamilton RB, Morse JR, Norgren R. 1986. Projections of thalamic gustatory and lingual areas in the monkey, *Macaca fascicularis*. *J. Comp. Neurol.* 244:213–28

Radna RJ, MacLean PD. 1981. Vagal elicitation of respiratory-type and other unit responses in basal limbic structures of squirrel monkeys. *Brain Res.* 213:45–61

Rathner JA, McAllen RM. 1999. Differential control of sympathetic drive to the rat tail artery and kidney by medullary premotor cell groups. *Brain Res.* 834:196–99

Rathner JA, Owens NC, McAllen RM. 2001. Cold-activated raphe-spinal neurons in rats. *J. Physiol.* 535:841–54

Ricardo JA, Koh ET. 1978. Anatomical evidence of direct projections from the nucleus of the solitary tract to the hypothalamus, amygdala, and other forebrain structures in the rat. *Brain Res.* 153:1–26

Roppolo JR, Nadelhaft I, De Groat WC. 1985. The organization of pudendal motoneurons and primary afferent projections in the spinal cord of the rhesus monkey revealed by horseradish peroxidase. *J. Comp. Neurol.* 234:475–88

Ross CA, Ruggiero DA, Reis DJ. 1985. Projections from the nucleus tractus solitarii to the rostral ventrolateral medulla. *J. Comp. Neurol.* 242:511–34

Rotto-Percelay DM, Wheeler JG, Osorio FA, Platt KB, Loewy AD. 1992. Transneuronal labeling of spinal interneurons and sympathetic preganglionic neurons after pseudorabies virus injections in the rat medial gastrocnemius muscle. *Brain Res.* 574:291–306

Sacca L, Perez G, Carteni G, Rengo F. 1977. Evaluation of the role of the sympathetic nervous system in the glucoregulatory response to insulin-induced hypoglycemia in the rat. *Endocrinology* 101:1016–22

Saper CB. 1982a. Convergence of autonomic and limbic connections in the insular cortex of the rat. *J. Comp. Neurol.* 210:163–73

Saper CB. 1982b. Reciprocal parabrachial-cortical projections in the rat. *Brain Res.* 242:33–40

Saper CB. 1985. Organization of cerebral cortical afferent systems in the rat. II. Hypothalamocortical projections. *J. Comp. Neurol.* 237:21–46

Saper CB. 1995. The central autonomic system. In *The Rat Nervous System*, ed. G Paxinos, pp. 155–85. San Diego: Academic

Saper CB. 2000. Pain as a visceral sensation. *Prog. Brain Res.* 122:237–43

Saper CB, Loewy AD, Swanson LW, Cowan WM. 1976. Direct hypothalamo-autonomic connections. *Brain Res.* 117:305–12

Sato A, Sato Y, Schmidt RF. 1997. The impact of somatosensory input on autonomic functions. *Rev. Physiol. Biochem. Pharmacol.* 130:1–328

Sato A, Schmidt RF. 1973. Somatosympathetic reflexes: afferent fibers, central pathways, discharge characteristics. *Physiol. Rev.* 53:916–47

Sato K, Nitta E. 2000. A case of ipsilateral ageusia, sensorineural hearing loss and facial sensorimotor disturbance due to pontine lesion. *Rinsho Shinkeigaku* 40:487–89

Schadt JC, Hasser EM. 2001. Defense reaction alters the response to blood loss in the conscious rabbit. *Am. J. Physiol. Regul. Integr. Comp. Physiol.* 280:R985–93

Schramm LP, Strack AM, Platt KB, Loewy AD. 1993. Peripheral and central pathways regulating the kidney: a study using pseudorabies virus. *Brain Res.* 616:251–62

Scott TR, Plata-Salaman CR. 1999. Taste in the monkey cortex. *Physiol. Behav.* 67:489–511

Scott TR, Yaxley S, Sienkiewicz ZJ, Rolls ET.

1986. Gustatory responses in the frontal opercular cortex of the alert cynomolgus monkey. *J. Neurophysiol.* 56:876–90

Selye H. 1975. Homeostasis and the reaction to stress: a discussion of Walter B. Cannon's contributions. In *The Life and Contributions of Walter Bradford Cannon 1871–1945*, ed. CM Brooks, K Koizumi, JO Pinkston, pp. 89–107. Brooklyn: State Univ. N.Y.

Shimoda O, Ikuta Y, Nishi M, Uneda C. 1998. Magnitude of skin vasomotor reflex represents the intensity of nociception under general anesthesia. *J. Auton. Nerv. Syst.* 71:183–89

Silver JR. 2000. Early autonomic dysreflexia. *Spinal Cord* 38:229–33

Smith JE, Jansen AS, Gilbey MP, Loewy AD. 1998. CNS cell groups projecting to sympathetic outflow of tail artery: neural circuits involved in heat loss in the rat. *Brain Res.* 786:153–64

Strack AM, Loewy AD. 1990. Pseudorabies virus: a highly specific transneuronal cell body marker in the sympathetic nervous system. *J. Neurosci.* 10:2139–47

Strack AM, Sawyer WB, Hughes JH, Platt KB, Loewy AD. 1989. A general pattern of CNS innervation of the sympathetic outflow demonstrated by transneuronal pseudorabies viral infections. *Brain Res.* 491:156–62

Strauther GR, Longo WE, Virgo KS, Johnson FE. 1999. Appendicitis in patients with previous spinal cord injury. *Am. J. Surg.* 178:403–5

Sugiura Y, Terui N, Hosoya Y. 1989. Difference in distribution of central terminals between visceral and somatic unmyelinated (C) primary afferent fibers. *J. Neurophysiol.* 62:834–37

Sugiura Y, Terui N, Hosoya Y, Tonosaki Y, Nishiyama K, Honda T. 1993. Quantitative analysis of central terminal projections of visceral and somatic unmyelinated (C) primary afferent fibers in the guinea pig. *J. Comp. Neurol.* 332:315–25

Sved AF, Cano G, Card JP. 2001. Neuroanatomical specificity of the circuits controlling sympathetic outflow to different targets. *Clin. Exp. Pharmacol. Physiol.* 28:115–19

Swanson LW, Sawchenko PE. 1983. Hypothalamic integration: organization of the paraventricular and supraoptic nuclei. *Annu. Rev. Neurosci.* 6:269–324

Tucker DC, Saper CB. 1985. Specificity of spinal projections from hypothalamic and brainstem areas which innervate sympathetic preganglionic neurons. *Brain Res.* 360:159–64

van den Pol AN. 1999. Hypothalamic hypocretin (orexin): robust innervation of the spinal cord. *J. Neurosci.* 19:3171–82

Weiss ML, Chowdhury SI. 1998. The renal afferent pathways in the rat: a pseudorabies virus study. *Brain Res.* 812:227–41

Willis WD, Al-Chaer ED, Quast MJ, Westlund KN. 1999. A visceral pain pathway in the dorsal column of the spinal cord. *Proc. Natl. Acad. Sci. USA* 96:7675–79

Woulfe JM, Flumerfelt BA, Hrycyshyn AW. 1990. Efferent connections of the A1 noradrenergic cell group: a DBH immunohistochemical and PHA-L anterograde tracing study. *Exp. Neurol.* 109:308–22

Yamaguchi S, Ito M, Ohshima N. 2001. Somatosensory nociceptive mechanical stimulation modulates systemic and mesenteric microvascular hemodynamics in anesthetized rats. *Auton. Neurosci.* 88:160–66

Yamamoto T, Matsuo R, Kawamura Y. 1980. Localization of cortical gustatory area in rats and its role in taste discrimination. *J. Neurophysiol.* 44:440–55

Yardley CP, Hilton SM. 1987. Vasodilatation in hind-limb skeletal muscle evoked as part of the defence reaction in the rat. *J. Auton. Nerv. Syst.* 19:127–36

Yasui Y, Saper CB, Cechetto DF. 1989. Calcitonin gene-related peptide immunoreactivity in the visceral sensory cortex, thalamus, and related pathways in the rat. *J. Comp. Neurol.* 290:487–501

Yaxley S, Rolls ET, Sienkiewicz ZJ. 1990. Gustatory responses of single neurons in the insula of the macaque monkey. *J. Neurophysiol.* 63:689–700

Zhang ET, Craig AD. 1997. Morphology and

distribution of spinothalamic lamina I neurons in the monkey. *J. Neurosci.* 17:3274–84

Zhang SP, Bandler R, Carrive P. 1990. Flight and immobility evoked by excitatory amino acid microinjection within distinct parts of the subtentorial midbrain periaqueductal gray of the cat. *Brain Res.* 520:73–82

Zhang YH, Hosono T, Yanase-Fujiwara M, Chen XM, Kanosue K. 1997. Effect of midbrain stimulations on thermoregulatory vasomotor responses in rats. *J. Physiol.* 503:177–86

Zhang YH, Lu J, Elmquist JK, Saper CB. 2000. Lipopolysaccharide activates specific populations of hypothalamic and brainstem neurons that project to the spinal cord. *J. Neurosci.* 20:6578–86

THE ROLE OF NOTCH IN PROMOTING GLIAL AND NEURAL STEM CELL FATES

Nicholas Gaiano[1,2] and Gord Fishell[1]

[1]Developmental Genetics Program, and Department of Cell Biology, Skirball Institute of Biomolecular Medicine, New York University School of Medicine, New York, NY 10016; email: fishell@saturn.med.nyu.edu
[2]Current address: Departments of Neurology and Neurosciences, Institute for Cell Engineering, Johns Hopkins University School of Medicine, Baltimore, Maryland 21287; email: ngaiano1@jhmi.edu

Key Words cell-fate specification, gliogenesis, radial glia, *Drosophila*, vertebrate

■ **Abstract** The Notch signaling pathway has long been known to influence cell fate in the developing nervous system. However, this pathway has generally been thought to inhibit the specification of certain cell types in favor of others, or to simply maintain a progenitor pool. Recently, this view has been challenged by numerous studies suggesting that Notch may play an instructive role in promoting glial development. This work has inspired a new look at the role of Notch signaling in specifying cell fate. It has also prompted further consideration of the emerging view that in some contexts glia may be multipotent progenitors. This review examines the role of Notch during gliogenesis in both fruit flies and vertebrates, as well as evidence in vertebrates that some glia may be stem cells.

INTRODUCTION

The Notch/Delta signaling pathway is highly conserved across species and is widely used during both vertebrate and invertebrate development to regulate cell fate in the developing embryo (Artavanis-Tsakonas et al. 1999, Lewis 1998). The Notch family of proteins are cell-surface receptors that are activated by the ligands Delta and Jagged (Serrate is the fly homologue of Jagged). Both the receptors and ligands are single-pass transmembrane proteins, which suggests that signaling through the Notch receptor requires cell-cell contact. Upon ligand binding the intracellular portion of the Notch receptor is cleaved and enters the nucleus, where it influences the expression of numerous transcription factors. A great deal of research has been focused on understanding how the Notch signal is transmitted and regulated on the molecular level (Mumm & Kopan 2000, Weinmaster 1997). In addition, many studies have examined the role of the Notch pathway during the development of nonneural tissues, such as the somites (Barrantes et al. 1999, Jiang et al. 2000), the limbs (Irvine & Vogt 1997, Vargesson et al. 1998), and the immune system (Anderson et al. 2001, Osborne & Miele 1999). For the purposes of this review, however, these issues are not considered further, and instead the role of Notch signaling on a cellular level during neural development is discussed.

Much of the initial understanding of the Notch pathway came from studies in worms and flies. In these systems, Notch was found to influence fate choices between cells with equivalent developmental potential. For example, in the worm vulva, although two cells have the potential to become a specialized cell type called the anchor cell, the Notch homologue lin-12 ensures that only one does, while the other becomes a ventral uterine precursor (Seydoux & Greenwald 1989, Sternberg & Horvitz 1989). Similarly, in flies, Notch influences the fate of cells in both the central nervous system (CNS) and peripheral nervous system (PNS). For instance, Notch is used to identify a single cell, among a small cluster of equivalent cells, to become either a neuroblast (also called neuroglioblast) in the CNS (Artavanis-Tsakonas et al. 1991), or a sensory organ precursor (SOP) in the PNS (Furukawa et al. 1992, Schweisguth 1995, Schweisguth & Posakony 1992). Subsequently, Notch signaling is used to regulate the acquisition of distinct fates by the daughter cells of neuroblasts and SOPs (Guo et al. 1996). Thus, in both the worm vulva and fly nervous system, Notch influences the decision between alternative cell fates.

The mechanism by which neuroblasts and SOPs are specified in the fly ectoderm has strongly influenced our view of the role of Notch in the vertebrate CNS. During vertebrate neural development, the canonical view has been that Notch signaling is used to maintain a pool of uncommitted precursors, while a subset of cells are selected to leave this pool and differentiate into neurons (Lewis 1996). This balance between progenitor maintenance and neuronal differentiation allows the continuous generation of neurons throughout development and permits temporal control over the specification of distinct neuronal fates. The common theme between neuroblast selection in flies and the selection of cells to undergo neuronal differentiation in the progenitor pool of vertebrates is that Notch activity inhibits the surrounding cells from becoming the primary cell type being specified.

The apparent tendency of Notch to inhibit differentiation has suggested that this pathway is an indirect regulator of cell fate, rather than a direct or "instructive" regulator. The widespread use of Notch signaling to influence the specification of many different cell fates during development has further suggested that Notch is unlikely to instructively influence these fates. Recently, however, numerous studies in vertebrates have suggested that rather than simply inhibiting neuronal differentiation and maintaining a neural progenitor state, Notch may, in some contexts, promote the acquisition of glial identity (Furukawa et al. 2000, Gaiano et al. 2000, Hojo et al. 2000, Morrison et al. 2000, Scheer et al. 2001). This work has found that Notch can actively direct cells toward certain fates and thereby has called for a re-evaluation of Notch's potential role as an instructive signal during development.

In this review the role of Notch during gliogenesis in both flies and vertebrates is examined. Though it is clear that Notch signaling influences gliogenesis across species, there is no uniform instructive role for Notch during this process. Depending upon the context, Notch can either promote or inhibit gliogenesis. It is interesting that in some cases Notch signaling in vertebrates promotes glial cell types that may retain progenitor character. This work suggests that, in certain contexts, Notch can maintain progenitor identity, consistent with the traditional view but that these progenitors acquire glial characteristics. In addition to an overview of the role of Notch during gliogenesis, the evidence is discussed that certain glial cell types, which are promoted by Notch signaling (i.e., radial glia, astrocytes, Müller glia), may be multipotent progenitors.

NOTCH DURING GLIOGENESIS IN FLIES

In the fruit fly, CNS and PNS progenitors are ectodermally derived (Campos-Ortega 1993, Jan & Jan 1994, Modolell 1997) (Figure 1a). The expression domains of both dorsal/ventral and anterior/posterior patterning genes define groups of 4–7 cells termed proneural clusters (Bhat 1999, Jan & Jan 1994). These cells express the proneural genes of the *achaete-scute* complex and *atonal*, which encode basic helix loop helix transcription factors that impart a neural ground state to ectodermal cells (Campos-Ortega 1993, Modolell 1997, Skeath & Doe 1996). Although all cells in a proneural cluster have the potential to give rise to neural cell types, normally only one cell in each is specified to become a neuroblast in the CNS, or an SOP in the PNS. This specification is achieved through cell-cell signaling, and is mediated by the Notch pathway. The cell that becomes the neuroblast or the SOP expresses the highest levels of the Notch ligand Delta, thus activating Notch in the surrounding cells, inhibiting their differentiation into neuroblasts or SOPs. This inhibition is achieved, at least in part, through the downregulation of proneural genes by Notch activity. In the absence of Notch signaling, all of the cells in a cluster continue to express proneural genes and become neuroblasts or SOPs, a circumstance that leads to the neurogenic phenotype characteristic of mutations in the Notch pathway (Jan & Jan 1994, Knust & Campos-Ortega 1989, Xu et al. 1990). The selection of neural progenitors is only the first of Notch's roles in generating neural cell diversity in the fly. Notch signaling is also used to regulate binary fate choices during the expansion of the neuroblast and SOP lineages (Doe & Skeath 1996, Guo et al. 1996). Among the many fate choices influenced by Notch in the fly nervous system, the role of this pathway during glial specification is focused on below.

The Fly CNS

Notch has been found to influence the generation of CNS glia in flies at several stages. First, as described above, the specification of a single neuroblast in each proneural cluster is controlled by Notch (Figure 1a). In specific CNS clusters, the cell that delaminates will give rise only to glia and is therefore called a glioblast (Jacobs et al. 1989). In Notch mutants, the inability to select a single "blast" cell from each proneural cluster results in the generation of supernumerary glioblasts. Consequently, extra glia are generated, which suggests that Notch represses glial

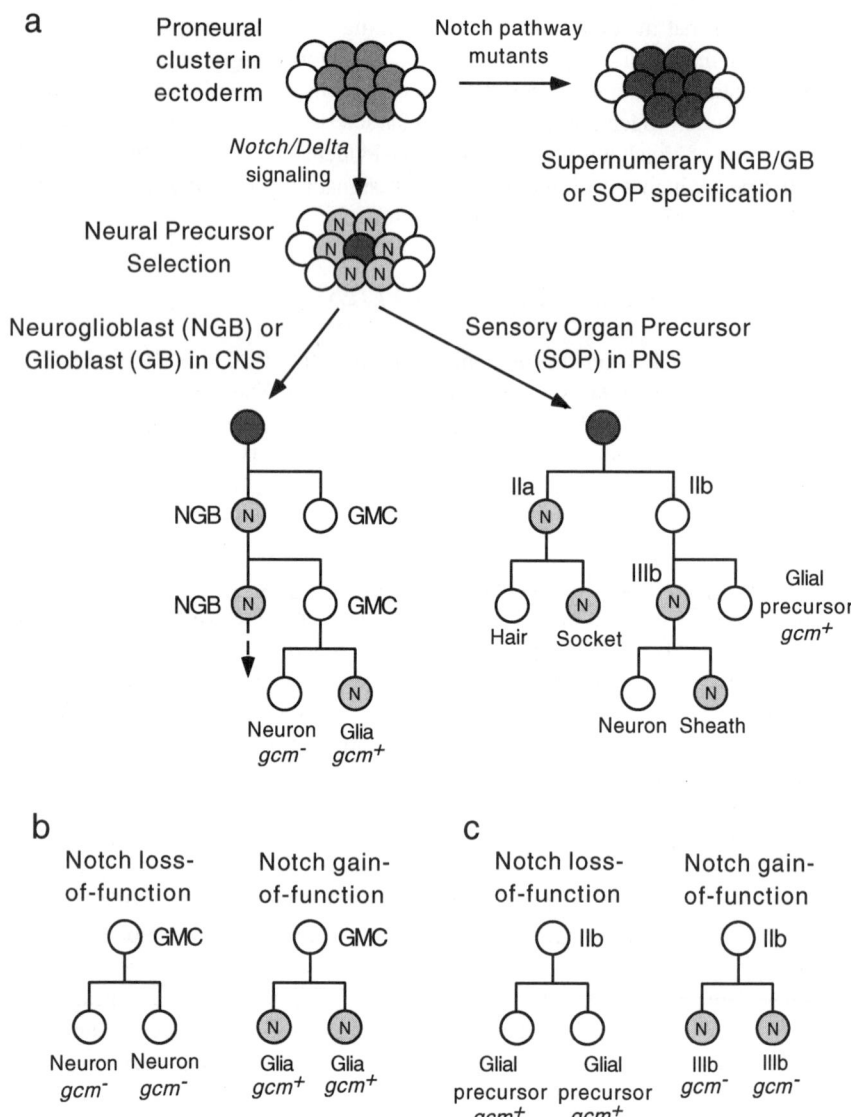

Figure 1 Role of Notch during cell-fate specification in the fly CNS and PNS. (*a*) Selection of neural precursors (*dark gray*) from proneural clusters (*medium gray*) using Notch signaling. Cells with "N" are receiving Notch signaling (*light gray*). *Left*, in the CNS the selected cell undergoes multiple rounds of self-renewing division, while also generating ganglion mother cells (GMCs). *Right*, in the PNS the SOP produces numerous cell types including a glial precursor. (*b*) Notch activity controls the GMC neuronal-glial fate choice through *glial cells missing* (*gcm*) (see Udolph et al. 2001). (*c*) Notch activity controls the IIIb-glial precursor fate choice through *gcm* (see Van De Bor & Giangrande 2001).

fate in this context. Strictly speaking this is true, but the role of Notch with respect to specifying glia is not specific because Notch limits the number of glioblasts or neuroblasts, depending on the proneural cluster, via the same mechanism. In support of a nonspecific function of Notch, it has also been found that Notch activity limits the delamination of the nonneural oenocyte precursors from ectoderm in the same manner (Hartenstein et al. 1992).

Recent work in the CNS has uncovered a more specific role for Notch during the decision to acquire glial identity (Udolph et al. 2001). After delamination from the ectoderm, most neuroblasts undergo a series of self-renewing divisions that also produce cells called ganglion mother cells (GMCs) (Figure 1a, *left*). GMCs divide again and typically give rise to one neuronal daughter and one glial daughter. In a recent study, Udolph et al. (2001) examined the role of Notch during this neuronal-glial fate choice and found Notch activity to be essential for this asymmetric division (Figure 1b). Specifically, this work examined the generation of the subperineurial glia (SPG) and showed that the loss of Notch function led to SPG loss and a concomitant increase in neuron number in this lineage.

To determine if Notch is sufficient to specify SPG fate, the authors examined the effect of increased Notch function, using either a mutation in Numb (a negative regulator of Notch) or by expression of a constitutively active form of Notch (ActN). Consistent with the loss-of-function result, increased Notch activity led to additional glia at the expense of neurons (Figure 1b). Both the loss-of-function and gain-of-function effects appeared to be mediated through the expression of *glial cells missing* (*gcm*), a gene believed to play an instructive role in the neuronal-glial fate choice throughout the embryo (Hosoya et al. 1995, Jones et al. 1995, Vincent et al. 1996). The authors examined the expression of *gcm* in the SPG lineage and found that *gcm* was normally expressed in the GMC daughter destined to become the SPG. In Notch mutants, *gcm* was not expressed in these cells (Figure 1b). In contrast, Notch activity in both daughters led to *gcm* expression in both, and extra glia at the expense of neurons. These data are consistent with the notion that Notch can play an instructive role during glial specification. However, it is important to note that in *gcm* mutants, activation of Notch did not promote SPG identity, which indicates that at least part of the "instructive" role of *Notch* in this context is mediated through its effect on *gcm* expression (Udolph et al. 2001).

The Fly PNS

The development of sensory organs in the fly PNS occurs in a stereotypic pattern after the specification of SOPs (Jan & Jan 1994) (Figure 1a, *right*). In the classically described lineage, each SOP divides into two cells, termed IIa and IIb. These cells each divide again to give rise to the hair and socket cells (IIa), and neuron and sheath cells (IIb). Analysis of the Notch pathway in the SOP lineage has proven quite useful for understanding the role of Notch during cell-fate specification. These studies utilized temperature-sensitive alleles, as well as mutations that either blocked or enhanced Notch signaling, to demonstrate that Notch has sequential roles during the expansion of the SOP lineage into four distinct daughter cells. Specifically, in the SOP division, Notch is needed for specification of the IIa cell, and in the second round of divisions Notch is needed for specification of the socket cell (from IIa) and the sheath cell (from IIb). Although Notch activity in the SOP lineage is clearly required to generate specific cellular fates, this work has not been widely interpreted to suggest that Notch plays an instructive role in this context. Depending on the timing of Notch activation, different cell fates are specified, which is inconsistent with the idea that Notch provides specific cell fate instruction.

More recently, the SOP lineage that gives rise to the mechanosensory bristle in the adult has been shown to include a fifth cell type, that of a glial precursor (GP) (Gho et al. 1999, Reddy & Rodrigues 1999, Van De Bor et al. 2000; Figure 1a, *right*). After the first SOP division to generate the IIa and IIb cells, IIb divides again to generate two cells, IIIb and a GP, which migrates away from the group and gives rise to numerous glia. IIIb then divides to give rise to the neuron and sheath cells formerly believed to be derived directly from IIb. The fact that the GP cell type has only recently been identified is likely a function of its migration away from the cluster, and the recent advent of time-lapse video microscopy.

Initial evidence that the Notch pathway might influence specification of the SOP GP came from the observation that Numb protein is segregated into this cell (Gho et al. 1999). As mentioned above, Numb antagonizes Notch signaling, which suggests that reduced Notch activity is required to acquire GP identity. Further work by Van De Bor & Giangrande (2001) has demonstrated this point convincingly. This study examined the effect of altering the Notch pathway in this lineage, using both loss-of-function and gain-of-function approaches. They found that a reduction in Notch signaling led to an increase in the number of GPs in SOP clusters, at the expense of other cell types, and that enhancing Notch activity led to a reduction in the number of GPs (Figure 1c).

Similar to the results described above for SPG development in the CNS, Notch was found to affect *gcm* expression in the SOP lineage (Van De Bor & Giangrande 2001). However, in contrast to the SPG lineage, where Notch positively regulated *gcm*, Notch negatively regulated this gene in the wing SOP. Therefore, in both lineages, the influence of Notch is mediated through regulation of *gcm* expression. Remarkably, however, Notch has opposite effects on *gcm* expression in these two contexts. This observation clearly suggests that attributing a uniform instructive function to Notch with respect to glial development in flies is not possible. Instead, Notch appears to behave in a context-dependent manner. The consequence of Notch activity is likely dictated by the developmental state of the cells in question, as well as other extrinsic cues they receive.

NOTCH DURING GLIOGENESIS IN VERTEBRATES

Over the course of the past decade four different Notch receptors, numerous forms of the ligands Delta and Jagged, and a variety of other Notch pathway members have been identified in vertebrates (Artavanis-Tsakonas et al. 1999, Lewis 1998,

Weinmaster 1997). During this time, it has become apparent that the Notch pathway functions in vertebrates in a manner similar to that observed in flies. Early evidence that Notch could inhibit neuronal differentiation in vertebrates came from in vitro work in the embryonic carcinoma cell line P19 cells. Under certain conditions these cells can be prompted to differentiate into neurons, astrocytes, or myoblasts. By introducing an active form of Notch (ActN) into P19 cells, Nye et al. (1994) showed that Notch could inhibit the differentiation of these cells into neurons and myoblasts. This inhibition was consistent with the action of Notch in invertebrates, as well as with an earlier study in frogs that suggested an inhibitory role for Notch in vertebrate cell-fate specification (Coffman et al. 1993). However, unexpectedly, in the P19 study ActN did not inhibit astrocyte differentiation. The authors concluded that glial fate, unlike the neuronal and myoblast fates, was refractory to Notch inhibition.

More recent work has suggested that the role of Notch during vertebrate gliogenesis is more complex than initially observed by Nye et al. (1994). Numerous studies have found that rather than simply not inhibiting gliogenesis, Notch may actively promote certain glial fates. Such fates include those of astrocytes, radial glia in the forebrain and cerebellum, Müller glia in the retina, and Schwann cells in the neural crest (each discussed below). It is interesting that in contrast, Notch has been found to inhibit oligodendrogliogenesis in the optic nerve. These studies suggest that, as in the fly nervous system, the role of Notch during vertebrate gliogenesis is not uniform.

Müller Glia

One of the first studies to suggest that Notch might actively influence glial fate in vertebrates was performed in the retina. Dorsky et al. (1995) found that the last cells to express Notch in the frog retina became Müller glia. The authors postulated that in retinal precursors Notch activation inhibited the acquisition of early born cell types, in favor of the later born Müller glia. When ActN was introduced into the retina however, even Müller glial fate was inhibited, and the ActN-expressing cells were quiescent with a neuroepithelial morphology. Taken together, these results suggested that prolonged expression of the Notch receptor led cells toward an eventual glial fate, but that the final acquisition of that fate required downregulation of Notch activity. In this case, Notch did not appear to be acting in an instructive manner, although the possibility that transient Notch activity specified a pre–Müller glial state, which could differentiate only after downregulation of Notch, could not be ruled out.

Subsequently, Bao et al. (1997) examined the role of Notch in the rat retina. In contrast to the frog study, however, this work found that cells expressing ActN were proliferatively active and acquired morphologies akin to Müller glia. More recent analysis by the same group has confirmed that ActN-expressing cells express Müller glial markers, and that expression of Hes-1, a downstream effector of Notch, can also promote Müller glial identity (Furukawa et al. 2000). The *Hes* genes are basic helix loop helix transcriptional repressors that may inhibit the activity of proneural transcription factors, such as Mash1 and Neurogenin2, which have recently been found to promote neurogenesis and block gliogenesis (Nieto et al. 2001, Sun et al. 2001). Consistent with the gain-of-function result described above, expression of a dominant negative form of Hes-1 blocked Müller glial fate, indicating that Notch signaling is necessary to attain this fate (Furukawa et al. 2000). Interestingly, the authors of this study suggested that the cell fate promoted by Notch signaling might represent "a hitherto undescribed population of persisting adult progenitor cells whose morphology and gene expression overlap considerably with those of Müller glia." This notion is supported by recent work indicating that Müller glia in the adult retina possess stem cell character (see below).

Recently the role of Hes-5 in the mouse retina has been examined (Hojo et al. 2000) and has proven consistent with the role of Hes-1 described above. Misexpression of Hes-5 promoted Müller glial fate, and loss of Hes-5 function led to a reduced number of Müller glia. Although the *Hes* genes are currently the primary known effectors of Notch signaling, the effects of ActN in the retina were not completely recapitulated by expression of Hes-1 or Hes-5. Unlike ActN, these *Hes* genes did not promote enhanced proliferation among retinal progenitors, which suggests that Notch activation promoted proliferation through other *Hes* genes, or yet to be identified Notch targets.

Generally consistent with the frog and rodent studies, Scheer et al. (2001) have found that expression of ActN in the zebrafish retina can also promote Müller glial identity. This study used the Gal4/UAS system to drive ActN expression, and found that ActN-expressing cells had one of two fates, that of Müller glia or of seemingly undifferentiated cells. The authors also found that expression of ActN reduced proliferation of these cells, which is consistent with the frog data, but in contrast to the rodent data. One unique finding of the zebrafish study was the observation that the Müller glial marker zrf-1 was expressed three days early as a result of Notch activation. This premature expression suggested that rather than passively guiding retinal progenitors toward Müller glial fate, ActN actively promoted that fate. In contrast, in the rat study, no premature glial marker expression was observed. This difference may reflect variability between species, or a lack of sufficiently early Müller glial markers for use in the rat. All told, however, it seems clear from this body of work that Notch activation plays a central role during the generation of glia in the retina, although the extent to which this is an active role remains to be sorted out.

Radial Glia

Contemporary with the retinal work described above were gain-of-function studies examining the role of Notch signaling in the mouse telencephalon (Chambers et al. 2001, Gaiano et al. 2000, Ishibashi et al. 1994). The first such efforts used a retroviral vector to drive expression of Hes-1 in the embryonic neocortex (Ishibashi et al. 1994). This study found that while most control infected cells migrated away from the proliferative ventricular zone (VZ), cells expressing Hes-1 did not. Instead those cells remained in the VZ and later were detected in the adult ependymal layer, an epithelial-like sheet of cells that lines the lateral ventricles and is considered the last vestige of the VZ. Although the ependymal fate of these cells was of limited interest at the time, it has become intriguing in light of more recent work suggesting that the ependyma may contain neural stem cells in the adult brain (see below). One unfortunate limitation of this study was the use of nuclear localized lacZ as the reporter, which prevented the authors from using morphology to identify cell type.

In a more recent study, our group examined the effects of ActN in the embryonic telencephalon (Gaiano et al. 2000). This study used retroviral vectors and the human placental alkaline phosphatase gene as the reporter to better visualize the morphology of infected cells. ActN was introduced into telencephalic progenitors in vivo at embryonic day (E)9.5 (prior to the onset of neurogenesis in this region) and was found to promote radial glial morphology and marker expression. Radial glia have their cell bodies in the VZ and extend a long radial process to the pial surface (Schmechel & Rakic 1979b). These cells have traditionally been thought to provide a migratory scaffold along which newly generated neurons migrate from the VZ to postmitotic areas (Rakic 1988, 1995). Thus, it seems plausible that these migratory neurons, expressing Notch ligands such as Delta, promote radial glial identity through the activation of Notch expressed along radial glial fibers (Figure 2). The recent observation that Delta-expressing cells are closely associated with radial glia supports this model (Campos et al. 2001). The promotion of radial glial identity by Notch activation suggests that the Hes-1 misexpression described above (Ishibashi et al. 1994) might also have promoted radial glial identity. While this may be true, it should not be assumed since ActN and Hes-1 do not have identical phenotypes in the rodent retina.

At first glance, the promotion of radial glial identity by ActN supports an instructive role for Notch in gliogenesis. Radial glia are one of the first cell types evident in the forebrain and as such are unlikely to represent a default state resulting from inhibition of all others cell types. Furthermore, ActN-infected cells were found to express the radial glial marker, brain lipid binding protein (BLBP), earlier and at higher levels than uninfected radial glia (Gaiano et al. 2000). This result is similar to the premature expression of zrf-1 observed in zebrafish retinal cells expressing ActN described above (Scheer et al. 2001). Such marker upregulation strongly suggests that Notch is actively promoting glial fate, rather than simply inhibiting other fates and indirectly leading to glial identity. Recent work examining the role of Notch in the cerebellum has suggested that Notch activation can promote Bergmann glial identity (R. Machold, D. Kittell, N. Gaiano, G. Fishell, in preparation). This result is not entirely surprising in that Bergmann glia are akin to the radial glia in the telencephalon, although Bergmann glia persist into adulthood.

When ActN-infected embryos were allowed to develop to adulthood, infected cells became astrocytes (Chambers et al. 2001, Gaiano et al. 2000), which is consistent with a known fate of radial glia (Schmechel & Rakic 1979b, Voigt 1989), and to a more limited extent ependymal cells. Interestingly, many infected cells were subependymal astrocytes, a cell type that in addition to ependymal cells

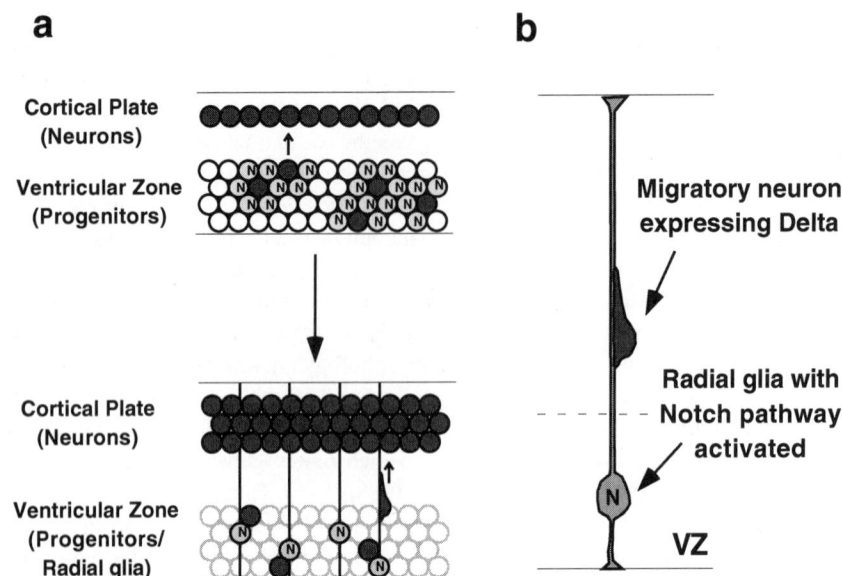

Figure 2 Model for generation and maintenance of radial glial identity during mouse brain development. (*a*) *Top*, some cells in the ventricular zone (VZ) express high levels of a Notch ligand, such as Delta (*dark gray*), as they prepare to leave the VZ. *Bottom*, adjacent cells have Notch activated in them (*light gray* with "N"), and attain radial glia morphology as development proceeds. (*b*) As Delta-expressing cells migrate out of the VZ to differentiate, they activate Notch in radial glia, thereby maintaining the radial glial scaffold (see Campos et al. 2001, Gaiano et al. 2000).

has been argued to possess stem cell character in the adult brain (Doetsch et al. 1999, Johansson et al. 1999, Laywell et al. 2000).

The promotion of radial glial identity by ActN embryonically, and of putative stem cell identity postnatally, suggests that radial glia may be the lineal precursors of adult neural stem cells. The idea that radial glia might be embryonic neural progenitors was proposed years ago (Alvarez-Buylla et al. 1990, Gray & Sanes 1992, Lendahl et al. 1990) and has recently gained substantial credence (see below). Consequently, the observation that Notch promotes radial glial identity may support, rather than contradict, the more traditional view that Notch promotes a progenitor state. Even so, the upregulation of BLBP indicates that the progenitor state promoted by Notch is likely to differ from that of the neuroepithelium as a whole.

Astrocytes

Our recent in vivo studies have found that, subsequent to promoting radial glial identity in the embryo, activation of Notch promoted astrocytic fate in the adult brain (Chambers et al. 2001, Gaiano et al. 2000). Although this in vivo work did not address whether Notch acted instructively to specify astrocyte identity, a recent in vitro study has suggested that Notch can "instructively restrict" CNS stem cells to become astrocytes (Tanigaki et al. 2001). This latter work used adult hippocampal progenitors (AHPs), which are neural stem cells derived from the rat hippocampus. Using either stable transfection or retroviral infection, the authors introduced activated forms of Notch1 and Notch3 into AHPs. Consequently, they found that astrocyte identity was promoted at the expense of neuronal and oligodendroglial identity. The authors then determined whether transient activation of Notch1 was sufficient to promote astrocyte fate. They fused ActN1 to the estrogen receptor to create a form that would be nuclear localized (and thereby active) only in the presence of 4-hydroxytamoxifen. Transient activation (36 h) of this form of ActN1 was found to bias AHPs toward astrocyte identity as assayed four days later. The authors then showed that Notch's ability to generate astrocytes appears to be independent of the astrocyte inducing properties of ciliary neurotrophic factor (CNTF) (Bonni et al. 1997, Johe et al. 1996).

This study clearly suggested that Notch can instructively promote astrocyte fate. However, the evidence that it restricts progenitors to this fate should be qualified. Specifically, in a clonal analysis of AHPs continuously expressing ActN1 or ActN3, the authors found 20% of the clones to be purely neuronal and 40%–50% to be mixed (possessing neurons and astrocytes) (Tanigaki et al. 2001). Although this experiment did find enhanced astrogliogenesis in ActN-infected clones, the presence of so many neurons in these clones was not consistent with Notch activation restricting cells to an astrocytic fate. Nevertheless, the observation that Notch biases AHPs toward astrocytic fate is clearly of interest, in particular to those trying to control the fate of neural stem cells in vitro.

Oligodendrocytes

Although Notch can promote Müller glial, radial glial, and astroglial fates in the mammalian brain, Wang et al. (1998) have found that Notch activation inhibits oligodendroglial differentiation. In particular, this study examined the development of oligodendrocytes in the rat optic nerve. Prior to myelination of retinal ganglion cell axons, the optic nerve contains oligodendrocyte precursor cells (OPCs). Although in vivo these cells are likely to give rise exclusively to oligodendrocytes, they have been found in vitro to be capable of generating astrocytes (Raff et al. 1983) and even neurons (Kondo & Raff 2000). The study by Wang et al. (1998) showed that OPCs express Notch1, and that the Notch ligand Jagged1 is expressed by retinal ganglion cells along their axons. Furthermore, Jagged1 expression was found to be downregulated coincident with the initiation of retinal ganglion cell activity and myelination. These data suggest a model of optic nerve development in which retinal ganglion axons use Notch signaling to delay myelination until they innervate their targets (Figure 3). In support of this model, the study further showed that Notch activation in vitro could inhibit the differentiation of OPCs. In a subsequent study, the authors found that the helix loop helix transcriptional

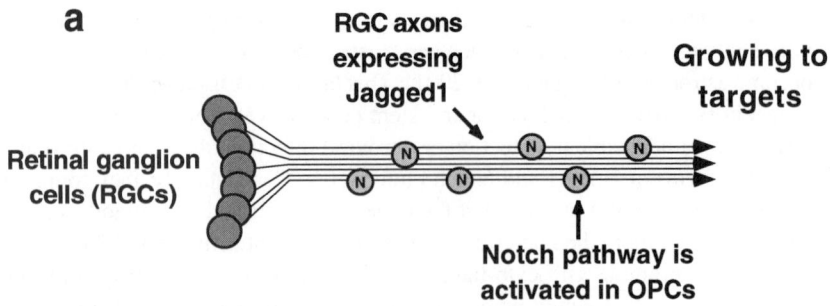

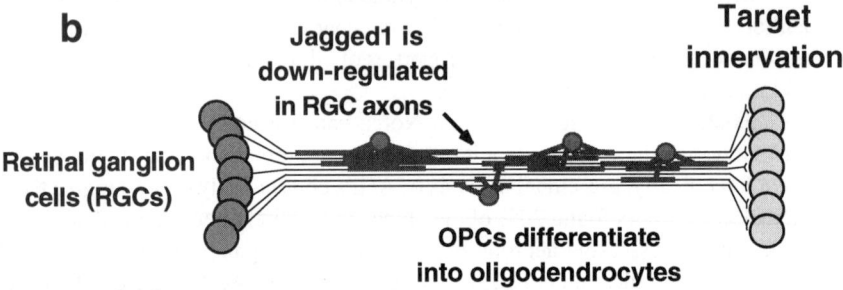

Figure 3 Model for optic nerve myelination as controlled by Notch-mediated inhibition of oligodendrocyte differentiation. (*a*) While retinal ganglion cells (RGCs) are growing to targets, their axons express the Notch ligand Jagged1. In the process they inhibit oligodendrocyte precursor cells (OPCs) from differentiating into oligodendrocytes and initiating myelinating. (*b*) After target innervation Jagged1 is downregulated in RGC axons and the OPCs proceed to differentiate (see Wang et al. 1998).

repressor Id2 can also inhibit OPC differentiation (Wang et al. 2001), although the role of Notch in this inhibition remains to be clarified.

The observations that Notch can promote Müller glial, radial glial, and astroglial fates in the mammalian CNS while inhibiting oligodendroglial differentiation suggests that the role of Notch during mammalian gliogenesis is not uniform. However, it is worth noting that oligodendrocytes are functionally unique among the CNS macroglia, and there is no reason to expect that Notch or any signaling pathway should behave similarly in such diverse cell types simply because they fall into the same broad class.

Schwann Cells

In the mammalian PNS the neural crest is a multipotent precursor population that gives rise to a wide variety of cell types including neurons and Schwann cells. Functionally, Schwann cells are the PNS equivalent of oligodendrocytes in that they myelinate peripheral axons. However, unlike the inhibitory role Notch appears to play during oligodendrocyte development (see above), a recent study has suggested that Notch irreversibly commits neural crest stem cells (NCSCs) to Schwann cell fate (Morrison et al. 2000). Initially, the authors found that expression of activated Notch in NCSCs in vivo inhibited neuronal differentiation. To examine the effect of activated Notch in the neural crest in vitro, the authors isolated NCSCs from the E14.5 rat sciatic nerve, and consistent with the in vivo data, neuronal differentiation was inhibited. In addition, the study found that both the rate and extent of Schwann cell differentiation was increased even after transient Notch activation. This result supported an instructive role for Notch in promoting Schwann cell fate. Furthermore, when transient Notch activation in NCSCs was followed by exposure to bone morphogenetic protein 2, a cue that promotes neuronal fate, the cells still became Schwann cells. Thus, transient Notch activation appeared not only to have instructed NCSCs toward Schwann cell fate, but to have done so irreversibly.

In an interesting twist, when Notch activation was performed coincident with exposure to bone morphogenetic protein 2, both Schwann cell and myofibroblast fate (a third cell type that can be derived from NCSCs in vitro) were promoted at the expense of neurons. This result showed that the influence of Notch could be modified by additional signals, underscoring the notion that even this seemingly instructive role of Notch is context dependent. Subsequent work has found that the response of NCSCs to Notch activation varies considerably depending on the age and location from which the cells are derived, further demonstrating the importance of the cellular "ground" state (S.J. Morrison, personal communication).

There are clear parallels between the studies of Notch's role in oligodendrocyte and Schwann cell development that make the contrasting outcomes challenging to understand. In both cases the precursor cells, rather than being derived from nascent progenitor pools, were derived from developing nerve tracts (OPCs from the optic nerve, and NCSCs from the sciatic nerve) (Morrison et al. 2000, Wang et al. 1998). Both the OPCs and NCSCs used in these studies have been found to be multipotent in vitro, although they are likely to give rise primarily to their respective myelinating cell types in vivo. Why then does Notch appear to have the opposite effect on these precursor populations? The simple answer may be that regardless of their similarities, OPCs and NCSCs are nonetheless different cell types, with distinct origins and intrinsic characters. The ability of Notch to promote gliogenesis in one cell type, while blocking it in a similar cell type, supports the premise that an instructive role for Notch in gliogenesis cannot be assigned without contextual consideration.

NOTCH, GLIA, AND STEM CELLS?

In vertebrates, the traditional view has been that Notch signaling inhibits differentiation and maintains cells as progenitors. In contrast, the newly emerging view is that Notch can positively promote certain glial fates, and as such may serve as an instructive signal. The maintenance of a progenitor state and the promotion of glial identity may seem mutually exclusive; however, recent studies examining the developmental potential of "differentiated" glia have suggested that some glial cell types can possess progenitor character (Doetsch et al. 1999, Fischer & Reh 2001, Laywell et al. 2000).

Several recent reports have found that radial glia may be multipotent neural progenitors in the embryo (Hartfuss et al. 2001, Malatesta et al. 2000, Noctor et al. 2001). One such study examined the potential of isolated cortical radial glia in vitro and found that in addition to generating glial daughters, these cells can also give rise to neurons (Malatesta et al. 2000). Noctor et al. (2001) provided further evidence that radial glia can generate neurons using time-lapse video observation of single-labeled radial glia in slice culture. The authors observed radial glia dividing and producing daughter cells that migrated to the cortical plate and expressed neuronal markers. In light of these data, the promotion of radial glial identity by ActN suggests that Notch may promote progenitor fate (Gaiano et al. 2000). Furthermore, recent analysis suggests that cells infected with ActN in vivo can display multipotent stem cell character when placed in vitro (N. Gaiano, S. Nery, M. Rutlin, F. Radtke, & G. Fishell, submitted). Whether all radial glia specified by Notch have stem cell character is unknown. Götz and colleagues have suggested that some radial glia are neurogenic, some are gliogenic, and others are multipotent (Malatesta et al. 2000). If at least some radial glia are embryonic neural stem cells, they are likely to be lineally related to stem cells present in the adult brain (see Figure 4). In the past, the observation that the radial glia scaffold transformed into astrocytes postnatally (Schmechel & Rakic 1979b, Voigt 1989) suggested that radial glia were committed glial cells that maintained a specialized morphology during development (Levitt et al. 1981, 1983; Schmechel & Rakic 1979a). More recently, however, several groups have suggested that astrocytes are capable of generating neurons and may be neural stem cells in the adult brain (Doetsch et al. 1999, Laywell et al. 2000). This work, together with the observation that Notch can promote astrocyte identity, supports the notion that in addition to specifying glial fate, Notch may be maintaining stem cell character. Further support for this idea has come from recent evidence that Müller glia, which are specified at least in part by Notch, may be multipotent stem cells in the adult retina. Fischer & Reh (2001) have found in the chick that subsequent to retinal damage, Müller glia can re-enter the cell cycle and give rise to new retinal neurons as well as additional Müller glia.

The idea that mature cell types, such as astrocytes and Müller glia, are stem cells may seem at odds with the traditional view that stem cells do not express markers of differentiated cell types. However, recent studies in a variety of tissues, including the nervous system, have suggested that this traditional view was misleading (Fuchs & Segre 2000). In the skin, for example, stem cells have been found to express markers previously thought to be present only in differentiated keratinocytes. Similarly, it has recently been suggested that both intestinal and hematopoietic stem cells possess molecular and/or morphological characteristics of differentiated cell types (Fuchs & Segre 2000).

While it is becoming increasingly believable that some glia may be stem cells in the adult vertebrate nervous system, there is certainly no reason to think that

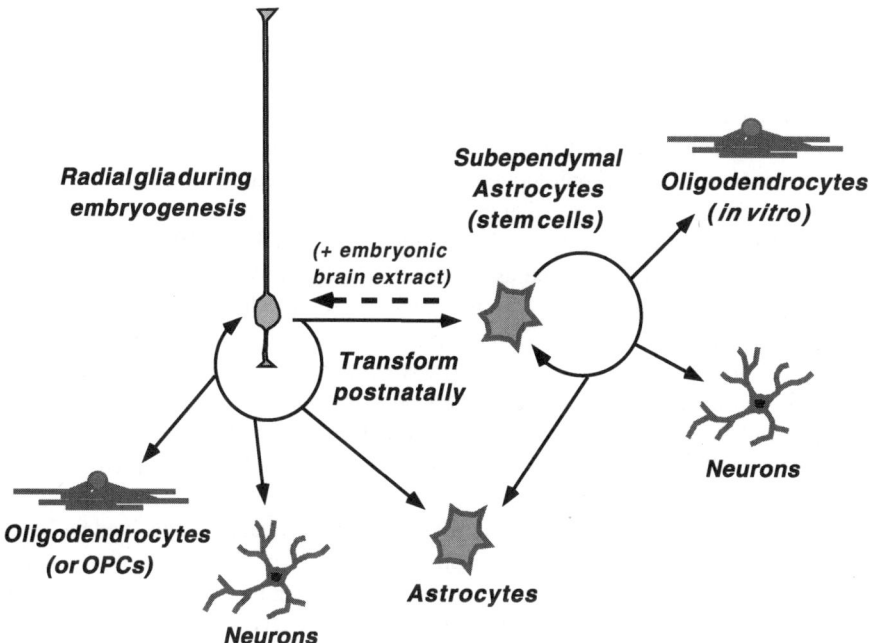

Figure 4 Model of radial glia and astrocytes as lineally related multipotent progenitors. During embryogenesis radial glia can give rise to both neurons and glia. As radial glia transform into astrocytes postnatally, some retain stem cell character in the form of specialized astrocytes present in the subependymal layer. These cells have been found to be capable of self-renewal as well as the generation of all three major CNS cell types in vitro (see Doetsch et al. 1999). When exposed to embryonic cues, astrocytes may reacquire an embryonic progenitor state (see Hunter & Hatten 1995), which suggests a reversible continuum between these cell types.

all glia are stem cells. Prior to any suspicions about their stem cell character, glia were known to play many fundamental supporting roles in the functioning of the mature nervous system. Even among those glial cell types recently suggested to possess stem cell character, only subsets are likely to be stem cells. For example, while both subependymal and dispersed astrocytes have been found to possess stem cell character early postnatally, dispersed astrocytes gradually lose this stem cell character (Laywell et al. 2000).

So what is the relationship between Notch, glia, and stem cells? The observations that Notch can induce early expression of glial markers in certain neural progenitors suggest that it is altering these cells. For example, the expression of BLBP in radial glia specified by ActN clearly distinguishes these cells from neuroepithelial progenitors present early in development (Gaiano et al. 2000). One possibility is that BLBP-expressing radial glia represent a more "mature" progenitor state, which is precociously promoted by expression of ActN. These cells may be the embryonic form of astrocytic stem cells present in the adult. The only difference between these cells might be their environment; the promotion of radial glial and astrocytic fates by ActN is the same phenomenon at different times. This notion is supported by the observation that astrocytes revert into radial glia in the presence of embryonic brain extract (Hunter & Hatten 1995) (see Figure 4). Interestingly, in primates radial glia express the canonical astrocytic marker, glial fibrillary acidic protein, during development (Levitt & Rakic 1980). The relationship between radial glia and astrocytes, with regard to their potential stem cell properties, has recently been discussed by Alvarez-Buylla and colleagues (1990).

On a final note, of the four glial cell types recently found to be specified by Notch, three have also been found to possess potential stem cell character: radial glia, astrocytes, and Müller glia. This begs the question, do the Schwann cells induced by Notch in NCSCs also possess such character? The studies by Morrison et al. (2000) suggest that this is unlikely. However, as the impact of Notch in NCSCs can vary depending on age and environment (S. J. Morrison, personal communication), perhaps under the right circumstances the Schwann cells specified by Notch activation might exhibit stem cell character as well. Supporting this notion, a recent study has shown that, in certain conditions, OPCs can be transformed into neural stem cells in vitro (Kondo & Raff 2000). Interestingly, like radial glia (in primates), Müller glia (in the context of retinal damage), and astrocytes, Schwann cells derived from NCSCs also express glial fibrillary acidic protein (Shah et al. 1994, Stemple & Anderson 1992), which suggests a commonality between these cell types.

CONCLUSIONS

In the past, Notch signaling has not been considered instructive because it was believed to have a nonspecific inhibitory effect on cellular differentiation. As such the Notch pathway was thought to passively influence cell fate by controlling the ability of progenitors to respond to instructive developmental cues. However, the recent data discussed in this review has suggested that Notch can play a more active role in directing cell fate. Specifically, many studies have found evidence of a role for Notch in promoting glial and perhaps stem cell identities. While intriguing, it is worth noting that this role is context dependent, both in vertebrates and in invertebrates, as Notch can either promote or inhibit gliogenesis, depending on the cell type being examined. Therefore, although in specific contexts the role of Notch might be termed instructive, it is currently not possible to define a uniform role for Notch with regard to glial fate.

The flurry of recent studies regarding the influence of Notch during gliogenesis has significantly altered our understanding of the role of this pathway during development. Further study promises to clarify the relationships between Notch signaling, gliogenesis, and the maintenance of stem cell populations both in the embryo and the adult. Particularly important will be identifying the molecular targets of the Notch pathway, as well as the manner in which Notch signaling interacts with other signaling pathways that influence cell fate. By better understanding the molecular circuitry downsteam of Notch, we will gain fundamental insight into the global regulation not only of glial specification, but also the maintenance of progenitor identity.

ACKNOWLEDGMENTS

The authors are grateful to members of the Fishell Lab and the Developmental Genetics Program at the Skirball Institute for continuous intellectual input. The authors have been funded by a postdoctoral fellowship from the American Cancer Society (PF4473 to N. G.), a Career Development Award from the Burroughs Wellcome Fund (to N. G.), a grant from the Children's Brain Tumor Foundation (to G. F.), and a grant from the NINDS (NS32993 to G. F.).

The *Annual Review of Neuroscience* is online at http://neuro.annualreviews.org

LITERATURE CITED

Alvarez-Buylla A, Theelen M, Nottebohm F. 1990. Proliferation "hot spots" in adult avian ventricular zone reveal radial cell division. *Neuron* 5:101–9

Anderson AC, Robey EA, Huang YH. 2001. Notch signaling in lymphocyte development. *Curr. Opin. Genet. Dev.* 11:554–60

Artavanis-Tsakonas S, Delidakis C, Fehon RG. 1991. The Notch locus and the cell biology of neuroblast segregation. *Annu. Rev. Cell Biol.* 7:427–52

Artavanis-Tsakonas S, Rand MD, Lake RJ. 1999. Notch signaling: cell fate control and signal integration in development. *Science* 284:770–76

Bao ZZ, Cepko CL. 1997. The expression and function of Notch pathway genes in the developing rat eye. *J. Neurosci.* 17:1425–34

Barrantes IB, Elia AJ, Wunsch K, De Angelis MH, Mak TW, et al. 1999. Interaction between Notch signalling and Lunatic fringe during somite boundary formation in the mouse. *Curr. Biol.* 9:470–80

Bhat KM. 1999. Segment polarity genes in neuroblast formation and identity specification during Drosophila neurogenesis. *Bioessays* 21:472–85

Bonni A, Sun Y, Nadal-Vicens M, Bhatt A, Frank DA, et al. 1997. Regulation of gliogenesis in the central nervous system by the JAK-STAT signaling pathway. *Science* 278:477–83

Campos LS, Duarte AJ, Branco T, Henrique D. 2001. mDll1 and mDll3 expression in the developing mouse brain: role in the establishment of the early cortex. *J. Neurosci. Res.* 64:590–98

Campos-Ortega JA. 1993. Mechanisms of early neurogenesis in Drosophila melanogaster. *J. Neurobiol.* 24:1305–27

Chambers CB, Peng Y, Nguyen H, Gaiano N, Fishell G, Nye JS. 2001. Spatiotemporal selectivity of response to Notch1 signals in mammalian forebrain precursors. *Development* 128:689–702

Coffman CR, Skoglund P, Harris WA, Kintner CR. 1993. Expression of an extracellular deletion of Xotch diverts cell fate in Xenopus embryos. *Cell* 73:659–71

Doe CQ, Skeath JB. 1996. Neurogenesis in the insect central nervous system. *Curr. Opin. Neurobiol.* 6:18–24

Doetsch F, Caille I, Lim DA, Garcia-Verdugo JM, Alvarez-Buylla A. 1999. Subventricular zone astrocytes are neural stem cells in the adult mammalian brain. *Cell* 97:703–16

Dorsky RI, Rapaport DH, Harris WA. 1995. Xotch inhibits cell differentiation in the Xenopus retina. *Neuron* 14:487–96

Fischer AJ, Reh TA. 2001. Muller glia are a potential source of neural regeneration in the postnatal chicken retina. *Nat. Neurosci.* 4:247–52

Fuchs E, Segre JA. 2000. Stem cells: a new lease on life. *Cell* 100:143–55

Furukawa T, Maruyama S, Kawaichi M, Honjo T. 1992. The Drosophila homolog of the immunoglobulin recombination signal-binding protein regulates peripheral nervous system development. *Cell* 69:1191–97

Furukawa T, Mukherjee S, Bao ZZ, Morrow EM, Cepko CL. 2000. rax, Hes1, and notch1 promote the formation of Muller glia by postnatal retinal progenitor cells. *Neuron* 26:383–94

Gaiano N, Nye JS, Fishell G. 2000. Radial glial identity is promoted by Notch1 signaling in the murine forebrain. *Neuron* 26:395–404

Gho M, Bellaiche Y, Schweisguth F. 1999. Revisiting the Drosophila microchaete lineage: a novel intrinsically asymmetric cell division generates a glial cell. *Development* 126:3573–84

Gray GE, Sanes JR. 1992. Lineage of radial glia in the chicken optic tectum. *Development* 114:271–83

Guo M, Jan LY, Jan YN. 1996. Control of daughter cell fates during asymmetric division: interaction of Numb and Notch. *Neuron* 17:27–41

Hartenstein AY, Rugendorff A, Tepass U, Hartenstein V. 1992. The function of the neurogenic genes during epithelial development in the Drosophila embryo. *Development* 116:1203–20

Hartfuss E, Galli R, Heins N, Gotz M. 2001. Characterization of CNS precursor subtypes and radial glia. *Dev. Biol.* 229:15–30

Hojo M, Ohtsuka T, Hashimoto N, Gradwohl G, Guillemot F, Kageyama R. 2000. Glial cell fate specification modulated by the bHLH gene Hes5 in mouse retina. *Development* 127:2515–22

Hosoya T, Takizawa K, Nitta K, Hotta Y. 1995. Glial cells missing: a binary switch between neuronal and glial determination in Drosophila. *Cell* 82:1025–36

Hunter KE, Hatten ME. 1995. Radial glial cell transformation to astrocytes is bidirectional: regulation by a diffusible factor in embryonic forebrain. *Proc. Natl. Acad. Sci. USA* 92:2061–65

Irvine KD, Vogt TF. 1997. Dorsal-ventral signaling in limb development. *Curr. Opin. Cell. Biol.* 9:867–76

Ishibashi M, Moriyoshi K, Sasai Y, Shiota K, Nakanishi S, Kageyama R. 1994. Persistent expression of helix-loop-helix factor HES-1 prevents mammalian neural differentiation in the central nervous system. *Embo. J.* 13:1799–805

Jacobs JR, Hiromi Y, Patel NH, Goodman CS. 1989. Lineage, migration, and morphogenesis of longitudinal glia in the Drosophila CNS as revealed by a molecular lineage marker. *Neuron* 2:1625–31

Jan YN, Jan LY. 1994. Genetic control of cell fate specification in Drosophila peripheral nervous system. *Annu. Rev. Genet.* 28:373–93

Jiang YJ, Aerne BL, Smithers L, Haddon C, Ish-Horowicz D, Lewis J. 2000. Notch signalling and the synchronization of the somite segmentation clock. *Nature* 408:475–79

Johansson CB, Momma S, Clarke DL, Risling M, Lendahl U, Frisen J. 1999. Identification of a neural stem cell in the adult mammalian central nervous system. *Cell* 96:25–34

Johe KK, Hazel TG, Muller T, Dugich-Djordjevic MM, McKay RD. 1996. Single factors direct the differentiation of stem cells from the fetal and adult central nervous system. *Genes. Dev.* 10:3129–40

Jones BW, Fetter RD, Tear G, Goodman CS. 1995. Glial cells missing: a genetic switch that controls glial versus neuronal fate. *Cell* 82:1013–23

Knust E, Campos-Ortega JA. 1989. The molecular genetics of early neurogenesis in Drosophila melanogaster. *Bioessays* 11:95–100

Kondo T, Raff M. 2000. Oligodendrocyte precursor cells reprogrammed to become multipotential CNS stem cells. *Science* 289: 1754–57

Laywell ED, Rakic P, Kukekov VG, Holland EC, Steindler DA. 2000. Identification of a multipotent astrocytic stem cell in the immature and adult mouse brain. *Proc. Natl. Acad. Sci. USA* 97:13,883–88

Lendahl U, Zimmerman LB, McKay RD. 1990. CNS stem cells express a new class of intermediate filament protein. *Cell* 60:585–95

Levitt P, Cooper ML, Rakic P. 1981. Coexistence of neuronal and glial precursor cells in the cerebral ventricular zone of the fetal monkey: an ultrastructural immunoperoxidase analysis. *J. Neurosci.* 1:27–39

Levitt P, Cooper ML, Rakic P. 1983. Early divergence and changing proportions of neuronal and glial precursor cells in the primate cerebral ventricular zone. *Dev. Biol.* 96:472–84

Levitt P, Rakic P. 1980. Immunoperoxidase localization of glial fibrillary acidic protein in radial glial cells and astrocytes of the developing rhesus monkey brain. *J. Comp. Neurol.* 193:815–40

Lewis J. 1996. Neurogenic genes and vertebrate neurogenesis. *Curr. Opin. Neurobiol.* 6:3–10

Lewis J. 1998. Notch signalling and the control of cell fate choices in vertebrates. *Semin. Cell. Dev. Biol.* 9:583–89

Malatesta P, Hartfuss E, Gotz M. 2000. Isolation of radial glial cells by fluorescent-activated cell sorting reveals a neuronal lineage. *Development* 127:5253–63

Modolell J. 1997. Patterning of the adult peripheral nervous system of Drosophila. *Perspect. Dev. Neurobiol.* 4:285–96

Morrison SJ, Perez SE, Qiao Z, Verdi JM, Hicks C, et al. 2000. Transient Notch activation initiates an irreversible switch from neurogenesis to gliogenesis by neural crest stem cells. *Cell* 101:499–510

Mumm JS, Kopan R. 2000. Notch signaling: from the outside in. *Dev. Biol.* 228:151–65

Nieto M, Schuurmans C, Britz O, Guillemot F. 2001. Neural bHLH genes control the neuronal versus glial fate decision in cortical progenitors. *Neuron* 29:401–13

Noctor SC, Flint AC, Weissman TA, Dammerman RS, Kriegstein AR. 2001. Neurons derived from radial glial cells establish radial units in neocortex. *Nature* 409:714–20

Nye JS, Kopan R, Axel R. 1994. An activated Notch suppresses neurogenesis and myogenesis but not gliogenesis in mammalian cells. *Development* 120:2421–30

Osborne B, Miele L. 1999. Notch and the immune system. *Immunity* 11:653–63

Raff MC, Miller RH, Noble M. 1983. A glial progenitor cell that develops in vitro into an astrocyte or an oligodendrocyte depending on culture medium. *Nature* 303:390–96

Rakic P. 1988. Specification of cerebral cortical areas. *Science* 241:170–76

Rakic P. 1995. Radial versus tangential migration of neuronal clones in the developing cerebral cortex. *Proc. Natl. Acad. Sci. USA* 92:11,323–27

Reddy GV, Rodrigues V. 1999. A glial cell arises from an additional division within the mechanosensory lineage during development of the microchaete on the Drosophila notum. *Development* 126:4617–22

Scheer N, Groth A, Hans S, Campos-Ortega JA. 2001. An instructive function for Notch in promoting gliogenesis in the zebrafish retina. *Development* 128:1099–107

Schmechel DE, Rakic P. 1979a. Arrested proliferation of radial glial cells during midgestation in rhesus monkey. *Nature* 277:303–5

Schmechel DE, Rakic P. 1979b. A Golgi study of radial glial cells in developing monkey telencephalon: morphogenesis and transformation into astrocytes. *Anat. Embryol. (Berl.)* 156:115–52

Schweisguth F. 1995. Suppressor of Hairless is required for signal reception during lateral inhibition in the Drosophila pupal notum. *Development* 121:1875–84

Schweisguth F, Posakony JW. 1992. Suppressor of Hairless, the Drosophila homolog of the mouse recombination signal-binding protein gene, controls sensory organ cell fates. *Cell* 69:1199–212

Seydoux G, Greenwald I. 1989. Cell autonomy of lin-12 function in a cell fate decision in C. elegans. *Cell* 57:1237–45

Shah NM, Marchionni MA, Isaacs I, Stroobant P, Anderson DJ. 1994. Glial growth factor restricts mammalian neural crest stem cells to a glial fate. *Cell* 77:349–60

Skeath JB, Doe CQ. 1996. The achaete-scute complex proneural genes contribute to neural precursor specification in the Drosophila CNS. *Curr. Biol.* 6:1146–52

Stemple DL, Anderson DJ. 1992. Isolation of a stem cell for neurons and glia from the mammalian neural crest. *Cell* 71:973–85

Sternberg PW, Horvitz HR. 1989. The combined action of two intercellular signaling pathways specifies three cell fates during vulval induction in C. elegans. *Cell* 58:679–93

Sun Y, Nadal-Vicens M, Misono S, Lin MZ, Zubiaga A, et al. 2001. Neurogenin promotes neurogenesis and inhibits glial differentiation by independent mechanisms. *Cell* 104:365–76

Tanigaki K, Nogaki F, Takahashi J, Tashiro K, Kurooka H, Honjo T. 2001. Notch1 and Notch3 instructively restrict bFGF-responsive multipotent neural progenitor cells to an astroglial fate. *Neuron* 29:45–55

Udolph G, Rath P, Chia W. 2001. A requirement for Notch in the genesis of a subset of glial cells in the Drosophila embryonic central nervous system which arise through asymmetric divisions. *Development* 128:1457–66

Van De Bor V, Giangrande A. 2001. Notch signaling represses the glial fate in fly PNS. *Development* 128:1381–90

Van De Bor V, Walther R, Giangrande A. 2000. Some fly sensory organs are gliogenic and require glide/gcm in a precursor that divides symmetrically and produces glial cells. *Development* 127:3735–43

Vargesson N, Patel K, Lewis J, Tickle C. 1998. Expression patterns of Notch1, Serrate1, Serrate2 and Delta1 in tissues of the developing chick limb. *Mech. Dev.* 77:197–99

Vincent S, Vonesch JL, Giangrande A. 1996. Glide directs glial fate commitment and cell fate switch between neurones and glia. *Development* 122:131–39

Voigt T. 1989. Development of glial cells in the cerebral wall of ferrets: direct tracing of their transformation from radial glia into astrocytes. *J. Comp. Neurol.* 289:74–88

Wang S, Sdrulla A, Johnson JE, Yokota Y, Barres BA. 2001. A role for the helix-loop-helix protein Id2 in the control of oligodendrocyte development. *Neuron* 29:603–14

Wang S, Sdrulla AD, diSibio G, Bush G, Nofziger D, et al. 1998. Notch receptor activation inhibits oligodendrocyte differentiation. *Neuron* 21:63–75

Weinmaster G. 1997. The ins and outs of notch signaling. *Mol. Cell. Neurosci.* 9:91–102

Xu T, Rebay I, Fleming RJ, Scottgale TN, Artavanis-Tsakonas S. 1990. The Notch locus and the genetic circuitry involved in early Drosophila neurogenesis. *Genes. Dev.* 4:464–75

MULTIPLE SCLEROSIS: Deeper Understanding of Its Pathogenesis Reveals New Targets for Therapy*

Lawrence Steinman,[1] Roland Martin,[2] Claude Bernard,[3] Paul Conlon,[4] and Jorge R. Oksenberg[5]

[1]*Department of Neurology and Neurological Sciences, Stanford University School of Medicine, Stanford, California 94305; email: steinman@stanford.edu*
[2]*Neuroimmunology Branch, National Institutes of Health, Bethesda, Maryland; email: MartinR@ninds.nih.gov*
[3]*Neuroimmunology Laboratory, Department of Biochemistry, La Trobe University, Bundoora, Victoria, 3083 Australia; email: C.Bernard@latrobe.edu.au*
[4]*Neurocrine Biosciences, Science Park Drive, San Diego, California; email: pconlon@neurocrine.com*
[5]*Department of Neurology, University of California, San Francisco, California; email: oksen@itsa.ucsf.edu*

Key Words microarray, proteomics, osteopontin, immunotherapy, genomics

■ **Abstract** Recent technological breakthroughs allowing for large-scale analysis of gene transcripts and large-scale monitoring of the immune response with protein chips are revealing new participants in the pathogenesis of multiple sclerosis. Some of these participants may be useful targets for therapy.

New Players in the Pathogenesis of MS

As multiple sclerosis (MS) tissue is analyzed with increasingly powerful technologies, a variety of new targets are being revealed. Large-scale sequencing of mRNA from MS brain plaques (Chabas et al. 2001), gene microarray analysis of transcripts from MS brain, and proteomic analysis of the immune response (Robinson et al. 2002) are revealing a far more complex picture than could have been imagined even a few years ago. These studies are helping us to understand the mechanism of action of the current group of pharmaceutical approaches to MS (Steinman 2001a,b) and to show why these interventions are so modest in modifying disease.

Multiple sclerosis often begins in early adulthood with an autoimmune inflammatory strike against components of the myelin sheath (Steinman 2001a). Paralysis, sensory disturbances, incoordination, and visual impairment are common features. The disease often starts with an attack lasting for days to weeks, followed

---

*The U.S. Government has the right to retain a nonexclusive, royalty-free license in and to any copyright covering this paper.

by remission lasting months to years. This relapsing remitting phase often lasts for five to ten years. About 30% of individuals with relapsing-remitting MS enter into a secondary chronic progressive state. This chronic progressive state is often characterized by the inability to walk, which leaves the MS patient wheelchair-bound. In the chronic progressive phase, distinct attacks are rare, and the disease progresses insidiously. In rare instances, clinical disability begins with this progressive phase, and in this case the disease is called primary progressive MS.

Recent evidence indicates that the earlier phase of disease characterized by distinct attacks followed by remission may be mediated by an autoimmune reaction. The subsequent chronic phase of disease is due to degeneration of both the myelin sheath, synthesized by oligodendroglial cells, as well as the underlying axon, which emanates from the neuronal cell body some distance away. Indeed it is axon loss in the spinal cord and spinal cord atrophy that correlate most strongly with paralysis and the inability to walk (Trapp et al. 1998, Loseff et al. 1996). We are just beginning to identify the pathways involved in the progression of disease from the relapse-remitting phase to the chronic progressive phase.

Worldwide, approximately one million individuals are afflicted with MS. Women with the disease outnumber men two to one. This bias toward females is seen in other diseases presumed to be autoimmune like rheumatoid arthritis, systemic lupus erythematosus, and thyroiditis. Genome-wide studies have revealed that susceptibility to MS is linked to genes in the major histocompatibility complex on chromosome 6 (Haines et al. 1996, Sawcer et al. 1996, Ebers et al. 1996). Alleles of certain class II genes, HLA DR and HLA DQ, confer the strongest degree of risk of MS. In addition to the well-established MHC association (HLA-DR2haplotype), full genome screens of families with multiple cases of MS also support a role for several additional unidentified genes, each with a modest effect. Other genes within the human leukocyte antigen (HLA) complex are involved in the pathogenesis of MS, including TNF-α, various components of the complement cascade, and myelin oligodendroglial glycoprotein. More recently, transcriptional profiling using gene microarrays and large-scale sequencing of transcripts from MS lesion material have revealed genes involved in the pathogenesis of acute disease, like immunoglobulin and interleukin 6 (Baranzini et al. 2000), as well as genes like osteopontin (Chabas et al. 2001) that are expressed and play a role in the transition from relapsing remitting to chronic MS (Figure 1, reprinted from Steinman et al. 2001a. See color insert).

THE AUTOIMMUNE ATTACK: A DESTRUCTIVE CASCADE OF INFLAMMATORY MOLECULES

T Cells Reactive to Inducible and Constituent Components of the Myelin Sheath are Present in Normal Blood

The ability to discriminate between a foreign pathogen and self-tissue is a key feature of the immune system. Paul Ehrlich's doctrine of horror autotoxicus appreciated the impact of an immune system gone awry in an autoimmune disease. We have learned recently from sequencing the human genome and the genome of various microbes that all biological organisms share many genes. Hence various proteins are used in a modular fashion to build structures whose intrinsic components may resemble each other in an extraordinary manner. The immune system, in recognizing a structure on a foreign microbe, may mistakenly attack self (horror autotoxicus) if the microbe and human share a common gene sequence encoding one of these conserved structural motifs. Indeed many microbial protein sequences share homologies with structures found on the myelin sheath, and this leads to an attack against myelin via a process called molecular mimicry. Relapses in MS are often triggered by common infections with viruses. Viruses such as herpes virus-6, influenza, measles, papilloma virus, and Epstein-Barr virus have genes encoding sequences mimicking those found in the major structural proteins of myelin. Indeed antibodies to components of the myelin sheath cross-react and bind sequences from these microbes (Warren et al. 1995, Wucherpfennig et al. 1997, Genain et al. 1999). T cells also recognize sequences from the myelin sheath that are shared with these microbial sequences (Wucherpfennig & Strominger 1995). Once an immune cell becomes activated, whether by a foreign microbe, a self-protein, or a microbial superantigen, it may penetrate the blood brain barrier.

How Lymphocytes Penetrate the Blood Brain Barrier

Penetration of the blood brain barrier by activated lymphocytes is a multistep process (Figure 1). There are specialized capillary endothelial cells in the CNS, which are nonfenestrated and connected through tight junctions. These capillary endothelial cells are induced to express vascular cellular adhesion molecule (V-CAM) and class II molecules of the major histocompatibility complex by gamma-interferon and tumor necrosis factor-alpha (TNF-α), which are released in the inflammatory response (Cannella & Raine 1995). Activated lymphocytes are able to diapedese (literally walk through) this barrier by virtue of adhesion molecules such as integrins, particularly alpha-4 integrin, which binds to V-CAM, and members of the immunoglobulin supergene family like CD4, which bind to major histocompatibility complex (MHC) class II. Any activated T cell expressing very late antigen (VLA)-4 may bind to adhesion molecules on the surface of inflamed endothelium and begin penetration of the endothelium, the first component of the blood brain barrier. Blockade of VLA-4 reverses clinical paralysis in acute experimental autoimmune encephalomyelitis (EAE), an animal model of MS, and prevents further relapses in the chronic model of this disease. In acute MS lesions, VLA-4 is found on T cells that collect in the perivascular lymphocyte cuff, a region around veins and capillaries limited by the extracellular matrix. V-CAM, along with class II MHC molecules, is found on inflamed endothelium. Clinical studies with a humanized version of the anti-VLA4 antibody are now in Phase III trials, after showing promise in Phase II studies in reducing the incidence of relapses of MS.

Once the activated lymphocytes have extravasated, they must still pass through a barrier of extracellular matrix comprised of type IV collagen before entering into the central nervous system. Alpha 1 integrin may play a role in binding to collagen type IV (Kern et al. 1994). Following contact with collagen, immune cells secrete enzymes such as matrix metalloproteases (MMP), which allow the activated lymphocytes to gain access to the white matter surrounding axons of the central nervous system. Matrix metalloproteases 2 and 9, also called gelatinase A and B, specifically degrade collagen type IV, which is found surrounding inflamed brain endothelium.

The MMPs are a family of structurally and functionally related enzymes involved in the degradation of the extracellular matrix as well as the proteolysis of myelin components in MS. MMPs contain a Zn^{2+} ion at their active site. MMPs also contain TNF-α convertase activity and induce the cleavage of TNF-α from a cell-bound form to a soluble form. Thus MMP inhibition can block TNF-α and thereby downregulate the induction of adhesion molecules. MMPs are inhibited by tissue inhibitors of matrix metalloproteases (TIMPs). TIMP-1 is present in the spinal fluid of MS patients. TIMP-1 is inducible by cytokines including TNF-α.

Both gelatinase A and B are detectable in the spinal fluid of MS patients, and gelatinase B immunoreactivity is seen in MS lesions (Gijbels et al. 1994, Brosnan & Raine 1996, Hartung 1995, Conlon et al. 1999). Gelatinase B is present in MS lesions in endothelial cells, pericytes, macrophages, and astrocytes. Myelin-specific T cell clones derived from MS patients also produce gelatinase B upon activation with antigen. The presence of gelatinase B in the perivascular infiltrate is associated with the disruption of the type IV collagen-positive basement membrane and is critical in the opening of the blood brain barrier. Once the blood brain barrier is breached inflammatory cells spread into the white matter of the CNS.

In MS the activated T cells, as well as natural killer cells, secrete gelatinases A and B, which allows penetration through the extracellular matrix surrounding blood vessels in the central nervous system. Beta-interferon is a potent inhibitor of gelatinase B activity, and this may account for its success as a therapy for MS. Inhibition of gelatinase may thus interfere with T cell migration into the CNS, as well as T cell secretion of TNF-α, a critical cytokine in the process of demyelination. It should be noted, however, that TNF-α also has neuroprotective functions (Probert et al. 1995, Akassaglou et al. 1997, Liu et al. 1998). TNF-α promotes proliferation of oligodendrocyte progenitors and promotes remyelination (Arnett et al. 2001). MMP inhibitors are under intense development for clinical trials in MS.

Specificity of the T and B Cell Responses

Despite our advances in sequencing the genome, the cause of MS is unknown. In the peripheral blood of normal individuals and of individuals with MS there is abundant evidence of T cells that have receptors directed to various components of the myelin sheath, including myelin basic protein, myelin oligodendroglial glycoprotein, and proteolipid protein (reviewed in Steinman et al. 1995). At the site of demyelination in the central nervous system of MS patients, peptides encompassing the immunodominant epitope of myelin basic protein, $MBP_{(83-99)}$, are bound to the DR2 molecule expressed inflammatory cells (Krogsgaard et al. 2000). Moreover, at the site of MS brain plaques in patients who are HLA DR2, there are T cell receptor rearrangements characteristic of T cell clones reactive to MBP$_{(83-99)}$ bound to HLA DR2 (Oksenberg et al. 1993). In addition, immunoglobulins with specificity for this portion of the MBP molecule have been eluted from MS brain plaques (Wucherpfennig et al. 1997, Genain et al. 1999). Recently, further evidence has been reported of clonal rearrangements of immunoglobulin genes in MS lesions (Baranzini et al. 1999). The specificity of these clonally rearranged immunoglobulins remains unknown.

Effector Molecules Involved in Myelin Damage and Progression of MS

Once immune cells have spread to the white matter of the CNS, the immune response is targeted to the entire supramolecular complex of myelin. Antibodies to various myelin proteins and lipids of the myelin sheath, as well as to molecules expressed in the CNS, are secreted by B cells that have migrated to the brain or from serum that has extravasated across the blood brain barrier (Steinman 2001a). The complement proteins are activated with membrane attack complexes that are composed of the terminal components of this cascade and that appear in the spinal fluid. T cells are targeted to certain key portions of proteins normally found in the myelin sheath, including myelin basic protein, myelin oligodendroglial glycoprotein, and proteolipid protein, as well as stress proteins like alpha B crystallin, found in the myelin sheath after activation by the inflammatory response. The T cells produce cytokines, notably lymphotoxin (LTα, also known as TNF-β) and TNF-α, and then influence macrophages, as well as microglial cells and astrocytes, to produce nitric oxide and osteopontin. LTα, TNF-α, and LTβ are all members of the TNF family. LTα is secreted as an LTα3 homotrimer. Like TNF-α it can bind to the p55 TNF receptor (p55 TNFR I) or the p75 TNF (p75 TNFRII).

The free radical nitric oxide (NO) is a major mediator in autoimmune diseases. NO is involved in the killing of oligodendroglial cells by microglia. Expression of induced nitric oxide synthase (iNOS) has been found in demyelinating lesions in MS. Both γ-interferon and TNF-α induce the transcription iNOS in astrocytes, microglia, and macrophages. iNOS catalyzes the synthesis of NO. The combined effect of antibody, complement, NO, and TNF-α damages myelin and induces the macrophage to phagocytose large pieces of the myelin sheath. In addition, macrophages and T cells produce osteopontin, which further induces Th1 cytokines including γ-interferon and IL-12 while downregulating Th2 cytokines like IL-10. Th1 cytokines may induce exacerbations of MS, though Th2 cytokines may reduce the extent of MS lesions (Bielekova et al. 2000, Kappos et al. 2000, Steinman 2001a, Chabas et al. 2001). This concerted attack of immune cells, the complement cascade, and inflammatory mediators including cytokines, osteopontin, and nitric oxide produces areas of demyelination, impairing electrical conduction along the axon and producing the pathophysiologic defect.

TNF-α, lymphotoxin, and other members of the TNF receptor family may play a key role in the pathogenesis of oligodendroglial damage. TNF is elevated in the spinal fluid in relapses of MS. TNF has been demonstrated in MS lesions. Myelin basic protein reactive T cells from MS patients that express the disease-associated HLA DRB1*15 allele produce increased levels of TNF-α. Experimental trials with altered peptide analogues of myelin basic protein, which downregulate the expression of TNF and upregulate Th2 cytokines, reveal a decrease in the size of new lesions in white matter on magnetic resonance scans (Kappos et al. 2000). Patients receiving the altered peptide ligand, who had active scans before treatment, showed reduced volume of enhancement on magnetic resonance (MR) scans after four months of treatment in 17 of 21 cases (Kappos et al. 2000). There was no increase in disease exacerbations and no worsening of disability in this 144-patient placebo-controlled double-blind trial. Local allergy was seen in 9% of patients, similar to what is seen in Copaxone-treated patients and perhaps consistent with a shift to Th2 (Kappos et al. 2000).

Bielekova and colleagues studied eight patients treated with the same APL (Bielekova et al. 2000). They reported exacerbations in three of these patients. There was an increase in MBP reactive T cells in one patient with MS who had an exacerbation, as well as another patient with both demyelinating central and peripheral nervous system disease, who had a relapse. In another patient who worsened, in this study, there was initial improvement in chronic symptoms and improvement on MR scans. After five months there was a relapse, but MBP reactive T cells had disappeared after treatment with the altered peptide (Bielekova et al. 2000).

In *Nature Immunology*, this laboratory summarized the two APL trials in this way:

> A particular attractive strategy has been to use autoantigenic peptides with modifications in TCR contact positions, that is, altered peptide ligands (APLs). APLs can mediate anergy, TCR antagonism or, most interestingly, bystander suppression. The latter mechanism refers to the induction of an APL-specific TH2-like cell population that cross-reacts with the native autoantigen and thus dampens immune responsiveness whenever autoantigen is released. Recently, an APL of the immunodominant MBP peptide (amino acids 83–99) was tested in phase II trials [(Bielekova et al. 2000, Kappos et al. 2000)]. Several interesting observations emerged. Unexpectedly, the APL induced allergic reactions in a substantial number of MS patients. In addition, at a high dose, some patients showed disease exacerbations that were mediated by APL-specific TH1 cells with cross-reactivity with the native MBP peptide [(Bielekova et al. 2000)]. In contrast, a lower dose showed a trend towards clinical benefit, probably via a TH2 shift [(Kappos et al. 2000)]. Although these studies show that our basic concepts about disease induction by specific autoantigens are probably correct, they also highlight that the correct dose of APL and its route of administration need further investigation [(Martin et al. 2001)]. Reprinted with permission from *Nature Immunology*.

Given the placebo-controlled double-blind study performed by Kappos and colleagues and the smaller open trial reported by Bielekova and colleagues, optimization of dosage and timing of administration may allow further trials of this promising approach—an approach that involves shifting the balance of cytokines from autoaggressive to suppressive (Genain & Zamvil 2000).

Similar also is the approved drug, Copaxone, which induces a shift toward Th2 cytokine production by T cells reactive to myelin (Duda et al. 2000). Copaxone reduces the frequency of relapses in early MS and decreases the degree of inflammatory activity in white matter on magnetic resonance images of white matter. Copaxone is associated with allergic responses in about 10% of patients as well (Kappos et al. 2000). Recently, Pedotti and colleagues have shown that myelin antigens can induce classic anaphylactic responses, though these responses can be easily contained with antihistamines (Pedotti et al. 2001). Clearly induction of Th2 shifts may be beneficial for autoimmunity but may then induce allergy to self.

New Targets Revealed in MS Pathology with High Throughput Sequencing of MS Brain Libraries

This laboratory recently published a large-scale analysis of transcripts in MS brain plaques (Chabas et al. 2001).[1] "High throughput sequencing of expressed sequence tags (EST), utilizing non-normalized cDNA brain libraries generated from MS brain lesions and control brain, has revealed a potential role for osteopontin in the progression of MS (Chabas et al. 2001). Using this protocol the mRNA populations present in the brain specimens are accurately represented, which enables the quantitative estimation of transcripts and comparisons between specimens" (Chabas et al. 2001) (Table 1). Molecular mining of two sequenced libraries and their comparison with a normal brain library, matched for size and tissue type and constructed with an identical protocol, revealed that osteopontin (OPN) transcripts were frequently detected and exclusive to the MS mRNA population but not to that of control brain (Table 1).

We sequenced over 11,000 clones from MS libraries 1 and 2 and control libraries, respectively, and focused analysis on genes present in both MS libraries but absent in the control library (Chabas et al. 2001). This yielded 423 genes, including 26 novel genes. From those, 54 genes showed a mean fold change of 2.5 or higher in MS libraries 1 and 2 (Table 1). Transcripts for B-crystallin, an inducible heat shock protein localized in the myelin sheath and known to be targeted by T cells in MS, were the most abundant transcripts unique to MS plaques (Table 1). The next five most abundant transcripts included those for prostaglandin D synthetase, prostatic binding protein, ribosomal protein L17, and OPN.

We also analyzed all genes present in each of the three cDNA libraries and found 330 (7 novel) genes. "Based on the clone count of each sequenced gene, a table

---

[1] We quote extensively from this source. Reprinted with permission from *Science*.

TABLE 1 Genes transcribed in MS brain plaques[a]

Gene description	MS1 abundance	MS2 abundance	Average clone count	Cellular function	Genomic location
Alpha B-crystallin	7	12	9.5	Cell structure/motility	11q22.3-q23.1
Prostaglandin D synthase	8	7	7.5	Cell signaling/cell communication	9q34.2-34.3
Prostatic binding protein	6	7	6.5	Cell signaling/cell communication	12q24.1*
Ribosomal protein L17	10	2	6	Gene/protein expression	18q
Osteopontin	8	3	5.5	Cell structure/motility/ signaling	4q21-q25
KIAA1376	6	3	4.5	Unclassified	5
Vimentin	4	5	4.5	Cell structure/motility	10p13
Cardiac gap junction protein	5	3	4	Cell signaling/cell communication	6q21-q23.2*
DNA-binding protein	4	4	4	Gene/protein expression	12q23-24.1*
G protein–coupled receptor	2	6	4	Cell signaling/cell communication	16p12
ATPase, Na/K transporting, alpha 2 (KIAA0778)	2	6	4	Cell signaling/cell communication	1q21-q23
MORF-related gene X	1	7	4	Gene/protein expression	Xq22*
KIAA0365	1	7	4	Unclassified	19p12
Gp130 associated protein GAM	6	1	3.5	Unclassified	19p13.3
Cathepsin D	6	1	3.5	Cell/organism defense	11p15.5
Dihydropyrimidinase-related protein-3	6	1	3.5	Metabolism	5q32
Stathmin	4	3	3.5	Cell division	1p36.1-p35*
Ribosomal protein L41	4	3	3.5	Unclassified	22q12
Human amyloid precursor-like protein 1	4	3	3.5	Unclassified	N/A
Small acidic protein	3	4	3.5	Unclassified	N/A
Selenoprotein W	3	4	3.5	Metabolism	19q13.3*
Tyrosine and tryptophan hydroxylase activator	2	5	3.5	Cell signaling/cell communication	22q12.3
KIAA0517 (brain)	4	2	3	Unclassified	4q28
PHR1 isoform 4 [Mus musculus]	3	3	3	Unclassified	N/A
Phosphoglycerate mutase, brain	2	4	3	Metabolism	10q25.3
Heat shock 60kD protein 1 (chaperonin)	2	4	3	Cell/organism defense	2
HUMCD59A Human lymphocytic antigen CD59/MEM43	2	4	3	Unclassified	11p13

(Continued)

TABLE 1 (*Continued*)

Gene description	MS1 abundance	MS2 abundance	Average clone count	Cellular function	Genomic location
Similar to Myc	2	4	3	Unclassified	N/A
Rac1 gene	2	4	3	Cell signaling/cell communication	Xq26.2-27.2
Septin 3	1	5	3	Cell division	22q13.1
Human translation initiation factor 5	1	5	3	Gene/protein expression	14q32*
Cysteine string protein [Bos taurus]	2	3	2.5	Unclassified	N/A
Cytochrome c oxidase subunit IV	4	1	2.5	Metabolism	16q24.1
Apolipoprotein D	4	1	2.5	Metabolism	3q26.2-qter
Cystatin C (cysteine proteinase inhibitor precursor)	4	1	2.5	Metabolism	20p11.2
Clathrin assembly protein lymphoid myeloid leukemia	4	1	2.5	Unclassified	11q14
Metallothionein-II pseudogene	4	1	2.5	Unclassified	4p11-q21
Beta-synuclein	3	2	2.5	Unclassified	5q35
Lactate dehydrogenase B	3	2	2.5	Metabolism	12p12.2-p12.1
FK-506 binding protein homologue (FKBP38)	3	2	2.5	Cell signaling/cell communication	19p12
NADH: ubiquinone oxidoreductase B22 subunit	3	2	2.5	Metabolism	8q13.3
KIAA0582 (brain)	3	2	2.5	Unclassified	2p12
Centromere autoantigen B	3	2	2.5	Unclassified	20p13
Basigin	3	2	2.5	Cell signaling/cell communication	19p13.3
S-adenosyl homocysteine hydrolase-like 1	3	2	2.5	Metabolism	1
Chaperonin containing TCP1, subunit 8 (theta)	2	3	2.5	Gene/protein expression	21q22.11
Transcription elongation factor B (SIII), polypeptide 1-lil	2	3	2.5	Gene/protein expression	5q31
Testis enhanced gene transcript	2	3	2.5	Cell division	12q12-q13
Quinoid dihydropteridine reductase	2	3	2.5	Metabolism	4p15.31
Non muscle myosin alkali light chain	2	3	2.5	Unclassified	12
Matrix Gla protein	2	3	2.5	Unclassified	12p13.1-p12.3
CGI-49 protein	2	3	2.5	Unclassified	1
Human EST H08032.1 (NID: g872854)	2	3	2.5	Unclassified	7q11.23-q21.1

[a]Only genes with a mean fold change of >2.5 are listed. N/A, mapping position is not known. *, genomic regions that reached nominal criteria of linkage in genome-wide screenings. Reprinted from *Science* (Chabas et al. 2001).

was constructed with transcripts showing an average fold difference (AFD) equal to or greater than ±2.00 between MS and control. Forty of these transcripts were divided into three levels, based on the consistency of differential expression across libraries. Some of these genes included myelin basic protein (MBP), heat shock protein 70 (HSP-70), glial fibrillary acidic protein (GFAP), and synaptobrevin. MBP transcripts displayed consistent high levels of expression in the three libraries, which suggests a very high turnover rate for this protein. Expression of HSP70-1, which is involved in myelin folding, was significantly elevated. Although not differentially expressed, GFAP was among the three most abundant species in all the libraries, consistent with a prominent glial (or astrocytic) response in the MS brains. Six genes belonging to the KIAA group of large-size cloned mRNAs showed differential expression. The decreased transcription of synaptobrevin may be of interest given that it belongs to a family of small integral membrane proteins specific for synaptic vesicles in neurons." Recent evidence indicates that axonal loss is one of the major components of pathology in MS (Pitt et al. 2000, Chabas 2001).

Given the known inflammatory role for OPN, we examined the cellular expression pattern of this protein in human MS plaques and in control tissue by immunohistochemistry. Within active MS plaques, OPN was found on microvascular endothelial cells and macrophages and in white matter adjacent to plaques. Reactive astrocytes and microglia also expressed OPN.

"The potential role of OPN in demyelinating disease was next tested using OPN-deficient mice. EAE was induced using myelin oligodendroglial peptide MOG 35-55 in complete Freund's adjuvant (CFA)in OPN−/− mice and OPN+/+ controls. EAE was observed in 100% of both OPN+/+ and OPN−/− mice with MOG 35-55" (Chabas et al. 2001). Despite this equivalence between the two groups in the incidence of disease, severity of disease was significantly reduced in all animals in the OPN−/− group, and these mice were totally protected from EAE-related death. Thus, OPN significantly influenced the course of progressive EAE induced by MOG 35-55.

"The rate of relapses and remissions was next tested. During the first 26 days, OPN−/− mice displayed a distinct evolution of EAE with a much higher percentage of mice having remissions compared to the controls. To examine whether different immune responses were involved in OPN−/− and OPN+/+ animals, we tested the profile of cytokine expression in these mice" (Chabas et al. 2001). Because "EAE is a T cell–mediated disease, we first analyzed the T cell proliferative response to the auto-antigen MOG 35-55 in the OPN−/− mice. T cells in OPN−/− mice showed a reduced proliferative response to MOG 35-55, compared with OPN+/+ T cells. In addition, IL-10 production was increased in T cells reactive to MOG 35-55 in OPN−/− mice that had developed EAE, compared with T cells in OPN+/+ mice. At the same time, IFN-γ and IL-12 production was diminished in the cultures of spleen cells stimulated with MOG" (Chabas et al. 2001).

Because IFN-γ and IL-12 are important proinflammatory cytokines in MS, the finding that in OPN−/− mice there is reduced production of these cytokines is consistent with the hypothesis that OPN may play a critical role in the modulation of Th1 immune responses in MS and EAE. Further, IL-10 has been associated with remission from EAE. In this context, the enhancement of myelin-specific IL-10 production in OPN−/− mice may account for the tendency of these mice to go into remission. Sustained expression of IL-10 may thus be an important factor in the reversal of relapsing MS, and its absence may allow the development of secondary progressive MS. OPN also has a more dramatic effect on IL-12 than TNF-α. Downregulating TNF-α may have deleterious consequences (Liu et al. 1998, Lenercept Study Group 1999), though OPN may target IL-12 and downregulate Th1 responses without influencing TNF-α.

The identification of γ-interferon, osteopontin, and IL-12 as potential mediators of progression in demyelinating disease may offer some explanation for the recent failures of inhibition of TNF-α in MS patients (Lenercept Study Group 1999, van Oosten et al. 1996). γ-interferon enhances major histocompatibility class II expression in the nervous system. IL-12 potentiates this effect. On the other hand TNF-α may inhibit IL-12 (Ma 2001, Steinman 2001d). TNF-α suppresses responses of both Th1 and Th2 cells (Cope et al. 1997), while anti-TNF-α may enhance T cell proliferation and cytokine production. These results may have contributed to the failure of anti-TNF in the clinic in MS. Inhibition of TNF-α may block important proinflammatory feedback loops and contribute to the worsening of MS. The FDA required that the manufacturer of Enbrel, a soluble TNF receptor-immunoglobulin Fc construct approved for use in rheumatoid arthritis, notify physicians that patients who have MS and are also being treated for rheumatoid arthritis should not receive this drug because of its worsening effect on MS. Alternatively, because there is strong evidence that anti-TNF-α is suppressive in various models of MS, these reagents may simply not have penetrated the blood brain barrier adequately (Steinman 2001d). However, in one model of EAE, TNF-α itself ameliorated disease (Liu et al. 1998).

"OPN may have pleiotropic functions in the pathogenesis of demyelinating disease. OPN production by glial cells may lead to the attraction of Th1 T cells, and its presence in glial and ependymal cells may allow inflammatory T cells to penetrate the brain. Finally, our data suggest that neurons may also secrete this proinflammatory molecule and participate in the autoimmune process. Potentially, neuronal OPN secretion could modulate inflammation and demyelination and influence the clinical severity of the disease. Consistent with this idea, a role for neurons in the pathophysiology of MS and EAE has recently been described (Pitt et al. 2000, Steinman 2000), and neurons are known to be capable of cytokine production" (Chabas et al. 2001).

CD44 is a known ligand of OPN that mediates a decrease of IL-10 production. As we have recently shown, OPN−/− mice produced elevated IL-10 during the course of EAE. We recently demonstrated that anti-CD44 antibodies prevented EAE (Brocke et al. 1999), which suggests that the proinflammatory effect of OPN in MS and EAE may be mediated by CD44. OPN is clearly situated at a number of checkpoints that would allow diverse activities in the course of autoimmune-mediated demyelination.

DEGENERATION OF THE MYELIN SHEATH AND THE AXON

During chronic MS, when exacerbations and remissions of MS are rare, there is evidence for axon loss and atrophy of the brain and spinal cord (Trapp et al. 1998). AMPA (α-amino-3-hydroxy-5-methyl-4-isoxazolepropionic acid)/kainate receptors, which mediate toxicity induced by the excitatory neurotransmitter glutamate, are present on oligodendroglial cells and neurons in EAE and in MS. During inflammation in both MS and its animal model, EAE, lymphocytes, brain microglia, and macrophages release excessive amounts of glutamate, which then activate AMPA receptors (Figure 1, see color figure). Blockade of these receptors with antagonists ameliorates EAE, including clinical relapses when treatment is begun after the onset of paralysis (Pitt et al. 2000, Smith et al. 2000). The blockade of AMPA/kainate receptors does not influence the immune response to myelin antigens, but it somehow protects oligodendroglial cells and axons from immune-mediated damage. Damage may be mediated by increased fluxes of calcium, which may cause necrotic damage to oligodendroglial cells and axons. The use of neuroprotective agents that block sub-types of glutamate receptors has been a prime direction in the development of new therapies for stroke and neurodegenerative conditions, and this approach may prove useful for treatment of the chronic degenerative phase of MS as well.

NEW THERAPEUTIC APPROACHES IN MS

Recognition of an inflammatory and a neurodegenerative phase of MS has allowed the targeting of therapies specific for various phases of MS. Thus, drugs like β-interferon interfere with lymphocyte migration to brain, and altered peptides and Copaxone alter the cytokine production by autoimmune T cells, while glutamate receptor antagonists block the insidious atrophy and death of oligodendroglial cells and the underlying axon. Current treatments with beta interferons and Copaxone only reduce the frequency of relapses by a third (Steinman 2001c). Other approaches use statins, which block the development of EAE and inhibit LFA-1 (Stanislaus et al. 2001), and antihistamines, which engage H1 receptors found in MS brain can block EAE (Pedotti et al. 2001); this may provide new approaches for previously approved drugs. Finally, it is now possible to reverse ongoing paralysis in the EAE model, by tolerizing the immune system via injection of DNA encoding myelin antigens along with DNA encoding the Th2 cytokine IL-4 (Garren et al. 2001). DNA vaccination has been taken into the clinic for infectious disease and cancer, and trials are now being organized to apply this approach to autoimmune diseases, including MS. It is hoped that these new discoveries will lead to improved treatments and ultimately to better control of MS, as well as a cure for this disease (Steinman 2001c).

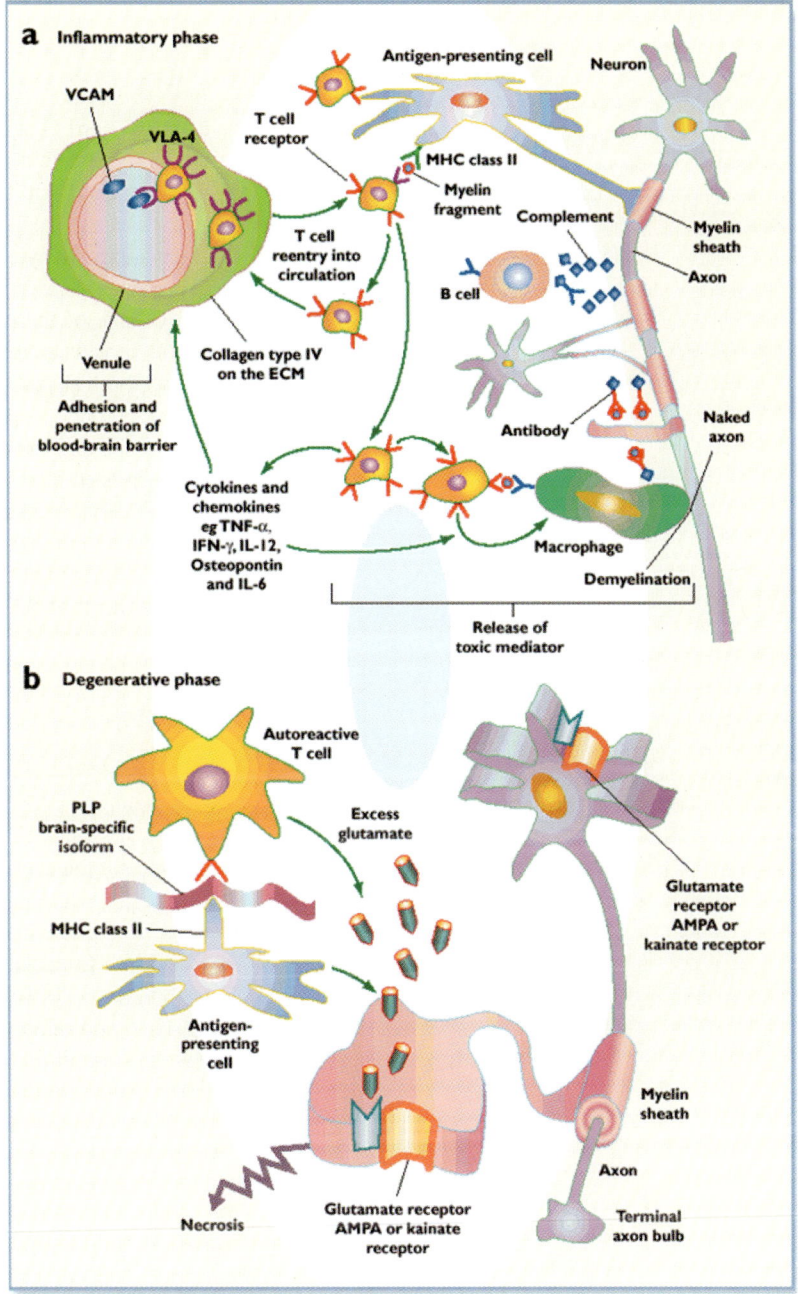

Figure 1 The two stages of the progression of MS: (*a*) the autoimmune attack and (*b*) neurodegeneration. Reprinted from *Nature Immunology*, Steinman 2001a.

ACKNOWLEDGMENTS

The support of the Phil N. Allen trust and the National Institutes of Health is gratefully acknowledged.

The *Annual Review of Neuroscience* is online at http://neuro.annualreviews.org

LITERATURE CITED

Akassoglou K, Probert L, Kontogeorgos G, Kollias G. 1997. Astrocyte-specific but not neuron-specific transmembrane TNF triggers inflammation and degeneration in the central nervous system of transgenic mice. *J. Immunol.* 158:438–45

Arnett H, Mason J, Suzuk K, Matsushima G, Ting J. 2001. TNFalpha promotes proliferation of oligodendrocyte progenitors and remyelination. *Nat. Neurosci.* 4:1116–22

Baranzini SE, Elfstrom C, Chang SY, Butunoi C, Murray R, et al. 2000. Transcriptional analysis of multiple sclerosis brain lesions reveals a complex pattern of cytokine expression. *J. Immunol.* 165(11):6576–82

Baranzini SE, Jeong MC, Butunoi C, Murray RS, Bernard CC, Oksenberg JR. 1999. B cell repertoire diversity and clonal expansion in multiple sclerosis brain lesions. *J. Immunol.* 163(9):5133–44

Bielekova B, Goodwin B, Richert N, Cortese I, Kondo T, et al. 2000. Encephalitogenic potential of the myelin basic protein peptide [amino acids 83-99] in multiple sclerosis: results of a phase II clinical trial with an altered peptide ligand. *Nat. Med.* 6:1167–75

Brocke S, Piercy C, Steinman L, Weissman IL, Veromaa T. 1999. Antibodies to CD44 and integrin alpha 4, but not L-selectin, prevent CNS inflammation and experimental encephalomyelitis by blocking secondary leukocyte recruitment. *Proc. Natl. Acad. Sci. USA* 96:6896–6901

Brosnan CF, Raine CS. 1996. Mechanisms of immune injury in multiple sclerosis. *Brain Pathol.* 6:243–57

Cannella B, Raine CS. 1995. The adhesion molecule and cytokine profile of multiple sclerosis lesions. *Ann. Neurol.* 37:424–35

Chabas D, Baranzini S, Mitchell D, Bernard CCA, Rittling S, et al. 2001. The influence of the pro-inflammatory cytokine, osteopontin, on autoimmune demyelinating disease. *Science* 294:1731–35

Conlon P, Oksenberg JR, Zhang J, Steinman L. 1999. The immunobiology of multiple sclerosis: an autoimmune disease of the central nervous system. *Neurobiol. Dis.* 6:149–66

Cope AP, Liblau RS, Yang XD, Congia M, Laudanna C, et al. 1997. Chronic TNF alters T cell responses by attenuating T cell receptor signaling. *J. Exp. Med.* 185:1573–84

Duda PW, Krieger JI, Schmied MC, Balentine C, Hafler DA. 2000. Human and murine CD4 T cell reactivity to a complex antigen: recognition of the synthetic random polypeptide glatiramer acetate. *J. Immunol.* 165(12):7300–7

Ebers GC, Kukay K, Bulman DE, Sadovnick AD, et al. 1996. A full genome search in multiple sclerosis. *Nat. Genet.* 13:472–76

Garren H, Ruiz PJ, Watkins TA, Fontoura P, Nguyen L-VT, et al. 2001. Combination of gene delivery and DNA vaccination to protect from and reverse Th1 autoimmune disease via deviation to the Th2 pathway. *Immunity* 15:15–22

Genain CP, Cannella B, Hauser SL, Raine CS. 1999. Identification of autoantibodies associated with myelin damage in multiple sclerosis. *Nat. Med.* 5:170–75

Genain CP, Zamvil SS. 2000. Specific immunotherapy: one size does not fit all. *Nat. Med.* 1098–100

Gijbels K, Galardy R, Steinman L. 1994. Reversal of EAE with a hydroxymate inhibitor of matrix metalloproteases. *J. Clin. Investig.* 94:2177–82

Haines JL, Ter-Minassian M, Bazyk A, Gusella JF, Kim DJ, et al. 1996. A complete genomic screen for multiple sclerosis underscores a role for the major histocompatibility complex. The Multiple Sclerosis Genetics Group. *Nat. Genet.* 13:469–71

Hartung HP. 1995. Pathogenesis of inflammatory demyelination: implications for therapy. *Curr. Opin. Neurol.* 8:191–99

Kappos L, Comi G, Panitch H, Oger J, Antel J, et al., and the APL in Relapsing MS Study Group. 2000. Induction of a non-encephalitogenic Th2 autoimmune response in multiple sclerosis after administration of an altered peptide ligand in a placebo controlled, randomized phase II trial. *Nat. Med.* 6(10):1176–82

Kern A, Brisewitz R, Bank I, Marcantonio EE. 1994. The role of the I domain in ligand binding of the human integrin $\alpha 1 \beta 1$. *J. Biol. Chem.* 269:22811–16

Krogsgaard M, Wucherpfennig KW, Canella B, Hansen BE, Svejgaard A, et al. 2000. Visualization of myelin basic protein T cell epitopes in multiple sclerosis lesions using a monoclonal antibody specific for the human histocompatibility leukocyte antigen DR2-MBP 85-99 complex. *J. Exp. Med.* 191:1395–1412

The Lenercept Multiple Sclerosis Study Group and The University of British Columbia MS/MRI Analysis Group. 1999. TNF neutralization in MS: results of a randomized, placebo-controlled multicenter study. *Neurology* 53(3):457–65

Liu J, Marino MW, Wong G, Grail D, Dunn A, et al. 1998. TNF is a potent anti-inflammatory cytokine in autoimmune-mediated demyelination. *Nat. Med.* 4(1):78–83

Loseff NA, Webb SL, O'Riordan JI, Page R, Wang L, et al. 1996. Spinal cord atrophy and disability in MS. A new reproducible and sensitive MR method with potential to monitor disease progression. *Brain* 119:701–8

Ma X. 2001. TNF-alpha and IL-12: a balancing act in macrophage functioning. *Microbes Infect.* 3:121–29

Martin R, Sturzebecher CS, McFarland HF. 2001. Immunotherapy of multiple sclerosis: Where are we? Where should we go? *Nat. Immunol.* 2:785–88

Oksenberg JR, Panzara MA, Begovich AB, Mitchell D, Erlich HA, et al. 1993. Selection for T cell receptor $V\beta$-$D\beta$-$J\beta$ gene rearrangements with specificity for a myelin basic protein peptide in brain lesions of multiple sclerosis. *Nature* 362:68–70

Pedotti R, Mitchell D, Wedemeyer J, Karpuj M, Chabas D, et al. 2001. An unexpected version of horror autotoxicus: anaphylactic shock to a self-peptide. *Nat. Immunol.* 2:216–22

Pitt D, Werner P, Raine C. 2000. Glutamate excitotoxicity as a mechanism in multiple sclerosis. *Nat. Med.* 6:67–70

Probert L, Akassoglou K, Pasparakis M, Kontogeorgos G, Kollias G. 1995. Spontaneous inflammatory demyelinating disease in transgenic mice showing central nervous system-specific expression of tumor necrosis factor alpha. *Proc. Natl. Acad. Sci. USA* 92:11294–98

Robinson WH, DiGennaro C, Hueber W, Haab B, Kamachi M, et al. 2002. Antigen arrays for multiplex characterization of autoantibody responses. *Nat. Med.* 8:295–301

Sawcer S, Jones HB, Feakes R, Gray J, Smaldon N, et al. 1996. A genome screen in multiple sclerosis reveals susceptibility loci on chromosome 6p21 and 17q22. *Nat. Genet.* 13:464–68

Smith T, Groom A, Zhu B, Turski L. 2000. Autoimmune encephalomyelitis ameliorated by AMPA antagonists. *Nat. Med.* 6:62–66

Stanislaus R, Singh A, Singh I. 2001. Lovastatin treatment decreases mononuclear cell infiltration into the CNS of Lewis rats with EAE. *J. Neurosci. Res.* 66:155–62

Steinman L. 2000. Multiple approaches to multiple sclerosis. *Nat. Med.* 6:15–16

Steinman L. 2001a. Multiple sclerosis: a two stage disease. *Nat. Immunol.* 2:762–65

Steinman L. 2001b. Gene microarrays and experimental demyelinating disease: a tool to enhance serendipity. *Brain* 124:1897–99

Steinman L. 2001c. Immunotherapy of multiple sclerosis: the end of the beginning. *Curr. Opin. Immunol.* 13(5):597–600

Steinman L. 2001d. Blockade of gamma interferon might be beneficial in MS. *Multiple Scler.* 7:275–76

Steinman L, Waisman A, Altmann A. 1995. Major T cell responses in multiple sclerosis. *Mol. Med. Today* 1:79–83

Trapp BD, Peterson J, Ransohoff RM, Rudick R, Mork S, Bo L. 1998. Axonal transection in MS lesions. *N. Engl. J. Med.* 338:278–85

van Oosten BW, Barkhof F, Truyen L, Boringa JB, Bertelsmann FW, et al. 1996. Increased MRI activity and immune activation in two multiple sclerosis patients treated with the monoclonal anti-tumor necrosis factor antibody cA2. *Neurology* 47:1531–34

Warren KG, Catz I, Steinman L. 1995. Fine specificity of the antibody response to myelin basic protein in the central nervous system in multiple sclerosis: the minimal B cell epitope and a model of its unique features. *Proc. Natl. Acad. Sci. USA* 92:11061–65

Wucherpfenig KW, Catz I, Hausmann S, Strominger JL, Steinman L, Warren KG. 1997. Recognition of the immunodominant myelin basic protein peptide by autoantibodies and HLA-DR2 restricted T cell clones from multiple sclerosis patients: identity of key contact residues in the B-cell and T-cell epitopes. *J. Clin. Investig.* 100:1114–22

Wucherpfennig KW, Strominger JL. 1995. Molecular mimicry in T-cell mediated autoimmunity: viral peptides activate T cell clones specific for myelin basic protein. *Cell* 80:695–705

WIRED FOR REPRODUCTION: Organization and Development of Sexually Dimorphic Circuits in the Mammalian Forebrain

Richard B. Simerly

Division of Neuroscience, Oregon Regional Primate Research Center, Oregon Health and Sciences University, Beaverton, Oregon 97006; email: simerlyr@ohsu.edu

Key Words hypothalamus, limbic, sexual differentiation, amygdala, hippocampus

■ **Abstract** Mammalian reproduction depends on the coordinated expression of behavior with precisely timed physiological events that are fundamentally different in males and females. An improved understanding of the neuroanatomical relationships between sexually dimorphic parts of the forebrain has contributed to a significant paradigm shift in how functional neural systems are approached experimentally. This review focuses on the organization of interconnected limbic-hypothalamic pathways that participate in the neural control of reproduction and summarizes what is known about the developmental neurobiology of these pathways. Sex steroid hormones such as estrogen and testosterone have much in common with neurotrophins and regulate cell death, neuronal migration, neurogenesis, and neurotransmitter plasticity. In addition, these hormones direct formation of sexually dimorphic circuits by influencing axonal guidance and synaptogenesis. The signaling events underlying the developmental activities of sex steroids involve interactions between nuclear hormone receptors and other transcriptional regulators, as well as interactions at multiple levels with neurotrophin and neurotransmitter signal transduction pathways.

INTRODUCTION

A principal goal of brain development is to produce the necessary neural architecture for integration of information from the external environment with internal cues that reflect important aspects of an animal's physiological state. This integration allows the elaboration of adaptive behavioral and physiological responses that are essential for an individual's survival, as well as for propagation of the species. From an evolutionary perspective, the most adaptive physiological responses are those that ensure successful reproduction. The long-term consequences of adaptive behavioral profiles that enhance survival are of little significance if an animal lacks the reproductive fitness necessary to pass on its genome. Moreover, the coordination of physiological events with behavior is a prerequisite to successful reproduction. For example, it is of no benefit to a mammalian species if females display appropriate solicitation behaviors and successfully copulate with conspecific males but have not ovulated. Males have similar requirements for physiological coordination; an individual that has mature sperm and is ready to impregnate a female will not get the chance if he displays agonistic behaviors. Thus, the future of a species often rests with the ability of its members to coordinate behavioral responses with physiological processes in response to sexually relevant cues. This coordination of behavior and physiology must also be reliable, which depends in part on how consistently the neural circuits underlying neuroendocrine integration are constructed and regulated.

Mammals reproduce sexually; males and females of a species display distinct patterns of copulatory behaviors and neuroendocrine physiology (Gerall & Givon 1992, Gorski & Jacobson 1981). This array of sex-specific behaviors and physiological responses is so vital to the success of mammalian species that robust developmental mechanisms have evolved to produce distinct yet complimentary neural systems that ensure the coordinated expression of reproductive function in male and female mammals. In this review key aspects of sexually dimorphic neural systems in the rodent forebrain are examined to consider developmental mechanisms that may be responsible for specifying sex-specific aspects of these neural pathways. Although the regions dealt with in detail play major roles in reproduction, it is important to note that significant sexual dimorphisms have been documented throughout the central nervous system, from the cerebral cortex to spinal motor neurons; therefore, the process of sexual differentiation of the brain should be viewed as a widespread series of developmental events with functional significance for diverse behaviors and physiological responses.

The central tenet of sexual differentiation is that the brain is bipotential but develops differently in males and females under the influence of sex steroid hormones during the perinatal period. In male rats, secretion of androgen from the differentiated testis produces two perinatal elevations in plasma testosterone, the first of which occurs on day 18 of gestation, and the second at approximately 2 h after birth (Corbier et al. 1992, Weisz & Ward 1980). The resulting difference in exposure of the brain to testosterone, or to its metabolites dihydrotestosterone and estradiol (Simpson et al. 1994), causes the brain to change its structure and function. Thus, the perinatal steroid environment determines whether male or female copulatory behavior is expressed, or whether the pituitary gland is able to mount a preovulatory surge of gonadotropin secretion. Before a significant effort was made to identify sex differences in brain architecture, it was suspected that the biological basis of these functional dimorphisms is hormonal modification of the brain, and work carried out during the past two decades has provided strong support for this notion. A large number of morphological and neurochemical sexual dimorphisms that are dependent on exposure to sex steroid hormones during the perinatal period have been documented in the mammalian brain (De Vries & Simerly 2002, Gorski 1996, Madeira & Lieberman 1995, McEwen 2001). Although it is likely that genetic background influences the degree to which various sexually dimorphic regions develop and may influence expression of sexually dimorphic traits that are independent of hormone action (Arnold 1997), it is clear that many brain dimorphisms can be completely reversed by hormone treatment alone.

Sexually Dimorphic Forebrain Pathways

The hypothalamus plays a critical role in coordinating expression of reproductive behaviors and physiological responses with environmental cues. Its close anatomical and physiological relationship with the pituitary gland provides an effective means for coordinating diverse homeostatic processes through neuroendocrine regulation of hormone secretion. The hypothalamus also shares strong connections with the limbic region of the forebrain so it can effectively coordinate neuroendocrine responses with sensory cues that regulate motivated behavior. The preoptic region of the hypothalamus was the historical focus of early studies on morphological sex differences, owing in part to its dominant role in the regulation of copulatory behavior and gonadotropin secretion (Gerall & Givon 1992, Larsson 1979, Pfaff 1980), but also because it contained a high density of neurons that were known to be targets for steroid hormones (see Simerly 1993 for summary). The modern era of sexual differentiation research was ushered in when Raisman and Field used electron microscopy to identify the first clear sex difference in neuronal connectivity (Raisman & Field 1971). Because it was generally believed that stereotypic patterns of behavior were dependent on the organization of neural connections, their finding that sexually dimorphic patterns of synaptology in the dorsal part of the medial preoptic area could be reversed by treatment with testosterone provided the first clear evidence that the developmental role of sex steroid hormones on behavior may have a structural basis. However, subsequent efforts were not successful to demonstrate that the observed sex difference in synaptic relationships were responsible for specific dimorphisms in behavior and gonadotropin secretion. Nevertheless, this work brought together the experimental paradigms used to study sexual differentiation of neuroendocrine physiology with modern neuroanatomical approaches to developmental plasticity.

The developmental actions of sex steroid hormones are not limited to relatively subtle alterations in synaptic organization. Soon after the publication of the paper by Raisman & Field, Gorski and colleagues reported a dramatic sexual dimorphism in the size of a group of cells in the medial preoptic area of the rat, which they designated the sexually dimorphic nucleus of the preoptic area (Gorski et al. 1978). Furthermore, they showed that perinatal exposure to testosterone or estrogen could cause a nucleus to form in females equal in size to that of males. This sensitivity to hormones declines after the first week of life (Gorski 1985). These observations had an enormous impact on the field because they demonstrated that sex steroid hormones can cause profound changes in regional anatomy, and they implied that testosterone and/or estrogen can specify cell number in hypothalamic nuclei. Because the number, size, or density of sexually dimorphic morphological features tended to be greater in males, a general principle emerged that steroid exposure during the perinatal period promoted neuronal development. However, the anteroventral periventricular nucleus (AVPV) of the preoptic region was found to be larger in female rodents, suggesting that sexual dimorphisms may also favor females (Bleier et al. 1982). The demonstration that the AVPV contained a greater number of dopaminergic neurons in females, which can be reduced to that of males by a single injection of testosterone, indicated that sex steroid hormones may actually facilitate loss of neurons in certain regions (Simerly et al. 1985).

THE MEDIAL PREOPTIC NUCLEUS   The sexually dimorphic nucleus of the preoptic area comprises neurons that are part of the medial preoptic nucleus (MPN), a nucleus known for its dominant role in expression of male sexual behavior (Gorski 1985, Larsson 1979). The MPN is a sexually dimorphic complex made up of three distinct subdivisions that can be distinguished on the basis of neurochemistry and cytoarchitecture (Simerly 1995b). The cell-dense central part of the MPN (MPNc) is the most dimorphic part of the nucleus with substantially more neurons in male rats than are found in females (Madeira et al. 1999). The medial part of the MPN (MPNm) is also larger in males and, like the MPNc, contains a high density of neurons that express large numbers of receptors for estrogen and androgen (Simerly et al. 1990). The sexually dimorphic nucleus identified by Gorski and collegues was not originally defined in anatomical terms but was later shown to correspond to the MPNc and subpopulations of cells in the MPNm. The observation that in males MPNc neurons display an accelerated decline in neuron number postnatally provided support for the notion that exposure to sex steroids enhanced their survival, which led to more neurons in the male MPNc (Dodson & Gorski 1993).

Each subdivision of the MPN shows a distinct pattern of connectivity: The MPNm sends its strongest projections to the periventricular zone of the hypothalamus, which is primarily involved in the control of hormone secretion from the anterior pituitary, while the MPNc sends its major projections to other sexually dimorphic forebrain nuclei (Simerly & Swanson 1988). Considering the widespread pattern of projections of the MPN, and the predominately bidirectional nature of these connections, it is clear why it proved so difficult to define discrete functional roles for the sexually dimorphic nucleus of the preoptic region or for the synaptic dimorphism studied by Raisman and Field. These regions do not function as centers but rather participate in interrelated functional neural systems that collectively integrate diverse aspects of reproductive function and contribute to multiple aspects of homeostasis. This problem is not unique to the study of neural mechanisms underlying reproduction but is being confronted by investigators studying how a variety of homeostatic functions are mediated by interrelated forebrain pathways (Kruk 1991, Sawchenko et al. 2000, Watts 2001), and by researchers interested in how distributed neural networks mediate cognition. Thus, the MPNc lies at the center of what can be viewed as a limbic-hypothalamic neural network of regions that develop under the influence of sex steroid hormones and collectively influence reproduction differently in males and females, as well as impacting other sexually dimorphic aspects of neuroendocrine function.

THE ANTEROVENTRAL PERIVENTRICULAR NUCLEUS (AVPV) Because gonadotropin secretion is perhaps the most significant sex difference in reproductive physiology, some of the earliest studies of sexual differentiation focused on the impact of sex steroid hormones on the phasic secretion of luteinizing hormone (LH), which initiates ovulation in female mammals (see Gerall & Givon 1992 for review). Treatment of ovariectomized adult female rats with estrogen causes a massive surge in LH secretion, yet similar treatments in males fail to induce a similar response. This sexually dimorphic response to hormone treatment can be reversed by castrating male rats at birth, and treatment of neonatal female rats with a single dose of testosterone results in permanent anovulatory sterility. Evidence from a variety of experimental approaches indicates that sex steroids act at the level of the preoptic region during postnatal life to organize the neural pathways controlling preovulatory gonadotropin secretion. The AVPV is a likely site of action because it plays a critical role in controlling the preovulatory LH surge and is sensitive to the developmental actions of sex steroid hormones (see Simerly 1998, Terasawa et al. 1980 for reviews). Consistent with its neuroendocrine role, neurons in the AVPV primarily innervate other parts of the periventricular zone and provide direct projections to gonadotropin releasing hormone (GnRH) neurons, as well as to tuberoinfundibular neurons (TIDA) in the arcuate nucleus that control secretion of prolactin (Gu & Simerly 1997). These projections appear to be more robust in females, which is not surprising given the larger volume of the AVPV in this sex, and is consistent with earlier ultrastructural studies that revealed a greater synaptic density in the arcuate nucleus of the hypothalamus in female rats (Matsumoto et al. 2000). Dendritic arborization of arcuate neurons also appears to be greater in female rats, although there are slightly more neurons in the nucleus in males (Leal et al. 1998).

The total number of neurons in the AVPV has not been determined in male and female rats, but cellular markers for dopaminergic neurons and peptidergic neurons (Simerly 1989, 1991) have identified subpopulations of AVPV neurons that are more numerous in females. It is surprising that neurons that contain the opioid peptide enkephalin are more abundant in male rats, which demonstrates that the developmental effects of sex steroids show considerable cell-type specificity, even within a single nucleus (Simerly 1991). Some of the strongest afferents to the AVPV are from other sexually dimorphic nuclei such as the MPNm/c. It is also heavily innervated by the accessory olfactory and septohippocampal pathways that innervate other sexually dimorphic parts of the hypothalamus (Simerly 1998). Thus, the AVPV is of interest not only because it illustrates the region and cell-type–specific activity of sex steroid hormones in regulating neuronal development, but also because it represents a site for sensory convergence at an interface between the limbic region of the telencephalon and neuroendocrine circuits.

PARALLEL LIMBIC-HYPOTHALAMIC SENSORY PATHWAYS In addition to the AVPV and MPN, several other forebrain regions are known to undergo sexual differentiation (Figure 1). Each of these regions contains high densities of neurons that express receptors for steroid hormones and are located in the hypothalamus or in limbic nuclei that have strong connections with the sexually dimorphic hypothalamic nuclei. Sexually dimorphic nuclei of the hypothalamus are either in the periventricular zone of the hypothalamus, as are the AVPV and arcuate nucleus, or are in the medial zone. Medial zone nuclei such as the MPN and ventrolateral part of the ventromedial hypothalamic nucleus (VMHvl) play central roles in mediating reproductive behavior (Meisel & Sachs 1994, Pfaff et al. 1994). The medial and central parts of the MPN share bidirectional projections with the VMHvl, and anatomical links between the medial and periventricular zone sexually dimorphic nuclei provide a possible substrate for coordination of copulatory behavior with gonadotropin secretion and associated autonomic responses. The major routes for sensory activation of this core circuitry of reproduction are via projections from the ventral subiculum of the hippocampal formation to the ventral part of the lateral septal nucleus and via the accessory olfactory pathway. Of

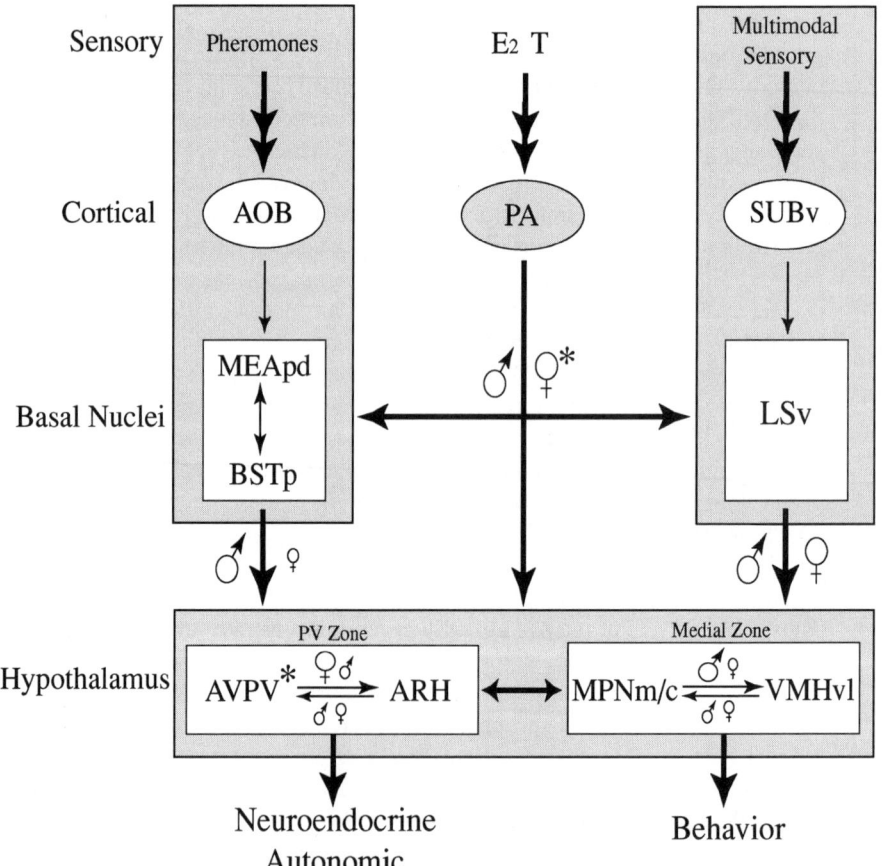

particular importance to reproduction are pheromonal cues from the vomeronasal organ relayed centrally by the accessory olfactory bulb, which is itself sexually dimorphic with more projection neurons in males (Guillamon & Segovia 1996, Simerly 1990) and which functions as primary olfactory cortex for pheromonal information (see Swanson & Petrovich 1998 for references and a discussion of this issue). This olfactory information is transmitted to the hypothalamus primarily by two sexually dimorphic nuclei: the medial nucleus of the amygdala and principal nucleus of the bed nuclei of the stria terminalis (BSTp). The posterodorsal part of the medial amygdaloid nucleus (MEApd) and BSTp are both larger in males (Guillamon & Segovia 1996), contain high densities of hormone-sensitive neurons (Shughrue et al. 1997, Simerly et al. 1990), and provide strong projections to sexually dimorphic nuclei in the periventricular and medial zones of the hypothalamus (Canteras et al. 1995, Gu et al. 2001, Simerly 1990). The dense projections of the MEApd and BSTp to the AVPV and arcuate nucleus of the hypothalamus

←──

Figure 1 Overview of sexually dimorphic pathways in the forebrain of the rat. The cell groups are organized to emphasize the major routes for sensory information impacting sexually dimorphic nuclei in the hypothalamus and to illustrate the relative representations of major pathways between forebrain sexually dimorphic cell groups in male and female animals. Two interconnected functional sexually dimorphic subcircuits in the hypothalamus mediate behavioral aspects of reproduction (MPNm/c and VMHvl in the medial zone of the hypothalamus) and associated neuroendocrine/autonomic responses (AVPV and ARH in the periventricular zone of the hypothalamus). Descending projections from the AVPV to the ARH are more robust in females, and descending projections from the MPNm/c to the VMH predominate in males. A descending pathway that is more robust in males conveys pheromonal information from the vomeronasal organ (which is also larger in males) to the hypothalamus via the accessory olfactory bulb (AOB) and nuclei in the amygdala and bed nuclei of the stria terminalis. A parallel descending pathway conveying multimodal sensory information to the hypothalamus via the subiculum and lateral septum appears to be represented similarly in male and female rats. The PA responds to fluctuations in circulating levels of estradiol (E2) and testosterone (T) and provides what is presumably a stimulatory input to both descending sensory pathways as well as to the hypothalamic effector nuclei. This pathway has roughly equal representation in males and females, but the asterisks denote one exception: The AVPV receives only a minor input from the PA in female rats. Four levels in a hypothetical functional hierarchy are proposed that correspond to a model advanced by Swanson for comparing functional neural systems in the mammalian forebrain (Swanson 2000). According to this scheme pathways that impact effector regions of the hypothalamus consist of a triple descending pathway composed of a glutamatergic cortical stimulatory projection and overlapping GABAergic projections from basal nuclei in a pattern analogous to the classical isocortical-striatal-pallidal model. See text for details and abbreviations.

(ARH) in the periventricular zone, and to the MPNm/c and VMHvl in the medial zone, are among the strongest descending projections of these sexually dimorphic nuclei. Thus, these limbic-hypothalamic pathways provide a particularly robust route for carrying pheromonal information to sexually dimorphic hypothalamic nuclei mediating reproduction.

Other sensory information that may impact the activity of sexually dimorphic nuclei is derived from multimodal association cortex and passes to the hypothalamus via glutamatergic projections from the ventral part of the subiculum and region CA1 of the hippocampal formation to the ventral part of the lateral septal nucleus (LSv) (Canteras & Swanson 1992). The LSv is a distinct part of the septal complex that, like the MEApd and BSTp, contains high densities of neurons that express sex steroid receptors and can be viewed functionally as a basal nucleus for the ventral subiculum and CA1. Its projections appear to be largely GABAergic (Risold & Swanson 1997a) and overlap with those of the MEApd and BSTp, which provides strong inputs to the AVPV and ARH, as well as to the MPNm/c and VMHvl (Risold & Swanson 1997b). However, no part of this septohippocampal-hypothalamic sensory pathway appears to be sexually dimorphic (F. Varoqueaux & R. B. Simerly, unpublished observations), although rostral and caudal parts of the lateral septal nucleus do receive a sexually dimorphic vasopressinergic innervation (De Vries & Miller 1998). Thus, the sexually dimorphic nuclei in the periventricular and medial zones of the hypothalamus receive completely overlapping inputs from two separate sensory pathways: Pheromonal information is conveyed along a sexually dimorphic pathway from the accessory olfactory bulb, and multimodal sensory information from isocortex is transmitted along a ventral subiculoseptal pathway that appears to be similar in males and females (Figure 1).

According to a recent model for telencephalic projections onto hypothalamic motor regions proposed by Swanson (2000), components of the accessory olfactory pathway, or ventral subiculoseptal pathway, can be viewed as being analogous to the cerebral cortex and basal ganglia. According to this model virtually all parts of the cerebral cortex provide a triple descending projection to brainstem regions involved in behavioral control. The cortex sends excitatory inputs to the basal nuclei such as the striatum and pallidum, which have descending inhibitory and disinhibitory projections, respectively. Thus, the major glutamatergic projection from the AOB (olfactory cortex) to the MEApd (striatum), and the massive GABAergic projections from the MEApd to the BSTp (pallidum), can be viewed as analogous to key aspects of basal ganglia circuitry. The MEApd and BSTp, in turn, send overlapping GABAergic projections to sexually dimorphic nuclei in the hypothalamus in much the same way as projections of striatal and pallidal neurons converge onto the substantia nigra. The parallel ventral subiculoseptal pathway also has a hierarchical organization with the ventral subiculum providing excitatory cortical inputs to the ventral part of the lateral septal nucleus, which may function as a basal nucleus by sending GABAergic projections to sexually dimorphic parts of the hypothalamus. The utility of this model for generating new approaches to understanding sensory integration and control of reproduction seems clear, but the accuracy of its predictions remains to be validated experimentally.

The accessory olfactory and ventral subiculoseptal pathways represent the major limbic-hypothalamic pathways impacting reproduction. The posterior nucleus of the amygdala (PA) (see Canteras et al. 1992a) is the only other part of the telencephalon that provides strong projections to sexually dimorphic hypothalamic nuclei. The PA is the caudal-most part of the amygdala and was previously considered to be a transition zone between the amygdala and hippocampal formation (see Swanson & Petrovich 1998 for review). An analysis of its projections, however, clearly indicates that this region should be included in the division of the amygdala that is primarily involved in relaying olfactory information to the hypothalamus because its only significant sensory afferents are indirect inputs from the accessory and main olfactory pathways. The PA appears to lack GABAergic neurons, which indicates that it provides primarily glutamatergic descending projections, and the pattern of its connections suggests that it occupies a cortical position in forebrain circuit hierarchies (Swanson 2000). Although virtually nothing is known with certainty about the functional role of the PA, its high density of neurons that express sex steroid receptors and massive projections to sexually dimorphic parts of the forebrain involved in reproduction suggest a key role in regulating sexually differentiated aspects of behavior and hormone secretion. Moreover, the specificity of its projections to sexually dimorphic forebrain regions might lead one to suspect that the PA itself is dimorphic and may provide differential projections to the hypothalamus in males and females, but this is not the case. A recent examination of PA morphology found that it is of approximately equal volume in male and female rats and lacks a clear sex difference in cell density (Gu et al. 2000). The descending projections of the PA are also remarkably similar in males and females. Thus, it provides massive inputs to the BSTp, MEApd, MPNm/c, and VMHvl that are similar in both male and female rats, but the projection to the AVPV is much more robust in males, similar to the inputs from the MEApd and BSTp (Gu et al. 2000, Hutton et al. 1998). Another unique aspect of the PA is the apparent lack of robust autoregulation of estrogen and androgen receptors by sex steroid hormones. In other parts of the forebrain, gonadectomy and steroid replacement have profound effects on cellular levels of sex steroid receptors, but in the PA expression of these receptors appears to be relatively stable, at least at the level of mRNA (Gu et al. 2000). One possible interpretation of this curious finding is that hormone-sensitive activity in the PA may accurately reflect dynamic changes in levels of circulating sex steroid hormones owing to this lack of autoregulation. In the future it will be important to establish whether neuronal activity in the PA changes as a function of sex steroid hormone levels in the blood, which would support a possible role for the PA as a stable central sensor of endocrine status.

Despite the robust innervation of sexually dimorphic nuclei in the hypothalamus by the BSTp and amgdala (MEApd and PA), neither the periventricular nor the medial zone dimorphic nuclei provide substantial return projections. Instead, feedback appears to be conveyed by the ventral premammillary nucleus (PMv), which is itself sexually dimorphic, contains high densities of receptors for sex steroids, and receives a strong input from the MPNc (Akesson & Micevych 1995, Simerly et al. 1990, Simerly & Swanson 1988). The PMv receives massive inputs from the MEApd, PA, and BSTp and sends equally strong return projections, as well as providing dense inputs to the LSv (Canteras et al. 1992b). Although the neurotransmitter in these pathways is unknown, the projections of the PMv represent the strongest hypothalamic projection to telencephalic regions that innervate sexually dimorphic nuclei. The function of this robust hypothalamic projection remains a mystery, but the PMv has been implicated in the neural regulation of reproductive behavior and gonadotropin secretion from the anterior pituitary (Akesson & Micevych 1995, Beltramino & Taleisnik 1985).

The organization of the sexually dimorphic forebrain pathways discussed above suggests that they function as part of a broader forebrain circuitry that conveys sensory information to effector regions in the hypothalamus and brainstem. These effector regions then integrate this information with visceral and endocrine signals that ultimately impact neuroendocrine, visceromotor, and behavioral responses. Sensory cues, such as pheromones, are transmitted along sexually dimorphic pathways that presumably do not impact the hypothalamus in the same way in male and female animals. Other sensory information, such as the multiple modalities relayed from the ventral subiculum and CA1 via the LSv, reaches the hypothalamus along pathways that are similar in both sexes, which provides a way for information from the environment to impact hypothalamic circuits similarly in males and females. Visual or main olfactory cues related to foraging behavior may represent an example of sensory information that impacts both sexes equally.

However, even sensory influences transmitted to the hypothalamus along monomorphic pathways may contribute to sexually dimorphic responses because the hypothalamic regions innervated are sexually differentiated. For example, the LSv provides strong inputs to both the AVPV and MPNm/c, which may process the afferent multimodal information differently in each sex. Sexually dimorphic pathways such as the accessory olfactory pathway provide more robust sensory inputs to hypothalamic nuclei in males, which indicates that there is greater convergence of this information onto hypothalamic neurons in target nuclei. This convergence is even more profound in target regions with fewer neurons in males, as is the case with the AVPV. Alternatively, descending projections from the LSv appear to be more divergent in males since there are more neurons in target nuclei such as the MPNm/c in males relative to that of females.

Although at present it is difficult to confidently predict the functional impact of sexually dimorphic patterns of sensory convergence and divergence on specific reproductive functions, it appears likely that the sexually dimorphic representations of these sensory routes and hypothalamic targets impose a sex-specific bias on information processing at nodal points in these circuits. The emerging appreciation of the sexually dimorphic organization of sensory pathways, and a detailed understanding of the cellular relationships that define the signaling balance encoded in patterns of sensory convergence and divergence onto hypothalamic circuits, is a prerequisite to an improved understanding of how these pathways function in the control of neuroendocrine physiology and behavior.

The recent clarification of anatomical relationships between sexually dimorphic parts of the forebrain and new theoretical proposals on information processing in cortico-hypothalamic pathways (Swanson 2000, Swanson & Petrovich 1998) has generated new insights into possible mechanisms underlying reproductive function. Investigators are currently applying a neural systems approach to this problem, and it will be useful to view these neural networks from the perspective of how descending forebrain pathways function generally. Thus, the hierarchical organization of other descending pathways, such as those from the cerebral cortex to the basal ganglia and brainstem motor systems, may have much in common with how descending sexually dimorphic pathways provide cortical input to neural systems mediating reproduction.

Development of Sexually Dimorphic Pathways

CELL NUMBER   Numbers of cells in brain regions and the connections between them are major determinants of the functional properties of forebrain neural circuits. Although it is clear that sex steroid hormones play an important role in determining the size of sexually dimorphic nuclei and can alter patterns of connectivity, the cellular mechanisms underlying the developmental actions of these hormones are only now beginning to be elucidated. Nevertheless it is clear that exposure to high levels of testosterone during the first week of life can dramatically alter the numbers of neurons that occupy sexually dimorphic nuclei in the adult. Cellular mechanisms proposed to explain how sex steroids accomplish such dramatic changes in cell number fall into three main categories: neurogenesis, neuronal migration, and cell death.

Of the potential mechanisms underlying hormonal control of cell number during development, a significant alteration in neurogenesis has found the least experimental support. Indeed, neuronal birthdating studies using thymidine or 5-bromodeoxyuridine (BrdU) labeling have provided evidence to the contrary. In the MPN, this is surprising because many of the neurons in the MPNc are born during the prenatal surge in testosterone secretion. Dodson & Gorski addressed this question by labeling neurons in the MPNc with thymidine and found that treating females with testosterone before or after the testosterone surge had the same effect on the number of labeled neurons in the MPNc of 30-day-old animals (Dodson et al. 1988), which suggests that hormone treatment does not enhance or prolong neurogenesis. Using the same experimental approach they also demonstrated that perinatal exposure to testosterone prevented the loss of neurons that normally occurs in the MPNc of females during the first ten postnatal days (Dodson & Gorski 1993). Using similar methods Arai and colleagues arrived at the conclusion that testosterone does not alter neurogenesis in the AVPV (Arai & Murakami 1994). Neurons in the AVPV are born between embryonic day 13 (E13) and E17. Exposure of female rats to high levels of testosterone on E14–16 did not produce a significant difference in the number of BrdU-labeled neurons in the AVPV of male, female, or androgen-treated female rats on E17 (Nishiuzuka et al. 1993). However, there was a significant reduction in the number of labeled neurons by E21 in males and in females exposed to T, indicating that the hormone exposure begins to cause a reduction in cell number after neurogenesis is complete. Although neurogenesis does not appear to be a major contributor to sexual differentiation of neuronal number in the MPN or AVPV, estrogen promotes numbers of newly generated neurons in the forebrain of avian species (Burek et al. 1995). Moreover, in mature rats estrogen appears to increase formation of new neurons in regions where neurogenesis continues well into adulthood, including the olfactory bulb (Kaba et al. 1988) and dentate gyrus of the hippocampal formation (Tanapat et al. 1999). There are more newborn neurons in the dentate gyrus of females relative to that of age-matched males, but this difference is transient and appears to be related to changes in circulating levels of estrogen that occur in the female during the estrous cycle.

Even less attention has been paid to whether sex differences in neuronal migration contribute to differences in neuronal number in sexually dimorphic nuclei in the mammalian forebrain. Thymidine labeling was used to examine the location of preoptic neurons that become postmitotic on E18 and are presumably destined to reside in the MPN (Jacobson et al. 1985). No significant differences were noted in the locations of labeled cells at various ages, so it was concluded that sexually dimorphic patterns of neuronal migration do not make a major contribution to the development of the sexually dimorphic nucleus of the preoptic area. Tobet and colleagues also addressed this question using BrdU labeling and subsequent evaluation of relative cell densities in ferrets killed at three different prenatal ages (Park et al. 1996). Consistent with the earlier findings in rats, they did not find evidence for sexually dimorphic patterns of migration in the preoptic region. Neither of these labeling methods is optimal for studying neuronal migration because it is difficult to determine when migrating neurons reach their targets and the course of individual cells cannot be traced. Optical recordings obtained by using time-lapse imaging of labeled cells provide a more direct appreciation for defining neuronal migratory routes, often with surprising results (O'Rourke et al. 1992). Tobet et al. (Henderson et al. 1999) used this approach on cultured brain slices to study migration in the preoptic region of the hypothalamus, and the results suggest that there is more medial to lateral migration in males than in females. The possibility that this observed difference in migratory orientation was related to differences in the orientation or density of radial glia was also examined, but again no significant sex differences were detected. The importance of glia in regulating patterns of neuronal migration has long been appreciated and steroid hormones clearly influence glial morphology in the hypothalamus (Garcia-Segura et al. 1996a). A sex difference in the morphology of astroglia was observed in the arcuate nucleus of neonatal rats, and this difference can be manipulated with testosterone treatment (Mong et al. 1999). However, such differences do not appear to be a general feature of sexually dimorphic regions because comparable differences were not found in the VMH.

Although the data to date do not support the notion that migration plays a major role in determining differences in neuronal number in sexually dimorphic nuclei, it remains possible that experimental limitations are largely responsible for this lack of information.

The major way that steroid hormones alter neuron number in sexually dimorphic regions is by influencing cell death. The earlier indirect assessments of hormone-induced changes in cell number have been confirmed by morphometric analyses utilizing a variety of differential cell-labeling procedures (Arai et al. 1996, Dodson & Gorski 1993, Forger & Breedlove 1987, Kay et al. 1999, Nunez et al. 2001, Yoshida et al. 2000). Thus, exposure to steroids during perinatal life enhances cell number in nuclei such as the MPN and BSTp, which are larger in males, and decreases the number of cells in the AVPV. It is less clear how steroids regulate cellular processes during development that lead to changes in cell number. Programmed cell death (PCD) occurs throughout the nervous system and is a major determinant of neuronal number (Burek & Oppenheim 1996, Vaux & Korsmeyer 1999). The notion is widely accepted that hormones influence the numbers of cells that reside in mature sexually dimorphic nuclei by altering the rate or duration of PCD, but this remains largely unproven. The problem of reliably determining that changes in cell number are due to PCD, as opposed to migration or necrotic mechanisms related to physiological insults, is difficult because cells that undergo PCD often are degraded quite rapidly, leading to underestimation of the amount of PCD that occurs in response to hormone exposure. An additional difficulty is that PCD is a dynamic process, and most experimental evaluations lack adequate temporal resolution to fully appreciate its impact on cell number; in addition, there are so many agents that are capable of initiating cell death independent of developmental programs that experimental results are often dependent on peculiar attributes of experimental model systems. The best genetic characterization of a cell-death process in the central nervous system is apoptosis (Yuan & Yankner 2000), which was first defined in morphological terms (Kerr et al. 1972) and was later linked to genetic pathways of PCD in the nematode *C. elegans* (Ellis & Horvitz 1986). Histochemical labeling of fragmented DNA with TUNEL (TdT-mediated dUTP-biotin nick end labeling) staining suggests that at least some of the neurons in sexually dimorphic nuclei die by apoptosis (Arai & Murakami 1994). Although TUNEL labeling is consistent with an apoptotic mechanism of cell death and certainly detects DNA fragmentation (Kerr et al. 1972), morphological features of dying neurons together with the demonstrated involvement of caspases are generally thought to be more reliable indicators of apoptosis (Zhang et al. 1998).

Caspases are members of a family of cysteine proteases that exist in cells as inactive zymogens, which are activated by proteolytic cleavage. Once activated, caspases cleave a broad spectrum of proteins within cells that ultimately result in cell death (Thornberry & Lazebnik 1998). Mutation of Caspase 3 or 9 in mice are generally lethal before birth but do result in reduced apoptosis in the central nervous system (Ranger et al. 2001). However, in certain brain regions and cell types, PCD persists during prenatal development in caspase knockouts, which suggests that the activity of caspases and the occurance of caspase-independent cell-death pathways may be a region-specific feature of early brain development (Oppenheim et al. 2001). Hormonal regulation of caspase expression during development of the mammalian forebrain has not been reported, nor has estrogen or testosterone been shown to alter caspase activation in sexually dimorphic nuclei, but recent preliminary findings suggest that estrogen may influence cell number in the AVPV through a caspase-dependent mechanism (Waters et al. 2000). Perhaps more problematic will be the definition of signaling pathways linking steroid hormones and apoptosis effector mechanisms that lead to alterations in apoptosis. Thus, although there is general agreement that sex steroid hormones regulate PCD, there is very little known about how these regulatory signals affect genetic pathways responsible for apoptosis. However, estrogen was recently reported to induce a neuroprotective activity in cultured fetal neurons that functions to inhibit caspase 6–mediated cell death through a receptor-dependent nongenomic pathway (Zhang et al. 2001). Given the variety of cellular signaling pathways that are apparently influenced by estrogen (see below), there may be multiple mechanisms for regulating the activity of effector caspases.

In addition to the caspases, several mammalian homologues of genes involved in PCD pathways have been identified (Nijhawan et al. 2000). Members of the antiapoptosis Bcl family such as Bcl2 and BclX interact with adapter proteins involved in activation of caspases to inhibit apoptosis and can themselves be regulated by proapoptotic family members like Bad and Bax that promote PCD (Conradt & Horvitz 1998). Thus, it appears that the ratio of antiapoptotic to proapoptotic proteins represents a critical molecular mechanism for determining whether a cell survives or undergoes apoptosis. Steroid hormones regulate cellular events most commonly by acting as ligand-activated transcription factors (Aranda & Pascual 2001, Zhang & Lazar 2000), so the possible hormonal regulation of antiapoptotic to proapoptotic proteins represents a likely mechanism for influencing cell death. Treatment of ovariectomized adult female rats with estrogen increased levels of bcl2 mRNA in the hypothalamus (Garcia-Segura et al. 2001), and estradiol also influenced bcl2 gene expression in the cerebral cortex of adult rats following experimentally induced ischemia (Wise et al. 2001). The rapidly expanding literature on the neuroprotective actions of estrogen in adult rats is complemented by numerous studies in vitro that more directly inform its possible roles during development (see Garcia-Segura et al. 2001, McEwen 2001, Wise et al. 2001 for reviews). For example, the ability of estrogen to enhance neuronal survival in primary cultures of hypothalamic and cortical neurons, or in cortical explants, can be blocked by estrogen receptor (ER) antagonists, which suggests a critical role for the ER in mediating the neurotrophic actions of estrogen (see Wise et al. 2001 for references). Stable transfection of PC12 cells with the alpha form of ER (ERα) increased the viability of estrogen-treated cultures following withdrawal of serum-containing medium (Gollapudi & Oblinger 1999a). ERα is presumably responsible for estrogen-induced increases in Bcl-xl mRNA and reduced expression of Bad mRNA in these cells (Gollapudi & Oblinger 1999b).

Although it remains to be shown definitively that the neuroprotective effects of estrogen in the brain are due to differential expression of apoptotic factors, deletion of Bcl2 or BclX leads to massive cell death in the CNS, whereas mice lacking the proapoptotic protein Bax have developmental defects due to failure of normal developmental cell death (Knudson & Korsmeyer 1997, White et al. 1998). Overexpression of Bcl2 has proven more informative and appears to function in a neuroprotective way following experimentally induced neuronal damage (Alkayed et al. 2001, Martinou et al. 1994, Parsadanian et al. 1998). The impact of these genetic manipulations on neuronal survival in sexually dimorphic nuclei has not been examined, but overexpression of Bcl2 appears to decrease programmed cell death in the spinal nucleus of the bulbocavernosus (SNB) in female mice. Nearly all of the motoneurons in the SNB of females normally die unless exposed to androgen during perinatal life, and this action of androgen appears to depend on expression of androgen receptors within the spinal motoneurons (Breedlove 1992, Forger et al. 1992). Overexpression of Bcl2 in neurons increased the number of motoneurons in the SNB by approximately 40% (Zup et al. 2001). The molecular mechanisms underlying the observed effects of Bcl2 on hormonally regulated neuroprotection are unknown, but the ability of estrogen to alter Bcl2 gene expression via protein-protein interactions with other transcription factors such as Sp-1 and Brn-3a (see Alkayed et al. 2001 for references), or the induction of Bcl2 expression by hormone-sensitive transcription factors such as the cAMP response element binding protein (CREB; see below), suggest possible regulatory mechanisms.

To date, analysis of PCD in the hawk moth represents the most completely characterized example of hormone-dependent neuronal survival and underscores some of the gaps in the mammalian literature (see Alkayed et al. 2001, Truman et al. 1992). During metamorphosis, motoneurons innervating specific proleg muscles in the hawk-moth larva die owing to a rapid fall in levels of the steroid hormone ecdysone, which signals the end of pupal life at adult emergence. These motoneurons express receptors for ecdysone and show segment-specific patterns of hormonally triggered PCD that are restricted to certain developmental stages. That this process occurs in isolated motoneuron cultures in vitro demonstrates cell autonomous regulation of neuronal survival (Zee & Weeks 2001). Given the conservation of PCD signaling mechanisms, it is likely that hormonal regulation of neuronal survival in mammalian systems is equally dependent on cell autonomous factors. Perinatal hormones appear to reduce cell death in some regions, such as the MPN and BSTp, while increasing the loss of cells in the AVPV. Moreover, these effects are cell-type–specific. For example, exposure to sex steroids perinatally decreased cell number and increased DNA fragmentation in the AVPV during the postnatal period but did not significantly alter the volume of the nucleus until much later (Davis et al. 1996). However, the same perinatal hormone treatment increases the number of enkephalinergic neurons that mature in the AVPV of rats, while decreasing numbers of dopaminergic or dynorphin-containing neurons (Simerly 1998). Thus, examinations of gross changes in the volume or neuron number of sexually dimorphic nuclei are likely to obscure cell-specific signaling events.

Sexual differentiation of distinct neuronal subpopulations in sexually dimorphic nuclei has been studied by using immunohistochemical markers (De Vries 1990), but visualization of neurotransmitter substances often requires severe colchicine pretreatment. Equally problematic is that neurochemical markers are often activationally regulated by sex steroids, which can confound interpretation of developmental events. Tyrosine hydroxylase (TH) has proven to be a reliable marker for dopaminergic neurons, is abundantly expressed in hypothalamic neurons, and is relatively resistant to acute regulation by sex steroid hormones. TH immunostaining was used to reveal that the AVPV contains a sexually dimorphic population of dopaminergic neurons that are more abundant in female rats and that sexually differentiate in response to perinatal sex steroids (see Simerly 1999 for review). This sex difference develops postnatally and can be completely reversed by exposing newborn animals to either testosterone or estrogen. Estrogen is as effective as testosterone in defeminizing the pattern of gonadotropin secretion, and the development of the sexually dimorphic dopaminergic neurons in the AVPV is dependent on the alpha form of the ER. The sex difference in dopamine neurons in the AVPV is conserved in C57Bl/6 mice, and most of the TH neurons transiently express the ERα during the first week of life (E.M. Waters & R.B. Simerly, unpublished observations). Mutation of the ERα prevents the loss of TH immunoreactive neurons normally seen in males, but TH neurons in the AVPV of Tfm mice, which have a naturally occurring mutation in the androgen receptor, develop normally, which indicates that sexual differentiation of these cells is independent of the androgen receptor (Simerly et al. 1997). Moreover, estrogen appears to act directly on the AVPV to specify the number of TH immunoreactive neurons that remain in adult animals because exposure of organotypic explants of the AVPV to either testosterone or estradiol for 24 hours under defined conditions results in a persistent loss of these cells (Ibanez et al. 1998). The effects of steroid exposure on TH immunoreactive neurons in the AVPV appear to be permanent, but hormone exposure is less effective if it is delayed until after 6 days in vitro, which suggests that estrogen exerts a cell-type– and receptor-specific action on the sexual differentiation of dopamine neurons in the AVPV during a restricted postnatal period. Although it is likely that estrogen can increase PCD in the AVPV (Arai et al. 1996), it is unclear whether it enhances death of dopaminergic neurons or simply effects a lasting change in neurotransmitter phenotype. An example of the latter pattern of development is found in the BST, where perinatal steroids appear to specify a subset of galanin-containing neurons to coexpress vasopressin (De Vries & Simerly 2002). Thus, in addition to morphological changes, the number of neurons in brain regions that control reproduction may undergo dramatic changes in gene expression during perinatal development and display sex-specific patterns of differentiation into mature phenotypes.

CONNECTIVITY The organization of neural connections between sexually dimorphic nuclei determines how information is transmitted through these pathways and therefore has a profound influence on their functional output. In the developing brain, axons must navigate through complex environments under the influence of local cues to reach their appropriate targets to accurately establish patterns of connectivity that determine the display of adaptive behavioral and physiological responses. Over 25 years ago Toran-Allerand and colleagues (Toran-Allerand 1976) demonstrated the neurotrophic effects of estrogen on neurite outgrowth from hypothalamic explants that were remarkably similar to the trophic action of nerve growth factor on neurite outgrowth from dorsal root ganglion neurons (Levi-Montalcini 1987). This similarity did not go unappreciated by Toran-Allerand who went on to show that estrogen synergized with insulin to promote neurite extension of both hypothalamic and cortical explants, and she correctly predicted a role for the insulin-like growth factor I (IGF-I) (Toran-Allerand et al. 1988). Although its role in sexual differentiation remains to be defined, IGF-I has a clear neuroprotective role that is dependent on estrogen (Garcia-Segura et al. 1996b). In addition, the IGF-I receptor is necessary for hormonal regulation of synaptic plasticity in the arcuate nucleus of the hypothalamus by estrogen (Cardona-Gomez et al. 2000). Common elements in the IGF-I signal transduction pathway and nongenomic mechanisms for estrogen signaling such as those mediated by the mitogen-activated protein kinase (MAPK) pathway (see below) may underlie many of the interactions between estrogen and IGF-I on neural development (Garcia-Segura et al. 2001). In addition to neurotrophic factors, other proteins involved in axonal growth and synaptogenesis, such as the growth-associated protein-43 (GAP-43) (Skene 1989), may participate in hormone-induced neurite outgrowth. Expression of GAP-43 is regulated by estrogen in the hypothalamus of postnatal and adult rats, and males have significantly higher levels of GAP-43 in the preoptic region during the postnatal period (Shughrue & Dorsa 1992, 1994).

SYNAPTOGENESIS   There is now compelling evidence that estrogen represents a powerful neurotrophic agent that promotes synaptogenesis in a variety of functional neural systems during development and in adulthood. Estrogen treatment increases the density of axodendritic contacts in the hypothalamus and hippocampus (Matsumoto et al. 2000, Woolley & McEwen 1992), and some of these effects are accompanied by changes in neuronal signaling (Woolley 1999, Woolley et al. 1997). In the arcuate nucleus of the hypothalamus, the number of both spine and somatic synapses in female rats is approximately twice that of males, but there do not appear to be differences in the incidence of synapses on dendritic shafts (Matsumoto et al. 2000). Whether the greater innervation of the arcuate nucleus by the AVPV contributes to this sex difference is unknown, but exposure to testosterone perinatally caused a significant reduction in the density of axospinous synapses by the second postnatal day (Mong et al. 2001). In contrast, the ventrolateral part of the VMH has more synapses in male rats, relative to that of females (Matsumoto et al. 2000), which may reflect its innervation by more neurons in the MPN and other sexually dimorphic cell groups. The sexually dimorphic synaptic pattern in the VMH is localized to its ventrolateral part, which contains most of the neurons that express estrogen receptors.

Estrogen receptors are coexpressed with neurotrophin receptors, such as the trkA and trkB receptors in forebrain neurons, and estrogen regulates expression of nerve growth factor (NGF) and brain-derived neurotrophic factor (BDNF) in the cortex, hippocampus, and basal forebrain. Estrogens and neurotrophins tend to exert reciprocal regulation of their receptors, and both the trkA and p75NTR genes contain estrogen response elements, which suggests that at least some of this regulation may be mediated directly by estrogen receptors at the level of gene transcription (see Toran-Allerand 1995 for review). Neurotrophins also appear to participate in the hormonally induced changes in dendritic spine density that occur in the hypothalamus and hippocampus. In primary cultures of embryonic hippocampal neurons, estrogen causes a twofold increase in spine density. BDNF and GABA appear to be involved in this induction of dendritic spines because suppression of BDNF expression by estrogen appears to reduce γ-aminobutyric acid (GABA) activity in inhibitory interneurons (Segal & Murphy 2001). The resulting increase in the activity of pyramidal neurons is thought to lead to formation of new dendritic spines, a process that appears to involve phosphorylation of transcription factor CREB (Murphy et al. 1998). CREB may also participate in the regulation of sexual differentiation by GABA. Treatment of newborn males with the GABA agonist muscimol increased CREB phosphorylation in sexually dimorphic nuclei, whereas only decreases in phospho-CREB were seen in females (Auger et al. 2001b). Thus, neuronal signaling initiated by exposure to excitatory GABA during postnatal development may exert a sex-specific activation of CREB with enduring consequences on neuronal survival and differentiation (McCarthy et al. 1997). That CREB phosphorylation is increased by estrogen in sexually dimorphic nuclei such as the AVPV, VMH, and BSTp (Auger et al. 2001a, Gu et al. 1996, Zhou et al. 1996), in vivo as it is in the hippocampal cultures, supports the notion that increased activation of CREB by estrogen promotes synaptogenesis in sexually dimorphic forebrain circuits. Estrogen may also signal through the estrogen receptor to influence spine formation locally through receptors expressed in or near the spine (Blaustein et al. 1992, McEwen et al. 2001, Milner et al. 2001). Taken together with direct imaging experiments that show dynamic interactions between dendritic spines and afferent axons during synaptogenesis (Dailey et al. 1994, Zhai et al. 2001, Ziv & Smith 1996), these observations suggest the possibility that steroid hormones influence synapse formation in sexually dimorphic nuclei by regulating the formation of dendritic spines, which then promote synapse formation with afferent axons.

AXONAL GUIDANCE   Little is known about whether sex steroids direct axons to their targets or simply determine the number of neurons in sexually dimorphic nuclei that provide inputs to other parts of the forebrain. In general, regions that contain more neurons in one sex have more neurons that provide inputs to other parts of the forebrain. Thus, there are more neurons in the MPN in males that project to the periaqueductal gray (Lisciotto & Morrell 1994) and more neurons in the AVPV of females that project to the arcuate nucleus of the hypothalamus (Gu &

Simerly 1997). On a similar note, the projection from the BSTp to the AVPV is approximately an order of magnitude more robust in male rats compared with the homologous pathway in females, a sex difference determined by perinatal exposure to testosterone (Gu et al. 2001, Hutton et al. 1998). However, the magnitude of this sex difference is surprising given that the AVPV is smaller and has fewer neurons in males. In most neural systems the number of projection neurons tends to be developmentally regulated so as to promote registration with target fields. Therefore, one might expect that, although there are approximately twice as many neurons in the BSTp of males, the reduction in the number of target neurons in the AVPV would produce a compensatory reduction in the density of inputs to the AVPV from the BSTp. Indeed the opposite is true: The magnitude of the sex difference in the projection from the BSTp to the AVPV is much greater than that of other BSTp terminal fields, which suggests that there is a considerable degree of site-specific regulation in the development of BSTp projections (Gu et al. 2001). Thus, the projection from the BSTp to the AVPV represents a direct neural pathway between two regions with divergent developmental histories: Exposure to sex steroids increases the number of cells in the BSTp, with robust projections to the AVPV, yet has the opposite effect on AVPV neurons, resulting in a massive convergence of information relayed by the BSTp onto the remaining AVPV neurons in males. It is intriguing to speculate that such target-dependent control of the development of connections between sexually dimorphic nuclei provides great flexibility in determining circuit architecture and that the functional outcome may be that the same sensory cue can have profoundly different effects on reproductive function in males and females.

Evidence from a variety of model systems indicates that axons respond to regionally specific contact-mediated guidance cues and are either attracted or repelled by diffusible factors that influence the behavior of individual growth cones as they seek their targets (see Goodman 1996, Goodman & Shatz 1993, Song & Poo 1999, Tessier-Lavigne & Goodman 1996 for reviews). The response of a particular axon depends on its complement of receptors for chemotropic molecules and on its sensitivity to activity-dependent developmental activities. Although it is common for axons to form exuberant terminal fields that are then restricted through regressive events such as those that are responsible for localizing retinal projections to specific parts of the tectum (Cowan et al. 1984, O'Leary 1992), the BSTp to AVPV pathway develops in a sexually dimorphic pattern that suggests a directed mechanism of axonal guidance (Hutton et al. 1998). Moreover, sex steroids act on the AVPV in a target-dependent way to direct development of BSTp inputs despite the fact that both the AVPV and BSTp express high levels of receptors for sex steroids both in vivo and in vitro (Hutton & Simerly 1997). Reconstitution of the BSTp to AVPV pathway in vitro by preparing heterochronic cocultures revealed that a high density of neurites extend from the BSTp to the AVPV explant only when the AVPV explant is derived from a male animal or from a female that was treated with testosterone perinatally (Ibanez et al. 2001). Only a few neurites extend from the BSTp explant toward the AVPV explant when the cocultures are prepared with an AVPV explant derived from a female rat, regardless of whether the BSTp explant is taken from a male or female rat. Therefore, even though many sexually dimorphic nuclei express steroid hormone receptors during development, the development of connections between them may be directed in a site-specific way, rather than by a concomitant action on several interconnected sexually dimorphic nuclei.

Because the cocultures are suspended in a collagen matrix, the target-dependent developmental activity specifying formation of BSTp to AVPV connections is likely to be a diffusible factor that acts on BSTp neurons. Moreover, when cocultures are prepared with AVPV explants derived from rats that are significantly younger or older than P10–P12, the time when BSTp to AVPV projections normally form in vivo, there is a marked reduction in the density of neurites that grow between the explants. This observation suggests that the diffusible factor is expressed in a temporally specific pattern. Thus, the sexual differentiation of the BSTp to AVPV pathway is likely a target-dependent developmental event mediated by the hormonal induction of chemotropic factors that act specifically on BSTp neurons during a defined postnatal critical period (Ibanez et al. 2001). Proof of this hypothesis awaits isolation of differentially expressed factors that display the appropriate developmental activity. Netrins, semaphorins, and slit proteins are certainly candidates (Brose & Tessier-Lavigne 2000, Raper 2000, Tessier-Lavigne & Goodman 1996), and members of the ephrin family of factors (O'Leary & Wilkinson 1999) may influence migration of BSTp axons through the preoptic region prior to reaching the AVPV. Although there is rapid progress in identifying molecular cues that direct formation of neural connections, none have been shown to be developmentally regulated by sex steroids. As outlined above, the projections of the AVPV are also sexually dimorphic, and it is unknown if the development of sexually dimorphic projections to GnRH or TIDA neurons are the result of a direct action of sex steroids on AVPV neurons or are induced by target-derived factors. GnRH neurons contain ERβ but do not express the ERα (Herbison & Pape 2001, Hrabovszky et al. 2001, Shivers et al. 1983), and a substantial number of TIDA neurons express ERα (Sar 1984). Therefore, estrogen may direct development of projections from the AVPV to these neuroendocrine neurons through either a neurotrophic action on AVPV neurons, a target-dependent mechanism, or a combination of the two. Alternatively, estrogen may act on GnRH neurons through an ER-independent mechanism or through transynaptic regulation of factors that specify synapse formation.

Sex Steroid Signaling During Development

One of the major advantages of studying developmental events regulated by sex steroid hormones is that much is known about the detailed signaling pathways utilized by these hormones to alter cellular processes. Thus, a key developmental factor in controlling sexual differentiation of forebrain pathways is the expression of receptors for estrogen and androgen by neurons in sexually dimorphic nuclei. Indeed all of the major sexually dimorphic nuclei in the mammalian forebrain contain high densities of neurons that express ERα, AR, and, possibly, the ERβ

(Shughrue et al. 1997, Simerly 1995b). The demonstration by Toran-Allerand and colleagues that the ability of estrogen to induce neurite outgrowth from brain explants appeared to be limited to tissue that contained high densities of estrogen receptors (Toran-Allerand et al. 1980) supports this concept. On a similar note, the failure of dopamine neurons in the AVPV to sexually differentiate in ERα knockout mice (Simerly et al. 1997) is a good example of how important these nuclear proteins can be for neuronal sexual differentiation. The major role of steroid receptors is to function as ligand-activated transcription factors that regulate diverse patterns of gene expression (see Aranda & Pascual 2001, Zhang & Lazar 2000 for reviews). These receptors bind specific sequences that function as *cis*-acting hormone response elements located within or near hormone responsive genes to influence promoter activity. The promoter regions of several genes known to effect neural development contain putative estrogen response elements, including brain-derived neurotrophic factor (Sohrabji et al. 1994). Coexpression of the ERα and ERβ in sexually dimorphic nuclei (Greco et al. 2001) adds considerable range to potential regulatory actions of estrogen because the two receptors appear to play complementary but not redundant roles in modulating gene expression, show distinct patterns of autoregulation, and have unique pharmacological properties (see Woolley 1999 for references).

Similarities between hormone response elements for different hormone receptors can lead to functional interactions. For example, thyroid hormone receptors can inhibit estrogen induction of transcription (Glass et al. 1988), which may account for similar interactions between estrogen and thyroid hormone in the modulation of behavior (see Pfaff et al. 2000 for review). Because thyroid hormone receptors are expressed abundantly in sexually dimorphic nuclei, their coexpression with ERs within individual neurons (Kia et al. 2001) may allow estrogen to directly suppress or enhance the profound developmental activities of thyroid hormone on neuronal development. Neonatal treatment with thyroid hormone results in elevated levels of choline acetyltransferase in male, but not female, rats (Westlind-Danielsson et al. 1991). The choline acetyltransferase gene contains a putative ERE (Miller et al. 1999), and the sexually dimorphic response to thyroid hormone supports the idea that sex steroids can alter brain development through interactions with thyroid hormone receptors. However, the same hormonal treatments did not alter the morphology of hippocampal neurons (Gould et al. 1991). More central to the action of sex steroid hormone receptors are coregulatory proteins that function as coactivators to increase transactivation by steroid hormone receptors, or they act as corepressors to lower transcriptional activity of target genes (McKenna et al. 1999, Shibata et al. 1997). Expression of these coregulator proteins in cell-type–specific patterns has a major impact on how cells respond to sex steroids and may allow different populations of neurons to display divergent responses to perinatal hormone exposure. Differential expression of nuclear receptor coregulators may also be responsible for specifying critical periods for the developmental activities of sex steroid hormones, such as the temporally specific loss of dopamine neurons in the AVPV or the target-dependent induction of afferents from the BSTp.

In addition to the direct genomic actions of nuclear hormone receptors, steroid hormones can regulate transcriptional events by altering expression of other transcription factors such as the protooncogene Fos (Insel 1990) or by coupling to second-messenger pathways (Kelly & Levin 2001). Thus, hormonal regulation of transcription factors or second-messenger signaling to the nucleus may mediate the induction of hormone-sensitive genes that lack conventional hormone response elements. Estrogen elicits a rapid and sustained phosphorylation of mitogen-activated protein kinase (MAPK) (Singer et al. 1999, Singh et al. 1999, Watters et al. 1997) as well as activates extracellular signal–regulated kinases (see Toran-Allerand et al. 1999 for review). In addition, rapid phosphorylation of Akt following exposure to estrogen may impact neuronal survival (Datta et al. 1999, Singh 2001) as well as contribute to the stimulation of CREB phosphorylation that promotes synaptogenesis in the hypothalamus and hippocampus. Thus, hormonal regulation of multiple signaling cascades with both rapid and sustained consequences provides a means of integrating and coordinating the action of sex steroids on diverse developmental events. A clear role for this type of developmental regulation in sexual differentiation is lacking, but estrogen induces a receptor-dependent increase in phosphorylation of CREB in the AVPV of adult female, but not male, rats (Gu et al. 1996; G. Gu & R.B. Simerly, unpublished data). One possible interpretation of this sex difference is that the cells that undergo CREB activation have been lost in males. However, neurons in the ventromedial nucleus of the hypothalamus show sexually dimorphic patterns of CREB phosphorylation (Auger et al. 2001a) despite similar numbers of cells in this nucleus (Madeira et al. 2001), which suggests that sexual differentiation may extend to differences in signal transduction pathways. An alternative interpretation is that the organization of local circuits in the VMH (Flanagan-Cato 2000), or sexually dimorphic afferents from other parts of the forebrain, are responsible for sex differences in transynaptic activation of CREB. CREB regulates gene transcription by acting directly on calcium/cAMP response elements but also through interactions with CREB binding protein (CBP) (Shaywitz & Greenberg 1999). In addition to CREB, CBP has binding sites for a wide variety of transcription factors including Fos and Jun (Vo & Goodman 2001). CBP can also bind estrogen and thyroid hormone receptors, which indicates that these protein-protein interactions represent a powerful means of integrating diverse molecular signals at the transcriptional level.

CONCLUSION

Despite many unresolved issues, it is now clear that steroid hormones effect permanent changes in the development of multiple interconnected regions of the mammalian forebrain that participate in the neural control of reproduction and influence other homeostatic functions as well. Estrogen and testosterone regulate most major developmental events including neurogenesis, neuronal migration, cell death, and neurotransmitter plasticity. In addition, sex steroid hormones specify sex-specific patterns of neuronal connectivity by affecting axonal guidance and synaptogenesis. The signaling events mediating these developmental activities interact at multiple levels with neurotrophin and neurotransmitter signal transduction pathways. In addition, sex steroid hormones signal to the nucleus through their ligand-activated receptors to influence a broad array of gene-expression events that contribute to the important developmental role of these hormones in specifying the architecture of forebrain pathways that are fundamental to propagation of mammalian species.

ACKNOWLEDGMENTS

I would like to acknowlege my many collegues for discussions that aided in the development of the ideas presented here and members of my laboratory for contributing their data and insight. I also thank Drs. Eva Polston and Bradley Cooke for helpful comments on the manuscript. Work in the author's laboratory is supported by grants from the NIH (NS37952, DK55819, and RR00163).

The *Annual Review of Neuroscience* is online at http://neuro.annualreviews.org

LITERATURE CITED

Akesson TR, Micevych PE. 1995. Sex steroid regulation of tachykinin peptides in neuronal circuitry mediating reproductive functions. See Micevych & Hammer Jr 1995, pp. 207–33

Alkayed NJ, Goto S, Sugo N, Joh HD, Klaus J, et al. 2001. Estrogen and Bcl-2: gene induction and effect of transgene in experimental stroke. *J. Neurosci.* 21:7543–50

Arai Y, Murakami S. 1994. Androgen enhances neuronal degeneration in the developing preoptic area: apoptosis in the anteroventral periventricular nucleus (AVPvN-POA). *Horm. Behav.* 28(4):313–19

Arai Y, Sekine Y, Murakami S. 1996. Estrogen and apoptosis in the developing sexually dimorphic preoptic area in female rats. *Neurosci. Res.* 25:403–7

Aranda A, Pascual A. 2001. Nuclear hormone receptors and gene expression. *Physiol. Rev.* 81:1269–304

Arnold AP. 1997. Sexual differentiation of the zebra finch song system: positive evidence, negative evidence, null hypotheses, and a paradigm shift. *J. Neurobiol.* 33:572–84

Auger AP, Hexter DP, McCarthy MM. 2001a. Sex difference in the phosphorylation of cAMP response element binding protein (CREB) in neonatal rat brain. *Brain Res.* 890:110–17

Auger AP, Perrot-Sinal TS, McCarthy MM. 2001b. Excitatory versus inhibitory GABA as a divergence point in steroid-mediated sexual differentiation of the brain. *Proc. Natl. Acad. Sci. USA* 98:8059–64

Beltramino C, Taleisnik S. 1985. Ventral premammillary nuclei mediate pheromonal-induced LH release stimuli in the rat. *Neuroendocrinology* 41:119–24

Blaustein JD, Lehman MN, Turcotte JC, Greene G. 1992. Estrogen receptors in dendrites and axon terminals in the guinea pig hypothalamus. *Endocrinology* 131:281–90

Bleier R, Byne W, Siggelkow I. 1982. Cytoarchitectonic sexual dimorphisms of the medial preoptic and anterior hypothalamic areas in guinea pig, rat, hamster, and mouse. *J. Comp. Neurol.* 212:118–30

Breedlove SM. 1992. Sexual dimorphism in the vertebrate nervous system. *J. Neurosci.* 12:4133–42

Brose K, Tessier-Lavigne M. 2000. Slit

proteins: key regulators of axon guidance, axonal branching, and cell migration. *Curr. Opin. Neurobiol.* 10:95–102

Burek MJ, Nordeen KW, Nordeen EJ. 1995. Estrogen promotes neuron addition to an avian song-control nucleus by regulating postmitotic events. *Dev. Brain Res.* 85(2):220–24

Burek MJ, Oppenheim RW. 1996. Programmed cell death in the developing nervous system. *Brain Pathol.* 6:427–46

Canteras NS, Simerly RB, Swanson LW. 1992a. Connections of the posterior nucleus of the amygdala. *J. Comp. Neurol.* 324:143–79

Canteras NS, Simerly RB, Swanson LW. 1992b. Projections of the ventral premamillary nucleus. *J. Comp. Neurol.* 324:195–212

Canteras NS, Simerly RB, Swanson LW. 1995. Organization of projections from the medial nucleus of the amygdala: a PHAL study in the rat. *J. Comp. Neurol.* 360(2):213–45

Canteras NS, Swanson LW. 1992. Projections of the ventral subiculum to the amygdala, septum, and hypothalamus: A PHAL anterograde tract-tracing study in the rat. *J. Comp. Neurol.* 324:180–94

Cardona-Gomez GP, Trejo JL, Fernandez AM, Garcia-Segura LM. 2000. Estrogen receptors and insulin-like growth factor-I receptors mediate estrogen-dependent synaptic plasticity. *Neuroreport* 11:1735–38

Conradt B, Horvitz HR. 1998. The C. elegans protein EGL-1 is required for programmed cell death and interacts with the Bcl-2-like protein CED-9. *Cell* 93:519–29

Corbier P, Edwards DA, Roffi J. 1992. The neonatal testosterone surge: a comparative study. *Arch. Int. Physiol. Biochim. Biophys.* 100:127–31

Cowan WM, Fawcett JW, O'Leary DD, Stanfield BB. 1984. Regressive events in neurogenesis. *Science* 225:1258–65

Dailey ME, Buchanan J, Bergles DE, Smith SJ. 1994. Mossy fiber growth and synaptogenesis in rat hippocampal slices in vitro. *J. Neurosci.* 14(3 Pt 1):1060–78

Datta SR, Brunet A, Greenberg ME. 1999. Cellular survival: a play in three Akts. *Genes Dev.* 13:2905–27

Davis EC, Shryne JE, Gorski RA. 1996. Structural sexual dimorphisms in the anteroventral periventricular nucleus of the rat hypothalamus are sensitive to gonadal steroids perinatally, but develop peripubertally. *Neuroendocrinology* 63:142–48

De Vries GJ. 1990. Sex differences in neurotransmitter systems. *J. Neuroendo.* 2:1–13

De Vries GJ, Miller MA. 1998. Anatomy and function of extrahypothalamic vasopressin systems in the brain. *Prog. Brain Res.* 119:3–20

De Vries GJ, Simerly RB. 2002. Anatomy, development, and function of sexually dimorphic neural circuits in the mammalian brain. See Pfaff et al. 2002

Dodson RE, Gorski RA. 1993. Testosterone propionate administration prevents the loss of neurons within the central part of the medial preoptic nucleus. *J. Neurobiol.* 24:80–88

Dodson RE, Shryne JE, Gorski RA. 1988. Hormonal modification of the number of total and late-arising neurons in the central part of the medial preoptic nucleus of the rat. *J. Comp. Neurol.* 275:623–29

Ellis HM, Horvitz HR. 1986. Genetic control of programmed cell death in the nematode C. elegans. *Cell* 44:817–29

Fahrbach S, Weeks JC. 2002. Hormonal regulation of neural and behavioral plasticity in insects. See Pfaff et al. 2002

Flanagan-Cato LM. 2000. Estrogen-induced remodeling of hypothalamic neural circuitry. *Front Neuroendocrinol.* 21:309–29

Forger NG, Breedlove SM. 1987. Motoneuronal death during human fetal development. *J. Comp. Neurol.* 264:118–22

Forger NG, Hodges LL, Roberts SL, Breedlove SM. 1992. Regulation of motoneuron death in the spinal nucleus of the bulbocavernosus. *J. Neurobiol.* 23(9):1192–203

Garcia-Segura LM, Azcoitia I, DonCarlos LL. 2001. Neuroprotection by estradiol. *Prog. Neurobiol.* 63:29–60

Garcia-Segura LM, Chowen JA, Naftolin F. 1996a. Endocrine glia: roles of glial cells in the brain actions of steroid and thyroid hormones and in the regulation of hormone secretion. *Front Neuroendocrinol.* 17:180–211

Garcia-Segura LM, Duenas M, Fernandez-Galaz MC, Chowen JA, Argente J, et al. 1996b. Interaction of the signalling pathways of insulin-like growth factor-I and sex steroids in the neuroendocrine hypothalamus. *Horm. Res.* 46:160–64

Gerall AA, Givon L. 1992. Early androgen and age-related modifications in female rat reproduction. In *Handbook of Behavioral Neurobiology*, ed. AA Gerall, H Moltz, IL Ward, pp. 313–54. New York: Plenum

Glass CK, Holloway JM, Devary OV, Rosenfeld MG. 1988. The thyroid hormone receptor binds with opposite transcriptional effects to a common sequence motif in thyroid hormone and estrogen response elements. *Cell* 54:313–23

Gollapudi L, Oblinger MM. 1999a. Estrogen and NGF synergistically protect terminally differentiated, ERalpha-transfected PC12 cells from apoptosis. *J. Neurosci. Res.* 56:471–81

Gollapudi L, Oblinger MM. 1999b. Stable transfection of PC12 cells with estrogen receptor (ERalpha): protective effects of estrogen on cell survival after serum deprivation. *J. Neurosci. Res.* 56:99–108

Goodman CS. 1996. Mechanisms and molecules that control growth cone guidance. *Annu. Rev. Neurosci.* 19:341–77

Goodman CS, Shatz CJ. 1993. Developmental mechanisms that generate precise patterns of neuronal connectivity. *Cell, Vol. 72/Neuron, Vol. 10 (Suppl.)* January:77–98

Gorski RA. 1985. The 13th J.A.F. Stevenson Memorial Lecture. Sexual differentiation of the brain: possible mechanisms and implications. *Can. J. Physiol. Pharmacol.* 63:577–94

Gorski RA. 1996. Gonadal hormones and the organization of brain structure and function. In *The Lifespan Development of Individuals Behavioral, Neurobiological, and Psychosocial Perspectives*, ed. D Magnusson, pp. 315–40. Cambridge: Cambridge Univ. Press

Gorski RA, Gordon JH, Shryne JE, Southam AM. 1978. Evidence for a morphological sex difference within the medial preoptic area of the rat brain. *Brain Res.* 148:333–46

Gorski RA, Jacobson CD. 1981. Sexual differentiation of the brain. In *Clinics in Andrology*, ed. SJ Kogan, ESE Hafez, pp. 109–34. The Hague: Martinus Nijhoff

Gould E, Woolley CS, McEwen BS. 1991. The hippocampal formation: morphological changes induced by thyroid, gonadal and adrenal hormones. *Psychoneuroendocrinology* 16:67–84

Greco B, Allegretto EA, Tetel MJ, Blaustein JD. 2001. Coexpression of ERβ with ERα and progestin receptor proteins in the female rat forebrain: effects of estradiol treatment. *Endocrinology* 142:5172–81

Gu G, Chen S, Coleman J, Simerly RB. 2000. The posterior nucleus of the amygdala: a unique component of sexually dimorphic forebrain circuits. *Soc. Neurosci. Abstr.* 26:592

Gu GB, Simerly RB. 1997. Target specific hormonal regulation of sexually dimorphic projections from the principal nucleus of the bed nuclei of the stria terminalis. *Soc. Neurosci. Abstr.* 23:341

Gu G, Rojo AA, Zee MC, Yu J, Simerly RB. 1996. Hormonal regulation of CREB phosphorylation in the anteroventral periventricular nucleus. *J. Neurosci.* 16(9):3034–44

Gu GB, Simerly RB. 1997. Projections of the sexually dimorphic anteroventral periventricular nucleus in the female rat. *J. Comp. Neurol.* 384:142–64

Guillamon A, Segovia S. 1996. Sexual dimorphism in the CNS and the role of steroids. In *CNS Neurotransmitters and Neuromodulators Neuroactive Steroids*, ed. TW Stone, pp. 127–52. Boca Raton: CRC

Henderson RG, Brown AE, Tobet SA. 1999. Sex differences in cell migration in the preoptic area/anterior hypothalamus of mice. *J. Neurobiol.* 41:252–66

Herbison AE, Pape JR. 2001. New evidence for estrogen receptors in gonadotropin-releasing hormone neurons. *Front Neuroendocrinol.* 22:292–308

Hrabovszky E, Steinhauser A, Barabas K,

Shughrue PJ, Petersen SL, et al. 2001. Estrogen receptor-beta immunoreactivity in luteinizing hormone-releasing hormone neurons of the rat brain. *Endocrinology* 142: 3261–64

Hutton LA, Gu GB, Simerly RB. 1998. Development of a sexually dimorphic projection from the bed nuclei of the stria terminalis to the anteroventral periventricular nucleus in the rat. *J. Neurosci.* 18(8):3003–13

Hutton LA, Simerly RB. 1997. Influence of sex steroids on expression of estrogen receptor mRNA in explant cultures of the principal nucleus of the bed nuclei of the stria terminalis. *Soc. Neurosci. Abstr.* 23:343

Ibanez MA, Zee J, Crabtree M, Simerly RB. 1998. Developmental critical period for sexual differentiation of dopaminergic neurons in the anteroventral periventricular nucleus (AVPV). *Soc. Neurosci.* 24:1546

Ibanez MA, Gu GB, Simerly RB. 2001. Target dependent sexual differentiation of a limbic-hypothalamic pathway. *J. Neurosci.* 21: 5652–59

Insel TR. 1990. Regional induction of c-fos-like protein in rat brain after estradiol administration. *Endocrinology* 126:1849–53

Jacobson CD, Davis FC, Gorski RA. 1985. Formation of the sexually dimorphic nucleus of the preoptic area: neuronal growth, migration and changes in cell number. *Dev. Brain Res.* 21:7–18

Kaba H, Rosser AE, Keverne EB. 1988. Hormonal enhancement of neurogenesis and its relationship to the duration of olfactory memory. *Neuroscience* 24:93–98

Kay JN, Hannigan P, Kelley DB. 1999. Trophic effects of androgen: development and hormonal regulation of neuron number in a sexually dimorphic vocal motor nucleus. *J. Neurobiol.* 40:375–85

Kelly MJ, Levin ER. 2001. Rapid actions of plasma membrane estrogen receptors. *Trends Endocrinol. Metab.* 12:152–56

Kerr JF, Wyllie AH, Currie AR. 1972. Apoptosis: a basic biological phenomenon with wide-ranging implications in tissue kinetics. *Br. J. Cancer* 26:239–57

Kia HK, Krebs CJ, Koibuchi N, Chin WW, Pfaff DW. 2001. Co-expression of estrogen and thyroid hormone receptors in individual hypothalamic neurons. *J. Comp. Neurol.* 437:286–95

Knobil E, Neill J. 1994. *The Physiology of Reproduction.* New York: Raven

Knudson CM, Korsmeyer SJ. 1997. Bcl-2 and Bax function independently to regulate cell death. *Nat. Genet.* 16:358–63

Kruk MR. 1991. Ethology and pharmacology of hypothalamic aggression in the rat. *Neurosci. Biobehav. Rev.* 15:527–38

Larsson K. 1979. Features of the neuroendocrine regulation of masculine sexual behavior. In *Endocrine Control of Sexual Behavior*, ed. C Beyer, pp. 77–163. New York: Raven

Leal S, Andrade JP, Paula-Barbosa MM, Madeira MD. 1998. Arcuate nucleus of the hypothalamus: effects of age and sex. *J. Comp. Neurol.* 401:65–88

Levi-Montalcini R. 1987. The nerve growth factor 35 years later. *Science* 237:1154–62

Lisciotto CA, Morrell JI. 1994. Sex differences in the distribution and projections of testosterone target neurons in the medial preoptic area and the bed nucleus of the stria terminalis of rats. *Horm. Behav.* 28:492–502

Madeira MD, Ferreira-Silva L, Paula-Barbosa MM. 2001. Influence of sex and estrus cycle on the sexual dimorphisms of the hypothalamic ventromedial nucleus: stereological evaluation and Golgi study. *J. Comp. Neurol.* 432:329–45

Madeira MD, Leal S, Paula-Barbosa MM. 1999. Stereological evaluation and Golgi study of the sexual dimorphisms in the volume, cell numbers, and cell size in the medial preoptic nucleus of the rat. *J. Neurocytol.* 28:131–48

Madeira MD, Lieberman AR. 1995. Sexual dimorphism in the mammalian limbic system. *Prog. Neurobiol.* 45:275–333

Martinou J-C, Dubois-Dauphin M, Staple JK, Rodriguez I, Frankowski H, et al. 1994. Overexpression of BCL-2 in transgenic mice protects neurons from naturally occurring cell death and experimental ischemia. *Neuron* 13: 1017–30

Matsumoto A, Sekine Y, Murakami S, Arai Y. 2000. Sexual differentiation of neuronal circuitry in the hypothalamus. In *Sexual Differentiation of the Brain*, ed. A Matsumoto, pp. 203–27. New York: CRC

McCarthy MM, Davis AM, Mong JA. 1997. Excitatory neurotransmission and sexual differentiation of the brain. *Brain Res. Bull* 44: 487–95

McEwen B, Akama K, Alves S, Brake WG, Bulloch K, et al. 2001. Tracking the estrogen receptor in neurons: implications for estrogen-induced synapse formation. *Proc. Natl. Acad. Sci. USA* 98:7093–100

McEwen BS. 2001. Invited review. Estrogens effects on the brain: multiple sites and molecular mechanisms. *J. Appl. Physiol.* 91:2785–801

McKenna NJ, Lanz RB, O'Malley BW. 1999. Nuclear receptor coregulators: cellular and molecular biology. *Endocr. Rev.* 20:321–44

Meisel RL, Sachs BD. 1994. The physiology of male sexual behavior. See Knobil & Neill 1994, pp. 3–105

Miller MM, Hyder SM, Assayag R, Panarella SR, Tousignant P, Franklin KB. 1999. Estrogen modulates spontaneous alternation and the cholinergic phenotype in the basal forebrain. *Neuroscience* 91:1143–53

Milner TA, McEwen BS, Hayashi S, Li CJ, Reagan LP, Alves SE. 2001. Ultrastructural evidence that hippocampal alpha estrogen receptors are located at extranuclear sites. *J. Comp. Neurol.* 429:355–71

Mong JA, Glaser E, McCarthy MM. 1999. Gonadal steroids promote glial differentiation and alter neuronal morphology in the developing hypothalamus in a regionally specific manner. *J. Neurosci.* 19:1464–72

Mong JA, Roberts RC, Kelly JJ, McCarthy MM. 2001. Gonadal steroids reduce the density of axospinous synapses in the developing rat arcuate nucleus: an electron microscopy analysis. *J. Comp. Neurol.* 432:259–67

Murphy DD, Cole NB, Segal M. 1998. Brain-derived neurotrophic factor mediates estradiol-induced dendritic spine formation in hippocampal neurons. *Proc. Natl. Acad. Sci. USA* 95:11412–17

Micevych PE, Hammer RP Jr, eds. 1995. *Neurobiological Effects of Sex Steroid Hormones*, Cambridge, UK: Cambridge Univ. Press

Nijhawan D, Honarpour N, Wang X. 2000. Apoptosis in neural development and disease. *Annu. Rev. Neurosci.* 23:73–87

Nishiuzuka M, Sumida H, Kano Y, Arai Y. 1993. Formation of neurons in the sexually dimorphic anteroventral periventricular nucleus of the preoptic area of the rat: effects of prenatal treatment with testosterone propionate. *J. Neuroendocrinol.* 5:569–73

Nunez JL, Lauschke DM, Juraska JM. 2001. Cell death in the development of the posterior cortex in male and female rats. *J. Comp. Neurol.* 436:32–41

O'Leary DD, Wilkinson DG. 1999. Eph receptors and ephrins in neural development. *Curr. Opin. Neurobiol.* 9:65–73

O'Leary DDM. 1992. Development of connectional diversity and specificity in the mammalian brain by the pruning of collateral projections. *Curr. Opin. Neurobiol.* 2:70–77

O'Rourke NA, Dailey ME, Smith SJ, McConnell SK. 1992. Diverse migratory pathways in the developing cerebral cortex. *Science* 258:299–304

Oppenheim RW, Flavell RA, Vinsant S, Prevette D, Kuan CY, Rakic P. 2001. Programmed cell death of developing mammalian neurons after genetic deletion of caspases. *J. Neurosci.* 21:4752–760

Park J-J, Baum MJ, Paredes RG, Tobet SA. 1996. Neurogenesis and cell migration into the sexually dimorphic preoptic area/anterior hypothalamus of the fetal ferret. *J. Neurobiol.* 30:315–28

Parsadanian AS, Cheng Y, Keller-Peck CR, Holtzman DM, Snider WD. 1998. Bcl-xL is an antiapoptotic regulator for postnatal CNS neurons. *J. Neurosci.* 18:1009–19

Pfaff D, Arnold A, Etgen A, Fahrbach S, Rubin R, eds. 2002. *Hormones, Brain and Behavior*. San Diego: Academic

Pfaff DW. 1980. *Estrogens and Brain Function.* New York: Springer-Verlag. 281 pp.

Pfaff DW, Schwartz-Giblin S, McCarthy MM, Kow L. 1994. Cellular mechanisms of female reproductive behaviors. See Knobil & Neill 1994, pp. 107–220

Pfaff DW, Vasudevan N, Kia HK, Zhu YS, Chan J, et al. 2000. Estrogens, brain and behavior: studies in fundamental neurobiology and observations related to women's health. *J. Steroid. Biochem. Mol. Biol.* 74:365–73

Raisman G, Field PM. 1971. Sexual dimorphism in the preoptic area of the rat. *Science* 173:731–33

Ranger AM, Malynn BA, Korsmeyer SJ. 2001. Mouse models of cell death. *Nat. Genet.* 28: 113–18

Raper JA. 2000. Semaphorins and their receptors in vertebrates and invertebrates. *Curr. Opin. Neurobiol.* 10:88–94

Risold PY, Swanson LW. 1997a. Chemoarchitecture of the rat lateral septal nucleus. *Brain Res. Rev.* 24:91–113

Risold PY, Swanson LW. 1997b. Connections of the rat lateral septal complex. *Brain Res. Rev.* 24:115–95

Sar M. 1984. Estradiol is concentrated in tyrosine hydroxylase-containing neurons of the hypothalamus. *Science* 223:938–40

Sawchenko PE, Li HY, Ericsson A. 2000. Circuits and mechanisms governing hypothalamic responses to stress: a tale of two paradigms. *Prog. Brain Res.* 122:61–78

Segal M, Murphy D. 2001. Estradiol induces formation of dendritic spines in hippocampal neurons: functional correlates. *Horm. Behav.* 40:156–59

Shaywitz AJ, Greenberg ME. 1999. CREB: a stimulus-induced transcription factor activated by a diverse array of extracellular signals. *Annu. Rev. Biochem.* 68:821–61

Shibata H, Spencer TE, Onate SA, Jenster G, Tsai SY, et al. 1997. Role of co-activators and co-repressors in the mechanism of steroid/thyroid receptor action. *Recent Prog. Horm. Res.* 52:141–64; discussion 64–65

Shivers BD, Harlan RE, Morrell JI, Pfaff DW. 1983. Absence of oestradiol concentration in cell nuclei of LHRH-immunoreactive neurones. *Nature* 304:345–47

Shughrue PJ, Dorsa DM. 1993. Estrogen modulates the growth-associated protein GAP-43 mRNA in the rat preoptic area and basal hypothalamus. *Neuroendocrinology* 57(3): 439–47

Shughrue PJ, Dorsa DM. 1994. The ontogeny of GAP-43 (neuromodulin) mRNA in postnatal rat brain: evidence for a sex dimorphism. *J. Comp. Neurol.* 340:174–84

Shughrue PJ, Lane MV, Merchenthaler I. 1997. Comparative distribution of estrogen receptor-α and -β mRNA in the rat central nervous system. *J. Comp. Neurol.* 388:507–25

Simerly RB. 1989. Hormonal control of the development and regulation of tyrosine hydroxylase expression within a sexually dimorphic population of dopaminergic cells in the hypothalamus. *Mol. Brain Res.* 6:297–310

Simerly RB. 1990. Hormonal control of neuropeptide gene expression in sexually dimorphic olfactory pathways. *Trends Neurosci.* 13:104–10

Simerly RB. 1991. Prodynorphin and proenkephalin gene expression in the anteroventral periventricular nucleus of the rat: sexual differentiation and hormonal regulation. *Mol. Cell. Neurosci.* 2:473–84

Simerly RB. 1993. Distribution and regulation of steroid hormone receptor gene expression in the central nervous system. In *Advances in Neurology, Vol. 59*, ed. FJ Seil, pp. 207–26. New York: Raven

Simerly RB. 1995a. Anatomical substrates of hypothalamic integration. In *The Rat Nervous System*, ed. G Paxinos, pp. 353–76. San Francisco: Academic

Simerly RB. 1995b. Hormonal regulation of limbic and hypothalamic pathways. See Micevych & Hammer Jr 1995, pp. 85–114

Simerly RB. 1998. Organization and regulation of sexually dimorphic neuroendocrine pathways. *Behav. Brain Res.* 92:195–203

Simerly RB. 1999. Development of sexually dimorphic forebrain pathways. In *Sexual Differentiation in the Brain*, ed. A Matsumoto. Boca Raton: CRC

Simerly RB, Chang C, Muramatsu M, Swanson LW. 1990. Distribution of androgen and estrogen receptor mRNA-containing cells in the rat brain: an in situ hybridization study. *J. Comp. Neurol.* 294:76–95

Simerly RB, Swanson LW. 1988. Projections of the medial preoptic nucleus: a Phaseolus vulgaris leucoagglutinin anterograde tract-tracing study in the rat. *J. Comp. Neurol.* 270: 209–42

Simerly RB, Swanson LW, Gorski RA. 1985. The distribution of monoaminergic cells and fibers in a periventricular nucleus involved in the control of gonadotropin release: immunohistochemical evidence for a dopaminergic sexual dimorphism. *Brain Res.* 330:55–64

Simerly RB, Zee MC, Pendleton JW, Lubahn DB, Korach KS. 1997. Estrogen receptor-dependent sexual differentiation of dopaminergic neurons in the preoptic region of the mouse. *Proc. Natl. Acad. Sci. USA* 94: 14077–82

Simpson ER, Mahendroo MS, Means GD, Kilgore MW, Hinshelwood MM, et al. 1994. Aromatase cytochrome P450, the enzyme responsible for estrogen biosynthesis. *Endocrine Rev.* 15:342–55

Singer CA, Figueroa-Masot XA, Batchelor RH, Dorsa DM. 1999. The mitogen-activated protein kinase pathway mediates estrogen neuroprotection after glutamate toxicity in primary cortical neurons. *J. Neurosci.* 19:2455–63

Singh M. 2001. Ovarian hormones elicit phosphorylation of Akt and extracellular-signal regulated kinase in explants of the cerebral cortex. *Endocrine* 14:407–15

Singh M, Sétáló G Jr, Guan X, Warren M, Toran-Allerand CD. 1999. Estrogen-induced activation of mitogen-activated protein kinase in cerebral cortical explants: convergence of estrogen and neurotrophin signaling pathways. *J. Neurosci.* 19:1179–88

Skene JH. 1989. Axonal growth-associated proteins. *Annu. Rev. Neurosci.* 12:127–56

Sohrabji F, Miranda RC, Toran-Allernad CD. 1994. Identification of a potential estrogen response element in the gene coding for brain derived neurotrophic factor (BDNF). *Soc. Neurosci. Abstr.* 20:1303

Song HJ, Poo MM. 1999. Signal transduction underlying growth cone guidance by diffusible factors. *Curr. Opin. Neurobiol.* 9:355–63

Swanson LW. 2000. Cerebral hemisphere regulation of motivated behavior(1). *Brain Res.* 886:113–64

Swanson LW, Petrovich GD. 1998. What is the amygdala? *Trends Neurosci.* 21:323–31

Tanapat P, Hastings NB, Reeves AJ, Gould E. 1999. Estrogen stimulates a transient increase in the number of new neurons in the dentate gyrus of the adult female rat. *J. Neurosci.* 19:5792–801

Terasawa E, Wiegand SJ, Bridson WE. 1980. A role for medial preoptic nucleus on afternoon of proestrus in female rats. *Am. J. Physiol.* 238:E533–29

Tessier-Lavigne M, Goodman CS. 1996. The molecular biology of axon guidance. *Science* 274:1123–33

Thornberry NA, Lazebnik Y. 1998. Caspases: enemies within. *Science* 281:1312–16

Toran-Allerand CD. 1976. Sex steroids and the development of the newborn mouse hypothalamus and prepoptic area in vitro: implications for sexual differentiation. *Brain Res.* 106:407–12

Toran-Allerand CD. 1995. Developmental interactions of estrogens with neurotrophins and their receptors. See Micevych & Hammer Jr 1995, pp. 391–411

Toran-Allerand CD, Ellis L, Pfenninger KH. 1988. Estrogen and insulin synergism in neurite growth enhancement in vitro: mediation of steroid effects by interactions with growth factors? *Dev. Brain Res.* 41:87–100

Toran-Allerand CD, Gerlach JL, McEwen BS. 1980. Autoradiographic localization of [3H] estradiol related to steroid responsiveness in cultures of the newborn mouse hypothalamus and preoptic area. *Brain Res.* 184:517–22

Toran-Allerand CD, Singh M, Sétáló G Jr. 1999. Novel mechanisms of estrogen action in the brain: new players in an old story. *Front. Neuroendocrinol.* 20:97–121

Truman JW, Thorn RS, Robinow S. 1992. Programmed neuronal death in insect development. *J. Neurobiol.* 23(9):1295–311

Vaux DL, Korsmeyer SJ. 1999. Cell death in development. *Cell* 96:245–54

Vo N, Goodman RH. 2001. CREB-binding protein and p300 in transcriptional regulation. *J. Biol. Chem.* 276:13505–8

Waters EM, Ibanez MA, Simerly RB. 2000. Estrogen causes sexual differentiation of dopaminergic neurons in the AVPV of the rat hypothalamus by inducing caspase mediated cell death. *Soc. Neurosci. Abstr.* 26:323

Watters JJ, Campbell JS, Cunningham MJ, Krebs EG, Dorsa DM. 1997. Rapid membrane effects of steroids in neuroblastoma cells: effects of estrogen on mitogen activated protein kinase signaling cascade and c-fos immediate early gene transcription. *Endocrinology* 138:4030–33

Watts AG. 2001. Neuropeptides and the integration of motor responses to dehydration. *Annu. Rev. Neurosci.* 24:357–84

Weisz J, Ward IL. 1980. Plasma testosterone and progesterone titers of pregnant rats, their male and female fetuses and neonatal offspring. *Endocrinology* 106:306–16

Westlind-Danielsson A, Gould E, McEwen BS. 1991. Thyroid hormone causes sexually distinct neurochemical and morphological alterations in rat septal-diagonal band neurons. *J. Neurochem.* 56:119–28

White FA, Keller-Peck CR, Knudson CM, Korsmeyer SJ, Snider WD. 1998. Widespread elimination of naturally occurring neuronal death in Bax-deficient mice. *J. Neurosci.* 18:1428–39

Wise PM, Dubal DB, Wilson ME, Rau SW, Liu Y. 2001. Estrogens: trophic and protective factors in the adult brain. *Front Neuroendocrinol.* 22:33–66

Woolley CS. 1999. Electrophysiological and cellular effects of estrogen on neuronal function. *Crit. Rev. Neurobiol.* 13:1–20

Woolley CS, McEwen BS. 1992. Estradiol mediates fluctuation in hippocampal synapse density during the estrous cycle in the adult rat. *J. Neurosci.* 12:2549–54

Woolley CS, Weiland NG, McEwen BS, Schwartzkroin PA. 1997. Estradiol increases the sensitivity of hippocampal CA1 pyramidal cells to NMDA receptor-mediated synaptic input: correlation with dendritic spine density. *J. Neurosci.* 17:1848–59

Yoshida M, Yuri K, Kizaki Z, Sawada T, Kawata M. 2000. The distributions of apoptotic cells in the medial preoptic areas of male and female neonatal rats. *Neurosci. Res.* 36:1–7

Yuan J, Yankner BA. 2000. Apoptosis in the nervous system. *Nature* 407:802–9

Zee MC, Weeks JC. 2001. Developmental change in the steroid hormone signal for cell-autonomous, segment-specific programmed cell death of a motoneuron. *Dev. Biol.* 235:45–61

Zhai RG, Vardinon-Friedman H, Cases-Langhoff C, Becker B, Gundelfinger ED, et al. 2001. Assembling the presynaptic active zone: a characterization of an active zone precursor vesicle. *Neuron* 29:131–43

Zhang J, Lazar MA. 2000. The mechanism of action of thyroid hormones. *Annu. Rev. Physiol.* 62:439–66

Zhang J, Liu X, Scherer DC, van Kaer L, Wang X, Xu M. 1998. Resistance to DNA fragmentation and chromatin condensation in mice lacking the DNA fragmentation factor 45. *Proc. Natl. Acad. Sci. USA* 95:12480–85

Zhang Y, Tounekti O, Akerman B, Goodyer CG, LeBlanc A. 2001. 17-beta-estradiol induces an inhibitor of active caspases. *J. Neurosci.* 21:RC176

Zhou Y, Watters JJ, Dorsa DM. 1996. Estrogen rapidly induces the phosphorylation of the cAMP response element binding protein in rat brain. *Endocrinology* 137:2163–66

Ziv NE, Smith SJ. 1996. Evidence for a role of dendritic filopodia in synaptogenesis and spine formation. *Neuron* 17:91–102

Zup SL, Bengston L, Tabor A, Forger NG. 2001. BCL-2 overexpression rescues motoneurons in the spinal nucleus of the bulbocavernosus of female mice. *Soc. Neurosci. (Abstr.)* 27: In press

CENTRAL NERVOUS SYSTEM DAMAGE, MONOCYTES AND MACROPHAGES, AND NEUROLOGICAL DISORDERS IN AIDS

Kenneth C. Williams[1] and William F. Hickey[2]

[1]*Department of Medicine, Harvard Medical School Division of Viral Pathogenesis Beth Israel Deaconess Medical Center, Boston, Massachusetts;*
email: kenneth_williams@hms.harvard.edu
[2]*Department of Pathology, Dartmouth Medical School, Dartmouth Hitchcock Medical Center, One Medical Center Drive, Lebanon, New Hampshire 03756;*
email: William.F.Hickey@Dartmouth.edu

Key Words human immunodeficiency virus, simian immunodeficiency virus, perivascular cells, microglia, HIV-associated dementia

■ **Abstract** This review focuses on the role of the extended macrophage/monocyte family in the central nervous system during HIV or SIV infection. The accumulated data, buttressed by recent experimental results, suggest that these cells play a central, pathogenic role in retroviral-associated CNS disease. While the immune system is able to combat the underlying retroviral infection, the accumulation and widespread activation of macrophages, microglia, and perivascular cells in the CNS are held in check. However, with the collapse of the immune system and the disappearance of the $CD4^+$ T cell population, productive infection reemerges, especially in CNS macrophages. These cells, as well as noninfected macrophages, are stimulated to high levels of activation. When members of this cell group become highly activated, they elaborate a wide spectrum of deleterious substances into the neural parenchyma. In the final phases of HIV or SIV infection, this chronic, widespread, and dramatic level of macrophage/ monocyte/microglial activation constitutes a self-sustaining state of macrophage dysregulation, which results in pathological alterations and the emergence of various neurological problems.

INTRODUCTION

In the two decades since the identification of acquired immunodeficiency syndrome (AIDS), we have learned that after the immune system, the nervous system suffers most significantly. Approximately 60% of human immunodeficiency virus type 1 (HIV-1)-infected people demonstrate some form of neurological dysfunction, and neuropathological changes are found in 80% to 90% of autopsied cases (Anders et al. 1986, Gray et al. 1991, Navia et al. 1986a, Petito et al. 1986). A variety of neurological disorders are attributable to HIV infection. Approximately

30%–40% of infected individuals develop a subcortical dementia—HIV-associated dementia, acronymically termed HAD (Kolson & Gonzalez-Scarano 2000; McArthur & Grant 1998; Navia et al. 1986a,b; Price et al. 1988). This clinical syndrome is characterized by deterioration of cognitive and motor function and behavioral changes (Price et al. 1988), and it usually has, as its pathological substrate, HIV-encephalitis (HIVE). Although HAD has been recognized for years, its pathogenesis remains obscure. Correlations have been noted between HAD and central nervous system (CNS) viral load (Wiley et al. 1998), neuronal dendritic pathology (Masliah et al. 1992, 1997), neuronal loss (Adle-Biassette et al. 1995; Everall et al. 1993, 1991; Gelbard et al. 1995; Ketzler et al. 1990; Petito & Roberts 1995; Wiley et al. 1991), and the accumulation of macrophages and multinucleated giant cells (MNGC) in the CNS (Budka 1986, Glass et al. 1995). The exact anatomical and physiological alterations giving rise to HAD are not defined but are presumed to be a result of damage to neuronal circuitry. Because HIV does not directly infect neurons, attempts to understand the neuronal injury and death have focused on indirect mechanisms involving viral proteins, the interaction of virus with chemokine receptors on neurons and astrocytes, and the release of proinflammatory molecules from macrophages and microglia. Recently, these soluble factors, elaborated by infected and/or activated brain macrophages and microglia, have been implicated as the most significant source of HIV-associated CNS disease (Gonzalez-Scarano & Baltuch 1999, Kaul et al. 2001, Langford & Masliah 2001, Power & Johnson 2001, Rausch & Davis 2001, Yong et al. 2001). In this review the term macrophage will be used for economy, but it should be understood to encompass multiple cell types derived from the bone marrow including perivascular cells, meningeal macrophages, choroid plexus macrophages, and phagocytic macrophages.

Many questions concerning HIV neuropathogenesis cannot be addressed in humans and are therefore studied in animal models. The simian immunodeficiency virus (SIV)-infected rhesus macaque is an ideal model in which to study AIDS pathogenesis (Desrosiers 1990, Hurtel et al. 1993, Lackner 1994, Sasseville & Lackner 1997, Sharer et al. 1991, Westmoreland et al. 1998a). Simian immunodeficiency viruses have extensive sequence homology, a genomic organization, and biological properties resembling HIV-1 and HIV-2 (Desrosiers 1990). The targets of SIV infection, like HIV-1, are monocytes/macrophages and lymphocytes (Desrosiers et al. 1991, Hurtel et al. 1993, Ringler et al. 1989). Similar to HIV infection in humans, approximately 30%–40% of simian immunodeficiency virus (SIV)-infected macaques develop encephalitis with formation of multinucleated giant cells (MNGC), while about 90% have CNS lesions (Lackner et al. 1991a, Sasseville & Lackner 1997). The reason only 30%–40% of retrovirally infected humans and primates develop encephalitis and/or mental deterioration is not known, but most probably these conditions involve changes in immune system function as it relates to HIV/SIV infection. Histological features typical of HIVE- and SIV-encephalitis (HIVE/SIVE) lesions are gliosis, microglial nodules, detectable virus in the CNS, perivascular macrophage accumulation, and the chronic presence of activated macrophages and multinucleated giant cells MNGC (Budka 1986, Budka et al. 1991, Kure et al. 1990a, Lackner 1994, Nielson et al. 1984, Price et al. 1988,

Sasseville & Lackner 1997, Wiley et al. 1986). Although the presence of virus in the CNS appears to be required for disease induction, the number of inflammatory macrophages and not viral load is the best indicator of both HAD (Glass et al. 1995) and neural damage in SIV-encephalitis (Zink et al. 1999). Moreover, though virus enters the CNS early after infection, symptoms of HAD usually do not occur until the immune system fails and AIDS develops. At that time there is an influx and accumulation of macrophages in the CNS (Gartner 2000, Gray et al. 1992, Lane et al. 1996). The products of chronically activated microglia and macrophages, believed to be central players in neurodegenerative changes, have been extensively reviewed (Gonzalez-Scarano & Baltuch 1999, Kaul et al. 2001, Langford & Masliah 2001, Power & Johnson 2001, Rausch & Davis 2001, Yong et al. 2001). In this review, we discuss the concept of macrophage dysregulation by reviewing observations on HIV infection in humans and SIV infection in nonhuman primates. Although HIV/SIV is required for the initiation CNS disease and for immune suppression, macrophage dysregulation might ultimately be responsible for neurologic dysfunction. Evidence is presented for the role of macrophages in response to lentivirus early after infection (during primary infection) and terminally in individuals with AIDS and encephalitis. Similar pathogenic mechanisms involving chronically activated members of the microglia/macrophage group may play a role in other neurological conditions including multiple sclerosis (MS), Alzheimer's disease, and Parkinson's disease, where viral etiologies are less certain (Cotter et al. 1999, Dickson et al. 1993, Gonzalez-Scarano & Baltuch 1999, Kusdra et al. 2000, Pulliam et al. 1997, Rausch & Davis 2001, Smits et al. 2000).

MONOCYTE/MACROPHAGE LINEAGE CELLS OF THE CNS

The CNS contains several macrophage populations, some of which are long-term residents while others are more transient (Hickey & Kimura 1988, Hickey et al. 1992, Lassmann et al. 1993). Normal monocyte/macrophage populations of the CNS include macrophages of the choroid plexus, macrophages of the meninges, microglia (the resident brain macrophage), perivascular cells (a CNS resident monocyte-like cell found around small parenchymal vessels), and the MNGC in encephalitis (Budka 1986, Dickson et al. 1991, Kure et al. 1990b, Hickey 1998). It has been proposed that members of this family actually regulate T cell responses in the CNS (Aloisi et al. 2000) and may govern the ability of T cells to enter the neural parenchyma itself (Tran et al. 1998). Though all of these cells are of bone marrow origin and therefore share myeloid markers, their biology including cell surface markers, turnover, and location within the CNS is different.

Early observations with HIV and recent work in the SIV model point to the perivascular cells, a subset of macrophages in perivascular locations, as the main cellular target of infection in the CNS (Davis et al. 1992, Hurtel et al. 1991, Lane et al. 1996, Price et al. 1988, Reinhart et al. 1997, Sharer et al. 1991, Williams et al. 2001b). Another possibility is that the resident brain macrophage, the microglial cell, is the main cell type infected (Dickson et al. 1993, Gonzalez-Scarano & Baltuch 1999, Kure et al. 1990b, Lee et al. 1993a, Sharpless et al. 1992, Watkins et al. 1990). The majority of in vitro analyses of brain macrophage infection and the release of soluble factors from these infected macrophages have therefore focused on the microglia (He et al. 1997, Jordan et al. 1991, Lee et al. 1993a, Sharpless et al. 1992, Strizki et al. 1996, Watkins et al. 1990). Other macrophage/monocyte subsets have not been as thoroughly analyzed but do merit close inspection. The identity of the macrophage populations infected in the CNS is important and relevant to questions concerning how virus enters the CNS, the role of the CNS as a viral reservoir, and the relevance of in vitro models of viral infection.

Evidence for the perivascular cell being a major target of lentiviral infection, and for the accumulation of activated, noninfected macrophages contributing to CNS neuropathology, comes from the rhesus macaque model. We have used immunohistochemistry and confocal microscopy to distinguish between microglia, perivascular cells, and MNGCs. Similar to what has been demonstrated in multiple sclerosis in humans and experimental allergic encephalomyelitis (EAE) in rodents, resident microglia can be distinguished from perivascular cells and recently recruited macrophages based on the expression of CD14 and CD45 (Ford et al. 1995, Sedgwick et al. 1991, Ulvestad et al. 1994b, Williams et al. 1992). CD14 and CD45 are easily detected on perivascular cells but not on parenchymal microglia in the normal and the SIV-infected CNS (Williams et al. 2001a). In situ hybridization for viral RNA followed by immunohistochemistry for CD14, or immunohistochemical double-labeling studies assessing viral protein (gp120 and p28) and CD14, demonstrate that the majority of viral RNA and gp120, p28-positive cells are CD14-positive perivascular cells (Williams et al. 2001b). MNGCs that are intensely positive for viral RNA and protein are also strongly CD14- and CD45-positive, which indicates that MNGC could arise from the fusion of perivascular macrophages (Williams et al. 2001b). Recent studies in humans confirm that CD14- or CD45-positive perivascular cells are one of, if not the major, CNS population infected by HIV (Fischer-Smith et al. 2001, Gartner et al. 2000; K. Williams, S. Corey, unpublished observations).

HIV AND SIV INFECTION OF THE CNS

HIV and SIV infection of the CNS happens early, occurring between a few days to a few weeks post infection (p.i.) of the periphery (Chakrabarti et al. 1991, Davis et al. 1992, Lackner et al. 1994). The mechanism by which virus enters the CNS is not defined. One possibility is that free or extracellular virus infects brain endothelial cells and thereby the virus gains access to the neural parenchyma. Alternatively, the virus may somehow enter the CNS directly without infecting intermediate cells (Banks et al. 2001). Weighing against both of these mechanisms somewhat are observations that, while viral RNA is readily isolated from the cerebral spinal fluid (CSF) of HIV/SIV-infected humans and macaques, and CSF viral loads appear to correlate with CNS disease (Price 2000, Zink & Clements 2000, Zink et al. 1999), cell free virus is difficult to detect in the CNS parenchyma (Nath & Geiger

1998, Parmentier et al. 1992, Wesselingh et al. 1993). Another possibility is that virus enters the CNS within an infected monocyte/macrophage destined to become some form of tissue-resident cell (Davis et al. 1992, Peluso et al. 1985, Williams & Blakemore 1990, Williams et al. 2001b). Because the CNS has been considered as immunologically privileged, or relatively so, it has been suggested that virus enters early after primary infection and thereafter replicates beyond the control of the peripheral immune system. Past reports of viral genome sequence analysis support the notion that CNS-specific or neurotropic forms of the virus exist (Hughes et al. 1997; Lane et al. 1995; Power et al. 1995, 1994; Strizki et al. 1996). Yet, recent studies question this idea, demonstrating that viral sequences within specific CNS regions match, phylogenetically, with sequences found in the bone marrow (Gartner 2000, Liu et al. 2000), which thereby supports the hypothesis that the virus could be carried into the CNS in hematogenously derived cells. Brain imaging and neuropathological analysis suggest early signs of neuronal injury coincident with seroconversion, underscoring the possibility that neuronal injury or death resulting from viral infection may begin soon after infection (Gonzalez et al. 2000, Gray et al. 1996). It is not known whether viral proteins, free virus, virally infected macrophages, or all of these are responsible for neuronal injury.

Because viral infection of the CNS is an early event post infection (Davis et al. 1992, Gray et al. 1996, Lackner et al. 1994), while the disease producing neuronal damage and drop out occurs late in the course of infection, HAD has been considered a chronic, neurodegenerative disease (Kolson & Gonzalez-Scarano 2000). Although neurological signs and symptoms—typically mild and transient—may be detected early post infection (Gray et al. 1992, 1996; Horn et al. 1998; Marcondes et al. 2001), the more severe, progressive CNS ailments typically occur over a far longer period of time, frequently signaling the development of AIDS (Lackner 1994, Masliah et al. 1994, Murray et al. 1992, Navia et al. 1986b, Wiley & Achim 1994, Zink et al. 1998). CNS infection and aseptic meningitis occur early after HIV or SIV infection, although productive infection of the CNS parenchyma does not; neither does neuronal damage leading to dysfunction, degeneration, and cell death typically occur early after infection. This later chronic, progressive CNS damage developing in HIV/SIV infection may be best tied to the levels of soluble factors released by HIV/SIV-infected and -activated macrophages and microglia, as well as to the density of these cells.

Early retroviral infection of the CNS is not sustained and productive but does result in monocyte/macrophage activation and enhanced entry of these cells. Evidence from serial sacrifice studies in SIV-infected rhesus macaques demonstrates the rapid appearance of virus in the CNS, predominantly within perivascular cuffs of macrophages (Lane et al. 1996, Williams et al. 2001b). The virus then seems to disappear for a variable period, only to reemerge with the collapse of the immune system. Concomitant with the reemergence of virus is the perivascular accumulation of activated cells of monocyte/macrophage lineage, some of which are infected (Glass et al. 1995, Lane et al. 1996, Williams et al. 2001b). Thus, the clinical onset of neurological disease, and its acceleration as immune function fails, may directly relate to the dysregulation and accumulation of highly activated macrophages—both retrovirally infected and noninfected—in the perivascular areas of the CNS parenchyma. Microglia in the vicinity would likewise be stimulated, and possibly infected, via their foot processes, which extend down to the perivascular space (Lassmann et al. 1991). Support for this hypothesis can be derived from several lines of evidence (Gartner 2000).

1.) During the asymptomatic period of infection, the CNS of HIV$^+$ humans and SIV$^+$ monkeys seem to contain neither retroviral RNA nor viral proteins. Yet, viral DNA (genomically integrated virus) can be detected via PCR (Bell et al. 1993, Donaldson et al. 1994, Gosztonyi et al. 1994, Williams et al. 2001b).

2.) Protection provided by the immune system prevents the development of neurological problems such as HAD. In both humans and monkeys, effective therapy with protease inhibitors, themselves not believed to cross the blood-brain barrier, has been associated with a decreased incidence of HAD and correction of conduction abnormalities in the CNS of monkeys (Brodt et al. 1997, Dore et al. 1999, Fox et al. 2000).

3.) Neurological dysfunction and pathological alterations in the CNS attributable to HIV or SIV do not appear until the immune system deteriorates, which signals the onset of AIDS.

4.) In humans the number of activated macrophages in the CNS correlates better with HAD than does viral load (Glass et al. 1995). Likewise in SIV$^+$ monkeys, the severity of CNS pathological alterations correlates well with the number of activated macrophages (Zink et al. 1999).

5.) Macrophages in the perivascular area of both humans and primates are the primary type of cell infected with HIV/SIV throughout the course of infection (Budka 1986, Budka et al. 1991, Chakrabarti et al. 1991, Lane et al. 1996, Reinhart et al. 1997, Williams et al. 2001b)

6.) In the SIV model, neuropathological changes are more readily induced by intravenous, rather than intracerebral, injection (Hurtrel et al. 1991). This suggests that damage to the CNS involves a process most effectively initiated outside the nervous system. This process likely includes infection of monocyte/macrophages or their bone marrow precursors, which subsequently traffic to the CNS.

In considering such a hypothesis it seems that at least a quantitative, and most probably a qualitative, alteration in the macrophage/monocyte populations in the CNS must occur concomitantly with the evolution of the CNS dysfunction. Certainly, there is evidence that at this phase of the infection the number of activated CNS macrophages increases (Stevenson & Gendelman 1994); likewise, the number of virally infected CNS macrophages goes up (Koenig et al. 1986, Stevenson & Gendelman 1994). However, it is more difficult to document that these cells are physiologically different. As the immune system fails, throughout the body monocyte/macrophage family members must assume an increasingly demanding defensive role in dealing with both pathogens and commensals. There would be many stimuli to monocyte/macrophage activation resulting from elevated levels of microbiological agents and their products, proinflammatory signals from damaged and/or infected tissue, and the emerging prevalence of retrovirally infected cells and viral proteins (Fauci 1988, Pantaleo et al. 1993, Stevenson & Gendelman 1994). The situation would most probably be the same in the CNS. Lymphocytes that had previously controlled the level of productively infected cells would disappear, and macrophages would be pressed into service. But in the CNS this might have special consequences. Macrophages in the perivascular areas becoming activated would probably stimulate both their retention at the site and the recruitment of additional ones. Because some of these very cells are productively infected by virus, a vicious cycle could be envisioned wherein macrophage activation was unremitting. In a number of well-defined conditions, this occurs at sites where the immune system cannot eliminate a specific stimulus. When a foreign body is present or when an intracellular pathogen cannot be killed (as in tuberculosis), highly activated members of the macrophage family fuse to form giant cells in an attempt to rid the body of what the immune system cannot destroy. MNGC form also when HIV/SIV infection of the nervous system enters the phase associated with neurological disease. Indeed, the presence of MNGC is a diagnostic hallmark of HIVE/SIVE (Budka 1986, Budka et al. 1991). Although other neuropathological disorders are characterized by the presence of myriad phagocytic macrophages (e.g., infarction), dense parenchymal infiltration by activated macrophages (e.g., multiple sclerosis), or widespread microglial reaction (e.g., typical viral encephalitis), MNGC formation in the CNS is rare, being almost pathognomonic for macrophage responses in retrovirally infected hosts.

The observations described above support the importance of macrophage accumulation, rather than viral infection itself, to CNS pathology and neurologic disease. Soluble substances released by virally infected or cytokine-stimulated monocyte/macrophages and microglia can activate or disrupt the blood brain barrier (BBB). They also stimulate macrophages in the CNS, augmenting their recruitment and retention. In vitro investigations with HIV-1 show that TNFα, more so than viral infection, induces monocyte/macrophage migration and accumulation in a BBB model (Schmidtmayerova et al. 1996a). Similarly, HIV infection appears to drive monocytes toward the activated macrophage phenotype, resulting in the release of matrix-metalloproteinases that degrade extracellular matrix integrity (Ghorpade et al. 2001). Brain-resident macrophages and microglia, as well as astrocytes in HIV- and SIV-infected CNS, synthesize chemokines including MIP-1α and MIP-1β, which are involved in stimulating monocyte traffic and retention (Conant et al. 1998, Persidsky et al. 1999, Schmidtmayerova et al. 1996a). Both of these chemokines have been demonstrated in the CNS of HIV-infected patients with HAD (Conant et al. 1998, Sanders et al. 1998, Schmidtmayerova et al. 1996a). Injured neurons themselves may participate in recruiting macrophages to the CNS by secreting fractalkine, thereby potentially contributing to a cycle of increasing neuron damage and macrophage recruitment (Tong et al. 2000). In addition to chemokines, matrix metalloproteinases have been demonstrated in HIV- and SIV-infected CNS and are made by MNGC and microglial nodules (Ghorpade et al. 2001, Sanders et al. 1998, Schmidtmayerova et al. 1996a). Lastly, viral infection itself has been demonstrated not only to elevate cytokines that are important in subsequent macrophage activation (IL-1 and IL-6) but also to result in prolonged long-term survival of these cells (Berman et al. 1994).

These collected observations point out mechanisms by which retroviral infection directly or indirectly results in monocyte/macrophage activation, stimulates their migration to and activation in the CNS, and enhances their retention and accumulation. The prolonged presence of activated members of this family, especially the perivascular cells, would produce deleterious changes to the CNS parenchyma and the BBB. Thus, it is possible to entertain the hypothesis that activated cells of this group contribute significantly to the deleterious neurological effects of HIV-1/SIV infection.

THE MOLECULAR BASIS OF MACROPHAGE TROPISM OF HIV AND SIV IN THE CNS

Molecular studies of viral sequences found in the brain underscore the importance of macrophages and macrophage-tropic viruses contributing to CNS disease. In fact, though virtually all pathogenic strains of SIV enter the CNS, macrophage tropism appears to be required for virus to remain and actively replicate (Desrosiers et al. 1991, Simon et al. 1992). HIV and SIV infect T lymphocytes and macrophages via CD4 and chemokine receptor–dependent mechanisms. Different strains of virus are described as being T cell- and/or macrophage-tropic based upon the cell types in which they grow best in vitro. HIV and SIV isolates from encephalitic brain preferentially infect macrophages in vitro and are thus described as macrophage-tropic (Desrosiers et al. 1991, Simon et al. 1992). In the rhesus macaque model, it is possible to use defined, molecular clones of SIV with restricted cell tropism. When animals are infected with SIVmac239, a pathogenic molecular clone that is considered lymphocyte tropic, 30% of the infected animals go on to produce macrophage-tropic viral variants and develop SIV-encephalitis with MNGC (Desrosiers et al. 1991, Kodama et al. 1993, Mori et al. 1992). Sequence analyses of virus in the CNS of these animals demonstrate subtle changes within the *env* gene (as few as 2 or 3 amino acids) that yield a so-called macrophage-tropic virus (Mori et al. 1992).

The notion of neurotropic viral sequences unique to the CNS has been studied but is not resolved. Studies in SIV suggest that in addition to *env* sequences that are important for macrophage tropism, sequences within the *nef* gene define neurovirulence (Mankowski et al. 1997). Several reports address the question of brain-specific and neurotropic sequences in demented versus nondemented patients with HIV (Di Stefano et al. 1996, Korber et al. 1994). It has been suggested that HIV preferentially infects brain microglia more efficiently than it does blood monocytes or monocyte-derived macrophages (Strizki et al. 1996); however, others have not found preferential replication by HIV in one population over another (Ghorpade et al. 1998). Macrophage tropism is necessary for virus to remain in the CNS and for the development of encephalitis. It is interesting to note

Gartner and colleagues have found that the gp160 of HIV recovered from patients with dementia demonstrate remarkable sequence similarity between isolates from subcortical regions of the brain, in particular deep white matter, and those from the bone marrow (Gartner et al. 2000). These data are counter to the notion of brain-specific sequences and point to the bone marrow as a likely source of virus entering the CNS terminally. It is important not only to have virus in the CNS but also to have actively replicating virus (Teo et al. 1997). Thus, both macrophage tropism and active replication of virus and the expression of viral proteins in CNS cells are important correlates of neurological disease in HIV and SIV infection.

THE ROLE OF MACROPHAGES IN THE COURSE OF VIRAL INFECTION OF THE CNS

Soon after viral infection in the periphery, at the time of sero-conversion signaling the onset of the anti-HIV immune response, HIV and SIV enter the CNS and can be detected in the CSF and in perivascular cell aggregates in symptomatic and asymptomatic humans (Chiodi et al. 1988, Gyorkey et al. 1987, Ho et al. 1985, Sinclair et al. 1994) and monkeys (Horn et al. 1998, Lackner et al. 1994, Lane et al. 1995, Smith et al. 1995, von Herrath et al. 1995). A minimal, productive viral infection of the CNS, as detected by in situ hybridization and immunohistochemistry, does occur in macrophages soon after infection by HIV/SIV (Chakrabarti et al. 1991, Davis et al. 1992, Williams et al. 2001b). When the asymptomatic phase of the disease arrives, productive infection is undetectable (Gosztonyi et al. 1994, Sinclair et al. 1994, Williams et al. 2001b). A report of an in situ hybridization and immunohistochemical double-labeling study for viral RNA, viral protein, and macrophage-associated molecules locates HIV in macrophages within perivascular aggregates (Davis et al. 1992, Gosztonyi et al. 1994). Likewise, studies of early events in the SIV model have demonstrated virus within perivascular cuffs as early as 3 days post infection and consistently by 7–14 days (Chakrabarti et al. 1991, Lackner et al. 1994, Williams et al. 2001b). The presence of SIV in the CNS in these studies is associated with the accumulation of perivascular macrophages wherein the virus is found (Chakrabarti et al. 1991, Lackner et al. 1994, Lane et al. 1996, Williams et al. 2001b). In fact, double-labeling studies for SIV RNA or protein, and markers of CNS perivascular cells (CD14 or CD45), demonstrate an accumulation of perivascular macrophages that are the elements principally infected in macaques (Williams et al. 2001b). In the terminal phases of the infection, an accumulation of p24-positive macrophages (CD14 and CD45 positive) has been demonstrated in humans (Fischer-Smith et al. 2001; Gartner et al. 2000, 1999).

In addition to the presence or accumulation of virus-bearing macrophages in perivascular sites early in infection, studies of CSF, serum, and CNS tissues demonstrate elevated levels of immunologically active or potentially deleterious factors bespeaking macrophage activation. These include quinolinate and β-2-microglobulin, IL-1, TNFα, IL-6 MMP-2, -7 and, -9, and prostoglandins (Achim et al. 1993; Conant et al. 1999; Gallo et al. 1989; Giulian et al. 1990;

Griffin et al. 1994; Heyes et al. 1991, 1992; Heyes & Lackner 1990; Jordan & Heyes 1993; Perrella et al. 1992; Pulliam et al. 1996; Sporer et al. 1998; Tyor et al. 1992; Wiley et al. 1992). When found in the CSF, their presence appears to correlate with CNS disease progression or severity (Conant et al. 1999, Griffin et al. 1994). Some of these factors have also been demonstrated in the CNS parenchyma (Achim et al. 1993, Pulliam et al. 1996, Schmidtmayerova et al. 1996a, Tyor et al. 1992). Furthermore, in the SIV model, high-resolution magnetic resonance spectroscopy (MRS) using snap frozen tissues and quantitative neuropathological measurements of synaptophysin, calbindin, benzodiazepine receptor (for macrophage activation) and N-acetylaspartate (NAA) demonstrate macrophage activation as well as early signs of neuronal injury within weeks of infection (Gonzalez et al. 2000). Neurophysiological studies confirm MRI signs of early neuronal injury concomitant with macrophage activation after primary infection of the CNS (Horn et al. 1998). It is interesting to note that indicators of neuronal damage by MRS have also been documented in non-AIDS-related neurologic diseases as diverse as multiple sclerosis, infarction, and Alzheimer's disease (Arnold et al. 1994, Meyerhoff et al. 1994, Shonk et al. 1995). All of these conditions are linked by the presence of highly activated monocytes, macrophages, and microglia in the nervous system parenchyma. In Alzheimer's disease amyloid is believed to be the unremitting stimulus to macrophage activation; in HIV/SIV it is most probably the virus, at least initially. In MS the stimulus is unknown.

Highly activated macrophages, as appear in the CNS in HIVE or SIVE, are potent agents in inflammatory reactions. They express a variety of immunologically active cell-surface molecules, and elaborate cytokines, reactive oxygen species, and neurotoxic substances. On the cell membrane MHC class I and II, CD4, B7-1 and B7-2, CD40, Fc receptors I–III, and chemokine receptors (CCR) 3, 5, and CXCR4 are all strongly expressed—some merely increased over basal levels, others expressed de novo (Aloisi et al. 2000; Becher et al. 2001; Berman et al. 1999; De Simone et al. 1995; Ghorpade et al. 2001; He et al. 1997; Hesselgesser & Horuk 1999; Jordan et al. 1991; Klein et al. 1999; Perry & Gordon 1987, 1988; Peudenier et al. 1991; Rottman et al. 1997; Sanders et al. 1998; Tan et al. 1999; Tanabe et al. 1997; Ulvestad et al. 1994a,c; Wei & Jonakait 1999; Westmoreland et al. 1998b; Williams et al. 1994). The cells secrete matrix metalloproteinases MMP-2, 7, and 9, as well as IL-1, IL-6, and TNFα, NO, L-cysteine, and Ntox (Adamson et al. 1996, Berman et al. 1999, Ghorpade et al. 2001, Giulian et al. 1993, Gottschall & Deb 1996, Gottschall et al. 1995, Lee et al. 1993b, Mollace et al. 2001, Pulliam et al. 1991, Seilhean et al. 1997, Tyor et al. 1992, Yeh et al. 2000). At a subcellular level, the cytokine/chemokine-rich environment serves to further stimulate macrophage/microglial cells by activating p38 (MAP-Kinase), which further stimulates these cells though it is damaging to neurons (Kaul et al. 2001, Kaul & Lipton 1999). These proinflammatory molecules are found in macrophage populations in the HIV- and SIV-infected CNS as well as other immune-mediated neurologic diseases including MS and Alzheimer's disease (Akiyama et al. 2000, Kaul et al. 2001). Extensive studies of the differential expression of these molecules on macrophage subpopulations in HIV and SIV have not been performed, but in general perivascular macrophages have higher basal levels of all of these, when present. Additionally, the expression of most of these molecules is usually higher on macrophages and microglia in the white matter as opposed to gray matter. Some studies point to the selective expression of some of these markers in the CNS (MHC class II and MMP-9, TNFα) as being differentially expressed in HAD versus non-HAD CNS or in SIV-infected macaques that are rapid progressors versus non-rapid progressors (Berman et al. 1999, Conant et al. 1999). In the context of HIV and SIV infection, virus might be the initiating element, but activated macrophages might be the common denominator of CNS injury.

CD8 LYMPHOCYTES IN HIV AND SIV ENCEPHALITIS

As noted previously, following the initial viremia after infection, the burst of viral replication is brought under control by $CD8^+$ T lymphocytes (Barouch & Letvin 2001, Schmitz et al. 1999). $CD4^+$ lymphocytes provide help both for the CD8 lymphocytes in controlling viremia and for B cells in producing antiretroviral antibodies (Altfeld & Rosenberg 2000; Rosenberg et al. 2000, 1997). If these mechanisms function as they usually do, the viral load drops and an asymptomatic phase of infection ensues (Schmitz et al. 1999). $CD4^+$ T cells are rarely found within the CNS parenchyma during primary HIV/SIV infection or terminally. $CD8^+$ T cells are a consistent, albeit minor, population of HIV/SIV-encephalitis lesions (Lackner et al. 1991b). HIV and SIV antigen-specific $CD8^+$ T lymphocytes have been demonstrated in the CSF and CNS parenchyma of humans and monkeys (von Herrath et al. 1995, Walker et al. 1987). It is interesting that after HIV infection, macrophages release MIP-1α/β, which can cause recruitment of CD8s (Schmidtmayerova et al. 1996a). Because $CD8^+$ T lymphocytes are found within HIV/SIV lesions and are often surrounded by macrophages with high levels of viral antigen on their surface, the significance of their presence is obscure. $CD8^+$ lymphocytes may be acting to directly control viral production or kill infected cells. Evidence from this laboratory, using SIV-gag tetramers, demonstrates a surprisingly high percentage of CD3/CD8-positive lymphocytes within the choroid plexus and meninges that are SIV-gag–specific by 21 days post infection (K. Williams, S. Corey, unpublished observations).

Fox and colleagues have been able to dissociate CNS tissues of macaques that were SIV infected and died with encephalitis and to recover CD8 lymphocytes that were SIV-antigen specific as demonstrated in bulk antigen stimulation assays (von Herrath et al. 1995). More recently, this group demonstrated that highly activated $CD8^+$ T lymphocytes in the CNS of SIV-infected macaques correlated with CNS dysfunction (Marcondes et al. 2001). HIV antigen–specific CD8 lymphocytes have also been demonstrated in the CSF of humans that are HIV infected (Walker et al. 1987). Our laboratory has used a recently described model wherein CD8-depleting monoclonal antibody (Schmitz et al. 1999) was used to deplete this lymphocyte phenotype in SIV-infected macaques. Of the animals treated with the anti-CD8 antibody, 50% exhibited a persistent depletion of the $CD8^+$ phenotype for longer than 28 days. Of the monkeys showing this persistent CD8 depletion, greater than 80% developed AIDS with SIV-encephalitis (Williams et al. 2001a; K. Williams, S. Corey, unpublished observations). This is in comparison to the baseline 30%–40% of humans who develop HIV-encephalitis and the 40%–50% of macaques that are rapid progressors developing SIV-encephalitis (McArthur et al. 1993, Westmoreland et al. 1998a). Histologically, the encephalitis in 4 of the 5 persistently CD8-depleted animals was more severe than the disease observed in an archive of 400 other monkeys that were SIV infected and died with AIDS and SIVE. They had massive infiltration of the perivascular area by macrophages, many of which were infected. Virtually every tissue section examined had MNGCs. Clinically, the persistently depleted animals became paralyzed at the onset of or following the development of AIDS (Williams, unpublished observations). These data underscore the CD8 lymphocyte contribution to the control of monocyte/macrophage entry, survival, persistence, and/or accumulation in the CNS. Though not fully documented, $CD8^+$ lymphocytes may actually control the level of infection occurring in monocyte/macrophages in the bone marrow or blood, as well as the traffic of these cells to the CNS. The ability of $CD8^+$ lymphocytes to directly lyse infected cells and to secrete factors that inhibit viral replication has been documented. Therefore, in retroviral infections of human and monkey, loss of the efficacy of the $CD8^+$ lymphocytes in controlling virus may permit the emergence of a dysfunctional, activated macrophage population that is neuropathogenic.

MACROPHAGES IN HIV AND SIV ENCEPHALITIS

Macrophages are the most common cellular component of inflammation in brains of humans and monkeys with HIV/SIV encephalitis (Glass et al. 1995). T lymphocytes are less common, but $CD8^+$ T lymphocytes are routinely found within the choroid plexus, meninges, and perivascular cuffs. Both infected and uninfected macrophages infiltrate the meninges, choroid plexus, perivascular spaces, and the brain parenchyma (Budka et al. 1991, Gendelman et al. 1997, Glass et al. 1995). There is little-to-no cell-free virus within the CNS of individuals with encephalitis; the majority of viral RNA, DNA, and protein is associated with macrophages and MNGC (Budka 1986, Budka et al. 1991, Petito et al. 1999). The number of activated macrophages, only some of which are virally infected, is a better predictor of HAD than viral load in humans and of CNS pathology in SIV-infected macaques (Glass et al. 1995, Zink et al. 1999). There is an evolving belief that macrophages are the most important targets of CNS infection and are the most probable means by which virus is carried into the CNS (Peluso et al. 1985; Persidsky et al. 1999, 1997; Williams et al. 2001b). Thus, special attention is again directed to the perivascular cells mentioned above.

Examination of rodent bone marrow, chimeras, and human marrow transplant recipients demonstrates that CNS perivascular cells continuously turn over, being replaced by bone marrow–derived monocyte/macrophages (Hickey & Kimura 1988, Hickey et al. 1992, Lassmann et al. 1993). Meningeal macrophages are the most rapidly replenished population. In contrast, parenchymal microglia are an extremely stable population with little to no turnover within the life span of the host

(Hickey & Kimura 1988, Hickey et al. 1992, Lassmann et al. 1993, Unger et al. 1993). These observations in part might therefore account for the observed kinetics of both CNS viral entry and clinical symptomatology early in infection. The early, transient meningitis probably heralds the arrival of infected macrophages in the meninges and the onset of the immune system's attempt to eliminate them. The emerging immunological control of the viral infection in macrophage lineage cells may explain the disappearance of virus from meninges and parenchyma during the asymptomatic stage, then its reappearance as productive macrophage infection reemerges in terminal AIDS. It is widely documented that in both HIV and SIV infection the majority of the CNS cells are not infected. Members of the extended monocyte/macrophage family are infected (Dickson et al. 1991, Koenig et al. 1986, Kure et al. 1990b). These data suggest that macrophage activation and infection, especially terminally, play a role in the pathogenesis of neurological disease.

The phenotype of the perivascular cell is similar to that of a minor monocyte population found in the peripheral blood. Though suggestive, it remains to be proven whether one represents the precursor form of the other. The perivascular cell may derive from distinct marrow precursors and enter the CNS to perform a unique set of functions (Walker 1999). The circulating monocyte population is defined by the level of CD14 expression and co-expression of CD16 and/or CD69 (Passlick et al. 1989). Notably, populations of CD14/CD16- and CD69-positive monocytes are expanded in the peripheral blood of HIV-infected patients (Thieblemont et al. 1995). Pulliam et al. identified a similar expansion of CD14/CD16 and CD14/CD69 monocytes in HIV-infected patients. Moreover, they noted a high correlation between an increase in this hematogenous CD14/CD69 population and the occurrence of HAD—noninfected individuals; HIV$^+$ AIDS patients without HAD did not exhibit the phenomenon (Pulliam et al. 1997). Similarly, others have observed an increase in the percentage of CD14/CD16-positive cells in the blood of HIV-infected individuals with HAD versus nondemented patients (Gartner et al. 2000). In monkeys we have noted a seemingly selective expansion of CD14/CD69-positive monocytes in the blood of SIV-infected macaques that develop SIV-encephalitis (K. Williams, unpublished observations). At the histological level, this population of monocyte/macrophages has been found in the perivascular space of SIV-infected macaques (K. Williams, unpublished observations) and humans (Fischer-Smith et al. 2001). It is not surprising that these CD14/CD16- or CD14/CD69-positive perivascular cells are HIV-infected and express p24 (Fischer-Smith et al. 2001, Gartner et al. 2000). It seems clear that HIV-associated macrophage secretion of cytokines, reactive oxygen species, matrix metalloproteases, Ntox, and other agents potentially damaging to neurons warrants further study.

CONTROL OF MACROPHAGE ACCUMULATION IN LENTIVIRAL ENCEPHALITIS

In this review we put forth one new and perhaps unifying hypothesis: In the phase of HIV/SIV infection that follows the virally induced failure of the immune system, dysregulation of the macrophage/monocyte populations occurs. The chronic presence of highly activated members of the macrophage family in various tissues of the body constitutes a pathogenic mechanism unto itself. Chronic macrophage dysregulation causes neurological disease. This has been proposed as a basic pathogenic mechanism in Alzheimer's disease (Akiyama et al. 2000). However, if this is true, then most retrovirally infected individuals must possess mechanisms to prevent macrophage dysregulation for a substantial period of time.

It is clear that infected and/or activated macrophages are in the CNS of humans and nonhuman primates with terminal AIDS, HIVE, HAD, SIVE, and other retroviral-associated neurological diseases. What is less clear is the mechanism by which monocyte/macrophages either traffic to, or accumulate in, the CNS during end-stage disease. There are CD14/CD16- or CD14/CD69-positive macrophages from the bone marrow carrying the macrophage-tropic strains programmed to enter the CNS under basal, nonactivation conditions. When they are activated, either by the evolving HIV-encephalitis or by intercurrent challenges and stimuli secondary to HIVE, there is an increased traffic of similar macrophages into the CNS. The mechanisms responsible for increased traffic are not known, and why these cells remain in the CNS during terminal AIDS has yet to be determined. Current data point to the role of immune activation, chemokine and cytokine production, and alterations in the immune system, controlling such traffic (Langford & Masliah 2001; Persidsky et al. 1999, 2000; Sasseville et al. 1994).

Numerous studies underscore the importance of adhesion molecules and chemokine expression in both early and end-stage disease. Research in humans and macaques has identified MCP-1, Rantes, and IP-10 in the CSF as an indicator of disease progression (Cinque et al. 1998, Conant et al. 1998, Kelder et al. 1998, Kolb et al. 1999). Vascular cell adhesion molecule-1 (VCAM-1) and intercellular adhesion molecule-1 (ICAM-1) have been detected on CNS endothelial cells in SIVE (Sasseville et al. 1992). Neuron-derived fractalkine may participate (Tong et al. 2000). It is less clear whether these molecules play a central role in monocyte/macrophage recruitment or are merely a result of endothelial activation. However, proteins of HIV-1 itself cause endothelial cell activation (Boven et al. 2000; Dhawan et al. 1997, 1994; Weiss et al. 1999). Macrophages of the type most relevant to HIV/SIV infection, the perivascular cells, enter the healthy CNS as part of normal physiology (Hickey et al. 1991). Likewise, when lymphocytes are activated in vitro and injected intravenously, these cells can be found in the CNS a few hours after infusion (Hickey et al. 1991). Thus, there clearly must exist a rapid entry mechanism that most likely is chemokine receptor–independent and does not require CNS chemokine expression. So what occurs in HIV- and SIV-infected CNS? Unlike MS, EAE, typical viral encephalitis and Alzheimer's disease, in HIV- and SIV-encephalitis macrophages appear to accumulate concomitantly with emerging dysfunction of the immune system. Thus, it appears that the appropriately functioning immune system in some way controls the traffic and accumulation of macrophages in the CNS. Perhaps it is as simple as macrophages attempting to perform the functions of the faltering immune system.

In summary, the mechanisms responsible for the accumulation of macrophages in the CNS as HIVE/SIVE emerge are unclear. Yet, the fact remains that they do accumulate. Whether this change is the result of loss of the immune system's regulation of the virus, or some intrinsic CNS alteration encouraging cell entry, or both, requires further investigation.

CONCLUSION

As shown in Figure 1, the progression of HIV or SIV infection in the CNS follows a protracted course. It is our contention that macrophage/monocyte cells that continuously enter the CNS as part of normal physiology are the central participants in the production of neurological disease. After an early infectious phase, the immune system controls macrophage infection. Yet, as the immune system slowly fails and $CD4^+$ T cell levels fall, the ability of both $CD8^+$ cells and B cells (which may require $CD4^+$ T cell help for effective functioning) become less able to prevent the emergence of productively infected cells. In the nervous system, the productively

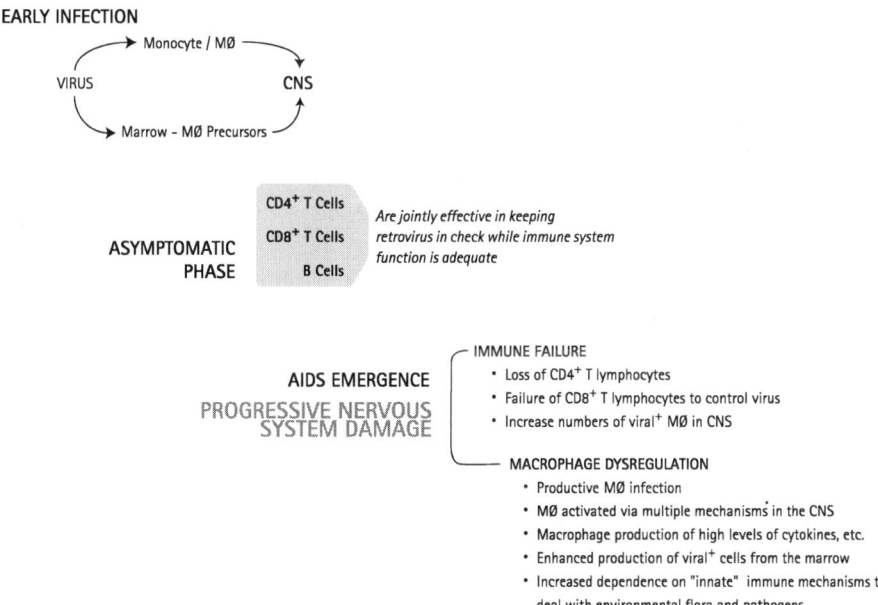

Figure 1 Schematic of early and asymptomatic stages of HIV/SIV infection and the development of AIDS and CNS pathology. The early stage of infection is characterized by monocyte and brain macrophage precursors, some of which are infected, entering the CNS and resulting in nonproductive infection of the CNS. During the asymptomatic phase components of the peripheral immune system, including CD4 and CD8 T lymphocytes and the humoral immune system, keep virus in check, and productive infection of the CNS is nondetectable. With the emergence of AIDS characterized by a loss of CD4 T lymphocytes and failure of CD8 T lymphocytes to control viral replication, there is an accumulation of activated, infected and noninfected macrophages in the CNS that correlates with neuropathology.

infected cells are predominantly members of the macrophage/monocyte family. This review suggests the importance of the accumulation of a specific type of monocyte/macrophage in the CNS as the principle contributors to the syndromes of neuro AIDS. The accumulation of these cells clearly is predictive of neurological dysfunction. In what becomes a self-amplifying cycle, infected macrophages activate other macrophages in the vicinity. The viral products, cytokines, and chemokines released from infected and/or activated macrophages and cells reacting to them attract and activate increasing numbers of monocytes. In the absence of immune system control of the retrovirus, the process proceeds unchecked. This cycle of attraction, activation, infection, and relentless stimulation of macrophages, without a mechanism to eliminate the offending stimulus, constitutes macrophage dysregulation. Strains of retrovirus effectively infecting specific macrophage populations would then be seen to stimulate this cycle in the CNS. The accumulation of highly activated, pathogenic macrophages in the tissues of the nervous system appears to be the substrate for a broad spectrum of retrovirally associated neurological disorders. The mechanisms by which unhealthy macrophages contribute to neurological signs in AIDS- and non–AIDS-related diseases warrant further study.

ACKNOWLEDGMENTS

This work was supported in part by Public Health Service grants NS27321 (WH), NS37654 (KW), and NS40237 (KW). We thank Drs. Norm Letvin and Woong-Ki Kim (Department of Medicine, Harvard Medical School) for discussion of the manuscript. We thank Lily Rappoli for graphic services.

The *Annual Review of Neuroscience* is online at http://neuro.annualreviews.org

LITERATURE CITED

Achim CL, Heyes MP, Wiley CA. 1993. Quantitation of human immunodeficiency virus, immune activation factors, and quinolinic acid in AIDS brains. *J. Clin. Invest.* 91:2769–75

Adamson DC, Wildemann B, Sasaki M, Glass JD, McArthur JC, et al. 1996. Immunologic NO synthase: elevation in severe AIDS dementia and induction by HIV-1 gp41. *Science* 274:1917–21

Adle-Biassette H, Levy Y, Colombel M, Poron F, Natchev S, et al. 1995. Neuronal apoptosis in HIV infection in adults. *Neuropathol. Appl. Neurobiol.* 21:218–27

Akiyama H, Barger S, Barnum S, Bradt B, Bauer J, et al. 2000. Inflammation and Alzheimer's disease. *Neurobiol. Aging* 21:383–421

Aloisi F, Ria F, Adorini L. 2000. Regulation of T-cell responses by CNS antigen-presenting cells: different roles for microglia and astrocytes. *Immunol. Today* 21:141–47

Altfeld M, Rosenberg ES. 2000. The role of CD4(+) T helper cells in the cytotoxic T lymphocyte response to HIV-1. *Curr. Opin. Immunol.* 12:375–80

Anders KH, Guerra WF, Tomiyasu V, Verity MA, Vinters HV. 1986. The neuropathology of AIDS. UCLA experience and review. *Am. J. Pathol.* 124:537–58

Arnold DL, Riess GT, Matthews PM, Francis GS, Collins DL, et al. 1994. Use of proton magnetic resonance spectroscopy for monitoring disease progression in multiple sclerosis. *Ann. Neurol.* 36:76–82

Banks WA, Freed EO, Wolf KM, Robinson SM, Franko M, Kumar VB. 2001. Transport of human immunodeficiency virus type 1 pseudoviruses across the blood-brain barrier: role of envelope proteins and adsorptive endocytosis. *J. Virol.* 75:4681–91

Barouch DH, Letvin NL. 2001. CD8+ cytotoxic T lymphocyte responses to lentiviruses and herpesviruses. *Curr. Opin. Immunol.* 13:479–82

Becher B, Durell BG, Miga AV, Hickey WF, Noelle RJ. 2001. The clinical course of experimental autoimmune encephalomyelitis and inflammation is controlled by the expression of CD40 within the central nervous system. *J. Exp. Med.* 193:967–74

Bell JE, Busuttil A, Ironside JW, Rebus S, Donaldson YK, et al. 1993. Human immunodeficiency virus and the brain: investigation of virus load and neuropathologic changes in pre-AIDS subjects. *J. Infect. Dis.* 168:818–24

Berman MA, Zaldivar F Jr, Imfeld KL, Kenney JS, Sandborg CI. 1994. HIV-1 infection of macrophages promotes long-term survival and sustained release of interleukins 1 alpha and 6. *AIDS Res. Hum. Retroviruses* 10:529–39

Berman NE, Marcario JK, Yong C, Raghavan R, Raymond LA, et al. 1999. Microglial activation and neurological symptoms in the SIV model of NeuroAIDS: association of MHC-II and MMP-9 expression with behavioral deficits and evoked potential changes. *Neurobiol. Dis.* 6:486–98

Boven LA, Middel J, Breij EC, Schotte D, Verhoef J, et al. 2000. Interactions between HIV-infected monocyte-derived macrophages and human brain microvascular endothelial cells result in increased expression of CC chemokines. *J. Neurovirol.* 6:382–89

Brodt HR, Kamps BS, Gute P, Knupp B, Staszewski S, Helm EB. 1997. Changing incidence of AIDS-defining illnesses in the era of antiretroviral combination therapy. *Aids* 11:1731–38

Budka H. 1986. Multinucleated giant cells in brain: a hallmark of the acquired immune deficiency syndrome (AIDS). *Acta Neuropathol.* 69:253–58

Budka H, Wiley CA, Kleihues P, Artigas J, Asbury AK, et al. 1991. HIV-associated disease of the nervous system: review of nomenclature and proposal for neuropathology-based terminology. *Brain Pathol.* 1:143–52

Chakrabarti L, Hurtrel M, Maire M, Vazeux R, Dormont D, et al. 1991. Early viral replication in the brain of SIV-infected rhesus monkeys. *Am. J. Pathol.* 139:1273–80

Chiodi F, Albert J, Olausson E, Hagberg L, Sönnerborg A, et al. 1988. Isolation frequency of human immunodeficiency virus from cerebrospinal fluid and blood of patients with varying severity of HIV infection. *AIDS Res. Hum. Retroviruses* 4:351–58

Cinque P, Vago L, Mengozzi M, Torri V, Ceresa D, et al. 1998. Elevated cerebrospinal fluid levels of monocyte chemotactic protein-1 correlate with HIV-1 encephalitis and local viral replication. *Aids* 12:1327–32

Conant K, Garzino-Demo A, Nath A, McArthur JC, Halliday W, et al. 1998. Induction of monocyte chemoattractant protein-1 in HIV-1 Tat-stimulated astrocytes and elevation in AIDS dementia. *Proc. Natl. Acad. Sci. USA* 95:3117–21

Conant K, McArthur JC, Griffin DE, Sjulson L, Wahl LM, Irani DN. 1999. Cerebrospinal fluid levels of MMP-2, 7, and 9 are elevated in association with human immunodeficiency virus dementia. *Ann. Neurol.* 46:391–98

Cotter RL, Burke WJ, Thomas VS, Potter JF, Zheng J, Gendelman HE. 1999. Insights into the neurodegenerative process of Alzheimer's disease: a role for mononuclear phagocyte-associated inflammation and neurotoxicity. *J. Leukoc. Biol.* 65:416–27

Davis LE, Hjelle BL, Miller VE, Palmer DL, Llewellyn AL, et al. 1992. Early viral brain invasion in iatrogenic human immunodeficiency virus infection. *Neurology* 42:1736–39

De Simone R, Giampaolo A, Giometto B, Gallo P, Levi G, et al. 1995. The costimulatory molecule B7 is expressed on human microglia in culture and in multiple sclerosis acute lesions. *J. Neuropathol. Exp. Neurol.* 54:175–87

Desrosiers RC. 1990. The simian immunodeficiency viruses. *Annu. Rev. Immunol.* 8:557–78

Desrosiers RC, Hansen-Moosa A, Mori K, Bouvier DP, King NW, et al. 1991. Macrophage-tropic variants of SIV are associated with specific AIDS-related lesions but are not essential for the development of AIDS. *Am. J. Pathol.* 139:29–35

Dhawan S, Puri RK, Kumar A, Duplan H, Masson JM, Aggarwal BB. 1997. Human immunodeficiency virus-1-tat protein induces the cell surface expression of endothelial leukocyte adhesion molecule-1, vascular cell adhesion molecule-1, and intercellular adhesion molecule-1 in human endothelial cells. *Blood* 90:1535–44

Dhawan S, Weeks BS, Soderland C, Schnaper HW, Toro LA, et al. 1994. HIV-1 infection alters monocyte interactions with human microvascular endothelial cells. *J. Immunol.* 154:422–32

Dickson DW, Lee SC, Mattiace LA, Yen SH, Brosnan C. 1993. Microglia and cytokines in neurological disease, with special reference to AIDS and Alzheimer's disease. *Glia* 7:75–83

Dickson DW, Mattiace LA, Kure K, Hutchins K, Lyman WD, Brosnan CF. 1991. Microglia in human disease, with an emphasis on acquired immune deficiency syndrome. *Lab. Invest.* 64:135–56

Di Stefano M, Wilt S, Gray F, Dubois-Dalcq M, Chiodi F. 1996. HIV type 1 V3 sequences and the development of dementia during AIDS. *AIDS Res. Hum. Retroviruses* 12:471–76

Donaldson YK, Bell JE, Ironside JW, Brettle RP, Robertson JR, et al. 1994. Redistribution of HIV outside the lymphoid system with onset of AIDS. *Lancet* 343:382–85

Dore GJ, Correll PK, Li Y, Kaldor JM, Cooper DA, Brew BJ. 1999. Changes to AIDS dementia complex in the era of highly active antiretroviral therapy. *Aids* 13:1249–53

Everall I, Luthert P, Lantos P. 1993. A review of neuronal damage in human immunodeficiency virus infection. *J. Neuropathol. Exp. Neurol.* 52:561–66

Everall IP, Luthert PJ, Lantos PL. 1991. Neuronal loss in the frontal cortex in HIV infection. *Lancet* 337:1119–21

Fauci AS. 1988. The human immunodeficiency virus: infectivity and mechanisms of pathogenesis. *Science* 239:617–22

Fischer-Smith T, Croul S, Sverstiuk AE, Capini C, L'Heureux D, et al. 2001. CNS invasion by CD14/CD16 positive peripheral blood-derived monocytes in HIV dementia: perivascular accumulation and reservoir of HIV infection. *J. Neuro. Virol.* 7:528–41

Ford AL, Goodsall AL, Hickey WF, Sedgwick JD. 1995. Normal adult microglia seperated from other CNS macrophages by flow cytometric sorting: phenotypic differences defined and direct ex vivo antigen presentation to myelin basic protein reactive CD4 cells compared. *J. Immunol.* 154:4309–21

Fox HS, Weed MR, Huitron-Resendiz S, Baig J, Horn TF, et al. 2000. Antiviral treatment normalizes neurophysiological but not movement abnormalities in simian immunodeficiency virus-infected monkeys. *J. Clin. Invest.* 106:37–45

Gallo P, Frei K, Rordorf C, Lazdins J, Tavolato B, Fontana A. 1989. Human immunodeficiency virus type 1 (HIV-1) infection of the central nervous system: an evaluation of cytokines in cerebrospinal fluid. *J. Neuroimmunol.* 23:109–16

Gartner S. 2000. HIV infection and dementia. *Science* 287:602–4

Gartner S, Liu Y, Hunter E, Sacktor N, Conant K, et al. 2000. *Role of systemic events in the development of HIV encephalitis and dementia*. Presented at 3rd Int. Symp. Neuro. Virol., San Francisco, CA

Gartner S, Liu Y, Tang XP, Hunter E, Sharma U, McArthur JC. 1999. *HIV dementia is associated with activated circulating monocytes.*

Presented at Infect. Nerv. Syst.: Host Pathogen Interact., Taos, NM

Gelbard HA, James HJ, Sharer LR, Perry SW, Saito Y, et al. 1995. Apoptotic neurons in brains from paediatric patients with HIV-1 encephalitis and progressive encephalopathy. *Neuropathol. Appl. Neurobiol.* 21:208–17

Gendelman HE, Persidsky Y, Ghorpade A, Limoges J, Stins M, et al. 1997. The neuropathogenesis of the AIDS dementia complex. *Aids* 11:S35–45

Ghorpade A, Nukuna A, Che M, Haggerty S, Persidsky Y, et al. 1998. Human immunodeficiency virus neurotropism: an analysis of viral replication and cytopathicity for divergent strains in monocytes and microglia. *J. Virol.* 72:3340–50

Ghorpade A, Persidskaia R, Suryadevara R, Che M, Liu XJ, et al. 2001. Mononuclear phagocyte differentiation, activation, and viral infection regulate matrix metalloproteinase expression: implications for human immunodeficiency virus type 1–associated dementia. *J. Virol.* 75:6572–83

Giulian D, Vaca K, Noonan CA. 1990. Secretion of neurotoxins by mononuclear phagocytes infected with HIV-1. *Science* 250:1593–96

Giulian D, Wendt E, Vaca K, Noonan CA. 1993. The envelope glycoprotein of human immunodeficiency virus type 1 stimulates release of neurotoxins from monocytes. *Proc. Natl. Acad. Sci. USA* 90:2769–73

Glass JD, Fedor H, Wesselingh SL, McArthur JC. 1995. Immunocytochemical quantitation of human immunodeficiency virus in the brain: correlations with dementia. *Ann. Neurol.* 38:755–62

Gonzalez RG, Cheng LL, Westmoreland SV, Sakaie KE, Becerra LR, et al. 2000. Early brain injury in the SIV-macaque model of AIDS. *Aids* 14:2841–49

Gonzalez-Scarano F, Baltuch G. 1999. Microglia as mediators of inflammatory and degenerative diseases. *Annu. Rev. Neurosci.* 22:219–40

Gosztonyi G, Artigas J, Lamperth L, Webster H deF. 1994. Human immunodeficiency virus (HIV) distribution in HIV encephalitis: study of 19 cases with combined use of in situ hybridization and immunocytochemistry. *J. Neuropathol. Exp. Neurol.* 53:521–34

Gottschall PE, Deb S. 1996. Regulation of matrix metalloproteinase expressions in astrocytes, microglia and neurons. *Neuroimmunomodulation* 3:69–75

Gottschall PE, Yu X, Bing B. 1995. Increased production of gelatinase B (matrix metalloproteinase-9) and interleukin-6 by activated rat microglia in culture. *J. Neurosci. Res.* 42:335–42

Gray F, Geny C, Lionnet F, Dournon E, Fenelon G, et al. 1991. Neuropathologic study of 135 adult cases of acquired immunodeficiency syndrome (AIDS). *Ann. Pathol.* 11:236–47

Gray F, Lescs MC, Keohane C, Paraire F, Marc B, et al. 1992. Early brain changes in HIV infection: neuropathological study of 11 HIV seropositive, non-AIDS cases. *J. Neuropathol. Exp. Neurol.* 51:177–85

Gray F, Scaravilli F, Everall I, Chretien F, An S, et al. 1996. Neuropathology of early HIV-1 infection. *Brain Pathol.* 6:1–15

Griffin DE, Wesselingh SL, McArthur JC. 1994. Elevated central nervous system prostaglandins in human immunodeficiency virus-associated dementia. *Ann. Neurol.* 35:592–97

Gyorkey F, Melnick JL, Gyorkey P. 1987. Human immunodeficiency virus in brain biopsies of patients with AIDS and progressive encephalopathy. *J. Infect. Dis.* 155:870–76

He J, Chen Y, Farzan M, Choe H, Ohagen A, et al. 1997. CCR3 and CCR5 are co-receptors for HIV-1 infection of microglia. *Nature* 385:645–49

Hesselgesser J, Horuk R. 1999. Chemokine and chemokine receptor expression in the central nervous system. *J. Neurovirol.* 5:13–26

Heyes MP, Brew BJ, Martin A, Price RW, Salazar AM, et al. 1991. Quinolinic acid in cerebrospinal fluid and serum in HIV-1 infection: relationship to clinical and neurological status. *Ann. Neurol.* 29:202–9

Heyes MP, Jordan EK, Lee K, Saito K, Frank JA, et al. 1992. Relationship of neurologic status in macaques infected with the simian immunodeficiency virus to cerebrospinal fluid quinolinic acid and kynurenic acid. *Brain Res.* 570:237–50

Heyes MP, Lackner A. 1990. Increased cerebrospinal fluid quinolinic acid and kynurenic acid in acute septicemia. *J. Neurochem.* 55:338–41

Hickey WF, Hsu BL, Kimura H. 1991. T Lymphocyte entry into the central nervous system. *J. Neurosci. Res.* 28:254–60

Hickey WF, Kimura H. 1988. Perivascular microglial cells of the CNS are bone marrow-derived and present antigen. *Science* 239:290–92

Hickey WF, Vass K, Lassmann H. 1992. Bone marrow derived elements in the central nervous system: an immunohistochemical and ultrastructural survey of rat chimeras. *J. Neuropathol. Exp. Neurol.* 5:246–56

Ho DD, Rota TR, Schooley RT, Kaplan JC, Allan JD, et al. 1985. Isolation of HTLV-III from cerebrospinal fluid and neural tissues of patients with neurologic syndromes related to the acquired immunodeficiency syndrome. *N. Engl. J. Med.* 313:1493–97

Horn TFW, Huitron-Resendiz S, Weed MR, Henriksen SJ, Fox HS. 1998. Early physiological abnormalities after simian immunodeficiency virus infection. *Proc. Natl. Acad. Sci. USA* 95:15072–77

Hughes ES, Bell JE, Simmonds P. 1997. Investigation of the dynamics of the spread of human immunodeficiency virus to brain and other tissues by evolutionary analysis of sequences from the p17gag and env genes. *J. Virol.* 71:1272–80

Hurtrel B, Chakrabarti L, Hurtrel M, Maire MA, Dormont D, Montagnier L. 1991. Early SIV encephalopathy. *J. Med. Primatol.* 20:159–66

Hurtrel B, Chakrabarti L, Hurtrel M, Montagnier L. 1993. Target cells during early SIV encephalopathy. *Res. Virol.* 144:41–46

Jordan CA, Watkins BA, Kufta C, Dubois-Dalcq M. 1991. Infection of brain microglial cells by human immunodeficiency virus type 1 is CD4 dependent. *J. Virol.* 65:736–42

Jordan EK, Heyes MP. 1993. Virus isolation and quinolinic acid in primary and chronic simian immunodeficiency virus infection. *Aids* 7:1173–79

Kaul M, Garden GA, Lipton SA. 2001. Pathways to neuronal injury and apoptosis in HIV-associated dementia. *Nature* 410:988–94

Kaul M, Lipton SA. 1999. Chemokines and activated macrophages in HIV gp120-induced neuronal apoptosis. *Proc. Natl. Acad. Sci. USA* 96:8212–16

Kelder W, McArthur JC, Nance-Sproson T, McClernon D, Griffin DE. 1998. Beta-chemokines MCP-1 and RANTES are selectively increased in cerebrospinal fluid of patients with human immunodeficiency virus-associated dementia. *Ann. Neurol.* 44:831–35

Ketzler S, Weis S, Haug H, Budka H. 1990. Loss of neurons in the frontal cortex in AIDS brains. *Acta Neuropathol. (Berl.)* 80:92–94

Klein RS, Williams KC, Alvarez-Hernandez X, Westmoreland S, Force T, et al. 1999. Chemokine receptor expression and signaling in macaque and human fetal neurons and astrocytes: implications for the neuropathogenesis of AIDS. *J. Immunol.* 163:1636–46

Kodama T, Mori K, Kawahara T, Ringler DJ, Desrosiers RC. 1993. Analysis of simian immunodeficiency virus sequence variation in tissues of rhesus macaques with AIDS. *J. Virol.* 67:6522–34

Koenig S, Gendelman HE, Orenstein JM, Dal Canto MC, Pezeshkpour GH, et al. 1986. Detection of AIDS virus in macrophages in brain tissue from AIDS patients with encephalopathy. *Science* 233:1089–93

Kolb SA, Sporer B, Lahrtz F, Koedel U, Pfister HW, Fontana A. 1999. Identification of a T cell chemotactic factor in the cerebrospinal fluid of HIV-1-infected individuals as interferon-gamma inducible protein 10. *J. Neuroimmunol.* 93:172–81

Kolson DL, Gonzalez-Scarano F. 2000. HIV and HIV dementia. *J. Clin. Invest.* 106:11–13

Korber BTM, Kunstman KJ, Patterson BK,

Furtado M, McEvilly MM, et al. 1994. Genetic differences between blood- and brain-derived viral sequences from human immunodeficiency virus type 1-infected patients: evidence of conserved elements in the V3 region of the envelope protein of brain-derived sequences. *J. Virol.* 68:7467–81

Kure K, Lyman WD, Weidenheim KM, Dickson DW. 1990a. Cellular localization of an HIV-1 antigen in subacute AIDS encephalitis using an improved double-labeling immunohistochemical method. *Am. J. Pathol.* 136:1085–92

Kure K, Weidenheim KM, Lyman WD, Dickson DW. 1990b. Morphology and distribution of HIV-1 gp41-positive microglia in subacute AIDS encephalitis. Pattern of involvement resembling a multisystem degeneration. *Acta Neuropathol.* 80:393–400

Kusdra L, Rempel H, Yaffe K, Pulliam L. 2000. Elevation of CD69+ monocyte/macrophages in patients with Alzheimer's disease. *Immunobiology* 202:26–33

Lackner AA. 1994. Pathology of simian immunodeficiency virus induced disease. In *Current Topics in Microbiology and Immunology: Simian Immunodeficiency Virus*, ed. RC Desrosiers, N Letvin, pp. 35–64. Berlin: Springer Verlag

Lackner AA, Dandekar S, Gardner MB. 1991a. Neurobiology of simian and feline immunodeficiency virus infections. *Brain Pathol.* 1:201–12

Lackner AA, Smith MO, Munn RJ, Martfeld DJ, Gardner MB, et al. 1991b. Localization of simian immunodeficiency virus in the central nervous system of rhesus monkeys. *Am. J. Pathol.* 139:609–21

Lackner AA, Vogel P, Ramos RA, Kluge JD, Marthas M. 1994. Early events in tissues during infection with pathogenic (SIVmac239) and nonpathogenic (SIVmac1A11) molecular clones of simian immunodeficiency virus. *Am. J. Pathol.* 145:428–39

Lane JH, Sasseville VG, Smith MO, Vogel P, Pauley DR, et al. 1996. Neuroinvasion by simian immunodeficiency virus coincides with increased numbers of perivascular macrophages/microglia and intrathecal immune activation. *J. Neurovirol.* 2:423–32

Lane TE, Buchmeier MJ, Watry DD, Jakubowski DB, Fox HS. 1995. Serial passage of microglial SIV results in selection of homogeneous env quasispecies in the brain. *Virology* 212:458–65

Langford D, Masliah E. 2001. Crosstalk between components of the blood brain barrier and cells of the CNS in microglial activation in AIDS. *Brain Pathol.* 11:306–12

Lassmann H, Schmied M, Vass K, Hickey WF. 1993. Bone marrow derived elements and resident microglia in brain inflammation. *Glia* 7:19–24

Lassmann H, Zimprich F, Vass K, Hickey WF. 1991. Microglial cells are a component of the perivascular glia limitans. *J. Neurosci. Res.* 28:236–43

Lee SC, Hatch WC, Liu W, Kress Y, Lyman WD, Dickson DW. 1993a. Productive infection of human fetal microglia by HIV-1. *Am. J. Pathol.* 143:1032–39

Lee SC, Liu W, Dickson DW, Brosnan CF, Berman JW. 1993b. Cytokine production by human fetal microglia and astrocytes. *J. Immunol.* 150:2659–67

Liu Y, Tang XP, McArthur JC, Scott J, Gartner S. 2000. Analysis of human immunodeficiency virus type 1 gp60 sequences from a patient with HIV dementia: evidence for monocyte trafficking into the brain. *J. Neurovirol.* 6 (Suppl.):S70–81

Mankowski JL, Flaherty MT, Spelman JP, Hauer DA, Didier PJ, et al. 1997. Pathogenesis of simian immunodeficiency virus encephalitis: viral determinants of neurovirulence. *J. Virol.* 71:6055–60

Marcondes MC, Burudi EM, Huitron-Resendiz S, Sanchez-Alavez M, Watry D, et al. 2001. Highly activated cd8(+) t cells in the brain correlate with early central nervous system dysfunction in simian immunodeficiency virus infection. *J. Immunol.* 167:5429–38

Masliah E, Achim CL, Ge N, De Teresa R, Wiley CA. 1994. Cellular neuropathology in HIV encephalitis. In *HIV, AIDS and the*

Brain, ed. RW Price, SW Perry, pp. 119–31. New York: Raven

Masliah E, Ge N, Morey M, DeTeresa R, Terry RD, Wiley CA. 1992. Cortical dendritic pathology in human immunodeficiency virus encephalitis. *Lab. Invest.* 66:285–91

Masliah E, Heaton RK, Marcotte TD, Ellis RJ, Wiley CA, et al. 1997. Dendritic injury is a pathological substrate for human immunodeficiency virus-related cognitive disorders. HNRC Group. The HIV Neurobehavioral Research Center. *Ann. Neurol.* 42:963–72

McArthur JC, Grant I. 1998. HIV neurocognitive disorders. In *The Neurology of AIDS*, ed. HE Gendleman, SA Lipton, L Epstein, S Swindells, pp. 499–523. New York: Chapmann and Hall

McArthur JC, Hoover DR, Bacellar H, Miller EN, Cohen BA, et al. 1993. Dementia in AIDS patients: incidence and risk factors. *Neurology* 43:2245–52

Meyerhoff DJ, MacKay S, Constans JM, Norman D, Van Dyke C, et al. 1994. Axonal injury and membrane alterations in Alzheimer's disease suggested by in vivo proton magnetic resonance spectroscopic imaging. *Ann. Neurol.* 36:40–47

Mollace V, Nottet HS, Clayette P, Turco MC, Muscoli C, et al. 2001. Oxidative stress and neuroAIDS: triggers, modulators and novel antioxidants. *Trends Neurosci.* 24:411–16

Mori K, Ringler DJ, Kodama T, Desrosiers RC. 1992. Complex determinants of macrophage tropism in *env* of simian immunodeficiency virus. *J. Virol.* 66:2067–75

Murray EA, Rausch DM, Lendvay J, Sharer LR, Eiden LE. 1992. Cognitive and motor impairments associated with SIV infection in rhesus monkeys. *Science* 255:1246–49

Nath A, Geiger J. 1998. Neurobiological aspects of human immunodeficiency virus infection: neurotoxic mechanisms. *Prog. Neurobiol.* 54:19–33

Navia BA, Cho ES, Petito CK, Price RW. 1986a. The AIDS dementia complex: II. Neuropathology. *Ann. Neurol.* 19:525–35

Navia BA, Jordan BD, Price RW. 1986b. The AIDS dementia complex: I. Clinical features. *Ann. Neurol.* 19:517–24

Nielson SL, Petito CK, Urmacher CD, Posner JB. 1984. Subacute encephalitis in acquired immune deficiency syndrome: a postmortem study. *Am. J. Clin. Pathol.* 82:678–82

Pantaleo G, Graziosi C, Demarest JF, Butini L, Montroni M, et al. 1993. HIV infection is active and progressive in lymphoid tissue during the clinically latent stage of disease. *Nature* 362:355–58

Parmentier HK, van Wichen DF, Meyling FH, Goudsmit J, Schuurman HJ. 1992. Epitopes of human immunodeficiency virus regulatory proteins tat, nef, and rev are expressed in normal human tissue. *Am. J. Pathol.* 141:1209–16

Passlick B, Flieger D, Ziegler-Heitbrock HW. 1989. Identification and characterization of a novel monocyte subpopulation in human peripheral blood. *Blood* 74:2527–34

Peluso R, Haase A, Stowring L, Edwards M, Ventura P. 1985. A trojan horse mechanism for the spread of visna virus in monocytes. *Virology* 147:231–36

Perrella O, Finelli L, Carrieri PB. 1992. The role of cytokines in AIDS-dementia complex. *Acta Neurol. (Napoli)* 14:342–44

Perry VH, Gordon S. 1987. Modulation of CD4 antigens on macrophages and microglia in rat brain. *J. Exp. Med.* 166:1138–43

Perry VH, Gordon S. 1988. Macrophages and microglia in the nervous system. *Trends Neurosci.* 11:273–75

Persidsky Y, Ghorpade A, Rasmussen J, Limoges J, Liu XJ, et al. 1999. Microglial and astrocyte chemokines regulate monocyte migration through the blood-brain barrier in human immunodeficiency virus-1 encephalitis. *Am. J. Pathol.* 155:1599–611

Persidsky Y, Stins M, Way D, Witte MH, Weinand M, et al. 1997. A model for monocyte migration through the blood-brain barrier during HIV-1 encephalitis. *J. Immunol.* 158:3499–510

Persidsky Y, Zheng J, Miller D, Gendelman HE. 2000. Mononuclear phagocytes mediate blood-brain barrier compromise and neuronal injury during HIV-1-associated dementia. *J. Leukoc. Biol.* 68:413–22

Petito CK, Cho ES, Lemann W, Navia BA, Price RW. 1986. Neuropathology of acquired immune deficiency syndrome (AIDS): an autopsy review. *J. Neuropathol. Exp. Neurol.* 45:635–46

Petito CK, Kerza-Kwiatecki AP, Gendelman HE, McCarthy M, Nath A, et al. 1999. Review: neuronal injury in HIV infection. *J. Neurovirol.* 5:327–41

Petito CK, Roberts B. 1995. Evidence of apoptotic cell death in HIV encephalitis. *Am. J. Pathol.* 146:1121–30

Peudenier S, Hery C, Ng KH, Tardieu M. 1991. HIV receptors within the brain: a study of CD4 and MHC-II on human neurons, astrocytes and microglial cells. *Res. Virol.* 142:145–49

Power C, Johnson RT. 2001. Neuroimmune and neurovirological aspects of human immunodeficiency virus infection. *Adv. Virus Res.* 56:389–433

Power C, McArthur JC, Johnson RT, Griffin DE, Glass JD, et al. 1994. Demented and nondemented patients with AIDS differ in brain-derived human immunodeficiency virus type 1 envelope sequences. *J. Virol.* 68:4643–49

Power C, McArthur JC, Johnson RT, Griffin DE, Glass JD, et al. 1995. Distinct HIV-1 env sequences are associated with neurotropism and neurovirulence. *Curr. Top. Microbiol. Immunol.* 202:89–104

Price RW. 2000. The two faces of HIV infection of cerebrospinal fluid. *Trends Microbiol.* 8:387–91

Price RW, Brew B, Sidtis J, Rosenblum M, Scheck AC, Cleary P. 1988. The brain in AIDS: central nervous system HIV-1 infection and AIDS dementia complex. *Science* 239:586–92

Pulliam L, Clarke JA, McGrath MS, Moore D, McGuire D. 1996. Monokine products as predictors of AIDS dementia. *Aids* 10:1495–500

Pulliam L, Gascon R, Stubblebine M, McGuire D, McGrath MS. 1997. Unique monocyte subset in patients with AIDS dementia. *Lancet* 349:692–95

Pulliam L, Herndier BG, Tang NM, McGrath MS. 1991. Human immunodeficiency virus-infected macrophages produce soluble factors that cause histological and neurochemical alterations in cultured human brains. *J. Clin. Invest.* 87:503–12

Rausch DM, Davis MR. 2001. HIV in the CNS: pathogenic relationships to systemic HIV disease and other CNS diseases. *J. Neurovirol.* 7:85–96

Reinhart TA, Rogan MJ, Huddleston D, Rausch DM, Eiden LE, Haase AT. 1997. Simian immunodeficiency virus burden in tissues and cellular compartments during clinical latency and AIDS. *J. Infect. Dis.* 176:1198–208

Ringler DJ, Wyand MS, Walsh DG, MacKey JJ, Chalifoux LV, et al. 1989. Cellular localization of simian immunodeficiency virus in lymphoid tissues: I. Immunohistochemistry and electron microscopy. *Am. J. Pathol.* 134:373–83

Rosenberg ES, Altfeld M, Poon SH, Phillips MN, Wilkes BM, et al. 2000. Immune control of HIV-1 after early treatment of acute infection. *Nature* 407:523–26

Rosenberg ES, Billingsley JM, Caliendo AM, Boswell SL, Sax PE, et al. 1997. Vigorous HIV-1-specific CD4+ T cell responses associated with control of viremia. *Science* 278:1447–50

Rottman JB, Ganley KP, Williams K, Wu L, Mackay CR, Ringler DJ. 1997. Cellular localization of the chemokine receptor CCR5. Correlation to cellular targets of HIV-1 infection. *Am. J. Pathol.* 151:1341–51

Sanders VJ, Pittman CA, White MG, Wang G, Wiley CA, Achim CL. 1998. Chemokines and receptors in HIV encephalitis. *AIDS* 12:1021–26

Sasseville VG, Lackner AA. 1997. Neuropathogenesis of simian immunodeficiency virus infection in macaque monkeys. *J. Neurovirol.* 3:1–9

Sasseville VG, Newman W, Brodie SJ, Hesterberg P, Pauley D, Ringler DJ. 1994. Monocyte adhesion to endothelium in simian immunodeficiency virus-induced AIDS encephalitis is mediated by vascular cell adhesion molecule-1/a4b1 integrin interactions. *Am. J. Pathol.* 144:27–40

Sasseville VG, Newman WA, Lackner AA, Smith MO, Lausen NC, et al. 1992. Elevated vascular cell adhesion molecule-1 in AIDS encephalitis induced by simian immunodeficiency virus. *Am. J. Pathol.* 141:1021–30

Schmidtmayerova H, Nottet HSLM, Nuovo G, Raabe T, Flanagan CR, et al. 1996a. Human immunodeficiency virus type 1 infection alters chemokine beta peptide expression in human monocytes: implications for recruitment of leukocytes into brain and lymph nodes. *Proc. Natl. Acad. Sci. USA* 93:700–4

Schmidtmayerova H, Nottet HSLM, Nuovo G, Raabe T, Flanagan CR, et al. 1996b. Human immunodeficiency virus type 1 infection alters chemokine b peptide expression in human monocytes: implications for recruitment of leukocytes into brain and lymph nodes. *Proc. Natl. Acad. Sci. USA* 93:700–4

Schmitz JE, Kuroda MJ, Santra S, Sasseville VG, Simon MA, et al. 1999. Control of viremia in simian immunodeficiency virus infection of CD8+ lymphocytes. *Science* 283:857–60

Sedgwick JD, Schender S, Imrich H, Dorries R, Butcher GW, ter Meulen V. 1991. Isolation and direct characterization of resident microglial cells from the normal and inflamed central nervous system. *Proc. Natl. Acad. Sci. USA* 88:7438–42

Seilhean D, Kobayashi K, He Y, Uchihara T, Rosenblum O, et al. 1997. Tumor necrosis factor-alpha, microglia and astrocytes in AIDS dementia complex. *Acta Neuropathol. (Berl.)* 93:508–17

Sharer LR, Michaels J, Murphey-Corb H, Hu F-S, Kuebler DJ, et al. 1991. Serial pathogenesis study of SIV brain infection. *J. Med. Primatol.* 20:211–17

Sharpless N, Gilbert D, Vandercam B, Zhou JM, Verdin E, et al. 1992. The restricted nature of HIV-1 tropism for cultured neural cells. *Virology* 191:813–25

Shonk TK, Moats RA, Gifford P, Michaelis T, Mandigo JC, et al. 1995. Probable Alzheimer disease: diagnosis with proton MR spectroscopy. *Radiology* 195:65–72

Simon MA, Chalifoux LV, Ringler DJ. 1992. Pathologic features of SIV-induced disease and the association of macrophage infection with disease evolution. *AIDS Res. Hum. Retroviruses* 8:327–37

Sinclair E, Gray F, Ciardi A, Scaravilli F. 1994. Immunohistochemical changes and PCR detection of HIV provirus DNA in brains of asymptomatic HIV-positive patients. *J. Neuropathol. Exp. Neurol.* 53:43–50

Smith MO, Heyes MP, Lackner AA. 1995. Early intrathecal events in rhesus macaques (*Macaca mulatta*) infected with pathogenic or nonpathogenic molecular clones of simian immunodeficiency virus. *Lab. Invest.* 72:547–58

Smits HA, Boven LA, Pereira CF, Verhoef J, Nottet HS. 2000. Role of macrophage activation in the pathogenesis of Alzheimer's disease and human immunodeficiency virus type 1-associated dementia. *Eur. J. Clin. Invest.* 30:526–35

Sporer B, Paul R, Koedel U, Grimm R, Wick M, et al. 1998. Presence of matrix metalloproteinase-9 activity in the cerebrospinal fluid of human immunodeficiency virus-infected patients. *J. Infect. Dis.* 178:854–57

Stevenson M, Gendelman HE. 1994. Cellular and viral determinants that regulate HIV-1 infection in macrophages. *J. Leukoc. Biol.* 56:278–88

Strizki JM, Albright AV, O'Conner M, Perrin L, Gonzalez-Scarano F. 1996. Infection of primary human microglia and monocyte-derived macrophages with human immunodeficiency virus type 1 isolates: evidence of differential tropism. *J. Virol.* 70:7654–62

Tan J, Town T, Saxe M, Paris D, Wu Y, Mullan M. 1999. Ligation of microglial CD40 results in p44/42 mitogen-activated protein kinase-dependent TNF-alpha production that is opposed by TGF-beta 1 and IL-10. *J. Immunol.* 163:6614–21

Tanabe S, Heesen M, Yoshizawa I, Berman MA, Luo Y, et al. 1997. Functional expression of

CXC-chemokine receptor-4/fusin on mouse microglial cells and astrocytes. *J. Immunol.* 1997:905–11

Teo I, Veryard C, Barnes H, An SF, Jones M, et al. 1997. Circular forms of unintegrated human immunodeficiency virus type 1 DNA and high levels of viral protein expression: association with dementia and multinucleated giant cells in the brains of patients with AIDS. *J. Virol.* 71:2928–33

Thieblemont N, Weiss L, Sadeghi HM, Estcourt C, Haeffner-Cavaillon N. 1995. CD14lowCD16high: a cytokine-producing monocyte subset which expands during human immunodeficiency virus infection. *Eur. J. Immunol.* 25:3418–24

Tong N, Perry SW, Zhang Q, James HJ, Guo H, et al. 2000. Neuronal fractalkine expression in HIV-1 encephalitis: roles for macrophage recruitment and neuroprotection in the central nervous system. *J. Immunol.* 164:1333–39

Tran EH, Hoekstra K, van Rooijen N, Dijkstra CD, Owens T. 1998. Immune invasion of the central nervous system parenchyma and experimental allergic encephalomyelitis, but not leukocyte extravasation from blood, are prevented in macrophage-depleted mice. *J. Immunol.* 161:3767–75

Tyor WR, Glass JD, Griffin JW, Becker PS, McArthur JC, et al. 1992. Cytokine expression in the brain during the acquired immunodeficiency syndrome. *Ann. Neurol.* 31:349–60

Ulvestad E, Williams K, Bo L, Trapp B, Antel J, Mork S. 1994a. HLA class II molecules (HLA-DR, DP, DQ) on cells in the human CNS studied in situ and in vitro. *Immunology* 82:535–41

Ulvestad E, Williams K, Mork S, Antel J, Nyland H. 1994b. Phenotypic differences between human monocytes/macrophages and microglial cells studied in situ and in vitro. *J. Neuropathol. Exp. Neurol.* 53:492–501

Ulvestad E, Williams K, Vedeler C, Antel J, Nyland H, et al. 1994c. Reactive microglia in multiple sclerosis lesions have an increased expression of receptors for the Fc part of IgG. *J. Neurol. Sci.* 121:125–31

Unger ER, Sung JH, Manivel JC, Chenggis ML, Blazar BR, Krivit W. 1993. Male donor-derived cells in the brains of female sex-mismatched bone marrow transplant recipients: a Y chromosome specific in situ hybridization study. *J. Neuropathol. Exp. Neurol.* 52:460–70

von Herrath M, Oldstone MBA, Fox HS. 1995. Simian immunodeficiency virus (SIV)-specific CTL in cerebrospinal fluid and brains of SIV-infected rhesus macaques. *J. Immunol.* 154:5582–89

Walker BS, Chakrabarti S, Moss B, Paradis T, Flynn T, et al. 1987. HIV specific cytotoxic T lymphocytes in seropositive individuals. *Nature* 328:345

Walker WS. 1999. Separate precursor cells for macrophages and microglia in mouse brain: immunophenotypic and immunoregulatory properties of the progeny. *J. Neuroimmunol.* 94:127–33

Watkins BA, Dorn HH, Kelly WB, Armstrong RC, Potts BJ, et al. 1990. Specific tropism of HIV-1 for microglial cells in primary human brain cultures. *Science* 249:549–53

Wei R, Jonakait GM. 1999. Neurotrophins and the anti-inflammatory agents interleukin-4 (IL-4), IL-10, IL-11 and transforming growth factor-beta1 (TGF-beta1) down-regulate T cell costimulatory molecules B7 and CD40 on cultured rat microglia. *J. Neuroimmunol.* 95:8–18

Weiss JM, Nath A, Major EO, Berman JW. 1999. HIV-1 Tat induces monocyte chemoattractant protein-1-mediated monocyte transmigration across a model of the human blood-brain barrier and up-regulates CCR5 expression on human monocytes. *J. Immunol.* 163:2953–59

Wesselingh SL, Power C, Glass JD, Tyor WR, McArthur JC, et al. 1993. Intracerebral cytokine messenger RNA expression in acquired immunodeficiency syndrome dementia. *Ann. Neurol.* 33:576–82

Westmoreland SV, Halpern E, Lackner AA. 1998a. Simian immunodeficiency virus encephalitis in rhesus macaques is associated with rapid disease progression. *J. Neurovirol.* 4:260–68

Westmoreland SV, Rottman JB, Williams KC, Lackner AA, Sasseville VG. 1998b. Chemokine receptor expression on resident and inflammatory cells in the brain of macaques with simian immunodeficiency virus encephalitis. *Am. J. Pathol.* 152:659–65

Wiley CA, Achim CL. 1994. Human immunodeficiency virus encephalitis is the pathological correlate of dementia in acquired immunodeficiency syndrome. *Ann. Neurol.* 36:673–76

Wiley CA, Achim CL, Schrier RD, Heyes MP, McCutchan JA, Grant I. 1992. Relationship of cerebrospinal fluid immune activation associated factors to HIV encephalitis. *Aids* 6:1299–307

Wiley CA, Masliah E, Morey M, Lemere C, DeTeresa R, et al. 1991. Neocortical damage during HIV infection. *Ann. Neurol.* 29:651–57

Wiley CA, Schrier RD, Nelson JA, Lampert PW, Oldstone MBA. 1986. Cellular localization of human immunodeficiency virus infection within the brains of acquired immunodeficiency syndrome patients. *Proc. Natl. Acad. Sci. USA* 83:7089–93

Wiley CA, Soontornniyomkij V, Radhakrishnan L, Masliah E, Mellors J, et al. 1998. Distribution of brain HIV load in AIDS. *Brain Pathol.* 8:277–84

Williams AE, Blakemore WF. 1990. Monocyte-mediated entry of pathogens into the central nervous system. *Neuropathol. Appl. Neurobiol.* 16:377–92

Williams KC, Alvarez X, Lackner AA. 2001a. Central nervous system perivascular cells are immunoregulatory cells that connect the CNS with the peripheral immune system. *Glia* 36:156–64

Williams KC, Barr-Or A, Ulvestad E, Olivier A, Antl JP, Yong WV. 1992. Biology of adult human microglia in culture: comparisons with peripheral blood monocytes and astrocytes. *J. Neuropath. Exp. Neurol.* 51:538–49

Williams KC, Corey S, Westmoreland SV, Pauley D, Knight H, et al. 2001b. Perivascular macrophages are the primary cell type productively infected by simian immunodeficiency virus in the brains of macaques: implications for the neuropathogenesis of AIDS. *J. Exp. Med.* 193:905–15

Williams K, Ulvestad E, Antel JP. 1994. B7/BB-1 expression on adult human microglia studied in vitro and in situ. *Eur. J. Immunol.* 24:3031–37

Yeh MW, Kaul M, Zheng J, Nottet HS, Thylin M, et al. 2000. Cytokine-stimulated, but not HIV-infected, human monocyte-derived macrophages produce neurotoxic levels of l-cysteine. *J. Immunol.* 164:4265–70

Yong VW, Power C, Forsyth P, Edwards DR. 2001. Metalloproteinases in biology and pathology of the nervous system. *Nat. Rev. Neurosci.* 2:502–11

Zink MC, Clements JE. 2000. The two faces of HIV infection of cerebrospinal fluid. *Trends Microbiol.* 8:390–91

Zink MC, Spelman JP, Robinson RB, Clements JE. 1998. SIV infection of macaques—modeling the progression to AIDS dementia. *J. Neurovirol.* 4:249–59

Zink MC, Suryanarayana K, Mankowski JL, Shen A, Piatak M Jr, et al. 1999. High viral load in the cerebrospinal fluid and brain correlates with severity of simian immunodeficiency virus encephalitis. *J. Virol.* 73:10480–88

LEARNING AND MEMORY FUNCTIONS OF THE BASAL GANGLIA

Mark G. Packard[1] and Barbara J. Knowlton[2]

[1]Department of Psychology, Yale University, New Haven, Connecticut 06520;
email: mark.packard@yale.edu
[2]Department of Psychology, University of California, Los Angeles,
California 90095-1563; email: knowlton@psych.ucla.edu

Key Words striatum, neostriatum, caudate nucleus, Parkinson's disease, Huntington's disease

Authors' note: In behavioral sections of the present review the term basal ganglia is often used to refer to the caudate nucleus and putmen (i.e., dorsal striatum). These structures are perhaps primary, but they are certainly not selective components of a group of subcortical structures that make up the basal ganglia. The broader term is used here simply in the interest of attracting the widest general readership of investigators interested in basal ganglia function.

■ **Abstract** Although the mammalian basal ganglia have long been implicated in motor behavior, it is generally recognized that the behavioral functions of this subcortical group of structures are not exclusively motoric in nature. Extensive evidence now indicates a role for the basal ganglia, in particular the dorsal striatum, in learning and memory. One prominent hypothesis is that this brain region mediates a form of learning in which stimulus-response (S-R) associations or habits are incrementally acquired. Support for this hypothesis is provided by numerous neurobehavioral studies in different mammalian species, including rats, monkeys, and humans. In rats and monkeys, localized brain lesion and pharmacological approaches have been used to examine the role of the basal ganglia in S-R learning. In humans, study of patients with neurodegenerative diseases that compromise the basal ganglia, as well as research using brain neuroimaging techniques, also provide evidence of a role for the basal ganglia in habit learning. Several of these studies have dissociated the role of the basal ganglia in S-R learning from those of a cognitive or declarative medial temporal lobe memory system that includes the hippocampus as a primary component. Evidence suggests that during learning, basal ganglia and medial temporal lobe memory systems are activated simultaneously and that in some learning situations competitive interference exists between these two systems.

INTRODUCTION

In the early 1900s, Kinnier Wilson coined the term extrapyramidal system to describe a mammalian basal ganglia system that interacts with brain-stem structures independently from the pyramidal tract to influence motor behavior. Wilson's emphasis on the role of the basal ganglia in motor behavior was driven by his early discoveries (1912, 1914) and those of Vogt (1911), revealing motor disorders in humans following damage to this brain region. The recognition that Parkinson's disease (a progressive neurodegenerative disease characterized by limb rigidity, tremors, and difficulty initiating movement) is fundamentally a disorder of basal ganglia function provided further evidence of the important role of this brain region in motor behavior. Decades of subsequent research on the behavioral functions of the mammalian basal ganglia have revealed a group of structures whose functions are diverse in nature, and it has long been recognized that behavioral classification of the basal ganglia strictly as a motor system is not tenable. Rather, evidence indicates that one nonmotor function of the basal ganglia involves participation in learning and memory. This view was partly espoused in early experimental research in animals investigating the effects of caudate nucleus lesions on performance of delayed response and alternation behavior (e.g., Battig et al. 1960; Butters & Rosvold 1968; Chorover & Gross 1963; Divac 1968, 1972; Divac & Oberg 1975; Gross et al. 1965; Kirkby 1969) and on performance of conditioned avoidance behavior (e.g., Allen & Davidson 1973, Kirkby & Polgar 1974, Neill & Grossman 1970, Prado-Alcala et al. 1975, Winocur 1974, Winocur & Mills 1969). The present review describes findings obtained primarily over the past decade supporting a role for the basal ganglia in mammalian learning and memory. Following a brief consideration of relevant anatomy and neurochemistry, the role of the basal ganglia in learning and memory is described based on findings from studies employing brain lesion and behavioral pharmacology approaches in experimental animals. It should be noted that there is also extensive research examining the role of the basal ganglia in learning and memory with electrophysiological techniques in behaving animals (e.g., Aosaki et al. 1994, Graybiel et al. 1994, Hikosaka et al. 1989, Jog et al. 1999, Mizumori et al. 2000, Rolls et al. 1983, Wiener 1993) and with neural computational modeling (e.g., Beiser et al. 1997, Berns & Sejnowski 1998, Gillies & Arbuthnott 2000). However, an adequate description of these two latter approaches is difficult given the prescribed limits of the present review, and therefore the reader is encouraged to examine this important additional literature. Following review of research in lower animals, the role of the basal ganglia in human learning and memory is described. This area of research involves neuropsychological studies of humans with neurological disorders that primarily compromise the basal ganglia (e.g., Parkinson's disease, Huntington's Chorea) and findings from experiments using brain neuroimaging techniques.

One description of the role of the basal ganglia in learning and memory in both lower animals and humans has been offered in the context of a multiple-systems approach to memory organization (e.g., Knowlton et al. 1996a, Mishkin & Petri 1984, Packard et al. 1989). According to this idea, the basal ganglia mediate a form of learning and memory in which stimulus-response (S-R) associations or habits are incrementally acquired. In several studies in which a neuroscientific approach has been used, the putative S-R habit mnemonic function of the basal ganglia has been dissociated from that of a cognitive medial temporal lobe memory system in which the hippocampus is a major component (Packard 2001). Therefore, the present review also includes a consideration of how relatively independent basal ganglia and medial temporal lobe memory systems may interact in learning and memory.

BASAL GANGLIA: BRIEF ANATOMICAL CONSIDERATIONS

In 1664 anatomist Thomas Willis termed a prominent subcortical region of the telencephalon *corpus striatum*. Neuronal tracing techniques developed by Nauta and colleagues in the mid-1950s allowed for elucidation of connectivity of the broadly defined corpus striatal region, and the term basal ganglia was adopted to refer to a group of subcortical structures that included as primary components the caudate nucleus and putamen (the caudate-putamen are fairly undifferentiated in the rat but are separated by the internal capsule in primates), the globus pallidus, and the claustrum. Heimer and colleagues (e.g., Heimer & Van Hoesen 1979) subsequently adopted the term ventral striatum to delineate the most ventral aspects of the striatum (i.e., nucleus accumbens and portions of the olfactory tubercle) from more dorsal regions (i.e., caudate nucleus or dorsal striatum). Thus, the core structures of the mammalian basal ganglia include the dorsal striatum, ventral striatum, and globus pallidus. In addition, the substantia nigra, ventral tegmental area, and the subthalamic nucleus may be considered associated basal ganglia structures via their reciprocal connections with the core structures (for reviews see Ohye et al. 1996, Parent 1986).

The present review is restricted to addressing the role of the mammalian dorsal striatum (caudate nucleus and putamen) in learning and memory. However, it should be noted that several lines of evidence also suggest a role for the ventral striatum (e.g., nucleus accumbens) in learning and memory. This hypothesis was originally offered by Mogenson and colleagues (1980), who proposed that projections to the ventral striatum from limbic brain regions provide an interface between motivational states and behavioral action. Consistent with this idea, the mnemonic functions of the nucleus accumbens have been associated with forms of memory that are mediated by limbic structures that target the ventral striatum (i.e., hippocampus and amygdala) and, importantly, are unaffected by damage to the dorsal striatum (for reviews see Cador et al. 1989, Setlow 1997).

The basal ganglia receive input from virtually all regions of the cerebral cortex, and these corticostriatal pathways are topographically organized (e.g., McGeorge & Faull 1989, Veening et al. 1980). The discovery of corticobasal ganglia loops delineates an important feature of the anatomical organization of the mammalian basal ganglia, and these pathways have been elegantly described in the monkey (for review see Alexander et al. 1986). Specific cortical regions project to the dorsal and ventral striatum, and pallidal output from the basal ganglia loops back into these same cortical regions via various thalamic nuclei. Evidence suggests that at least five parallel corticobasal ganglia loops exist (for review see Kimura & Graybiel 1995). With regards to learning and memory functions, one interesting recent hypothesis is that fronto-cortical-striatal loops are used by the basal ganglia to essentially train the cortex to produce learned motor responses in the presence of a particular pattern of sensory information (Wise et al. 1996). However, it is important to note that, although basal ganglia output is clearly looped via the globus pallidus and thalamus back to specific cortical sites, pallidal and nigral outputs also directly project to downstream brain-stem structures that allow for rapid access to spinal control of motor responses.

The basal ganglia also receive a prominent projection from the thalamus, and intralaminar thalamic nuclei are recognized as the primary origin of the thalamostriatal pathway. However, projections to the neostriatum originating in various other thalamic nuclei have also been identified. Overlapping territories exist between thalamic regions innervated by the output nuclei of the basal ganglia and thalamostriatal projection neurons, which suggests the presence of feedback circuitry between these two brain regions (for review see Mengual et al. 1999).

BASAL GANGLIA: BRIEF NEUROCHEMICAL CONSIDERATIONS

Neurochemically the basal ganglia is characterized by a prominent input from midbrain dopaminergic pathways originating in the substantia nigra and ventral tegmental area, primarily innervating the dorsal and ventral striatum, respectively (for reviews see Graybiel 1990, Gerfen & Wilson 1996). Corticostriatal, thalamostriatal, and afferent projections from limbic structures including the amygdala and hippocampus utilize excitatory amino acid neurotransmission and are predominantly glutamatergic (Fonnum et al. 1981). Medium spiny output neurons of the neostriatum, composing approximately 90% of striatal neurons, use gamma-amino butyric acid as a neurotransmitter. An additional prominent component of basal ganglia neurochemistry is a large population of cholinergic interneurons (Lynch et al. 1972). For each of these systems, the full complement of receptor subtype families that have been indented for each neurotransmitter are present in varying densities and distribution patterns in the basal ganglia. Finally, several neuropeptides are also localized in the basal ganglia, including various opioids, cholecystokinin, substance P, somatostatin, and neurotensin.

NEUROANATOMICAL AND NEUROCHEMICAL ORGANIZATION OF THE BASAL GANGLIA: PATCH AND MATRIX

In the mid-1980s, a series of important findings demonstrated the existence of two neural compartments in the mammalian neostriatum that are neurochemically and anatomically differentiated and are commonly termed the striatal patch and matrix (for review see Gerfen 1992). Neurochemically, the patch compartments of the striatum (also termed striosomes) are characterized by low levels of acetylcholine and high levels of various opiates and substance P. In contrast, the matrix compartment is characterized by cholinergic and somatostatin-containing neurons. Both striatal compartments receive dopaminergic input, although dopamine pathways originating in the ventral tegmental area and substantia nigra appear to primarily innervate the patch and matrix, respectively. Anatomically, corticostriatal and thalamostriatal projections are closely associated with the striatal matrix, while projections from limbic structures to the striatum (e.g., hippocampus, amygdala) appear to primarily innervate striatal patches. Investigation of the functional significance of the neurochemical and anatomical differentiation observed between these two striatal compartments represents an evolving area of basal-ganglia research. With regards to the role of the basal ganglia in learning and memory, one hypothesis is that the striatal matrix primarily mediates the mnemonic functions of the dorsal striatum (White 1989a).

ROLE OF THE BASAL GANGLIA IN LEARNING AND MEMORY: A THEORETICAL FORMULATION

As previously mentioned, one account of the role of the basal ganglia in learning and memory posits that the caudate nucleus mediates a form of learning and memory in which S-R associations or habits (e.g., Thorndike 1933, Hull 1943) are incrementally acquired (Knowlton et al. 1996a, Mishkin & Petri 1984, Packard et al. 1989; for a similar early proposal based on human studies see Phillips & Carr 1987). This hypothesis was originally proposed largely on the basis of evidence that lesions of the monkey putamen impair simultaneous visual discrimination learning (e.g., Buerger et al. 1974). In this task, animals are presented concurrently with two objects (i.e., stimuli), and selection of the same object within a given pair (i.e., response) is followed by food reward. According to S-R learning theory, the satisfying or annoying nature of the reinforcer simply serves to strengthen or weaken learning and is not itself represented in the associations formed (i.e., Thorndike's 1933 Law of Effect).

The hypothesis that the basal ganglia mediate S-R habit learning has gained support from studies in rats (e.g., Graybiel 1998; Jog et al. 1999; Kesner et al. 1993; McDonald & White 1993, 1994; Packard et al. 1989; Packard & McGaugh 1992, 1996; Packard & Teather 1997, 1998; White 1997), monkeys (e.g., Fernandez-Ruiz et al. 2001, Kimura 1995; Teng et al. 2000), and humans (e.g., Butters et al. 1994; Heindel et al. 1988; Knowlton et al. 1996a; Martone et al. 1984). In particular, several of these studies have used dissociation methodology to provide support for the hypothesis that the basal ganglia and hippocampus are parts of independent memory systems that mediate the acquisition of S-R habits and cognitive (e.g., Tolman 1932) forms of memory, respectively.

ROLE OF THE BASAL GANGLIA IN LEARNING AND MEMORY: LESION STUDIES

Beginning in the early 1960s, numerous studies conducted in experimental animals used brain-lesion techniques to implicate the mammalian basal ganglia in performance of delayed alternation and response tasks (e.g., Battig et al. 1960; Butters & Rosvold 1968; Chorover & Gross 1963; Divac 1968, 1972; Divac & Oberg 1975; Gross et al. 1965; Kirkby 1969). Several investigators subsequently demonstrated that lesions of the basal ganglia impair acquisition of various types of conditioned avoidance behavior in rats (e.g., Allen & Davidson 1973, Kirkby & Polgar 1974, Neill & Grossman 1971, Prado-Alcala et al. 1975, Winocur 1974, Winocur & Mills 1969), providing further evidence of a potential mnemonic role for this brain region (for early reviews on putative learning and memory functions of the caudate nucleus see Chozick 1983, Oberg & Divac 1979). However, the conclusion that the effects of a pretraining lesion on task acquisition result from an impairment of learning and memory per se must always be offered cautiously, as such lesions may potentially disrupt nonmnemonic functions (e.g., sensory, motivational, and/or motoric) that contribute to task performance. One strategy for dissociating lesion effects on mnemonic versus nonmnemonic factors is to employ pairs of learning tasks that share the same motivational, sensory, and motoric characteristics. The first use of dissociation methodology to directly test the hypothesized selective role of the basal ganglia (dorsal striatum) in S-R habit learning involved a study in which two eight-arm radial maze tasks were used (Packard et al. 1989). In the standard win-shift version of the radial maze task introduced by Olton & Samuelson (1976), rats obtain food rewards by visiting each arm of the maze once within a daily training session, and re-entries into maze arms that were previously visited are scored as errors. In a newly developed win-stay version of the radial maze task, rats obtained food rewards by visiting four randomly selected and illuminated maze arms twice within a daily training session, and visits to unlit maze arms are scored as errors. Performance in the win-shift task requires rats to remember those arms that have been previously visited within a daily training session, and this task is essentially a prototypical test of spatial working memory (e.g., Olton & Papas 1979) and/or may involve the use of a Tolmanian (1932) cognitive mapping strategy (e.g., O'Keefe & Nadel 1978). In contrast, acquisition of the win-stay task requires rats to learn to approach lit maze arms, and this task is essentially a simultaneous visual discrimination that may involve acquisition of a Hullian (1943) S-R habit. When rats are trained on these two tasks following electrolytic lesions of the dorsal striatum, a dissociation is observed; dorsal striatal lesions impair acquisition of the win-stay task and do not affect acquisition of the win-shift task. Interestingly, lesions of the hippocampal system (fimbria-fornix) produce the opposite pattern of results, providing evidence of a double dissociation of the mnemonic functions of the dorsal striatum and hippocampal system (Packard et al. 1989). These findings were subsequently replicated in a study employing neurotoxic dorsal striatal and hippocampal lesions (McDonald & White 1993). Importantly, when rats are well trained in the win-stay radial maze task and subsequently exposed to reinforcer devaluation (via pairing of the food reward with nauseating lithium chloride injections), they continue to approach illuminated maze arms (Sage & Knowlton 2000). This finding suggests that representation of the food reward is not guiding learned behavior in this task; rather, performance of the caudate-dependent win-stay task involves acquisition of an S-R (light-approach) and not a stimulus-stimulus (light-food) association.

An additional study utilized two water maze tasks to investigate the selective role of the basal ganglia in S-R memory (Packard & McGaugh 1992). In these tasks, analogous to those originally introduced by Morris (1984), two rubber balls protruding above the water surface served as cues. One ball (correct) was located on top of a platform that could be used to escape the water, and the other ball (incorrect) was located on top of a thin rod and thus did not provide escape. The two balls also differed in visual appearance (i.e., vertical versus horizontal, black versus white stripes). In a cognitive version of the task, the correct platform was located in the same spatial location on every trial, but the visual appearance of the ball varied. Therefore, this version of the task requires rats to learn to approach the correct ball on the basis of spatial location, and not visual pattern. In an S-R habit version of the task, the correct platform was located in different spatial locations across trials, but the visual pattern was consistent. Therefore, this task could be acquired by learning an approach response to the visual cue. Lesions of the dorsal striatum impair acquisition of the S-R habit task, without affecting acquisition of the spatial task (Packard & McGaugh 1992). In another version of the water maze task, rats were trained to swim to a visibly cued escape platform located in a constant spatial location (McDonald & White 1994). Following acquisition of the task, the nature of the learned behavior was probed by moving the visible platform to a novel spatial location. On the probe trial, half of the control rats swam to the old platform location (indicating the use of spatial memory), and half swam to the visibly cued platform in the new location (indicating the use of S-R memory). In contrast, all of the rats with dorsal striatal lesions swam to the old spatial location, indicating an impairment in S-R learning. Again, it is of interest to note that, in both of the water maze studies described above, lesions of the hippocampal system (fimbria-fornix) produced the opposite pattern of results (i.e., a selective impairment of spatial memory).

Numerous other studies have used irreversible pretraining lesions of the basal ganglia to demonstrate impairments in the acquisition of learning tasks that theoretically could be acquired by an S-R habit memory system. For example, in rats, caudate nucleus lesions impair the acquisition of two-way active avoidance behavior (e.g., Green et al. 1967, Kirkby & Kimble 1968, Neill & Grossman 1971), simultaneous tactile discriminations (Colombo et al. 1989), simple straight-alley runway behavior (Kirkby et al. 1981, Salinas et al. 1998), invariant reference memory in a four-arms baited, four-arms unbaited radial maze task (Colombo et al. 1989, Packard & White 1990), conditional visual (Reading et al. 1991, Winocur & Estes 1998), and auditory (Adams et al. 2001) discrimination learning.

In addition to studies employing irreversible brain lesions, recent research using intracaudate infusion of drugs that produce temporary neural inactivation (i.e., tetrodotoxin or lidocaine) also reveal a role for the basal ganglia in learning and memory (Lorenzi et al. 1995, Packard & McGaugh 1996). In one study (Packard & McGaugh 1996), a plus-maze task that figured prominently in the historic debate between S-R and cognitive learning theorists (for review see Restle 1957) was used to examine the role of the caudate nucleus in response learning. Potegal (1972) originally proposed that the caudate nucleus mediates egocentric response learning, and several experimental findings support this idea (e.g., Abraham et al. 1983, Brasted et al. 1997, Cook & Kesner 1988, Kesner et al. 1993, Mitchell & Hall 1988, Robbins & Brown 1990, Thompson et al. 1980). In the aforementioned plus-maze study (Packard & McGaugh 1996), rats were trained in a daily session to obtain food from a consistently baited goal box (west) and were trained to approach this maze arm from the same start box on each trial (south). Following seven days of training (i.e., on day eight), rats were given a probe trial in which they were placed in the start box opposite to that used during training (north). On the probe trial, rats that entered the west arm (i.e., the spatial location where food was located during training) were designated place learners, and rats that entered the east arm (i.e., made the same body turn response that had been reinforced during training) were designated response learners. Prior to the probe trial, rats received intradorsolateral caudate injections of the local anesthetic lidocaine, in order to examine the role of this brain region in the expression of previously learned behavior. On the day-eight probe trial, rats receiving vehicle or lidocaine injections into the dorsolateral caudate were predominantly place learners, providing further evidence that the functional integrity of the dorsolateral caudate is not necessary for expression of spatial or place learning. However, with extended training in the cross-maze, intact rats switch from the use of place learning to a response-learning tendency (Hicks 1964, Ritchie et al. 1950). Therefore, the rats were trained for an additional seven days, given a second probe trial on day 16, and again received intracerebral injections of lidocaine prior to the probe trial. On this second probe trial, rats receiving vehicle injections into the dorsolateral caudate were predominantly response learners, revealing a switch from the use of place to response learning with extended training. In contrast, rats receiving intradorsolateral caudate injections of lidocaine prior to the second probe trial exhibited place learning, demonstrating a blockade of the expression of response learning. This finding also indicates that when the shift from the use of place to response learning occurs the place representation can be brought back into use by blockade of the caudate nucleus response-learning system. Consistent with the double dissociations between basal ganglia and hippocampal mnemonic function observed in studies using irreversible lesions, intradorsal hippocampal infusions of lidocaine selectively impaired the expression of place learning (Packard & McGaugh 1996).

Finally, recent evidence also indicates that pretraining neurotoxic lesions of monkey basal ganglia (i.e., ventrocaudal neostriatum) impair concurrent visual-discrimination learning and leave visual-recognition memory intact (Fernandez-Ruiz et al. 2001). Similarly, conjoint damage to the hippocampus and ventrocaudal neostriatum (but not hippocampus alone) impairs concurrent and pattern-discrimination learning (Teng et al. 2000). Therefore, the selective impairment of S-R habit learning that has been revealed following lesion damage to the basal ganglia in rats is also observed following neostriatal damage in monkeys. Taken together, these findings suggest that the mnemonic functions of the basal ganglia generalize across mammalian species, an idea that is also supported by research on humans (e.g., Knowlton et al. 1996a).

BASAL GANGLIA, LEARNING, AND MEMORY: FUNCTIONAL HETEROGENEITY

Several early studies on the role of the basal ganglia in learning and memory were guided in part by anatomical evidence demonstrating the existence of the corticostriatal pathways. One hypothesis that has continued to garner support is that the mnemonic function of the caudate nucleus is organized based on the nature of the topographical cortical input this structure receives. For example, in experimental animals lesions of either the frontal cortex or the medial region of the caudate nucleus to which it projects produce similar impairments in performance of delayed alternation and response tasks (e.g., Divac 1972; Kolb 1977; Rosvold 1968, 1972). Moreover, lesions of regions of the caudate nucleus that receive visual or olfactory input selectively impair conditioned emotional responding based on visual or olfactory stimuli, respectively (Viaud & White, 1989; see Pisa & Cyr 1990, Winocur 1974 for additional examples of regional specificity of the dorsal striatum in learning).

One suggestion with regard to functional heterogeneity of the mnemonic role of the basal ganglia is that S-R habit learning may selectively involve lateral regions of the dorsal striatum, whereas the medial dorsal striatum may mediate a cognitive form of memory that appears similar to that typically associated with the hippocampus. This idea is based in part on evidence that lesions of the medial, but not lateral, dorsal striatum, impair the use of spatial navigation in a hidden-platform water maze task (e.g., Devan & White 1999, Furtado & Mazurek 1996, Whishaw et al. 1987) and bias rats toward the use of S-R memory in a water maze task in which competing place- and cue-learning preferences are simultaneously assessed (Devan et al. 1999, Devan & White 1999). It should be noted that in some of these water maze studies, medial dorsal striatal lesions did not completely prevent spatial learning (Devan et al. 1999, Devan & White 1999, Whishaw et al. 1987). Moreover, there is behavioral evidence that appears inconsistent with the idea that medial dorsal striatal lesions produce mnemonic deficits identical to those produced by hippocampal system lesions. For example, large lesions of the dorsal striatum that include lateral and medial regions (Packard et al. 1989, 1992), as well as lesions restricted to the medial dorsal striatum (Sakamoto & Okaichi 2001), do not impair acquisition of hippocampus-dependent win-shift behavior in the radial maze. Moreover, in the previously described spatial and S-R visual-discrimination water maze tasks (Packard & McGaugh 1992), medial dorsal striatal lesions selectively impaired acquisition of the S-R task, whereas fimbria-fornix lesions produced the opposite effect. Similarly, lesions of the medial caudate impair the S-R component of a sequential learning task, without affecting the use of spatial information, whereas hippocampal lesions produce the opposite effect in this task (DeCoteau & Kesner 2000). Finally, recent evidence suggests that separate lesions of the medial or lateral dorsal striatum each impair acquisition of an S-R auditory conditional-response association task (Adams et al. 2001). In sum, findings of a number of studies suggest the existence of functional heterogeneity within the dorsal striatum in learning. However, further research is clearly necessary to determine whether such heterogeneity is limited to modality-specific S-R memory, or whether medial regions of dorsal striatum may in part mediate more cognitive forms of information processing.

ROLE OF THE BASAL GANGLIA IN LEARNING AND MEMORY: POSTTRAINING DRUG STUDIES

An additional experimental approach that has been used to assess the mnemonic functions of the basal ganglia involves posttraining manipulations of this structure. Studies using localized intracerebral posttraining treatments have contributed significantly to our understanding of the neuroanatomical and neurochemical bases of memory. Early experimental findings (Breen & McGaugh 1961, Burnham 1904, Duncan 1949, Zubin & Barrera 1941) supported the hypothesis (Hebb 1949, Muller & Pilzecker 1900) that memory is in a labile state immediately following a training experience, and over time the information is consolidated into a more permanent state. Consistent with consolidation theory, a critical feature of these treatments is that they are time dependent, that is, they are most effective when administered shortly after training and lose effectiveness as the training-treatment interval is increased. The time-dependent nature of posttraining treatments also indicates that the effects on retention are not due to a proactive influence on motivational, sensory, or motoric processes (for reviews see McGaugh 1966, 1989, 2000).

Early research utilized posttraining electrical stimulation to implicate the basal ganglia in memory processes (for review see Kesner & Wilburn 1974). Subsequent research using posttraining pharmacological treatments has implicated dopaminergic, cholinergic, and glutamatergic neurotransmission in dorsal striatal memory processes. With regard to dopamine, studies employing 6-hydroxydopamine lesions to deplete striatal dopamine initially suggested a role for the nigrostriatal dopamine pathway in the mnemonic functions of the basal ganglia (e.g., Neill et al. 1974, White 1988, Zis et al. 1974; see also Major & White 1978).

Memory enhancement produced by posttraining intracaudate infusion of the indirect catecholamine agonist amphetamine in conditioned emotional-response tasks provides direct evidence of a role for striatal dopamine in memory consolidation (Carr & White 1984, Viaud & White 1989). Posttraining intracaudate injections of various dopamine receptor agonists (i.e., amphetamine, D1 receptor, and D2 receptor agents) also enhance memory in the win-stay radial maze task and have no effect in the win-shift task (Packard & White 1991). Similarly, posttraining intracaudate amphetamine injections enhance memory in an S-R visible platform water maze task, but they have no effect on memory in a hidden platform (spatial) task in the same apparatus (Packard et al. 1994, Packard & Teather 1998). In both of these latter studies, posttraining intradorsal hippocampal infusions of dopamine agonists produced the opposite pattern, a selective enhancement spatial memory.

In addition to evidence indicating a role for striatal dopamine in memory, extensive evidence from rats trained in various conditioned avoidance behaviors indicates a role for cholinergic mechanisms within the basal ganglia in memory consolidation (e.g., Deadwyler et al. 1972; Haycock et al. 1973; Neill & Grossman 1971; Packard et al. 1996; Prado-Alcala et al. 1972, 1981; Prado-Alcala & Cobos-Zapian 1977, 1979). Specifically, posttraining intradorsal striatal infusions of muscarinic cholinergic receptor agonist and antagonist drugs typically enhance and impair memory, respectively. Recent studies have also implicated glutamatergic function in dorsal striatal memory processes. Posttraining intracaudate infusions of glutamate and the NMDA receptor antagonist AP5 enhance and impair memory in an S-R visible platform water maze task, respectively (Packard & Teather 1997, 1999). Evidence also suggests a role for metabotropic glutamate receptors in caudate-dependent memory in the visible platform task (Packard et al. 2001). As observed following administration of dopaminergic drugs, the effects of posttraining intracaudate infusion administration of glutamate, AP5, and mGluR agents are task dependent; similar infusions have no effect on memory in a hidden platform water maze task. Finally, a potential role for GABAergic transmission in basal ganglia memory processes was initially suggested by findings indicating that posttraining infusions of the GABA receptor antagonist picrotoxin into the substantia nigra impairs memory (Kim & Routtenberg 1976), and recent research has demonstrated memory-impairing effects of picrotoxin infused directly into the dorsal striatum (Salado-Castillo et al. 1996).

In view of the evidence implicating dopamine, acetylcholine, glutamate, and GABA in the mnemonic functions of the basal ganglia, important questions remain concerning the neural mechanisms that integrate the action of these neurotransmitters. According to one hypothesis (White 1989a, White et al. 1994), glutamatergic corticostriatal projections provide the dorsal striatum with sensory information underlying the formation of S-R associations, whereas GABAergic output to the globus pallidus mediates the motor or response element of habit memory. Dopaminergic input to dorsal striatum is hypothesized to provide a reinforcing signal that effectively stamps together S-R associations. Evidence indicating that 6-hydroxydopamine lesions of the nigrostriatal dopamine pathway block the memory-enhancing effects of posttraining intracaudate infusion of cholinergic agents suggests the existence of acetylcholine-dopamine interactions in striatal memory processes (White et al. 1994).

It is important to note that the putative reinforcing action of dopamine on the formation of S-R habits in the dorsal striatum is differentiated from the rewarding properties of this neurotransmitter. Extensive evidence suggests that affective or rewarding properties of dopamine are mediated by release of this transmitter in the ventral striatum (e.g., via ventral tegmental area projections to the nucleus accumbens). The proposed role of the ventral tegmental-accumbens dopamine pathway in stimulus-reward learning (e.g., Shultz et al. 1997, Sutton & Beninger 1999, Wickens 1990), in which affective information may be represented in memory, may be different than the reinforcing action of this neurotransmitter in S-R habit learning, in which affective information is not represented in the underlying associative structure (for a discussion of theoretical distinctions between the concepts of reward and reinforcement see White 1989b). This suggestion is consistent with evidence indicating that 6-hydroxydopamine lesions of the dorsal striatum do not block the primary rewarding affective properties of dopamine agonists as measured in conditioned place preference behavior or impair the acquisition of conditioned reinforcement tasks that putatively involve stimulus-reward learning (e.g., Cador et al. 1989).

Further research is necessary to elucidate the synaptic and cellular mechanisms by which dopamine, acetylcholine, and glutamate influence dorsal striatal memory processes. Interestingly, each of these transmitter systems has been implicated in various forms of synaptic plasticity [i.e., long-term potentiation (LTP) and long-term depression (LTD)] that have been identified in the basal ganglia (e.g., Calabresi et al. 1992, Centonze et al. 1999, Charpier & Deniau 1997, Garcia-Munoz et al. 1992, Lovinger et al. 1993, Kerr & Wickens 2001), although the relationship between neostriatal LTP and/or LTD and the mnemonic functions of this brain region is currently unknown.

ROLE OF THE BASAL GANGLIA IN HUMAN LEARNING AND MEMORY

Investigation of the role of the basal ganglia in human learning and memory has provided some convergence with data from experimental animals. The fact that the cerebral cortex is highly developed in humans (especially frontal lobe) suggests that corticostriatal loops may subserve even more complex functions in humans. Patients with basal ganglia disorders exhibit impairments in a number of cognitive tasks (for review see Glosser 2001). Cognitive impairment is a primary feature of Huntington's disease, which involves cell loss in the caudate and putamen. Patients with Parkinson's disease also exhibit a variety of cognitive deficits, though not typically as severe. In Parkinson's disease, cell death in the substantia nigra leads to a loss of dopaminergic input to the caudate and putamen. In both of these patient groups, motor deficits are the most obvious consequences of basal ganglia disease. However, as has been described in experimental animals, the demonstration of nonmotor deficits following basal ganglia dysfunction also suggests a broader role for this brain region in human behavior, one that clearly includes learning and memory functions.

PROBABILISTIC CLASSIFICATION LEARNING

The deficits in habit learning that have been demonstrated in experimental animals with damage or dysfunction in the caudate nucleus appear to have an analog in humans with basal ganglia dysfunction. However, the maze tasks that have been used with rats may not readily lend themselves to studies of human learning and memory. For instance, although the caudate-dependent win-stay radial maze task is learned gradually and incrementally across many trials in rats (McDonald & White 1993, Packard et al. 1989), it seems likely that humans would surmise that the illuminated maze arms are selectively baited within a few trials. In order to tap into a human habit learning system, it may be necessary to circumvent the use of explicit or declarative memory. For example, by using a learning task in which cues and outcomes are probabilistically related, explicit memory for what has occurred on each particular trial is not as useful as a general sense of the relationship between cues and outcomes gleaned across numerous training trials.

One version of a probabilistic classification task involves a weather prediction game (Knowlton et al. 1994). There are four cues in the task (i.e., cards with geometric shapes), and these cues predict one of the outcomes approximately 60–85% of the time. Subjects are told that on each trial they will be seeing a set of cues on a computer screen and that their task is to guess whether the cues predict sun or rain. If the subjects make a correct response, they hear a high tone and see a smiling face on the screen, and if their response is incorrect they hear a low tone and see a frowning face on the screen. Although subjects often feel as if they are simply guessing, they nevertheless generally exhibit learning in this task over 50–100 trials, as evidenced by a tendency to choose the more highly associated outcome. Evidence suggests that damage to the basal ganglia results in a deficit in this probabilistic classification task, as patients with Huntington's disease and Parkinson's disease are impaired in task acquisition (Knowlton et al. 1996a). Although the locus of cell loss is different in the two diseases—Parkinson's disease affects the dopaminergic input to the striatum and Huntington's disease affects cells in the striatum itself—it appears that in either disorder the circuitry that is required for learning the stimulus outcome associations is disrupted. The deficit is particularly noteworthy in patients with Parkinson's disease because these patients are able to show evidence of normal declarative memory for the training episode.

It is important to note that in various caudate-dependent learning tasks, rats with lesions of the hippocampal system exhibit intact (or in some cases facilitated) acquisition. Therefore, if learning underlying the probabilistic classification task is analogous to caudate-dependent S-R learning, then it should be acquired normally by patients with damage to structures in the medial temporal lobe that are critical for declarative memory. In fact, it does appear that temporal lobe amnesic patients perform normally on this task (Knowlton et al. 1994, Reber et al. 1996). Thus, a double dissociation between declarative memory and habit learning is observed when patients with Parkinson's disease are compared to amnesic patients. This double dissociation parallels the findings with experimental animals, which suggests that habit learning may be a mnemonic function of the basal ganglia that is conserved across species.

MOTOR SKILL AND PERCEPTUAL-MOTOR LEARNING

In addition to the probabilistic classification task, patients with basal ganglia dysfunction are impaired on other tasks in which procedures or habits may be acquired. Patients with basal ganglia dysfunction exhibit deficits in motor skill learning, primarily for open loop motor skills that are not under the direct control of visual feedback (Gabrieli et al. 1997, Harrington et al. 1990). Although the existence of motor performance deficits can complicate the interpretation of these findings, impaired acquisition has been reported in learning paradigms in which motor performance problems are unlikely to be the primary cause of a failure to show learning. For example, patients with Huntington's disease exhibit reduced weight–biasing effects (Heindel et al. 1991). In this study, subjects lifted a set of weights and then later judged the heaviness of a new set of weights. Subjects' prior experience with the weights affected their judgments; if they had initially lifted heavier weights, they judged the test weight to be lighter than if they had initially lifted lighter weights. This biasing effect may occur because the motor program for lifting the target weight is influenced by previous experience (i.e., if the subject has just been lifting heavy weights, the motor system may be prepared to lift heavy weights). Therefore, a test weight may appear lighter if the motor system is prepared to lift heavy weights, whereas the same weight may seem heavier if the motor system has been prepared to lift lighter weights. This biasing effect does not require declarative memory for experience with the weights because patients with Alzheimer's disease exhibit bias to the same extent as control subjects. The deficit exhibited by patients with Huntington's disease may occur because of a difficulty adapting motor programs based on experience.

Patients with Huntington's disease also exhibit deficits in perceptual-motor skill learning. For example, in the prism adaptation task, subjects wear prism goggles that effectively shift their visual world. Subjects initially make reaching errors while wearing prism goggles, but with practice, they are able to reduce the error.

When subjects remove the goggles there is even a transient error in reaching as the perceptual-motor system readapts to the normal visual world. Patients with Huntington's disease do not adapt as well as control subjects, while patients with Alzheimer's disease are able to adapt normally despite their declarative memory problems (Paulsen et al. 1993). Thus, it appears that changes in motor behavior based on perceptual input depend on the neostriatum, and not on cortical and medial temporal lobe regions affected in Alzheimer's disease.

An additional set of tasks that appear to tap into basal ganglia mnemonic processing involves sequence learning (e.g., Laforce & Doyon 2001, Willingham et al. 1996). In the serial reaction-time task (Nissen & Bullemer 1987), for instance, subjects see a series of stimuli such as asterisks appearing on a computer screen, and their task is to press the key directly below each asterisk as it appears. Unbeknownst to the subject, the asterisks appear according to a fixed sequence. As subjects practice the task, their reaction times decrease (i.e., they press the keys faster in response to the asterisks). In normal subjects, much of this learning is specific to the sequence, and this can be demonstrated by switching from the fixed sequence to randomly appearing asterisks. Although this change is not readily apparent to subjects, their reaction times slow down significantly when this shift occurs. Subjects are not generally able to recognize the sequence after training and may deny that there was a fixed sequence, even though they exhibit sensitivity to the sequence through their performance. There are several reports of poor performance by patients with basal ganglia disorders on the serial reaction-time task. Patients with Huntington's disease, for instance, do not exhibit decreases in reaction time when the fixed sequence is switched to one that is random (Willingham & Koroshetz 1993). One complicating factor is that the movement disorder exhibited by these patients results in their initial performance being much slower and variable than that of controls. Although the slower performance of the patients with Huntington's disease could on the one hand give them more room to improve, it may also be the case that their difficulty with simply performing the task overwhelms any sequence-specific learning that might be present. Nevertheless, the data are consistent with the idea that the basal ganglia are important for learning a visuomotor sequence. In contrast, as is the case with other implicit learning tasks, amnesic patients show normal sequence learning (Nissen & Bullemer 1987, Reber & Squire 1994).

The data from patients with Parkinson's disease on sequence learning are less clear. Although several studies have observed impaired sequence learning in Parkinson's patients, other studies have not (Helmuth et al. 2000, Smith et al. 2001, Sommer et al. 1999). There are several factors that may contribute to this inconsistent pattern of results. First, subjects in the patient groups appear to have differed in the severity of the disease. As Parkinson's disease primarily involves degeneration of projections to the putamen (Morrish et al. 1996), it is possible that patients with early Parkinson's disease may not exhibit a deficit if sequence learning depends on the integrity of the caudate nucleus. Second, the sequences used in different experiments may differ in important ways. For example, some sequences are not balanced for first-order dependencies. If A, B, C, and D represent the four locations in which asterisks appear, the sequence DCBACBDCAB is formed such that position D is always followed by position C. In contrast, the sequence DCBA-CADABCDB does not have this property. For each position, all other positions are equally likely to follow. Sequence-specific speed up would thus indicate that subjects are learning more than just the dependencies of single items. Sequences can be constructed that are even more complex, with second-order dependencies balanced for all groups. Given that there are several aspects of the sequence that can be learned, it may be that patients with Parkinson's disease are impaired on only some types of sequence learning. If so, it would help investigators focus on the type of information processing mediated by the basal ganglia. For example, although the serial reaction-time task measures motor response speed, it appears that learning is more abstract than a sequence of movements. Subjects exhibit good transfer when switched to another effector (such as the other hand) or if the locations are mapped onto a different set of keys. However, if the locations are changed, performance suffers greatly, even if the same pattern of movements occurs that had occurred during training (Willingham et al. 2000). Thus, it appears that what may be acquired is a sequence of locations that one should respond to, rather than a motor sequence.

ROLE OF BASAL GANGLIA IN LEARNING AND MEMORY: HUMAN NEUROIMAGING STUDIES

In addition to neuropsychological studies, there are several human neuroimaging studies indicating that the basal ganglia is involved in learning skills and habits. Activation in the caudate nucleus has been observed while subjects are learning the skill of reading mirror-reversed text (Dong et al. 2000). Interestingly, the striatal involvement may only be present during learning, with highly skilled performance primarily activating cortical circuitry (Poldrack & Gabrieli 2001). Several neuroimaging studies also have demonstrated activation of the basal ganglia during learning of the serial reaction-time task (Doyon et al. 1996, Rauch et al. 1997). In these studies, activation in the caudate nucleus is observed while subjects are performing the serial reaction-time task with a fixed sequence, compared to blocks of trials for which locations occur randomly. Moreover, activation in the caudate nucleus is associated with performance of an implicitly learned sequence, but not when subjects are explicitly told the sequence beforehand and can therefore consciously anticipate the location of the upcoming stimulus. In another study, positron emission tomography (PET) was used to measure dopamine release while participants played a video game (Koepp et al. 1998). As subjects played the game and improved their performance, there was decreased binding of radiolabeled raclopride (a dopamine antagonist) in dorsal and ventral striatum in comparison to a control condition. These data suggest that endogenous dopamine release increased in the neostriatum during practice, consistent with research in lower animals demonstrating a role for striatal dopamine in S-R habit learning (e.g., Packard & White 1991).

INTERACTIONS BETWEEN BASAL GANGLIA AND MEDIAL TEMPORAL LOBE MEMORY SYSTEMS: TEMPORAL ASPECTS

As previously described, numerous studies have dissociated the S-R habit memory function of the basal ganglia from those of a declarative medial temporal lobe memory system that includes the hippocampus as a primary component. Within the context of a multiple-systems approach to memory organization, one important question concerns the nature of potential interactions between different memory systems. In considering this issue, one can start with the observation that during learning basal ganglia and hippocampal memory systems appear to be activated simultaneously and in parallel (McDonald & White 1994, Packard & McGaugh 1996). Recall that, with extended training in a plus maze, expression of response-learning tendencies comes to overshadow previously dominant place learning (Hicks 1964, Packard & McGaugh 1996, Ritchie et al. 1950). This shift indicates that in a learning task for which both memory systems can provide an adequate solution, the hippocampal system mediates a rapid form of learning that initially controls behavior but that eventually cedes behavioral control to a caudate memory system mediating a more slowly developing S-R form of learning. Consistent with this suggestion, rats that are overtrained in the caudate-dependent win-stay radial maze task appear to perform based on S-R associations because reinforcer devaluation does not alter response accuracy or latency (Sage & Knowlton 2000). However, early in training before asymptotic performance has been reached, rats do show longer latencies to run down cued arms if the reinforcer has been devalued, which suggests that a representation of the food is present at early stages of learning. In view of evidence that performance on the hippocampus-dependent win-shift radial arm maze task is affected by reinforcer devaluation (Sage & Knowlton 2000), it appears that the reinforcer is represented in what has been learned by the hippocampus about which arms have been visited.

Taken together, these findings raise the possibility that postposttraining injections of a memory-enhancing agent into the dorsal striatum or hippocampus during early training might influence the time course of the shift in the use of these two structures to guide learned behavior. In an experiment designed to address this implication, rats received postposttraining intradorsal hippocampal or intradorsolateral caudate infusion of glutamate during early time points (on days 3–5) of plus-maze training (Packard 1999). Rats receiving vehicle injections predominantly displayed place learning on an early (day 8) probe trial and response learning on a later (day 16) probe trial. However, rats receiving postposttraining intrahippocampal infusions of glutamate predominantly displayed place learning on both the early and late probe trials, which suggests that infusion of glutamate into the hippocampus strengthened a spatial representation, effectively blocking the shift to response learning that occurs with extended training. In contrast, rats given postposttraining glutamate infusions into the caudate-putamen predominantly displayed response learning on both the early and late probe trials, which suggests that infusion of glutamate into the caudate-putamen accelerated the shift to response learning that occurs in control rats only following extended training. Therefore, manipulation of a neurotransmitter system relevant to the mnemonic processes mediated by both the basal ganglia and hippocampus can bias animals toward the use of a specific memory system.

Although plus-maze studies in experimental animals clearly demonstrate that with extended training there is a shift from the use of cognitive to habit memory, this phenomenon has been observed in animals in which both memory systems are functional at the initiation of training. However, it is of interest to note that in patients with Parkinson's disease, deficits on a probabilistic classification task are most apparent early in training. After extended training (more than about 150 trials), Parkinson's patients do tend to approach the performance of control subjects. It is likely that in this situation, subjects are eventually able to gain sufficient declarative knowledge of the task structure to further improve their performance and may begin to make optimal choices for some cue patterns rather than simply probability match. Patients with Parkinson's disease would presumably gain declarative knowledge of the task along with control subjects. Patients with amnesia, as well as control subjects, would not be able to acquire this declarative knowledge and thus would be relatively impaired later in training. Indeed, this pattern has been observed when amnesic patients are given extended training on this task (Knowlton et al. 1994).

INTERACTIONS BETWEEN BASAL GANGLIA AND MEDIAL TEMPORAL LOBE MEMORY SYSTEMS: COMPETITIVE ASPECTS

As simultaneous and parallel activation of basal ganglia and medial temporal lobe memory systems occurs during learning, one form of interaction between these systems appears to be competitive or interfering in nature. Sherry & Schacter (1987) hypothesized that the presence of functional incompatibility, in which an existing memory system is unable to provide an adequate solution in a situation involving novel information or task demands, may have driven natural selection processes that ultimately resulted in the evolution of multiple memory systems. One can envision a type of racehorse model in which both systems undergo learning-related changes, with the system that comes up with the most valid and reinforced response enjoying a strengthening of its control on behavior.

In experimental settings, competitive interference between different memory systems may be potentially revealed in studies in which pretraining lesions of a given system result in enhanced acquisition of a task relative to brain-intact animals. For example, in the caudate-dependent win-stay radial maze task, it is conceivable that spatial information processed by the hippocampal system, which provides the rat with information concerning those maze arms in which food has already been retrieved, may interfere with the task requirement of revisiting maze arms in which food was recently removed. Consistent with this hypothesis, lesions of the hippocampal system facilitate acquisition of caudate-dependent win-stay radial maze behavior (Packard et al. 1989, McDonald & White 1993). In addition, pretraining hippocampal system lesions (Matthews & Best 1995) and postposttraining neural inactivation of the dorsal hippocampus (Schroeder et al. 2002) facilitate acquisition of caudate-dependent response learning. The enhancing effect of hippocampal lesions on acquisition of caudate-dependent two-way active avoidance behavior has also been interpreted as resulting from the removal of spatial information processing that would tend to interfere with the task requirement of returning to a spatial location in which electrical shock has recently been administered (O'Keefe & Nadel 1978). In some learning situations, interference with hippocampal memory processes by the caudate nucleus may also occur. Consistent with this suggestion, lesions of the caudate nucleus facilitate acquisition of a spatial Y-maze discrimination task (Mitchell & Hall 1988), perhaps by disrupting the use of a potentially interfering response-learning strategy.

Investigation of potential neurochemical mechanisms that mediate the interaction between basal ganglia and temporal lobe memory systems is at an early stage (Gold et al. 2001, Packard 1999). As previously described, in a plus-maze task that can be acquired using either caudate-dependent response learning or hippocampus-dependent place learning, postposttraining intracerebral infusions of glutamate can bias animals toward the use of a particular memory system (Packard 1999). Other findings suggest that dorsal striatal cholinergic function may also influence the relative efficiency and use of hippocampal and caudate memory systems. For example, increases in acetylcholinetransferase activity in the dorsolateral caudate are negatively correlated with accuracy in a hippocampus-dependent working memory task (Colombo & Gallagher 1998). In addition, in vivo microdialysis reveals that during acquisition of a plus-maze task, a higher ratio of acetylcholine release in the hippocampus relative to the dorsal striatum is associated with the use of place learning on a subsequent probe trial (C. McIntyre, C. K. Marriot, P.E. Gold, submitted). Remarkably, the ratio of acetylcholine release in the hippocampus relative to the dorsal striatum prior to initial training predicts whether rats will later employ place or response learning in the plus maze, which suggests that relative levels of cholinergic activity may in part determine individual differences in the use of these two memory systems (C. McIntyre, C. K. Marriot, P. E. Gold, submitted).

Consideration of memory processes that may be active in the probabilistic classification task suggests that interference between basal ganglia and medial temporal lobe memory systems is also likely to occur in humans. For example, in the probabilistic classification task, episodic memory for a particular trial in which a lower probability outcome occurs would contradict the response based on an S-R habit that has developed across training. For the very first few responses, the subject may select an outcome based on their declarative memory of the reinforcement received when one of the cues present had appeared shortly before. However, as the cue-response habit strengthens, the striatal memory system could guide behavior more accurately with potentially less effort. With extended training, there may be enough exposure to individual trials to allow declarative knowledge of the task structure to be acquired. At this point, subjects may begin to choose the most associated outcome and begin to probability maximize. Consistent with this suggestion is evidence indicating that patients with medial temporal lobe damage are relatively impaired following extended training in this task (Knowlton et al. 1994). In addition, in some cases of caudate-dependent sequence learning, explicit knowledge of the sequence can impair acquisition as measured by reaction time, which suggests that effortful retrieval of explicit knowledge can interfere with the performance of the implicitly learned sequence.

Direct evidence for the idea that basal ganglia and medial temporal lobe memory systems may compete in some learning situations is provided by human neuroimaging studies employing the probabilistic classification task. During learning of the weather prediction task, activation is present in the caudate nucleus, and there is a concomitant decrease in medial temporal lobe activation relative to the activation present when the subject performs a low-level baseline task such as indicating if more than two cards are present (Poldrack et al. 1999). Furthermore, a recent fMRI study using a design in which activation can be measured on individual trials has shown that activation in the caudate nucleus and medial temporal lobe is negatively correlated within subjects (Poldrack et al. 2001). This study also demonstrates that at the beginning of learning it appears that subjects rely on medial temporal lobe structures; this dependence rapidly declines with training and with an increase in dependence on the striatum.

At present, the factors that may influence the interaction between basal ganglia and medial temporal lobe memory systems are not well understood. The use of various reinforcement/training parameters (e.g., correction versus noncorrection, spaced versus massed trials) has been shown to influence the relative use of hippocampus-dependent place learning and caudate-dependent response learning in the plus-maze task (for review see Restle 1957), and these parameters might affect the interaction between these two memory systems in other situations as well. The nature of the visual environment (e.g., heterogenous versus homogenous surrounds) also influences the type of learning observed in various maze tasks and appears to bias the brain toward the use of hippocampus-dependent or caudate-dependent learning. In the caudate-dependent win-stay radial maze task, for instance, the removal of extra-maze cues enhances task acquisition in brain-intact rats, an effect that is strikingly similar to the effects of damaging the hippocampal system and training rats in this task in the presence of abundant extra-maze cues (Packard & White 1987).

In addition to experimental factors, the neural basis of competitive interference between basal ganglia and medial temporal lobe memory systems is unknown. There are reportedly direct connections between the entorhinal cortex and the neostriatum in the rat that, when stimulated, have been shown to mediate a long phase of inhibition after initial excitation (Finch et al. 1995). Interestingly, there is a report that neuronal activity in the human caudate nucleus shows a similar phenomenon while the subject is performing a declarative memory task. In this study, disease, as well as human neuroimaging studies, have provided some support for the role of the basal ganglia in S-R habit learning. Finally, evidence suggests that in a given learning situation basal ganglia and medial temporal lobe systems are activated simultaneously, and recent studies have begun to examine the nature of the interaction between these brain regions in learning and memory.

The *Annual Review of Neuroscience* is online at http://neuro.annualreviews.org

LITERATURE CITED

Abdullaev YG, Melnichuk KV. 1997. Cognitive operations in the human caudate nucleus. *Neurosci. Lett.* 234:151–55

Abraham L, Potegal M, Miller S. 1983. Evidence for caudate nucleus involvement in an egocentric spatial task: return form passive transport. *Physiol. Psychol.* 11:11–17

Adams S, Kesner RP, Ragozzino ME. 2001. Role of the medial and lateral caudate-putamen in mediating an auditory conditional response association. *Neurobiol. Learn. Mem.* 76:106–16

Alexander GE, DeLong MR, Strick PL. 1986. Parallel organization of functionally segregated circuits linking basal ganglia and cortex. *Annu. Rev. Neurosci.* 9:357–81

Allen JD, Davidson CS. 1973. Effects of caudate lesions on signaled and nonsignaled Sidman avoidance in the rat. *Behav. Biol.* 8:239–50

Aosaki T, Graybiel AM, Kimura M. 1994. Effect of the nigrostriatal dopamine system on acquired neural responses in the striatum of behaving monkeys. *Science* 265:412–15

Battig K, Rosvold HE, Mishkin M. 1960. Comparison of the effects of frontal and caudate lesions on delayed response and alternation in monkeys. *J. Comp. Physiol. Psych.* 53:400–4

Beiser DG, Hua SE, Houk JC. 1997. Network models of the basal ganglia. *Curr. Opin. Neurobiol.* 7:185–90

Berns GS, Sejnowski TJ. 1998. A computational model of how the basal ganglia produce sequences. *J. Cogn. Neurosci.* 10:108–21

Brasted PJ, Humby T, Dunnett SB, Robbins TW. 1997. Unilateral lesions of the dorsal striatum in rats disrupt responding in egocentric space. *J. Neurosci.* 17:8919–26

Breen RA, McGaugh JL. 1961. Facilitation of maze learning with post-trial injections of picrotoxin. *J. Comp. Physiol. Psychol.* 54:495–501

Buerger AA, Gross CG, Rocha-Miranda CE. 1974. Effects of ventral putamen lesions on discrimination learning by monkeys. *J. Comp. Physiol. Psychol.* 86:440–46

Burnham WH. 1904. Retroactive amnesia: illustrative cases and a tentative explanation. *Am. J. Psychol.* 14:382–96

Butters N, Rosvold HE. 1968. Effect of caudate and septal lesions on resistance to extinction and delayed alternation. *J. Comp. Physiol. Psychol.* 65:397–403

Butters N, Salmon D, Heindel WC. 1994. Specificity of the memory deficits associated with basal ganglia dysfunction. *Rev. Neurol.* 150:580–87

Cador M, Robbins TW, Everitt BJ. 1989. Involvement of the amygdala in stimulus-reward associations: interaction with the ventral striatum. *Neuroscience* 30:77–86

Cahill L, McGaugh JL. 1998. Mechanisms of emotional arousal and lasting declarative memory. *Trends Neurosci.* 21:294–99

Calabresi P, Maj R, Pisani A, Mercuri NB, Bernardi G. 1992. Long-term synaptic depression in the rat striatum: physiological and pharmacological characterization. *J. Neurosci.* 12:4224–33

Carr GD, White NM. 1984. The relationship between stereotypy and memory improvement produced by amphetamine. *Psychopharmacology* 82:203–9

Centonze D, Gubellini P, Bernardi G, Calabresi P. 1999. Permissive role of interneurons in corticostriatal synaptic plasticity. *Brain Res. Rev.* 31:1–5

Charpier S, Deniau JM. 1997. In vivo activity-dependent plasticity at cortico-striatal connections: evidence for physiological long-term potentiation. *Proc. Natl. Acad. Sci. USA* 94:7036–40

Chorover SL, Gross CG. 1963. Caudate nucleus lesions: behavioral effects in the rat. *Science* 141:826–27

Chozick BS. 1983. The behavioral effects of lesions of the corpus striatum: a review. *Int. J. Neurosci.* 19:143–59

Colombo PJ, Davis HP, Volpe BT. 1989. Allocentric spatial and tactile memory impairments in rats with dorsal caudate lesions are affected by preoperative behavioral training. *Behav. Neurosci.* 103:1242–50

Colombo PJ, Gallagher M. 1998. Individual differences in spatial memory and striatal ChAT activity among young and aged rats. *Neurobiol. Learn. Mem.* 70:314–27

Cook D, Kesner RP. 1988. Caudate nucleus and memory for egocentric localization. *Behav. Neural Biol.* 49:332–43

Deadwyler SA, Montgomery D, Wyers EJ. 1972. Passive avoidance and carbachol excitation of the caudate nucleus. *Physiol. Behav.* 8:631–35

DeCoteau WE, Kesner RP. 2000. A double dissociation between the rat hippocampus and medial caudoputamen in processing two forms of knowledge. *Behav. Neurosci.* 114:1096–108

Desmond JE, Gabrieli JDE, Glover GH. 1998. Dissociation of frontal and cerebellar activity in a cognitive task: evidence for a distinction between selection and search. *Neuroimage* 7:368–78

Devan BD, McDonald RJ, White NM. 1999. Effects of medial and lateral caudate-putamen lesions on place- and cue-guided behaviors in the water maze: relation to thigmotaxis. *Behav. Brain Res.* 100:5–14

Devan BD, White NM. 1999. Parallel information processing in the dorsal striatum: relation to hippocampal function. *J. Neurosci.* 19:2789–98

Divac I. 1968. Functions of the caudate nucleus. *Acta Neurobiol. Exp. (Warsz)* 28:107–20

Divac I. 1972. Neostriatum and functions of the prefrontal cortex. *Acta. Neurobiol. Exp.* 32:461–77

Divac I, Oberg RGE. 1975. Dissociative effects of selective lesions in the caudate nucleus of cats and rats. *Acta Neuro. Exp.* 35:647–59

Divac I, Rosvold HE, Szwarcbart MK. 1967. Behavioral effects of selective ablation of the caudate nucleus. *J. Comp. Physiol. Psychol.* 63:183–90

Dong Y, Fukuyama H, Honda M, Okada T, Hanakawa T, et al. 2000. Essential role of the right superior parietal cortex in Japanese kana mirror reading: an fMRI study. *Brain* 123:790–99

Doyon J, Owen AM, Petrides M, Sziklas V, Evans AC. 1996. Functional anatomy of visuomotor skill learning in human subjects examined with positron emission tomography. *Eur. J. Neurosci.* 8:637–48

Duncan CP. 1949. The retroactive effect of electroshock on learning. *J. Comp. Physiol. Psychol.* 42:32–44

Fernandez-Ruiz J, Wang J, Aigner TG, Mishkin M. 2001. Visual habit formation in monkeys with neurotoxic lesions of the ventrocaudal neostriatum. *Proc. Natl. Acad. Sci. USA* 98:4196–201

Finch DM, Gigg J, Tan AM, Kosoyan OP. 1995. Neurophysiology and neuropharmacology of projections from entorhinal cortex to striatum in the rat. *Brain Res.* 670:233–47

Fonnum F, Storm-Mathisen J, Divac I. 1981. Biochemical evidence for glutamate as neurotransmitter in corticostriatal and corticothalamic fibres in rat brain. *Neuroscience* 6:863–73

Furtado JCS, Mazurek MF. 1996. Behavioral characterization of quinolinate-induced lesions of the medial striatum: relevance for Huntington's disease. *Exp. Neurol.* 138:158–68

Gabrieli JD, Stebbins GT, Singh J, Willingham DB, Goetz CG. 1997. Intact mirror-tracing and impaired rotary-pursuit skill learning in patients with Huntington's disease: evidence for dissociable memory systems in skill learning. *Neuropsychologia* 1:272–81

Garcia-Munoz M, Young SJ, Groves PM. 1992. Presynaptic long-term changes in excitability of the corticostriatal pathway. *Neuroreport* 3:357–60

Gerfen CR. 1992. The neostriatal mosiac: multiple levels of compartmental organization in the basal ganglia. *Annu. Rev. Neurosci.* 15:285–320

Gerfen CR, Wilson CJ. 1996. The basal ganglia. In *Handbook of Chemical Neuroanatomy, Integrated Systems of the CNS, Part III*, ed. LW Swanson, A Bjorkland, T Hokfelt, 12:371–468. New York: Elsevier

Gillies A, Arbuthnott G. 2000. Computational models of the basal ganglia. *Mov. Disord.* 15:762–70

Glosser G. 2001. Neurobehavioral aspects of movement disorders. *Neurologic Clin.* 19:535–51

Gold PE, McIntyre C, McNay E, Stefani M, Korol DL. 2001. Neurochemical referees of dueling memory systems. In *Memory Consolidation: Essays in Honor of James L. McGaugh*, ed. PE Gold, WT Greenough, pp. 219–48. Washington, DC: Am. Psychol. Assoc.

Graybiel AM. 1990. Neurotransmitters and neuromodulators in the basal ganglia. *Trends Neurosci.* 13:244–54

Graybiel AM. 1998. The basal ganglia and chunking of action repertories. *Neurobiol. Learn. Mem.* 70:119–36

Graybiel AM, Aosaki T, Flaherty AW, Kimura M. 1994. The basal ganglia and adaptive motor control. *Science* 265:1826–31

Green RH, Beatty WW, Schwartbaum JS. 1967. Comparative effects of septo-hippocampal and caudate lesions on avoidance behavior in rats. *J. Comp. Physiol. Psychol.* 64:444–52

Gross CG, Chorover SL, Cohen SM. 1965. Caudate, cortical hippocampal, and dorsal thalamic lesions in rats: alternation and Hebb-Williams maze performance. *Neuropsychologia* 3:53–68

Harrington DL, Haaland KY, Yeo RA, Marder E. 1990. Procedural memory in Parkinson's disease: impaired motor but not visuoperceptual learning. *J. Clin. Exp. Neuropsychol.* 12:323–39

Haycock JW, Deadwyler SA, Sideroff SI, McGaugh JL. 1973. Retrograde amnesia and cholinergic systems in the caudate-putamen complex and dorsal hippocampus of the rat. *Exp. Neurol.* 41:201–13

Hebb DO. 1949. *The Organization of Behavior*. New York: Wiley

Heimer L, van Hoesen G. 1979. Ventral striatum. In *The Neostriatum*, ed. I Divac, RGE Oberg, pp. 147–58. New York: Pergamon

Heindel WC, Butters N, Salmon DP. 1988. Impaired learning of a motor skill in patients with Huntington's disease. *Behav. Neurosci.* 102:141–50

Heindel WC, Salmon DP, Butters N. 1991. The biasing of weight judgments in Alzheimer's and Huntington's disease: a priming or programming phenomenon? *J. Clin. Exp. Neuropsychol.* 13:189–203

Helmuth LL, Mayr U, Daum I. 2000. Sequence learning in Parkinson's disease: a comparison of spatial-attention and number-response sequences. *Neuropsychologia* 38:1443–51

Hicks LH. 1964. Effects of overtraining on acquisition and reversal of place and response learning. *Psychol. Rep.* 15:459–62

Hikosaka O, Sakamoto M, Usui S. 1989. Functional properties of monkey caudate neurons. III. Activities related to expectation of target and reward. *J. Neurophysiol.* 61:814–32

Hull CL. 1943. *Principles of Behavior*. New York: Appleton-Century Crofts

Jog MS, Kubota Y, Connolly CI, Hillegaart V, Graybiel AM. 1999. Building neural representations of habits. *Science* 286:1745–49

Kerr JND, Wickens JR. 2001. Dopamine D-1/D-5 receptor activation is required for long-term potentiation in the rats neostriatum in vitro. *J. Neurophysiol.* 85:117–24

Kesner RP, Bolland BL, Dakis M. 1993. Memory for spatial locations, motor responses, and objects: triple dissociation among the hippocampus, caudate nucleus, and extrastriate visual cortex. *Exp. Brain Res.* 93:462–70

Kesner RP, Wilburn MW. 1974. A review of electrical stimulation of the brain in context of learning and retention. *Behav. Biol.* 10:259–93

Kim HJ, Routtenberg A. 1976. Retention disruption following post-trial picrotoxin injection into the substantia nigra. *Brain Res.* 113:620–25

Kim JJ, Lee H, Han JS, Packard MG. 2001. Amygdala is critical for stress-induced modulation of hippocampal LTP and learning. *J. Neurosci.* 21:5222–28

Kimura M. 1995. Role of the basal ganglia in behavioral learning. *Neurosci. Res.* 22:353–58

Kimura A, Graybiel AM. 1995. *Functions of the Cortico-Basal Ganglia Loop*, ed. M Kimura, AM Graybiel. Tokyo/New York: Springer

Kirkby RJ. 1969. Caudate nucleus lesions impair spontaneous alternation. *Percept. Mot. Skills* 29:550

Kirkby RJ, Polgar S. 1974. Active avoidance in the laboratory rats following lesions of the dorsal or ventral caudate nucleus. *Physiol. Psychol.* 2:301–6

Kirkby RJ, Polgar S, Coyle IR. 1981. Caudate nucleus lesions impair the ability of rats to learn a simple straight-alley task. *Percept. Mot. Skills* 52:499–502

Kirkby RJ, Kimble DP. 1968. Avoidance and escape behavior following striatal lesions in the rat. *Exp. Neurol.* 20:215–27

Knowlton BJ, Mangels JA, Squire LR. 1996a. A neostriatal habit learning system in humans. *Science* 273:1399–402

Knowlton BJ, Squire LR, Gluck MA. 1994. Probabilistic category learning in amnesia. *Learn. Mem.* 1:106–20

Knowlton BJ, Squire LR, Paulsen JS, Swerdlow N, Swenson M, et al. 1996b. Dissociations within nondeclarative memory in Huntington's disease. *Neuropsychologia* 10:1–11

Koepp MJ, Gunn RN, Lawrence AD, Cunningham VJ, Dagher A, et al. 1998. Evidence for striatal dopamine release during a video game. *Nature* 393:266–68

Kolb B. 1977. Studies on the caudate-putamen and the dorsomedial thalamic nucleus of the rat: implications for mammalian frontal-lobe functions. *Physiol. Behav.* 18:237–44

Laforce R Jr, Doyon J. 2001. Distinct contribution of the striatum and cerebellum to motor learning. *Brain Cogn.* 45:189–211

Lawrence AD, Watkins LHA, Sahakian BJ, Hodges JR, Robbins TW. 2000. Visual object and visuospatial cognition in Huntington's disease: implications for information processing in corticostriatal circuits. *Brain* 123:1349–64

Lorenzi CA, Baldi E, Bucherelli C, Tassoni G. 1995. Time-dependent deficits of rat memory consolidation induced by tetrodotoxin injections into the caudate-putamen, nucleus accumbens, and globus pallidus. *Neurobiol. Learn. Mem.* 63:87–93

Lovinger DM, Tyler EC, Marritt A. 1993. Short and long term depression in the rat neostriatum. *J. Neurophysiol.* 70:1937–49

Lynch GS, Lucas PA, Deadwyler SA. 1972. The demonstration of acetylcholinesterase containing neurones within the caudate nucleus of the rat. *Brain Res.* 45:617–21

Major R, White NM. 1978. Memory facilitation produced by self-stimulation reinforcement mediated by the nigro-neostriatal bundle. *Physiol. Behav.* 20:723–33

Marie RM, Barre L, Dupuy B, Viader F, Defer G, et al. 1999. Relationships between striatal dopamine denervation and frontal executive tests in Parkinson's disease. *Neurosci. Lett.* 260:77–80

Martone M, Butters N, Payne J, Becker J, Sax DS. 1984. Dissociations between skill learning and verbal recognition in amnesia and dementia. *Arch. Neurol.* 41:965–70

Matthews DB, Best PJ. 1995. Fimbria/fornix lesions facilitate the learning of a nonspatial response task. *Psychol. Bull. Rev.* 2:113–16

McDonald RJ, White NM. 1993. A triple dissociation of memory systems: hippocampus, amygdala, and dorsal striatum. *Behav. Neurosci.* 107:3–22

McDonald RJ, White NM. 1994. Parallel information processing in the water maze:

evidence for independent memory systems involving the dorsal striatum and hippocampus. *Behav. Neural Biol.* 61:260–70

McGaugh JL. 1966. Time-dependent processes in memory storage. *Science* 153:1351–58

McGaugh JL. 1989. Dissociating learning and performance: drug and hormone enhancement of memory storage. *Brain Res. Bull.* 23:339–45

McGaugh JL. 2000. Memory: a century of consolidation. *Science* 287:248–251

McGeorge AJ, Faull RLM. 1989. The organization of the projection from the cerebral cortex to the striatum in the rat. *Neuroscience* 29:503–37

Mengual E, de las Heras S, Erro E, Lanciego JL, Gimenez-Amaya JM. 1999. Thalamic interaction between the input and output systems of the basal ganglia. *J. Chem. Neuroanat.* 16:187–200

Middleton FA, Strick PL. 1996. The temporal lobe is a target of output from the basal ganglia. *Proc. Natl. Acad. Sci. USA* 93:8683–87

Mishkin M, Petri HL. 1984. Memories and habits: some implications for the analysis of learning and retention. In *Neuropsychology of Memory*, ed. LR Squire, N. Butters, pp. 287–96. New York: Guilford

Mitchell JA, Hall G. 1988. Caudate-putamen lesions in the rat may impair or potentiate maze learning depending upon availability and relevance of response cues. *Q. J. Exp. Psychol.* 40B(3):243–58

Mizumori JY, Ragozzino KE, Cooper BG. 2000. Location and head direction representation in the dorsal striatum of rats. *Psychobiology* 28:441–62

Mogenson GJ, Jones DL, Yim CY. 1980. From motivation to action: functional interface between the limbic system and the motor system. *Prog. Neurobiol.* 14:69–97

Morris RGM. 1984. Development of a water-maze procedure for studying spatial representation in the rat. *J. Neurosci. Methods* 11:47–60

Morrish PK, Sawle GV, Brooks DJ. 1996. Regional changes in [18F] dopa metabolism in the striatum in Parkinson's disease. *Brain* 119:2097–103

Muller GE, Pilzecker A. 1900. Experimentelle Beitrage zur Lehre vom Gedachtnis. *Z. Psychol.* (Suppl 1)

Neill DB, Grossman SP. 1971. Behavioral effects of lesions or cholinergic blockade of the dorsal and ventral caudate of rats. *J. Comp. Physiol. Psychol.* 71:311–17

Neill DB, Boggan WO, Grossman SP. 1974. Impairment of avoidance performance by intrastriatal administration of 6-hydroxydopamine. *Pharm. Biochem. Behav.* 2:97–103

Nissen MJ, Bullemer P. 1987. Attentional requirements of learning: evidence from performance measures. *Cogn. Psychol.* 19:1–32

O'Keefe J, Nadel L. 1978. *The Hippocampus as a Cognitive Map*. Oxford, UK: Oxford Univ. Press

Oberg RGE, Divac I. 1979. "Cognitive" functions of the neostriatum. In *The Neostriatum*, ed. I Divac, RGE Oberg, pp. 291–313. New York: Pergamon

Ohye C, Kimura M, McKenzie JS. 1996. *The Basal Ganglia V*, ed. C Ohye, M Kimura, JS McKenzie. New York: Plenum

Olton DS, Papas BC. 1979. Spatial memory and hippocampal function. *Neuropsychologia* 17:669–82

Olton DS, Samuelson RJ. 1976. Remembrance of places passed: spatial memory in rats. *J. Exp. Psychol: Animal Behav. Proc.* 2:97–115

Packard MG. 1999. Glutamate infused post-training into the hippocampus or caudate-putamen differentially strengthens place and response learning. *Proc. Natl. Acad. Sci. USA* 96:12881–86

Packard MG. 2001. On the neurobiology of multiple memory systems: Tolman versus Hull, system interactions, and the emotion-memory link. *Cogn. Process.* 2:3–24

Packard MG, Cahill L. 2001. Affective modulation of multiple memory systems. *Curr. Opin. Neurobiol.* 11:752–56

Packard MG, Cahill L, McGaugh JL. 1994. Amygdala modulation of hippocampal-dependent and caudate nucleus-dependent memory processes. *Proc. Natl. Acad. Sci. USA* 91:8477–81

Packard MG, Hirsh R, White NM. 1989. Differential effects of fornix and caudate nucleus lesions on two radial maze tasks: evidence for multiple memory systems. *J. Neurosci.* 9:1465–72

Packard MG, Introini-Collison IB, McGaugh JL. 1996. Stria terminalis lesions attenuate memory enhancement produced by intra-caudate nucleus injections of oxotremorine. *Neurobiol. Learn. Mem.* 65:278–82

Packard MG, McGaugh JL. 1992. Double dissociation of fornix and caudate nucleus lesions on acquisition of two water maze tasks: further evidence for multiple memory systems. *Behav. Neurosci.* 106:439–46

Packard MG, McGaugh JL. 1996. Inactivation of the hippocampus or caudate nucleus with lidocaine differentially affects expression of place and response learning. *Neurobiol. Learn. Mem.* 65:65–72

Packard MG, Teather LA. 1997. Double dissociation of hippocampal and dorsal striatal memory systems by post-training intracerebral injections of 2-amino-phosphonopentanoic acid. *Behav. Neurosci.* 111:543–51

Packard MG, Teather LA. 1998. Amygdala modulation of multiple memory systems: hippocampus and caudate-putamen. *Neurobiol. Learn. Mem.* 69:163–203

Packard MG, Teather LA. 1999. Dissociation of multiple memory systems by posttraining intracerebral injections of glutamate. *Psychobiology* 27:40–50

Packard MG, Vecchioli SF, Schroeder JP, Gasbarri A. 2001. Task-dependent role for dorsal striatum metabotropic glutamate receptors in memory. *Learn. Mem.* 8:96–103

Packard MG, White NM. 1987. Differential roles of the hippocampus and caudate nucleus in memory: selective mediation of "cognitive" and "associative" learning. *Soc. Neurosci. Abs.* 13:1005

Packard MG, White NM. 1990. Lesions of the caudate nucleus selectively impair acquisition of "reference memory" in the radial maze. *Behav. Neural Biol.* 53:39–50

Packard MG, White NM. 1991. Dissociation of hippocampus and caudate nucleus memory systems by posttraining intracerebral injection of dopamine agonists. *Behav. Neurosci.* 105:295–306

Packard MG, Williams CL, Cahill L, McGaugh JL. 1995. The anatomy of a memory modulatory system: from periphery to brain. In *Neurobehavioral Plasticity: Learning, Development, and Response to Brain Insults*, ed. NE Spear, LP Spear, ML Woodruff, pp. 149–83. Hillsdale, NJ: Erlbaum

Packard MG, Winocur G, White NM. 1992. The caudate nucleus and acquisition of win-shift radial maze behavior: effect of exposure to the reinforcer during maze adaptation. *Psychobiology* 20:127–32

Parent A. 1986. *Comparative Neurobiology of the Basal Ganglia*. New York: Wiley

Paulsen JS, Butters N, Salmon DP, Heindel WC, Swenson MR. 1993. Prism adaptation in Alzheimer's and Huntington's disease. *Neuropsychologia* 7:73–81

Phillips AG, Carr GD. 1987. Cognition and the basal ganglia: a possible substrate for procedural knowledge. *Can. J. Neurol. Sci.* 14:381–85

Pisa M, Cyr J. 1990. Regionally selective roles of the rat's striatum in modality specific discrimination learning and forelimb reaching. *Behav. Brain Res.* 37:281–92

Poldrack RA, Clark J, Pare-Blagoev J, Shohamy D, Creso Moyano J, et al. 2001. Interactive memory systems in the human brain. *Nature* 414:546–50

Poldrack RA, Gabrieli JDE. 2001. Characterizing the neural mechanisms of skill learning and repetition priming: evidence from mirror reading. *Brain* 124:67–82

Poldrack RA, Prabhakaran V, Seger C, Gabrieli JDE. 1999. Striatal activation during cognitive skill learning. *Neuropsychologia* 13:564–74

Potegal M. 1972. The caudate nucleus egocentric localization system. *Acta Neurobiol. Exp. (Warsz).* 32:479–94

Prado-Alcala RA, Cobos-Zapiain GC. 1977. Learning deficits induced by cholinergic blockade of the caudate nucleus as a function of experience. *Brain Res.* 138:190–96

Prado-Alcala RA, Cobos-Zapiain GG. 1979. Improvement of learned behavior through cholinergic stimulation of the caudate nucleus. *Neurosci. Lett.* 14:253–58

Prado-Alcala RA, Grinberg-Zylberbaun J, Alvarez-Leefmans J, Gomez A, Singer S, Brust-Carmona H. 1972. A possible caudate-cholinergic mechanism in two instrumental conditioned responses. *Psychopharmacology* 25:339–46

Prado-Alcala RA, Grinberg ZJ, Arditti ZL, Garcia MM, Prieto HG, et al. 1975. Learning deficits produced by chronic and reversible lesions of the corpus striatum in rats. *Physiol. Behav.* 15:283–87

Prado-Alcala RA, Signoret L, Figueroa M. 1981. Time-dependent retention deficits induced by post-training injections of atropine into the caudate nucleus. *Pharm. Biochem. Behav.* 15:633–36

Rauch SL, Whalen PJ, Savage CR, Curran T, Kendrick A, et al. 1997. Striatal recruitment during an implicit sequence learning task as measured by functional magnetic resonance imaging. *Hum. Brain Mapp.* 5:124–32

Reading PJ, Dunnett SB, Robbins TW. 1991. Dissociable roles of the ventral, medial and lateral striatum on the acquisition and performance of a complex visual stimulus response habit. *Behav. Brain Res.* 45:147–61

Reber PJ, Knowlton BJ, Squire LR. 1996. Dissociable properties of memory systems: differences in the flexibility of declarative and nondeclarative knowledge. *Behav. Neurosci.* 110:859–69

Reber PJ, Squire LR. 1994. Parallel brain systems for learning with and without awareness. *Learn. Mem.* 1:217–29

Restle F. 1957. Discrimination of cues in mazes: a resolution of the place vs. response controversy. *Psychol. Rev.* 64:217–28

Ritchie BF, Aeschliman B, Pierce P. 1950. Studies in spatial learning: VIII. Place performance and acquisition of place dispositions. *J. Comp. Physiol. Psychol.* 43:73–85

Robbins TW, Brown VJ. 1990. The role of the striatum in the mental chronometry of action: a theoretical review. *Rev. Neurosci.* 2:181–213

Rolls ET, Thorpe SJ, Maddison SP. 1983. Responses of striatal neurons in the behaving monkey. 1. Head of the caudate nucleus. *Behav. Brain Res.* 7:179–210

Rosvold HE. 1968. The prefrontal cortex and caudate nucleus: a system for effecting correction in response mechanisms. In *Mind as Tissue*, ed. C Rupp, pp. 21–38 New York: Harper Row

Rosvold HE. 1972. The frontal lobe system: cortical-subcortical interrelationships. *Acta Neurobiol. Exp.* 32:439–60

Sage JR, Knowlton BJ. 2000. Effects of US devaluation on win-stay and win-shift radial arm maze performance in rats. *Behav. Neurosci.* 114:295–306

Sakamoto T, Okaichi H. 2001. Use of win-stay and win-shift strategies in place and cue tasks by medial caudate putamen (MCPu) lesioned rats. *Neurobiol. Learn. Mem.* 76:192–208

Salado-Castillo R, Diaz del Guante MA, Alvarado R, Quirarte GL, Prado-Alcala RA. 1996. Effects of regional GABAergic blockade of the striatum on memory consolidation. *Neurobiol. Learn. Mem.* 66:102–8

Salinas JA, White NM. 1998. Contributions of the hippocampus, amygdala, and dorsal striatum to the response elicited by reward reduction. *Behav. Neurosci.* 112:812–26

Schroeder JP, Wingard JC, Packard MG. 2002. Post-training reversible inactivation of hippocampus reveals interference between memory systems. *Hippocampus* 12:280–84

Schultz W, Dayan P, Montague PR. 1997. A neural substrate of prediction and reward. *Science* 275:1593–99

Schultz W. 2000. Multiple reward signals in the brain. *Nat. Rev. Neurosci.* 1:199–208

Setlow B. 1997. The nucleus accumbens and learning and memory. *J. Neurosci. Res.* 49:515–21

Sherry DF, Schacter DL. 1987. The evolution

of multiple memory systems. *Psychol. Rev.* 94:439–54

Smith J, Siegert RJ, McDowall J. 2001. Preserved implicit learning on both the serial reaction time task and artificial grammar in patients with Parkinson's disease. *Brain Cogn.* 45:378–91

Sommer M, Grafman J, Clark K, Hallett M. 1999. Learning in Parkinson's disease: eyeblink conditioning, declarative learning and procedural learning. *J. Neurol. Neurosurg. Psychiatry* 67:27–34

Sutton MA, Beninger RJ. 1999. Psychopharmacology of conditioned reward: evidence for a rewarding signal at D1-like receptors. *Psychopharmacology* 144:95–110

Teng E, Stefanacci L, Squire LR, Zola SM. 2000. Contrasting effects on discrimination learning after hippocampal lesions and conjoint hippocampal-caudate lesions in monkeys. *J. Neurosci.* 20:3853–63

Thompson WG, Guilford MO, Hicks LH. 1980. Effects of caudate and cortical lesions on place and response learning in rats. *Physiol. Psychol.* 8:473–79

Thorndike EL. 1933. A proof of the law of effect. *Science* 77:173–75

Tolman EC. 1932. *Purposive Behavior in Animals and Men.* New York: Appleton-Century Crofts

Veening JG, Cornelissen FM, Lieven JM. 1980. The topical organization of the afferents to the caudatoputamen of the rat. A horseradish peroxidase study. *Neuroscience* 5:1253–68

Viaud MD, White NM. 1989. Dissociation of visual and olfactory conditioning in the neostriatum of rats. *Behav. Brain Res.* 32:31–42

Vogt C. 1911. Quelques considerations generales sur le syndrome du corps strie. *J. Psychol. Neurol. (Leipzig)* 18:479–88

Whishaw IQ, Mittlemann G, Bunch ST, Dunnett SB. 1987. Impairments in the acquisition, retention, and selection of spatial navigation strategies after medial caudate-putamen lesions in rats. *Behav. Brain Res.* 24:125–38

White NM, Major R. 1978. Effect of pimozide on the improvement in learning produced by self-stimulation and water reinforcement. *Pharm. Biochem. Behav.* 8:565–71

White NM. 1988. Effect of nigrostriatal dopamine depletion on the post-training, memory improving action of amphetamine. *Life Sci.* 43:7–12

White NM. 1989a. A functional hypothesis concerning the striatal matrix and patches: mediation of S-R memory and reward. *Life Sci.* 45:1943–57

White NM. 1989b. Reward or reinforcement: What's the difference? *Neurosci. Biobehav. Rev.* 13:181–86

White NM. 1997. Mnemonic functions of the basal ganglia. *Curr. Opin. Neurobiol.* 7:164–69

White NM, Viaud M, Packard MG. 1994. Dopaminergic-cholinergic function in neostriatal memory function: role of nigro-striatal terminals. In *Strategies for Studying Brain Disorders, Vol. 2, Schizophrenia, Movement Disorders, and Age-Related Cognitive Disorders,* ed. T Palomo, T Archer, R Beninger, pp. 299–312, Madrid: Editorial Complutense

Wickens JR. 1990. Striatal dopamine in motor activation and reward-mediated learning. Steps towards a unifying model. *J. Neural Transm.* 80:9–31

Wiener SI. 1993. Spatial and behavioral correlates of striatal neurons in rats performing a self-initiated navigation task. *J. Neurosci.* 13:3802–17

Willingham DB, Koroshetz WJ. 1993. Evidence for dissociable motor skills in Huntington's disease patients. *Psychobiology* 21:173–82

Willingham DB, Koroshetz WJ, Peterson EW. 1996. Motor skills have diverse neural bases: spared and impaired skill acquisition in Huntington's disease. *Neuropsychologia* 10:315–21

Willingham DB, Wells LA, Farrell JM, Stemwedel ME. 2000. Implicit motor sequence learning is represented in response locations. *Mem. Cogn.* 28:366–75

Willis T. 1664. *Cerebri Anatome, cui Accessit Nervorum Descriptio et Usus.* London: Martin & Allestry

Wilson SAK. 1912. Progressive lenticular degeneration: a familiar nervous disease associated with cirrhosis of the liver. *Brain* 34:295–509

Wilson SAK. 1914. An experimental research into the anatomy of the corpus striatum. *Brain* 36:427–92

Winocur G. 1974. Functional dissociation within the caudate nucleus of rats. *J. Comp. Physiol. Psychol.* 86:432–39

Winocur G, Estes G. 1998. Prefrontal cortex and caudate nucleus in conditional associative learning. *Behav. Neurosci.* 112:89–101

Winocur G, Mills JA. 1969. Effects of caudate lesions on avoidance behavior in rats. *J. Comp. Physiol. Psychol.* 65:552–57

Wise SP, Murray EA, Gerfen CR. 1996. The frontal cortex-basal ganglia system in primates. *Crit. Rev. Neurobiol.* 10:317–56

Zis AP, Fibiger HC, Phillips AG. 1974. Reversal by l-dopa of impaired learning due to destruction of the dopaminergic nigro-striatal projection. *Science* 185:960–63

Zubin J, Barrera SE. 1941. Effect of electric convulsive therapy on memory. *Proc. Soc. Exp. Biol. Med.* 48:596–97

SUBJECT INDEX

A

AIDS
 neurogenesis
  macrophages/monocytes'
   role in, 537–52
Alcoholism, 1, 3
AMPA receptors, 103–19
 delivery to synapses
  LTP and, 113–17
  role of AMPA receptor
   phosphorylation in,
   116–17
 endocytosis and LTD,
  107–13
 molecular interactions of,
  104–7
AMPA receptor trafficking,
 103–19
 general mechanisms,
  117–19
 LTD and, 103, 107–13
 LTP and, 103, 113–17
  AMPA receptor
   phosphorylation and,
   116–17
 synaptic plasticity and,
  103–19
Amytrophic lateral schlerosis,
 6
Appetite
 hypocretins and, 283,
  302–4
Arm movement
 posterior parietal cortex
  and, 189–215
  compensation for eye
   movements,
   202–3
  gain fields, 203–6
  intentional maps and,
   189–215

movement decisions,
 208–9
neuronal prosthetics and,
 211–14
reach planning in
 eye-centered
  coordinates, 201–2
saccade planning in
 eye-centered
  coordinates, 200–1
Arousal
 hypocretins and, 283,
  298–302
Asperger's syndrome, 20–25
Astrocytes
 Notch signaling pathway
  and, 480–81
Attention, 192–94, 209–10
 posterior parital cortex and,
  209–10
 versus intention, 192–94
Auditory system
 development, 51–86
 axon development, 69–71
 cochlear ganglion neuron
  development, 52–68
 development of central
  projections in, 68–86
 ganglion cell axons and
  cochlear nucleus neurons
   contacts between, 71–74
  innervation of inner ear,
   52–68
 trophic regulation of CNS
  targets and, 68–86
Autism, 1–7, 20–25
 chromosomal aberrations,
  21
 genetic linkages, 21–25
Autonomic nervous system
 autonomic pattern

generation, 433, 447–61
 functional organization
  of neuronal pools,
  453–58
 integration across
  multiple levels, 458–60
 organotopic organization
  of neuronal pools,
  450–53
 pattern of autonomic
  response, 448–50
central, 433–61
peripheral, 433
See also Central autonomic
 nervous system
Axon pathfinding, 251–73
 LIM code and, 251, 264–73

B

Basal ganglia
 learning and memory
  functions, 563–85
  functional heterogeneity,
   571–72
  human neuroimaging
   studies, 578
  lesion studies, 568–71
  neuroanatomical and
   neurochemical aspects,
   565–67
  posttraining drug
   studies, 572–74
  task switching and,
   583–84
  theory, 567–68
Bipolar disorder (BPD), 1–7,
 15–20, 25
 genetic linkages, 16–20
Blood brain barrier
 autoimmune attack and,
  493–94

595

C

Cell-fate specification, 471–87
  Notch signaling pathway and, 471–87
    astrocytes, 480–81
    Müller glia, 477–78, 480
    oligodentrocytes, 481–82
    radial glia, 478–80
    Schwann cells, 482–83
    stem cells, 483–87
Cell lines
  astrocyte, 389
  embryonic stem cells and teratocarcinomas, 394–98
    developmental pathway to neural cells, 396–98
    human, 398
    neural differentiation of, 395–96
  neural, 381–402
  neurospheres, 389–94
    directed differentiation of, 392–93
    monolayer cultures of progenitors and, 391
    transdifferentiation of, 393–94
  oncogene expression–derived, 383, 385–86
    C117 cells, 385–86
    HiB5 cells, 385–86
  pluripotent, 382–83
  PNS-derived lines, 387–89
  RN33B cells, 387
  ST14A cells, 387
  tumor-derived, 383–85
Central autonomic nervous system, 433–61
  autonomic pattern generation, 433, 447–61
    functional organization of neuronal pools and, 453–58
    integration across multiple levels, 458–60
    organotopic organization of neuronal pools and, 450–53
    patterns of autonomic response, 448–50
  visceral sensory information to conscious perception and, 433–46, 460–61
    cranial nerve visceral system, 437–43
    spinal visceral afferent system, 443–46
Cerebral cortex
  dendrite development in, 127–44
Channelopathies, 6
Cochlea
  afferent activation of, 59–61
  cochlear ganglion cells, 51–68
  outer and inner hair cells of, 52, 66–68
Cochlear ganglion, 51–68
  development of, 52–68
    afferent innervation types and typography, 59–61
    fiber growth mechanisms, 62–63
    ganglion neuron survival, 64–66
    neuron proliferation, 54–57
    neurotrophins and neurotrophin receptors and, 64–66
    pathfinding in mutant mammals, 63
    polarity, 58–59
    proneural genes, 57–58
    synaptogenesis between hair cells and afferent fibers, 66–68
    Type I and Type II spiral sensory neurons, 52
Cochlear nuclei (CN), 51–53, 71–86
  basic organization, 53
  ganglion cell innervation and, 51, 71–86
  influence of cochlear nerve on development of, 78–86
  synaptic activity and, 51, 71–86
Cranial nerve visceral sensory system, 437–43

D

Dementia
  HIV-associated, 537–52
Dendrite development
  dendritic stability and microtubules and, 142–43
  molecular control of cortical, 127–44
    mRNAs and proteins, 139–40
    regulation of actin dynamics, 140–42
    Rho GTPases and, 140–42
  regulation of apical dendrite orientation, 129–31
    semaphorin signaling and, 129–31
  regulation of dendritic growth and branching
    neurotrophic factors and, 131–33
    Notch signaling and, 133–34, 144
  regulation of dendritic spine development, 134–39
    activity-dependent plasticity, 137–38
    Golgi studies, 134–36
    mechanisms of spine motility, 138–39

mental retardation and, 139
time-lapse studies, 136–37
Dendrite morphology, 127–29
Dendritic spines
Down sydrome and, 139
fragile X syndrome and, 326–29
Depression
major depressive disorder, 2, 20
narcolepsy and, 302
See also Bipolar disorder
Down syndrome
dendritic spine defects and, 139
Drosophila
gliogenesis in, 471–87
CNS, 473–75
Notch signaling pathway and, 471–76
PNS, 474–76

E

Ear
See Auditory system development; Cochlea
Effective connectivity, 221, 225–27
Embryology
auditory system development, 51–86
Emotion
impact on autonomic function, 461
role of visceral sensory system, 461
See also Visceral sensory information
Encephalitis
HIV/SIV, 548–49
lentiviral, 549–50
Endocytosis
AMPA receptor trafficking and, 107–13

interacting proteins, 109–13
intracellular signaling pathways, 108–9
Estrogen
synaptogenesis and, 507, 523–24
Eye movement, 189–215
posterior parietal cortex and, 189–215
compensation for eye movements, 202–3
movement decisions, 208–9
neural prosthetics and, 211–14
saccade planning, 200–1

F

Fragile X syndrome, 6, 20, 22, 139, 315–30
cause of cognitive impairment in, 321–29
dendritic spine defects and, 139
FMR1 expression and, 317–19
prevalence, 321
repeat expansion in, 319–21
treatment, 329–30
Functional integration, 222–23, 225–48
Functional magnetic resonance (fMRI)
advances in paradigm design, 154–55
basic principles, 153–54
language studies, 151–83
inferior frontal lobe, 151, 156–69, 182–83
right hemisphere, 151, 174–83
temporal lobe, 151, 169–74, 182–83
See also Functional neuroimaging

Functional neuroimaging, 221–48
distributed circuitry and, 221–48
effective connectivity and, 221, 225–27
functional integration and, 222–23, 225–48
functional specializaton and, 222–25, 242–48
information theory and, 221, 227–33
predictive coding and, 221, 227–30, 233–48
See also Functional magnetic resonance
Functional specialization, 222–25, 242–48

G

Genetics
basic terms, 40–42
See also Human genome project
Glia
astroglia, 480–81
Müller glia, 477–78, 480
Notch signaling pathway and, 471–87
oligodendroglia, 480–81
radial glia, 471, 478–80
Gliogenesis
Notch signaling pathway and, 471–87
atrocytes, 480–81
Drosophila, 471–76
Müller glia, 477–78, 480
oligodendrocytes, 480–81
radial glia, 471, 478–80
Schwann cells, 482–83
vertebrate, 471–73, 476–87

H

Haplotype mapping, 1, 26–30
genetic basis of mental

illness and, 1, 26–30
Hippocampus, 507, 528
Horror autotoxicus, 492–93
Human genome project, 1–42
 genetic basis of mental
  illness and, 1–42
  autism, 1–7, 20–25
  bipolar disorder, 1–7,
   15–20, 25
  schizophrenia, 1–15, 25
 history of, 30–40
Human immunodeficiency
 virus (HIV), 537–52
 brain macrophages/
  monocytes in
  neuropathogenesis,
  537–49
  HIV/SIV encephalitis,
   548–49
  infection of CNS,
   540–44
  microphage tropism in
   CNS, 544–45
 HIV-associated dementia,
  537–52
Huntington's disease, 6
Hypocretins, 283–307
 appetite and, 283, 302–4
 arousal and, 298–302
 deficiency in animals
  narcolepsy and, 294–95
 deficiency in humans
  narcolepsy and, 296–97
 depression and, 302
 discovery of, 286
 gene and peptide structure,
  287
 neuroendocrinology of,
  304–5
 neurotransmission, 287–89
 sleep regulation and,
  283–307
Hypothalamus
 hypocretins and, 283–307
 reproductive behaviors
  and, 509–17
 parasympathetic and

sympathetic
 preganglionic neuron
 innervation and, 451–61
sleep regulation and,
 383–307

I

Inferior frontal gyrus, 151,
 156–69
 nonlanguage tasks, 166–68
 role in semantic
  processing, 151, 156–69
 phonology, 165–66
 priming effects, 157–59
 selective attention to
  meaning, 159–61
 syntax, 162–64
Information theory, 221,
 227–33, 239–48
Intention
 default plans, 194–98
 dynamic evolution of
  intention-related activity,
  198–99
 multisensory integration
  and coordinate
  transformations, 200–8
 versus attention, 192–94
Intentional maps, 189–215
 posterior parietal cortex
  and, 189–215
 dynamic evolution of
  intention-related
  activity, 198–99
 movement decisions,
  208–9
 multisensory integration
  and coordinate
  transformations,
  200–8

L

Language
 See Semantic processing
Learning
 posterior parietal cortex
  and, 211–14

neural prosthetics and,
 213
Learning and memory
 basal ganglia functions in,
  563–85
 functional heterogeneity,
  571–72
 human neuroimaging
  studies, 578
 lesion studies, 568–71
 motor skill and
  perceptual-motor
  learning, 576–78
 posttraining studies,
  572–74
 probabalistic
  classification, 575–76
 task switching and,
  583–84
 theory, 567–68
 See also Memory
LIM homeodomain
 code, 251, 264–73
 transcription factors, 251,
  264–73
Linkage analysis
 autism and, 21–25
 bipolar disorder and, 16–20
 schizophrenia and, 10–15
Logic
 right hemisphere and,
  177–78, 181
Long-term depression (LTD)
 AMPA receptor trafficking
  and, 103, 107–13
 endocytosis, 107–8
 interacting proteins and,
  109–13
 intracellular signaling
  pathways and, 108–9
 synaptic delivery and,
  115–17
Long-term potentiation (LTP)
 AMPA receptor trafficking
  and, 103–4, 113–17
 electrophysiological
  tagging to montitor

synaptic delivery, 114–15
optical detection of, 113–14
subcellular steady-state distribution of AMPA receptors, 113
synaptic delivery of endogenous receptors, 115–16

M
Manic depression
See Bipolar disorder
Major depressive disorder, 2, 20
Medial preoptic nucleus
sexual differentiation and, 510
Memory
LTD
AMPA receptor trafficking and, 103, 107–13
LTP
AMPA receptor trafficking and, 103–4, 113–17
See also Learning and Memory
Mendelian inheritance
defined, 40
Mental illnesses
disability adjusted life year, 2–3
genetic basis of, 1–42
autism, 1–7, 20–25
bipolar disorder, 1–7, 15–20, 25
schizophrenia, 1–15, 25, 409–27
lifetime prevalence
See also Neuropsychiatric disorders
Mental retardation, 315, 321–29
fragile X syndrome and, 321–29

Metaphor
right hemisphere and comprehension of meaning, 174–76, 181
Motor neuron
development, 251–73
LIM homeodomain transcription factors and, 251, 264–73
subtype specification, 251–73
Multiple sclerosis (MS), 491–502
autoimmune attack
blood brain barrier and, 493–94
effector molecules, 495
horror autotoxicus, 492–94
myelin sheath, 491–95
osteopontin and, 497–501
T and B cell response specificity, 494–95
genes transcribed in MS brain plaques, 497–501
new therapeutic approaches, 502
Myelin sheath
autoimmunity, 491–95
See also Multiple sclerosis

N
Narcolepsy, 283–302
canine, 293–95
depression and, 302
diagnosis and treatment, 292–93
genetics, 291–92
hypocretin deficiency and, 283, 294–307
canine, 294–95
human, 296–97
symptoms, 290–91
Neural cell lines, 381–402
astrocyte, 389
embryonic stem cells and

teratocarcinomas, 394–98
developmental pathway to neural cells, 396–98
human, 398
neural differentiation of, 395–96
neurospheres, 389–94
directed differentiation of, 392–93
monolayer cultures of progenitors and, 391
transdifferentiation of, 393–94
oncogene expression-derived, 383, 385–86
C117 cells, 385–86
HiB5 cells, 385–86
PNS-derived lines, 387–89
RN33B cells, 387
ST14A cells, 387
tumor-derived, 383–85
See also Neural stem cells
Neural prosthetics, 189, 211–14
posterior parietal cortex and, 211–14
Neural stem cells, 381–402
available cell lines, 381–402
defined, 381
transplantation research and, 401–2
See also Neural cell lines
Neuropsychiatric disorders, 409–27
schizophrenia
as a neurodevelopmental disorder, 409–27
neurodevelopmental models of, 418–20, 423–27
See also Depression; Mental illness
Neurotrophins
regulation of dendritic growth and branching

and, 127, 131–33
Nitric oxide
  myelin damage and, 495–96
Notch
  regulation of dendritic growth and branching and, 133–34, 144
Notch signaling pathway, 471–87
  cell-fate specification and, 471–87
  gliogenesis and, 471–87
    atrocytes, 480–81
    Drosophila, 471–76
    Müller glia, 477–78, 480
    oligodendrocytes, 480–81
    radial glia, 471, 478–80
    Schwann cells, 482–83
    vertebrate, 471–73, 476–87

O

Occlusion
  visual processing, 350–63
  motion processing, 360–63
Oligodendrocyte precursor cells, 394
Orexins
  See Hypocretins
Osteopontin
  multiple sclerosis progression and, 491, 497, 500–1

P

Parasympathetic nervous system, 433, 437–43
Parkinson's disease, 6
Perception
  contextual influences on, 339–73
  Gestalt psychology and, 341–42
  neurobiology of, 342–73
  neurophysiology of vision and, 342
  reductionist psychology of, 340–41
Periaqueductal gray matter, 454–55, 459
Phenotype
  defined, 40
Phonology
  inferior frontal gyrus and, 165–66
Polygenic
  neuropsychological disorders, 409–27
Posterior parietal cortex, 189–215
  intentional maps and, 189–215
    arm movement, 189–215
    dynamic evolution of intention-related activity, 198–99
    eye movement, 189–215
    learning and adaptation and, 211–14
    movement decisions, 208–9
    multisensory integration and coordinate transformations, 200–8
    neural prosthetics, 189, 211–14
    sensory motor integration, 189–215
Predictive coding, 221, 227–30, 233–48
Prosody, 179–81
  right hemisphere and, 179–81
Psychosis, 409–27
  See also Neuropsychiatric disorders, Schizophrenia

R

Retrovirus-associated CNS disease, 537–52
  See also Human immunodeficiency virus (HIV)
Rett syndrome (also Rett's syndrome), 20–25, 139
  dendritic spine defects and, 139
  MeCP2 mutations and, 25
Rho GTPases
  regulation of actin dynamics and, 127, 140–42
RNA binding
  fragile X syndrome and, 315
Rostral medullary raphe, 454

S

Schizophrenia, 1–15, 25, 409–27
  associated illnesses, 7–8
  candidate genes, 9–10
  concordance rate among monozygotic twins, 8
  dorsolateral prefrontal cortex and, 15
  genetic susceptibility, 1–15
  linkage studies, 10–15
  as a neurodevelopmental disorder, 411–27
    genetic and environmental factors, 409–11
  neurodevelopmental models of, 418–20, 423–27
    cascade effects of altered development, 425–27
    gene-environmental interactions, 424–27
    genetic predisposition, 424–27
    steady-state outcome, 425–27
  pathogenesis, 421–23
Schwann cells
  Notch signaling pathway and, 482–83

Semantic processing, 151, 156–69
 functional MRI studies, 151–83
 organization of categories and objects in temporal lobe, 151, 169–74, 182–83
  body parts, 172
  living vs. nonliving, 169–72
  word-specific effects, 172–73
 role of inferior frontal gyrus in, 151, 156–69
  phonology, 165–66
  priming effects, 157–59
  selective attention to meaning, 159–61
  syntax, 162–64
 role of right hemisphere in comprehension of meaning, 151, 174–83
  cohesion and repair, 178–79, 181
  linguistic context, 176–77, 181
  metaphor, 174–76, 181
  prosody, 179–81
  reason and logic, 177–78, 181
Semantic system
 functional MRI studies, 151–83
 organization of categories and objects in temporal lobe, 151, 159–74, 182–83
 role of left inferior frontal lobe, 151, 156–69, 182–83
 role of right hemisphere in comprehension of meaning, 151, 174–83
Semaphorins
 regulation of apical dendrite orientation and, 129–31
Sensory-motor integration
 posterior parietal cortex and, 189–215
Sex steroids, 507–29
 axonal guidance and, 507, 524–26, 528–29
 synaptogenesis and, 507, 523–24, 528–29
Sexual differentiation, 507–29
 sex steroid signaling during development, 526–28
 sexually dimorphic forebrain pathways, 509–26
  development of, 517–26
Simian immunodeficiency virus (SIV)
 brain macrophages/monocytes in neuropathogenesis of, 537–49
  infection of CNS, 540–44
  encephalitis, 548–49
Single nucleotide polymorphisms, 1, 26–30
Sleep disorders
 hypocretins and, 283–98
 narcolepsy, 283–98
Sleep regulation
 hypcretins and, 283–307
Spatial representation
 intentional maps and, 189–215
Speech
 See Semantic processing; Semantic system
Spinal visceral afferent system, 443–46
Spinocerebellar ataxias, 6
Stem cells
 cell-fate specification
  Notch signaling pathway and, 483–87
Sympathetic nervous system, 433, 438–39, 443–61
Synaptic plasticity
 AMPA receptor trafficking and, 103–19
  LTD and, 103, 107–13
  LTP and, 103–4, 113–17
  synaptic delivery and, 115–17
 fragile X syndrome and, 315, 327–29
Syntax
 inferior frontal gyrus and, 162–64, 166

T

Transcription factors, 251–73
 neuronal identity and
  apterous, 259, 261
  bHLH, 255–57
  Isl1, 259, 262
  LIM homeodomain, 251, 264–73
  lin-11, 259–61
  mec-3, 259, 261
Trinucleotide repeat
 fragile X syndrome and, 315–30

V

Ventrolateral medulla, 453–54
Vertebrate gliogenesis
 Müller glia, 477
 Notch signaling pathway and, 471–73, 476–87
Visceral sensory information, 433–46
 to conscious perception, 433–46
 cranial nerve visceral system, 437–43
 spinal visceral afferent system, 443–46
Visual cortex
 contextual influences on

602 SUBJECT INDEX

perception and, 339–73
Visual processing
  boundary assignment, 347
  contextual influences on, 339–73
    color appearance, 369–72
    neuronal correlates, 369–73
  depth-ordering, 339, 347
  figure-ground interpretation, 339, 347, 350–52
  filling in, 339, 347–68
  blind spot, 347–49
  modal and amodal completion, 352–60
  surface occlusion, 350–63
  visual gaps, 349–50
Volterra kernals, 225–27

Cumulative Indexes

CONTRIBUTING AUTHORS, VOLUMES 16–25

A
Abarbanel HDI, 24:263–97
Abraham WC, 19:437–62
Allada R, 24:1091–119
Allendoerfer KL, 17:185–218
Albright TD, 25:339–79
Amara SG, 16:73–93
Andersen RA, 20:299–326; 25:189–220
Anderson DJ, 16:129–58
Angleson JK, 22:1–10
Armstrong RC, 16:17–29
Arnold AP, 20:455–77
Ashley CT Jr, 18:77–99

B
Bal T, 20:185–215
Baltuch G, 22:219–40
Bandtlow CE, 16:565–95
Banker G, 17:267–310
Barbe MF, 20:1–24
Barde Y-A, 19:289–317
Bargmann CI, 16:47–71; 21:279–308
Barres BA, 23:579–612
Basbaum AI, 23:777–811
Bate M, 19:545–75
Baylor DA, 24:779–805
Bear MF, 19:437–62
Bernard C, 25:491–505
Bernard CCA, 17:247–65
Best PJ, 24:459–86
Betz WJ, 22:1–10
Bi G-Q, 24:139–66
Biel M, 17:399–418
Bonhoeffer T, 24:1071–89
Bookheimer S, 25:151–88
Bothwell M, 18:223–53
Bottjer SW, 20:455–77

Bouchard TJ Jr, 21:1–24
Boussaoud D, 20:25–42
Bowe MA, 18:443–62
Bowers BJ, 24:845–67
Bradley DC, 20:299–326
Brainard MS, 18:19–43
Broadie K, 19:545–75
Buck LB, 19:517–44
Bullock TH, 16:1–15
Buneo CA, 25:189–220
Buonomano D, 21:149–86
Burns ME, 24:779–805

C
Callaerts P, 20:479–528
Callaway EM, 21:47–74
Caminiti R, 20:25–42
Cannon SC, 19:141–64
Caron MG, 16:299–321
Carr CE, 16:223–43
Caterina MJ, 24:487–517
Cawthon R, 16:183–205
Chemelli RM, 24:429–58
Chen C, 20:157–84
Chen K, 22:197–217
Chiba A, 19:545–75
Chiu C-YP, 16:159–82
Choi DW, 21:347–75
Christie BR, 19:165–86
Ciaranello AL, 18:101–28
Ciaranello RD, 18:101–28
Clapham DE, 17:441–64
Cleveland DW, 19:187–217
Cochilla AJ, 22:1–10
Cohen JD, 24:167–202
Colamarino SA, 18:497–529
Colbert CM, 19:165–86
Colby CL, 22:319–50
Collinge J, 24:519–50

Conlon P, 25:491–505
Connor JA, 18:319–57
Corey DP, 20:563–90
Cowan WM, 23:343–91; 24:551–600; 25:1–50
Craig AM, 17:267–310
Cumming BG, 24:203–38
Curran T, 24:1005–39

D
Dacey DM, 23:743–75
Darnell RB, 24:239–62
Dasen JS, 24:327–55
Daw NW, 16:207–22
Deadwyler SA, 20:217–44
DeAngelis GC, 24:203–38
DeArmond SJ, 17:311–39
deCharms RC, 23:613–47
Deckwerth TL, 16:31–46
DeLuca NA, 19:265–87
DePaulo JR, 20:351–69
Desimone R, 18:193–222
Dickinson A, 23:473–500
Dijkhuizen P, 25:127–49
Donoghue JP, 23:393–415
Douglas R, 18:255–81
Doupe AJ, 22:567–631
Dreyfuss G, 20:269–98
Dubnau J, 21:407–44
du Lac S, 18:409–41
Duncan J, 18:193–222
Dunwiddie TV, 24:31–55

E
Eagleson KL, 20:1–24
Eatock RA, 23:285–314
Edwards RH, 20:125–56
Eisen JS, 17:1–30
Elfvin L-G, 16:471–507

603

Emeson RB, 19:27–52
Emery P, 24:1091–119

F
Fallon JR, 18:443–62
Ferster D, 23:441–71
Fiez JA, 16:509–30
Fink DJ, 19:265–87
Fischbach GD, 20:425–54
Fischbeck KH, 19:79–107
Fischer U, 20:269–98
Fishell G, 25:471–90
Fisher LJ, 18:159–92
Fitch RH, 20:327–49
Flanagan JG, 21:309–45
Flockerzi V, 17:399–418
Fox K, 16:207–22
Francis NJ, 22:541–66
Frankland PW, 21:127–48
Friedrich RW, 24:263–97
Friston K, 25:221–50
Fritzsch B, 25:51–101
Froehner SC, 16:347–68
Fuchs AF, 24:981–1004

G
Gage FH, 18:159–92
Gaiano N, 25:471–90
Garbers DL, 23:417–39
García-Añoveros J, 20:563–90
Gehring WJ, 20:479–528
Geppert M, 21:75–95
Ghosh A, 25:127–49
Gibson AD, 23:417–39
Gingrich JA, 16:299–321
Gingrich JR, 21:377–405
Glass JD, 19:1–26
Glorioso JC, 19:265–87
Gluck MA, 16:667–706
Goedert M, 24:1121–59
Goins WF, 19:265–87
Goldberg JL, 23:579–612
Goldberg ME, 22:319–50
Goldstein LSB, 23:39–72
González-Scarano F, 22:219–40

Goodman CS, 19:341–77
Gottlieb DI, 25:381–407
Granger R, 16:667–706
Gray CM, 18:555–86
Greenberg ME, 19:463–89
Griffin JW, 21:187–226
Grimwood PD, 23:649–711
Grote E, 16:95–127
Gudermann T, 20:395–423

H
Halder G, 20:479–528
Hampson RE, 20:217–44
Harris KM, 17:341–71
Harter DH, 23:343–91
Hatten ME, 18:385–408; 22:511–39
Hatton GI, 20:371–93
Hawkins RD, 16:625–65
Heinemann S, 17:31–108
Heintz N, 18:385–408
Hemmati-Brivanlou A, 20:43–60
Hendry SHC, 23:127–53
Herrup K, 20:61–90
Hickey WF, 25:537–62
Highstein SM, 17:465–88
Hildebrand JG, 20:591–629
Hlavin ML, 21:97–125
Ho TW, 21:187–226
Hofmann F, 17:399–418
Hökfelt T, 16:471–507
Hollmann M, 17:31–108
Honarpour N, 23:73–87
Howe CL, 24:1217–81
Huang EJ, 24:677–736
Hyman SE, 25:1–50

I
Ip NY, 19:491–515

J
Jahn R, 17:216–46
Jamison KR, 20:351–69
Jan L, 23:531–56
Jan LY, 20:91–123

Jan Y-N, 20:91–123; 23:531–56
Jessell TM, 22:261–94
Johnson EM Jr, 16:31–46
Johnson PB, 20:25–42
Johnson RT, 19:1–26
Johnston D, 19:165–86
Jones EG, 22:49–103; 23:1–37
Joyner AL, 24:869–96
Julius D, 24:487–517

K
Kamiguchi H, 21:97–125
Kandel ER, 16:625–65; 23:343–91
Kapfhammer JP, 16:565–95
Kaplan JM, 21:279–308
Karschin A, 23:89–125
Kastner S, 23:315–41
Kater SB, 17:341–71
Katz LC, 22:295–318
Kauer JS, 24:963–79
Kelly RB, 16:95–127
Keshishian H, 19:545–75
Keynes R, 17:109–32
Kida S, 21:127–48
Kimmel CB, 16:707–32
King DP, 23:713–42
Knowlton BJ, 25:563–93
Knudsen EI, 18:19–43
Kogan JH, 21:127–48
Koh JY, 21:347–75
Kopnisky KL, 25:1–50
Korsmeyer SJ, 20:245–67
Krumlauf R, 17:109–32
Krupa DJ, 17:519–49
Kuemerle B, 20:61–90
Kuhar MJ, 16:73–93
Kuhl PK, 22:567–631

L
Landis DMD, 17:133–51
Landis SC, 22:541–66
Laurent G, 24:263–97
LeDoux JE, 23:155–84
Lee KJ, 22:261–94

Lee MK, 19:187–217
Lee VM-Y, 24:1121–59
Lemke G, 24:87–105
Lemmon V, 21:97–125
Lester HA, 23:89–125
Levitt P, 20:1–24; 25:409–32
Lewin GR, 19:289–317
Lewis DA, 25:409–32
Lichtman JW, 22:389–442
Linden DJ, 18:319–57
Lindh B, 16:471–507
Lisberger SG, 18:409–41
Liu A, 24:869–96
Liu Y, 20:125–56
Lo DC, 22:295–318
Logothetis NK, 19:577–621
Lu B, 23:531–56

M

MacDermott AB, 22:443–85
Macdonald RL, 17:569–602
MacKinnon DF, 20:351–69
Madison DV, 17:153–83
Magee JC, 19:165–86
Mahowald M, 18:255–81
Malenka RC, 23:185–215; 25:103–26
Malicki DM, 18:283–317
Malinow R, 25:103–26
Mandel G, 16:323–45
Marder E, 21:25–45
Maren S, 24:897–931
Martin R, 25:491–505
Martin SJ, 23:649–711
Martinez S, 21:445–77
Masino SA, 24:31–55
Mason P, 24:737–77
Masland RH, 23:249–84
Matthews G, 19:219–33
Maunsell JHR, 16:369–402
McAllister AK, 22:295–318
McCormick DA, 20:185–215
McEwen BS, 22:105–22
McGue M, 21:1–24
McKhann GM, 21:187–226
McKinnon D, 16:323–45
McNamara JO, 22:175–95

Mead C, 18:255–81
Meaney MJ, 24:1161–92
Melton D, 20:43–60
Menzel R, 19:379–404
Merigan WH, 16:369–402
Merry DE, 20:245–67
Merzenich MM, 21:149–86
Michael WM, 20:269–98
Mignot E, 25:283–313
Miller A, 17:247–65
Miller EK, 24:167–202
Miller S, 20:327–49
Miller KD, 23:441–71
Minai A, 24:459–86
Miyashita Y, 16:245–63
Mobley WC, 24:1217–81
Mogil JS, 23:777–811
Mombaerts P, 22:487–509
Montminy MR, 16:17–29
Morris RGM, 23:649–711
Moschovakis AK, 17:465–88
Mueller BK, 22:351–88
Müller U, 19:379–404
Musunuru K, 24:239–62

N

Nakielny S, 20:269–98
Nestler EJ, 18:463–95
Newsome WT, 21:227–77
Nicola SM, 23:185–215
Nijhawan D, 23:73–87
Nishizuka Y, 17:551–67

O

Ochsner KN, 16:159–82
O'Donnell WT, 25:315–38
Oksenberg JR, 17:247–65; 25:491–505
Olanow CW, 22:123–44
O'Leary DDM, 17:419–39
Olsen RW, 17:569–602
Olshausen BA, 24:1193–1215
Orr HT, 23:217–47

P

Packard MG, 25:563–93
Parker AJ, 21:227–77

Paulson HL, 19:79–107
Pearson KG, 16:265–97
Petersen SE, 16:509–30
Pfaff SL, 25:251–81
Polleux F, 25:127–49
Poo M-M, 24:139–66
Price DL, 21:479–505
Prusiner SB, 17:311–39
Puelles L, 21:445–77

Q

Quinn WG, 24:1283–309

R

Rabinovich MI, 24:263–97
Radcliffe RA, 24:845–67
Raisman G, 20:529–62
Ranganathan R, 18:283–317
Raviola E, 23:249–84
Ray J, 18:159–92
Raymond JL, 18:409–41
Read HL, 23:501–29
Reeke GN Jr, 16:597–623
Reichardt LF, 24:677–736
Reid RC, 23:127–53
Reyes A, 24:653–75
Rice DS, 24:1005–39
Ridd M, 22:197–217
Robinson FR, 24:981–1004
Roder J, 21:377–405
Role LW, 22:443–85
Romo R, 24:107–37
Rosbash M, 24:1091–119
Rosen KM, 20:425–54
Rosenfeld MG, 24:327–55
Roses AD, 19:53–77
Ross ME, 24:1041–70
Rubel EW, 25:51–101
Rubenstein JLR, 21:445–77
Rubin GM, 17:373–97
Rubin LL, 22:11–28

S

Sala C, 24:1–29
Salinas E, 24:107–37
Sanes JN, 23:393–415
Sanes JR, 22:389–442

Saper CB, 25:433–69
Sargent PB, 16:403–43
Schacter DL, 16:159–82
Schall JD, 22:241–59
Schlaggar BL, 17:419–39
Schöneberg T, 20:395–423
Schreiner CE, 23:501–29
Schultz G, 20:395–423
Schultz W, 23:473–500
Schuman EM, 17:153–83; 24:299–325
Schwab ME, 16:565–95
Scott MP, 24:385–428
Searle JR, 23:557–78
Segal RA, 19:463–89
Sejnowski TJ, 18:409–41
Self DW, 18:463–95
Selkoe DJ, 17:489–517
Selverston AI, 16:531–46
Shatz CJ, 17:185–218
Sheinberg DL, 19:577–621
Sheng M, 24:1–29
Shepherd GM, 20:591–629
Shih JC, 22:197–217
Shimamura K, 21:445–77
Shirasaki R, 25:251–81
Shooter EM, 24:601–29
Siegelbaum SA, 16:625–65; 19:235–63; 22:443–85
Silva AJ, 21:127–48
Simerly RB, 25:507–36
Simoncelli EP, 24:1193–215
Simpson L, 19:27–52
Singer W, 18:555–86
Sinton CM, 24:429–58
Sisodia SS, 21:479–505
Snipes GJ, 18:45–75
Snyder LH, 20:299–326
Sofroniew MV, 24:1217–81
Soriano P, 18:1–18
Sporns O, 16:597–623
Squire LR, 16:547–63
Staddon JM, 22:11–28
Stein PSG, 16:207–22

Steindler DA, 16:445–70
Steinman L, 17:247–65; 25:491–505
Steward O, 24:299–325
Stoner GR, 25:339–79
Stopfer M, 24:263–97
Strittmatter WJ, 19:53–77
Südhof TC, 17:219–46; 21:75–95; 24:933–62
Surmeier DJ, 23:185–215
Suter U, 18:45–75
Sutter ML, 23:501–29

T
Taheri S, 25:283–313
Takahashi JS, 18:531–53; 23:713–42; 24:1091–119
Tallal P, 20:327–49
Tanaka C, 17:551–67
Tanaka K, 19:109–39
Tanji J, 24:631–51
Tatton WG, 22:123–44
Tennissen AM, 24:807–43
Tessier-Lavigne M, 18:497–529
Thompson KG, 22:241–59
Thompson RF, 17:519–49
Tonegawa S, 20:157–84
Trojanowski JQ, 24:1121–59
Tully T, 21:407–44
Tuttle R, 17:419–39

U
Ungerleider LG, 23:315–41

V
Vanderhaeghen P, 21:309–45
Viskochil D, 16:183–205
Volkovskii A, 24:263–97

W
Waddell S, 24:1283–309
Walsh CA, 24:1041–70

Wandell BA, 22:145–73
Wang X, 23:73–87
Warren ST, 18:77–99; 25:315–38
Watts AG, 24:357–84
Wechsler-Reya R, 24:385–428
Wehner JM, 24:845–67
Weinberger NM, 18:129–58
White AM, 24:459–86
White FJ, 19:405–36
White J, 24:963–79
White R, 16:183–205
Whitford KL, 25:127–49
Whitney KD, 22:175–95
Williams KC, 25:537–62
Willie JT, 24:429–58
Wise RA, 19:319–40
Wise SP, 20:25–42
Wolpaw JR, 24:807–43
Wong ROL, 22:29–47

X
Xing J, 20:299–326

Y
Yamasaki M, 21:97–125
Yanagisawa M, 24:429–58
Yancopoulos GD, 19:491–515
Yang Z, 23:39–72
Yu L, 23:777–811
Yuste R, 24:1071–89

Z
Zador A, 23:613–47
Zagotta WN, 19:235–63
Zeitzer JM, 25:283–313
Zeki S, 24:57–86
Zipursky SL, 17:373–97
Zoghbi HY, 23:217–47
Zola-Morgan S, 16:547–63; 18:359–83
Zuker CS, 18:283–317

CHAPTER TITLES, VOLUMES 16–25

Affect and Emotion

Neurobiology of Pavlovian Fear Conditioning	S Maren	24:897–931

Auditory System

Adaptation in Hair Cells	RA Eatock	23:285–314
Modular Organization of Frequency Integration in Primary Auditory Cortex	CE Schreiner, HL Read, ML Sutter	23:501–29
Auditory System Development: Primary Auditory Neurons and Their Targets	EW Rubel, B Fritzsch	25:51–101

Autonomic Nervous System

The Chemical Neuroanatomy of Sympathetic Ganglia	L-G Elfvin, B Lindh, T Hökfelt	16:471–507
The Central Autonomic Nervous System: Conscious Visceral Perception and Autonomic Pattern Generation	CB Saper	25:433–69

Basal Ganglia

Dopaminergic Modulation of Neuronal Excitability in the Striatum	SM Nicola, DJ Surmeier, RC Malenka	23:185–215
Learning and Memory Functions of the Basal Ganglia	MG Packard, BJ Knowlton	25:563–93

Behavioral Neuroscience

Emotion Circuits in the Brain	JE LeDoux	23:155–84
Quantitative Genetics and Mouse Behavior	JM Wehner, RA Radcliffe, BJ Bowers	24:845–67

Cerebellum

The Role of the Cerebellum in Voluntary Eye Movements	FR Robinson, AF Fuchs	24:981–1004

Cerebral Cortex

The Role of NMDA Receptors in Information Processing	NW Daw, PSG Stein, K Fox	16:207–22

Inferior Temporal Cortex: Where Visual Perception Meets Memory	Y Miyashita	16:245–63
Localization of Brain Function: The Legacy of Franz Joseph Gall (1758–1828)	S Zola-Morgan	18:359–83
Visual Feature Integration and the Temporal Correlation Hypothesis	W Singer, CM Gray	18:555–86
Inferotemporal Cortex and Object Vision	K Tanaka	19:109–39
Patterning and Specification of the Cerebral Cortex	P Levitt, MF Barbe, KL Eagleson	20:1–24
Sleep and Arousal: Thalamocortical Mechanisms	DA McCormick, T Bal	20:185–215
Multimodal Representation of Space in the Posterior Parietal Cortex and Its Use in Planning Movements	RA Andersen, LH Snyder, DC Bradley, J Xing	20:299–326
Local Circuits in Primary Visual Cortex of the Macaque Monkey	EM Callaway	21:47–74
Cortical and Subcortical Contributions to Activity-Dependent Plasticity in Primate Somatosensory Cortex	EG Jones	23:1–37
Touch and Go: Decision-Making Mechanisms in Somatosensation	R Romo, E Salinas	24:107–37
An Integrative Theory of Prefrontal Cortex Function	EK Miller, JD Cohen	24:167–202
Sequential Organization of Multiple Movements: Involvement of Cortical Motor Areas	J Tanji	24:631–51
Molecular Control of Cortical Dendrite Development	KL Whitford, P Dijkhuizen, F Polleux, A Ghosh	25:127–49
Intentional Maps in Posterior Parietal Cortex	RA Andersen, CA Buneo	25:189–20

Circadian and Other Rhythms

Molecular Neurobiology and Genetics of Circadian Rhythms in Mammals	JS Takahashi	18:531–53
Molecular Genetics of Circadian Rhythms in Mammals	DP King, JS Takahashi	23:713–42

Clinical Neuroscience

The Neurofibromatosis Type 1 Gene	D Viskochil, R White, R Cawthon	16:183–205
The Epigenetics of Multiple Sclerosis: Clues to Etiology and a Rationale for Immune Theory	L Steinman, A Miller, CCA Bernard, JR Oksenberg	17:247–65

Normal and Abnormal Biology of the β-Amyloid Precursor Protein	DJ Selkoe	17:489–517
Biology and Genetics of Hereditary Motor and Sensory Neuropathies	U Suter, GJ Snipes	18:45–75
Triplet Repeat Expansion Mutations: The Example of Fragile X Syndrome	ST Warren, CT Ashley Jr	18:77–99
The Neurobiology of Infantile Autism	AL Ciaranello, RD Ciaranello	18:101–28
Molecular Mechanisms of Drug Reinforcement and Addiction	DW Self, EJ Nestler	18:463–95
Human Immunodeficiency Virus and the Brain	JD Glass, RT Johnson	19:1–26
Apolipoprotein E and Alzheimer's Disease	WJ Strittmatter, AD Roses	19:53–77
Trinucleotide Repeats in Neurogenetic Disorders	HL Paulson, KH Fischbeck	19:79–107
Sodium Channel Defects in Myotonia and Periodic Paralysis	SC Cannon	19:141–64
Addictive Drugs and Brain Stimulation Reward	RA Wise	19:319–40
The Role of Vesicular Transport Proteins in Synaptic Transmission and Neural Degeneration	Y Liu, RH Edwards	20:125–56
Genetics of Manic Depressive Illness	DF MacKinnon, KR Jamison, JR DePaulo	20:351–69
Human Autoimmune Neuropathies	TW Ho, GM McKhann, JW Griffin	21:187–226
Zinc and Brain Injury	DW Choi, JY Koh	21:347–75
Mutant Genes in Familial Alzheimer's Disease and Transgenic Models	DL Price, SS Sisodia	21:479–505
Etiology and Pathogenesis of Parkinson's Disease	CW Olanow, WG Tatton	22:123–44
Microglia as Mediators of Inflammatory and Degenerative Diseases	F González-Scarano, G Baltuch	22:219–40
Glutamine Repeats and Neurodegeneration	HY Zoghbi, HT Orr	23:217–47
The Emergence of Modern Neuroscience: Some Implications for Neurology and Psychiatry	WM Cowan, DH Harter, ER Kandel	23:343–91
Paraneoplastic Neurologic Disease Antigens: RNA-Binding Proteins and Signaling Proteins in Neuronal Degeneration	K Musunuru, RB Darnell	24:239–62
The Developmental Biology of Brain Tumors	R Wechsler-Reya, MP Scott	24:385–428
Prion Diseases of Humans and Animals: Their Causes and Molecular Basis	J Collinge	24:519–50
Neurodegenerative Tauopathies	VM-Y Lee, M Goedert, JQ Trojanowski	24:1121–59

Nerve Growth Factor Signaling, Neuroprotection, and Neural Repair	MV Sofroniew, CL Howe, WC Mobley	24:1217–81
The Human Genome Project and Its Impact on Psychiatry	WM Cowan, KL Kopnisky, SE Hyman	25:1–50
Beyond Phrenology: What Can Neuroimaging Tell Us About Distributed Circuitry?	K Friston	25:221–50
A Decade of Molecular Studies of Fragile X Syndrome	WT O'Donnell, ST Warren	25:315–38
Schizophrenia as a Disorder of Neurodevelopment	DA Lewis, P Levitt	25:409–32
Multiple Sclerosis: Deeper Understanding of Its Pathogenesis Reveals New Targets for Therapy	L Steinman, R Martin, C Bernard, P Conlon, JR Oksenberg	25:491–505

Comparative Neuroscience

Processing of Temporal Information in the Brain	CE Carr	16:223–43
Patterning the Brain of the Zebrafish Embryo	CB Kimmel	16:707–32

Computational Approaches

Behaviorally Based Modeling and Computational Approaches to Neuroscience	GN Reeke Jr, O Sporns	16:597–623
Computational Models of the Neural Bases of Learning and Memory	MA Gluck, R Granger	16:667–706
From Biophysics to Models of Network Function	E Marder	21:25–45
Odor Encoding as an Active, Dynamical Process: Experiments, Computation and Theory	G Laurent, M Stopfer, RW Friedrich, MI Rabinovich, A Volkovskii, HDI Abarbanel	24:263–97
Natural Image Statistics and Neural Representation	EP Simoncelli, BA Olshausen	24:1193–1215

Cytoskeleton and Axonal Transport

Neuronal Polarity	AM Craig, G Banker	17:267–310
Neuronal Intermediate Filaments	MK Lee, DW Cleveland	19:187–217

Neurodegenerative Tauopathies	VM-Y Lee, M Goedert, JQ Trojanowski	24:1121–59

Developmental Neurobiology

Molecular Mechanisms of Developmental Neuronal Death	EM Johnson Jr, TL Deckwerth	16:31–46
Molecular Control of Cell Fate in the Neural Crest: The Sympathoadrenal Lineage	DJ Anderson	16:129–58
Inhibitors of Neurite Growth	ME Schwab, JP Kapfhammer, CE Bandtlow	16:565–95
Development of Motoneuronal Phenotype	JS Eisen	17:1–30
Hox Genes and Regionalization of the Nervous System	R Keynes, R Krumlauf	17:109–32
The Subplate, A Transient Neocortical Structure: Its Role in the Development of Connections between Thalamus and Cortex	KL Allendoerfer, CJ Shatz	17:185–218
Determination of Neuronal Cell Fate: Lessons From the R7 Neuron of *Drosophila*	SL Zipursky, GM Rubin	17:373–97
Specification of Neocortical Areas and Thalamocortical Connections	DDM O'Leary, BL Schlaggar, R Tuttle	17:419–39
Creating a Unified Representation of Visual and Auditory Space in the Brain	EI Knudsen, MS Brainard	18:19–43
Isolation, Characterization, and Use of Stem Cells from the CNS	FH Gage, J Ray, LJ Fisher	18:159–92
Mechanisms of Neural Patterning and Specification in the Developing Cerebellum	ME Hatten, N Heintz	18:385–408
The Role of the Floor Plate in Axon Guidance	SA Colamarino, M Tessier-Lavigne	18:497–529
Mechanisms and Molecules that Control Growth Cone Guidance	CS Goodman	19:341–77
The *Drosophila* Neuromuscular Junction: A Model System for Studying Synaptic Development and Function	H Keshishian, K Broadie, A Chiba, M Bate	19:545–75
Vertebrate Neural Induction	A Hemmati-Brivanlou, D Melton	20:43–60
Pax-6 in Development and Evolution	P Callaerts, G Halder, WJ Gehring	20:479–528
Adhesion Molecules and Inherited Diseases of the Human Nervous System	ML Hlavin, H Kamiguchi, M Yamasaki, V Lemmon	21:97–125

Regionalization of the Prosencephalic Neural Plate	JLR Rubenstein, K Shimamura, S Martinez, L Puelles	21:445–77
Retinal Waves and Visual System Development	ROL Wong	22:29–47
The Specification of Dorsal Cell Fates in the Vertebrate Central Nervous System	KJ Lee, TM Jessell	22:261–94
Growth Cone Guidance: First Steps Towards a Deeper Understanding	BK Mueller	22:351–88
Development of the Vertebrate Neuromuscular Junction	JR Sanes, JW Lichtman	22:389–442
Central Nervous System Neuronal Migration	ME Hatten	22:511–39
Cellular and Molecular Determinants of Sympathetic Neuron Development	NJ Francis, SC Landis	22:541–66
Control of Cell Divisions in the Nervous System: Symmetry and Asymmetry	B Lu, L Jan, Y-N Jan	23:531–56
The Relationship Between Neuronal Survival and Regeneration	JL Goldberg, BA Barres	23:579–612
Glial Control of Neuronal Development	G Lemke	24:87–105
Signaling and Transcriptional Mechanisms in Pituitary Development	MG Rosenfeld, JS Dasen	24:327–55
The Developmental Biology of Brain Tumors	R Wechsler-Reya, MP Scott	24:385–428
Early Days of the Nerve Growth Factor Proteins	EM Shooter	24:601–29
Neurotrophins: Roles in Neuronal Development and Function	EJ Huang, LF Reichardt	24:677–736
Early Anterior/Posterior Patterning of the Midbrain and Cerebellum	A Liu, AL Joyner	24:869–96
Role of the Reelin Signaling Pathway in Central Nervous System Development	DS Rice, T Curran	24:1005–39
Human Brain Malformations and Their Lessons for Neuronal Migration	ME Ross, CA Walsh	24:1041–70
Maternal Care, Gene Expression, and the Transmission of Individual Differences in Stress Reactivity Across Generations	MJ Meaney	24:1161–92
Nerve Growth Factor Signaling, Neuroprotection, and Neural Repair	MV Sofroniew, CL Howe, WC Mobley	24:1217–81
Molecular Control of Cortical Dendrite Development	KL Whitford, P Dijkhuizen, F Polleux, A Ghosh	25:127–49
Transcriptional Codes and the Control of Neuronal Identity	R Shirasaki, SL Pfaff	25:251–81
Large-Scale Sources of Neural Stem Cells	DI Gottlieb	25:381–407
The Role of Notch in Promoting Glial and Neural Stem Cell Fates	N Gaiano, G Fishell	25:471–90

Glia, Schwann Cells, and Extracellular Matrix

Glial Boundaries in the Developing Nervous System	DA Steindler	16:445–70
The Early Reactions of Non-Neuronal Cells to Brain Injury	DMD Landis	17:133–51
Glial Control of Neuronal Development	G Lemke	24:87–105
The Role of Notch in Promoting Glial and Neural Stem Cell Fates	N Gaiano, G Fishell	25:471–90
Central Nervous System Damage, Monocytes and Macrophages, and Neurological Disorders in AIDS	KC Williams, WF Hickey	25:537–62

Higher Functions

Monoamine Oxidase: From Genes to Behavior	JC Shih, K Chen, M Ridd	22:197–217
Mechanisms of Visual Attention in the Human Cortex	S Kastner, LG Ungerleider	23:315–41
Neuronal Coding of Prediction Errors	W Schultz, A Dickinson	23:473–500
Consciousness	JR Searle	23:557–78

Hippocampus

Spatial Processing in the Brain: The Activity of Hippocampal Place Cells	PJ Best, AM White, A Minai	24:459–86

History of Neuroscience

Spatial Processing in the Brain: The Activity of Hippocampal Place Cells	PJ Best, AM White, A Minai	24:459–86
Viktor Hamburger and Rita Levi-Montalcini: The Path to the Discovery of Nerve Growth Factor	WM Cowan	24:551–600
Early Days of the Nerve Growth Factor Proteins	EM Shooter	24:601–29
Flies, Genes, and Learning	S Waddell, WG Quinn	24:1283–309
The Human Genome Project and Its Impact on Psychiatry	WM Cowan, KL Kopnisky, SE Hyman	25:1–50

Ion Channels

Molecular Basis for Ca^{2+} Channel Diversity	F Hofmann, M Biel, V Flockerzi	17:399–418
Direct G Protein Activation of Ion Channels?	DE Clapham	17:441–64
Structure and Function of Cyclic Nucleotide-Gated Channels	WN Zagotta, SA Siegelbaum	19:235–63

Cloned Potassium Channels from Eukaryotes and Prokaryotes	LY Jan, YN Jan	20:91–123
Gain of Function Mutants: Ion Channels and G Protein-Coupled Receptors	HA Lester, A Karschin	23:89–125

Language

The Processing of Single Words Studied with Positron Emission Tomography	SE Petersen, JA Fiez	16:509–30
Birdsong and Human Speech: Common Themes and Mechanisms	AJ Doupe, PK Kuhl	22:567–631
Functional MRI of Language: New Approaches to Understanding the Cortical Organization of Semantic Processing	S Bookheimer	25:151–88

Learning and Memory

Implicit Memory: A Selective Review	DL Schacter, C-YP Chiu, KN Ochsner	16:159–82
Neuroanatomy of Memory	S Zola-Morgan, LR Squire	16:547–63
Organization of Memory Traces in the Mammalian Brain	RF Thompson, DJ Krupa	17:519–49
Learning and Memory in Honeybees: From Behavior to Neural Substrates	R Menzel, U Müller	19:379–404
Neurobiology of Speech Perception	RH Fitch, S Miller, P Tallal	20:327–49
CREB and Memory	AJ Silva, JH Kogan, PW Frankland, S Kida	21:127–48
Gene Discovery in *Drosophila*: New Insights for Learning and Memory	J Dubnau, T Tully	21:407–44
Synaptic Plasticity and Memory: An Evaluation of the Hypothesis	SJ Martin, PD Grimwood, RGM Morris	23:649–711
An Integrative Theory of Prefrontal Cortex Function	EK Miller, JD Cohen	24:167–202
Spatial Processing in the Brain: The Activity of Hippocampal Place Cells	PJ Best, AM White, A Minai	24:459–86
Neurobiology of Pavlovian Fear Conditioning	SA Maren	24:897–931
Learning and Memory Functions of the Basal Ganglia	MG Packard, BJ Knowlton	25:563–93

Miscellaneous

An Urge to Explain the Incomprehensible: Geoffrey Harris and the Discovery Control of the Pituitary Gland	G Raisman	20:529–62
The Cell Biology of the Blood-Brain Barrier	LL Rubin, JM Staddon	22:11–28

Making Brain Connections: Neuroanatomy and the Work of TPS Powell, 1923–1966	EG Jones	22:49–103
The Human Genome Project and Its Impact on Psychiatry	WM Cowan, KL Kopnisky, SE Hyman	25:1–50
Wired for Reproduction: Organization and Development of Sexually Dimorphic Circuits in the Mammalian Forebrain	RB Simerly	25:507–36

Molecular Neuroscience

Neurotransmitter Transporters: Recent Progress	SG Amara, MJ Kuhar	16:73–93
Molecular Basis of Neural-Specific Gene Expression	G Mandel, D McKinnon	16:323–45
Prion Diseases and Neurodegeneration	SB Prusiner, SJ DeArmond	17:311–39
The Protein Kinase C Family for Neuronal Signaling	C Tanaka, Y Nishizuka	17:551–67
Functional Interactions of Neurotrophins and Neurotrophin Receptors	M Bothwell	18:223–53
The Role of Agrin in Synapse Formation	MA Bowe, JR Fallon	18:443–62
RNA Editing	L Simpson, RB Emeson	19:27–52
Physiology of the Neurotrophins	GR Lewin, Y-A Barde	19:289–317
Intracellular Signaling Pathways Activated by Neurotrophic Factors	RA Segal, ME Greenberg	19:463–89
The Neurotrophins and CNTF: Two Families of Collaborative Neurotrophic Factors	NY Ip, GD Yancopoulos	19:491–515
RNA Transport	S Nakielny, U Fischer, WM Michael, G Dreyfuss	20:269–98
ARIA: A Neuromuscular Junction Neuregulin	GD Fischbach, KM Rosen	20:425–54
Rab3 and Synaptotagmin: The Yin and Yang of Synaptic Membrane Fusion	M Geppert, TC Südhof	21:75–95
Inducible Gene Expression in the Nervous System of Transgenic Mice	JR Gingrich, J Roder	21:377–405
Microtubule-Based Transport Systems in Neurons: The Roles of Kinesins and Dyneins	LSB Goldstein, Z Yang	23:39–72
Apoptosis in Neural Development and Disease	D Nijhawan, N Honarpour, X Wang	23:73–87
PDZ Domains and the Organization of Supramolecular Complexes	C Sala, MH Sheng	24:1–29
Protein Synthesis at Synaptic Sites on Dendrites	O Steward, E Schuman	24:299–325
Signaling and Transcriptional Mechanisms in Pituitary Development	MG Rosenfeld, JS Dasen	24:327–55

Prion Diseases of Humans and Animals: Their Causes and Molecular Basis	J Collinge	24:519–50
Early Days of the Nerve Growth Factor Proteins	EM Shooter	24:601–29
Neurotrophins: Roles in Neuronal Development and Function	EJ Huang, LF Reichardt	24:677–736
α-Latrotoxin and its Receptors: Neurexins and CIRL/Latrophilins	TC Südhof	24:933–62
AMPA Receptor Trafficking and Synaptic Plasticity	R Malinow, RC Malenka	25:103–26
Transcriptional Codes and the Control of Neuronal Identity	R Shirasaki, SL Pfaff	25:251–81

Motor Systems

The Anatomy and Physiology of Primate Neurons that Control Rapid Eye Movements	AK Moschovakis, SM Highstein	17:465–88
Learning and Memory in the Vestibulo-Ocular Reflex	S du Lac, JL Raymond, TJ Sejnowski, SG Lisberger	18:409–41
Premotor and Parietal Cortex: Corticocortical Connectivity and Combinatorial Computations	SP Wise, D Boussaoud, PB Johnson, R Caminiti	20:25–42
Plasticity and Primary Motor Cortex	JN Sanes, JP Donoghue	23:393–415
Neuropeptides and the Integration of Motor Responses to Dehydration	AG Watts	24:357–84
Sequential Organization of Multiple Movements: Involvement of Cortical Motor Areas	J Tanji	24:631–51
The Role of the Cerebellum in Voluntary Eye Movements	FR Robinson, AF Fuchs	24:981–1004

Neural Membranes

Protein Targeting in the Neuron	RB Kelly, E Grote	16:95–127

Neural Networks

Modeling of Neural Circuits: What Have We Learned?	AI Selverston	16:531–46
Neuromorphic Analogue VLSI	R Douglas, M Mahowald, C Mead	18:255–81
The Significance of Neuronal Ensemble Codes During Behavior and Cognition	SA Deadwyler, RE Hampson	20:217–44
Neural Representation and the Cortical Code	RC deCharms, A Zador	23:613–47

Neuroethology

Developmental Plasticity in Neural Circuits for a Learned Behavior	SW Bottjer, AP Arnold	20:455–77

Neurogenetics

Genetic and Cellular Analysis of Behavior in *C. elegans*	CI Bargmann	16:47–71
Gene Targeting in ES Cells	P Soriano	18:1–18
The Compartmentalization of the Cerebellum	K Herrup, B Kuemerle	20:61–90
Bcl-2 Gene Family in the Nervous System	DE Merry, SJ Korsmeyer	20:245–67
Genetic and Environmental Influences on Human Behavioral Differences	M McGue, TJ Bouchard Jr.	21:1–24
Signal Transduction in the *Caenorhabditis elegans* Nervous System	CI Bargmann, JM Kaplan	21:279–308
Quantitative Genetics and Mouse Behavior	JM Wehner, RA Radcliffe, BJ Bowers	24:845–67
Human Brain Malformations and Their Lessons for Neuronal Migration	ME Ross, CA Walsh	24:1041–70
Flies, Genes, and Learning	S Waddell, WG Quinn	24:1283–1309

Neuronal Membranes

Influence of Dendritic Conductances on the Input-Output Properties of Neurons	A Reyes	24:653–75

Neuronal Plasticity

Learning to Modulate Transmitter Release: Themes and Variations in Synaptic Plasticity	RD Hawkins, ER Kandel, SA Siegelbaum	16:625–65
Dynamic Regulation of Receptive Fields and Maps in the Adult Sensory Cortex	NM Weinberger	18:129–58
Long-Term Synaptic Depression	DJ Linden, JA Connor	18:319–57
Active Properties of Neuronal Dendrites	D Johnston, JC Magee, CM Colbert, BR Christie	19:165–86
Molecular Genetic Analysis of Synaptic Plasticity, Activity-Dependent Neural Development, Learning, and Memory in the Mammalian Brain	C Chen, S Tonegawa	20:157–84
Function-Related Plasticity in Hypothalamus	GI Hatton	20:371–93
Cortical Plasticity: From Synapses to Maps	D Buonomano, MM Merzenich	21:149–86
Stress and Hippocampal Plasticity	BS McEwen	22:105–22
Neurotrophins and Synaptic Plasticity	AK McAllister, LC Katz, DC Lo	22:295–318

Synaptic Modification by Correlated Activity: Hebb's Postulate Revisited	G-Q Bi, M-M Poo	24:139–66
Activity-Dependent Spinal Cord Plasticity in Health and Disease	JR Wolpaw, AM Tennissen	24:807–43
Morphological Changes in Dendritic Spines Associated with Long-Term Synaptic Plasticity	R Yuste, T Bonhoeffer	24:1071–89

Neuropeptides

Odor Encoding as an Active, Dynamical Process: Experiments, Computation and Theory	G Laurent, M Stopfer, RW Friedrich, MI Rabinovich, A Volkovskii, HDI Abarbanel	24:263–97
Neuropeptides and the Integration of Motor Responses to Dehydration	AG Watts	24:357–84
To Eat or Sleep? The Role of Orexin in Coordination of Feeding and Arousal	JT Willie, RM Chemelli, CM Sinton, M Yanagisawa	24:429–58
Imaging and Coding in the Olfactory System	JS Kauer, J White	24:963–79

Neuroscience Techniques

Common Principles of Motor Control in Vertebrates and Invertebrates	KG Pearson	16:265–97
Gene Transfer to Neurons Using Herpes Simplex Virus-Based Vectors	DJ Fink, NA DeLuca, WF Goins, JC Glorioso	19:265–87
Imaging and Coding in the Olfactory System	JS Kauer, J White	24:963–79

Neurotrophins

| Early Days of the Nerve Growth Factor Proteins | EM Shooter | 24:601–29 |
| Neurotrophins: Roles in Neuronal Development and Function | EJ Huang, LF Reichardt | 24:677–736 |

Olfaction/Taste

| Information Coding in the Vertebrate Olfactory System | LB Buck | 19:517–44 |
| Mechanisms of Olfactory Discrimination: Converging Evidence for Common Principles Across Phyla | JG Hildebrand, GM Shepherd | 20:591–629 |

Molecular Biology of Odorant Receptors in Vertebrates	P Mombaerts	22:487–509
Guanylyl Cyclases as a Family of Putative Odorant Receptors	AD Gibson, DL Garbers	23:417–39
Odor Encoding as an Active, Dynamical Process: Experiments, Computation and Theory	G Laurent, M Stopfer, RW Friedrich, MI Rabinovich, A Volkovskii, HDI Abarbanel	24:263–97
Imaging and Coding in the Olfactory System	JS Kauer, J White	24:963–79

Pain

Pain Genes?: Natural Variation and Transgenic Mutants	JS Mogil, L Yu, AI Basbaum	23:777–811
The Vanilloid Receptor: A Molecular Gateway to the Pain Pathway	MJ Caterina, D Julius	24:487–517
Contributions of the Medullary Raphe and Ventromedial Reticular Region to Pain Modulation and Other Homeostatic Functions	P Mason	24:737–777

Prefatory Chapter

Integrative Systems Research on the Brain: Resurgence and New Opportunities	TH Bullock	16:1–15

Receptors and Receptor Subtypes

Recent Advances in the Molecular Biology of Dopamine Receptors	JA Gingrich, MG Caron	16:299–321
Regulation of Ion Channel Distribution at Synapses	SC Froehner	16:347–68
The Diversity of Neuronal Nicotinic Acetylcholine Receptors	PB Sargent	16:403–43
Cloned Glutamate Receptors	M Hollmann, S Heinemann	17:31–108
$GABA_A$ Receptor Channels	RL Macdonald, RW Olsen	17:569–602
Functional and Structural Complexity of Signal Transduction via G-Protein-Coupled Receptors	T Gudermann, T Schöeneberg, G Schultz	20:395–423
The Ephrins and Eph Receptors in Neural Development	JG Flanagan, P Vanderhaeghen	21:309–45

The Vanilloid Receptor: A Molecular Gateway to the Pain Pathway	MJ Caterina, D Julius	24:487–517
Role of the Reelin Signaling Pathway in Central Nervous System Development	DS Rice, T Curran	24:1005–39
AMPA Receptor Trafficking and Synaptic Plasticity	R Malinow, RC Malenka	25:103–26

Sleep and Sleep Disorders

To Eat or Sleep? The Role of Orexin in Coordination of Feeding and Arousal	JT Willie, RM Chemelli, CM Sinton, M Yanagisawa	24:429–58
The Role of Hypocretins (Orexins) in Sleep Regulation and Narcolepsy	S Taheri, JM Zeitzer, E Mignot	25:283–313

Somatosensory System

Touch and Go: Decision-Making Mechanisms in Somatosensation	R Romo, E Salinas	24:107–37

Synapses/Synaptic Transmission

Transsynaptic Control of Gene Expression	RC Armstrong, MR Montminy	16:17–29
Synaptic Vesicles and Exocytosis	R Jahn, TC Südhof	17:216–46
Dendritic Spines: Cellular Specializations Imparting Both Stability and Flexibility to Synaptic Function	KM Harris, SB Kater	17:341–71
Neurotransmitter Release	G Matthews	19:219–33
Synaptic Regulation of Mesocorticolimbic Dopamine Neurons	FJ White	19:405–36
Long-Term Depression in Hippocampus	MF Bear, WC Abraham	19:437–62
Monitoring Secretory Membrane with FM1-43 Fluorescence	AJ Cochilla, JK Angleson, WJ Betz	22:1–10
Autoimmunity and Neurological Disease: Antibody Modulation of Synaptic Transmission	KD Whitney, JO McNamara	22:175–95
Presynaptic Ionotropic Receptors and the Control of Transmitter Release	AB MacDermott, LW Role, SA Siegelbaum	22:443–85
PDZ Domains and the Organization of Supramolecular Complexes	C Sala, MH Sheng	24:1–29
The Role and Regulation of Adenosine in the Central Nervous System	TV Dunwiddie, SA Masino	24:31–55

Synaptic Modification by Correlated Activity: Hebb's Postulate Revisited	G-Q Bi, M-M Poo	24:139–66
Influence of Dendritic Conductances on the Input-Output Properties of Neurons	A Reyes	24:653–75
α-Latrotoxin and its Receptors: Neurexins and CIRL/Latrophilins	TC Südhof	24:933–62

Transmitter Biochemistry

Nitric Oxide and Synaptic Function	EM Schuman, DV Madison	17:153–83

Vision

The Molecules of Mechanosensation	J García-Añoveros, DP Corey	20:563–90
Localization and Globalization in Conscious Vision	S Zeki	24:57–86
The Physiology of Stereopsis	BG Cumming, GC DeAngelis	24:203–38
Activation, Deactivation, and Adaptation in Vertebrate Photoreceptor Cells	ME Burns, DA Baylor	24:779–805
Natural Image Statistics and Neural Representation	EP Simoncelli, BA Olshausen	24:1193–1215
Contextual Influences on Visual Processing	TD Albright, GR Stoner	25:339–79

Visual System

How Parallel Are the Primate Visual Pathways?	WH Merigan, JHR Maunsell	16:369–402
Neural Mechanisms of Selective Visual Attention	R Desimone, J Duncan	18:193–222
Signal Transduction in *Drosophila* Photoreceptors	R Ranganathan, DM Malicki, CS Zuker	18:283–317
Visual Object Recognition	NK Logothetis, DL Sheinberg	19:577–621
Sense and the Single Neuron: Probing the Physiology of Perception	AJ Parker, WT Newsome	21:227–77
Computational Neuroimaging of Human Visual Cortex	BA Wandell	22:145–73
Neural Selection and Control of Visually Guided Eye Movements	JD Schall, KG Thompson	22:241–59
Space and Attention in Parietal Cortex	CL Colby, ME Goldberg	22:319–50
The Koniocellular Pathway in Primate Vision	SHC Hendry, RC Reid	23:127–53
Confronting Complexity: Strategies for Understanding the Microcircuitry of the Retina	RH Masland, E Raviola	23:249–84

Neural Mechanisms of Orientation Selectivity in the Visual Cortex	D Ferster, KD Miller	23:441–71
Parallel Pathways for Spectral Coding in Primate Retina	DM Dacey	23:743–75
The Role and Regulation of Adenosine in the Central Nervous System	TV Dunwiddie, SA Masino	24:31–55
Localization and Globalization in Conscious Vision	S Zeki	24:57–86
The Physiology of Stereopsis	BG Cumming, GC DeAngelis	24:203–38
Activation, Deactivation, and Adaptation in Vertebrate Photoreceptor Cells	ME Burns, DA Baylor	24:779–805
Natural Image Statistics and Neural Representation	EP Simoncelli, BA Olshausen	24:1193–1215